USED

P9-AGL-381

INVITATION TO BIOLOGY

INVITATION TO BIOLOGY

By Helena Curtis

WORTH PUBLISHERS, INC.

INVITATION TO BIOLOGY

BY HELENA CURTIS

ILLUSTRATIONS BY SHIRLEY BATY

PICTURE EDITOR: ANNE FELDMAN

DESIGNED BY MALCOLM GREAR DESIGNERS

TYPE SET IN PALATINO BY WOLF COMPOSITION

LIBRARY OF CONGRESS CATALOG CARD NUMBER 70–181459

ISBN NUMBER 0–87901–020–7

MANUFACTURED IN THE UNITED STATES OF AMERICA

THE PAPER USED IN THIS BOOK IS RECYCLED (70% POST CONSUMER WASTE)

FIRST PRINTING MARCH 1972

WORTH PUBLISHERS, INC.

70 FIFTH AVENUE

NEW YORK, NEW YORK 10011

This book is dedicated to Caroline Rogers

PREFACE

Invitation to Biology is, in many respects, a short version of *Biology*. We were encouraged in this undertaking not only by the wide acceptance of *Biology* but also by the diversity of the colleges and universities at which it was adopted and, most of all—at least for me personally—by students who wrote spontaneously to say how much they had enjoyed it. This book, like *Biology*, is dedicated to the proposition that man, by nature, is curious and that learning is a pleasurable process.

In the course of working on this book, we soon discovered that we could not simply condense or eliminate parts of *Biology* but that we had to replan and, largely, rewrite the entire text. With the exception of the Introduction, which has been shortened but is otherwise unchanged, there is hardly a paragraph that has not been rethought, rewritten, and, dare I say it, considerably improved!

Moreover, we had even more relentlessly to face the question of what to put in, and, much more difficult, what to leave out. What is modern biology *really*? What do students *have* to know? What will they remember 10 years from now? We do not pretend to have complete answers to these questions, but we have certainly been constantly concerned with them.

In view of these larger considerations, those familiar with *Biology* might be interested in the ways in which this text differs from its partial parent.

First—and this was the most difficult decision—I had to decide over again what to do about the chemistry, which is the part students like least and certainly forget the fastest. I seriously considered virtually eliminating it and found some support for that solution. However, I finally decided that to minimize the chemistry in this way would be to distort our present knowledge of biology and really cheat the newcomer to the field. Although I do not agree with Albert Lehninger when he says "Biology *is* chemistry," I think it is a useful thing to say. Certainly the great triumph of twentieth century biology is the gaining of widespread acceptance for the fact that all living structures and processes have a nonvitalistic—in short, a physiochemical—basis. This acceptance was not won by proclaiming this idea—which had been done at least a couple of centuries before—but by showing how it worked in detail. Similarly, I think it is important for students at least to glimpse these workings. As a consequence, although the presentation of cellular physiology in this text is less extensive than in *Biology*, some parts of it are done in more detail. Also, for the same reason—after weighing the problem in the same way—chemical structurals are presented at other places throughout the book. The point of these structurals is simply to emphasize once again that life processes do indeed have a chemical basis; no one should ever memorize the chemical structures of the cytokinins, for example, but presenting these is a quick reminder that growth and differentiation are the results of chemical interactions and that the configurations of chemicals do indeed determine their activities in living systems.

Genetics, the last section of Part I, follows the text of *Biology* quite closely. As in *Biology*, the last chapter, which is largely about work now going on, is optional. From time to time I have a quiet reverie about students reading it *anyway*, but in these days of academic and nonacademic pressures, this probably is not true.

In Part II, the section on plants has greatly benefited from the years of work that went into my collaboration with Peter Raven in writing *Biology of Plants*. The animal physiology section bears little resemblance to its counterpart in *Biology*. Here I have elected to concentrate more on mammalian physiology and, in particular, human physiology. This decision seems appropriate in view of the fact that students reading this book may not study any more life science. The animal physiology section contains a picture essay on human development which we find enormously exciting and believe students will also.

Part III, which deals with ecology and evolution, contains much of the material on social behavior of animals which was presented in a separate section in the previous text. It includes a discussion of the Hardy-Weinberg law, which some teachers may choose to omit. However, we felt that at a time the scientists in the field are trying so hard to put ecology and evolution on a mathematical basis, at least some idea of this basis was necessary. The text concludes with a discussion of the present ecological crisis, a problem which is coming to concern all inhabitants of the planet but in which biologists, quite rightly, feel they have special responsibilities. Our intention in this chapter was not to harangue or exort—which is being done amply at the present time—but to try to trace some of the underlying biological causes of our present dilemma and so, perhaps, help students to anticipate and prevent future catastrophes as well as to understand the present ones.

At the end of each section of the book is a glossary, which is not where you would ordinarily expect to find one, for which we offer this word of explanation. We worked hard on these glossaries and, in the course of writing and revising them, began to appreciate the usefulness of short, sharp definitions—not so much of anatomical terms, which are easy, but of processes and abstractions, which are far more difficult. So we have placed the glossaries where the student cannot fail to stumble over them from time to time and so be encouraged to use them as an integral part of the text, which they are.

After the glossaries, there are short lists of books for further reading. We have not included lists of *Scientific American* reprints, simply on the assumption that these excellent supplementary materials are widely known, widely available, and constantly increasing in number. We would like to urge that, if it is at all possible, students be encouraged to explore on their own, journals such as *Scientific American*, *Science*, and *American Scientist*. In this way, they will have the pleasure of discovering something for themselves and of seeing the process of the acquisition of scientific knowledge as it takes place.

In the preparation of this text, I am particularly indebted to a number of people. Winslow Briggs of Harvard University, Marc Lappé of the University of California, Berkeley, and Richard Lewontin of the University of Chicago, who read and criticized the entire text; David Shappirio of the University of Michigan, whose sharp eye dissected Part I and most of Part II; Michael Menaker of the University of Texas, who helped in planning the animal physiology section; and Roberts Rugh of College of Physicians and Surgeons of Columbia University who, in addition to supplying the superb photographs, was our advisor for the picture essay on human development. Walter L. Meagher helped to initiate and plan the entire project. Jean Ely, an editor, who wields a ball-point pen as if it were a scalpel, both cruel and curative, not only corrected syntax and ministered to the lame and halt among my sentences, but poked and queried me on matters of logic and coherence in a manner that would have brought to an end all but the longest and best of friendships, which ours is. And the people at Worth Publishers who held the whole thing together from beginning to end, with patience, good humor, and wit.

Among others who have made major contributions to the preparation of the book are:

John M. Anderson, Cornell University; William C. Bessler, Mankato State College; Walter F. Bodmer, University of Oxford; Robert Franke, Iowa State University; Ursula Goodenough, Harvard University; James Jamieson, Rockefeller University; C. Benjamin Melecca, Ohio State University; J. Richard McIntosh, University of Colorado, Boulder; Reginald Noble, Bowling Green State University; Ivan G. Palmblad, Utah State University; Valerie Pasztor, McGill University; Keith Porter, University of Colorado; Elmer Palmatier, University of Rhode Island; Richard F. Thompson, University of California, Irvine; William Vanderkloot, S.U.N.Y., Stony Brook; Richard H. Wilson, University of Texas; Stephen Vogel, Duke University.

Finally, we must thank all of the many teachers who wrote to us about *Biology* with criticisms and suggestions. They helped us greatly in preparing the present text and we earnestly hope they will continue. There is no one whose opinion we value more.

HELENA CURTIS
East Hampton, New York
February, 1972

CONTENTS IN BRIEF

TABLE OF CONTENTS

1

"Afterwards, on becoming very intimate with Fitz-Roy [the captain of the Beagle*], I heard that I had run a very narrow risk of being rejected on account of the shape of my nose! He . . . was convinced that he could judge of a man's character by the outline of his features; and he doubted whether anyone with my nose could possess sufficient energy and determination for the voyage. But I think he was afterwards well satisfied that my nose had spoken falsely."* (*Charles Darwin,* The Voyage of the Beagle)

Introduction

In 1831, a young Englishman, Charles Darwin, sailed from Devonport on what was to prove the most consequential voyage in the history of biology. Not yet twenty-three, Darwin had already abandoned a proposed career in medicine—he describes himself as fleeing a surgical theater in which an operation was being performed on an unanesthetized child—and was a reluctant candidate for the clergy, a profession deemed more suitable for the younger son of an English gentleman. An indifferent student, Darwin was an ardent hunter and fisherman, a collector of beetles, mollusks, and shells, and an amateur botanist and geologist. When the captain of the surveying ship *Beagle,* himself only a little older than Darwin, offered passage and a berth in his own cabin to any young man who would volunteer to go without pay as a naturalist, Darwin eagerly seized the opportunity to escape from Cambridge. This voyage, which lasted five years, shaped the course of Darwin's future work. He returned to an inherited fortune, an estate in the English countryside, and a lifetime of work and study.

THE ROAD TO EVOLUTIONARY THEORY

That Darwin was the founder of the modern theory of evolution is well known. In order to understand the meaning of his theory, however, it is useful to look briefly at the intellectual climate in which it was formulated.

The Ladder of Life

Aristotle, the first great biologist, believed that all living things could be arranged in a hierarchy. This hierarchy became known as the *Scala Naturae,* or ladder of nature, in which the simplest creatures had a humble position on the bottommost rung, man occupied the top, and all other organisms had their proper places between. European biologists of more modern times also believed in such a natural hierarchy. But whereas to Aristotle, living organisms had always existed, the later Europeans, in harmony with the teachings of the Scriptures, believed that all living things were the products of a divine creation. In either case, the concept that prevailed for 2,000 years was that all the kinds, or species (*species* simply means "kinds"), of animals had come into existence in their present form. Even those who believed in spontaneous generation (toads forming from the mud and snakes from a lady's hair dropped in a rain barrel) did not believe that any species had an historical relationship to any other one—any common ancestry, so to speak.

One of the more famous biologists who believed in original creation was Carolus Linnaeus of Sweden, the great eighteenth century systematist who established a method of classifying the plants and animals. All the time that Linnaeus was at work on his encyclopedic *Systema Naturae* and *Species Plantarum,* explorers of the New World were continuing to return to Europe with new species of plants and animals and even new kinds of human beings. Linnaeus revised edition after edition to accommodate these findings, but he did not change his opinions on the fixity of species. He was sometimes criticized by his contemporaries, who felt that his classifications were too artificial. Few, however, took issue with his belief in a special creation.

Preevolutionists

Among the first to suggest that species might undergo some changes in the course of time was the French scientist Georges-Louis Leclerc de Buffon (1707–1788). Buffon believed that these changes took place by a process of degeneration. He suggested that, in addition to the numerous creatures that were the product of the special creation at the beginning of the world, "There are lesser families conceived by Nature and produced by Time." In fact, as he summed it up, ". . . improvement and degeneration are the same thing, for both imply an alteration of the original constitution." If you are familiar with classical philosophy, you will see that Buffon was influenced by the Platonic concept of the ideal, or true form, of which all worldly expressions are merely imperfect copies.

Another early doubter of the fixity of species was Erasmus Darwin (1731–1802), Charles Darwin's grandfather. Erasmus Darwin was a physician, a gentleman naturalist, and a prolific and discursive writer, often in verse, on both botany and zoology. Erasmus Darwin suggested, largely in asides and footnotes, that species have historical connections with one another, that competition plays a role in the development of different species, that animals may change in response to their environment, and that their offspring may inherit these changes. For instance, a polar bear, he maintained, is an "ordinary" bear which has become modified by living in the Arctic and which passes these modifications along to its cubs. These ideas were never clearly formulated; what Erasmus Darwin thought is of interest largely because of its possible effects on Charles Darwin, although the latter, who was born after his grandfather died, did not hold his grandfather's views in high esteem.

2
Carolus Linnaeus, Swedish professor, physician, and naturalist, who initiated the system of classification of living organisms which we still use today. Linnaeus believed that all species, every kind of living thing, had been created simultaneously and that by classifying them, he was revealing the grand pattern of creation.

The Age of the Earth

It was the geologists far more than the biologists who paved the way for evolutionary theory. One of the most influential of these was James Hutton (1726–1797). Hutton proposed that the earth had been molded not by sudden, violent events but by slow and gradual processes—wind, weather, and the flow of water—the same processes that can be seen at work in the world today. This theory of Hutton's, which was known as uniformitarianism, was important for two reasons. First, it implied that the earth has a long history. This was a new idea. Christian theologians by counting the successive generations since Adam (as recorded in the Bible), had calculated the maximum lifespan of the world at about 6,000 years. No one had ever thought in terms of a longer period. And 6,000 years is not enough time for any form of evolution to have taken place. Second, it stated that change is the *normal* course of events, as opposed to the concept of a normally static system interrupted by an occasional unusual event.

The Fossil Record

During the late eighteenth century, there was a revival of interest in fossils. In previous centuries, fossils had been collected as curiosities, but they had generally been regarded either as accidents of nature—stones that somehow looked like shells—or as evidence of great natural catastrophes, such as proof of Noah's Flood. The English surveyor William Smith (1769–1839) was the first to make a systematic study of fossils. Whenever his work took him down into a mine or along canals or across country, he carefully noted the order of the different layers of rock, which are called strata, and collected the fossils from each layer. He eventually established that each stratum, no matter where he came across it in England, contained a characteristic group of specimens and that these fossils were actually the best means of identifying a particular stratum. (The use of fossils to identify strata is still widely practiced.) Smith did not interpret his findings, but the implica-

3

The first ideas about evolution came from the work of geologists who discovered that different layers, or strata, of the earth's surface contain different types of fossils. This view of the Grand Canyon clearly shows such strata, now seen as chapters in evolutionary history.

tion that the earth had been formed layer by layer over the course of time was an unavoidable one.

Like Hutton's world, the world seen and reported by William Smith was clearly a very ancient one. A revolution in geology was beginning; earth science was becoming a study of time and change rather than a mere cataloging of types of rocks. As a consequence, the history of the earth became inseparable from the history of living organisms, as revealed in the fossil record.

Catastrophism

Although the way was being prepared, the time was not yet ripe for a parallel revolution in biology. The dominating force in European science at the end of the eighteenth century was Georges Cuvier (1769–1832). Cuvier was the founder of paleontology, the scientific study of the fossil record. An expert in anatomy and zoology, he applied his special knowledge of the way in which animals are constructed to the study of fossil animals, and he was able to make brilliant deductions about the form of an entire animal from a few fragments of bone. We think of paleontology and evolution as so closely connected that it is surprising to learn that Cuvier was a staunch and powerful opponent of evolutionary theories. He recognized the fact that many species that had once existed no longer did. (In fact, according to modern estimates, considerably less than 1 percent of all species that have ever lived are represented on the earth today.) He explained the extinction of species by postulating a series of catastrophes. After each catastrophe, the most recent of which was the Flood, the species that remained alive repopulated the world.

The proponents of catastrophism were of two schools of thought: the deluvianists held that all the great upheavals which had destroyed extinct species were floods, and the vulcanists believed that the world had periodically been inundated with lava. According to both theories, however, the fossil record is simply the remains of those species of once living things which the violence of nature had eliminated. Time and nature, the catastrophists believed, act not to create but to eliminate.

THE THEORIES OF LAMARCK

The first scientist to work out a systematic theory of evolution was Jean Baptiste Lamarck (1744–1829). "This justly celebrated naturalist," as Darwin himself referred to him, boldly proposed in 1801 that all species, including man, are descended from other species. Lamarck, unlike most of the other zoologists of his time, was particularly interested in the one-celled organisms and other invertebrates, and it is undoubtedly his long study of these "simpler" forms of life that led him to think of living things in terms of constantly increasing complexity, each form leading from the other.

Like Cuvier and others, Lamarck noted that the older the rocks, the simpler the forms of life they contain, but unlike Cuvier, he interpreted this as meaning that the higher forms had risen from the simpler forms by a kind of progression. According to his hypothesis, this progression, or "evolution," to use the modern term, is dependent on two main forces. The first is the inheritance of acquired characteristics. Organs in animals become stronger or weaker, more or less important, through use or disuse, and these changes, according to Lamarck's theory, are transmitted from the parents to the progeny. His most famous example, and the one that Cuvier used most often to ridicule him, was that of the giraffe which stretched its neck longer and longer to reach the leaves on the higher trees and transmitted this longer neck to its offspring, who again stretched their necks, and so on.

The second important factor in Lamarck's theory of evolution was a universal creative principle, an unconscious striving upward on the *Scala Naturae* that moved every living creature toward greater complexity. Every amoeba was on its way to man. Some might get waylaid—the orangutan, for instance, by being caught in an unfavorable environment had been diverted off its course—but the will was always present. Life in its simplest forms was constantly emerging by spontaneous generation to fill the void left at the bottom of the ladder. In Lamarck's formulation, the ladder of life of the ancients had been transformed into a steadily ascending escalator powered by a universal will.

Lamarck's concept of the inheritance of acquired characteristics is an attractive one. Darwin borrowed from it consciously and unconsciously, and in fact, successive editions of *The Origin of Species* show that he was maneuvered into this particular Lamarckian point of view by some of his critics and by his own inability (owing to the primitive state of the science of genetics at that time) to explain some of the ways in which animals changed. Lamarck's theories persisted into the twentieth century in the work of the Russian biologist Lysenko.

Lamarck's contemporaries did not object to his ideas about the inheritance of acquired characteristics, as we would today with our more advanced knowledge of the mechanics of inheritance, nor did they criticize his belief in a metaphysical force, which was actually a common element in all the theories of the time. But these vague, unprovable postulates provided a very shaky foundation for so radical a proposal. Lamarck's championship of evolution was damaging not only to his own career but to the concept of evolution itself. The result of his theories was that both scientists and the public became even less prepared for an evolutionary doctrine.

DARWIN'S THEORY

The Earth Has a History

The person who most influenced Darwin, it is generally agreed, was Charles Lyell (1797–1875), a geologist who was Darwin's senior by only ten years. One of the few books that Darwin took with him on his voyage was the first volume of Lyell's newly published *Principles of Geology*, and the second volume was sent to him while he was on the *Beagle*. On the basis of his own observations and those of his predecessors, Lyell opposed the theory of catastrophes. Instead, he produced new evidence in support of Hutton's earlier theory of uniformitarianism. According to Lyell, the slow, steady, and cumulative effect of natural forces had produced continuous change in the course of the earth's history. Since this process is demonstrably slow, its results being barely visible in a single lifetime, it must have been going on for a very long time. In his earlier works, Lyell did not discuss the biological implications of his theory, but apparently they were clear to Darwin. If the earth had a long continuous history and if no forces other than outside physical agencies were needed to explain the events as they were recorded in the geologic record, might not living organisms have had a similar history? What Darwin's theory needed, as he knew very well, was time, and it was time that Lyell gave him.

The Voyage of the Beagle

This, then, was the intellectual equipment with which Charles Darwin set sail from Devonport. As the *Beagle* moved down the Atlantic coast of South America, through the Straits of Magellan, and up the Pacific coast, Darwin traveled the interior, fished, hunted, and rode horseback. He explored the rich fossil beds of South America (with the theories of Lyell fresh in his mind) and collected specimens of the many new kinds of plant and animal life he encountered. One of the impressions brought home to him most strongly by his long, slow trip down the coast and up again was the constantly changing varieties of organisms that he encountered. The birds and other animals on the west coast, for example, were very different from those on the east coast, and even as he moved slowly up the western coast, one species would give way to another one.

Most interesting to Darwin were the animals and plants that inhabited a small barren group of islands, the Galapagos, that lie some 580 miles off the coast of Ecuador. The Galapagos were named after the islands' most striking inhabitants, the tortoises (*galápagos* in Spanish), some of which weigh as much as 200 pounds or more. Each island

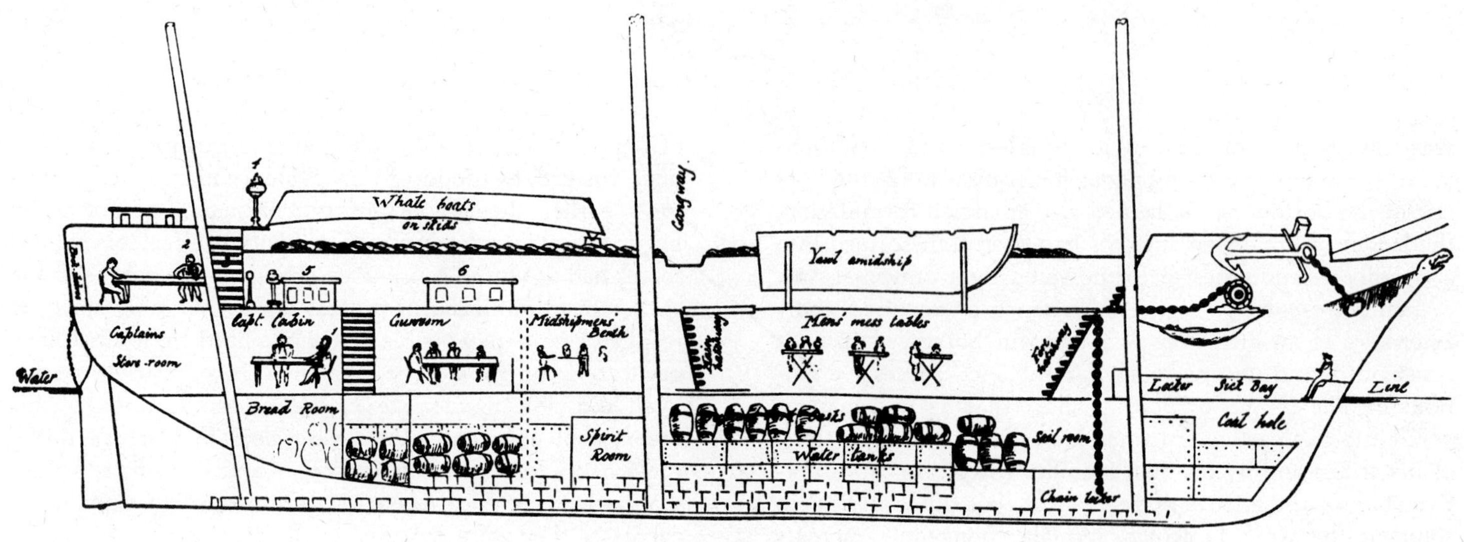

4

Cutaway view of the Beagle. Only 90 feet in length, this "good little vessel" set sail on its five-year voyage with 74 people aboard. Darwin shared the poop cabin with a midshipman and 22 chronometers belonging to Captain Fitzroy, who had a passion for exactness. His sleeping space was so confined that he had to remove a drawer from a locker to make room for his feet.

5

One of the tortoises after which the Galapagos are named. Darwin learned that the inhabitants of the Galapagos were able to distinguish among the tortoises from the different islands.

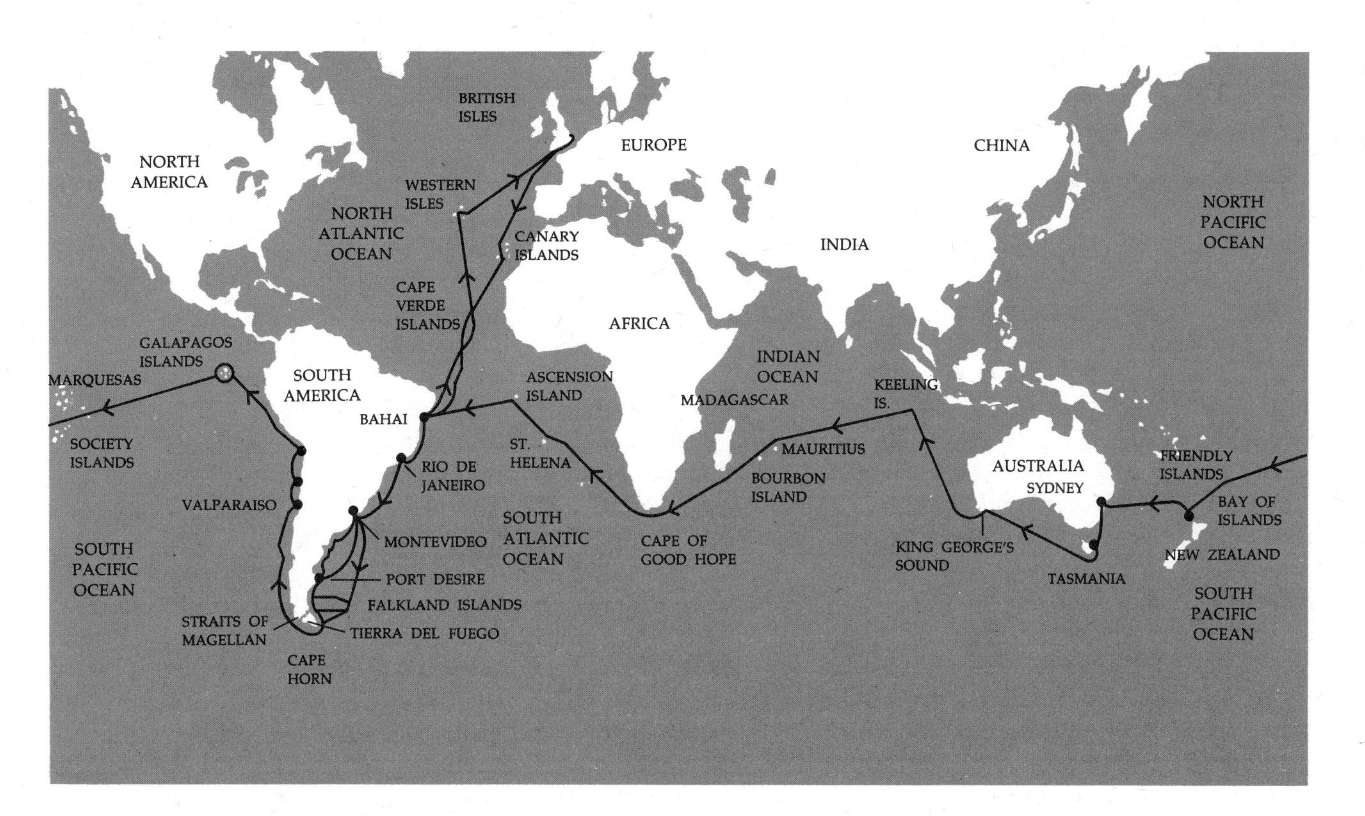

6
The Beagle's *voyage.*

has its own type of tortoise; the fishermen who frequented the islands and hunted the tortoises for food could readily tell which island any particular tortoise had come from. Then there was a group of finchlike birds, 13 species in all, which differed from one another in body size, and particularly in the type of food they ate. In fact, although still clearly finches, they had taken on many characteristics seen only in completely different types of birds on the mainland. The woodpeckerlike finch, for example, had taken on the woodpecker's role in routing insects out of the bark of trees. It is not fully equipped for this, however, lacking the long tongue with which the woodpecker flicks out insects from under the bark. In compensation, the woodpecker finch carries with it a small stick to pry the insects loose.

From his knowledge of geology, Darwin knew that these islands, clearly of volcanic origin, were much younger than the mainland. Yet the plants and animals of the islands were different from those of the mainland, and in fact, the inhabitants of different islands in the archipelago differed from one another. Were the living things on each island the product of a separate special creation? "One might really fancy," Darwin mused at a later date, "that from an original paucity of birds in this archipelago one species had been taken and modified for different ends." For years after his return, this problem continued, in his own word, to "haunt" him.

Development of the Theory

Not long after Darwin's return, he came across a book by the Reverend Thomas Malthus that had first appeared in 1798. In this book, Malthus warned, as economists have warned frequently ever since, that the human population was increasing so rapidly that it not only would soon outstrip the food supply but would leave "standing room only" on the earth. Darwin saw that this conclusion—that food supply and other factors hold populations in check —is true for all species, not just the human one. For example, a single breeding pair of elephants, which are the slowest breeders of all animals, could produce 19 million elephants in 750 years, yet the average number of elephants generally remains the same over the years. Where there might have been 19 million elephants in theory, there are, in fact, only two. The process by which the two survivors are "chosen" was termed by Darwin *natural selection*. He saw it as a process analogous to the type of selection exercised by breeders of cattle, horses, or dogs—which, as a country squire, he was very familiar with. In the case of artificial selection, man chooses variants for breeding on the basis of characteristics which seem to him to be desirable. In the case of natural selection, environmental conditions are the principal active force, operating on the variations continually produced in all species to "favor" some variants and "discourage" or eliminate others.

Where do the variations come from? According to Darwin's theory, variations occur absolutely at random. They are not produced by the environment, by a "creative force," or by the unconscious striving of the organism. In themselves, they have no direction; *direction is imposed entirely by natural selection.* A variation that gives an animal even a slight advantage makes that animal more likely to leave surviving offspring. Thus, to return to Lamarck's giraffe, an animal with a slightly longer neck has an advantage in feeding and so is apt to leave more offspring than one with a shorter neck. If the longer neck is an inherited trait, some of these offspring will also have long necks, and since the animals with longer necks are always favored, the next generation will have even longer necks. Finally, the population of short-necked giraffes will give way to a population of long-necked ones.

As you can see, the essential difference between Darwin's formulation and that of any of his predecessors is the central role he gave to the process of variation. Others had thought of variations as mere disturbances in the overall design, whereas Darwin saw that variations are the real fabric of the evolutionary process. Species arise, he saw, when differences among individuals within a group are gradually converted into differences between groups as the groups become separated in space and time.

The Origin of Species, which Darwin pondered for more than 20 years before its publication in 1859, is, in his own words, "one long argument." No experiments are performed. No new information is revealed. Fact after fact, observation after observation, culled from the remotest Pacific island to the neighbor's pasture, is recorded, analyzed, and commented upon. Every objection is anticipated and countered. Because the process of evolution is so slow, Darwin did not believe that direct proof of his theory was possible. However, as we shall see, the twentieth century has produced clear evidence of evolution in progress. No scientist now doubts that species have originated in the past and are still originating, that species have become extinct in the past and are still becoming extinct, and that all living things today have an ancestral species in the past.

The real difficulty in accepting Darwin's theory has always been that it seems to diminish man's significance. The new astronomy had made it clear that the earth is not the center of the universe or even of our own solar system. Now the new biology was asking man to accept the fact that he is the product of a random process and that, as far as science can show him, he is not created for any special purpose or as a part of any universal design. We are still dealing with this problem today.

Importance of the Theory in Modern Biology

What is the importance of this theory to modern biologists, many of whom are concerned with such areas of investigation as the chemistry of heredity—a phrase that would have been meaningless in Darwin's time—or the interpretation of subcellular structures newly revealed by the electron microscope or the tracing of a radioactive substance through the components of an ecosystem? In the words of Ernst Mayr of Harvard University:*

> The theory of evolution is quite rightly called the greatest unifying theory in biology. The diversity of organisms, similarities and differences between kinds of organisms, patterns of distribution and behavior, adaptation and interaction, all this was merely a bewildering chaos of facts until given meaning by the evolutionary theory. There is no area of biology in which that theory has not served as an ordering principle.

As Mayr indicates, because biologists now accept Darwin's theory as fact, they are able to make generalizations and connections of a sort that would otherwise be impossible. It is not necessary for them to examine every type of living organism in order to draw conclusions about structures and functions common to all living things. And, just as evolutionary theory gives meaning to the similarities among organisms, it also makes possible an understanding of the differences among them and how these differences arose.

Perhaps most important, an evolutionary perspective makes biology a much more interesting subject. Without such a unifying principle, biology can be only a descriptive science, a catalog of parts. Because of evolution, biologists may—indeed must—assume that biological structures and functions have a survival value; as you will see, the process of evolution is so inexorable that only that which is biologically useful can long survive. Hence one is permitted constantly to ask "why?". Why are flowers brightly colored? Why do dogs turn in circles before they lie down to sleep? Why are proteins manufactured in the cell only on certain structures (ribosomes), and why are these structures made the way they are? There are many questions such as these that remain unanswered—and undoubtedly many more good questions that have not yet even been asked. You may be among the ones to ask them.

* Ernst Mayr, *Animal Species and Evolution,* Harvard University Press, Cambridge, Mass., 1963.

Finally, for those of us concerned with the present environmental crisis—and who among us can afford not to be?—a knowledge of the processes of evolution helps in our understanding of the roots of our present dilemma. The interrelationships of the living things of an ecosystem are the end result of evolutionary processes which have built up a network of intricate and often, to us, invisible adaptations and dependencies. We have been late in discovering that pulling a single thread in this "web of life" can bring about changes in its entire structure.

Man, like all the other species that have originated and will originate, arose in and as a part of an ecological system. The human species, however, has now reached a point of ecological dominance without precedence in evolutionary history. Thus, of all the species, man has a power to disrupt and destroy. But also, of all the species, man has the capacity to accumulate knowledge to act with foresight. Perhaps we shall find a way to use this capacity more effectively in our relationships with the living world of which we are a part. This book is dedicated to that proposition.

QUESTIONS

1. How is a theory useful in science? Could a theory be harmful to scientific progress? Do you know of any modern theories opposed to Darwin's?

2. What is the essential difference between Darwin's theory of evolution and that of Lamarck?

3. The chief enemy of an English species of snail is the song thrush. Snails that grow on woodland floors have dark shells, whereas those that grow on grass have yellow shells, which are less clearly visible against the lighter background. Explain, in terms of Darwinian principles.

4. The phrase "chance and necessity" has been applied to the process by which species develop. Relate this to the fact that the snails growing on grass do not have green shells.

SUGGESTIONS FOR FURTHER READING

DARWIN, CHARLES: *The Origin of Species by Means of Natural Selection, or The Preservation of Favored Races in the Struggle for Life,* Doubleday & Company, Inc., Garden City, N.Y., 1960.*

Darwin's "long argument." Every student of biology should, at the very least, browse through this book to catch its special flavor and to begin to understand its extraordinary force.

DARWIN, CHARLES: *The Voyage of the Beagle,* Natural History Library, Doubleday & Company, Inc., Garden City, N.Y., 1962.*

Darwin's own chronicle of the expedition on which he made the discoveries and observations that eventually led him to his theory of evolution. The sensitive eager young Darwin that emerges from these pages is very unlike the image many of us have formed of him from his later portraits.

MOOREHEAD, ALAN: *Darwin and the Beagle,* Harper & Row, Publishers, Inc., New York, 1969.

A delightful narrative of Darwin's journey, beautifully illustrated with contemporary or near-contemporary drawings, paintings, and lithographs.

TOULMIN, STEPHEN, and JUNE GOODFIELD: *The Discovery of Time,* Harper & Row, Publishers, Inc., New York, 1965.*

The historical development of our concepts of time as they relate to nature, human nature, and human society.

* Available in paperback.

PART I

Cells

SECTION 1

The Nature of Living Things

1–1
Birth of a star. This is a nebula—a cloud of gases condensing in outer space—in the constellation Sagittarius, the archer. Our solar system had its beginnings in an event such as this about 5 billion years ago.

SECTION 1

The Nature of Living Things

We inhabit a planet on the edge of a galaxy which contains 100 billion stars. The star that is our sun was born almost 4.5 billion years ago. Like other stars, it formed from one of the swirling clouds of hydrogen gas that even today drift through interstellar space. The cloud condensed gradually; the hydrogen atoms were pulled toward one another by the force of gravity, falling into the center of the cloud and gathering speed as they fell. As the cluster grew denser, the atoms moved faster and faster. More and more atoms collided with each other, and the gas in the cloud became hotter and hotter. The collisions became increasingly violent as the temperature rose, until the atoms collided with such force that their nuclei fused, forming helium and releasing nuclear energy (in a process very similar to the explosive reaction in an H-bomb). Energy from this thermonuclear reaction, still going on at the heart of the sun, is radiated from its glowing surface. It is this energy, captured in the cells of green plants, on which all life on earth depends.

The planets, according to current theory, formed from the remaining gas and the ice and dust moving around the newborn star. At first, in the planets-to-be, particles were collected merely at random, but as each mass grew larger, other particles began to be attracted by the gravity of the central mass. The whirling dust and forming spheres continued to revolve around the sun until finally each planet had swept its own path clean, picking up loose matter like a giant snowball. The orbit nearest the sun was swept clean by Mercury, the next by Venus, the third by earth, the fourth by Mars, and so on out to Neptune. Pluto, the most distant of the nine planets, is generally believed to have started as a satellite of Neptune and to have drifted apart from the parent planet in the early days of the solar system.

THE ORIGIN OF LIFE

Why on Earth

In our solar system, earth among all the planets is most favored for the production of life. For one thing, earth is neither too close nor too distant from the sun. At very low temperatures, the chemical reactions on which life—any form of life—depends must virtually cease. At high temperatures, compounds are too unstable for life to form or survive. The planets beyond Mars—Jupiter, Saturn, Uranus, Neptune, and Pluto—are too cold. Mercury, the closest to the sun, is too hot, and so, apparently, is Venus. Earth's size and density are also greatly in its favor. Planets much smaller than earth do not have enough gravitational pull to hold an adequate atmosphere, and any planet much larger than earth may hold so dense an atmosphere that radiations from the sun cannot reach its surface. Of the eight other planets in our solar system, life can theoretically exist only on Mars and Venus. And even there, conditions seem to be too inhospitable to support any but the most primitive organisms. In the galaxy of which our solar

1–2
The planet earth, as seen from the moon. Of all the planets in this solar system, only earth is known to harbor life. All the life on earth exists on a thin film on the earth's surface known as the biosphere.

system is a part, however, the astronomers estimate that there are a billion other planets capable of supporting life. And there are uncounted billions of galaxies in the universe!

How Life Began

The earliest fossils found so far are about 3.2 billion years old. Sometime between the time earth formed and the date of these earliest fossils—an interval of more than 1 billion years—life began. The raw materials from which it was formed were found in the atmosphere of the young earth: carbon, oxygen, hydrogen, and nitrogen. These four elements, present in the gases of the early atmosphere, make up about 98 percent of the tissues of all living things. The remaining 2 percent are essential minerals, such as sulfur and phosphorus, which were present in the hardening crust of the earth's surface.

Through the thin atmosphere, the rays of the sun beat down on the harsh, bare surface of the young earth, bombarding it with light, heat, and ultraviolet radiation. Water vapor cooled in the upper atmosphere, fell on the crust of the earth as rain, and steamed up again, driven by the sun's heat. Violent rainstorms, accompanied by lightning, released electrical energy. Radioactive substances in the earth's crust emitted their energy, and molten rock and boiling water erupted from beneath the earth's surface. The energy in this vast crucible broke apart the simple gases of the atmosphere and reformed them into more complicated molecules. In the laboratory, under conditions simulating those of the earth's primitive atmosphere, it has been shown that such molecules—amino acids (the building blocks of proteins) and other compounds associated with living things—do form when gases are exposed to an energy source such as ultraviolet rays or electric discharges.

As biochemists reconstruct these events, the compounds that were formed in the atmosphere tended to be washed out by the driving rains and to collect in the oceans, which grew larger as the earth cooled. As a consequence, the ocean became an increasingly rich mixture of organic molecules. Some organic molecules have a tendency to aggregate in groups; in the primitive ocean, these groups probably took the form of droplets, similar to the droplets formed by oil in water. Such droplets of organic molecules appear to have been the forerunners of primitive cells, the first forms of life.

Life and Energy

The energy that was required to produce the first organic molecules came, then, from the heat, ultraviolet rays, and electrical disturbances that abounded in the primitive at-

1–3
Bolts of lightning in the steam boiling up from a volcanic crater. Such sources of energy would have been present on the primitive earth and might have contributed to the formation of organic molecules. This photograph, taken in 1963, shows the birth of the island of Surtsey off the coast of Iceland.

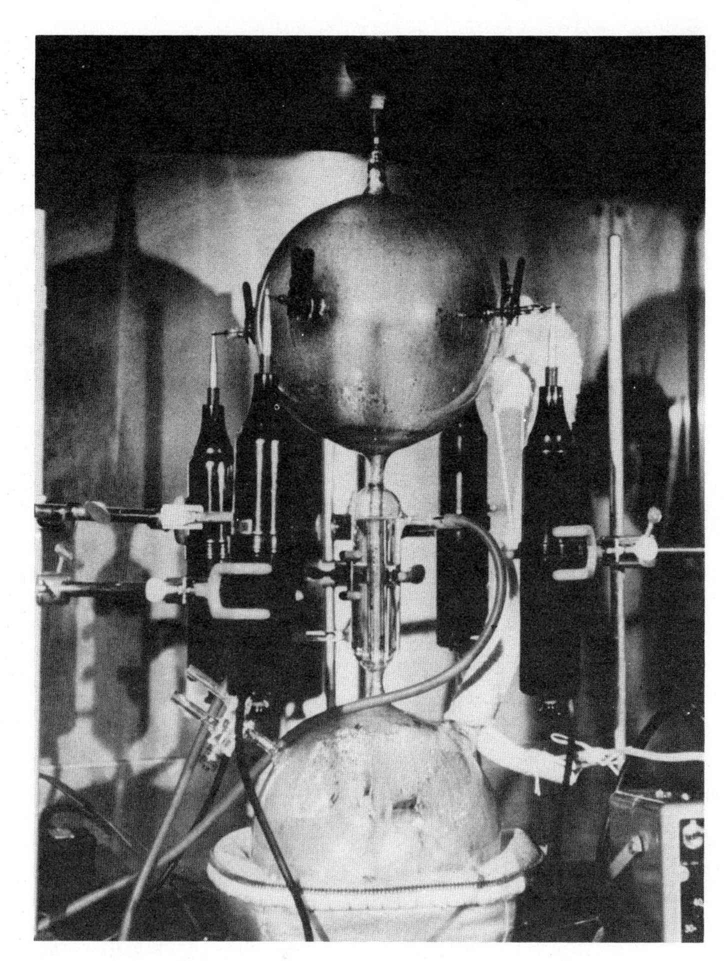

1–4
Conditions believed to exist on the primitive earth are simulated in this apparatus. Methane and ammonia are continuously circulated between a lower "ocean," which is heated, and an upper "atmosphere," through which an electric discharge is transmitted. At the end of 24 hours, 95 percent of the methane is converted to amino acids, the building blocks of proteins, and other organic molecules.

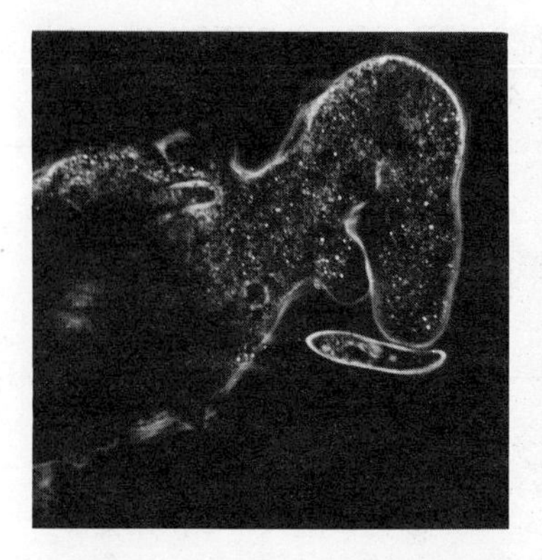
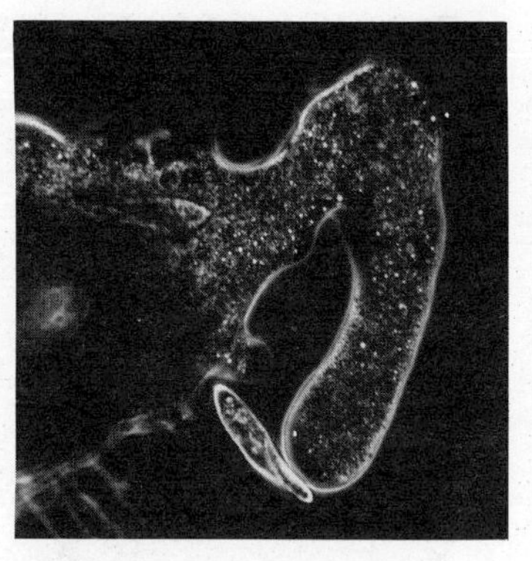
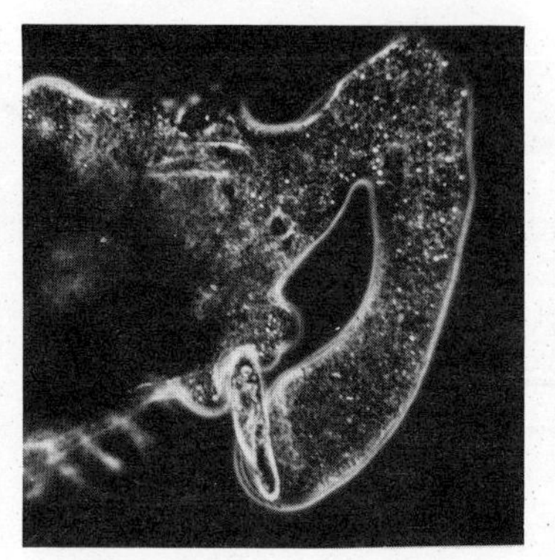
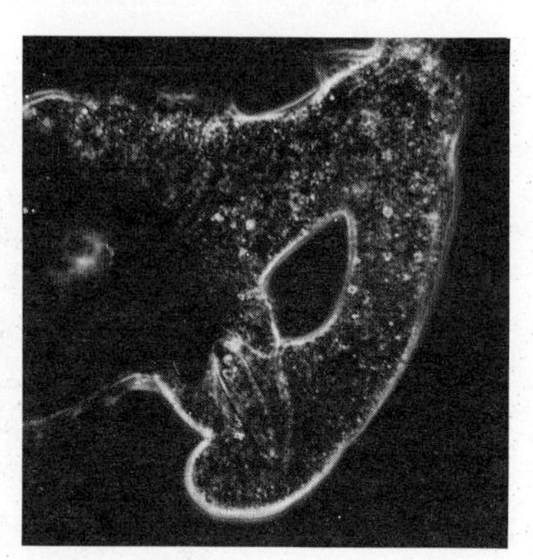

1–5
A modern heterotroph, the giant amoeba Chaos chaos, *captures a paramecium. Although seemingly disorganized, as its name would indicate,* Chaos chaos *is able to sense its prey, move toward it, and send out a pseudopod ("false foot") of the right size and shape to envelop it.*

mosphere. And these organic molecules became, in turn, the source of energy for the earliest forms of life. The primitive cells or cell-like structures were able to use these compounds, which were abundant in the "primordial soup," to satisfy their energy requirements, just as we use sugars, starches, proteins, and fats to satisfy our energy requirements.

These early cells were *heterotrophs*, a category of organisms which today includes all living things classified as animals or fungi and many of the one-celled organisms, the protists. *Hetero* comes from the Greek word meaning "other," and *troph* comes from *trophos*, "one that feeds." A heterotrophic organism is one that is dependent on others—that is, on an outside source of organic molecules—for its energy. Modern heterotrophs, such as ourselves, get their energy-supplying molecules from other organisms. We break these molecules down within our cells and use their energy. We can also release useful energy from organic molecules by burning them in the form of coal or oil or wood.

As the primitive heterotrophs increased in number, they began to use up the complex molecules on which their existence depended—and which had taken millions of years to accumulate. Organic molecules in free solution (not inside a cell) became more and more scarce. Competition began. Under the pressure of this competition, cells that could make efficient use of the limited energy sources now available were more likely to survive than cells that could not. In the course of time, by the long, slow process of weeding out the less fit, cells evolved that were able to make their own energy-rich molecules out of simple materials. Such cells are called *autotrophs*, "self-feeders." Without the evolution of autotrophs, life on earth would soon have come to an end.

The molecules on which the earliest heterotrophs depended were first formed in the atmosphere by the energy of the sun and later incorporated by the heterotrophs. The most successful of the autotrophs were those which evolved a process for making use directly of the sun's energy. This process is *photosynthesis*. Photosynthesis is the means by which the radiant energy of the sun is transformed into an energy source for both autotrophs and modern heterotrophs. With only very minor and unimportant exceptions, all the chemical energy that powers living systems today enters the living world through the process of photosynthesis.

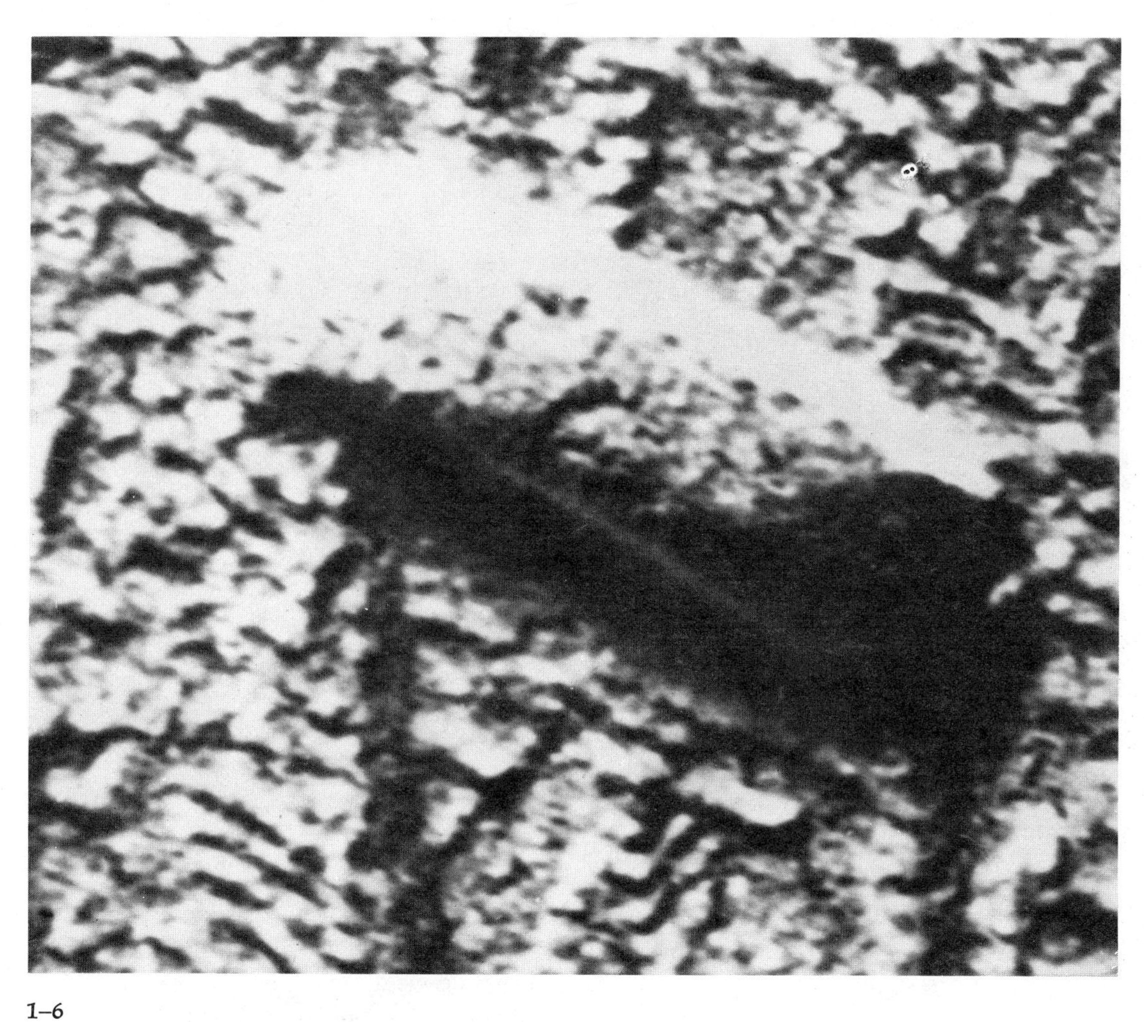

1–6

A fossil of the oldest known bacterium, which lived some 3.2 billion years ago. Eobacterium isolatum *("the isolated dawn bacterium"), as it is called, appears as a raised capsule shape on this microscopic fragment of ancient rock. The magnification in this electron micrograph is 260,000 times.*

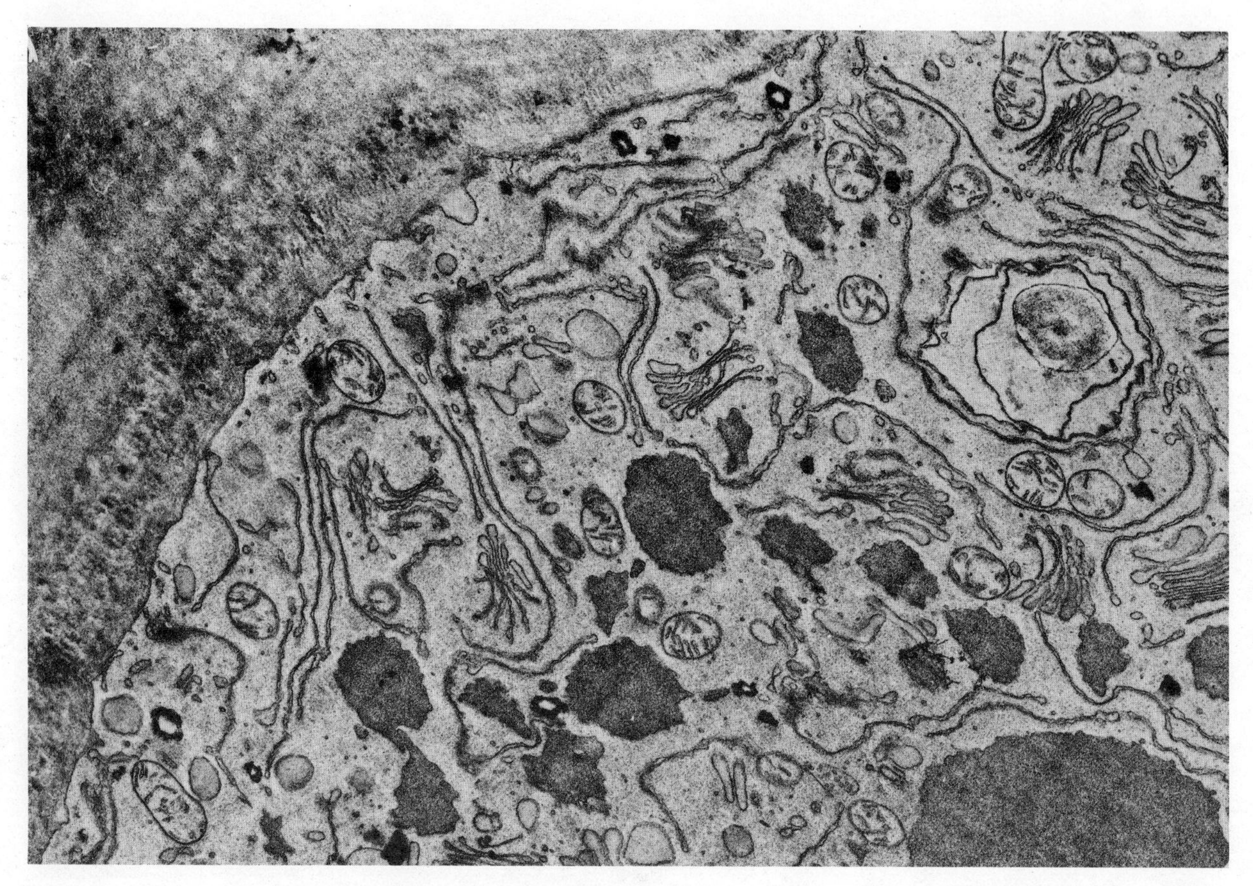

1–7
Living things characteristically have a complex and highly organized structure. This is a "simple" cell from a barley root as seen by the electron microscope. The cell wall is in the upper left corner of the picture. Notice the many different kinds of organelles, the sites of the cell's varied activities.

WHAT IS LIFE?

What do we mean when we speak of "living"? How do we recognize whether or not something is alive? These may seem like abstract questions, but actually they are even now being weighed not only by scientists interested in the origin of life some 3 billion years past but by those concerned with the problem of whether or not life now exists on the other planets of our solar system. None of these scientists has provided us with any single, simple way to draw a sharp line between the living and the nonliving, but they have defined the special characteristics which, taken together, identify living matter—the symptoms by which we know that life is present.*

The first characteristic of living things is that they are organized in specific structural and functional patterns, which are not encountered elsewhere. Even the earliest living things had definite sizes and structures, which is the chief way in which they can be recognized in ancient microfossils. More modern cells are enormously complex, with a nucleus and many smaller structures, or organelles, each of which has a specific function. Modern plants and animals are made up of millions and even billions of such cells organized into tissues, organs, and systems, each geared to carry out a particular function. Organization is one of the most important properties by which living matter can be identified.

The second characteristic is closely related to the first: living systems maintain a chemical composition quite different from that of their surroundings. The atoms present in living matter are the same as those in their environment, but they are present in different proportions and arranged in different ways. And although living systems constantly exchange certain materials with their environment—any single atom may be present only briefly in a living structure—they maintain this characteristic chemical composition. This important property is called *homeostasis*, "staying the same."

Third, living things have the capacity to take in energy from their environment and to transform and use it to make and remake molecules and to form structures. Many transform the chemical energy to kinetic energy, the energy of motion. Some use it to send electrical signals, and some even convert it to light energy and glow in the dark!

Fourth, living systems can respond to stimuli. Bacteria move toward or away from particular chemicals; green plants bend toward light; mealworms congregate where it is damp; cats pounce on small moving objects. This prop-

* This listing is borrowed, in part, from Albert Lehninger's new text, *Biochemistry*, Worth Publishers, Inc., New York, 1970.

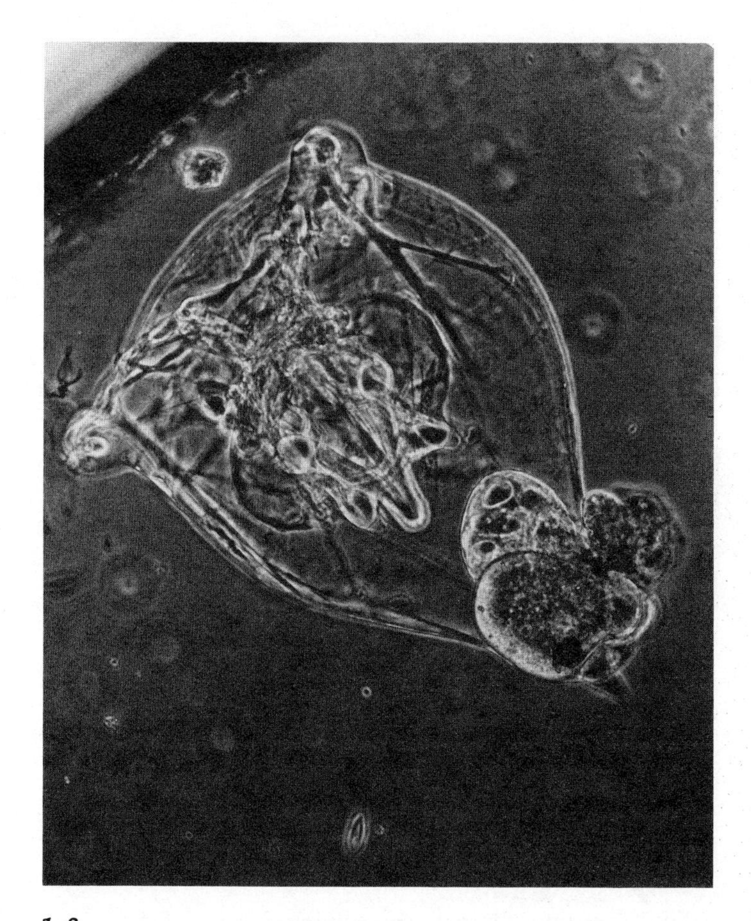

1–8
Living things, although they exchange materials with their environment, maintain a composition quite unlike that of their surroundings. Even this tiny, apparently fragile animal, a rotifer, is distinctly different from the water in which it lives.

1–9
Living things respond to stimuli. Here a seedling turns toward the light.

erty is generally known as *irritability*, and although different organisms respond to widely varying stimuli, the capacity to respond is a fundamental and almost universal characteristic of life.

Fifth, and most remarkably, living things have the capacity to reproduce themselves with astonishing fidelity (and yet, as we shall see, with just enough variation to provide the material on which evolution depends). This reproduction process has been going on now for more than 3 billion years. All organisms living today can trace their ancestry back to the first, primitive cells. (We shall be presenting evidence to support this statement in the course of this text.)

Finally, living things have what the French Nobel laureate Jacques Monod chooses to call teleonomy,* which is perhaps best defined as "purposefulness." By purposefulness, we do not mean that each living thing has a sense of purpose; an individual giraffe does not grow its neck longer in order to reach the high branches, nor does a flower emit a sweet odor in order to attract bees, nor does a pancreas cell secrete insulin in order to mobilize sugar. Yet giraffes do have long necks, flowers do emit sweet scents, and the pancreatic secretion does mobilize sugar, and the reason they do these particular things is that, in the course of evolution, these activities and the structures that make them possible have proved to have survival value for the individual and the species. As a consequence, it is appropriate to ask questions about the function of any structure or activity of any living thing in terms of its usefulness; what is this for, we can appropriately inquire about even the smallest molecule, the most microscopic structure.

Of course, this appearance of having been designed for a purpose is a property that is shared by man-made objects. Every component of a man-made object, even one that is merely decorative or amusing, has a purpose. Moreover, it is at least theoretically possible to design machines that possess the other properties which we think of as belonging to living things—a spaceship, for example, that runs on solar energy, replaces its component parts using materials from the atmosphere, and perhaps even reproduces itself! However, here it is easy to make an essential distinction: The information required to construct such a machine must come from outside the machine, that is, from the mind of man. The information by which living things organize their "purposeful" structures and functions, exchange materials with their environment, transform energy, respond to stimuli, and reproduce is all contained within the organisms themselves.

THE VITAL FORCE

How is it that living systems have these extraordinary properties, not truly shared by inanimate objects? In the seventeenth century, two entirely different answers to this question were proposed. One group of philosophers, called the vitalists, contended that living systems contain a vital

* Jaques Monod, *Chance and Necessity*, Alfred A. Knopf, Inc., New York, 1971.

force, a force that cannot be duplicated or even comprehended by man. Further, they believed that there are two entirely different kinds of chemistry: the chemistry of living things (organic chemistry) and the chemistry of the nonliving world (inorganic chemistry). In other words, according to the vitalists, the materials of which living organisms are composed do not conform to the same natural "laws" as those which govern the materials making up nonliving systems. The opposing group, the mechanists, believed that living organisms are essentially machines, which operate by the same mechanical, physical, and chemical laws as nonliving systems.

Today it is clear that organic chemistry "obeys the rules" of inorganic chemistry. The difference between a collection of chemicals and a living organism, with all its remarkable capacities, is not a matter of materials but one of organization. When we move from one level of organization to another, we move to something qualitatively different.

Consider, for example, protons, neutrons, and electrons. In isolation, they have certain well-defined physical properties. If they are put together—that is, organized—in a particular way, they form a carbon atom, which has quite different properties from those of any of the subatomic units that make it up. Put together in another way, and in different proportions, these same units form oxygen or uranium or any one of more than a hundred known elements. A proton represents one level of organization. An atom represents a second level.

At a third level of organization, atoms are grouped together into molecules. Hydrogen atoms and oxygen atoms combined form water molecules, and water molecules have properties that are entirely different from those of hydrogen and oxygen. Further, the behavior of a collection of molecules is different from the sum total of the activities of the separate molecules. At what level do those properties we recognize as living first appear? This is a surprisingly easy question to answer. At the cellular level of organization. A single cell exhibits all the properties that we have taken to describe living things.

First of all, a cell is highly organized in its structure and function. Second, although it contains the same chemical elements as are found in the external environment, it contains them in different proportions and grouped in different ways. It exchanges materials with its environment, yet maintains its highly characteristic chemical identity. Third, the important energy exchanges which provide the power sources for living organisms take place, as you will see in Chapters 2–5 and 2–6, at the level of the individual cell. Fourth, many cells, including single-celled organisms, special pigment-containing cells of plants, and nerve cells of animals, have remarkable capacities for responding in very precise and useful (to the organism) ways to specific stimuli. Fifth, cells reproduce themselves, and even the reproduction of many-celled organisms takes place through the agency of individual cells—the sperm and the egg. Finally, each component of the individual cell has a specific purpose or function, as does each of the individual cells of the many-celled organism.

We are going to begin this book, therefore, where life begins—with the single cell. We shall explore first the cell's component parts and their functions, and we shall conclude, in Section 3, with an investigation of the way in which the cell stores its essential information and transmits it to other cells in the course of reproduction. Part II of the book will be concerned with another level of organization, the multicellular plants and animals, and with the special properties, characteristics, and capacities that emerge at this level. Part III will examine the ways in which organisms are organized into populations and communities and how these populations and communities shape and are shaped by the forces of ecology and evolution.

SUMMARY

The sun and its planets were formed 4.5 billion years ago—the sun probably from the condensation and contraction of a hydrogen gas cloud, and the planets as accumulations of interstellar debris. Of the nine planets in this solar system, only earth is known to support life, but there are probably other planets in the galaxy with some form of life.

The primitive atmosphere of earth held the chief raw materials of living matter—hydrogen, oxygen, carbon, and nitrogen—combined in water vapor and gases. The energy required to break apart the simple gases in the atmosphere and re-form them into more complex molecules was supplied by heat, lightning, and high-energy radiation from the sun. Laboratory experiments have shown that the types of molecules that are characteristic of living organisms—that is, organic molecules—can be formed under such conditions. These molecules gradually accumulated to form cells, the first living things.

The earliest cells were probably heterotrophs, that is, organisms which depend on outside sources of organic molecules for their energy. Organisms that can use the sun's energy directly to make organic molecules are known as autotrophs. The most familiar of the autotrophs are the green plants.

Living systems are characterized by their possession of certain properties:

1. Living matter is highly organized and so is much more complex in its structure than nonliving matter.
2. Living matter has a chemical composition quite different in its proportions from that of its surroundings. Although it exchanges materials with its surroundings, it maintains this characteristic chemical composition.
3. Living things extract energy from their environment and use it to maintain their organization and their characteristic composition and also to move, reproduce, and carry out other functions. The ultimate source of this energy for almost all living systems is the sun.
4. Organisms respond to stimuli. This capacity is known as irritability.
5. Organisms reproduce themselves. Through this process of self-replication, modern organisms are linked with those that first appeared more than 3 billion years ago.
6. Organisms exhibit an apparent sense of purpose, and each component part and each activity has a specific survival value to the organism. This characteristic is a result of the process of evolution.

Modern biology holds that living systems are composed of the same types of substances as nonliving systems and that they obey the same "rules" of physics and chemistry. The difference between living and nonliving systems lies in their degree of organization. The first level of organization at which characteristics clearly associated with living systems appear is that of the cell.

QUESTIONS

1. Would you consider yourself a mechanist or a vitalist? Before you answer this question, think of how you define emotional states, such as love or hate, or abstract attributes, such as genius or creativity.

2. How would you define a law in science? How do natural laws differ from man-made, or social, laws?

3. In the case of microfossils, it is sometimes difficult to be sure that the specimen is not a flaw or bubble in the ancient rock. What criteria would you use in deciding that the specimen did, in fact, represent the remains of something once living?

GLOSSARY

AUTOTROPH (**auto**-trowf) [Gk. *autos,* self, + *trophos,* feeder]: An organism that is able to synthesize the nutritive substances it requires from inorganic substances.

HETEROTROPH (**het**-ero-trowf) [Gk *heteros,* other, + *trophos,* feeder]: An organism which cannot manufacture organic compounds and so must feed on complex organic food materials that have originated in other plants and animals.

HOMEOSTASIS (home-e-o-**stay**-sis) [Gk. *homos,* same, + *stasis,* standing]: The maintaining of a relatively stable internal physiological environment or equilibrium in an organism, population, or ecosystem.

IRRITABILITY: The ability to respond to stimuli or changes in environment; a general property of all living organisms.

PHOTOSYNTHESIS (photo-**sin**-thus-sis) [Gk. *photos,* light, + *syn,* together, + *tithenai,* to place]: The conversion of light energy to chemical energy; the production of carbohydrate from carbon dioxide in the presence of chlorophyll, using light energy.

SUGGESTIONS FOR FURTHER READING

HENDERSON, LAWRENCE J.: *The Fitness of the Environment,* Beacon Press, Boston, 1958.*

This book, first published in 1913, still wins the admiration of modern scientists because of the "great questions" that it dares to ask.

JASTROW, ROBERT: *Red Giants and White Dwarfs,* Harper & Row, Publishers, Inc., New York, 1967.

A short, handsomely illustrated book, covering "the evolution of stars, planets, and life."

OPARIN, A. I.: *Origin of Life,* Dover Publications, Inc., New York, 1938.*

Oparin, a Russian biochemist, was the first to argue that life arose spontaneously in the oceans of the primitive earth. Although his concepts have been somewhat modified in detail, they form the basis for the present scientific theories on the origin of living things.

SHERRINGTON, CHARLES S.: *Man on His Nature,* Cambridge University Press, New York, 1951.

"Life in Little," Chapter 3 of this delightful book, is an exquisite exposition of the living state. Sherrington's method of approach is to contrast his essentially modern view of life with that of an outstanding seventeenth-century vitalist.

* Available in paperback.

SECTION 2

The Cell

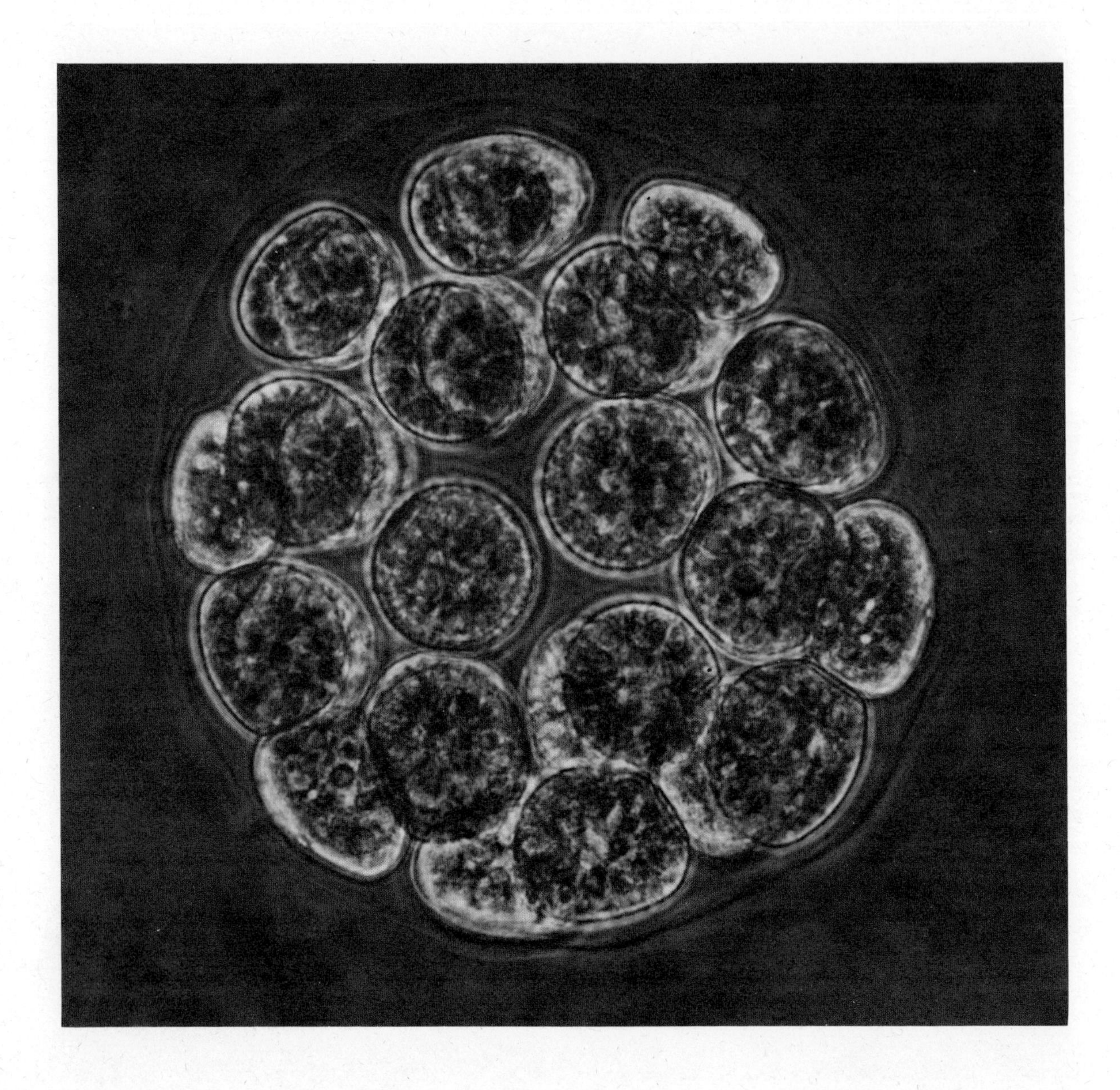

2–1
This simple organism, called Pandorina, *is made up of 32 cells, most of which are visible here, held together by a jellylike substance, whose outlines you can also see. Each of these cells is photosynthetic and can survive independent of the others. Each cell has two flagella, none of which is visible in this picture, probably because the flagella move too quickly. The flagella all point outward and beat in synchrony, so that the colony rolls through the water like a ball. To reproduce, each cell divides, producing a new cell inside, and then the parent colony breaks apart.*

CHAPTER 2–1

The Many Kinds of Cells

The concept that the cell is the basic functional unit of all living things is one of the great unifying ideas of biology. The cell doctrine, as it is called, is a fairly recent development. Individual cells were first observed in the seventeenth century by amateur microscopists. One of these, Antony van Leeuwenhoek, focused his homemade lenses on a drop of water, on the contents of his own mouth, and on his feces and found them teeming with life—"many thousands of living creatures, seen all alive in a drop of water. . . . This was for me, among all the marvels that I have discovered in nature, the most marvelous of all." But for Leeuwenhoek, these marvelous creatures were simply small animals. It was not until whole living things were dissected and looked at under the microscope that the true role of cells in nature could be understood.

Another of the seventeenth-century microscopists, Robert Hooke, noticed that cork and other plant tissues are made up of small cavities separated by walls. He called these cavities cells, meaning "little rooms." The word did not take on its present meaning, however, until 1839, when the German biologist Theodor Schwann proposed that all living organisms are composed of cells. For Hooke, cells were empty containers; for Schwann and the biologists who have followed him, cells are the building blocks of living organisms. Organisms are composed either of one cell or of many cells, and there is no life apart from the life of cells.

In 1858, the cell doctrine took on a broader significance when the great pathologist Rudolf Virchow generalized that cells can arise only from preexisting cells: "Where a cell exists, there must have been a preexisting cell, just as the animal arises only from an animal and the plant only from a plant. . . . Throughout the whole series of living forms, whether entire animal or plant organisms or their component parts, there rules an eternal law of continuous development."

WHAT IS A CELL?

Cells are units of living matter, including and surrounded by an outer membrane. Largely because of the special properties of this membrane, cells maintain a chemical composition different from that of the medium in which they live, such as blood or lymph in the case of animal cells, sap in the case of plant cells, or fresh or salt water. The living material in the cell, sometimes called the *protoplasm*, consists of the cytoplasm and the nucleus. In the *cytoplasm*, under a high-powered microscope, a variety of small structures—the *organelles* ("little organs")—can be seen. The *nucleus* in all except some very primitive cells (bacteria and blue-green algae) is surrounded by another membrane, the nuclear envelope. The material within the nuclear envelope is called the *nucleoplasm*. All cells except the bacteria and blue-green algae contain these same basic structures.

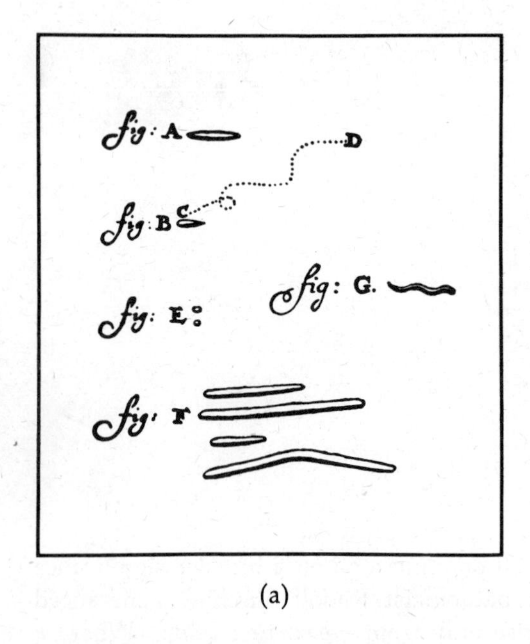

(a)

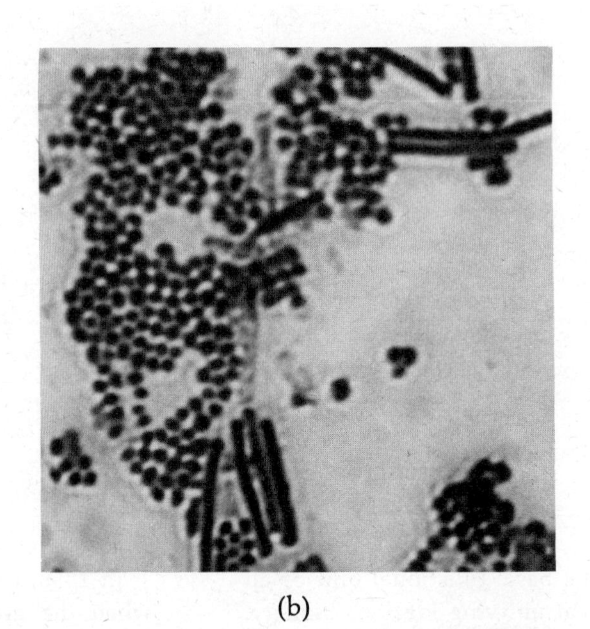

(b)

2–2
Antony van Leeuwenhoek (1632–1723), a cloth merchant, made the first microscope lenses with enough magnification to show individual cells. He spent most of his leisure time at microscopy, and his observations include the first visual records we have of protozoans, yeasts, bacteria, red blood cells, and sperm cells. Leeuwenhoek's drawing, made in 1683, of bacteria found in his mouth is shown on the left (a). *To the right is a modern photomicrograph of some of the same organisms* (b).

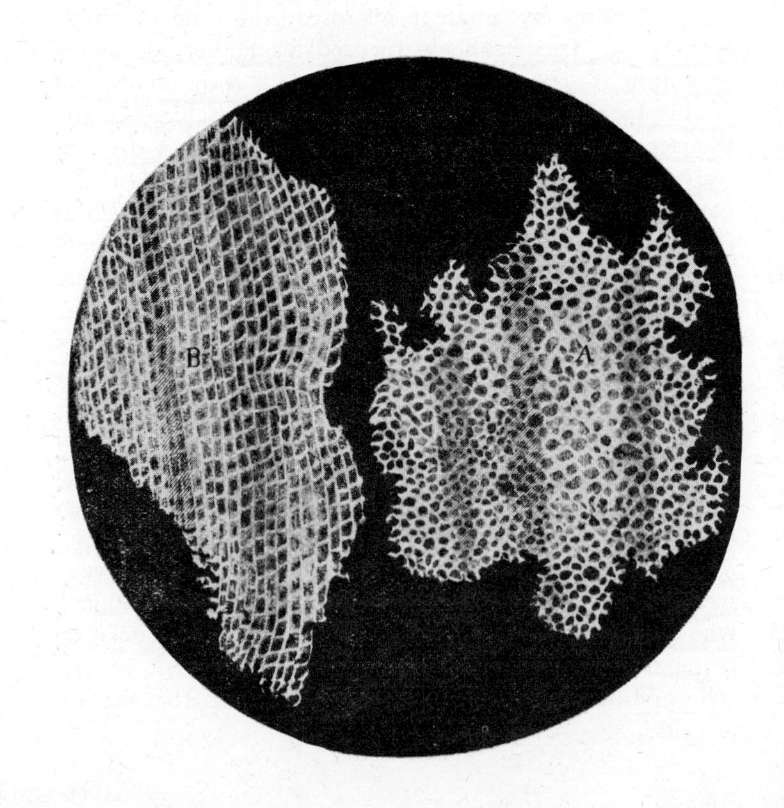

2–3
Robert Hooke's drawings of two slices of a piece of cork, reproduced from his Micrographia (1665). *Hooke was the first to use the word "cells" to describe these tiny compartments into which living organisms are organized.*

Cells are functional units as well as structural ones. Each cell, in at least some stage of its existence, carries out its own energy transformations. In the case of the green plant cell, the energy enters the cell in the form of light; energy enters other, nonphotosynthetic cells in chemical forms, often as a type of sugar. Using this energy, a cell is able to maintain and repair itself, to make a variety of complex molecules, such as enzymes, by which it carries out its particular functions, and to grow and divide.

Sizes of Cells

Cells vary greatly in size. The unicellular alga *Acetabularia* is usually 1 to 2 inches long. Egg cells, which contain large amounts of storage material, may be several inches in diameter. The majority of cells, however, can be seen only with the aid of a microscope. The units of measurement generally used in microscopy are microns, nanometers, and angstroms:

1 centimeter (cm) = 1/100 meter = 0.4 inch
1 millimeter (mm) = 1/1,000 meter = 1/10 cm
1 micron (μ)* = 1/1,000,000 meter = 1/10,000 cm
1 nanometer (nm) = 1,000,000,000 meter = 1/10,000,000 cm
1 angstrom (Å) = 1/10,000,000,000 meter = 1/100,000,000 cm

or,

$$10^{10}\ \text{Å} = 10^{9}\ \text{nm} = 10^{6}\ \mu = 10^{3}\ \text{mm} = 10^{2}\ \text{cm} = 1\ \text{meter}$$

The human egg cell, or ovum, is 1/10 millimeter in diameter, right on the edge of visibility for the unaided eye. Most plant cells are about half this diameter. Animal cells are generally smaller, typically only about 1/100 millimeter, or 10 microns, in diameter. The largest bacterial cells are about 5 microns in length, and the smallest are beyond the limits of light resolution.

* μ, the Greek letter corresponding to our m, is pronounced "mew."

The largest of all cells, the ostrich egg, is about 500,000 times longer than the smallest cell, a bacterium. These two extremes apparently represent a maximum and a minimum. Within this range, the size of each particular kind of cell seems to be the result of a compromise; on the one hand, the cell must be large enough to contain the structures and materials it requires to maintain its "life-style," and on the other hand, it must be small enough for materials to move readily between its surface and the inner areas of its cytoplasm. Gases such as oxygen and carbon dioxide move in and out of cells by diffusion, and as we shall see in the following chapter, diffusion is efficient only over small distances. Some cells that are large, such as the giant amoeba *Chaos chaos*, are very thin, thus the distance from the membrane to the innermost point of the cytoplasm is small.

Maximum size is also determined by the need of cells to have the greatest possible surface area across which to exchange materials with the environment. Figure 2–4 shows why, as volume increases, surface area decreases proportionately. Therefore, the smaller the cell, the greater will be its surface area in proportion to its size. As you will see in Section 6, when the egg cell begins dividing to form an embryo, it divides several times before any significant growth—that is, formation of new protoplasm—takes place; this has the effect of reducing the volume of each additional cell to a more efficient size.

Kinds of Cells

There is a great diversity among cells. Some cells, such as *Chaos chaos* and *Paramecium*, are separate living organisms which can carry out for themselves all the functions and activities necessary for their own survival and for the survival of their species. Such one-celled organisms are known as *protists*.* Protists that can photosynthesize are called *algae*, and those that cannot are called *protozoans*

* In this text we follow the classification system proposed by R. H. Whittaker (Science, **163**: 150–160, 1969). It is described further in Section 4.

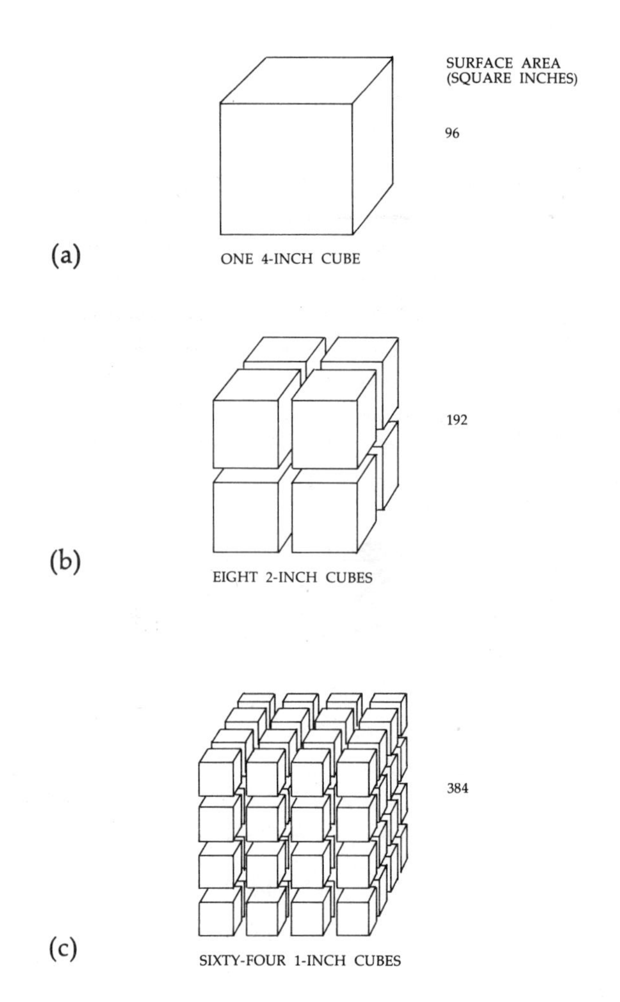

2–4

How surface area increases proportionately with decreasing volume. The total volume of each of the sets of cubes in (b) *and* (c) *is the same as that of the 4-inch cube at the top, but the total surface area of* (b) *is twice as great and that of* (c) *is four times greater.*

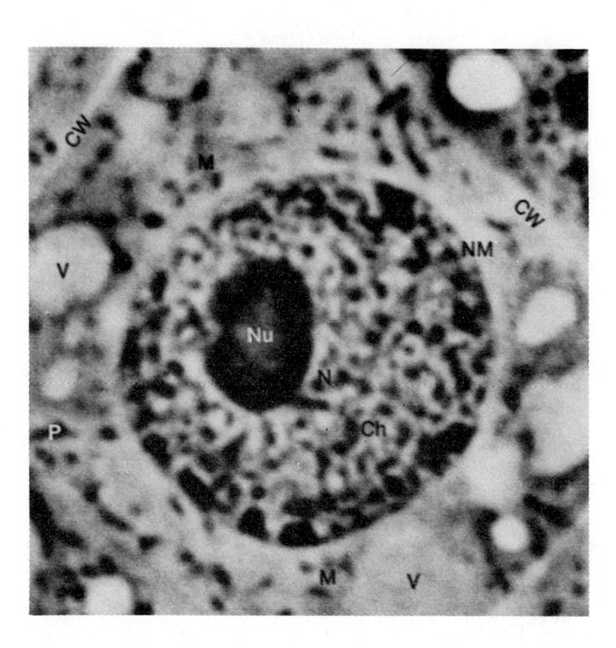

Cell as seen by the light microscope. The large central mass (N) *is the nucleus and the dark body within it* (Nu)*, the nucleolus. This cell is magnified 645 times.*

VIEWING THE MICROSCOPIC WORLD

The unaided human eye has a resolving power of about 1/10 millimeter, or 100 microns. This means that if you look at two lines that are less than 100 microns apart, they merge into a single line. Similarly, two dots less than 100 microns apart look like a single blurry dot. To separate structures closer than this, optical instruments such as microscopes are used. The best light microscope has a resolving power of 0.2 micron, or 200 nanometers, or 2,000 angstroms, and so improves on the naked eye about 500 times. It is theoretically impossible to build a light microscope that will do better than this.

Notice that resolving power and magnification are two different things; if you take a picture through the best light microscope of two lines that are less than 0.2 micron, or 200 nanometers, apart, you can enlarge that photograph indefinitely but the two lines will continue to blur together. By using more powerful lenses, you can increase magnification, but this will not improve resolution.

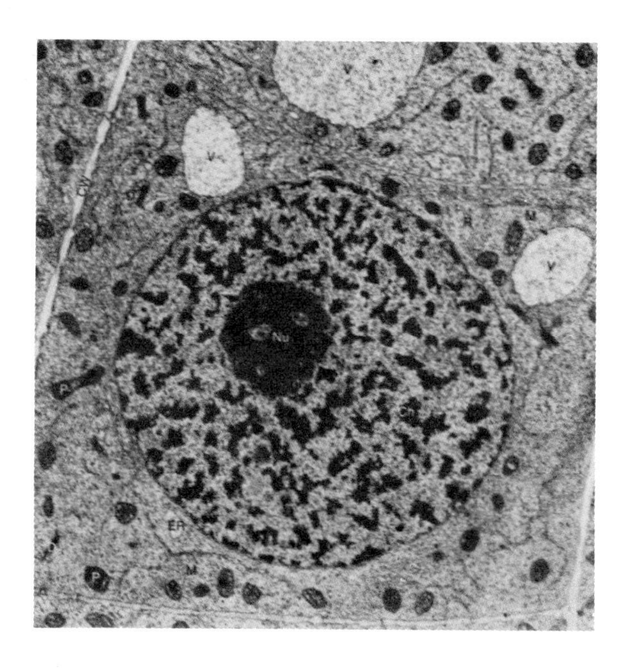

Cell at same magnification in an electron micrograph. Notice the difference in resolution.

Head of a fruit fly as recorded by the scanning electron microscope and magnified 75 times.

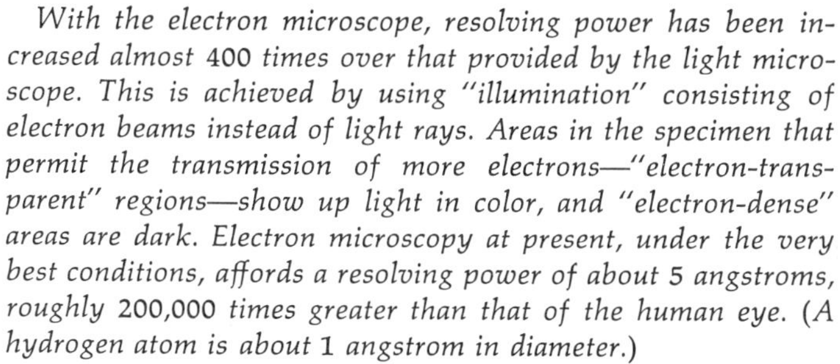

With the electron microscope, resolving power has been increased almost 400 times over that provided by the light microscope. This is achieved by using "illumination" consisting of electron beams instead of light rays. Areas in the specimen that permit the transmission of more electrons—"electron-transparent" regions—show up light in color, and "electron-dense" areas are dark. Electron microscopy at present, under the very best conditions, affords a resolving power of about 5 angstroms, roughly 200,000 times greater than that of the human eye. (A hydrogen atom is about 1 angstrom in diameter.)

Electrons, which have a very small mass, can pass through specimens only when the specimens are exceedingly thin. As a consequence, living materials have to be killed, fixed, embedded in hard materials, and sliced by special cutting instruments before they can be examined by transmission electron microscopy. Thus only dead cells can be studied. In the earlier days of electron microscopy, it was difficult to determine whether or not the specimens had been changed by the preparation methods.

Recently, electron microscopy has been expanded in its scope by the development of the scanning electron microscope. In scanning electron microscopy, the electrons whose imprints are recorded come from the surface of the specimen. The electron beam is focused into a fine probe, which is used to scan the specimen. Complete scanning takes a few seconds to a few minutes. As a result of the electron bombardment from the probe, the specimen emits low-energy secondary electrons. Variations in the surface of the specimen alter the number of secondary electrons emitted. Holes and fissures appear dark, and knobs and ridges are light. The secondary electrons are collected, amplified, and transmitted to a screen, which is scanned in synchrony with the electron probe. (The pictures of the surface of the moon transmitted back to earth from the Apollo flights were made in an analogous way.)

It is possible to examine living tissue by this technique, but the best resolution is obtained when the specimens are coated with a very thin layer of metal.

("first animals"). Although protists are single cells, they are not "simple." They have been evolving for millenia, and each species is exquisitely adapted to its small corner of the environment. Protists contain a variety of special structures which are found nowhere else in nature.

Other cells make up the tissues of multicellular organisms, including the fungi, the plants, and the animals. Such cells are called *somatic* ("body") cells. The fact that tissues are made up of separate functional units instead of being continuous is an advantage to the organism in terms of its survival. Cellular tissues are better able to withstand injury and to repair themselves than continuous tissue would be. For example, even when as much as two-thirds of a human liver is removed surgically, it can regenerate fully if the remaining tissue is healthy, simply by the production of new liver cells. Similarly, if a cut is made in the human skin, cells will migrate from nearby areas to fill in the wound. Moreover, the cellular type of organization means that, in the course of development, certain cells can differentiate—become different from one another—and specialize. Somatic cells often are specialists, in the sense that an engineer that designs only suspension bridges or a chef that makes only salads is a specialist. Like specialists in human societies, these cells function in coordination with other cells. Groups of similar cells make up *tissues,* and groups of tissues organized so that they carry out a common function are known as *organs*. Such specialist cells are dependent upon other cells for their existence, just as an engineer in our society is dependent upon the farmer who grows his food and on the mechanic who repairs his automobile.

Similarities among Cells

Despite the fact that somatic cells differ in many ways from each other and from single-celled organisms, there are remarkable similarities among all cells. These similarities are the subject of all the following chapters in this section. The most striking similarity, as you will find, concerns the chemical composition of cells. Although you may think of chemicals as lifeless, you will see that these "lifeless" chemicals actually make up the very fabric of living material. The revelation of the intricate details of life at the molecular level has been the great triumph of twentieth-century biology. Cells are also very similar in their structures, the membranes and organelles by which they organize and compartmentalize their activities. Finally, as we shall see at the end of the section, cells have similar needs for energy, and the same pathways for energy exchanges and transformations are found throughout the cellular world. The striking resemblances between cells of very different types—between an amoeba, for example, and a white cell from your own bloodstream—seem surprising at first. However, when we remember that all cells, regardless of their immediate origin, are linked together by the long thread of evolutionary history, it is easy to understand why each is, in essence, merely one variation of a basic, universal theme.

In the chapters that follow, we are going to focus on the similarities among cells in terms of what they are made of, how they are organized, and what they must do to stay alive. But first, as an introduction to the cell itself, let us look at some representative cells, beginning with two one-celled organisms, such as those that Leeuwenhoek saw, and concluding with one cell from a plant body and one from an animal.

A SINGLE-CELLED PLANT: *CHLAMYDOMONAS*

If you examine a drop of water from the sunlit surface of a pond or the bright corner of a home aquarium, you will be likely to encounter a number of different kinds of cells. One type will probably be *Chlamydomonas.** *Chlamydomonas,* which is only a single cell, is so small that you will have difficulty seeing it clearly under an ordinary micro-

* Pronounced "clammy-doe-moan-us."

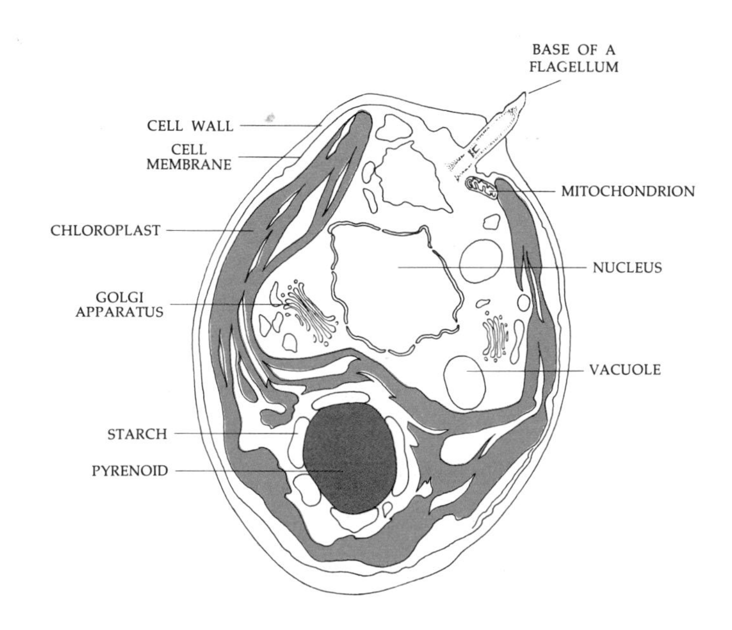

scope. It is easy to recognize, however, because it moves very rapidly and with a characteristic darting motion. This movement results from the beating of the two *flagella* ("whips") that protrude from its anterior (front) end. It is a bright grass-green color; this color is due to the presence of *chlorophyll*, a pigment involved in photosynthesis. *Chlamydomonas* is of special interest because it is believed to be a modern representative of the ancient cell line from which the land plants developed.

Figure 2–5 shows an electron micrograph of a single *Chlamydomonas*. Surrounding the cell, and appearing as a light shadow in the micrograph, is a tough but flexible *cell wall* made of cellulose, the material that also forms the cell walls of grasses and other plants. The possession of this type of wall is one of the reasons that *Chlamydomonas* is believed to be linked evolutionarily to the land plants. Water and materials dissolved in water can pass back and forth readily through the cellulose meshwork. Within the

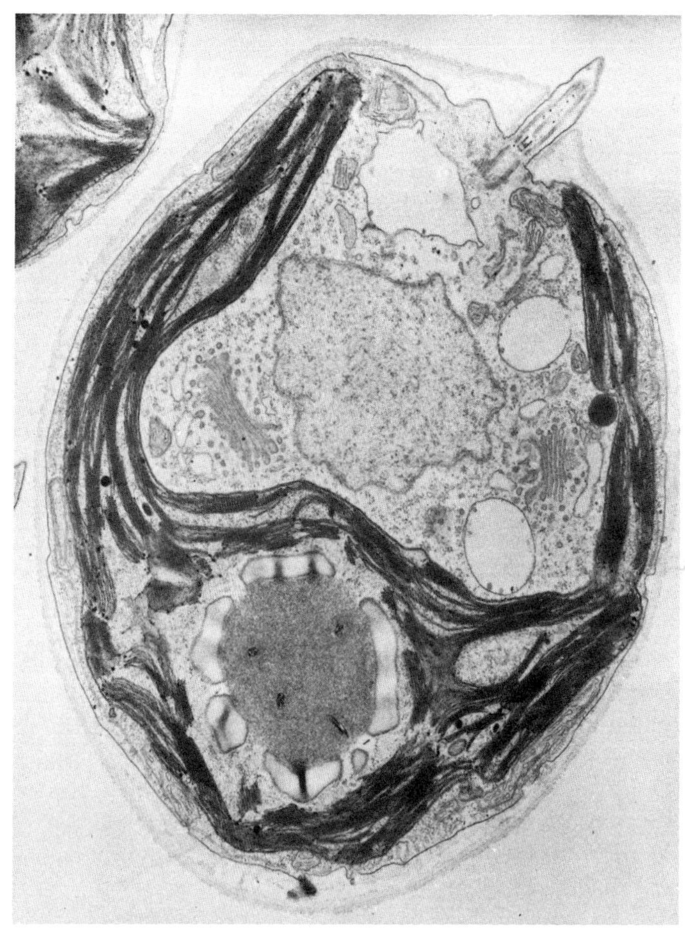

2–5

Electron micrograph of Chlamydomonas. *The drawing accompanying it will help you identify the structures. At the top, on the right, you can see a portion of one of the organism's two flagella. A single chloroplast fills most of the cell. Within the chloroplast starch is stored. Around the outside of the cell are both a cell membrane and a cell wall. The short straight line at the bottom of the picture provides a reference of size. Mu (μ) is the abbreviation for micron, which is 1/10,000 centimeter. This cell, for instance, is about 10 microns in diameter. The same system is used to indicate distances on a road map.*

cell wall is the *cell membrane*. As we mentioned previously, it is because of this function of the membrane that organisms are able to maintain a chemical composition unlike that of their environment—which is, as you will remember, one of the characteristics of living organisms.

In the upper part of the picture is a portion of a flagellum. Note that you can see that the base of the flagellum (which is called the basal body) is embedded in the cytoplasm and that the flagellum protrudes through the cell wall. The other flagellum cannot be seen at all in this particular specimen, which represents only a very thin slice of the original cell.

The most prominent structure within the cell is the *chloroplast*, whose irregular shape fills almost half of the cell body. The many dark strands within the chloroplast are layers of membranes containing chlorophyll. Photosynthesis takes place within these membranes. The function of the pyrenoid body is not known, but it is often seen surrounded by starch granules, as it is in this picture, and for this reason is believed to be involved in making starch from the sugars that are formed in the chloroplasts by photosynthesis. Plants commonly store their food as starch.

The nucleus is surrounded by a double membrane, the nuclear envelope, which shows up quite clearly in the picture. The nucleus carries all the genetic "information" of the cell; it "tells" the cell how to make and repair its various structures and how to carry out its functions.

On either side of the nucleus is a *Golgi body*. These are packaging centers, where materials made within the cell are assembled and packed into membrane-surrounded sacs, in which they are transported through the cell body. For example, an important function of the Golgi body in plant cells appears to be the assembling of materials for building the cell wall.

Also visible within the cell are smaller bodies known as *mitochondria*. Mitochondria are the powerhouses of the cell. Within the mitochondrion, the energy present in the sugars produced by photosynthesis is converted to a more readily available chemical form. You will notice that many of the mitochondria in this picture are clustered near the basal body of the flagellum. It is usual to find more of these organelles in areas of the cell which have high energy requirements.

Chlamydomonas lives in fresh water. Because it requires sunshine for photosynthesis, it is typically found near the surface. Photosynthesis provides all the energy it requires, although it can probably also take small food molecules in through its membrane.

Chlamydomonas is, in a sense, immortal. Under favorable conditions, each cell divides frequently. Within the cell wall, the cell divides, splitting down the middle to form two cells which are mirror images of one another. The chloroplast divides at this time, and the nucleus undergoes a very special kind of division, which will be described in detail in Chapter 3–1. (This special kind of division, called mitosis, ensures that each new cell gets the same genetic information as that contained in the nucleus of the original cell.) Finally, new cell walls begin to form, the old cell wall breaks down, and two tiny green cells dart out into the sunlit water, each a small but completely self-sufficient entity.

A PROTIST: *PARAMECIUM*

Another drop of aquarium water, preferably from a spot where debris and bacteria have collected, is likely to contain some examples of *Paramecium*. Paramecia, because of their shape, were the "slipper animacules" of the early microscopists. They are commonly found in fresh water and are completely harmless to man.

Unlike *Chlamydomonas*, paramecia are heterotrophs, depending on other living organisms as their food sources. Their usual diet consists of bacteria. A single paramecium can consume 5,000 bacteria in a day. A groove in the center of the cell funnels the bacteria and other food particles toward an indentation, called a gullet, in the cell membrane. After a mass of food particles has accumulated in the

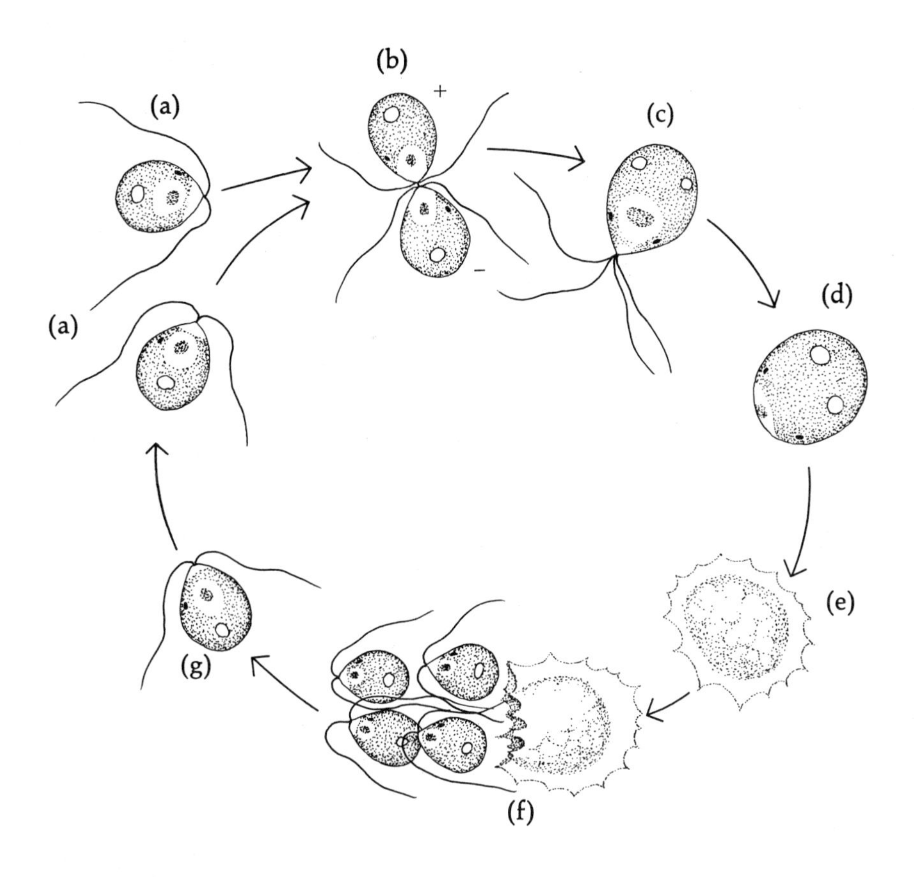

2–6
Chlamydomonas *leads a solitary existence except when it reproduces sexually. At this time, cells of different mating strains (marked + and −) are attracted to one another. At first, they stick together by their flagella* (b). *Next their cytoplasms fuse* (c), *and finally, their nuclei unite. The result is a single cell, the zygote* (d), *which loses its flagella, sinks to the bottom, and develops a thick, protective coat in which it can survive through periods of drought or cold* (e). *When conditions are favorable once again, the zygote germinates and divides, producing four new cells of* Chlamydomonas (f) *which will probably divide by mitosis but which might enter into another sexual cycle.*

gullet, the membrane begins to balloon inward, until finally it has formed a little food-filled sac surrounded by membrane. Such a membrane-surrounded cavity is called a *vacuole*. The vacuole pinches off from the cell membrane, which then seals itself as other particles begin to collect. This process, known as phagocytosis ("cell-eating"), is seen in many other one-celled organisms, such as *Chaos chaos*. (Figure 1–5.) White blood cells in our own bloodstream also pick up bacteria and other microscopic invaders and "devour" them in this way. The food vacuoles move through the body of the paramecium, fusing with other vacuoles that contain digestive enzymes. These enzymes break down the food into molecules, which are released through the membrane of the vacuole into the protoplasm. The debris remaining in the vacuole is released through the cytopyge.

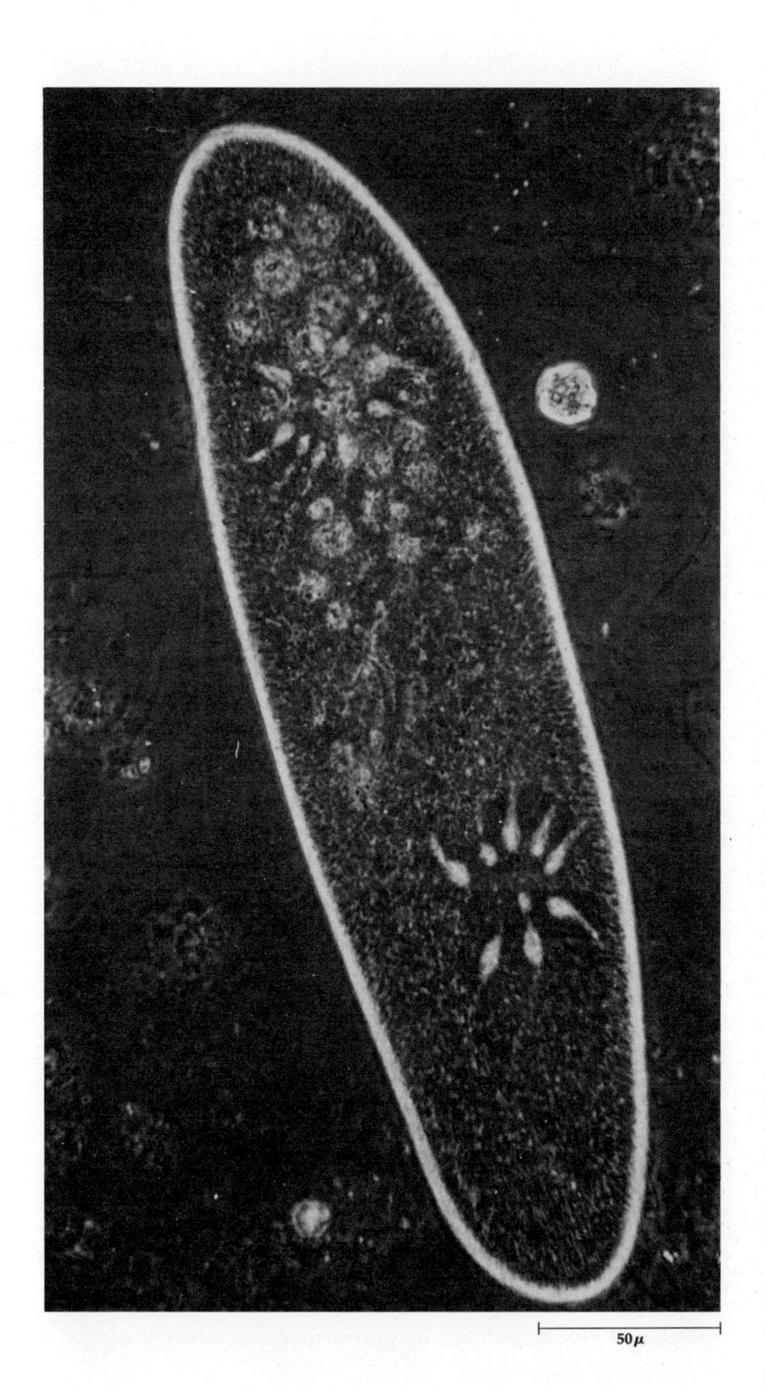

2–7

A paramecium at low magnification. In the photograph, two contractile vacuoles are visible. These rosettelike structures beat rhythmically, pumping excess water out of the cell body. Several food vacuoles can be seen around one of the contractile vacuoles. In the drawing, you can also see the "gullet" into which food particles are driven by the beating cilia. Paramecia and other ciliated protozoans differ from all other cells in that they have a macronucleus and one or more micronuclei. These nuclei are duplicates of one another and appear to be necessary for the control of the unusually large and complex cell structure typical of this group. The body of the protist is completely covered by cilia, although only a relatively few are shown here.

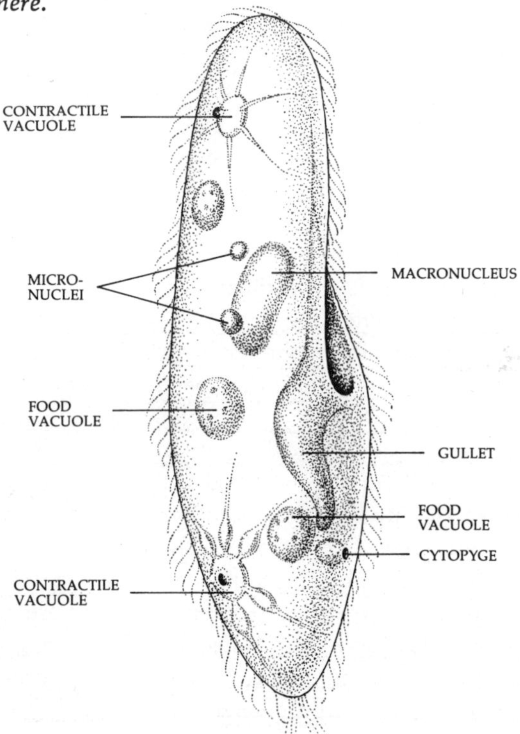

2–8
(a) *Electron micrograph of a food vacuole of a paramecium. The vacuole is full of bacteria. You can clearly see the surrounding membrane.* (b) *Edge of a vacuole. The membrane runs across the center of the picture. You can see some small dense vesicles outside of the membrane. It is believed that these contain digestive enzymes which are discharged into the food vacuole. In the lower right-hand corner of the picture are two mitochondria. Inside the vacuole it is apparent that digestion of the bacteria has begun.*

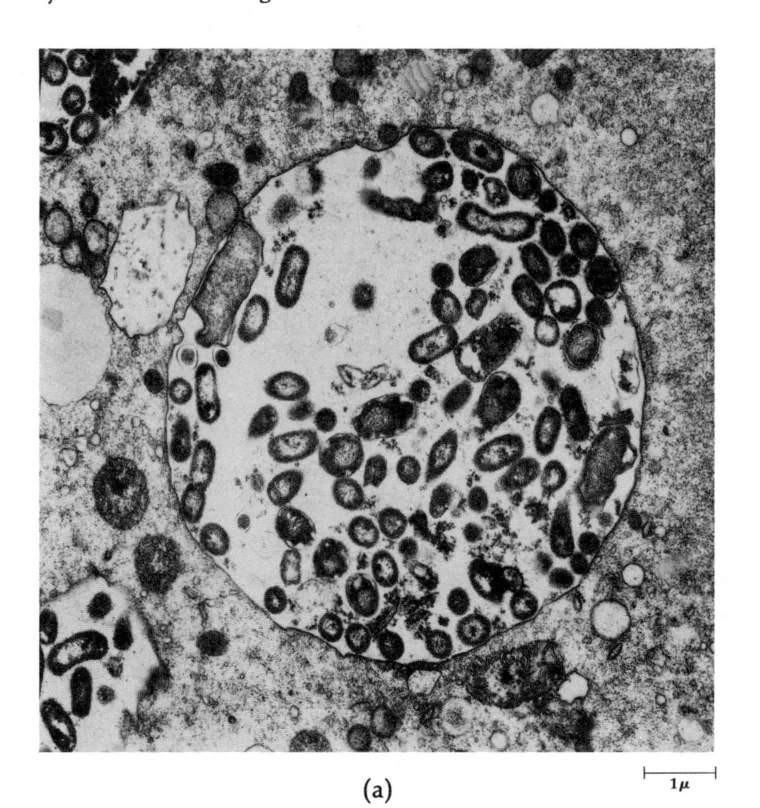

(a)

0.5μ

(b)

Movement in Paramecia

Paramecia are covered with short hairlike structures called *cilia*, a name which comes from the Latin word for "eyelash." There is no clear distinction between cilia and flagella, although "cilia" is the term generally applied to shorter structures occurring in greater numbers. *Paramecium caudatum** has some 2,500 cilia, which arise from rows of small cylinder-shaped basal bodies lying beneath the cell membrane. Their beating is coordinated. They do not all stroke at the same time, like the oars of a well-

* Biologists use a binomial ("two-name") system for classifying organisms. The first part of the name refers to the genus to which the organism belongs, and the second part refers to the species, a subdivision of the genus classification. In this name, for example, "*caudatum*" is a particular kind of paramecium, distinguished from all other paramecia by certain specific characteristics, such as size, shape, and number of cilia.

2–9

Because of the oblique beat of its body cilia, a paramecium rotates as it swims forward.

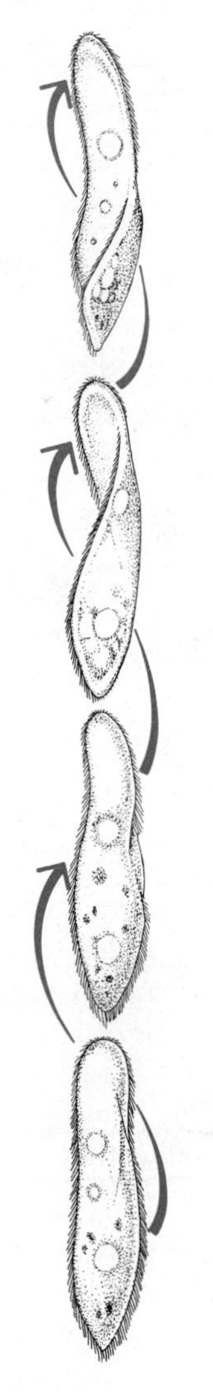

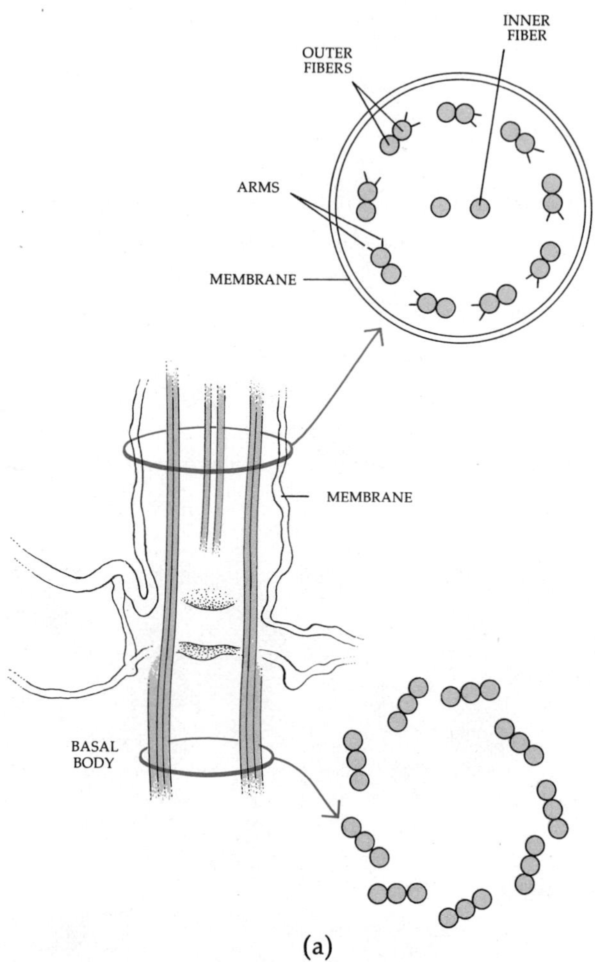

2–10

(a) *A cilium. All cilia and flagella, whether they are found on one-celled organisms or on the surfaces of cells within our own bodies, have this same basic structure, which consists of an outer ring of nine pairs of fibers surrounding two additional fibers in the center. The basal bodies from which they arise have nine outer triplets and no fibers in the center.* (b) *Part of a cilium of a paramecium. The membrane of the cilium is continuous with the cell membrane, and the basal body lies beneath its surface. Note the two large mitochondria.* (c) *Cross section of cilia showing many repetitions of the characteristic structure.*

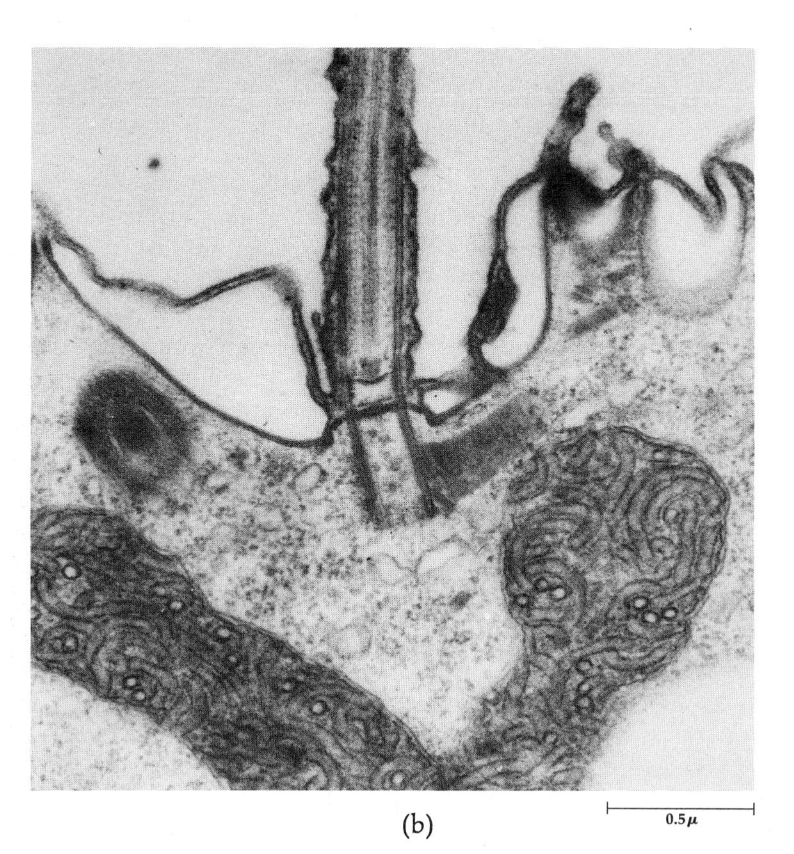

(b)

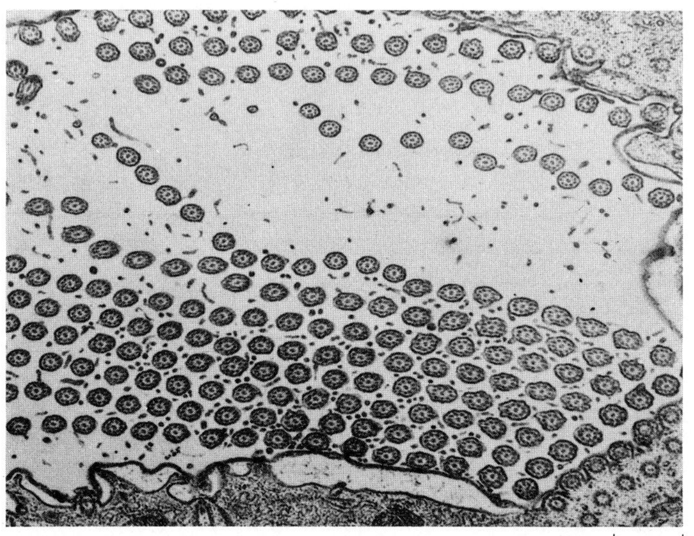

(c)

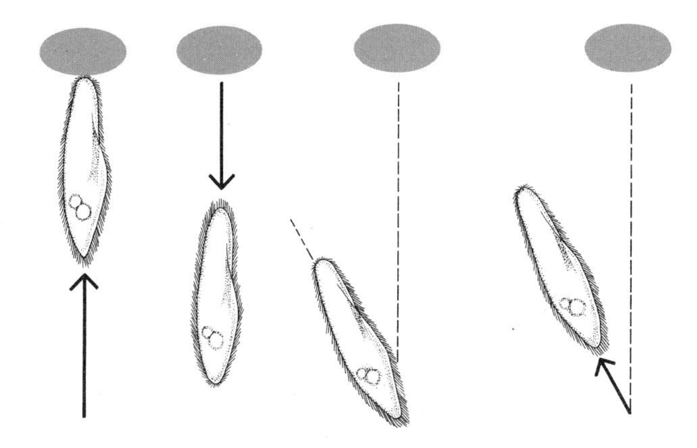

2–11
Paramecia respond to stimuli. Here a paramecium senses an irritating substance, stops short, backs up, turns 30°, and starts forward again in a new direction.

trained crew, but their stroking motion runs in waves down the body of the cell; in motion, they look like blades of wheat in a field blown by the wind. These waves travel at a slightly oblique angle, causing the paramecium to rotate as it swims.

In open water, the paramecium swims swiftly, at least on a microscopic scale, but it slows down in the presence of bacteria or food particles. When it is swimming slowly, the cilia around its oral groove beat more strongly than do those on the rest of its body. This creates currents in the water which draw in nearby particles and cause them to flow down the funnel-like groove to the "mouth."

A paramecium is capable of quite complex behavior. It can move faster or slower, depending on environmental cues. It can avoid obstacles, extremes of heat or cold, or irritating substances (see Figure 2–11). It can even reject certain food particles—with a flick of a cilium.

Like *Chlamydomonas*, paramecia continue to multiply as long as they are alive and healthy. In fact, as someone once calculated, if all the progeny of a single paramecium survived (assuming a division rate of once a day), in 113 days there would be a mass of paramecia equal to the volume of the earth.

THE CELLS OF PLANTS AND ANIMALS

The cells such as those observed by Hooke and Schwann that make up the tissues of plants and animals resemble single-celled organisms in many ways. They take in energy from their environment and convert it to forms that they can use. They make and unmake complex chemicals and structures, virtually the same chemicals and the same structures as those made by the single-celled organisms. However, they are dependent upon and contribute to the life of the whole organism, which continues to function because of the interrelated activities of its many cells. Such a cell lives, at most, only as long as the organism, and many cells, such as a human skin cell or the cell at the tip of a growing plant root, have a life-span of only a few days.

The Cell of a Plant: A Leaf Cell

Figure 2–12 is an electron micrograph of a cell from a leaf of a tobacco plant. This cell is of a general type known as parenchyma cells. Parenchyma cells, which are characteristically large with many sides and thin walls, are the "all-purpose" cells of higher plants. Parenchyma cells which contain chloroplasts, such as this cell, are photosynthetic cells. You can see that it resembles *Chlamydomonas* in several ways. First, it is surrounded by an outer cell wall. Almost all plant cells have such walls, and as the micrograph indicates, these walls are glued by pectin (the material that makes jellies jell) to the walls of adjoining cells. Within the cell wall is a cell membrane, which governs the passage of material in and out of the cell. As in *Chlamydomonas,* almost all the cytoplasm is taken up with chlorophyll-packed membranes. These membranes are contained in numerous chloroplasts rather than in a single, large chloroplast, such as we saw in *Chlamydomonas.* Between the chloroplasts can be seen a few mitochondria. This cell's energy requirements are relatively low compared with those of the constantly moving *Chlamydomonas.*

A large proportion of the body of the cell is taken up by a central vacuole. A membrane surrounds the vacuole, within which are salts and various small molecules dissolved in water. The water exerts pressure against the protoplasm, which, in turn, exerts pressure against the cell wall, keeping it distended. Most cells contain vacuoles—a few were visible in the electron micrograph of *Chlamydomonas*—but the large, single, central vacuole is characteristic of the cells of the higher plants.

Much of the sugar produced in the chloroplasts is exported to other, nonphotosynthetic cells in the plant body, which depend upon the photosynthetic cells for their sustenance. The photosynthetic cells, in turn, depend upon these other cells. Wax-coated, transparent epidermal cells protect the photosynthetic cell from drying out; stomatal cells make tiny openings in the leaf that supply the leaf with carbon dioxide and oxygen; root cells anchor the plant body in the earth and absorb water and minerals needed by all the cells; and cells of the vascular system form tubes that carry the water and minerals to the photosynthetic leaf cell.

A cell such as this one dies when the plant dies. At harvest, when the tobacco plant is cut down, the cell's water supply will be cut off. Its accumulated sugars and other materials will go up in smoke as the leaves are burned. In the case of an edible plant, some of these materials would be passed on to the cow, human, or other consumer.

So the cells of higher plants, while they resemble *Chlamydomonas* in many of their basic structures and functions, differ from this tiny one-celled organism in that they are part of an integrated, interdependent cellular community whose continued existence depends on the smooth functioning of each cell type.

The Cell of an Animal: An Intestinal Cell

Figure 2–14 shows a group of cells from the lining of the intestinal tract of a mouse. The cells from your intestinal tract would look almost exactly like these. These cells, like the leaf cell we discussed previously, have a specific function in the life of the organism. They absorb sugars and

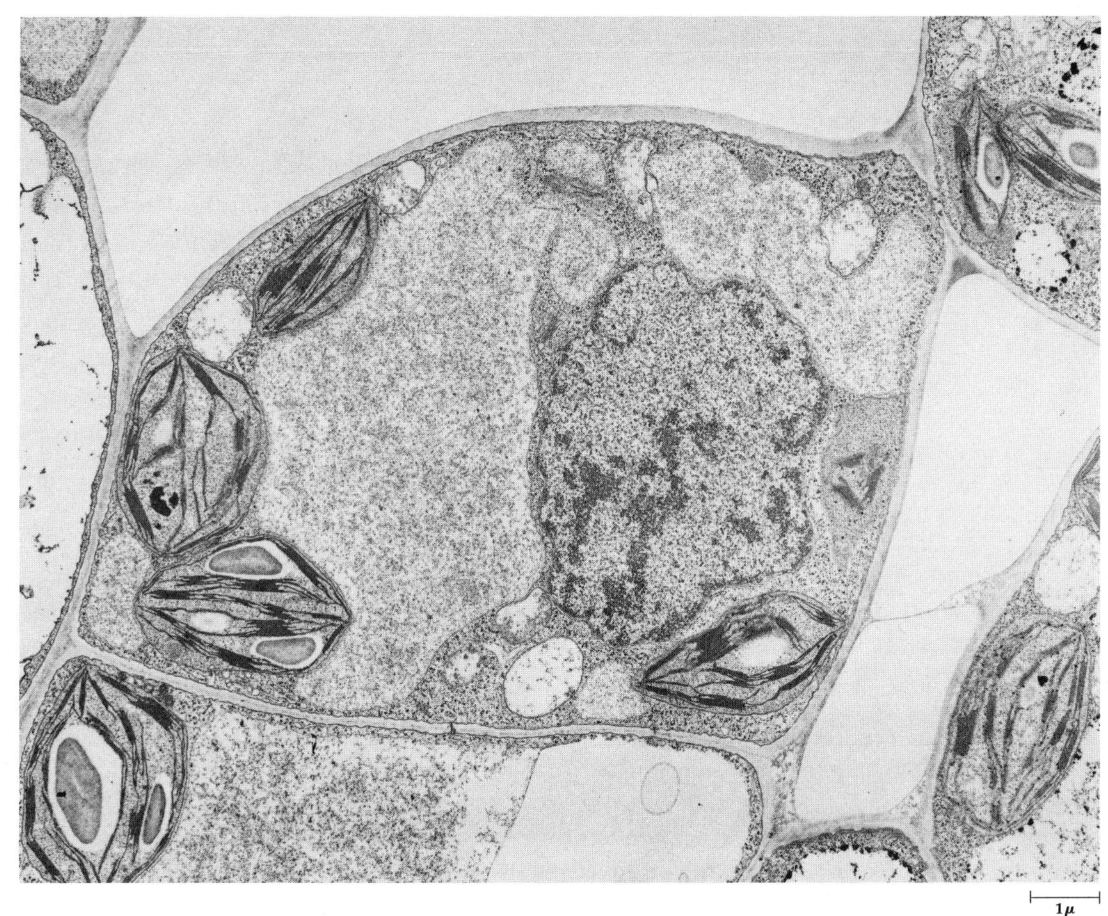

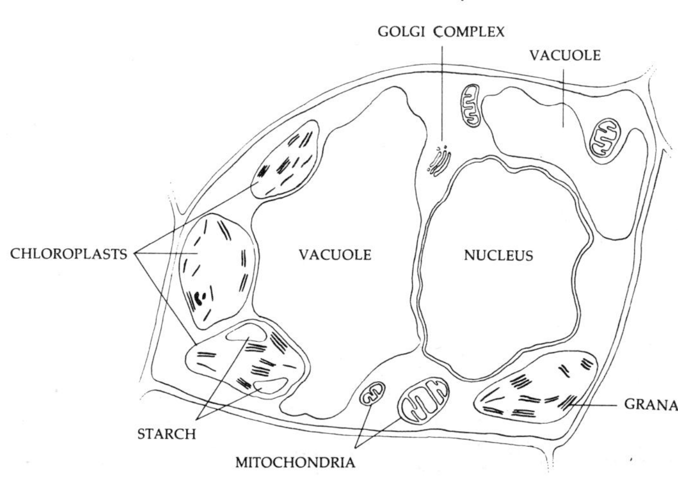

2–12
Photosynthetic cell from the leaf of a tobacco plant. A young plant cell, such as the one from a tobacco leaf shown here, usually contains several vacuoles in different parts of the cell. These vacuoles will swell as the cell reaches maturity, stretching the flexible cell wall, until most of the interior of the cell is taken up by a single large central vacuole. In the mature cell, the cellulose cell wall becomes thick and inflexible. The chloroplasts are the largest bodies in the cytoplasm. Within the chloroplasts, you can see several starch grains, the form in which reserve food supplies are stored.

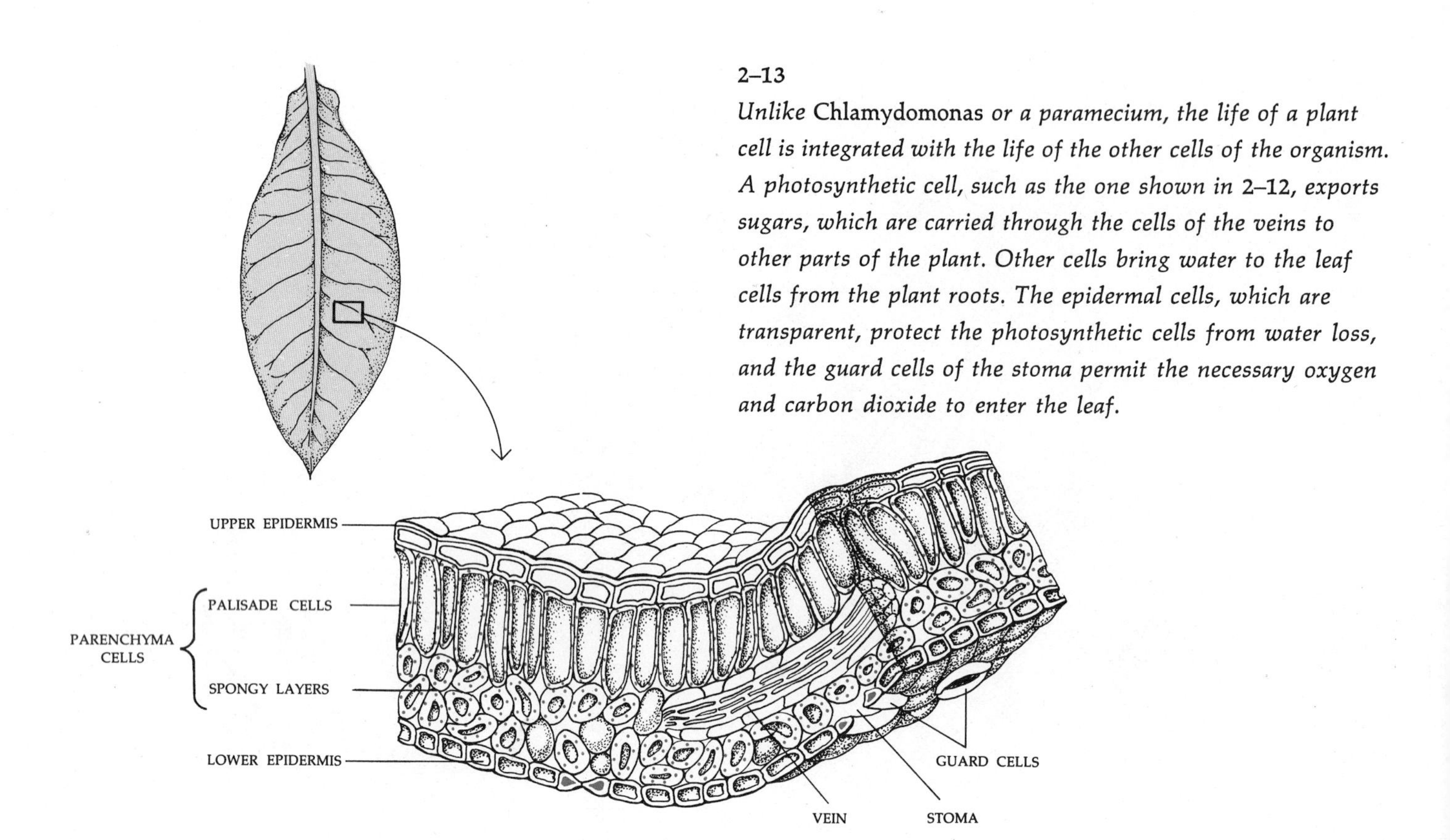

2–13
Unlike Chlamydomonas *or a paramecium, the life of a plant cell is integrated with the life of the other cells of the organism. A photosynthetic cell, such as the one shown in* 2–12*, exports sugars, which are carried through the cells of the veins to other parts of the plant. Other cells bring water to the leaf cells from the plant roots. The epidermal cells, which are transparent, protect the photosynthetic cells from water loss, and the guard cells of the stoma permit the necessary oxygen and carbon dioxide to enter the leaf.*

other energy-containing molecules from the material moving down the digestive tract. By the time the food eaten by a mouse or man reaches the small intestine, it has been acted upon by many digestive enzymes and has been broken down to such an extent that it can be taken up through the intestinal cell membrane.

The cells that line the intestine are epithelial cells, a type of cell that covers the internal and external surfaces of organisms. However, the intestinal epithelial cells do not form a smooth surface, such as the surface of your skin (which is also made up of epithelial cells); the cell layers form fingerlike projections, called intestinal villi, which greatly increase the surface area of the intestine. Furthermore, on the outer surface of each cell are projections similar in design but much smaller in scale, the microvilli. Most of the food that is taken up by the body, and all the needed minerals and vitamins as well, pass through these microvilli.

Each cell has some 3,000 individual microvilli and a square millimeter of intestine may have as many as 200 million. Tight junctions connect adjacent cells and prevent materials passing over the intestinal surface from entering between the cells.

Also clearly visible within each cell is a portion of endoplasmic reticulum. Endoplasmic reticulum is present in all

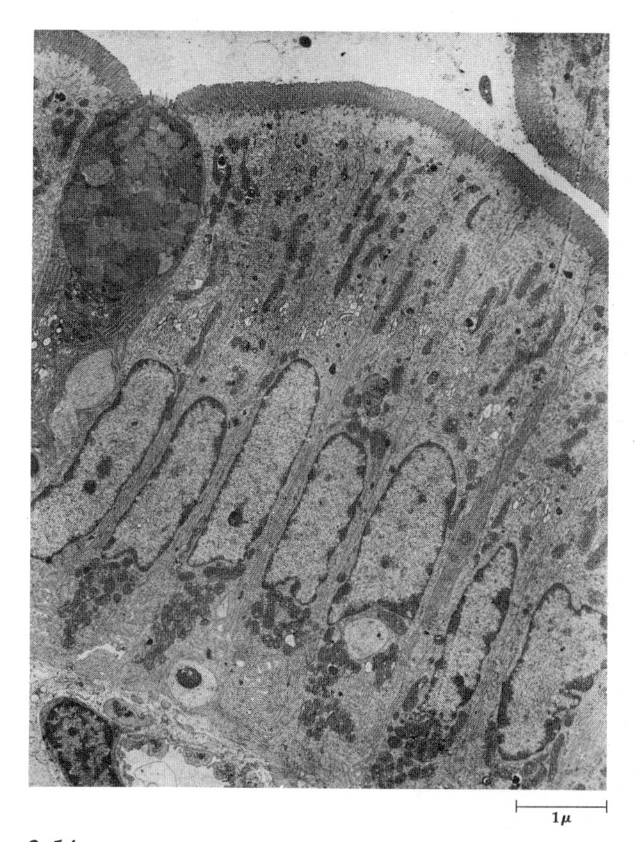

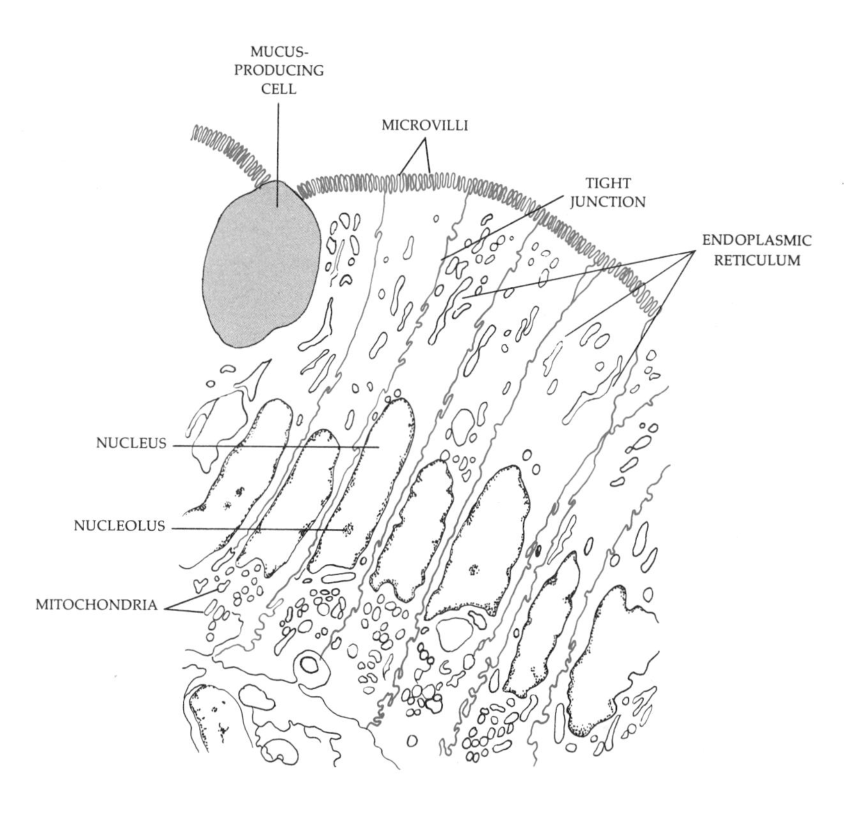

2–14

Cells from the lining of the intestine of a mouse. The microvilli on the outer surface of the cell form a "brush border" that greatly increases the capacity of the cell to absorb molecules into the cell body.

cells, although in varying amounts. It is a network of membranes which serves to separate portions of the cell into compartments, or "laboratories," where chemical reactions are carried out. In general, the more active a cell is in synthesizing materials—such as digestive enzymes or hormones—the more endoplasmic reticulum it will contain. (For a spectacular example of endoplasmic reticulum, see page 82c.) The most prominent body within each cell is the nucleus. Because these cells are in a resting (non-dividing) stage, only the nucleoli are clearly visible within the nuclei.

Like the leaf cell, the intestinal cell functions as part of a large cellular community. Intermingled with these ab-

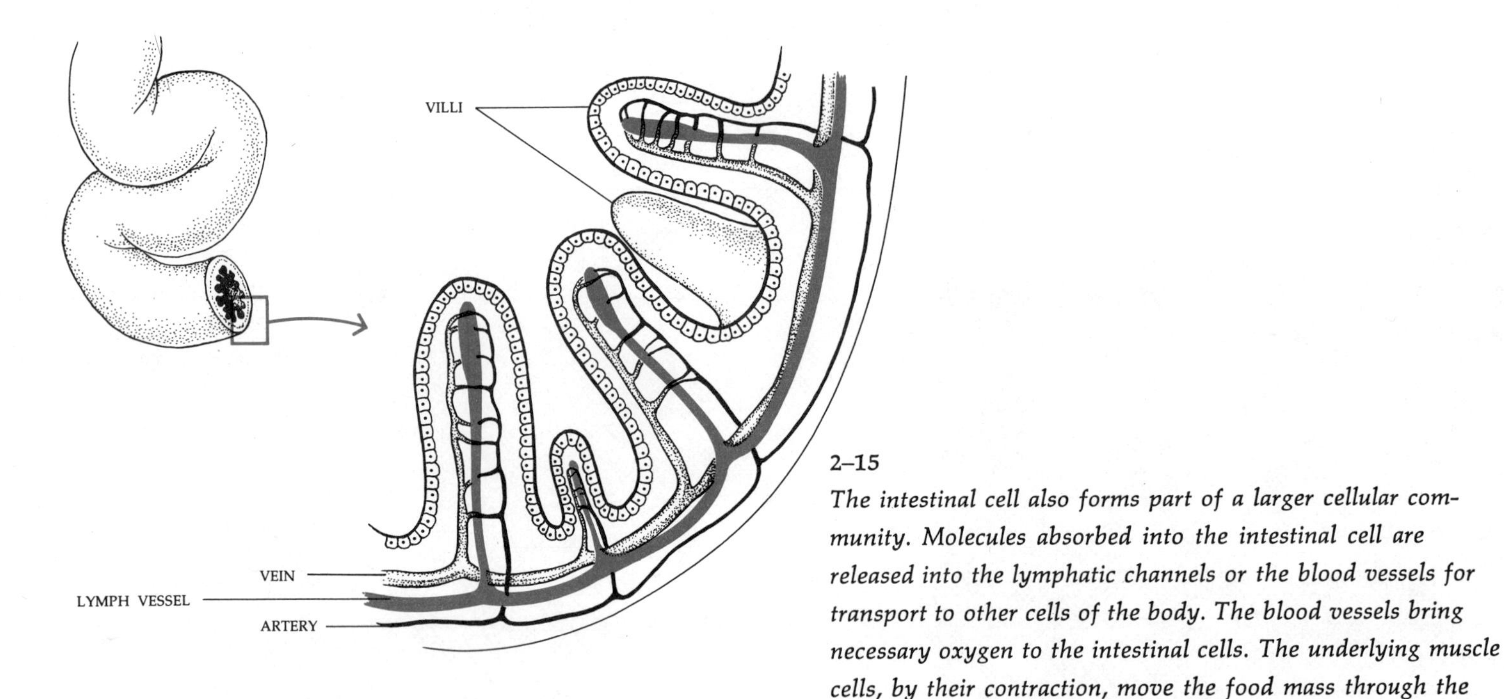

2–15
The intestinal cell also forms part of a larger cellular community. Molecules absorbed into the intestinal cell are released into the lymphatic channels or the blood vessels for transport to other cells of the body. The blood vessels bring necessary oxygen to the intestinal cells. The underlying muscle cells, by their contraction, move the food mass through the intestine.

sorptive cells, for instance, are cells that secrete mucus, which lubricates the intestinal tract. Within a few microns of each cell is a tiny blood vessel, a capillary, which brings oxygen in red blood cells and carries off carbon dioxide. Most of the food molecules absorbed by the cell are also carried off by the bloodstream to nourish the other cells of the body. Fat globules, which are too large for the small blood vessels, are carried off by the lymph, the watery fluid that bathes the body. The lymph contains amoebalike white blood cells that serve as scavengers, protecting the intestinal and other cells from bacteria and other invaders. At the base of the villi are sheets of muscle cells, which move the intestine in wavelike contractions, so that the food passes along the intestinal tract over the absorptive surface of the epithelial cells.

Each individual intestinal cell has a life of only two or three days. After this time, it is sloughed off and replaced by another, identical cell. These cells are formed from a layer of cells immediately below the epithelial surface which divide regularly, just as *Chlamydomonas* and *Paramecium* do. After division, one of the pair of new cells moves toward the surface of the intestine, becomes elongated, and develops microvilli—performing, for its brief life-span, the specialized job of absorbing food for the body. The other cell remains undifferentiated—that is, it does not develop any of the special features of its "sister" cell—but continues to divide, constantly supplying new cells, each with the same brief life-span. Like the paramecium, the intestinal cell is a dynamic functional unit with all the characteristics of living matter. However, while the paramecium leads a solitary and self-sufficient existence, the life—and death—of the intestinal cell is entirely subordinated to that of the community, the whole organism, of which it forms a part.

SUMMARY

The basic unit of life is the cell. All living things are composed of cells. One-celled organisms, such as *Chlamydomonas* and *Paramecium,* carry out all the activities necessary to their own survival. Somatic cells, the cells which make up the tissues of plants and animals, are specialized. Specialized cells carry out specific functions necessary to the survival of the organism, which is the community of cells in which they live. They depend on other cells in the organism for particular services, such as protection or the supply of water or oxygen or energy sources. At some stage in their existence, all cells, despite differences in appearance and in specialized function, are structurally similar; that is, every cell is essentially a unit of protoplasm surrounded by a membrane and containing a nucleus. All photosynthetic cells (with the exception of photosynthetic bacteria and blue-green algae) have chloroplasts which contain membranes packed with chlorophyll, the pigment involved in photosynthesis. All cells also contain mitochondria, where sugars are broken down to supply the cell's quick-energy requirements; endoplasmic reticulum, a membrane system that divides the cytoplasm into functional compartments. Golgi bodies are also commonly found in both plant and animal cells. All cells are functionally similar, in that they are able to carry out energy conversions, maintain chemical homeostasis, and make and remake cellular structures.

For a pictorial summary of the principal cell organelles in an animal cell and a plant cell, see Figure 2–16 on pages 46 and 47.

QUESTIONS

1. Which of the cells described in this chapter are autotrophs? Which are heterotrophs? How do the different types of heterotrophs each get the organic molecules they require?

2. The fact that the cilia of all organisms, from those of *Chlamydomonas* and *Paramecium* to those of *Homo sapiens,* have the same structure is used as evidence of evolutionary relationships among these diverse groups. Do you agree? Explain.

2–16(a)
The principal cell organelles of an animal cell from the liver of a rat.

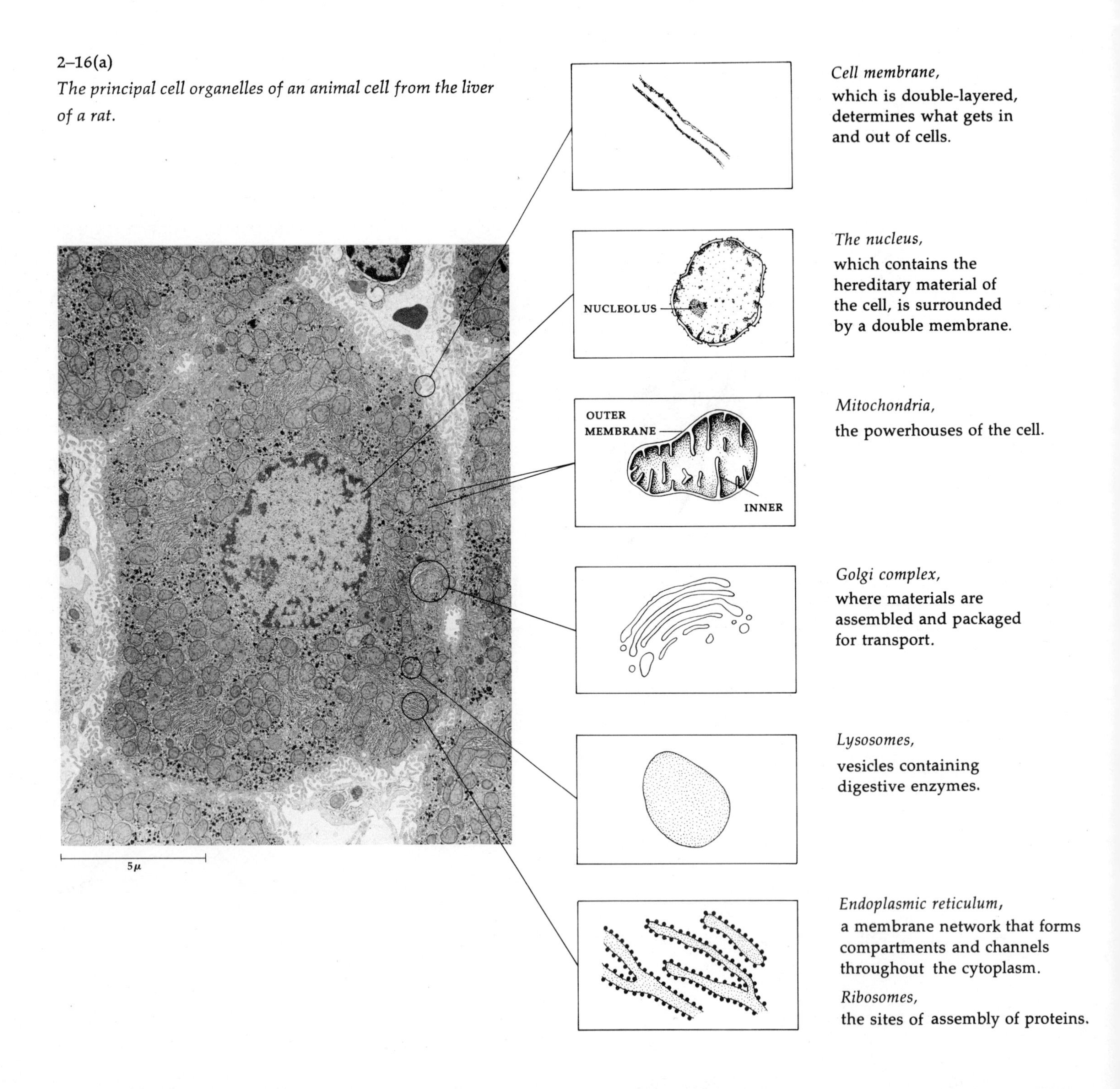

Cell membrane,
which is double-layered, determines what gets in and out of cells.

The nucleus,
which contains the hereditary material of the cell, is surrounded by a double membrane.

Mitochondria,
the powerhouses of the cell.

Golgi complex,
where materials are assembled and packaged for transport.

Lysosomes,
vesicles containing digestive enzymes.

Endoplasmic reticulum,
a membrane network that forms compartments and channels throughout the cytoplasm.

Ribosomes,
the sites of assembly of proteins.

2–16(b)

The principal cell organelles of a plant cell from a blade of timothy grass. (Myron C. Ledbetter.)

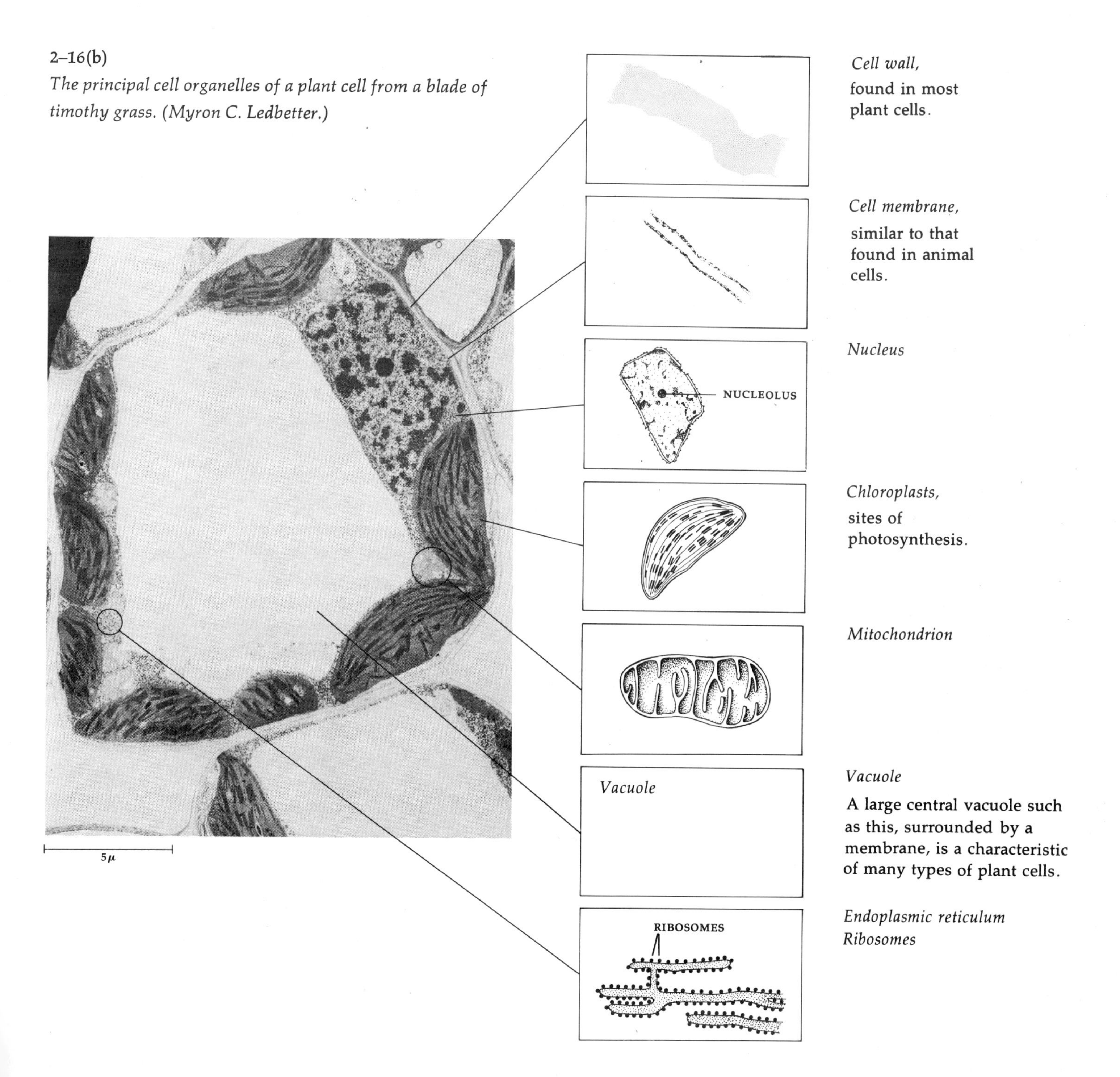

Cell wall, found in most plant cells.

Cell membrane, similar to that found in animal cells.

Nucleus

Chloroplasts, sites of photosynthesis.

Mitochondrion

Vacuole
A large central vacuole such as this, surrounded by a membrane, is a characteristic of many types of plant cells.

Endoplasmic reticulum
Ribosomes

2–17
Leaves of clover after rain.

CHAPTER 2–2

The Composition of Living Matter: Water

LIFE AND WATER

Life on this planet began in water, and today, almost wherever water is found, life is present, too. There are one-celled organisms that eke out their entire existence in no more water than that which can cling to a grain of sand. Some species of algae are found only on the melting undersurfaces of polar ice floes. Some species of bacteria and certain blue-green algae can tolerate the near-boiling water of hot springs. In the desert, plants race through an entire life cycle—seed to flower to seed—following a single rainfall. In the jungle, the water cupped in the leaves of a tropical plant forms a microcosm in which myriad small organisms are born, spawn, and die. We are interested in whether the soil of Mars and the dense atmosphere surrounding Venus contain water principally because we want to know if life is there. On our planet, and probably on others where life exists, life and water have been companions since life first began.

Water makes up 50 to 95 percent of the weight of any functioning living system and an equal or greater proportion of the weight of any living cell. It is a common liquid, by far the most common. Three-quarters of the surface of the earth is covered by water. In fact, if the earth's surface were absolutely smooth, all of it would be 1½ miles under water. But do not mistake "common" for "ordinary"; water is not in the least an ordinary liquid. It is, in fact, extraordinary in comparison with other liquids. If it were not, it is most probable that life could never have evolved.

THE EXTRAORDINARY PROPERTIES OF WATER

Surface Tension

Look, for example, at water dripping from a faucet. Each drop clings to the rim and dangles for a moment by a thread of water; then just as it breaks loose, tugged by gravity, its outer surface is drawn taut, enclosing the entire sphere as it falls free. Take a needle or a razor blade and place it gently on the surface of a glass of water. Although the metal is denser than water, it floats! Look at a pond in spring or summer; you will see water boatmen and other insects skating on its surface as if it were smooth and solid.

The phenomena just described are all the result of surface tension. Surface tension is caused by the cohesiveness of water molecules. The only liquid with greater surface tension than that of water is mercury. Atoms of mercury have so great an attraction for one another that they will not adhere to anything else. (Adhesion is the holding together of unlike substances, and cohesion is the holding together of like substances. Water both coheres and adheres strongly.)

Capillary Action

If you hold two dry glass slides together and dip one corner in water, the interaction of cohesion and adhesion will cause water to spread between the two slides. This is capillary action. Capillary action similarly causes water to rise

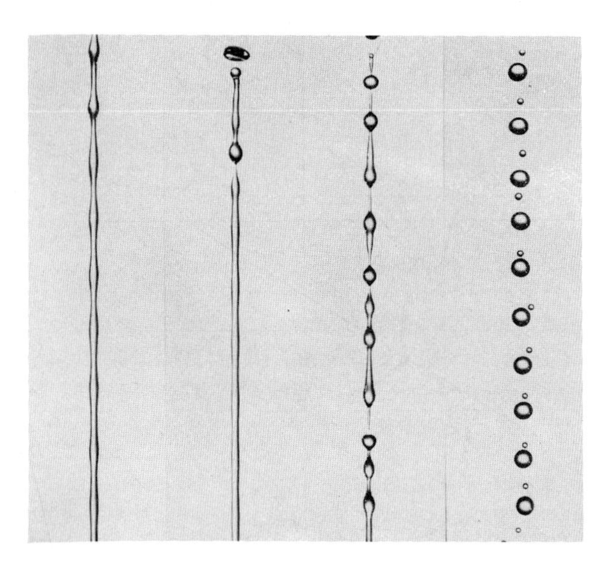

2–18
The formation of water droplets.

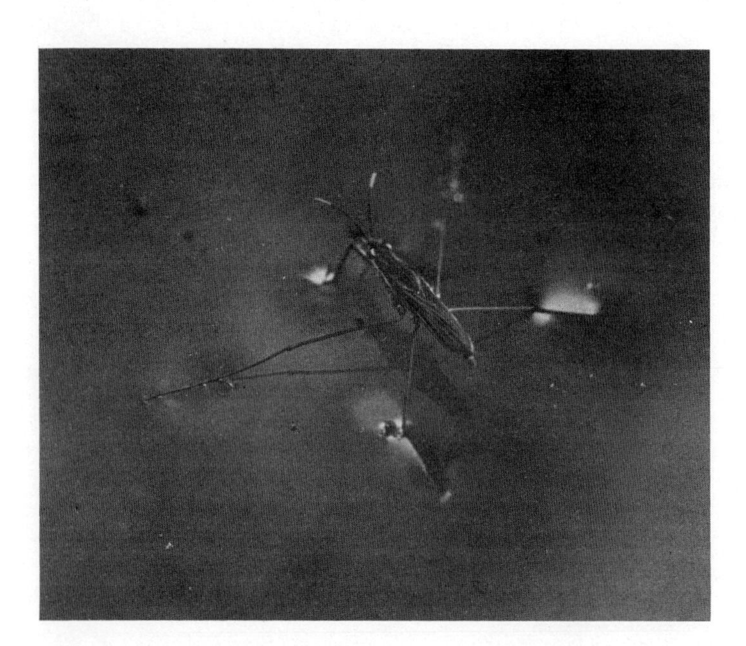

2–19
A water strider's legs dimple the surface of a pond.

in glass tubes of very fine bore, to creep up a piece of blotting paper, or to move slowly through the micropores of the soil and so become available to the roots of plants.

Imbibition

Imbibition ("drinking up") is the movement of water molecules into substances such as wood or gelatin, which swell or increase in volume as a result of adhesion between them and the water molecules. The pressures developed by imbibition can be astonishingly large. It is said that stone for the ancient Egyptian pyramids was quarried by driving wooden pegs into holes drilled in the rock face and then soaking the pegs with water. As the wood swelled, a force sufficient to break the slab of stone was created. Seeds imbibe water as they begin to germinate, often swelling to many times their original size as a result.

Water and Heat

The amount of heat a given amount of a substance requires for a given increase in temperature is its *specific heat*. One calorie of heat, by definition, is the amount of heat required to raise the temperature of 1 gram (1 cubic centimeter) of water 1°C. The specific heat of water is about twice the specific heat of oil or alcohol; that is, approximately 0.5 calorie will raise 1 gram of oil or alcohol 1°C. It is four times the specific heat of air or aluminum and nine times that of iron. Only liquid ammonia has a higher specific heat. In other words, it is relatively difficult to raise the temperature of water.

What does this high specific heat mean in biological terms? It means that for a given rate of heat input, the temperature of water will rise more slowly than the temperature of almost any other material. Conversely, the temperature will drop more slowly as heat is released. Because so much heat input or heat loss is required to raise or lower the temperature of water, organisms that live in the oceans live in an environment in which the temperature is relatively constant. This constancy of temperature is important because, as we shall see, biological reactions take

2–20
The germination of seeds begins with changes in the seed coat that permit a massive uptake of water. In these acorns, photographed on the forest floor, the embryonic roots have penetrated the seed coat and emerged through the tough outer layer of the fruit.

place only within a narrow temperature range. Similarly, land plants and animals, because of their high water content, are more easily able to maintain a relatively constant internal temperature.

Water also has a high *heat of vaporization.* Vaporization—or evaporation, as it is more commonly called—is the change from a liquid to a gas. It takes more than 500 calories to change a gram of liquid water into vapor, five times as much as for ether and almost twice as much as for ammonia.

Thus water, in comparison with other liquids, has both great stability and great fluidity. Both of these properties are essential to the role of water as the principal component of every living cell and of the medium in which each cell, from *Chlamydomonas* to a human ovum, lives its entire life.

THE STRUCTURE OF WATER

Why does water have these unusual properties? To answer this question, we must look at how a water molecule is put together. As you know, each water molecule is made up of two atoms of hydrogen and one of oxygen—H_2O. The hydrogen atom is the simplest of all atoms. Its nucleus contains a single positively charged particle, a proton, and outside the nucleus is a single negatively charged particle, an electron:

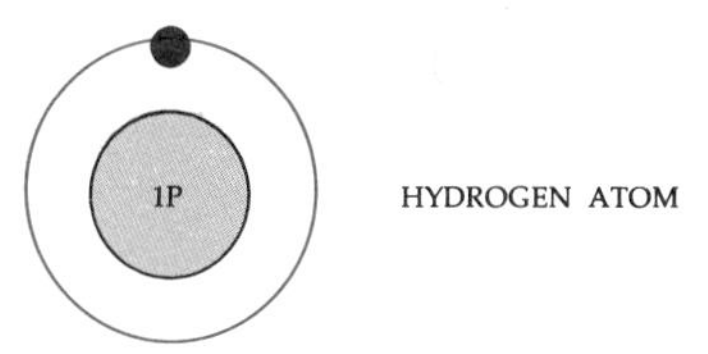

The electron moves in orbit around the proton. The path of movement of the electron is called the electron shell. The proton and the electron have equal charges, so that they cancel one another out. If, however, the hydrogen atom loses its single orbiting electron, it will have a charge of +1, becoming H^+. Such charged atoms or groups of atoms are referred to as ions. (A hydrogen ion consists of a single proton and is often referred to simply as a proton.)

The oxygen atom is a little more complicated. In its nucleus are eight protons, and outside the nucleus are eight electrons:

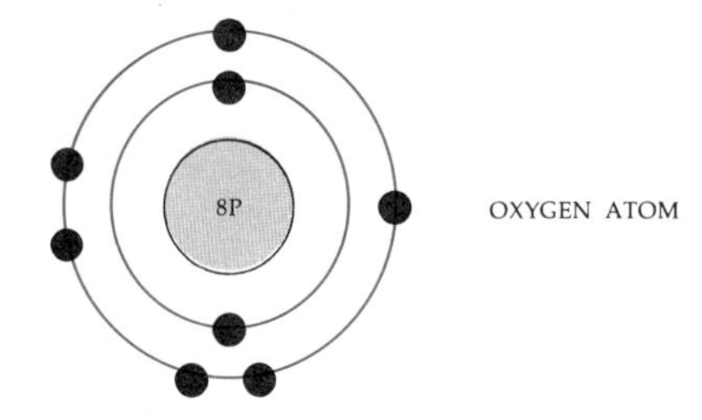

The inner shell has two electrons. (An inner shell can never have more than two atoms, according to the "rules" of atomic structure.) There are six electrons in the outer shell when oxygen is electrically neutral, but according to another "rule," this outer shell is not complete unless it has eight electrons. For reasons which are not clearly understood, atoms "seek" to complete their shells, which they often do by sharing electrons with other atoms.

In order to complete its outer shell, an oxygen atom requires two electrons. A hydrogen atom requires one. Thus, when one oxygen atom shares electrons with two hydrogen atoms, each completes its outer shell:

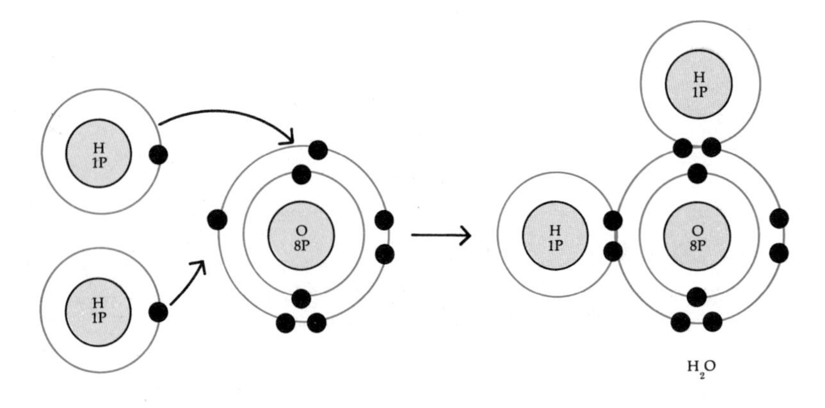

This electron sharing binds the three atoms together. Bonds such as this, which involve the sharing of electrons, are known as *covalent bonds*. Most molecules important in biological systems are held together by covalent bonds.

The Water Molecule

The water molecule is electrically neutral, having an equal number of electrons (−) and protons (+). However, the single electrons of the two hydrogen atoms, which are shared with the oxygen atom, are more strongly attracted to the oxygen nucleus and are therefore located nearer to it than to the hydrogen nuclei. As a consequence, the molecule has two local positive charges, carried by the hydrogen atoms, and two local negative charges, carried by the oxygen atom.

The water molecule, in terms of its electrical charges, is thus four-cornered, with two "positive" corners and two "negative" ones:

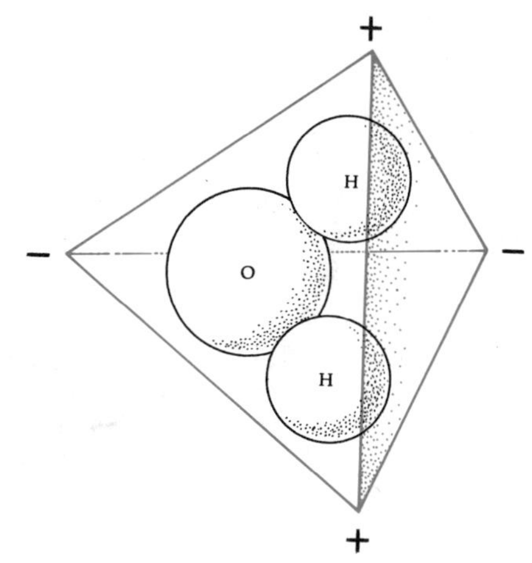

A molecule with zones of negative and positive charge is known as polar, by analogy with a magnet, which has a negative pole and a positive pole.

When a positively charged hydrogen atom of a water molecule comes into juxtaposition with an atom carrying a sufficiently strong negative charge—such as the oxygen atom of another water molecule—the force of the attraction forms a bond between them, which is known as a *hydrogen bond*. A hydrogen bond forms between the negative "corner" of one water molecule and the positive "corner" of another:

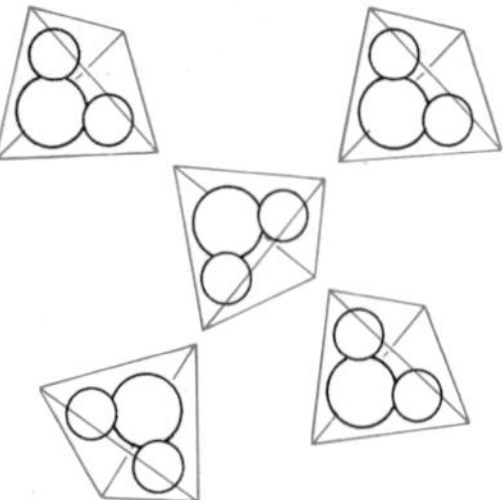

Every water molecule can therefore establish hydrogen bonds with four other molecules:

Liquid water is made up of water molecules bound together in this way. Here is a "picture" of liquid water as analyzed by a computer:

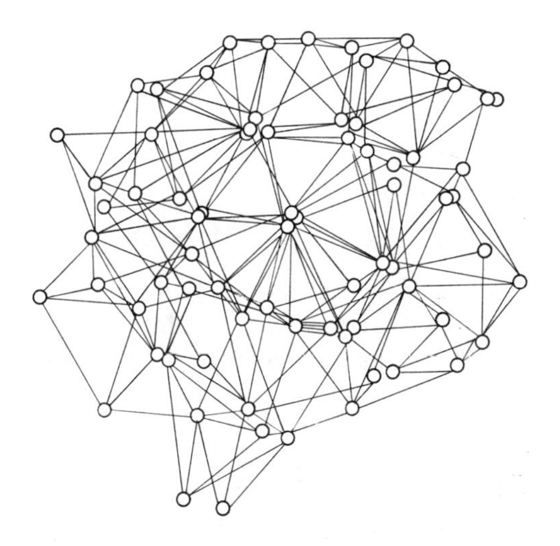

Any single hydrogen bond has only an exceedingly short lifetime; on an average, each such bond lasts about 10^{-11} second, which is 1/100,000,000,000 second. All together, however, they have considerable strength. It is because of these multiple, shifting hydrogen bonds that liquid water is both fluid and stable.

These attractions between water molecules are responsible for its surface tension and also for its capillary action, which takes place as a result of the molecules clinging to one another as well as to the surface along which they are creeping. The movement of water through the xylem cells in plants takes place in large part because of this attraction of the water molecules for one another, as we shall see in Chapter 5–5.

Hydrogen Bonds and Heat

The high specific heat of water is also a consequence of hydrogen bonding. Heat is a form of energy, the kinetic energy, or energy of movement, of molecules. In order for the kinetic energy of molecules of water to increase sufficiently for the temperature to rise 1°C, it is necessary to rupture a number of the hydrogen bonds holding the molecules together. If you put an iron skillet over a gas flame, the skillet will soon be red hot as a result of the transfer to the metal molecules of the heat energy produced by the gas flame. However, a pot of water will take much longer to heat up. This is because much more heat is used in breaking the hydrogen bonds holding the water molecules together and so a relatively small amount is available to increase molecular movement. (Note that heat and temperature are not the same. A lake may have a lower temperature than that of the bird flying over it, but the lake contains more heat because it has more molecular motion. Temperature is measured in degrees and reflects average energy of the molecules. Heat is measured in calories, which reflect both molecular movement and volume.)

Hydrogen bonding is also responsible for water's high heat of vaporization. Vaporization comes about because the rapidly moving molecules of a liquid break loose from the surface and enter the air. The hotter the liquid, the more rapid the movement of its molecules and, hence, the more rapid the rate of evaporation; but so long as a liquid is exposed to air that is below 100 percent humidity, evaporation will take place and will continue to take place right down to the last drop.

2–21
Deserts, hot in the daytime, become very cold at night. These drastic fluctuations in temperature occur because the air contains little or no water vapor.

In order for a water molecule to break loose from its fellow molecules—that is, to vaporize—the hydrogen bonds have to be broken. This requires heat energy. As we noted previously, more than 500 calories are needed to change 1 gram of liquid water into vapor (the exact amount depends on the temperature of the water). The same amount of heat would raise the temperature of 500 grams of water 1°C. As a consequence, when water evaporates, as from the surface of your skin or a leaf, it absorbs a great deal of heat from the immediate environment. Thus evaporation has a cooling effect. Evaporation from the surface of a land-dwelling plant or animal is one of the principal ways in which these organisms "unload" excess heat.

Water as a Solvent

The polarity of the water molecule is responsible for its powers of adhesion, which we described previously. It is attracted to both negatively and positively charged surfaces. Also, water's polarity makes it a good solvent for any other polar molecules—most biological molecules are polar—and for negatively or positively charged ions. Water molecules tend to pull apart molecules such as NaCl into their constituent ions. A number of such ions are important in the functioning of biological systems.

Ionization: Acids and Bases

Water molecules also have a slight tendency to ionize, separating into H^+ and OH^-. In any given volume of pure water, a very small but constant number of water molecules will be dissociated in this form. The number is constant because the tendency of water to dissociate is exactly offset by the tendency of the ions to reunite; thus even as some are ionizing, an equal number of others are forming bonds. This can be expressed as:

$$HOH \rightleftharpoons H^+ + OH^-$$

The arrows indicate that the reaction can go in either direction. The fact that the arrow pointing toward HOH is longer indicates that the reaction is more likely to go in that direction. (In any sample of water, only a small fraction exists in ionized form.)

In pure water, the number of H^+ ions exactly equals the number of OH^- ions. (This is necessarily the case since neither ion can be formed without the other when only H_2O molecules are present.) A solution acquires the properties we recognize as acid when the number of H^+ ions exceeds the number of OH^- ions; conversely, a solution is basic when the OH^- ions exceed the H^+ ions.

Hydrochloric acid (HCl) is an example of a common acid. It is strong acid, meaning that it tends to be almost completely ionized into H^+ and Cl^-, releasing additional H^+ ions into the solution. Sodium hydroxide (NaOH) is a com-

2–22
Ammonia is very similar to water in its chemical structure, and biologists have speculated about whether it might substitute for water in life processes (a). *The ammonia molecule* (b) *is made up of hydrogen atoms covalently bonded to nitrogen, which, like the oxygen in the water molecule, retains a slight negative charge. But because there are three hydrogens with slight positive charges to one nitrogen, ammonia does not have the cohesive power of water and evaporates much more quickly. Perhaps this is why no form of life based on ammonia has been found, although* NH_3 *was very common in the primitive atmosphere.*

"Ammonia! Ammonia!"

(a)

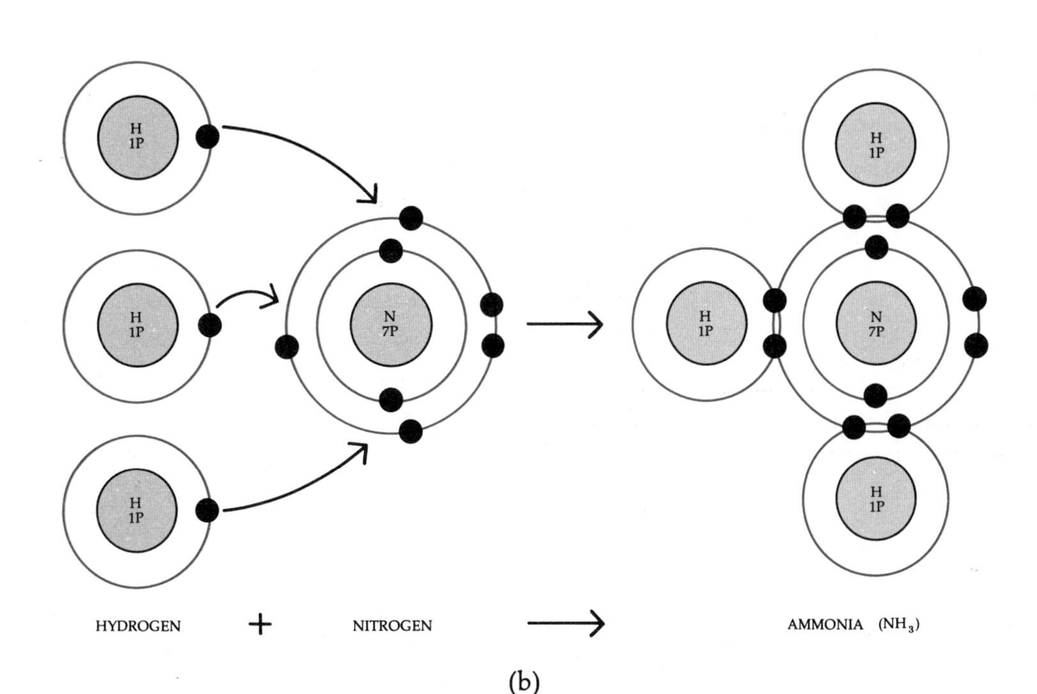

(b)

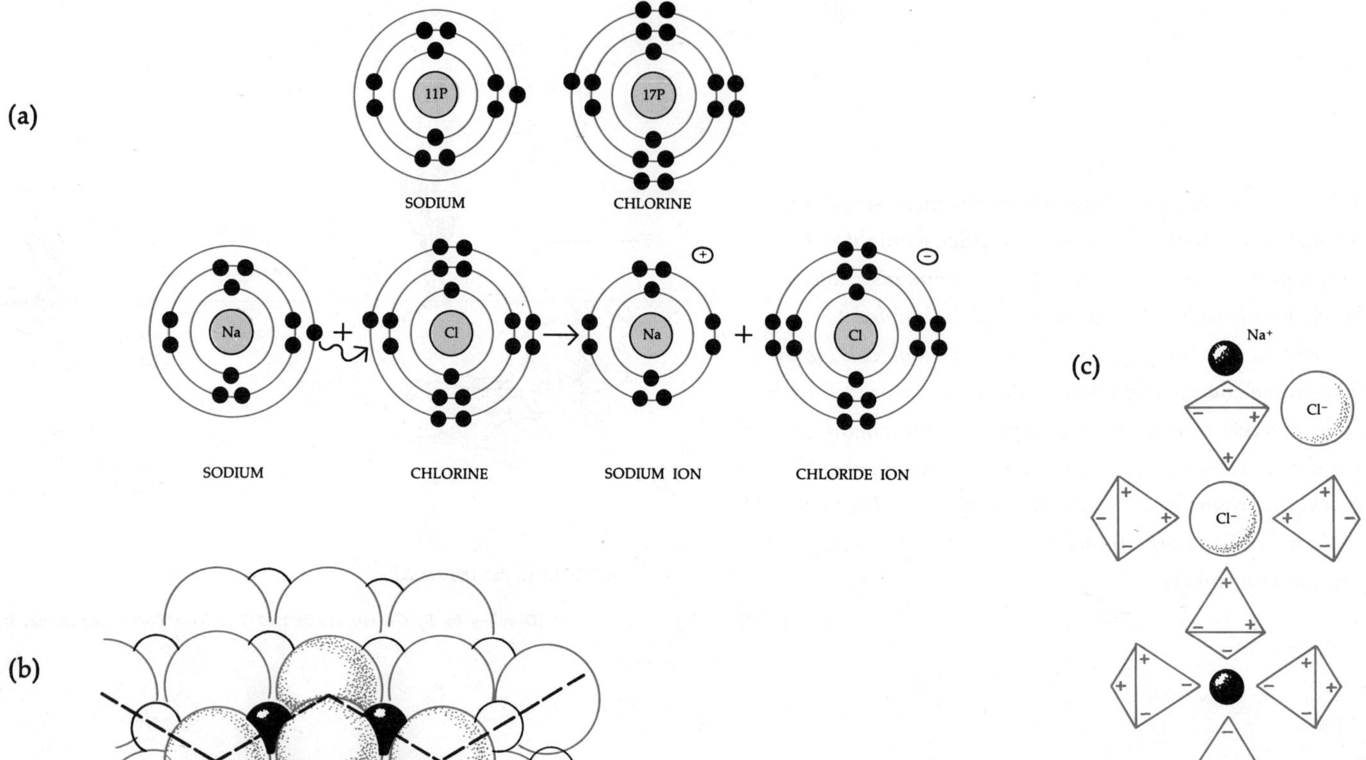

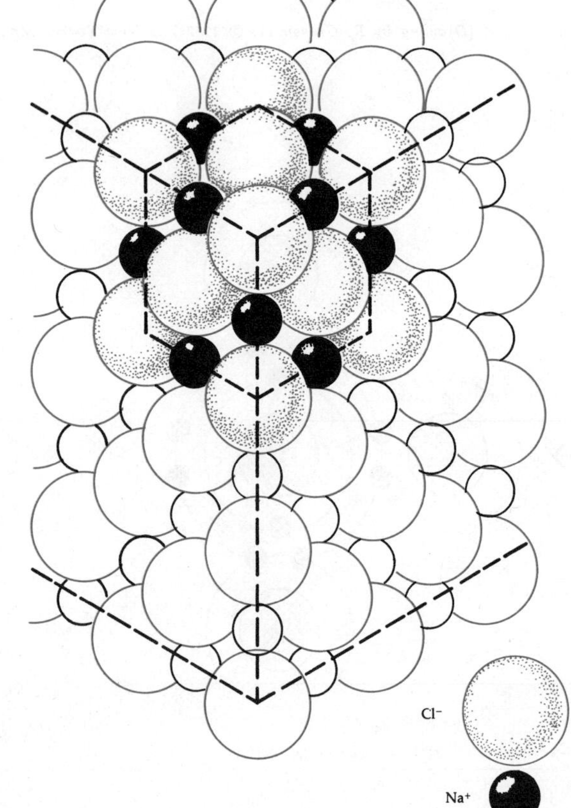

2–23

(a) *Sodium, which has only one electron in its outer shell, needs to lose this electron to become stable. Chlorine, which has seven electrons in its outer shell, needs to gain one. When sodium and chlorine interact, sodium loses its single electron and chlorine gains it; following this transaction, sodium has a positive charge and chlorine a negative one. Such charged atoms are called* ions. (b) *Oppositely charged ions attract one another. Table salt is crystalline* NaCl, *a latticework of alternating* Na^+ *and* Cl^- *ions held together by their opposite charges. Such bonds between oppositely charged ions are known as* ionic bonds. (c) *Ions are held in solution in water because of the polarity of the water molecule; water molecules will cluster around, and so segregate, the positively or negatively charged ions.*

mon strong base; it exists entirely as Na^+ and OH^-. Weak acids and weak bases are those which ionize only slightly.

Chemists define degrees of acidity by means of the pH scale. For example, 1/10,000,000 mole* of hydrogen ions, which is 10^{-7} mole, is referred to simply as pH 7 (see Table 2–1). At pH 7, the concentrations of free H^+ and OH^- ions are exactly the same; this is a "neutral" state. Any pH below 7 is acidic, and any pH above 7 is basic.

Table 2–1

The pH Scale

	Concentration of H^+ Ions (moles per liter)		*pH*	*Concentration of OH^- Ions (moles per liter)*	
Acidic	1.0	10^{0}	0	10^{-14}	
	0.1	10^{-1}	1	10^{-13}	
	0.01	10^{-2}	2	10^{-12}	
	0.001	10^{-3}	3	10^{-11}	
	0.0001	10^{-4}	4	10^{-10}	
	0.00001	10^{-5}	5	10^{-9}	
	0.000001	10^{-6}	6	10^{-8}	
Neutral	0.0000001	10^{-7}	7	10^{-7}	Neutral
Basic		10^{-8}	8	10^{-6}	0.000001
		10^{-9}	9	10^{-5}	0.00001
		10^{-10}	10	10^{-4}	0.0001
		10^{-11}	11	10^{-3}	0.001
		10^{-12}	12	10^{-2}	0.01
		10^{-13}	13	10^{-1}	0.1
		10^{-14}	14	10^{0}	1.0

WATER MOVEMENT

Many biological processes involve water movement. Water moves in living systems in three ways: by bulk flow, by diffusion, and by osmosis. Of these three, bulk flow is the most familiar.

Bulk Flow

Bulk flow is the overall movement of water (or some other liquid). It occurs in response to differences in the potential energy of water, usually referred to as *water potential.*

A simple example of water that has potential energy is water at the top of a hill. As this water runs downhill, its potential energy can be converted to mechanical energy by a water mill or to electrical energy by a hydroelectric turbine.

Pressure is another source of water potential. If we put water into a rubber bulb and squeeze the bulb, this water, like the water at the top of a hill, also has water potential, and it will move to an area of less water potential. Can we make the water that is running downhill run uphill by means of pressure? Obviously, we can. But only so long as the water potential produced by the pressure exceeds the water potential produced by gravity. Water moves from an area where water potential is greater to an area where water potential is less, regardless of the reason for the water potential. Your heart pumps blood to your brain against gravity by creating a greater water potential.

The concept of water potential is a useful one because it enables physiologists to predict the way in which water will move under various combinations of circumstances. Measurements of water potential are usually made in terms of the hydrostatic pressure required to stop the movement of water under the particular circumstances. This pressure is usually expressed in units of atmosphere. One atmosphere is the average pressure of the air at sea level.

* The mole is the principal unit of measure used for defining quantities of substances involved in chemical reactions. The number of particles in 1 mole of any substance (whether atoms, ions, or molecules) is always exactly the same: 6.023×10^{23}. For example, 1 mole of hydrogen ions contains 6.023×10^{23} ions, 1 mole of hydrogen contains 6.023×10^{23} hydrogen atoms, and 1 mole of water contains 6.023×10^{23} water molecules.

2–24
Bulk flow describes the overall movement of water from a position of higher water potential to one of lower water potential, as over this dam.

Diffusion

Diffusion is a familiar phenomenon. If you sprinkle a few drops of perfume in the corner of a room, the scent will eventually permeate the entire room even if the air is still. If you put a few drops of dye in one end of a rectangular glass tank full of water, the dye molecules will slowly distribute themselves evenly throughout the tank. (The process may take a day or more.) Why do the dye molecules move apart?

Most of us, being human, tend to explain these movements in human terms, that is, in terms of a crowded population which, in order to find "elbowroom," moves out to the suburbs as a result of population pressure. This is not true of diffusion processes. If you could observe the individual dye molecules in the tank, you would see that each one of them moves at random and moves individually; looking at any single molecule—at either its rate of motion or its direction of motion—gives you no clue at all about where the molecule is located with respect to the others. So how do the molecules get from one side of the tank to the other? In your imagination, take a thin cross section of the tank, running from top to bottom. Dye molecules will move in and out of the section, some moving in one direction, some moving in the other. But you will see more dye molecules moving from the side of greater concentration. Why? Simply because there are more dye molecules at that end of the tank. Consequently, if there are more dye molecules on the left, the overall movement will be from left to right, even though there is an equal probability that any one molecule of dye will move from right to left. Similarly, if you could see the movement of the individual water molecules in the tank, you would see that their overall movement is from right to left.

What happens when all the molecules are distributed evenly throughout the tank? The even distribution does not affect the behavior of the molecules as individuals; they still move at random. And, since the movements are random, just as many molecules go to the left as to the right.

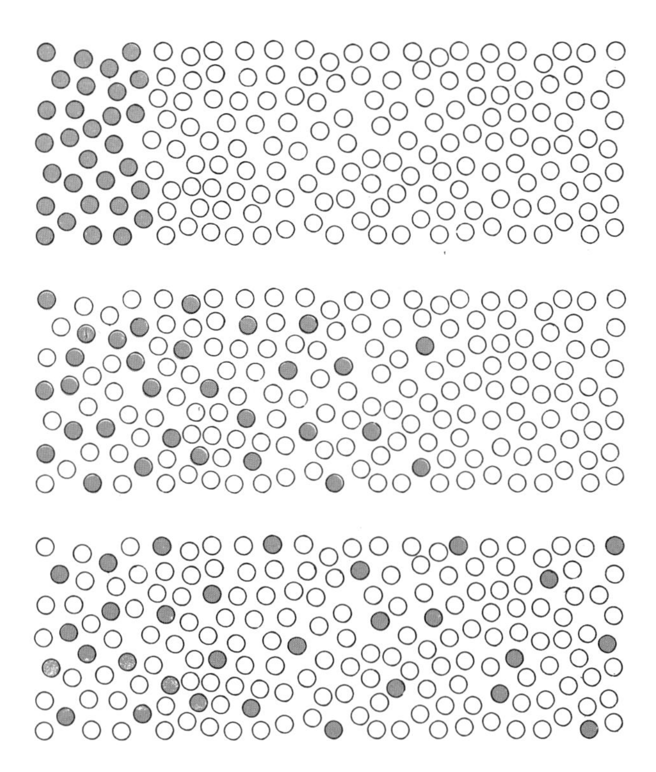

2–25
Diagram of diffusion process. Diffusion, which is the random movement of molecules, results in an even distribution of molecules. Notice that as one type of molecule (indicated by color) diffuses to the right, the other diffuses in the opposite direction.

But because there are now as many molecules of dye and as many molecules of water on one side of the tank as on the other, there is no overall direction of motion, although, provided the temperature has not changed, there is as much overall motion as before.

Substances that are moving from a region of greater concentration of their own molecules to a region of lesser concentration are said to be moving *along the gradient.* A substance moving in the opposite direction, toward a greater concentration of its own molecules, moves *against the gradient,* which is analogous to pushing something uphill. When the molecules have reached a state of equal distribution, that is, when there is no more gradient, they are said to be in *dynamic equilibrium.* In our imaginary tank, there are two gradients; the dye molecules are moving along one of them, and the water molecules are moving along the other.

A system in which diffusion is occurring is a system that possesses water potential. Since water, like other substances, will move from a region of greater concentration to a region of lesser concentration, the area of the tank in which there is pure water has a greater water potential than the area containing water plus dye or some other solute.

The essential characteristics of diffusion are (1) that each molecule moves as an individual and (2) that these movements are random. The net result of diffusion is that the diffusing substance becomes evenly distributed.

Cells and Diffusion

A principal way in which substances move in cells is by diffusion. (This process may be hastened by cell streaming, the movement of cytoplasm in cells, which is readily visible in *Amoeba* and in growing plant cells.) This dependence upon diffusion is one of the factors limiting cell size. As you can probably deduce by studying the simple diagram (Figure 2–25), the process becomes increasingly slower and less efficient as the distance "covered" by the diffusing molecules increases.

Water molecules, oxygen, carbon dioxide, and a few other simple molecules diffuse freely across cell membranes —as if the membranes had holes or pores in them, as they probably do. Larger molecules and positively charged atoms or molecules (the latter apparently stick on their way through) are stopped by the outer cell membrane and also by membranes within the cell, such as the endoplasmic reticulum, the nuclear envelope, and the membranes surrounding the chloroplasts and the mitochondria. These substances move across cell membranes by a process known as *active transport*, the same energy-requiring process as that by which the intestinal epithelial cell imports food molecules. How the cell membrane performs this important function is not known, but according to most theories, the membrane must have specific molecules on it that, under certain conditions, "recognize" other molecules and escort them into the cell. By active transport, substances can be moved against the diffusion gradient. Movement against the gradient requires the expenditure of energy.

Osmosis

In order to move in and out of living cells, water must move through cell membranes. Cell membranes, like many other kinds of membranes, permit the free passage of water but inhibit or retard the passage of most materials dissolved in the water (solutes). The movement of water through such a membrane is known as osmosis. In the absence of other forces (such as pressure), the movement of the water in osmosis will typically be from a region of lesser solute concentration (and therefore of greater water concentration) to a region of greater solute concentration (fewer water molecules). The presence of solute decreases the water potential and so causes the movement of water from a region of greater to a region of lesser water potential—from the "wetter" to the "drier" side of the membrane.

Osmosis can be measured by a simple device known as an osmometer. An example of an osmometer is shown in Figure 2–26. The beaker contains distilled water, and with-

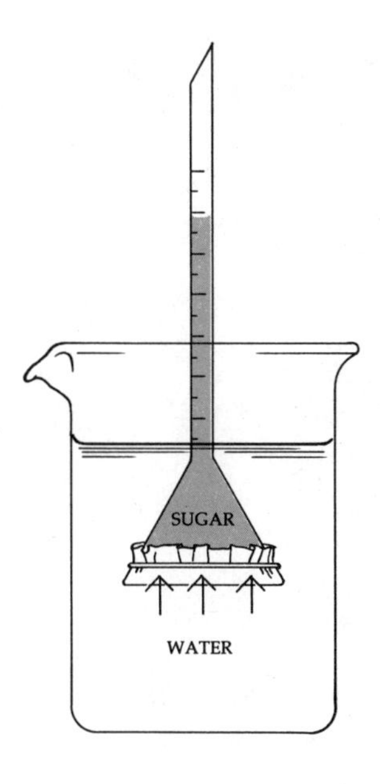

2–26

A simple osmometer. The funnel contains sugar and water, and the beaker contains distilled water. The membrane permits the passage of water but not sugar. The movement of water into the solution will cause the water to rise in the funnel until the water potentials on both sides of the membrane are equal. The amount that the water rises in the funnel indicates the osmotic potential of the solution.

in the funnel is water plus a solute such as sugar or table salt. Across the mouth of the funnel is what is called a selectively permeable membrane; such a membrane is freely permeable to water and not permeable to solute.

The movement of the water into the solution causes it to rise in the funnel until equilibrium is reached, that is, until the water potential on both sides of the membrane is equal (the amount of hydrostatic pressure—in this case, the force of gravity—exerted on the rising water in the funnel equals the osmotic pressure exerted by the concentration of the solution in the funnel).

The movement of water is affected less by what is dissolved in it than by how much—how many molecules or ions of solute—it contains. The word "isotonic" was coined to describe solutions that have equal osmotic concentrations. There is no net movement of water across a membrane separating two solutions that are isotonic to each other, unless, of course, hydrostatic pressure is exerted on one side. In comparing solutions of different concentration, the solution that has less solute and therefore a lower osmotic concentration is known as *hypotonic* and the one that has more solute and more osmotic concentration is known as *hypertonic*. Since solutes decrease water potential, a hypotonic solution has a higher water potential than a hypertonic one. In osmosis, water molecules move through a selectively permeable membrane into a hypertonic solution until the water potential is equal on both sides of the membrane. Water potential caused by the presence of dissolved substance is known as osmotic potential.

Cells and Osmosis

Cell membranes are selectively permeable. In freshwater protists, such as the paramecium, the interior of the cell is hypertonic to the surrounding water and so water tends to move into the cell by osmosis. If too much water were to move into the cell, it could dilute the cell contents to the point of interfering with function and could even lead eventually to rupture of the cell membrane. This is prevented by the contractile vacuole of the organism, which pumps water out.

Plant cells grow by taking up water. Because plant cells are usually hypertonic to their surrounding solution, water tends to move into them. This movement of water into the cell creates pressure within the cell against the cell wall—just as there would be pressure in the funnel of the osmometer if the water were not permitted to rise. The pressure causes the cell wall to expand and the cell to grow. Some 90 percent of the growth of a plant cell is a direct result of water uptake.

If the water potential on both sides of the plant cell membrane becomes equal, the net movement of water ceases. (We say "net movement" here because water molecules continue to diffuse back and forth across the membrane. However, these movements are in equilibrium; that is, as many water molecules are going in as are coming out.

However, as the plant cell matures, the cell wall stops growing. Moreover, mature plant cells, as we saw in Chapter 2–1, have large central vacuoles. These vacuoles often contain solutions of salts and other materials. (In citrus fruits, for example, they contain the acids which give the fruits their characteristic sour taste.) Because of these concentrated solutions, water continues to "try" to move into the cells. In the mature cell, however, the cell wall does not expand further and equilibrium is not reached. As a consequence, the cell wall remains under constant pressure. This water pressure exerted on the cell wall from inside is known as *turgor*. Turgor keeps the cell wall stiff and the plant body crisp.

If, however, a plant cell is placed in a strong sugar or salt solution—a hypertonic solution—water moves out and the cell body and the vacuole shrink within the cell wall, causing the cell membrane to pull away from the wall. This phenomenon is known as *plasmolysis*.

One-celled organisms that live in salt water are usually isotonic with the medium which they inhabit. Similarly, the cells of higher animals are isotonic with the blood and lymph that constitute the watery medium in which they live.

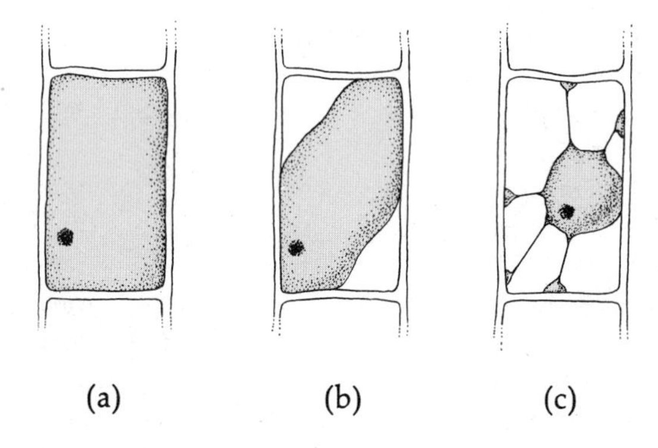

2–27
Plasmolysis in a leaf epidermal cell. Under normal conditions, the cell body fills the space within the cell walls (a). *When the cell is put in a sucrose solution, water passes out of the cell and the cell membrane contracts slightly* (b). *When the cell is immersed in a stronger solution, it loses large amounts of water and contracts still further* (c). *However, the process can be reversed if the cell is then transferred to pure water.*

WATER AND LIFE

Thus, water plays many and varied roles in relation to living things. Life first began in the warm primitive seas, and many types of modern living things spend their entire lives in the oceans or fresh waters of our planet. Every living cell, even the cell of a desert plant or animal, lives in water, and water, for most cells, is the source of the nutrients, minerals, vitamins, and other substances on which the cell depends. Finally, the bulk of the substance within a living cell is again water, which bathes and buoys the cell organelles, dissolves other vital chemicals, and often, as we shall see, takes a crucial part in cellular reactions.

There are two essential ingredients in the biosphere, the thin film on the surface of the planet in which life can exist. One of these is energy, supplied by the rays of the sun, and the other is liquid water. Wherever these are found, life is there, too.

SUMMARY

One of the chief components of cells is water. Water, which is the most common liquid in the biosphere, has a number of remarkable properties, which are responsible for its "fitness" for its role in living systems. These include a high surface tension, unusual capacities for cohesion and adhesion, and a high specific heat. These properties are a consequence of the chemical structure of the molecule. Water is made up of two hydrogen atoms and one oxygen atom held together by covalent bonds. (Covalent bonds, which involve the sharing of electrons between atoms, are the most common bonds in organic chemistry.)

As a result of the arrangement of the electrons, the water molecule is polar, carrying two weak negative charges and two weak positive charges. As a consequence, weak bonds form between water molecules. Such bonds, which link a positively charged hydrogen atom that is part of one molecule to a negatively charged oxygen atom that is part of another, are known as hydrogen bonds. Each water molecule can form hydrogen bonds with four other molecules, and although each bond is weak and constantly shifting, the total strength of the bonds holding the molecules together is very great. These strong intermolecular attractions account for the high surface tension of water, its high specific heat, and the fact that it evaporates relatively slowly.

The polarity of the molecule is responsible for water's capacity for adhesion and also for its capacity as a solvent for ions or other polar molecules.

Water moves by bulk flow, diffusion, and osmosis. The direction in which it moves is determined by water potential; that is, water movement takes place from where the water potential is greater to where it is less. Bulk flow is the overall movement of water, as when water flows downhill. Diffusion involves the random movement of molecules. Osmosis is the movement of water through a membrane which permits the passage of water but inhibits the movement of solutes (a selectively permeable membrane). In the absence of other forces, the movement of water in osmosis is from a region of lesser solute concentration (a hypotonic medium) to one of greater solute concentration (a hypertonic medium).

QUESTIONS

1. Sketch the water molecule and label the areas of positive charge and negative charge. What are the major consequences of the polarity of the water molecule? How are these effects important to living systems?

2. Imagine a pouch with a selectively permeable membrane containing a saltwater solution. It is immersed in a dish of fresh water. Which way will the water move? If you add salt to the water in the dish, how will this affect the water movement? What living systems exist under analogous conditions? How do you think they maintain water balance?

CHAPTER 2–3

The Composition of Living Matter: Some Organic Compounds

Almost a hundred different elements are present in the earth's crust, in the waters that cover much of the earth's surface, and in the thin veil of atmosphere surrounding it. Of these elements, only six make up some 99 percent of all living tissue. These six are carbon, hydrogen, nitrogen, oxygen, phosphorus, and sulfur—conveniently remembered as CHNOPS.

Table 2–2
Atomic Composition of Three Representative Organisms

Element	*Man*	*Alfalfa*	*Bacteria*
Carbon	19.37	11.34	12.14
Hydrogen	9.31	8.72	9.94
Nitrogen	5.14	0.825	3.04
Oxygen	62.81	77.90	73.68
Phosphorus	0.63	0.71	0.60
Sulfur	0.64	0.10	0.32
CHNOPS total:	97.90	99.60	99.72

CHNOPS combine in different ways to make up the molecules found in living systems. Here again, although many combinations are possible, only a few different kinds of molecules are found in large quantities in living systems —five, to be exact. One of these is water. The other four are sugars, lipids (including fats and waxes), amino acids (which combine to form proteins), and nucleotides. Only the first three of these will be discussed in this chapter; we shall take up the nucleotides in later chapters, when we discuss energy exchanges and genetic systems, in which nucleotides play key roles.

Each of these four molecules contains carbon; in fact, the modern definition of an organic compound is that it contains carbon atoms. All also contain hydrogen and oxygen. In addition, amino acids contain nitrogen and sulfur, and nucleotides contain nitrogen and phosphorus.

Interactions of these six atoms and the five types of molecules which they combine to form are responsible, in large part, for the special properties that we associate with living things.

THE CARBON ATOM

Just as knowledge of the structure of the water molecule helps us understand water's unusual characteristics, so we can better understand the central role of the carbon atom in the chemistry of living matter if we look at the way in which it is put together. A carbon atom has six protons and six electrons, two in the inner shell and four in the outer shell:

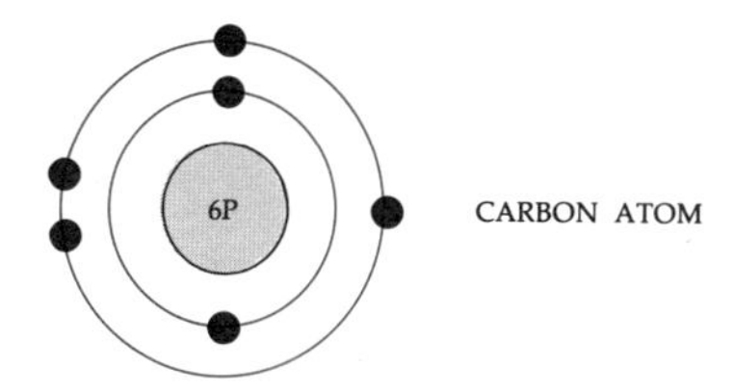

CARBON ATOM

This arrangement means that carbon does not tend either to gain or to lose electrons and become ionized but that, instead, it usually forms covalent bonds. Also, because of the four electrons in its outer shell, each carbon atom can form four covalent bonds simultaneously. For example, one carbon atom can combine with four hydrogen atoms to form methane gas (CH_4):

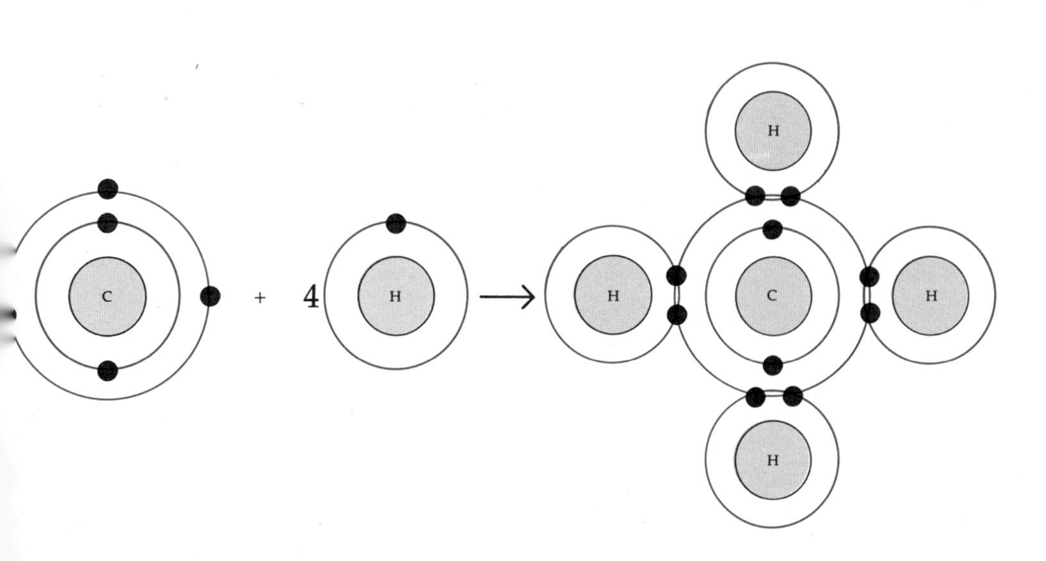

METHANE

Even more important, in terms of its biological role, carbon atoms can form bonds with each other. Butane gas, for instance, consists of a chain of 4 carbon atoms with 10 hydrogen atoms attached:

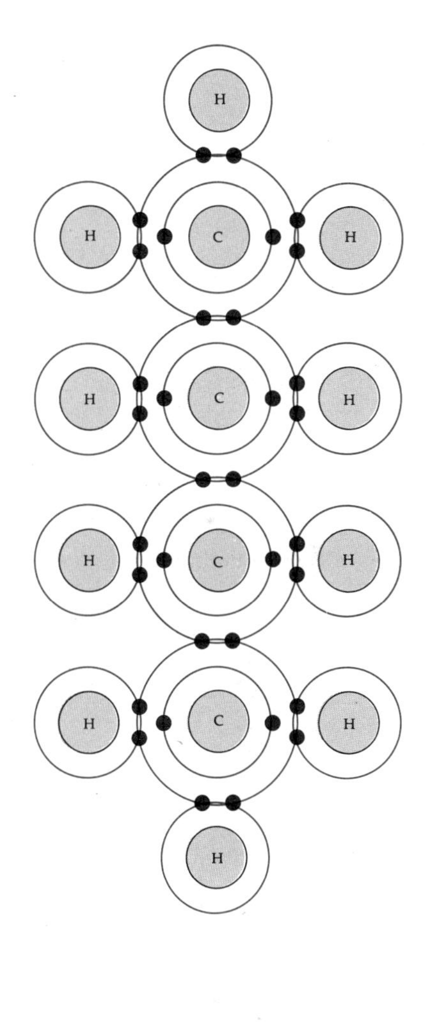

Molecules that consist only of hydrogen and carbon, such as methane and butane, are called *hydrocarbons.* In the primitive atmosphere, it is believed, carbon atoms were present in the form of such hydrocarbon gases. In the modern atmosphere, carbon is present as carbon dioxide. It is brought into the living world by the process of photosynthesis, in which the sun's energy is used to combine carbon, hydrogen, and oxygen into organic compounds.

SUGARS

Sugars, which are formed in the chloroplast by photosynthesis, serve two extremely important functions in the cell: (1) they are the principal energy source for most living systems, and (2) they are the basic materials from which many other kinds of molecules are built.

Sugars are composed of carbon, hydrogen, and oxygen atoms, and are therefore known collectively as *carbohydrates.* In each single sugar, the proportion is 1 carbon atom to 2 hydrogen atoms to 1 oxygen atom, as indicated by the shorthand symbol CH_2O. There are three kinds of sugars: (1) monosaccharides ("single sugars"), such as glucose, ribose, and fructose; (2) disaccharides ("two sugars"), such as sucrose, which is cane sugar, maltose, and lactose; and (3) polysaccharides ("many sugars"), such as cellulose and starch.

A Monosaccharide: Glucose

Glucose is the principal sugar of man and other vertebrates. It is in this form that sugar is generally transported through the animal body; a hospitalized patient receiving intravenous feeding is receiving glucose dissolved in water.

Glucose molecules also, as we shall see, are the structural units of many of the common polysaccharides including starch and cellulose.

MOLECULE MODELS

There are many different ways to describe a molecule. Glucose, for example, has 6 carbon atoms, 12 hydrogen atoms, and 6 oxygen atoms, so $C_6H_{12}O_6$ is one way to describe it. If we want to emphasize the fact that it is a carbohydrate, we can make this point by describing sugar as $6CH_2O$, which, of course, says just the same thing in a slightly different way. However, fructose, another simple sugar, also contains 6 carbons, 12 hydrogens, and 6 oxygens and has a similar structure—a chain of carbon atoms to which hydrogen and oxygen atoms are attached. The symbol = in the drawings below indicates a double bond. Double bonds involve the sharing of four rather than two electrons. The differences between the two molecules are a result of which carbons the other atoms are attached to. The molecules can therefore be described more precisely by drawing the carbon chains:

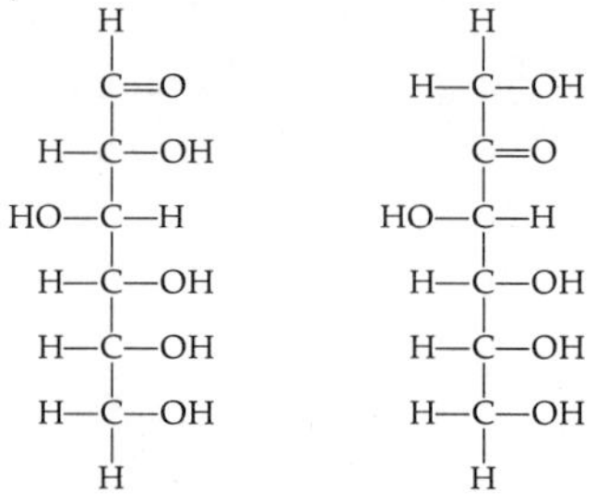

```
        H                  H
        |                  |
        C=O             H—C—OH
        |                  |
     H—C—OH               C=O
        |                  |
    HO—C—H             HO—C—H
        |                  |
     H—C—OH             H—C—OH
        |                  |
     H—C—OH             H—C—OH
        |                  |
     H—C—OH             H—C—OH
        |                  |
        H                  H
     GLUCOSE            FRUCTOSE
```

When glucose and fructose are in solution, however, the chains tend to form rings:

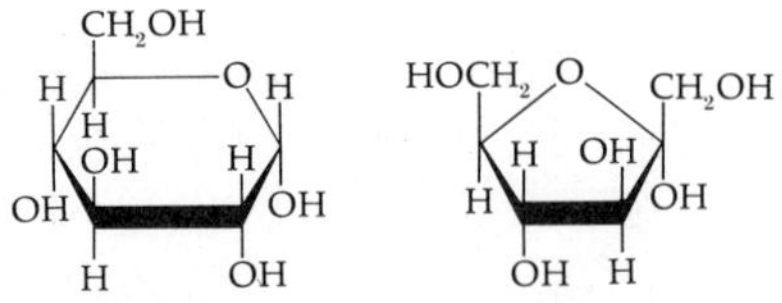

GLUCOSE $C_6H_{12}O_6$

FRUCTOSE $C_6H_{12}O_6$

The lower edge of the ring is made darker and thicker to emphasize the three-dimensional nature of the structure. By convention, the carbon atoms in the ring are "understood" and not labeled in this type of diagram. The position of each carbon atom in the ring is numbered:

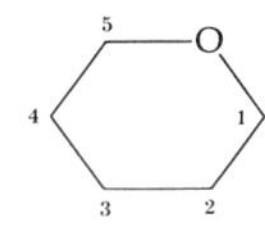

Another way of depicting molecules is by the "ball-and-stick" method, in which the balls are symbols of atoms and the sticks represent the bonds between them. In these models, for instance, carbon atoms are large black spheres, oxygen atoms large white spheres, and hydrogen atoms small black spheres:

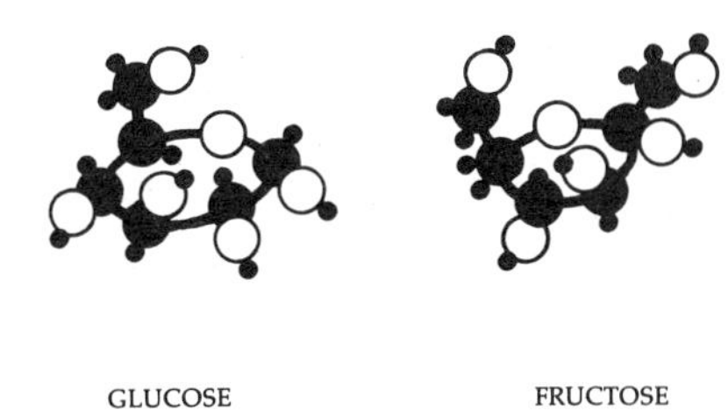

GLUCOSE FRUCTOSE

These models emphasize three-dimensional shape more than the simpler ring models do. They also make it possible to show very clearly the lengths of the bonds (that is, the relative distances between the different atoms) and the angles of the bonds. For instance, the water molecule looks like this in the ball-and-stick model:

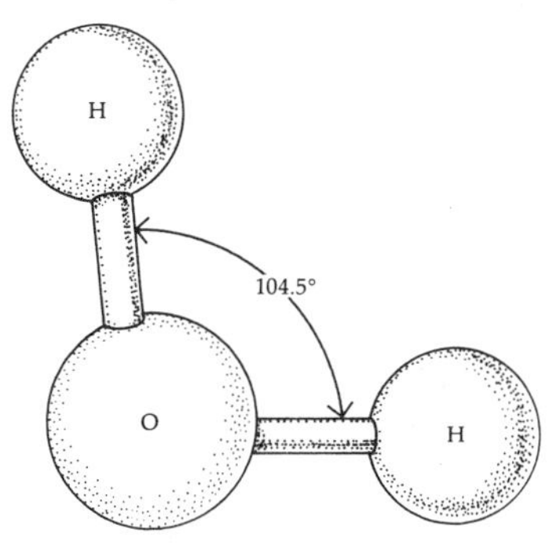

Finally, molecules can be diagramed by space-filling models:

FRUCTOSE

GLUCOSE

Actually, molecules do not fill space in the same way that we think of a table or a rock as filling space. The atoms that make up molecules consist mostly of empty space. As described by a baseball fan, if the outer shell of electrons in an atom were the size of the Astrodome that covers the Houston baseball stadium, the nucleus would be a ping-pong ball in the center of the stadium! What "fills" the space in these models are the paths of the circulating electrons, which set up areas of charge around the atoms. One molecule "sees" another molecule in terms of these areas of charge. Biochemical reactions are very specific; for instance, an enzyme that digests a glucose molecule will not have any effect on a fructose molecule because of the differences in shape. All the intricate biochemistry that goes on in the cell is based on this ability of molecules to "recognize" one another.

It is important to remember that none of these ways of showing the molecules attempts to tell what the molecule actually might look like if we could see it. Rather, they are models. A model, like a theory, is a way of organizing a particular set of scientific data.

Disaccharides

Disaccharides consist of two sugar molecules coupled together. Many common sugars are found in nature in this form. Maltose (malt sugar) consists of two glucose units chemically linked to one another. Lactose, a sugar which occurs only in milk, is made up of glucose combined with another monosaccharide, galactose. Sugar is transported through the blood of insects as another type of disaccharide, trehalose. Sucrose, or cane sugar, is the sugar most common in plants and is our usual table sugar. Sucrose is made up of the monosaccharides glucose and fructose.

Disaccharides are formed by the removal of a molecule of water from the two monosaccharide molecules:

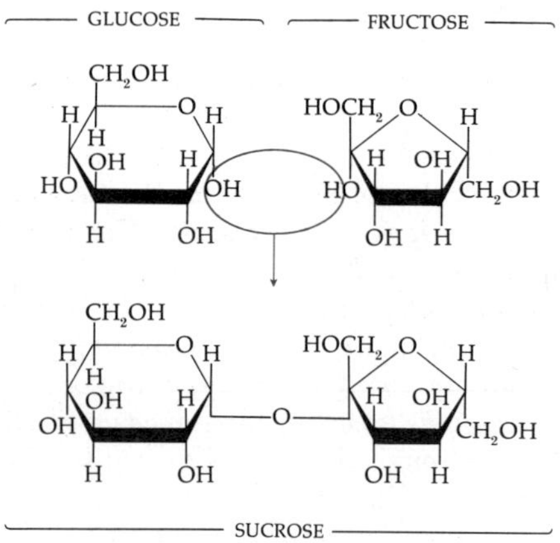

GLUCOSE — FRUCTOSE

SUCROSE

The addition of a molecule of water splits the sucrose molecule back into glucose and fructose. This splitting is known as *hydrolysis*, from *hydro*, meaning "water," and *lysis*, "breaking apart."

Polysaccharides

Polysaccharides are made up of monosaccharides linked together in long chains. They constitute a storage form for sugar. Starch is the principal storage polysaccharide in higher plants, and glycogen in higher animals and in fungi. Both are built up of many glucose units. The differences between them are in the length of the polysaccharide chains (starch is made up of longer chains) and in the ways in which the glucose molecules are linked. These sugars must be hydrolyzed to monosaccharides before they can be used as energy sources or transported through living systems.

Polysaccharides also play structural roles. In plants, the principal structural polysaccharide is cellulose. Although cellulose is made up of the same building materials as starch, the arrangement of its long-chain molecules makes it rigid, and so its biological role is very different. Also, because the bonds linking the glucose units in cellulose are different from those in starch or glycogen, cellulose cannot be broken apart by the same enzymes. Once monosaccharides are incorporated into the plant cell wall in the form of cellulose, they are no longer generally available as an energy source. Only some fungi, bacteria, and protists and a very few animals (silverfish, for example) possess enzyme systems capable of breaking down cellulose. Certain organisms, such as cows and other ruminants, termites, and cockroaches, are able to utilize cellulose as a source of energy, but only because of the microorganisms that inhabit their digestive tracts.

Sugar as Energy

In photosynthesis, carbon dioxide and water are combined to produce sugar:

Carbon dioxide + water + energy → sugar + oxygen

When the cell breaks down the sugar molecule, this process is reversed:

Sugar + oxygen → carbon dioxide + water + energy

(a)

(b)

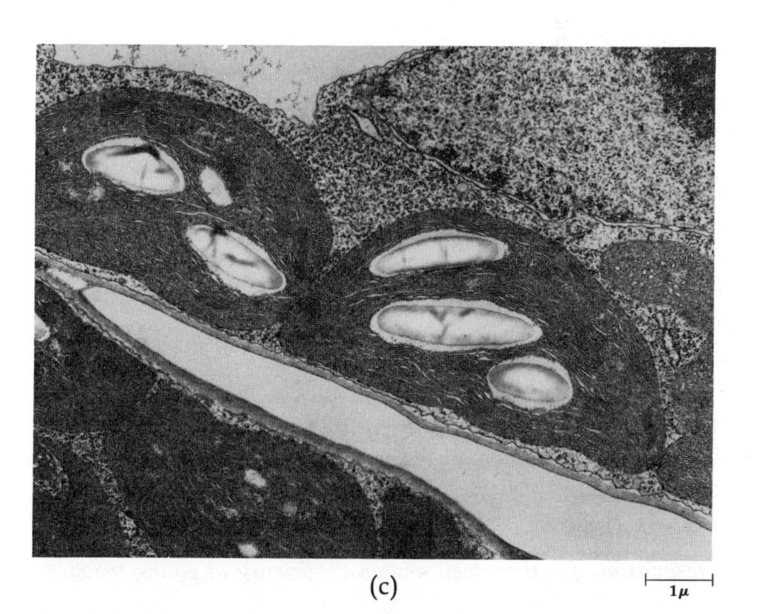

(c)

2–28

In plants, sugars are stored in the form of starch. Starch is composed of two different types of polysaccharides, amylose (a) *and amylopectin* (b). *A single molecule of amylose may contain* 1,000 *or more glucose units in a long unbranched chain, which winds to form a helix. A molecule of amylopectin is made up of about* 48 *to* 60 *glucose units arranged in shorter, branched chains. Starch molecules, perhaps because of their helical nature, tend to cluster into granules.* (c) *In this electron micrograph of chloroplasts in a tomato leaf, the large pale objects within the chloroplasts are starch grains. (Micrograph courtesy of Myron C. Ledbetter.)*

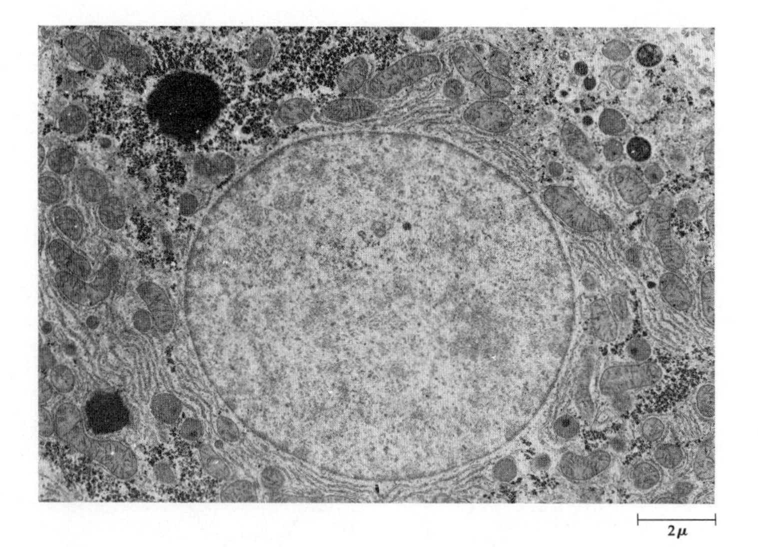

2–29
Glycogen, which is the common storage form for sugar in vertebrates, resembles amylopectin in its general structure except that each molecule contains only 16 to 24 glucose units. The dark granules in this liver cell—such as those surrounding the fat droplet at the upper left—are glycogen. When needed, the glycogen can be converted to glucose and released into the bloodstream.

CH$_2$OH O H H OH H H H OH — O — CH$_2$OH O H H OH H H H OH

(a)

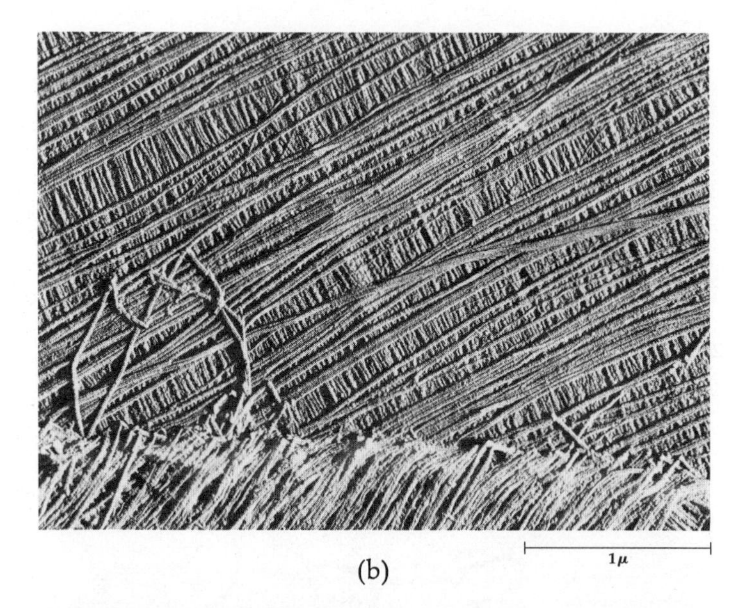

(b)

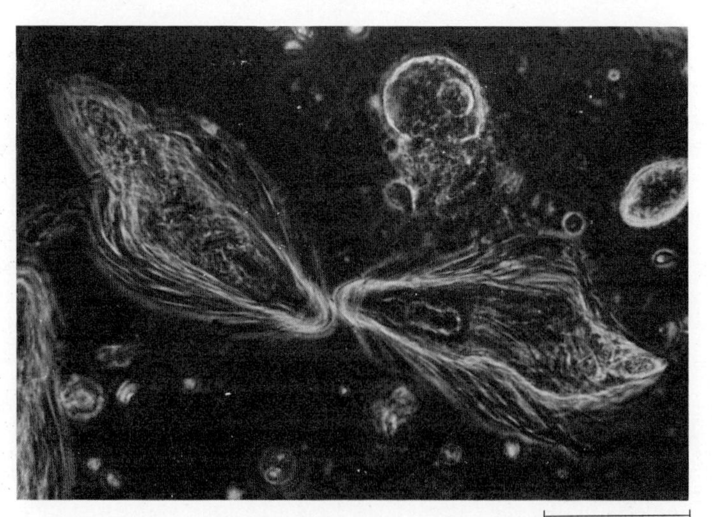

(c)

2–30
Cellulose is the principal structural material in plants. It consists of very long, straight molecules, each containing as many as 2,500 glucose units (a). *The molecules are bound together in microfibrils, each of which is composed of hundreds of cellulose molecules. These microfibrils are organized, in turn, into fibers. The cell wall of a plant cell consists of successive layers of these cellulose fibers* (b). *Because the bonds linking the glucose units in cellulose are different from those linking the glucose units in starch or glycogen, very few organisms can digest them and so utilize the sugar molecules. One such organism is the protist* Trichonympha, *of which two are shown here* (c). Trichonympha *lives in the gut of termites and is entirely responsible for their well-known proficiency at eating wood.*

2–31
Cicada molting. The shells, or exoskeletons, of insects are made of chitin, another type of polysaccharide. Some types of insects, after molting, recycle their sugars by thriftily eating their discarded exoskeletons.

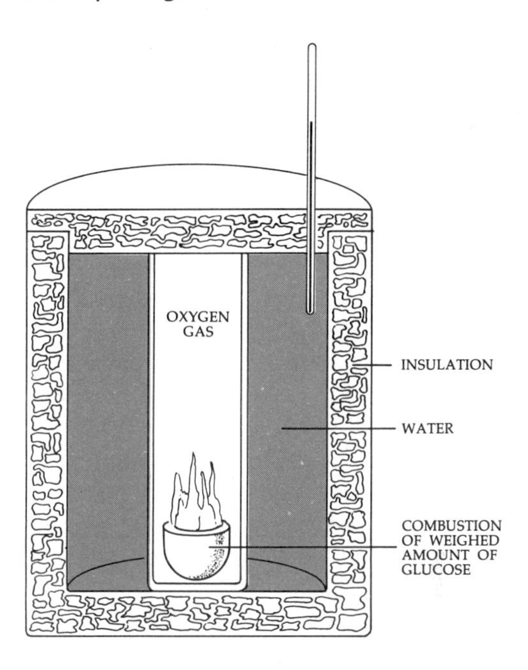

2–32
A calorimeter. A known quantity of glucose, or some other material, is burned; the rise in the temperature of the water indicates the number of calories present in the molecule.

Organic compounds, such as wood, gas, oil, and even paper, also release their stored energy when they are burned. Sugar, too, can be burned to release energy. In these cases, however, the energy stored in the molecule by photosynthesis is converted to heat energy. In the cell, much of the stored energy is converted to other forms of chemical energy and there is comparatively little release of heat, which would not be useful to the cell and, in any large amount, would certainly be harmful. The way in which the cell breaks down the sugar molecule and repackages its energy in a different chemical form is described in Chapter 2–5.

The amount of energy a compound contains is measured in terms of the number of calories released when the compound is burned or oxidized. In such a reaction, chemical bonds are broken and new bonds form. Energy is released because the new bonds have less energy than the old bonds — that is, they are not as strong as the old bonds. There is more energy in the bond holding a hydrogen atom to a carbon atom than in the bond holding an oxygen atom to a carbon atom; so when H—C bonds are broken and O—C bonds form, there is some energy "left over," which is released as heat.

Energy changes taking place during chemical reactions can be measured very precisely by means of a calorimeter (Figure 2–32). A known amount of glucose or some other substance is placed in the crucible and burned. The heat released causes the temperature of the surrounding water to rise. From the known weight of the water and the rise in its temperature, the heat of combustion of the glucose can be calculated in terms of calories. A calorie, remember, is the amount of heat required to raise the temperature of 1 gram of water 1°C.

A given amount of glucose yields the same number of calories when it is converted to carbon dioxide and water, whether this reaction takes place inside or outside the cell.

Sugar as a Starting Material

In addition to serving as the building blocks of polysaccharides, glucose and other monosaccharides also serve as starter materials for the construction of more complex molecules. As you know, sugar taken in the diet is converted to fat for storage if it is not used as energy. The change from sugar to fat involves a partial breakdown and then a rebuilding of the molecule. These processes take place within the cell. The building blocks of protein and of nucleic acids are also synthesized within the cell, frequently using glucose or another monosaccharide as a starting material.

LIPIDS

Lipids are fatty or oily substances. They have two principal distinguishing characteristics: (1) they are nonpolar and so are generally insoluble in water, and (2) they contain a larger proportion of carbon-hydrogen bonds than do other organic compounds and, as a consequence, store a larger amount of energy than do other organic compounds. Fats, on the average, yield about twice as many calories as an equivalent amount of carbohydrate or protein. These two characteristics determine their roles as structural materials and as energy reserves.

Fats

Lipids are stored in the cell chiefly in the form of fat, which cells synthesize from sugars. A fat molecule consists of three molecules of a fatty acid joined to one glycerol molecule. As with the disaccharides and polysaccharides, each bond is formed by the removal of a molecule of water, as shown at the bottom of the page.

The physical nature of the fat is determined by the chain length of the fatty acid and by whether the acid is saturated or unsaturated. In saturated fatty acids, such as stearic acid, every carbon atom in the chain (except the last one) holds two hydrogen atoms, which completes the bonding possibilities of the carbon atom. Unsaturated fatty acids, such as oleic acid, contain carbon atoms joined by double bonds. Such carbon atoms are able to form additional bonds with other atoms (hence the term "unsaturated").

Unsaturated fats, which tend to be oily liquids, are more common in plants than in animals; examples are olive oil, peanut oil, and corn oil. Animal fats, such as lard, contain saturated fatty acids and usually have higher melting temperatures.

Phospholipids

Closely related to the fats are the phospholipids—compounds in which the glycerol molecule is attached to only

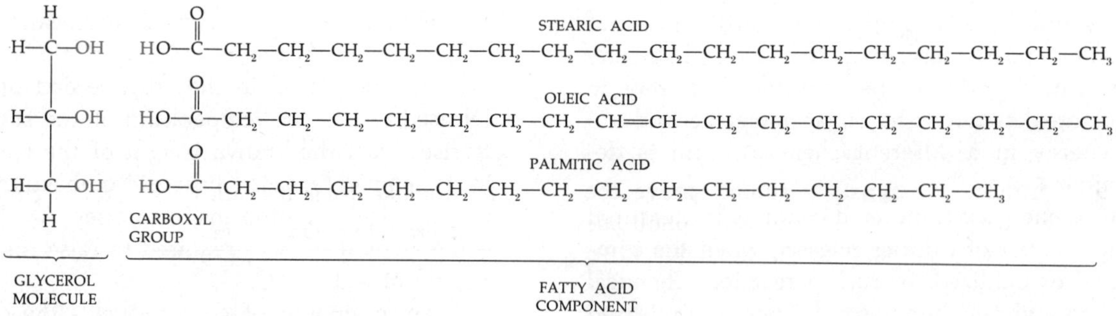

STEARIC ACID

$$H{-}\overset{H}{\underset{|}{\overset{|}{C}}}{-}OH \qquad HO{-}\overset{O}{\overset{\|}{C}}{-}CH_2{-}CH_2{-}CH_2{-}CH_2{-}CH_2{-}CH_2{-}CH_2{-}CH_2{-}CH_2{-}CH_2{-}CH_2{-}CH_2{-}CH_2{-}CH_2{-}CH_2{-}CH_2{-}CH_3$$

OLEIC ACID

$$H{-}\underset{|}{\overset{|}{C}}{-}OH \qquad HO{-}\overset{O}{\overset{\|}{C}}{-}CH_2{-}CH_2{-}CH_2{-}CH_2{-}CH_2{-}CH_2{-}CH_2{-}CH{=}CH{-}CH_2{-}CH_2{-}CH_2{-}CH_2{-}CH_2{-}CH_2{-}CH_2{-}CH_3$$

PALMITIC ACID

$$H{-}\underset{H}{\underset{|}{\overset{|}{C}}}{-}OH \qquad HO{-}\overset{O}{\overset{\|}{C}}{-}CH_2{-}CH_2{-}CH_2{-}CH_2{-}CH_2{-}CH_2{-}CH_2{-}CH_2{-}CH_2{-}CH_2{-}CH_2{-}CH_2{-}CH_2{-}CH_2{-}CH_3$$

CARBOXYL GROUP

GLYCEROL MOLECULE

FATTY ACID COMPONENT

```
                    |
                    O
                    |
                 O=P—O⁻
                    |
 H        H         O
1|       2|        3|
H—C———————C—————————CH2          POLAR HEAD
  |       |
  O       O
  |       |
  C=O     C=O
  |       |
  CH2     CH2
  |       |
  CH2     CH2
  |       |
  CH2     CH2
  |       |
  CH2     CH2
  |       |
  CH2     CH2
  |       |
  CH2     CH2
  |       |
  CH2     CH2
  |       |
  CH2     CH
  |       ||
  CH2     CH                     NONPOLAR TAILS
  |       |
  CH2     CH2
  |       |
  CH2     CH2
  |       |
  CH2     CH2
  |       |
  CH2     CH2
  |       |
  CH2     CH2
  |       |
  CH2     CH2
  |       |
  CH2     CH2
  |       |
  CH3     CH3
```

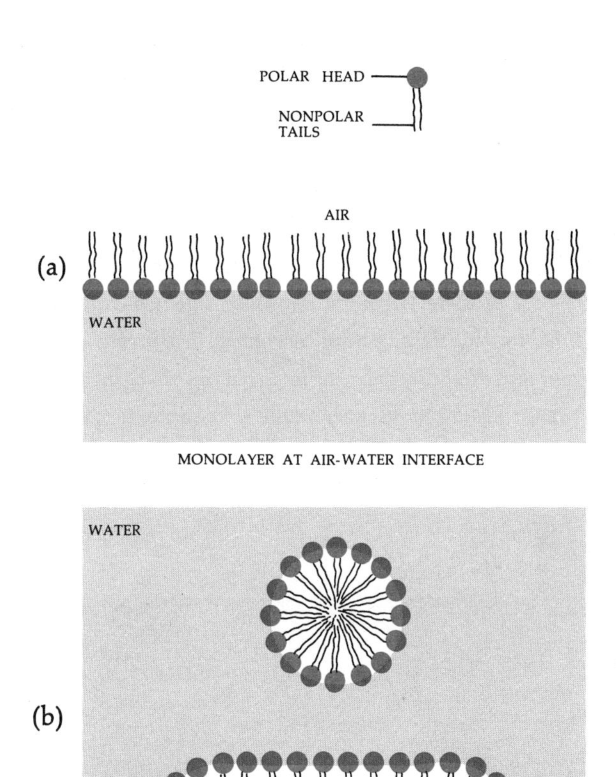

two fatty acids and the third space is occupied by a molecule containing phosphorus as shown above. The phosphate end of the phospholipid molecule is polar and therefore soluble in water, whereas the fatty acids are not. When phospholipids are added to water, they tend to form a film along its surface, with their polar heads under the water and the insoluble fatty acid chains protruding above the surface. In the watery interior of the cell, phospholipids tend to align themselves in rows, with the insoluble fatty acids oriented toward one another and the phosphate ends directed outward. Such configurations, which are assumed spontaneously by these molecules, are probably involved in cellular structures, particularly membranes.

2–33
In the phospholipid molecule, the glycerol-phosphate combination is soluble in water but the fatty acids are insoluble. As a consequence, when placed in water, the molecules tend to form a film on the water surface, with their hydrophilic heads beneath the surface and their hydrophobic tails projecting above it (a). *In the cell, they tend to cluster, forming micelles, with their soluble heads pointing outward* (b). *Self-assembly systems such as this, which depend upon attractions between molecules with different charges, are important in the forming of cellular structures.*

2–34

For reasons which are not clear, diets high in saturated fats, which are largely animal fats, are associated with heart disease. Studies of different populations in seven countries revealed that, with little variation, coronary death rate rose with the amount of saturated fats in the daily diet. In the United States, railroad employees made up the study group.

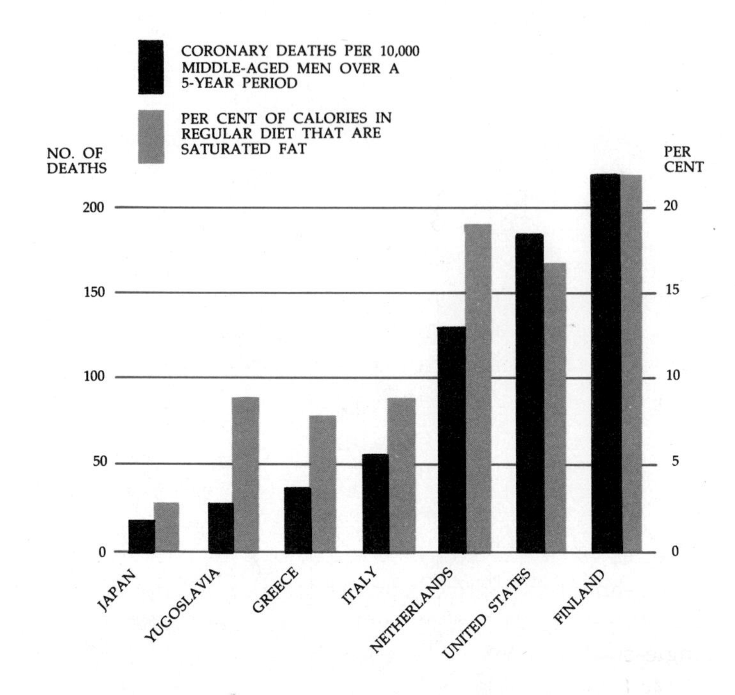

2–35

Waxes are also made of fatty acids. This electron micrograph shows waxy deposits in the upper surface of a eucalyptus leaf. Beneath these deposits is the cuticle, a wax-containing varnishlike layer, and beneath the cuticle is a layer of transparent epidermal cells. Biosynthesis of waxes, which protect exposed plant surfaces from water loss, is a property of all groups of higher plants.

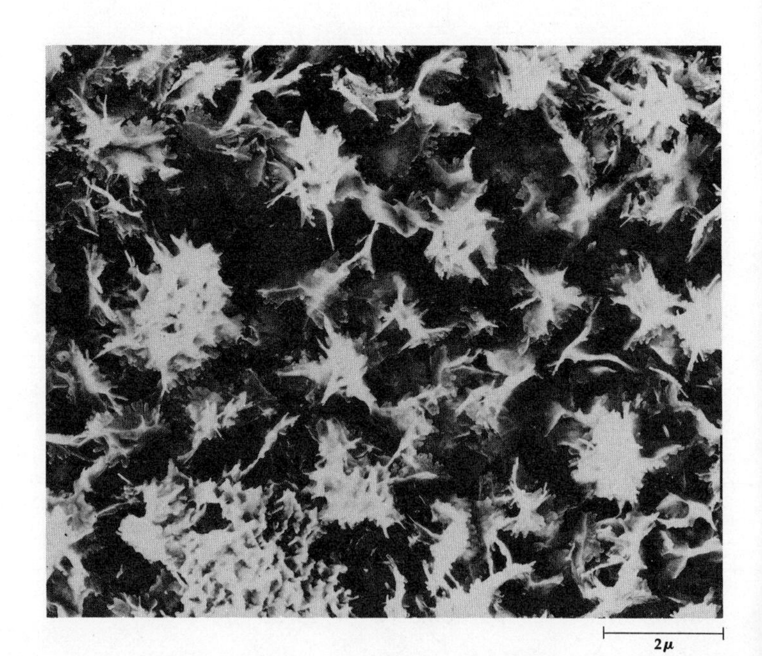

PROTEINS

Proteins, the substances that play such important and complicated roles in living systems, are made up of one or more chains of nitrogen-containing molecules known as amino acids. The sequence in which the amino acids are arranged in these chains determines the biological character of the protein molecule; even one small variation in this sequence may alter or destroy its function. Twenty different kinds of amino acids are found in proteins, and since protein molecules are large, often containing several hundred amino acids, the number of different amino acid sequences, and therefore the possible variety of protein molecules, is enormous—about as enormous as the number of different sentences that can be written with a 26-letter alphabet. The single-celled bacterium *Escherichia coli*,* for example, has 600 to 800 different kinds of proteins at any one time, and the cell of a higher plant or animal probably has several times that number. In a single complex organism such as man, there are thousands, possibly hundreds of thousands, of different proteins, each with a special job to do and each, by its unique chemical nature, specifically fitted for that job.

Amino Acids

Amino acids are made up of carbon, hydrogen, and oxygen, as are sugars. All of them also contain nitrogen. Every amino acid contains an amino group (NH_2) and a carboxyl group (COOH) bonded to a carbon atom:

$$H_2N-\underset{\substack{|\\H}}{\overset{\substack{R\\|}}{C}}-\underset{\substack{\|\\O}}{C}-OH$$

This is the basic structure of the molecule, and it is the same in all amino acids. The "R" stands for the rest of the molecule, which is different in each different kind of amino acid (Figure 2–36).

* Essha-ricky-uh coal-eye, frequently abbreviated as *E. coli*.

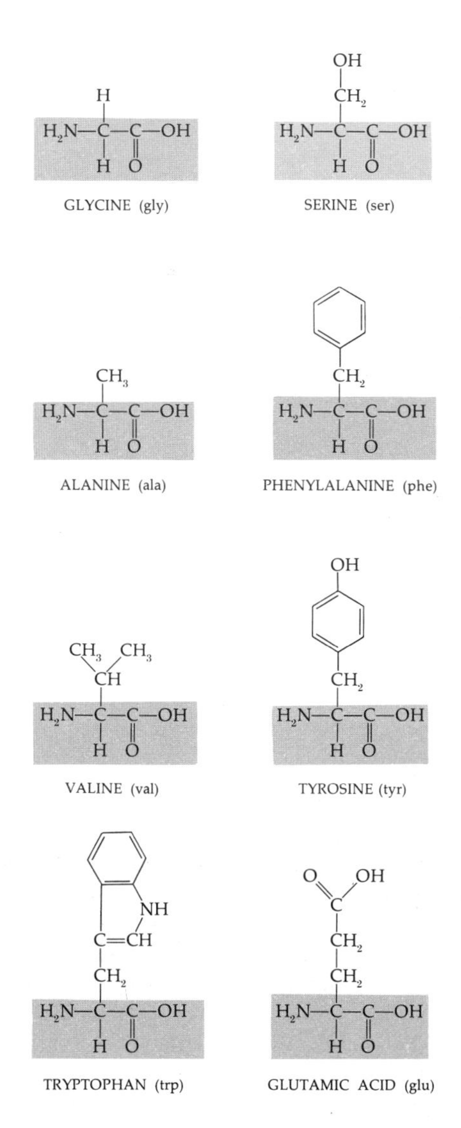

2–36

Eight of the 20 different kinds of amino acids found in proteins. As you can see, their basic structures are the same, but they differ in their R groups. The other 12 amino acids—not shown here—are glutamine (gln), arginine (arg), cysteine (cys), histidine (his), leucine (leu), asparagine (asn), methionine (met), isoleucine (ile), aspartic acid (asp), lysine (lys), threonine (thr), and proline (pro).

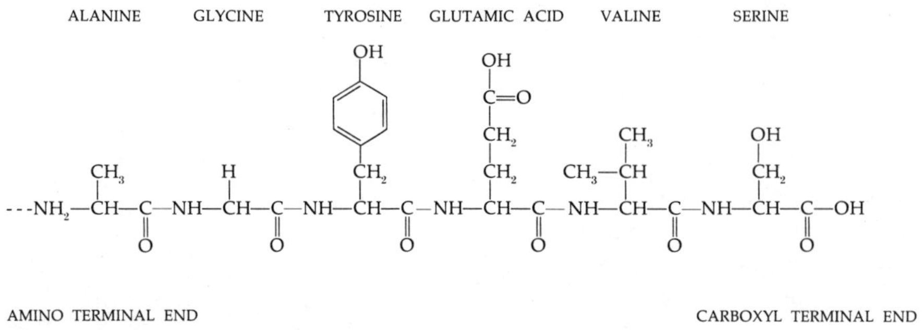

ALANINE GLYCINE TYROSINE GLUTAMIC ACID VALINE SERINE

$$
\cdots NH_2\!-\!\underset{\displaystyle CH_3}{CH}\!-\!\underset{\displaystyle \|\atop O}{C}\!-\!NH\!-\!\underset{\displaystyle H}{CH}\!-\!\underset{\displaystyle \|\atop O}{C}\!-\!NH\!-\!\underset{\displaystyle CH_2\!-\!C_6H_4\!-\!OH}{CH}\!-\!\underset{\displaystyle \|\atop O}{C}\!-\!NH\!-\!\underset{\displaystyle CH_2\!-\!CH_2\!-\!C(\!=\!O)\!-\!OH}{CH}\!-\!\underset{\displaystyle \|\atop O}{C}\!-\!NH\!-\!\underset{\displaystyle CH_3\!-\!CH\!-\!CH_3}{CH}\!-\!\underset{\displaystyle \|\atop O}{C}\!-\!NH\!-\!\underset{\displaystyle CH_2\!-\!OH}{CH}\!-\!\underset{\displaystyle \|\atop O}{C}\!-\!OH
$$

AMINO TERMINAL END CARBOXYL TERMINAL END

2–37

Polypeptides are chains of amino acids linked together by peptide bonds (shown in color), with the amino group of one acid joining the carboxyl group of its neighbor. The polypeptide chain shown here contains six different amino acids, but some chains may contain as many as 300 linked amino acids.

AMINO ACIDS AND NITROGEN

Like fats, amino acids are formed within living cells using sugars as starter materials. But while fats are made up only of carbon, hydrogen, and oxygen atoms, all available in the sugar and water of the cell, amino acids also contain nitrogen. Most of the nitrogen in the biosphere exists in the form of gas in the atmosphere. Only a few organisms, all microscopic, are able to incorporate nitrogen from the air into compounds—nitrites and nitrates—that can be used by living systems. Hence the amount of nitrogen available to the living world is very small.

Plants incorporate the nitrogen in the nitrites and nitrates into carbon-hydrogen compounds to form amino acids. Animals are able to synthesize some of their amino acids, using ammonia as a nitrogen source, but they cannot synthesize others, the so-called essential amino acids, and so must obtain them either directly or indirectly from plants. For adult human beings, the essential amino acids are lysine, tryptophan, threonine, methionine, phenylalanine, leucine, valine, and isoleucine.

Until recently, agricultural scientists concerned with the world's hungry concentrated on developing plants with a high caloric yield. Now, increasing recognition of the role of plants in supplying amino acids to the animal world has led to emphasis on the development of high-protein strains of food plants.

The amino "head" of one amino acid can be linked to the carboxyl "tail" of another by the removal of a molecule of water. (The "tail" of one amino acid loses OH, and the "head" of the next loses an H.) The linkage that is formed is known as a peptide bond (see Figure 2–37), and the molecule that is formed by the linking of many amino acids is called a polypeptide.

As you can see, in order to assemble amino acids into proteins, a cell must have not only a large enough quantity of amino acids but also all the different kinds—just as a typesetter, even though he may have a large supply of letters, cannot set a sentence if he is lacking in e's.

The Levels of Protein Organization

Polypeptides are assembled in the cell on the special cell structures called ribosomes. The head of one amino acid is linked to the tail of another, like a line of boxcars, always in a particular sequence. This linear assembly is known as the *primary structure* of the protein.

As the chain is assembled, it tends to take the form of a helix. The helix is held in shape because of hydrogen bonds that form between every fourth amino acid. The helix is known as the *secondary structure* of the protein.

2–38
The primary structure of the protein ribonuclease, which was synthesized in 1969. Ribonuclease, which contains 124 amino acids, was the first enzyme to be assembled in a laboratory. The dark lines are disulfide bonds. Disulfide bonds form between cysteine subunits; if these subunits were not in exactly the right place, the molecule would not assume this primary structure.

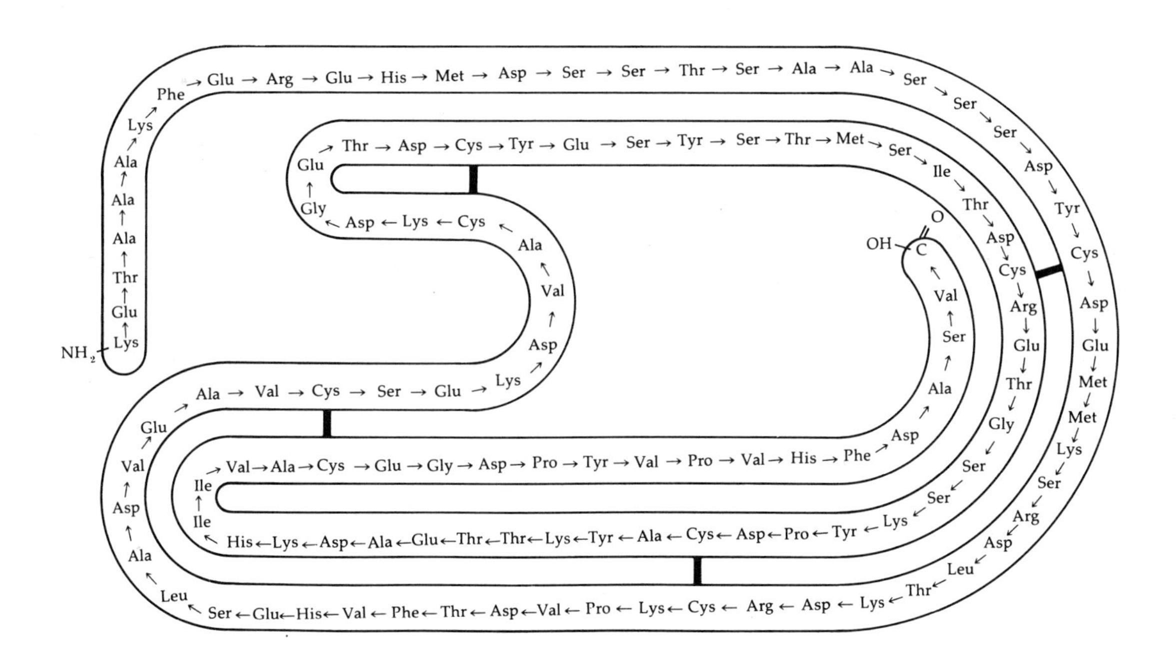

Some proteins, the fibrous proteins, tend to form ropes and sheets; these are the proteins which serve structural functions in the body. Hair, silk, wool, fingernails, and horns are all structural proteins. The most common structural protein in our own bodies is collagen, which is a principal component of cartilage, bone, and tendons. Collagen is made by specialized cells called fibroblasts, which spin out the long protein molecules of which collagen is composed. Three of these molecules wrap around each other to form a cablelike fiber. Collagen consists of collections of such fibers.

In other proteins, the long helix folds back on itself to make a complex *tertiary structure*. Such proteins are called globular proteins; enzymes are an example. These globular structures are also assumed spontaneously; they form as a result of disulfide bonds and of attractions between amino acids with different charges in the polypeptide chain.

Often two or more globular proteins will twine around each other in a *quaternary* (fourth) *structure*. Hemoglobin, for example, the oxygen-carrying molecule of the blood, is made up of four intertwined protein chains, each about 150 amino acids long. In every protein, the primary structure determines the secondary, tertiary, and quaternary structures.

Enzymes

Enzymes are globular proteins that are specialized to serve as catalysts, which are substances that accelerate the rate of a chemical reaction, remaining unchanged in the process. Because they remain unaltered, catalysts may be used over and over again, and so they are typically effective in very small amounts.

In the laboratory, the rates of chemical reactions are usually accelerated by the application of heat, which increases the force and frequency of collision among molecules. But in a cell, hundreds of different reactions are going on at the same time, and heat would speed up all these reactions indiscriminately. Moreover, heat would melt the lipids and would have other generally destructive ef-

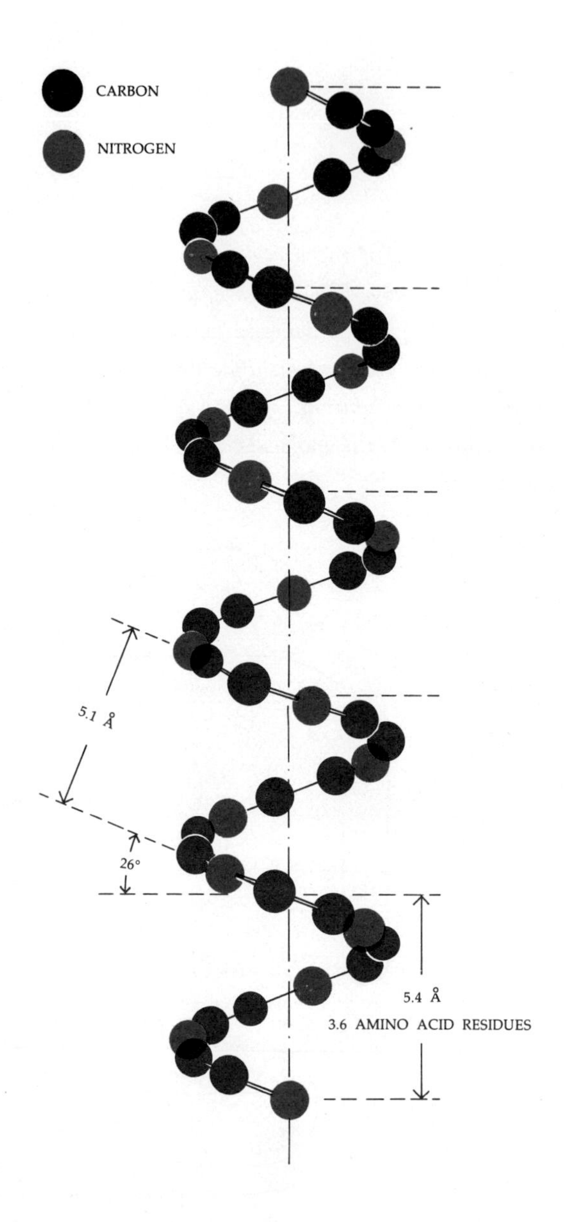

2–39a
The alpha helix arrangement is the most common secondary structure of polypeptide chains. The configuration is very regular in its geometry, with a turn occurring every 3.6 amino acids. The successive turns of the helix are held together by hydrogen bonds.

2–39b
Linus Pauling, who discovered the alpha helical structure of proteins and the role of the hydrogen bond in maintaining it.

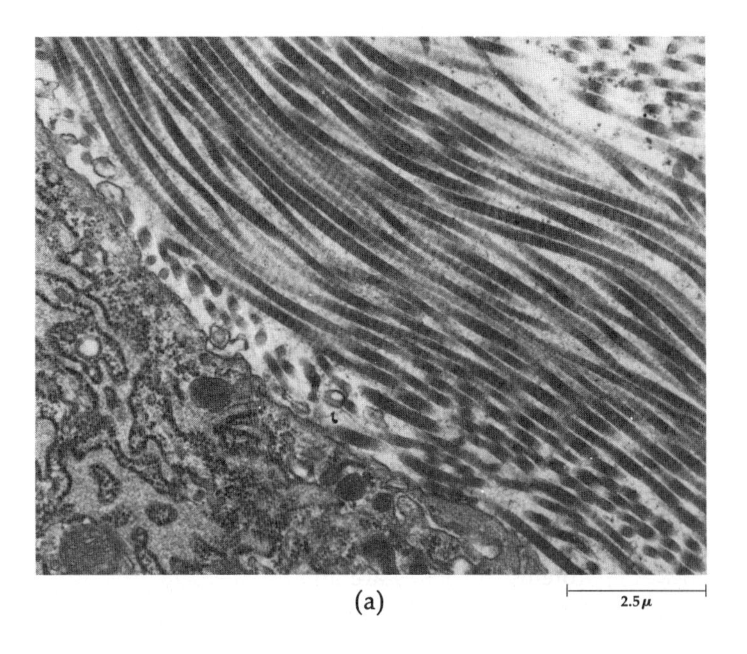

(a)

(b)

2–40
(a) *The structural protein collagen makes up about one-third of the protein in the human body and is probably the most abundant protein in the animal kingdom. Collagen is composed of long polypeptide chains woven together to form fibrils, such as those shown here. These fibers are a major constituent of skin, tendon, ligament, cartilage, and bone.* (b) *Keratin is a structural protein found in hair, wool, horn, nails, skin and feathers. These are cross sections of human hairs as seen by the scanning electron microscope.*

fects on the cell. Because of enzymes, cells are able to carry out chemical reactions at great speed and at comparatively low temperatures. A single enzyme molecule may catalyze as many as several hundred thousand reactions in a second!

The chemical, or chemicals, upon which an enzyme acts is known as its *substrate*. (In the reaction diagramed in Figure 2–41, sucrose is the substrate.) The site on the surface of the globular enzyme molecule into which the substrate fits is the *active site*. Only a few amino acids are involved in any particular active site. Some of these may be adjacent to one another in the primary structure, but often they are brought into proximity to one another by the intricate foldings of the chains involved in the tertiary structure. Over a thousand enzymes are now known, each of them capable of catalyzing some specific chemical reaction.

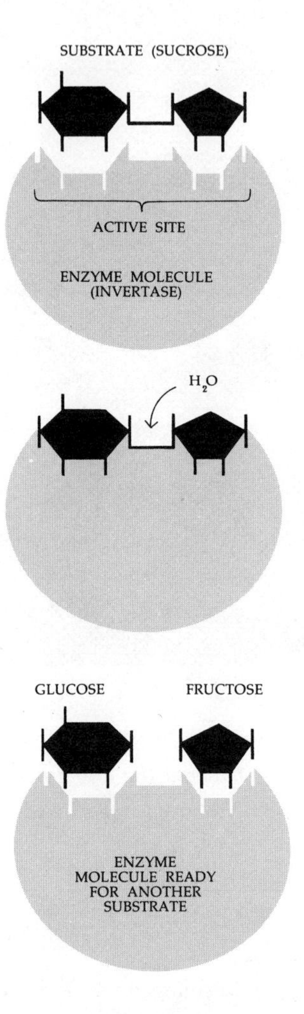

2–41
A model of enzyme action. Sucrose is hydrolyzed to yield a molecule of glucose and a molecule of fructose. The enzyme involved in this reaction is specific for this process; as you can see, the active site of the enzyme fits the opposing surface of the sucrose molecule. For example, a molecule composed of two subunits of glucose would not be affected by this enzyme.

SUMMARY

Living matter is composed of only a few of the naturally occurring elements. The bulk of living matter is water. Most of the rest is made up of organic compounds. Sugars, lipids, and proteins are three of the principal organic compounds in cells. (Nucleotides, to be discussed later, constitute the fourth.)

Sugars serve as a primary source of chemical energy for living systems and as important structural elements in cells. The simplest sugars are the monosaccharides ("single sugars"), such as glucose and fructose. Monosaccharides can be combined to form disaccharides ("two sugars"), such as sucrose, and polysaccharides (chains of many submolecules of sugar), such as starch, glycogen, and cellulose. These molecules can be broken apart again by hydrolysis, the addition of a water molecule.

Lipids are another source of energy and also of structural material for cells. Compounds in this group, which includes fats, waxes, and phospholipids, are generally insoluble in water.

Proteins are very large molecules composed of long chains of amino acids; these are known as polypeptide

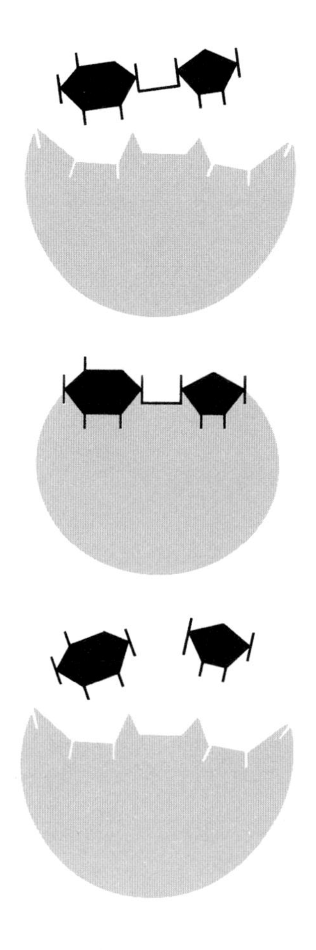

2–42
A new theory of enzyme action. Recent studies by D. E. Koshland at the University of California, Berkeley, indicate that substrates induce changes in shape in their enzymes. According to this theory, the enzyme is active only where it is in the particular shape induced by the substrate. Thus, according to Koshland's model, the substrate must be able to alter the molecule as well as to fit the active site. However, Koshland's theory is similar to the traditional one (illustrated in Figure 2–41) *in its emphasis on the importance of a tight "lock-and-key" fit between the substrate and the active site.*

chains. There are 20 different amino acids in proteins, and from these, an enormous number of different protein molecules are built. Fibrous proteins serve as structural elements, and globular proteins play important dynamic and metabolic roles. The principal levels of protein organization are (1) primary structure, the amino acid sequence; (2) secondary structure, the coiling or spiraling of the polypeptide chain; (3) tertiary structure, the folding of the coiled chain into various shapes; and (4) quaternary structure, the folding of two or more protein molecules around each other.

Because of enzymes, which are globular proteins, cells are able to catalyze reactions at high speed and at body temperature. The specificity of enzymes is due to the "lock-and-key" fit of the active site on the enzyme with the substrate molecule.

PICTURE ESSAY

The Architecture of the Cell

Atomic particles, such as electrons and protons, combine to form atoms, atoms combine to make molecules, and molecules combine to make macromolecules, such as proteins and polysaccharides. These macromolecules, in turn, come together in special ways, at another level of organization, to form the structures within the cell. Some of these structures have been discovered only recently, by use of the electron microscope. Many of them are not fully understood, either in terms of their chemical structure—just how the molecules fit together and assemble themselves or become assembled—or in terms of their function. One fact that is becoming clear, however, is that the cell is a low-budget architect, using the same basic materials—the same architectural modules—over and over again, in different ways for different purposes. In these pages, we are going to look first at some of the structural features common to all cells and then at some specialized structures.

Membranes

Cellular membranes are composed of lipids and proteins. At one time it was thought that every cellular membrane is a three-ply structure, like a sandwich, with the "filling" of the sandwich composed of a double layer of lipid molecules, their tails pointed inward, and the outer layers made up of protein, either in polypeptide chains (a) *or in globules* (b)*:*

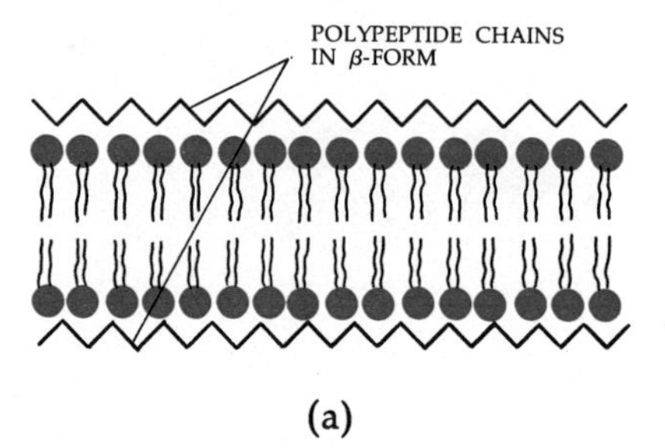

(a)

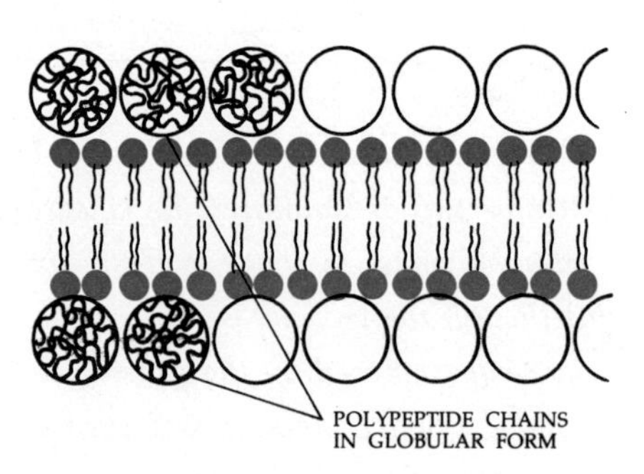

(b)

More recent studies indicate that the arrangement of the lipid and protein components of membranes may be more complex and may vary from cell to cell. Here, for instance, is another possible model, in which polypeptide chains coat clusters of lipids:

Although the exact arrangement of the membrane molecules is not known at this time, it is becoming increasingly clear that membranes play an important role in the life of the cell. All cells are surrounded by an outer membrane. This micrograph shows the outer membrane of a single-celled plant (Chlorella) *which is quite similar to* Chlamydomonas. *The cell wall has been torn away, revealing, in the center of the micrograph, the surface of the cell membrane. The function of the little particles on the membrane surface is not known. The three-dimensional quality of the electron micrograph is due to a technique known as freeze-etching, in which the specimen is frozen and split apart, rather than sliced.*

Endoplasmic reticulum, which fills the upper half of this micrograph, is a system of membranes which separate the cell into channels and compartments and provide surfaces on which chemical activities take place. The round objects that stud the membrane surfaces are ribosomes. Ribosomes are known to play an important part in the assembly of amino acids into proteins, but the exact way in which they function in this important cellular operation is not yet known. This cell is from a pancreas, an organ extremely active in the synthesis of digestive enzymes, which are proteins.

At the bottom of this same micrograph is a mitochondrion, an organelle composed chiefly of membranes. The mitochondrion is the organelle in which food molecules are broken down and converted to readily available chemical energy. Many of these reactions take place on the surface of these inner membranes.

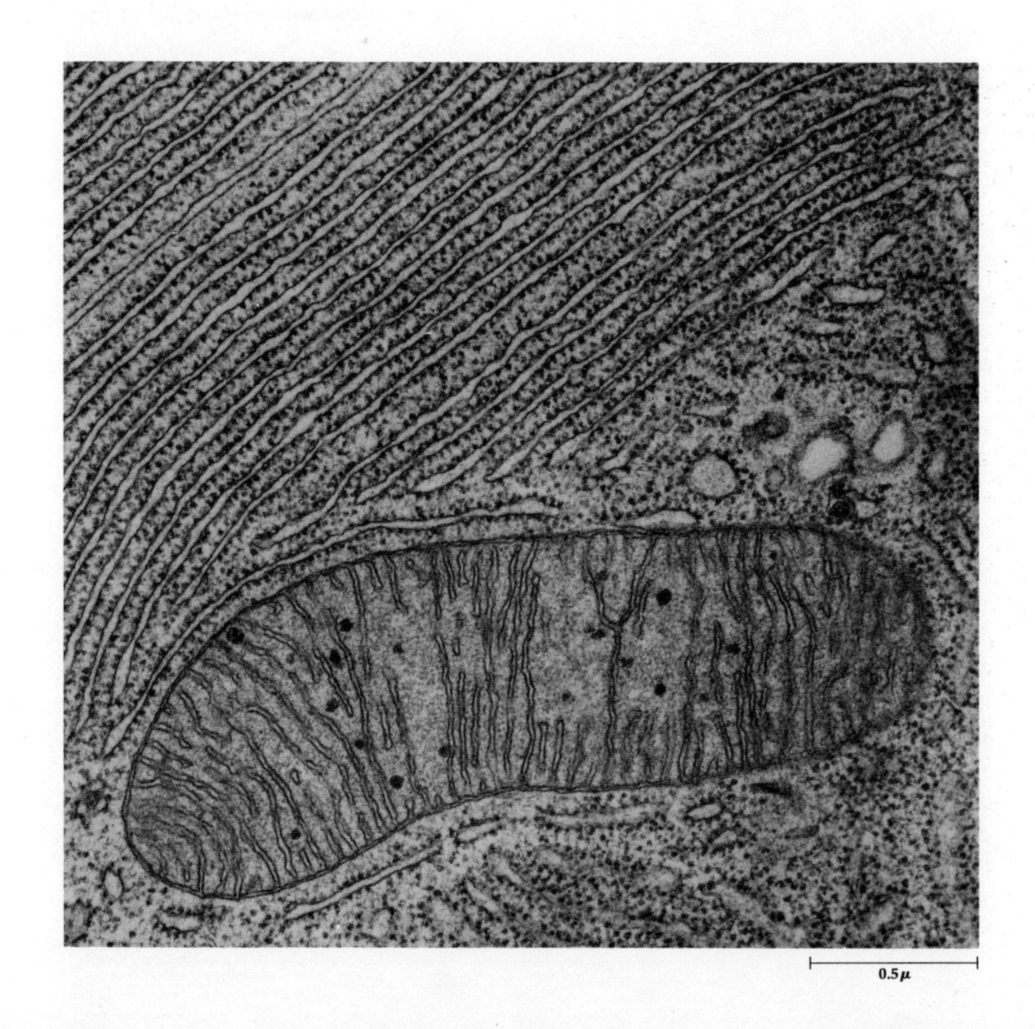

Golgi bodies are also composed of special arrangements of membranes. Compounds formed within the cell are packaged within membrane-enclosed vesicles at the Golgi bodies. The containers then move to the cell surface and release their contents. For instance, in the case of plant cells, the contents of the Golgi containers are used in the construction of the plant cell walls. This Golgi body is from Chlamydomonas.

In this cell from a barley root, you can see numerous Golgi bodies and many strands of endoplasmic reticulum. Numerous mitochondria—the small round bodies with internal membranes—are visible. The large body partially visible at the right is the nucleus, the repository of the cell's genetic information. The outer cell wall is separated from the cytoplasm by the cell membrane. It is visible only as a thin irregular line at the bottom left of the micrograph.

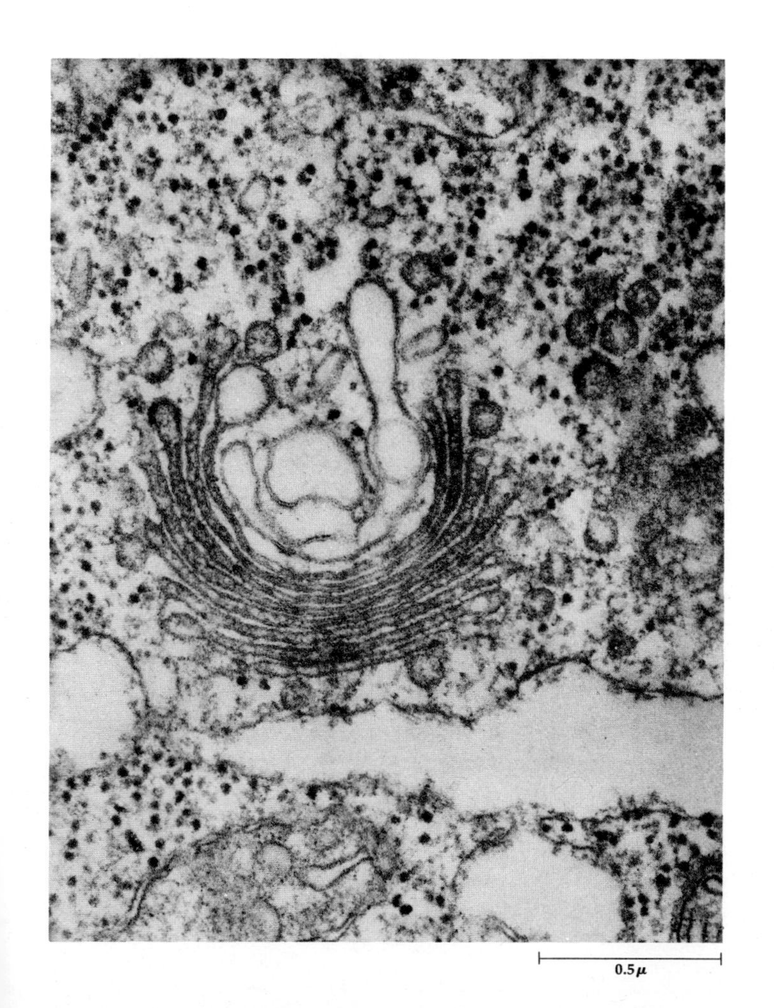

0.5μ

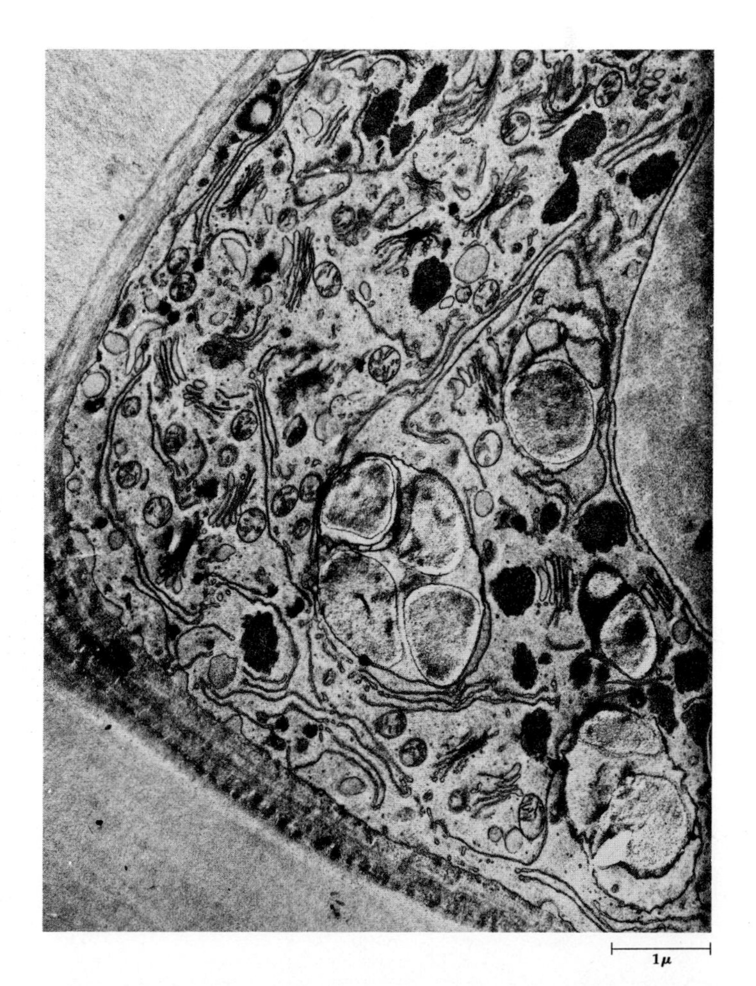

1μ

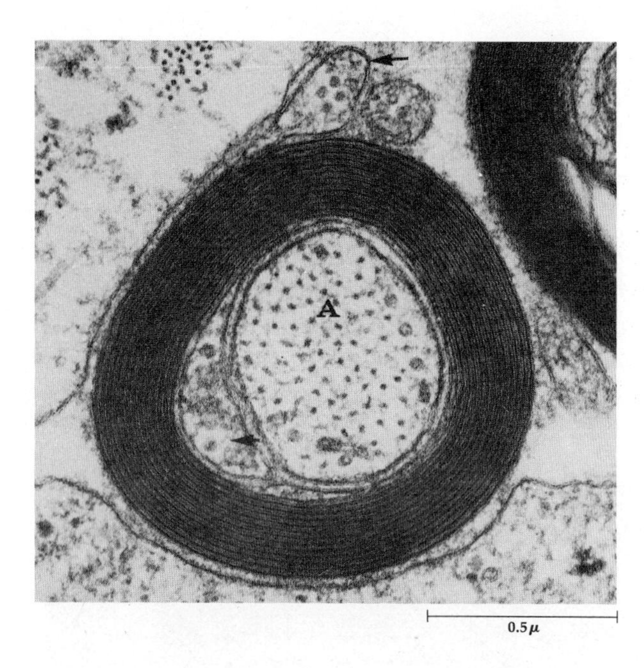

The most extravagant use of membranes in the living cell is seen in Schwann cells, a very specialized type of cell associated with nerve cells. Nerve cells characteristically have long cytoplasmic extensions through which the nerve impulses pass. Schwann cells, as they grow, wrap their cell membrane around and around these extensions, insulating them as with electrician's tape:

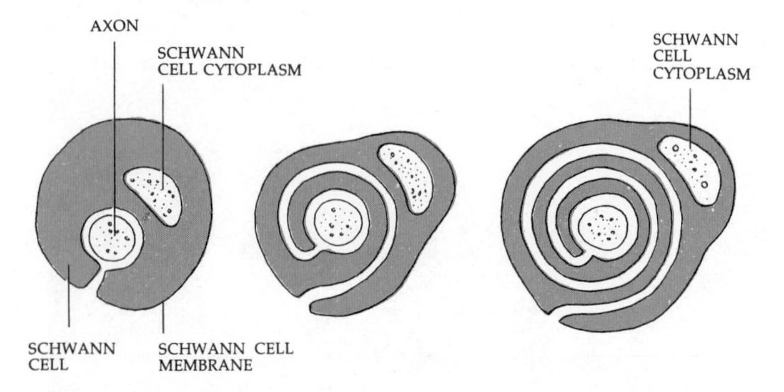

The dark area in this micrograph is a spirally wrapped membrane. The central area (marked A *for axon) is the nerve cell. The cytoplasm of the Schwann cell is indicated by the arrows.*

Tubules

Another universal structure within cells is the microtubule. Microtubules are very slender (only about 200 *angstroms in diameter) and so long that it is almost impossible to trace their whole length in the cell. They are made up of proteins.*

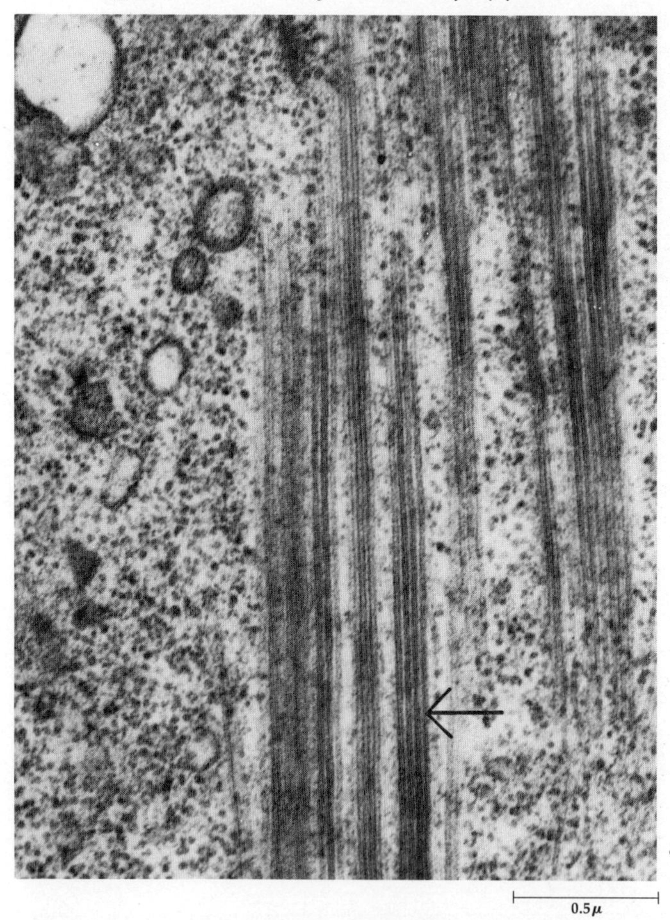

These are bundles of microtubules from the cytoplasm of a paramecium. Such bundles are characteristically found in the area around the gullet, and it has been suggested that they help to guide the food vacuoles away from this area.

This picture shows the same bundles in cross section.

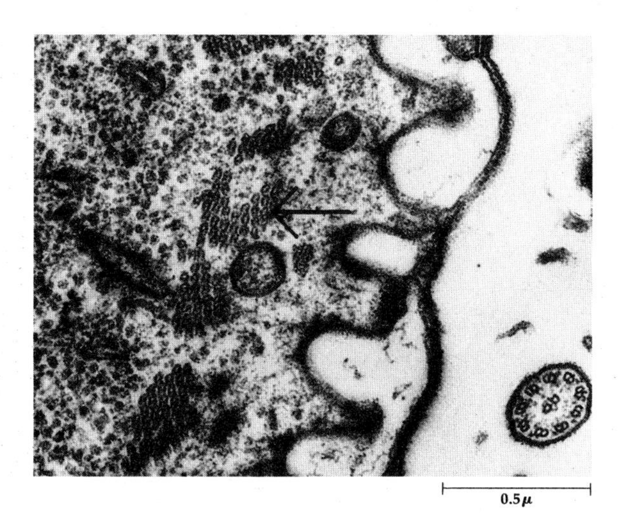

This is a model of the way in which protein subunits assemble themselves to form a microtubule:

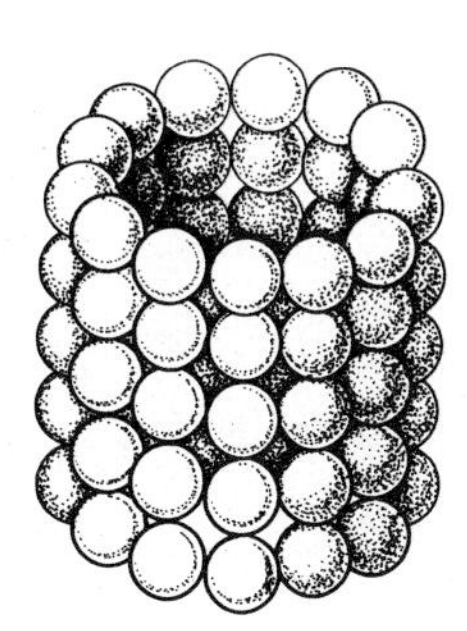

Microtubules in another arrangement make up the apparatus known as the spindle, which forms when cells divide and appears to be involved with movement of the chromosomes. You will see several examples of these in Section 3.

The 9-plus-2 structures found in cilia are arrays of microtubules. This display of cilia in cross section is from the one-celled organism Trichonympha *(Figure 2–29).*

The existence of all these individual structures has been known for some time, but it is only very recently that the electron microscopist has been able to demonstrate that they are all architecturally related in the "master plan" of the living cell.

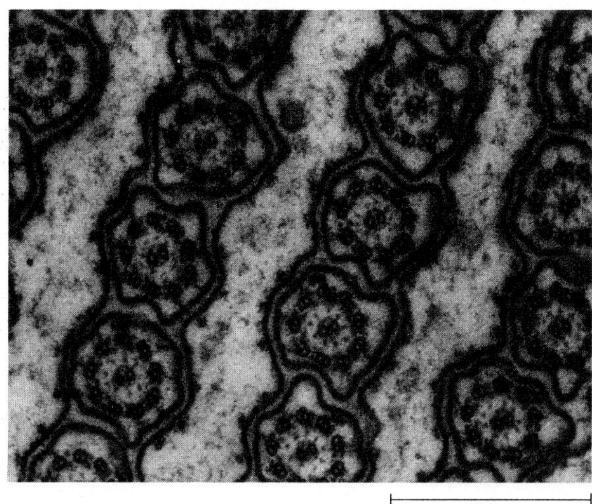

Some Specialized Cells

In the previous pages we discussed structures common to all cells. Specialized cells often have structures which represent unusual combinations of the same relatively few kinds of molecules we described in Chapter 2–3.

One of the best studied of all specialized cells is the muscle cell, shown at right. Muscle cells have the special function of contracting. They are made up largely of fibrils. A single muscle of the sort found in your arms and legs, for instance, is made up of many hundreds of thousands of muscle cells, arranged to form a kind of living cable. Each cell, in turn, contains some 1,000 to 2,000 smaller strands, or fibrils, in parallel array. The fibrils, which are contracting elements, are composed of units of alternating thick (myosin) and thin (actin) filaments.

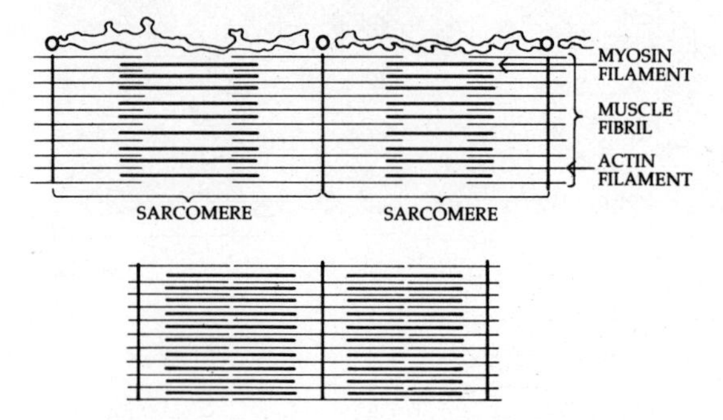

The arrangement of filaments is shown in the diagrams. Notice that the filaments do not become shorter as the fibril contracts (like a rubber band, for instance) but instead slide together.

Both myosin and actin are proteins. Myosin also acts as an enzyme, releasing energy for the muscle contraction. The "arms" extending from the inner tubules of cilia have a similar enzymatic function.

Red blood cells of humans and other mammals lose most of their organelles, including their nuclei, when they mature, so that they become little more than masses of hemoglobin enclosed in a membrane. As you can see below, they are flexible and can bend to pass through small blood vessels. The four red blood vessels shown here—the dark shapes, two on the left and two on the right—are carrying oxygen to the kidney.

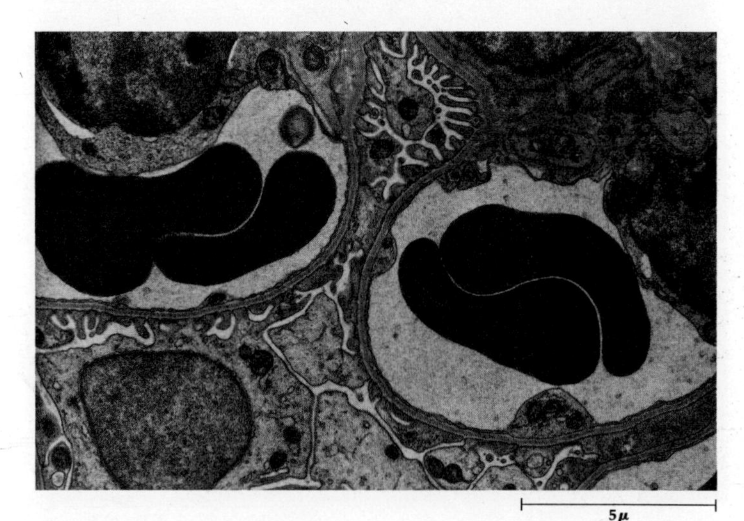

This is a bone cell, or osteocyte. When this cell was younger, it produced bone, which now surrounds it. (The large black area of the micrograph is the bone itself.) A network of canals which connects it to other bone cells supplies nutriments and oxygen. The smaller dark area at the top of the cell is glycogen, the storage form of sugar in animals. Several mitochondria and some endoplasmic reticulum are visible in the lower part of the picture.

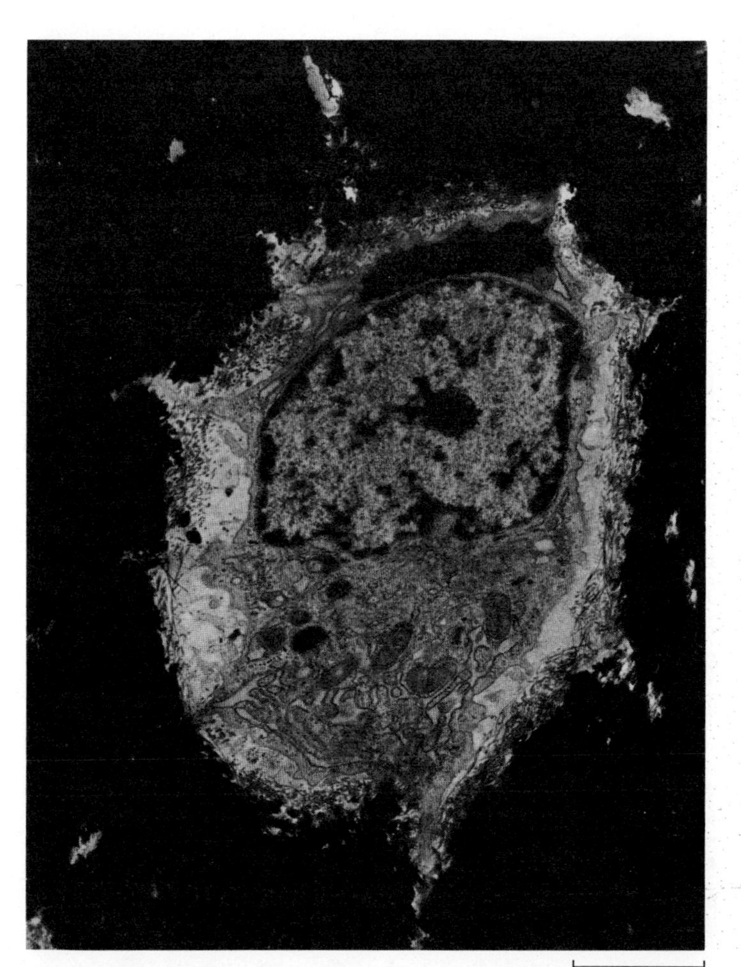

Here are two highly specialized plant cells. These cells, known as vessel elements, are stacked one on top of the other to form the water-conducting elements of the xylem. At maturity, the cells are dead and all that remains, as shown here, is the sturdy outer wall. This wall is mostly cellulose, reinforced with a stiffening material called lignin. To obtain this micrograph, the vessels were cut lengthwise, so that half of the cylinder is seen. The view is from the inside. The thicker line indicated by the arrow is the result of the junction of the two cells. It is the remains of the two cell walls, which perforated as the cells matured, forming one continuous vessel.

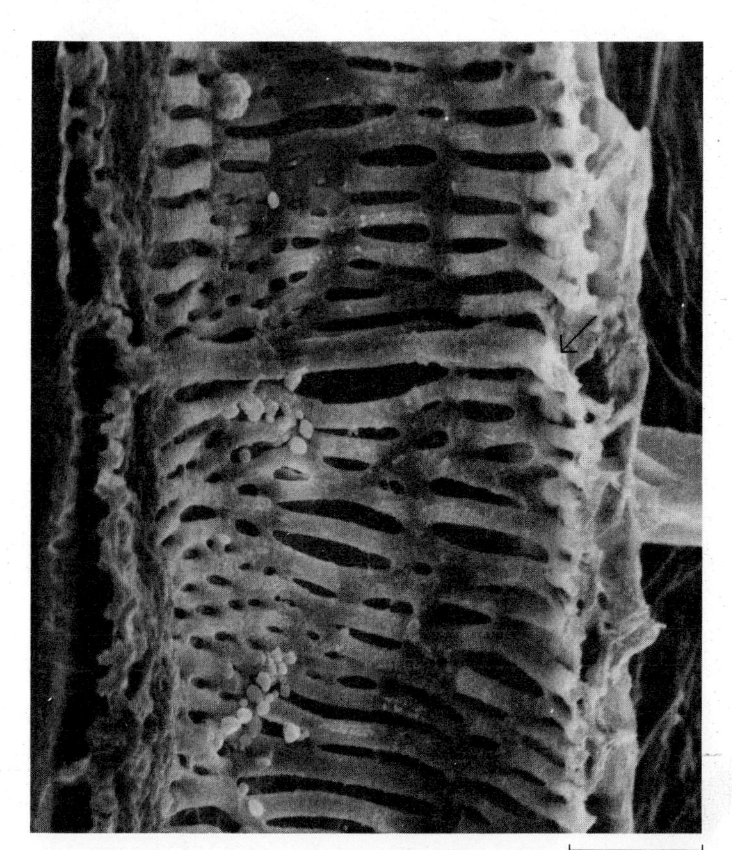

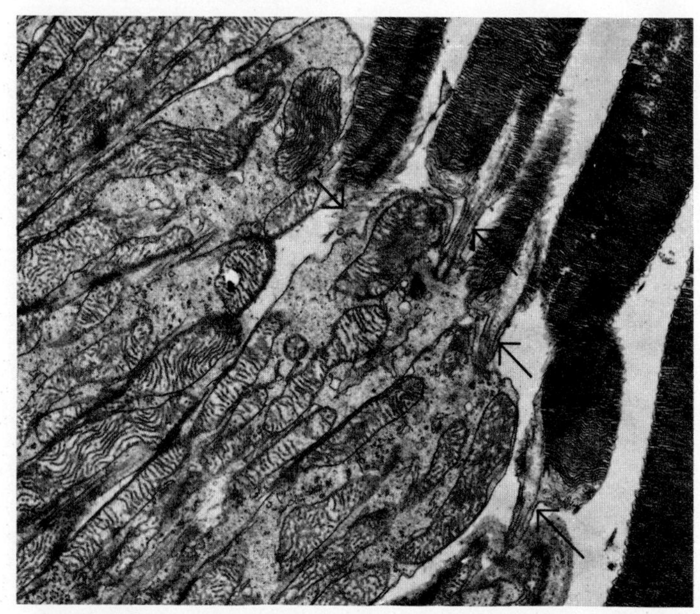

Among the most spectacular cells in the animal world are those involved in light perception. This group of rod cells is from the retina of a kangaroo rat, but the rod cells in your own eyes would look almost identical. The cells are arranged in the retina so that the dark, brushlike portion of the cell is facing toward the back of the eyeball; the light must pass through the cell body to reach it. The brushlike portion consists of a stack of membranes, piled up one on top of the other like poker chips. The visual pigment is built into these membranes. In the lower part of the cell, new pigment molecules and other substances required by the light receptor area are synthesized and transported through a narrow stalk (indicated by arrows) separating the two portions of the cell to the membrane-packed portion, where the actual work of this highly specialized cell is done. A cross-section of this stalk reveals, surprisingly, that it has the internal structure of a cilium, lacking only the two central fibers. The rod cell offers an example of how relatively simple and common structures—membranes and cilia—can be used in new ways to carry out new and specialized functions.

JOURNEY INTO A CELL

Arching over the roof of each nasal cavity is a patch of special tissue, the olfactory epithelium, which is responsible for our sense of smell (olfaction). The tissue is composed of three types of cells: supporting, basal, and olfactory.

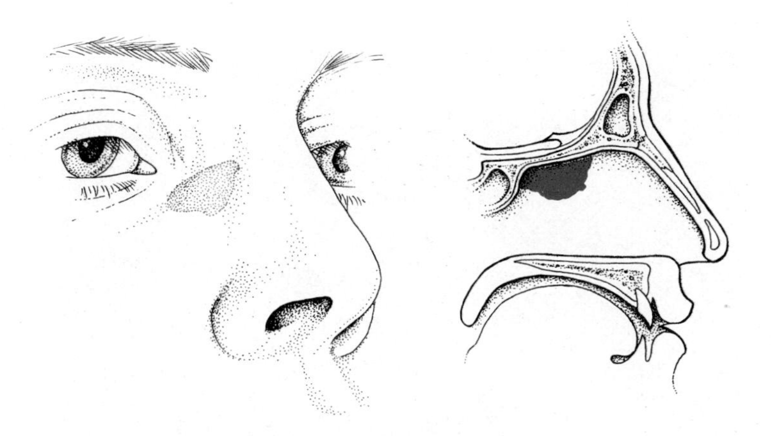

The supporting cells are tall and columnar, wider near the surface than they are deep within the tissue. Their outer surfaces are covered with microvilli, similar to the microvilli found on the surface of intestinal cells.

The basal cells, triangular in shape, are found along the deepest layer of the epithelium. Their function is unknown. They may give rise to new supporting cells when these are needed.

The olfactory cell is the sensory receptor. Its cell body is long and narrow, like that of the supporting cell. The part of the cell nearer the surface, which is very slender and delicate, passes through crevices between the supporting cells and is held upright by them.

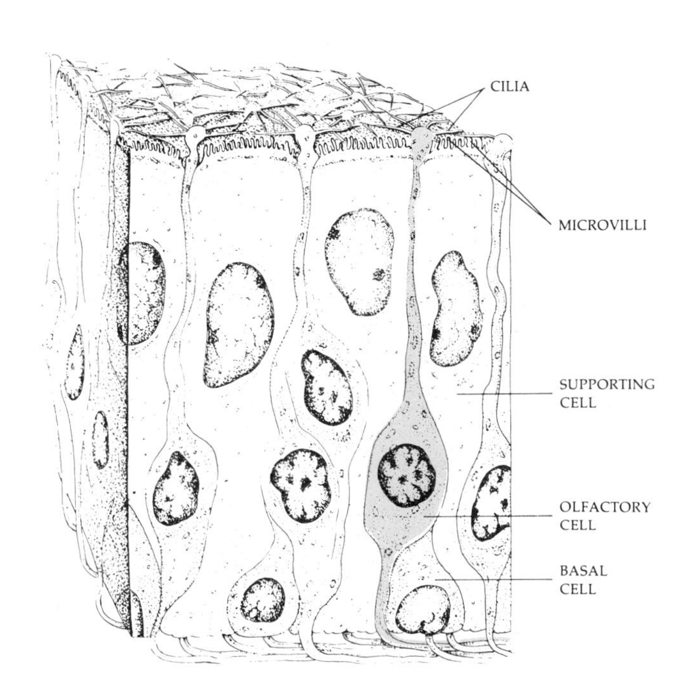

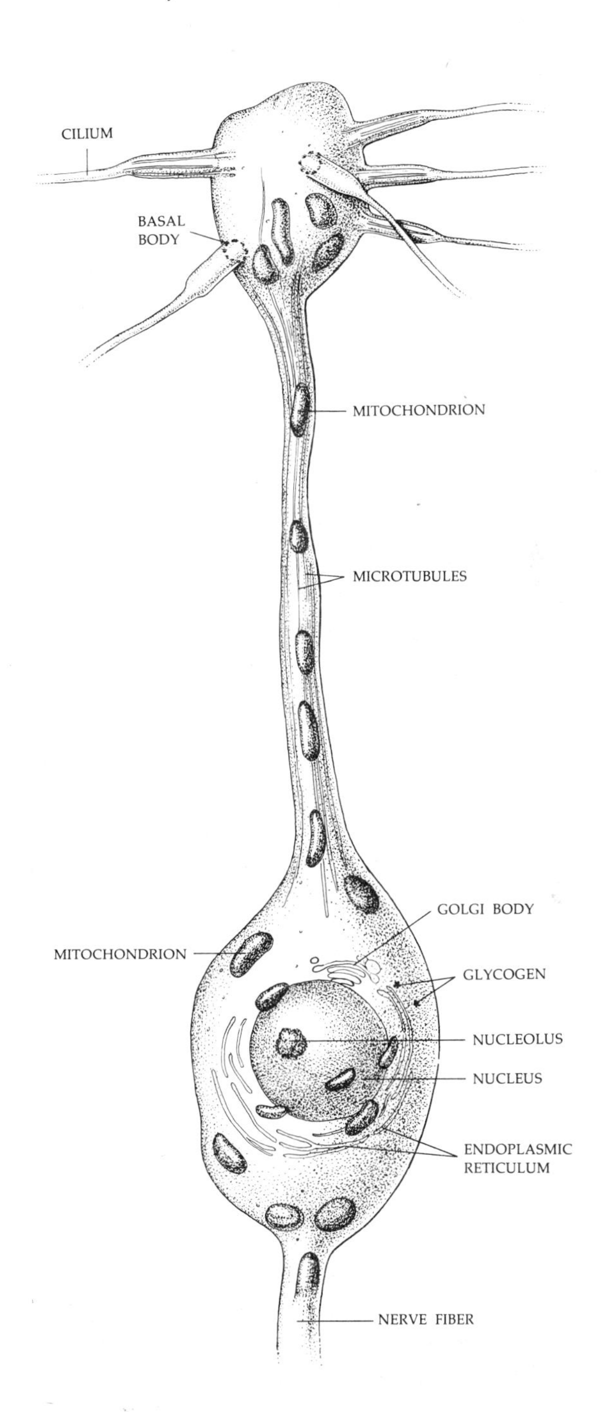

Cilia protrude from the upper surface of the olfactory cell. These are believed to be the odor receptors, although the way in which they function is not known. The shafts of the cilia close to the cell contain the usual 9-plus-2 tubules but the narrow portions contain only two. This outermost part of the cell is connected with the cell body by a long stalk containing microtubules arising from the deeper portion. From the deeper portion of the cell body, a nerve fiber extends through the underlying tissue which transports the sensory message to the brain.

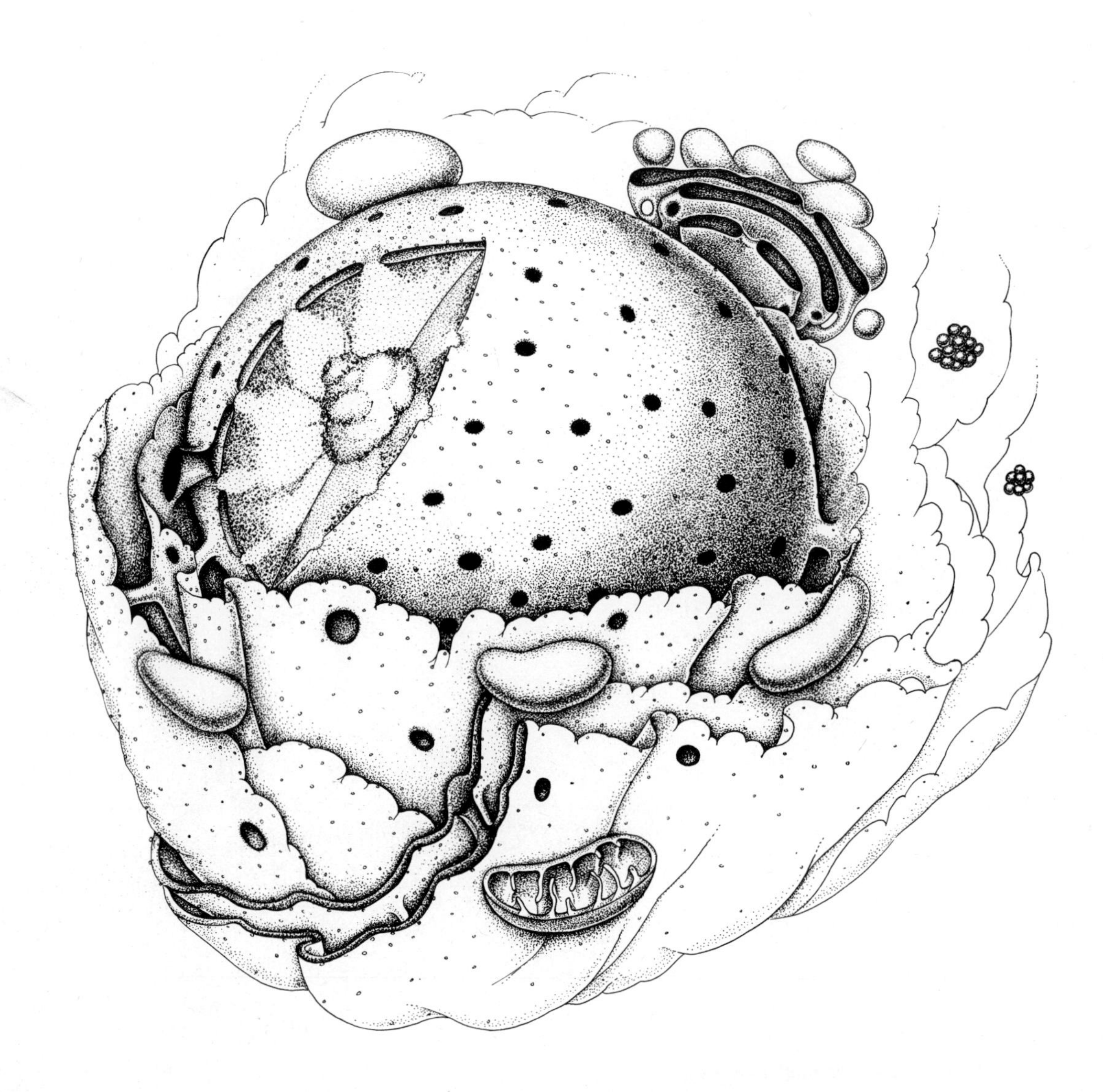

A portion of the interior of the olfactory cell. The dominant body is the nucleus. It is surrounded by a nuclear envelope, which is covered with nuclear pores. (Despite their name, these pores are not actually open but are covered with a thin diaphragm.) Some of the nuclear membrane has been cut away, revealing the underlying nucleolus. Continuous with the nuclear envelope is the endoplasmic reticulum, shown under the nucleus and extending to the right. Above the nucleus is a Golgi body and, to the right of the Golgi body, deposits of glycogen. Many ribosomes are visible, attached to the outer surface of the endoplasmic reticulum and of the nuclear envelope. There are also a number of mitochondria, one of which, at the bottom right, is cut open to show its inner structure.

CHAPTER 2–4

Energy, Light, and Life

Life here on earth depends upon the flow of energy from thermonuclear reactions taking place at the heart of the sun. The energy in the sunlight that strikes the earth becomes, through a series of operations performed by the plant and animal cells themselves, the energy that drives all the processes of life. Essentially, these operations consist of changing energy from one form to another, that is, transforming the radiant energy from the sun into the chemical energy used by living organisms to power the energy-requiring activities of the cell. This energy conversion takes place in two major stages, which together form a complete cycle. By the process of photosynthesis, the green plant cell uses the energy of sunlight to produce glucose and oxygen from carbon dioxide and water. The process of respiration, which takes place in both plant and animal cells, uses the products of photosynthesis—oxygen and glucose—to release this energy in a form that can be used by the cell for its work. The waste products of respiration are carbon dioxide and water.

Before we look at how the cell carries out these energy transformations, let us first look at the broader, related questions of the role of energy in the natural world and of the nature of light, the source of this energy.

2–43
The energy on which life depends enters the biosphere in the form of sunlight. This photograph, taken from Apollo 7, shows the morning sun reflected off the waters of the Atlantic Ocean and the Gulf of Mexico. The land mass in the center of the photo is the Florida peninsula.

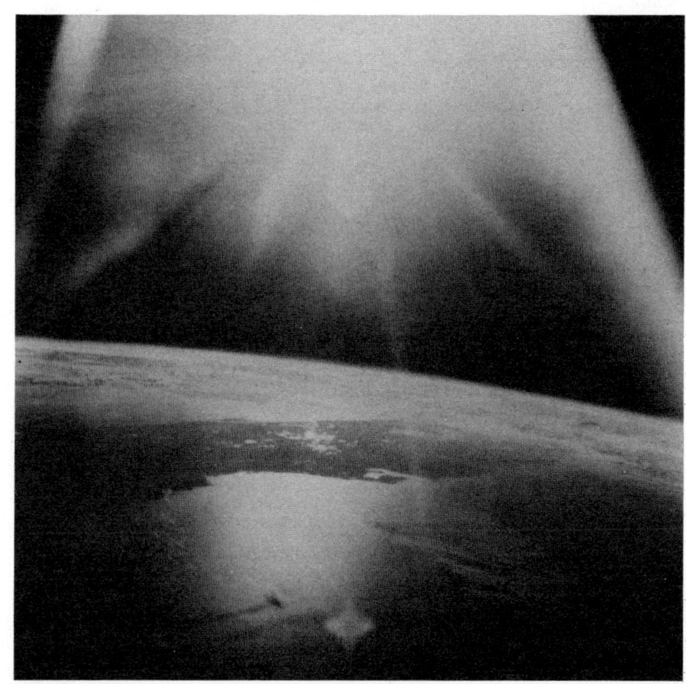

ENERGY IN BIOLOGICAL SYSTEMS

Certain laws apply to energy in both physical and biological systems. These laws are usually referred to as the laws

(a)

(b)

2–44

Energy transformations. (a) *Green leaves transform light energy to chemical energy. Organisms use this chemical energy in various ways. Hummingbirds* (b), *for example, convert large amounts of chemical energy to kinetic energy. Their energy requirements are so high that, by a special adaptation, their temperatures drop while they are at rest, which conserves stored chemical energy.* (c) *Certain organisms, such as these luminescent mushrooms, transform some of their chemical energy back into light energy, thereby glowing in the dark.* (d) *The nest of the mallee bird is a hole* 15 *feet in diameter filled with rotting vegetation. The heat produced as the chemical energy of the vegetation is converted to heat provides the warmth necessary to incubate the eggs.*

(c)

(d)

2–45
This tiny fuel cell can convert the chemical energy of glucose directly into electrical energy. Thus it could be powered by plant saps, for example, or by blood or body fluids. One aim of its developers, who are scientists at the Yerkes Primate Center in Georgia, is to produce this cell in such a form that it can be planted in the body and its electrical energy used to regulate the heartbeat.

of thermodynamics because heat (*thermo*) is the form in which energy changes are usually measured in the laboratory. The first law of thermodynamics states: *Energy can be changed from one form to another, but it cannot be created or destroyed.* The total energy of any system plus that of its surroundings thus remains constant, regardless of the physical or chemical changes it may undergo. Light is a form of energy, as is electricity. Light can be changed to electrical energy, and electrical energy can be changed to light (for example, by letting it flow through the tungsten wire in a light bulb). Energy can be stored in chemical bonds. When we burn carbohydrates, such as wood or paper, we release much of this energy in the form of heat. Warm-blooded animals "burn" (oxidize) sugars in maintaining their body temperature. Living systems are experts in energy conversions.

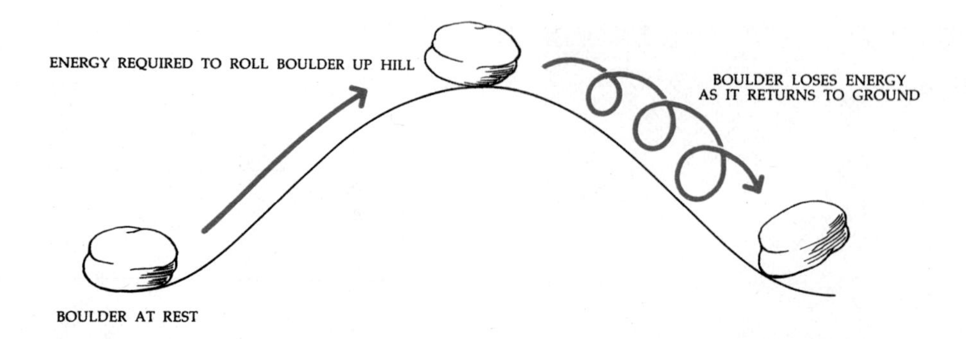

2–46
When a boulder is pushed uphill, kinetic energy is converted to potential energy and heat. When the boulder rolls downhill, potential energy is converted to kinetic energy and heat (which is also a form of energy).

Motion is another kind of energy. A gasoline engine, for example, produces energy in the form of motion. In order to do this, it must use up energy in another form—the chemical energy stored in the gasoline. By burning gasoline as fuel, the engine changes the chemical energy to heat and then changes the heat to mechanical movements.

Energy can be stored in various other forms. A boulder on a hill is an example of stored energy. The energy required to push the boulder up the hill is "stored" in the boulder as potential energy. When the boulder rolls downhill, the energy is released as kinetic (motion) energy.

Some useful energy is always converted to heat. When a boulder rolls downhill, its potential energy is converted to kinetic energy and also to heat energy due to friction. In a gasoline engine, about 75 percent of the energy originally present in the fuel is dissipated to the surroundings in the form of heat. Similarly, an electric motor converts only 25 to 50 percent of the electrical energy into mechanical energy, the rest going into heat. Heat, you will recall, is simply the random motion of atoms or molecules. The higher the heat, the greater the motion.

When we say that energy is "lost" as heat, what we actually mean is that it is no longer available to do work. Energy stored in a stick of dynamite can do work; that part of the energy that is released as heat at the time of the explosion is no longer available for work. The energy in the boulder at the top of the hill is also available to do work. If it is difficult to visualize a rolling boulder as having the capacity to do work, remember that boulders are not conventionally harnessed for this purpose. Substitute water, for which the first law of thermodynamics applies equally well. Suppose that, instead of the boulder, we transport water to the top of our imaginary hill and let it rush down again. We know from experience that the water could be used to turn a series of paddle wheels and that such wheels could be used to grind corn or do other work useful in human terms.

This brings us to the *second law of thermodynamics.* When stated in its simplest form, this law says that *in all natural processes, energy in a form to do work is eventually converted to heat energy, which is dissipated out into the surroundings.* This second law, unlike the first one, shows the direction that natural processes take. As a consequence, it is sometimes known as "time's arrow."

Now let us look at these two laws as they apply to biological systems. The cell is a specialist in energy exchange, interconverting electrical energy, light energy, kinetic energy, and chemical energy and also shifting the energy from one type of chemical bond to another, more convenient form. Most of the attributes that we would select as characteristic of living things are forms of energy exchange. Living organisms constantly take in useful energy from outside sources and release it into the environment in

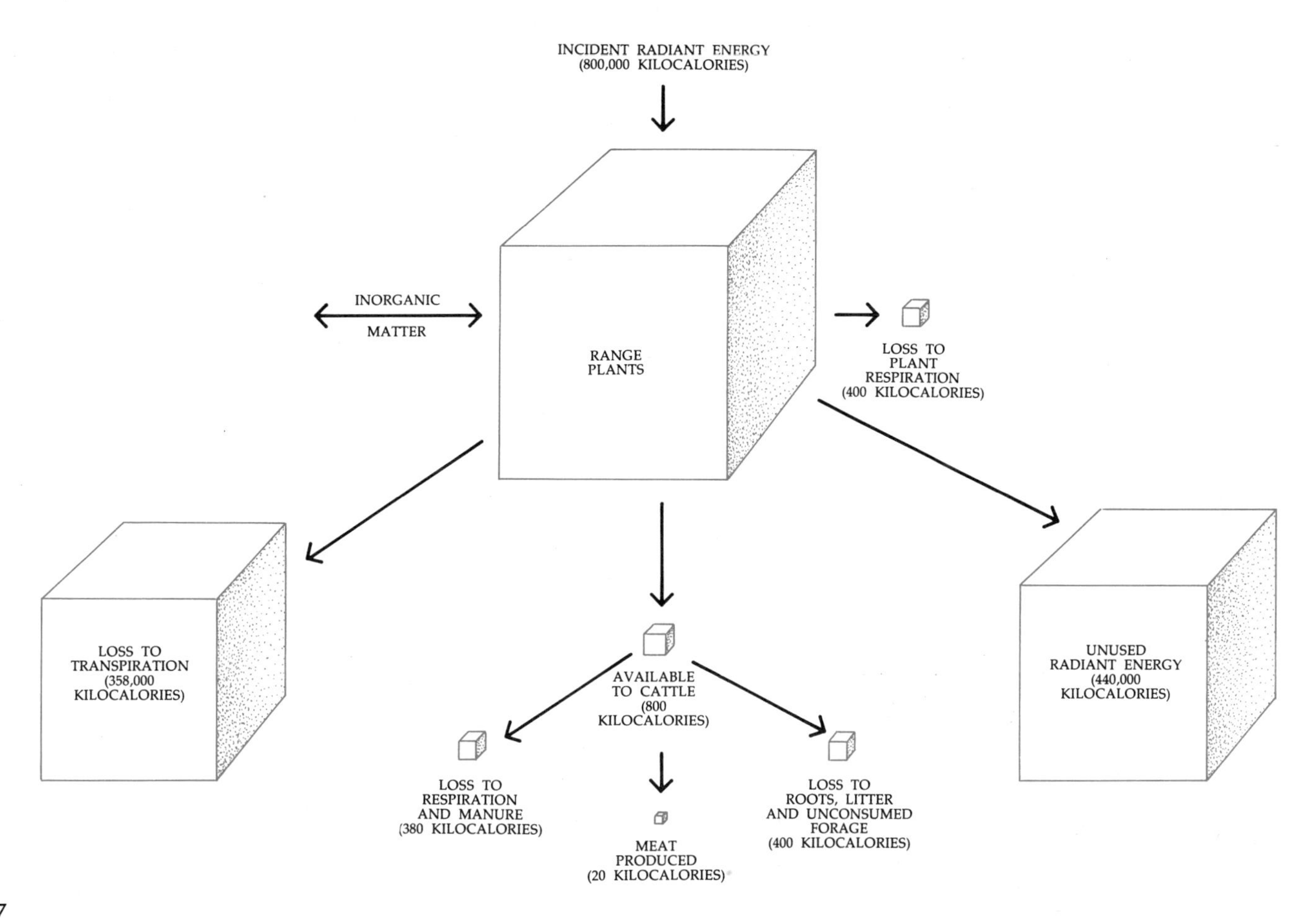

2–47

A biological example of the second law of thermodynamics. How energy is expended from the time it leaves the sun until it reaches the ultimate consumer is shown here for an eight-month period of meat production on an acre of rangeland in California. The top cube represents the 800,000 kilocalories of radiant energy available per square yard from October to May. Only 0.1 percent of this energy, 800 kilocalories, becomes available as food for the cattle grazing the rangeland. Of this amount, no more than 2.5 percent, in the form of 50 pounds of meat on the hoof, is available for human consumption at the end of the eight months.

a less useful form. The ultimate outside source is, of course, the sun. Life can continue to operate at this vast perpetual deficit only because of the tremendous stores of energy flowing into the living world every minute of every day.

The amount of energy delivered by the sun is 13×10^{23} calories per year—the number 13 followed by 23 zeros. It is a difficult quantity to imagine. For example, the amount of energy striking the earth every day is the equivalent of 1 million Hiroshima-sized bombs. About one-third of this energy is immediately reflected back as light (as it is from the moon). Much of it is absorbed by the earth and converted to heat. Some of this absorbed heat energy serves to evaporate the waters of the ocean, producing the clouds that, in turn, produce the rainfall on the land. Solar energy, in combination with other factors, is also responsible for the movements of air and of water that help set patterns of climate. A small proportion of this energy—less than 1 percent—is converted to chemical energy in a form that can be used by living systems. This fraction of the sun's energy is responsible for virtually all life on this planet.

Energy and the Electron

Atoms, you will remember, are made up of a central nucleus that contains protons (positively charged particles) and an outer cloud of electrons, which are negatively charged particles moving in orbitals around the nucleus. Electrons are found at certain fixed distances from the nucleus, arranged in electron shells, which have different energy levels. Energy conversions in chemical systems involve the movement of electrons from one energy level to another.

To understand this a little better, return to the analogy of the boulder. A boulder sitting still on flat ground has no energy. If you push it up a hill, it gains potential energy. So long as it sits on the peak of the hill, it neither gains nor loses energy. If it is permitted to roll down the hill, some of the potential energy is converted into kinetic energy and the rest is converted into heat. Similarly, as we pointed out in Chapter 2–2, water on a hilltop has potential energy, unlike water at the bottom of the hill. Energy has been expended to get the water up there—whether you carried it or an engine pumped it or the sun's energy converted it to water vapor which eventually precipitated as snow or rain. If you trapped the water up there—in a bucket, say, or by a dam or as snow—it would have potential, unreleased energy. If you released it, it could come down the hill, turn a waterwheel, converting some of its energy to another form. When it reached the bottom, all its potential energy would be spent.

The electron is like the boulder or the water on the hilltop in that an input of energy can raise it to a higher level. So long as it remains at this level, it possesses potential energy. When it returns to a lower level, its potential energy is released. The fall of electrons from higher to lower levels is the source of energy for biological work. Photosynthesis is a way of using light energy to push electrons to a higher level.

Oxidation-Reduction Reactions

The movement "uphill" or "downhill" of an electron often involves passing the electron from one atom or molecule to another atom or molecule. The passing of an electron from one molecule to another is known as an oxidation-reduction reaction. The loss of an electron is known as *oxidation*, and the compound that loses the electron is said to be oxidized. The reason electron loss is called oxidation is that many substances—such as carbohydrates—will lose electrons only when oxygen is available to accept them. Thus you can stop a fire from burning by smothering it (that is, cutting off its oxygen), and similarly, you can destroy the energy-yielding process of an animal by suffocating it, which cuts off the oxygen needed for cells to break down carbon compounds and convert their chemical energy to some other form of energy.

Reduction is, conversely, the gain of an electron. Oxidation and reduction take place simultaneously because an electron that is lost by one atom is accepted by another.

Often an electron travels in company with a proton—in short, as part of a hydrogen atom. In that case, oxidation involves the removal of the hydrogen ion (proton) and its electron from one substance and reduction involves the transfer of both a hydrogen ion and an electron to another substance. The reduction of oxygen—the addition of hydrogen atoms—thus results in the formation of water. The reduction of carbon dioxide (which occurs in photosynthesis) can result in the formation of carbohydrate (CH_2O) from carbon dioxide, hydrogen ions, and electrons. The oxidation of carbohydrate, whether it takes place in the cell or by burning wood or gas in a flame, yields carbon dioxide and water and releases the energy put into the molecule during photosynthesis. The water is formed when oxygen

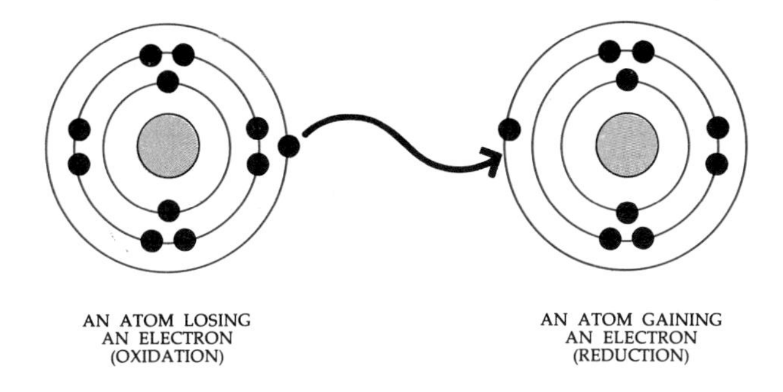

2–48
When an atom (or a molecule) loses an electron, it is said to be oxidized. When it gains an electron, it is reduced.

atoms accept the electrons (and hydrogen ions) removed from the carbohydrate.

As we noted previously, oxidation and reduction always take place simultaneously. However, if a reaction results in a net increase in chemical bond energy, it is referred to as a reduction reaction; for example, the chief result of photosynthesis is the *reduction* of carbon. Conversely, if there is a net decrease in chemical bond energy (with a release of energy as heat or light), the reaction is often referred to as an oxidation process. Sugar is *oxidized* to carbon dioxide and water.

LIGHT AND LIFE

Almost 300 years ago, the English physicist Sir Isaac Newton (1642–1727) separated visible light into a spectrum of colors by letting it pass through a prism. By this experiment, Newton showed that white light is actually made up of a number of different colors, ranging from violet at one end of the spectrum to red at the other. Their separation is possible because light of different colors is bent at different angles in passing through the prism. Newton believed that light was a stream of particles (or, as he termed them, "corpuscles") because, in part, of its tendency to travel in a straight line.

In the nineteenth century, through the genius of James Clerk Maxwell (1831–1879), it came to be known that what we experience as light is in truth a very small part of a vast continuous spectrum of radiation, the electromagnetic spectrum. As Maxwell showed, all the radiations included in this spectrum travel in waves. The wavelengths—that is, the distances from one peak to the next—range from those of gamma-rays, which are measured in angstroms, to those of low-frequency radio waves, which are measured in miles. Within the spectrum of visible light, red light has the longest wavelength, violet the shortest. Another feature that these radiations have in common is that, in a vacuum, they all travel at the same speed—186,300 miles (300,000 kilometers) per second.

By 1900, it had become clear, however, that the wave theory of light was not adequate. The key observation, a very simple one, was made in 1888: When a zinc plate is exposed to ultraviolet light, it acquires a positive charge. The metal, it was soon deduced, becomes positively charged because the light energy dislodges the electrons, forcing them out of the metal atoms. Subsequently it was discovered that this photoelectric effect, as it is called, can be produced in all metals. Every metal has a critical wavelength

2–49
When white light passes through a prism, as Sir Isaac Newton showed almost 300 years ago, it is sorted into a rainbowlike spectrum. By this experiment, Newton showed that white light is actually a mixture of different colors, ranging from violet at one end of the spectrum to red at the other.

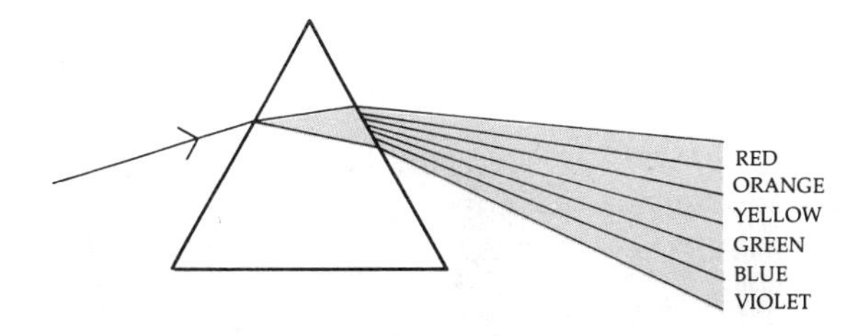

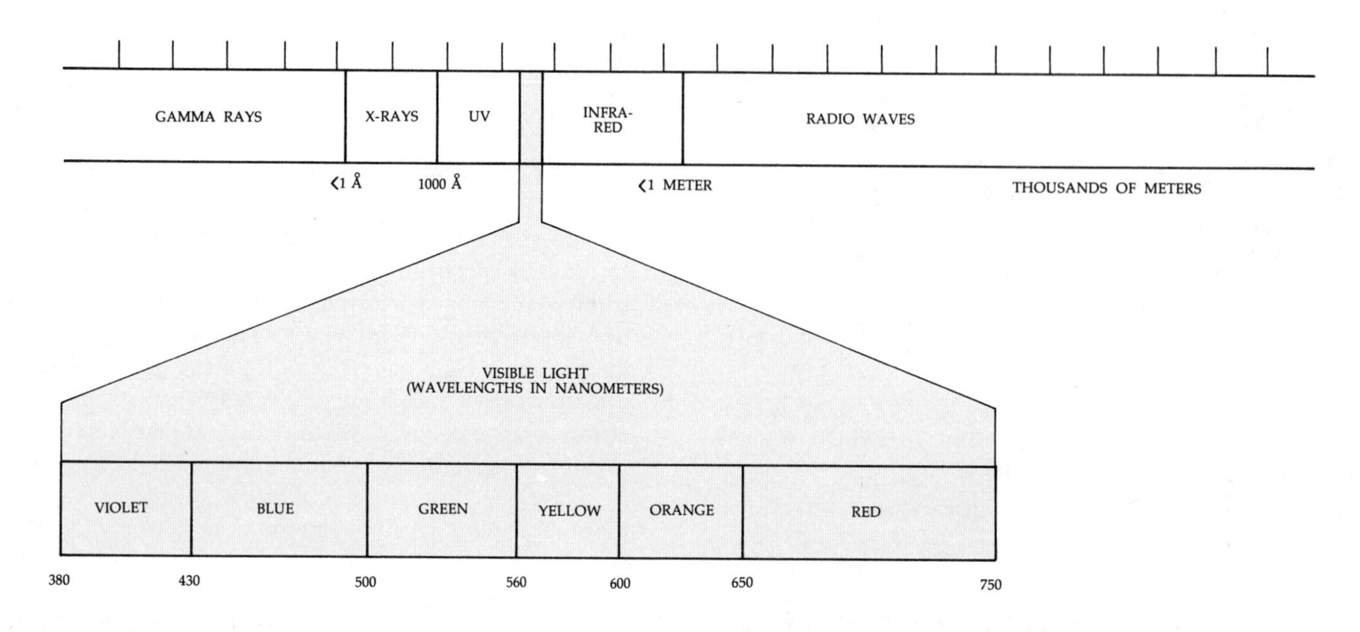

2–50

It is now known that the spectrum of visible light shown in Figure 2–49 represents only a very small portion of the vast electromagnetic spectrum.

for the effect; the light (visible or invisible) must be of that wavelength or a shorter wavelength for the effect to occur.

With some metals, such as sodium, potassium, and selenium, the critical wavelength is within the spectrum of visible light, and as a consequence, visible light striking the metal can set up a moving stream of electrons (such a stream is, of course, an electric current). The electric eyes that open doors for you at supermarkets or airlines terminals, burglar alarms, exposure meters, and television cameras all operate on this principle of turning light energy into electrical energy.

Wave or Particle?

Now here is the problem. The wave theory of light would lead you to predict that the brighter the light—that is, the stronger the beam—the greater the force with which the electrons would be dislodged. But as we have already seen, whether or not light can eject the electrons of a particular metal depends not on the brightness of the light but on its wavelength. A very weak beam of the critical wavelength or a shorter wavelength is effective, while a stronger beam of a longer wavelength is not. Furthermore, as was shown in 1902, increasing the brightness of the light increases the number of electrons dislodged but not the velocity at which they are ejected from the metal. To increase the velocity, one must use a shorter wavelength of light. Nor is it necessary for energy to be accumulated in the metal. With even a dim beam of a critical wavelength, an electron may be emitted the instant the light hits the metal.

To explain such phenomena, the particle theory of light was proposed by Albert Einstein in 1905. According to this theory, light is composed of particles of energy called *photons*. The energy of a photon is not the same for all kinds of light but is, in fact, inversely proportional to the wavelength—the longer the wavelength, the lower the energy. Photons of violet light, for example, have almost twice the energy of photons of red light, the longest visible wavelength.

The wave theory of light permits physicists to describe certain aspects of its behavior mathematically, and the photon theory permits another set of mathematical calculations and predictions. These two theories are no longer regarded as opposed to one another; rather, they are complementary, in the sense that both are required for a complete description of the phenomenon we know as light.

The Fitness of Light

Light, as Maxwell showed, is only a tiny band in a continuous spectrum. From the physicist's point of view, the difference between light and darkness—so dramatic to the human eye—is only a few nanometers of wavelength. There is no qualitative difference at all marking the borders of the light spectrum. Why does this particular small group of radiations, rather than some other, bathe our world in radiance, make the leaves grow and the flowers burst forth, cause the mating of fireflies and paolo worms, and when reflecting off the surface of the moon, excite the imagination of poets and lovers? Why is it that this tiny portion of the electromagnetic spectrum is responsible for vision, for the rhythmic, day-night regulation of many biological activities, for the bending of plants toward the light, and also for photosynthesis, on which all life depends? Is it an amazing coincidence that all these biological activities are dependent on these same wavelengths?

George Wald of Harvard, one of the greatest living experts on the subject of light and life, says no. He thinks that if life exists elsewhere in the universe, it is probably dependent on this same fragment of the vast spectrum. Wald bases this conjecture on two points. First, living things, as we have seen, are composed of large, complicated molecules held in special configurations and relationships to one another by hydrogen bonds and other weak bonds. Radiation of even slightly higher energies than the energy of violet light breaks these bonds and so disrupts the structure and function of the molecules. Radiations with wavelengths below 200 nanometers drive electrons out of atoms; hence they are called ionizing radiations. Light of wavelengths longer than those of the visible band is absorbed by water, which makes up the great bulk of all living things. When it does reach molecules, its lower energies cause them to

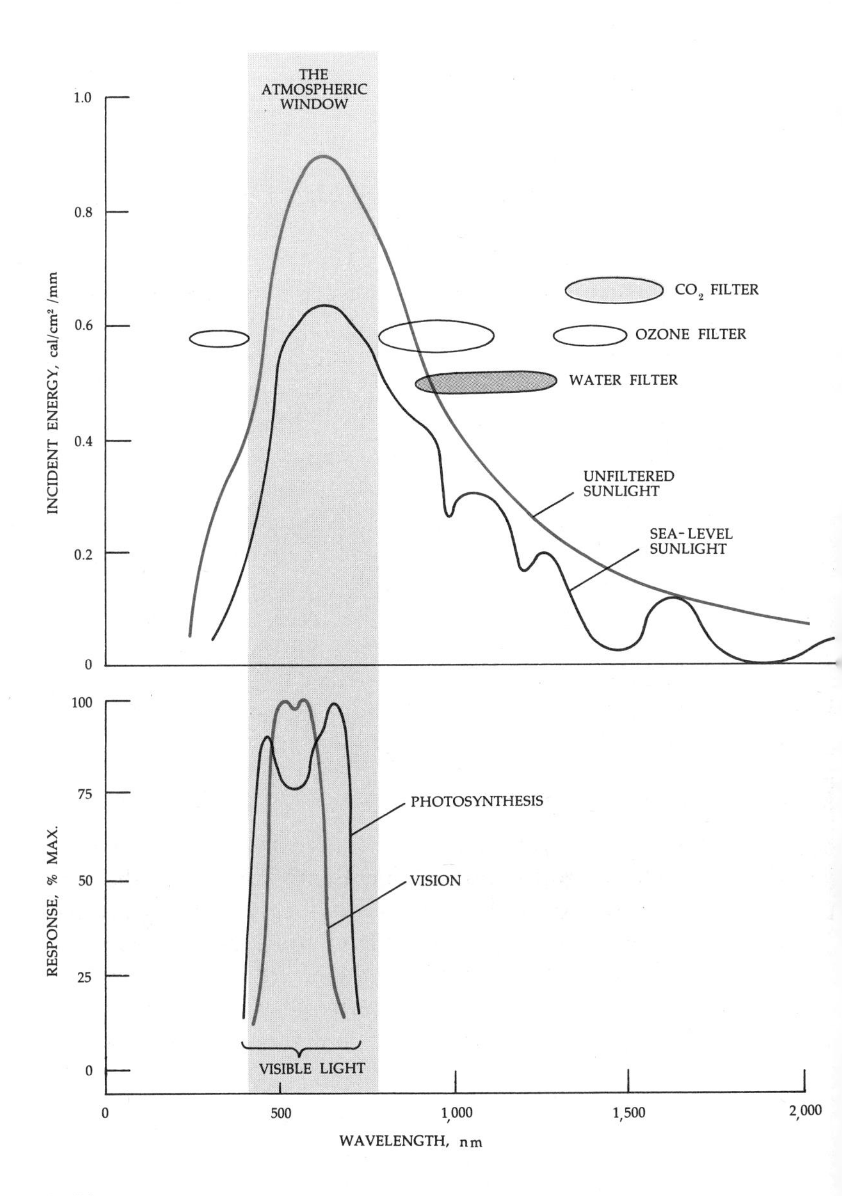

2–51

Sunlight is filtered by the atmosphere. The radiation reaching the earth through this "atmospheric window" is most abundant in the wavelengths involved in biological reactions such as photosynthesis and vision.

increase their motion (increasing heat) but do not trigger changes in their structure. Only those radiations within the range of visible light have the property of exciting molecules—that is, of raising electrons from one energy level to another—and so of producing biological changes.

The second reason for the visible band of the electromagnetic spectrum being "chosen" by living things is that it, above all, is what is available. Most of the radiation reaching the earth from the sun is within this range. Higher-energy wavelengths are screened out by the oxygen and ozone high in the atmosphere. Much infrared radiation is screened out by water vapor and carbon dioxide before it reaches the earth's surface.

This is an example of what has been termed "the fitness of the environment"; the suitability of the environment for life and that of life for the physical world are exquisitely interrelated. If they were not, life could not, of course, exist.

The Role of Pigments

In order for light energy to be used by living systems it must first be absorbed. Pigments are substances that absorb light energy. They are always easy to recognize because they are colored. They have to be. A black pigment is black because it absorbs light of all wavelengths. Red pigment is red because it absorbs all wavelengths *except red,* that is, it reflects or transmits red; a red pigment viewed under blue light will appear black because it has nothing to reflect or transmit. Chlorophyll, the pigment involved in photosynthesis, absorbs violet, blue, and red light and sends back green.

Chlorophyll *a,* the key compound in photosynthesis, is a porphyrin. The porphyrin ring is composed of four nitrogen-containing rings and their connecting carbon atoms. In the center of the four rings is an atom of magnesium. Attached to the porphyrin part of the molecule is a long hydrocarbon "tail." Chlorophyll *b* differs from chlorophyll *a* only in having a CHO group in the position at which chlorophyll *a* has a CH_3. Yet chlorophyll *b,* which is also found in plants, is not essential for photosynthesis. In fact, of all the pigments known, only chlorophyll *a* and bacteriochlorophyll and chlorobium chlorophyll (both of which are found in a few types of bacteria) can transform light energy to chemical energy.

Other plant pigments, carotenes, which are orange-red, absorb green and other lights and pass their energy on to chlorophyll *a.* (Chlorophyll *b,* which absorbs a slightly different spectrum of light than chlorophyll *a,* also performs this function in the plant cell.) Carotenes are responsible for the red and yellow colors of fruits and flowers; in the green leaf, the carotene color is masked by that of the chlorophyll and becomes evident only when the leaf dies and the chlorophyll fades. A principal carotene, beta-carotene, is a lipid:

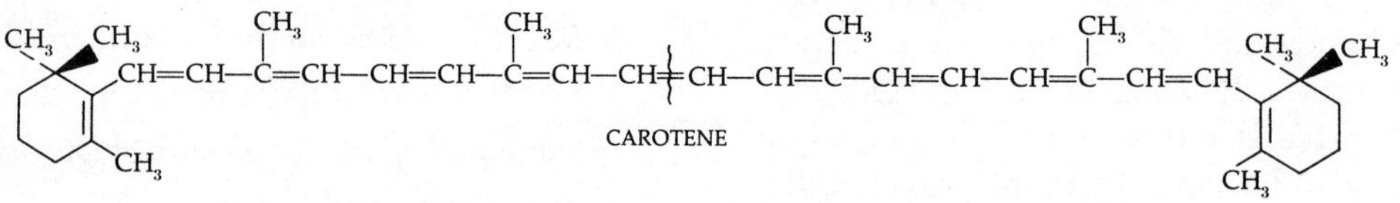

CH=CH—C(CH₃)=CH—CH=CH—C(CH₃)=CH—CH=CH—CH=C(CH₃)—CH=CH—CH=C(CH₃)—CH=CH

CAROTENE

If the carotene molecule is split in half at the point shown (by the addition of a molecule of water), the result is a compound known as vitamin A, another compound for which the animal world depends on the plant world:

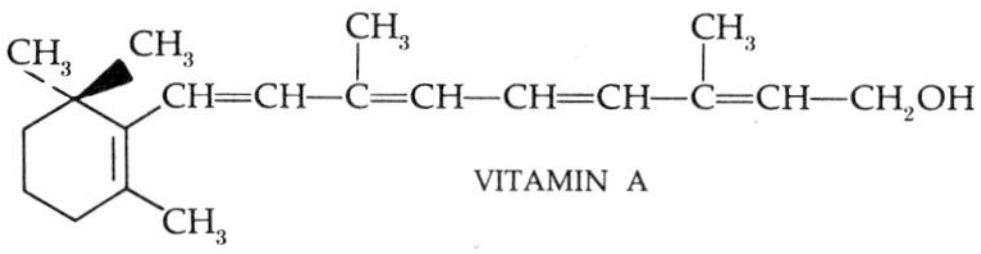

$$\text{(ring with } CH_3, CH_3, CH_3\text{)}-CH=CH-\overset{CH_3}{\overset{|}{C}}=CH-CH=CH-\overset{CH_3}{\overset{|}{C}}=CH-CH_2OH$$

VITAMIN A

Oxidation of vitamin A yields retinene:

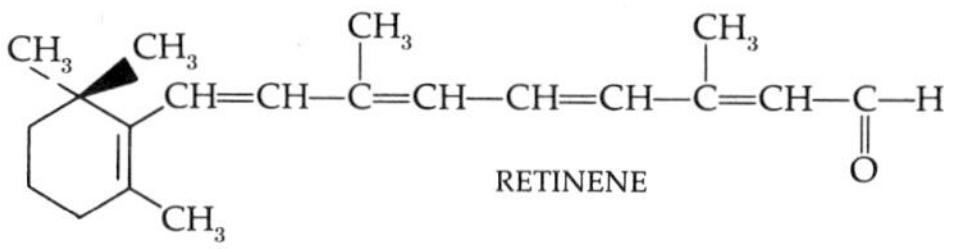

$$\text{(ring with } CH_3, CH_3, CH_3\text{)}-CH=CH-\overset{CH_3}{\overset{|}{C}}=CH-CH=CH-\overset{CH_3}{\overset{|}{C}}=CH-\underset{\underset{O}{\|}}{C}-H$$

RETINENE

The stacked membranes of the photoreceptor cells shown on page 82i contain coiled molecules of retinene. When these molecules are exposed to the light, they unfold and so set in motion the events that trigger the sensation of vision that is transferred to the brain. You are able to read this page because of the coiling and uncoiling of thousands of these molecules.

PHOTOSYNTHESIS AND RESPIRATION

In the process of photosynthesis, electrons are boosted "uphill." As they move downhill, in a series of oxidation-reduction reactions, their energy is converted to forms of chemical energy that can be widely used by all kinds of cells. These forms of chemical energy supply the necessary power for the biosynthesis of new compounds, for the transport of materials across cell borders, for movement, for cell division, and for the many other functions that make up the life processes of the individual cells. In the next chapter, we shall describe how the sugar glucose is broken down to yield chemical energy packaged in the form of a key molecule known as ATP.

SUMMARY

Living systems convert energy from one form to another in order to carry out the functions essential to their maintenance, growth, and reproduction. The sun is the ultimate source of this energy.

The laws of thermodynamics govern exchanges of energy in biological systems. The first law states that energy can be transformed from one form to another but cannot be created or destroyed. The second states that energy in a form to do work is eventually converted to heat energy, which is dissipated out into the surroundings. In other words, the supply of useful energy is constantly diminishing and must be replaced.

The exchanges of energy in living cells usually involve the movement of electrons from one energy level to another. Such movements are known as oxidation-reduction reactions. An atom or molecule that loses electrons is oxidized; one that gains electrons is reduced.

Light energy enters the living world by means of pigments. Among the biologically important pigments are chlorophyll *a*, which is essential in photosynthesis, chlorophyll *b* and carotene, which are accessories in photosynthesis, and retinene, derived from carotene and involved in vision.

QUESTIONS

1. "Entropy" is a word meaning disorder or randomness. According to the second law of thermodynamics, entropy always tends to increase. Can you explain why this is so? Do cells "break" this law?

2. In what ways do you think the laws of thermodynamics are relevant to present-day environmental problems?

3. One of the scientists concerned with developing the glucose fuel cell (Figure 2–45), has suggested half-jokingly that it might be possible to design a lawn mower powered by the sugars in the grass clippings. Can you think of any similar uses of energy conversions?

CHAPTER 2–5

Cellular Respiration: Fuel for the Living Cell

One of the most important chemicals in all living systems is ATP. ATP (adenosine triphosphate) is the major link between the radiant energy from the sun and the vital processes of the cell.

Cells with chloroplasts use solar energy to make sugar. These sugars, as we have seen, serve as a storage form of energy. This energy is made available for the moment-to-moment transactions of the cell by the process of cellular respiration, in which energy stored in sugar is converted to ATP. Sugar is like money in the bank. Before you can spend banked money to buy a hotdog or ride a bus, you have to cash a check and convert your "stored" money to another form, such as dollar bills or small change. ATP is equivalent to this immediately spendable form of money. Almost every energy-requiring transaction in the cell involves ATP, just as, alas, a large number of our daily transactions involve dollars and cents. For this reason, ATP is often referred to as the "universal currency" of the cell.

How does ATP perform this function? Again, to understand function we must look at structure. ATP is composed of three types of subunits. One of these is a compound known as adenine, which is a nitrogen base:

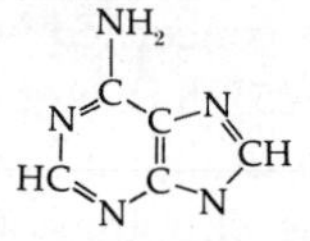

ADENINE

Adenine is a compound which we shall be mentioning frequently in Section 3 because it is one of the principal components of the genetic material. (The use of adenine for these two quite different purposes is another example of the economy with which the cell operates.)

The second subunit is a five-carbon sugar, ribose, similar in basic design to the sugars described in Chapter 2–3:

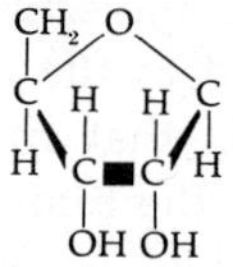

RIBOSE

The third subunit is phosphate, which is an atom of phosphorus combined with four oxygen atoms. Adenine plus the sugar plus one phosphate group make up a compound known as a *nucleotide,* which is one of the basic and important many-purpose chemical combinations in the cell. In ATP, as the name implies, there are three phosphates:

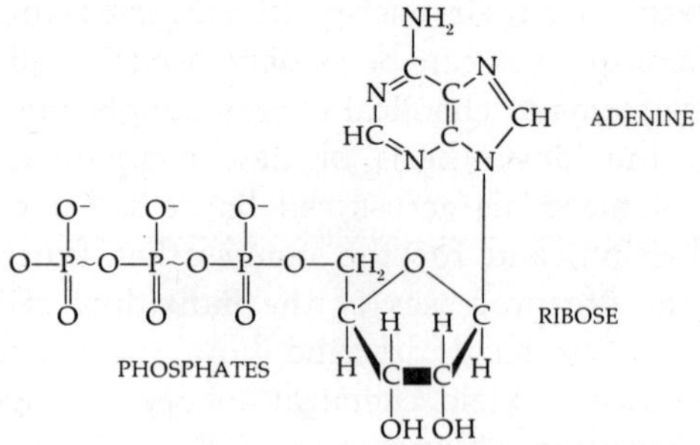

ADENINE

PHOSPHATES

RIBOSE

The symbol ~ indicates a so-called "high-energy" bond. "High-energy bond" is a somewhat confusing term. It does not mean a strong bond, such as the covalent bond between carbon and hydrogen. A high-energy bond is one that yields its energy readily.

One of the principal characteristics of phosphate groups—and the reason for their playing an important role in chemical exchanges—is their capacity to form high-energy bonds. In ATP, two such bonds link the three phosphates together. Usually, in the course of cellular transactions, only one bond is broken. With the loss of one phosphate group, the ATP molecule becomes ADP, adenosine diphosphate. The ADP molecule is then "recharged"—regaining the phosphate group and again becoming ATP—with the energy obtained from the oxidation of sugar or some other organic compound. This process is known as respiration because it requires oxygen for completion.

THE THREE STAGES OF CELLULAR RESPIRATION

Respiration is complicated in detail but simple in its overall design. The glucose molecule—the form in which a major end product of photosynthesis reaches the animal cell—is split, the hydrogen atoms (protons and electrons) are removed from the carbon atoms and combined with oxygen atoms, and the energy released by this process is used to convert ADP to ATP. Only about 40 percent of the energy is dissipated as heat, which is extremely efficient compared with an automobile engine, for example, in which about 75 percent of the energy released by the combustion of gasoline is "lost" as heat.

Respiration occurs in three stages: glycolysis, the Krebs cycle, and the electron transport chain. The first stage takes place in the cytoplasm, and the other two in the mitochondrion. In the first two stages, the hydrogen atoms are removed from carbon and passed to compounds known as *electron acceptors*. The electron acceptors involved in glycolysis and the Krebs cycle, NAD and FAD, have nucleotides as their structural base, just as ATP does. Both NAD and FAD can accept two hydrogen atoms. Another type of electron acceptor is involved in the electron transport chain. In this third stage of respiration, all the hydrogen atoms removed from the carbon during the first two stages are passed "downhill" along a series of compounds known as cytochromes. Cytochromes, while quite different from NAD and FAD, are also based on a familiar chemical structure. They resemble hemoglobin. Each has an iron-containing porphyrin ring and a protein "tail." In the mitochondrion, cytochromes are linked to the enzymes involved with the regeneration of ATP from ADP. The cytochromes in the electron transport chain, which pass the electrons rapidly from one to another, are usually referred to as *electron carriers*.

Glycolysis

In glycolysis, the six-carbon glucose molecule is split into 2 three-carbon molecules of a compound known as pyruvate:

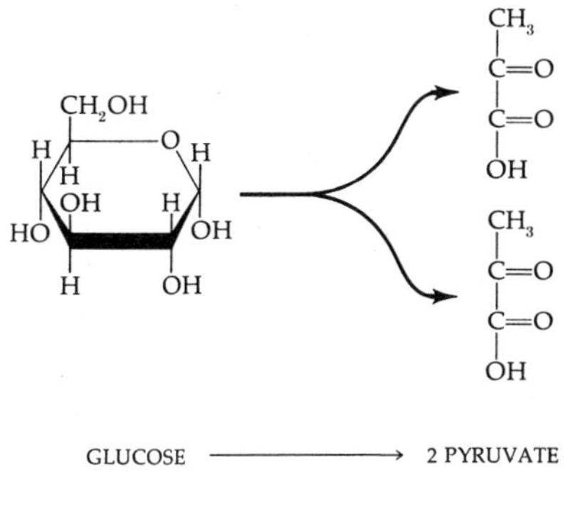

GLUCOSE ⟶ 2 PYRUVATE

This splitting takes place in nine separate steps, during which two molecules of ATP are formed from ADP and four hydrogen atoms are passed to NAD. Do not try to memorize these steps, but follow them closely, merely as one example of what a cell can do.

Step 1. Glycolysis begins with an input of energy, the result of ATP splitting to form ADP. The ATP terminal phosphate group is transferred to the 6 position of the glucose molecule. Some of the energy released from ATP is used to drive the reaction, and some is stored in the chemical bond attaching the phosphate to the sugar molecule. The reaction is catalyzed by a specific enzyme (hexokinase).

Step 2. The molecule is reorganized, again with the help of a particular enzyme. The six-sided ring characteristic of glucose becomes a five-sided fructose molecule. As you know, glucose and fructose both have the same number of atoms—$C_6H_{12}O_6$—and differ only in the arrangement of these atoms.

Step 3. This step is similar to step 1 and results in the attachment of a phosphate in the 1 position of the fructose molecule, producing fructose 1, 6-diphosphate, that is, fructose with phosphates in the 1 and 6 positions. This step also requires a special enzyme. (Enzymes are identifiable by the suffix "ase.")

Step 4. The molecule is split into 2 three-carbon molecules. One of these, glyceraldehyde phosphate, is acted upon by the next enzyme in the sequence. The molecule on the left is converted to glyceraldehyde phosphate by the enzyme isomerase.

Step 5. Glyceraldehyde phosphate molecules are oxidized —that is, hydrogen atoms with their electrons are removed —and NAD_{ox} (NAD oxidized) is transformed to NAD_{red} (NAD reduced). At the same time, using the energy yielded by the oxidation reaction, phosphate molecules are attached to the 3 position of each of the glyceraldehyde molecules in preparation for the formation of two molecules of ATP.

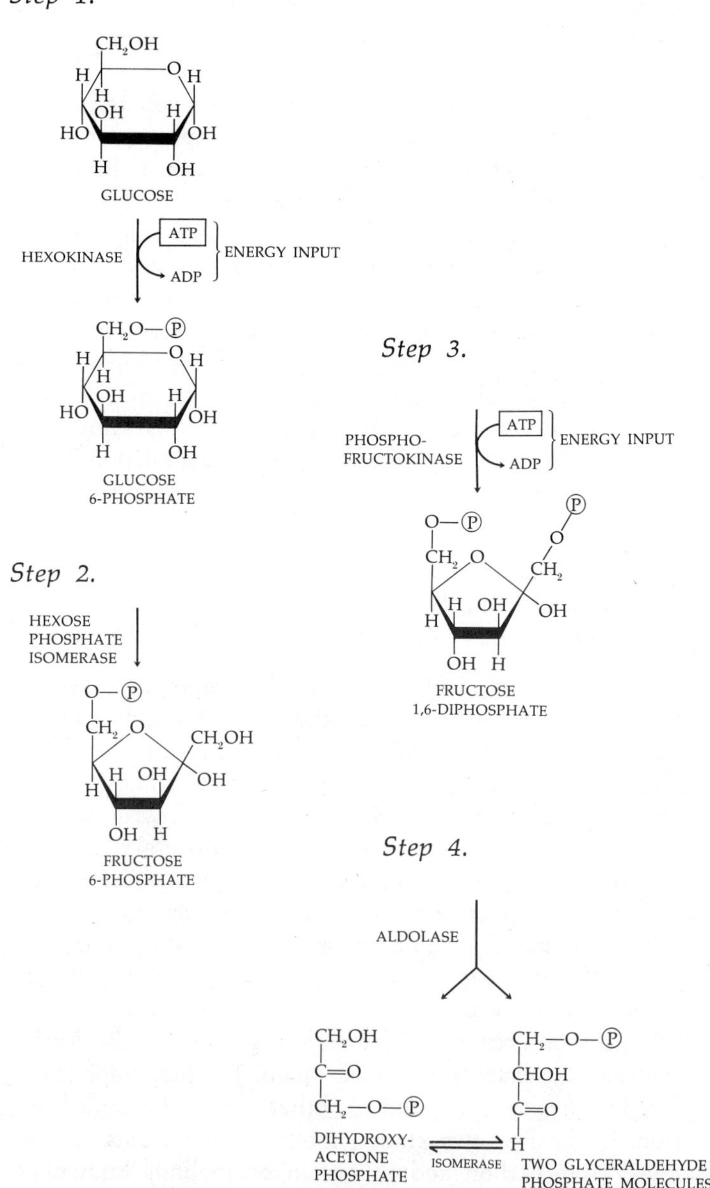

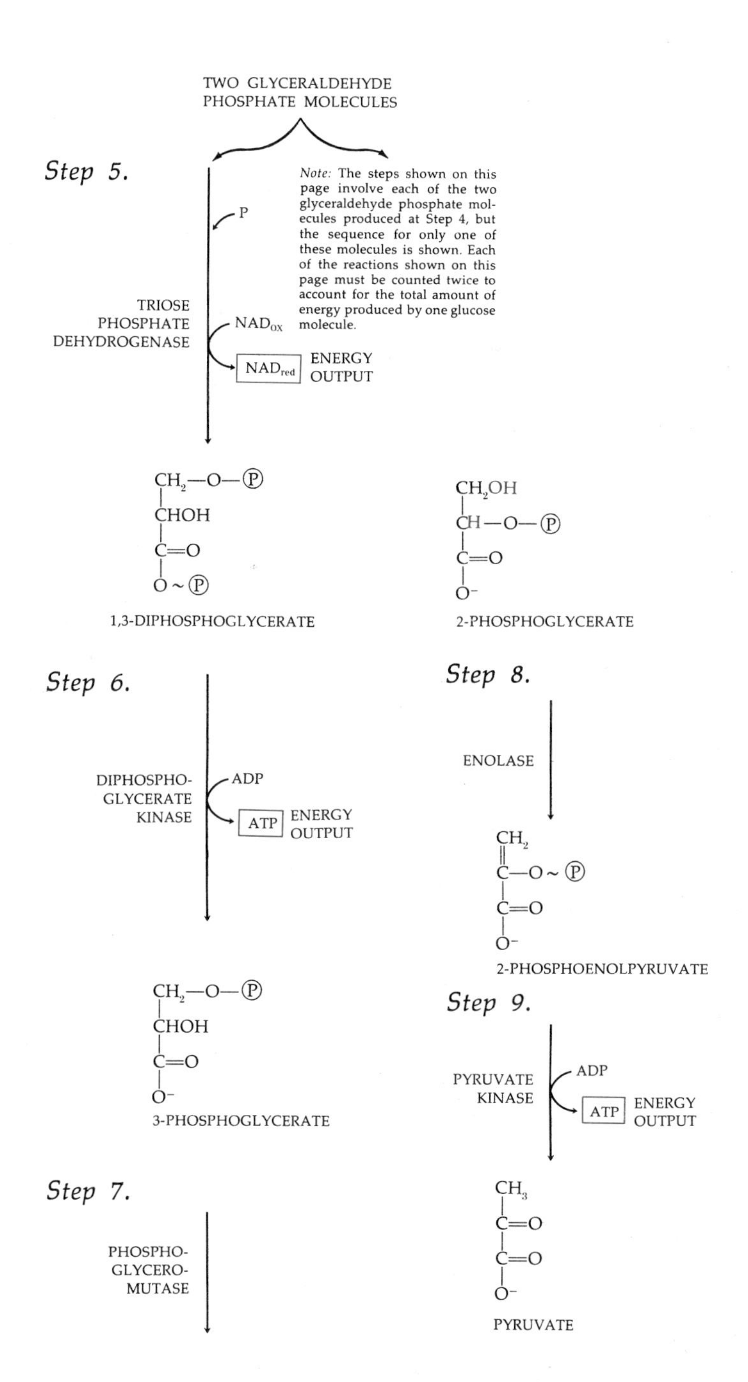

Step 6. A phosphate is released from the 1 position and transferred to ADP, forming ATP (a total of two molecules of ATP per molecule of glucose).

Step 7. The remaining phosphate group is transferred from the 3 position to the 2 position.

Step 8. In this step, a molecule of water is removed from the three-carbon compound. Following this internal rearrangement of the molecule, a high-energy bond is formed.

Step 9. The remaining phosphate is transferred to a molecule of ADP, forming another molecule of ATP (again, a total of two molecules of ATP per molecule of glucose).

The complete sequence begins with one molecule of glucose. Energy is put into the reaction at Steps 1 and 3 when a molecule of ATP gives up the energy stored in one of its three phosphate bonds and becomes ADP. From this point to the end of the glycolytic sequence, energy is released. At Step 5, a molecule of NAD_{ox} takes energy from the system and becomes NAD_{red}. At Steps 6 and 9, molecules of ADP take energy from the system, form additional phosphate bonds, and become ATP. Thus one glucose molecule, using energy from the phosphate bonds of two ATP molecules to initiate the glycolytic reaction, produces two NAD_{red} from two NAD_{ox} molecules and four ATP from four ADP molecules. Two molecules of pyruvate remain, and these two molecules still contain a large amount of the energy stored in the original glucose molecule. In a sequence of reactions known as the Krebs cycle and the electron transport chain, energy is released from the pyruvate molecules and packaged in the form of ATP so that it becomes available for other biochemical uses in the cell.

The glycolytic sequence does not require oxygen and is believed to have evolved before photosynthesis made oxygen available in the atmosphere.

Completing the Breakdown of Glucose

When oxygen is available, the two pyruvate molecules are broken down completely to carbon dioxide and water, releasing the remaining energy stored in the bonds. These stages—the Krebs cycle and the electron transport chain—take place within the mitochondrion.

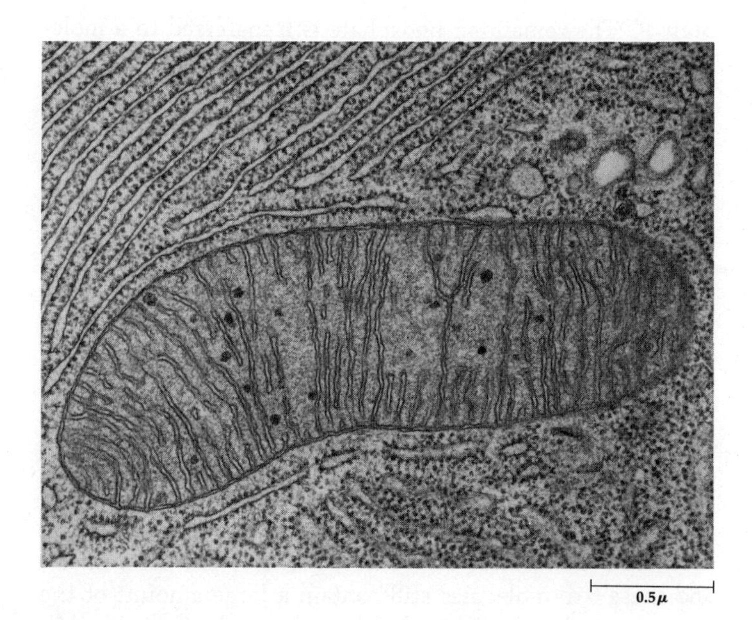

2–52
Mitochondria are surrounded by two membranes. The inner membrane, as you can see clearly in the electron micrograph, folds inward to make a series of shelves, or cristae. The enzymes and electron carriers involved in the final stages of cellular respiration are built into these internal membranes.

In preparation for these final stages, each of the three-carbon pyruvate molecules is oxidized, each yielding a molecule of carbon dioxide and a two-carbon fragment, known as an acetyl group, which is combined with a carrier compound called coenzyme A (CoA):

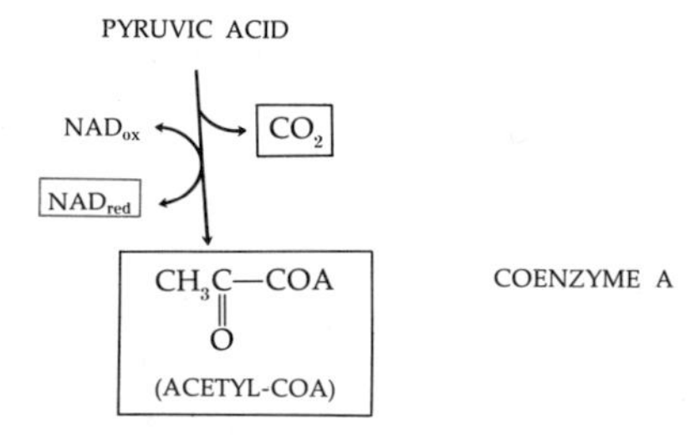

This is an energy-yielding reaction, in which a molecule of NAD_{red} is produced from NAD_{ox}.

Fats and amino acids can also be used as energy sources after first being converted to acetyl CoA, which then, like the acetyl CoA formed from sugars, enters the Krebs cycle.

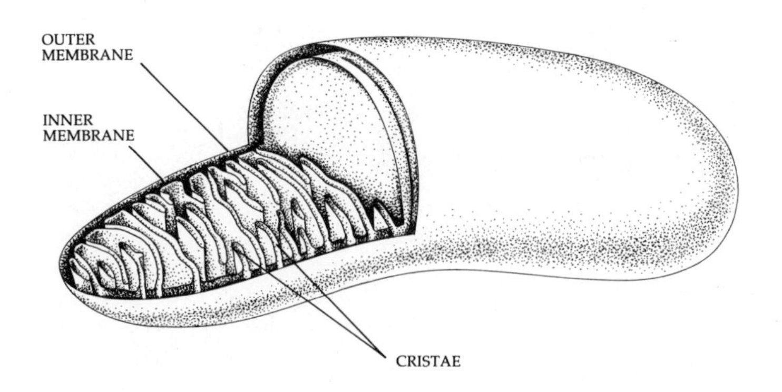

Krebs Cycle

In the Krebs cycle, the acetyl group (CH_3CO) is released from CoA and joined to a four-carbon compound, oxaloacetic acid. A six-carbon compound, citric acid, is thereby produced. In the course of the Krebs cycle, citric acid is oxidized to form carbon dioxide and oxaloacetic acid. The energy stored in the molecule is used to form three molecules of NAD_{red} from NAD_{ox}, one molecule of FAD_{red} from FAD_{ox}, and one ATP from ADP. The sugar molecule has now been completely oxidized.

The oxaloacetic acid is formed in the last step of the Krebs cycle, and so the sequence can begin again when another acetyl group combines with the oxaloacetic acid molecule. Cycling systems, such as this one, permit the cell to work with great efficiency and economy. A chemist in a laboratory usually produces only a very small amount of the desired end product in proportion to waste products or by-products of the reaction, which is one of the reasons our industrial pollution problem is so acute. In the cell, however, the waste products (carbon dioxide and water, in the case of the Krebs cycle) are usually few and relatively easy to dispose of.

FERMENTATION

Some cells, such as yeast cells, are able to live without oxygen. If oxygen is not present, yeast cells convert the pyruvate—the product of glycolysis—to alcohol. When yeast cells are provided with a supply of sugar, as they are in grapes and other fruit, and with no air, they produce alcohol. But yeast cells, like other living systems, have only a limited tolerance for alcohol, and when a concentration of about 12 percent is reached, they die and the process stops. Animal cells in the absence of oxygen convert pyruvate to lactic acid. Lactic acid also can be a cell poison. When it accumulates in muscle cells, for instance, it produces symptoms associated with muscle fatigue.

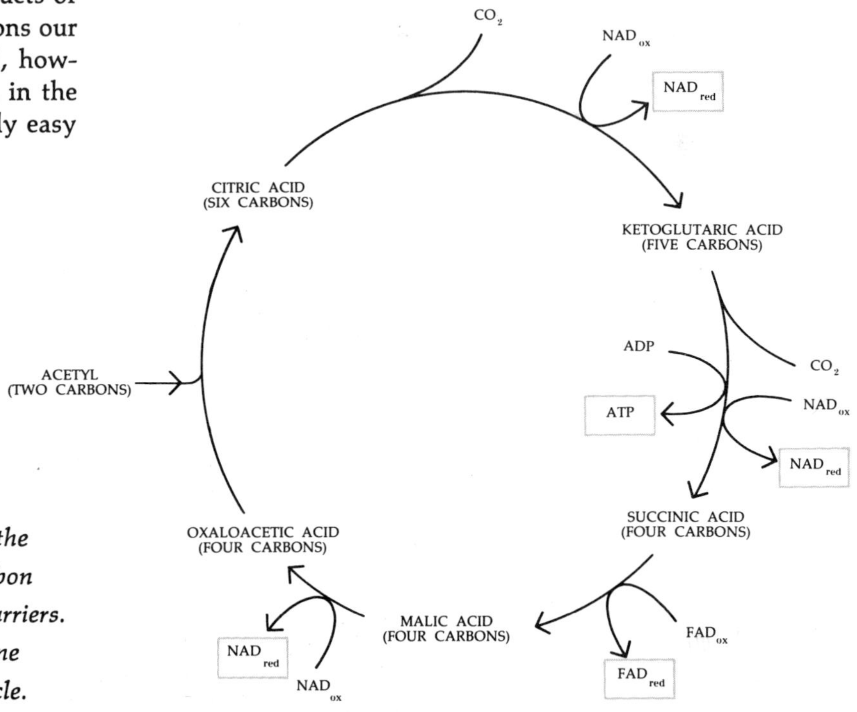

2–53

Summary of the Krebs cycle. In the course of the cycle, the carbons donated by the acetyl group are oxidized to carbon dioxide and the hydrogen atoms are passed to electron carriers. One molecule of ATP, three molecules of NAD_{red}, and one molecule of FAD_{red} represent the energy yield of the cycle.

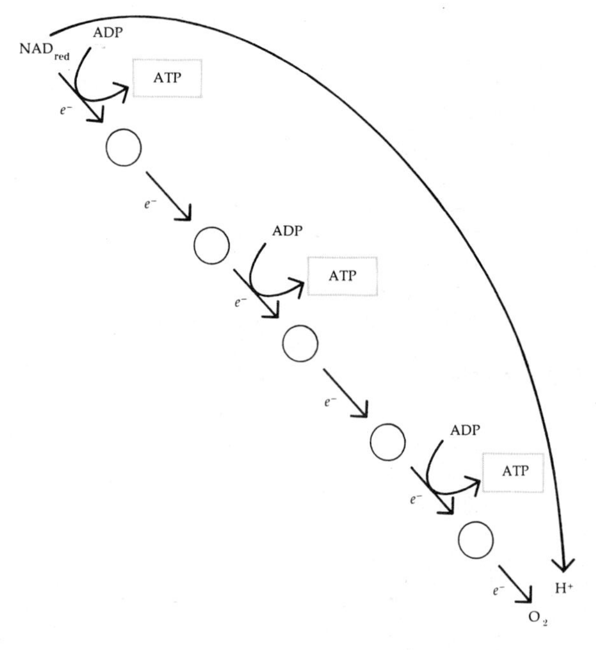

2–54

The electron transport chain somewhat resembles a series of waterwheels. Water as it flows from a point of higher water potential to one of lower water potential releases energy, which can be harnessed by waterwheels and so used to grind corn or perform other work functions. As electrons flow along the electron transport chain from a point of higher energy to a point of lower energy, the released energy is harnessed by the cytochromes and their coupling enzymes to produce ATP from ADP. The ATP is then used for the work of the cell.

Electron Transport Chain

In the electron transport chain, the electrons accepted by NAD and FAD during the preceding reactions are passed to a series of electron carriers. As the electrons are passed "downhill" along this electron transport chain, the energy released is transferred to molecules of ATP. At the end of the chain, the electrons are accepted by protons (hydrogen atoms) and combine with oxygen to produce water. Each arrow in Figure 2–55 indicates a pair of electrons. Each time one pair of electrons passes from NAD to oxygen, three molecules of ATP are formed from ADP and phosphate. Each time a pair of electrons passes from FAD, which holds them at a slightly lower energy level than NAD, two molecules of ATP are formed. As the balance sheet, Table 2–3, shows, the complete yield from a single molecule of glucose is 38 molecules of ATP.

Table 2–3

Summary of Energy Yield from One Molecule of Glucose

Glycolysis:		2 ATP	
	2 NAD$_{red}$ ⟶	6 ATP	⟶ 8 ATP
Pyruvate ⟶ acetyl CoA:	1 NAD$_{red}$ ⟶	3 ATP	(× 2) ⟶ 6 ATP
Krebs cycle:		1 ATP	
	3 NAD$_{red}$ ⟶	9 ATP	(× 2) ⟶ 24 ATP
	1 FAD$_{red}$ ⟶	2 ATP	

RESPIRATION AND THE CELL

Cellular respiration is a process carried out by virtually every kind of cell, from "simple" one-celled organisms, such as an amoeba or a paramecium, to the cells of muscle, bone, and epithelium—all the cells that make up the human body. And in all these cells, each so different from the other in appearance, the same array of enzymes and the same

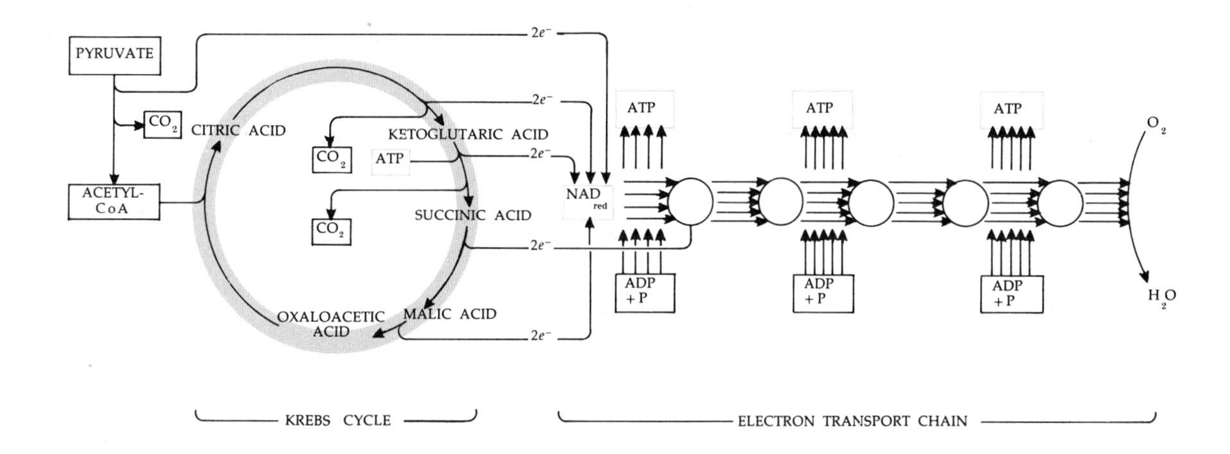

2–55
Summary of reactions which take place in the mitochondrion. In the Krebs cycle, the acetyl group (the remnant of the glucose molecule) is oxidized and the electron acceptors, NAD and FAD, are reduced. NAD and FAD then transfer their electrons to the series of cytochromes (indicated by the circles) that make up the electron transport chain. As these cytochromes pass the electrons "downhill," the energy released is used to make ATP from ADP.

structures—the folds of the mitochondrion—are involved.

Glycolysis was the first of the enzymatic reactions of the cell to be studied—largely because of its great importance to the wine industry—and the discovery of the significance of this process in the life of the cell and the working out of the details of its many complex steps is one of the great triumphs of modern biochemistry. In fact, one of the important challenges still facing biochemists concerns the nature of the link between the cytochromes and other carriers of the electron transport chain and the enzymes involved in making ATP from ADP in this last and crucial phase of respiration.

Respiration, a process carried out by nearly all living cells, offers a good example of the efficiency and order with which the cell carries out its many complex processes.

SUMMARY

The chief source of energy in nonphotosynthetic cells comes from the breakdown of glucose. As the glucose is broken down in a series of small enzymatic steps, the energy in the molecule is packaged in the form of "high-energy" bonds in molecules of ATP.

The first stage in the breakdown of glucose is glycolysis, in which the six-carbon glucose molecule is split into 2 three-carbon molecules of pyruvate and two new molecules of ATP and two of NAD_{red} are formed. This reaction takes place in the cytoplasm of the cell.

In the presence of oxygen, the three-carbon pyruvate molecules are broken down to two-carbon acetyl groups, which then enter the Krebs cycle. In the Krebs cycle, which takes place in the mitochondrion, the two-carbon acetyl group is broken apart in a series of reactions to carbon dioxide. In the course of the oxidation of the acetyl group, four electron acceptors (three NAD and one FAD) are reduced and another molecule of ATP is formed.

The third stage of breakdown of the glucose molecule is the electron transport chain, which involves a series of electron carriers and enzymes embedded in the inner membranes of the mitochondrion. Along this series of electron carriers, the high-energy electrons accepted by NAD_{red} and FAD_{red} during the Krebs cycle pass downhill to oxygen. Each time a pair of electrons passes down the electron transport chain, ATP is formed from ADP and phosphate. In the course of the breakdown of the glucose molecule, 38 molecules of ATP are formed, most of them in the mitochondrion.

QUESTION

Explain the advantages to the cell of a cycling system, such as the Krebs cycle.

CHAPTER 2–6

Photosynthesis

In plants, photosynthesis takes place only within those special organelles known as *chloroplasts*, which we saw in *Chlamydomonas* and in the leaf cell in Chapter 2–1. A single cell of a leaf may contain 40 to 50 chloroplasts, and a square millimeter, some 500,000. Chloroplasts move within the leaf cell, orienting their surfaces, which are often lens-shaped, so that they catch the light.

Chloroplasts, like mitochondria, are surrounded by two membranes. Within the membrane is a watery material, called the stroma, which is different in composition from the watery material that surrounds the organelles in the cytoplasm. In the light microscope under high power, it is possible to see little spots of green within the chloroplasts. These were called grana (grains) by the earlier microscopists, and this term is still in use. Under the electron microscope, however, it can be seen that the grana are part of an elaborate system of membranes called the lamellae. These membranes are organized in parallel pairs. The pairs of membranes are joined at each end, forming a disk-shaped structure, or thylakoid (from *thylakos*, the Greek word for "sac"). The grana are stacks of thylakoids.

All the chlorophyll is found packed into the lamellae together with other compounds involved in photosynthesis. This functional group of components is known as a *pigment system*. The pigment systems of all plants contain chlorophyll *a* and, in addition, chlorophyll *b* and carotene. Also packed in these pigment systems are the electron carriers, enzymes, and other molecules associated with the capture of light energy.

Photosynthesis occurs in two distinct steps: the first step, the "light reactions," takes place in the pigment systems within the membranes of the chloroplast. The second step, the so-called "dark reactions," takes place in the stroma. The dark reactions do not have to proceed in the dark, but neither do they require light—although, as we shall see, they require the products of the light reactions.

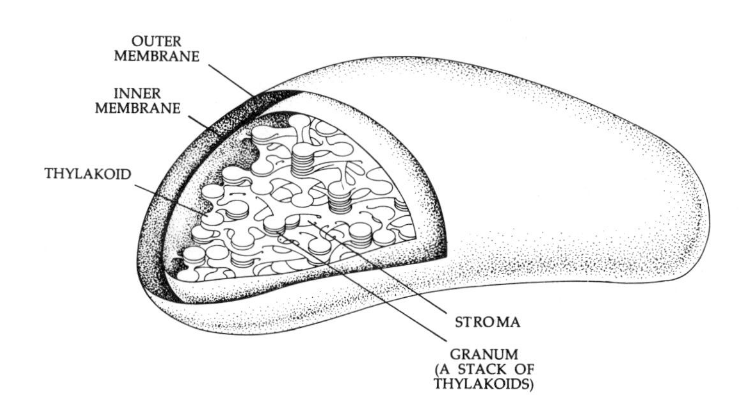

2–56

The drawing is a three-dimensional interpretation of chloroplast structure.

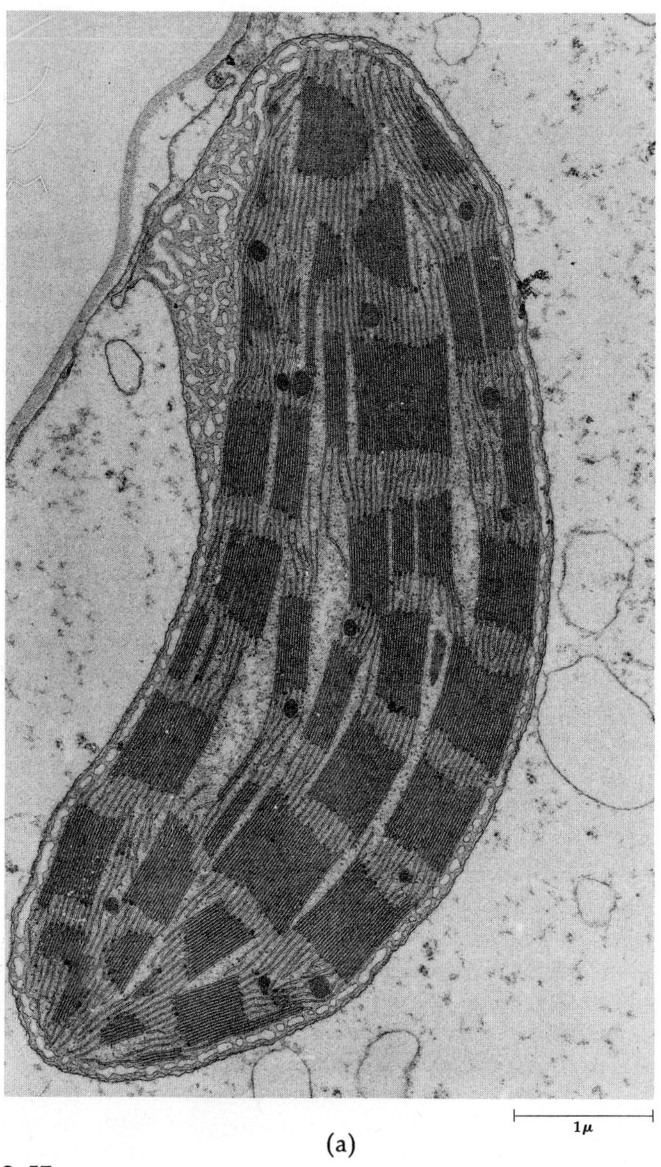

(a)

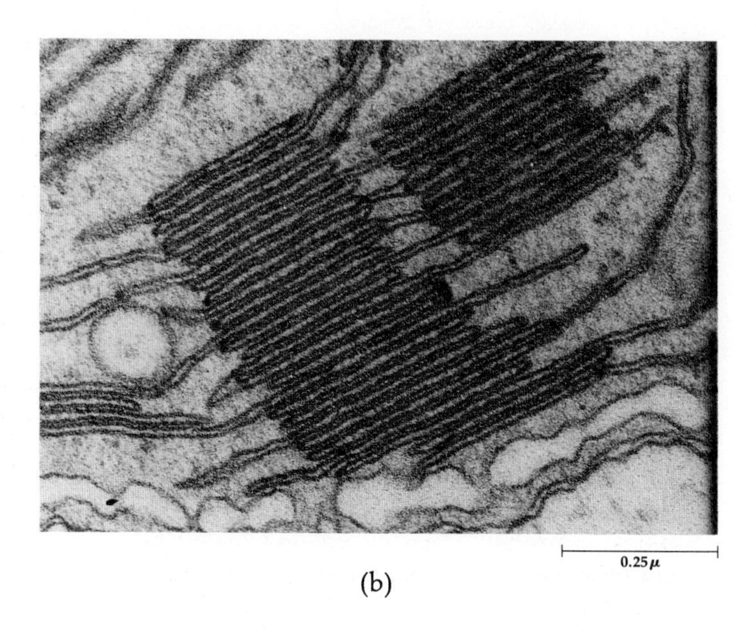

(b)

2–57

(a) *Chloroplasts, like mitochondria, are surrounded by a double membrane and have a system of internal membranes. The chloroplast membranes, called lamellae, are organized so that they form stacks, or grana. Three such stacks of membranes are shown at higher magnification in* (b).

THE LIGHT REACTIONS

Photosynthesis begins when light energy strikes a chlorophyll molecule and boosts an electron to a higher energy level. If light strikes an isolated molecule of chlorophyll, laboratory experiments show, the energy is immediately sent back as light, in the form of fluorescence, as the electron drops back to its original energy level. However, in the pigment system, the electron is passed downhill in a series of stages. In the course of this passage, the energy put in when the electron was boosted uphill is converted to chemical energy in the form of ATP and reduced electron carriers.

The exact way in which the electron is "captured" is not known, but recently a model has been devised for photosynthesis in plants which appears to satisfy all the known facts. According to this model, there are two separate inputs of energy involved, that is, two separate "uphill" boosts. These two uphill paths are designated photochemical system I and photochemical system II. They act sequentially, with photochemical system II first in the sequence and photochemical system I second. (Photosystem I is designated number I because this system, which does not produce oxygen, apparently evolved first.)

When light energy is trapped in pigment system II, it causes the lysis (breaking apart) of water. The energy of light pulls electrons from a water molecule, and the rest of the molecule "falls apart":

$$H_2O \xrightarrow{\text{light energy}} \tfrac{1}{2}O_2 + 2H^+ + 2e^-$$

The oxygen is released as gas. ($\tfrac{1}{2}O_2$ is a designation for one oxygen atom, that is, half of a molecule of oxygen gas.) The electrons are boosted uphill to the electron acceptor for photosystem II. From this electron acceptor, the electrons pass "downhill" along a series of cytochromes and other electron carriers. As they pass downhill, part of the energy is converted into ATP. The stream of electrons flows into the pigment assembly of photosystem I. Here the light energy absorbed by pigment system I gives the electrons a second boost, raising them to the electron acceptor of system I, an even higher energy level. From this acceptor, the electrons are passed downhill again. In the course of this passage, the electron acceptor NADP is reduced. (NADP is a close chemical relative of NAD.)

This process is sometimes known as the Z scheme since its three principal stages—uphill to the electron acceptor of system II, downhill to the reaction center of system I, and uphill again to the electron acceptor of system I—form the letter Z. (Although, as usually diagramed, it might better be called the N scheme.)

System I can also operate independent of system II, in what is known as cyclic electron flow. In this process, after the light energy received in system I pushes the electrons uphill to the electron acceptor of system I, the electrons pass downhill again along a chain of electron carriers to the system I pigments. ATP is formed, but no water is split, no $NADP_{red}$ is formed, and no oxygen is released. Primitive organisms, such as the photosynthetic bacteria, have only system I.

The end products of the first phase of photosynthesis, the light stage, are ATP and $NADP_{red}$.

PHOTOSYNTHESIS AND THE COMING OF OXYGEN

The first photosynthetic organism probably appeared more than 3 billion years ago. Before the evolution of photosynthesis, the physical characteristics of the planet itself were by far the most powerful forces in shaping the course of natural selection. But with the evolution of photosynthesis, organisms began to change the face of the planet, and they have continued to do so, at an ever-increasing rate, up to the present day.

One of the earliest and most important effects of photosynthesis was the change it brought about in the atmosphere of the earth. At first, this atmosphere probably consisted chiefly of methane (CH_4), *with minor amounts of hydrogen, nitrogen, ammonia* (NH_3), *and water vapor. Subsequently, as the interior of the earth assumed its present structure, methane and ammonia were broken down and their hydrogen atoms, which are very light, were released into space as hydrogen gas* (H_2). *Thus, carbon dioxide and nitrogen replaced methane and ammonia as significant components of the atmosphere.*

This is the atmosphere in which the first cells evolved. These earliest forms of life were, of course, adjusted to living in an environment without free oxygen. In fact, such anaerobic ("without air") organisms cannot survive in the presence of free oxygen.

With the evolution of photosynthesis, the composition of the atmosphere began slowly to change. Increasing amounts of atmospheric carbon dioxide were consumed as photosynthetic organisms multiplied, and free oxygen, a by-product of photosynthesis, began to accumulate. In the struggle for survival under these changing conditions, cell species arose for which oxygen was not a poison but a requirement of existence.

Oxygen-consuming organisms have an advantage over those that do not use oxygen: a higher yield of energy can be extracted per molecule from the oxidation ("burning") of carbon-containing compounds than from anaerobic fermentation processes. Energy resulting from the use of oxygen by cells made possible the development of increasingly active, increasingly complex organisms. Without oxygen, the higher forms of life that now exist on earth could not have evolved.

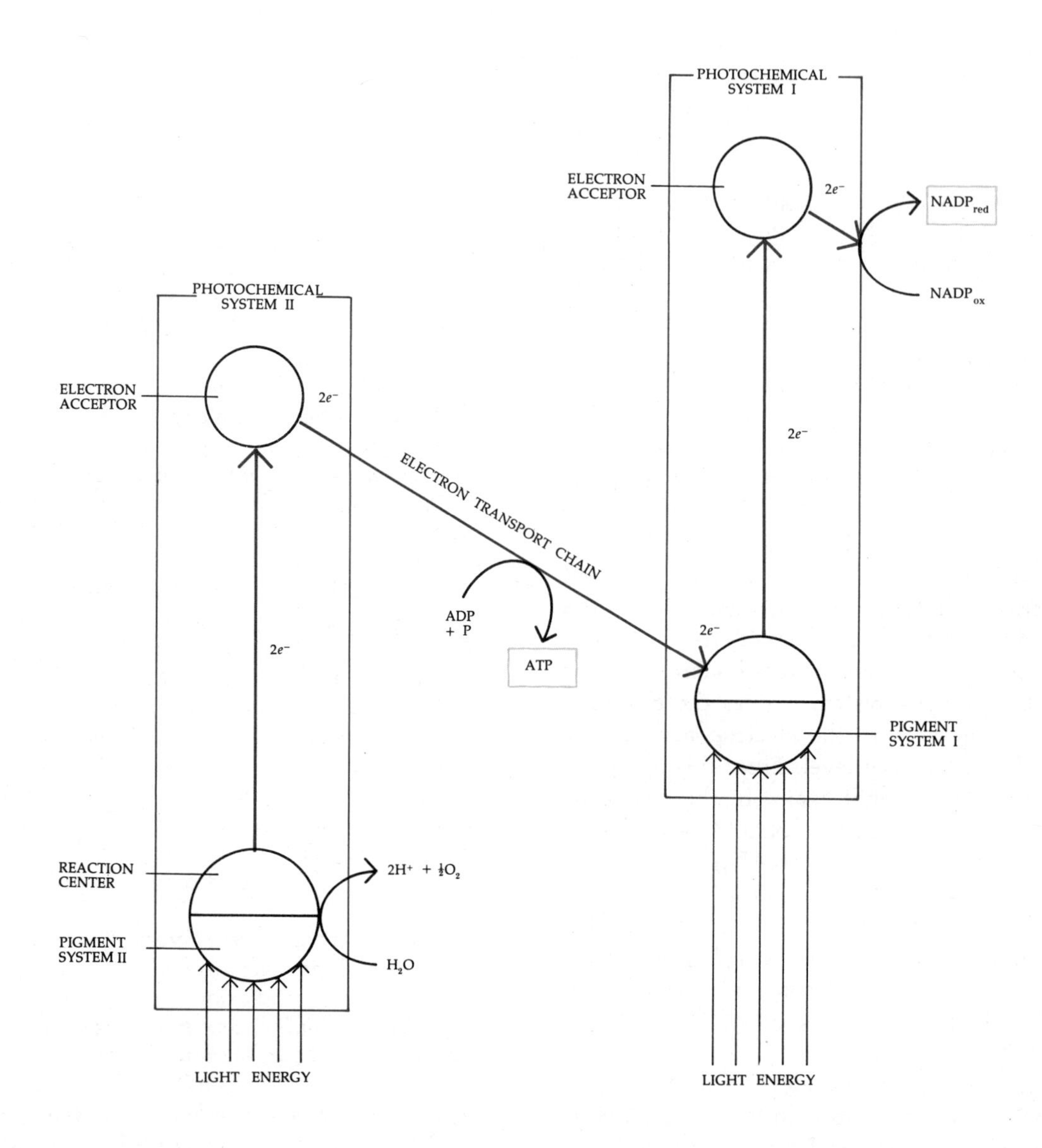

2–58

Light energy caught in the pigment system of photochemical system II lyses water and boosts electrons (e^-) *uphill to an electron acceptor. The electrons are then passed along a chain of electron-acceptor molecules to a lower energy level, the reaction center of pigment system I. As they pass along the electron transport chain, some of their energy is packaged in the form of ATP. At the reaction center of pigment system I, the light energy absorbed by pigment system I boosts the electrons to a second electron acceptor. From this acceptor, they are passed to* $NADP_{ox}$ *to form* $NAPD_{red}$*. ATP and* $NADP_{red}$ *represent the net gain from the two stages of the light reaction.*

THE DARK REACTIONS

Some of the ATP molecules produced in the first stage of photosynthesis are used as a direct energy source by the photosynthetic cell in which they are formed. Most of the ATP, however, and the high-energy electrons held by $NADP_{red}$ are used in the formation of sugar from carbon dioxide. The major biochemical pathway is called the Calvin cycle. This process involves linking carbon atoms into chains and also reducing the carbon atoms—that is, combining them with electrons (and, in this case, hydrogen ions). These reactions take place in the stroma and do not require light, although they require the energy-packed compounds produced by the light.

The Calvin Cycle

The cycle begins with ribulose diphosphate (RuDP), a five-carbon sugar with two phosphates attached. Do not try to memorize these steps.

Step 1. This compound is combined with carbon dioxide and water and immediately splits to form two molecules of a three-carbon compound, phosphoglycerate. For simplicity, only one of the two is shown here.

Step 2. The phosphoglycerate reacts with ATP, forming ADP and diphosphoglycerate (glycerate with two phosphates). Some of the energy from ATP is used to power this reaction, and some is stored in the new phosphate bond.

Step 3. Using energy released when $NADP_{red}$ is oxidized and the second phosphate bond is broken, the diphosphoglycerate undergoes a chemical change to glyceraldehyde phosphate, a three-carbon sugar-phosphate.

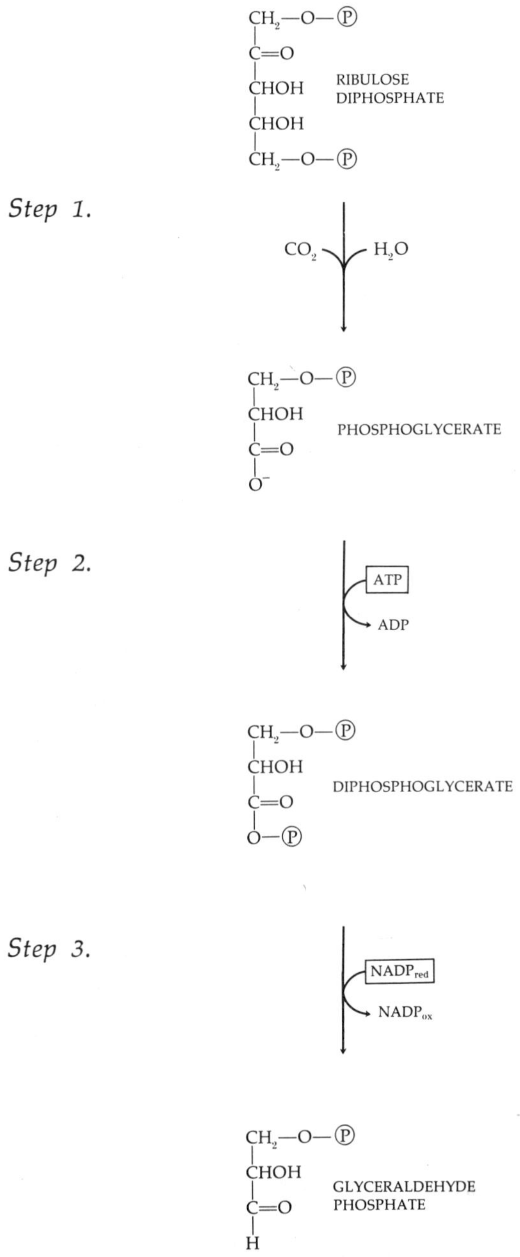

CH_2—O—Ⓟ
C=O
CHOH
CHOH
CH_2—O—Ⓟ
RIBULOSE DIPHOSPHATE

Step 1.
CO_2 H_2O

CH_2—O—Ⓟ
CHOH
C=O
O^-
PHOSPHOGLYCERATE

Step 2.
ATP
ADP

CH_2—O—Ⓟ
CHOH
C=O
O—Ⓟ
DIPHOSPHOGLYCERATE

Step 3.
$NADP_{red}$
$NADP_{ox}$

CH_2—O—Ⓟ
CHOH
C=O
H
GLYCERALDEHYDE PHOSPHATE

Since each cycle actually involves three molecules of RuDP (with three molecules of carbon dioxide and three molecules of water), six molecules of glyceraldehyde phosphate are formed. Five of the six are recombined by a series of enzymatic steps to produce three more molecules of RuDP. The RuDP then accepts three molecules of carbon dioxide and three molecules of water, and the cycle begins again.

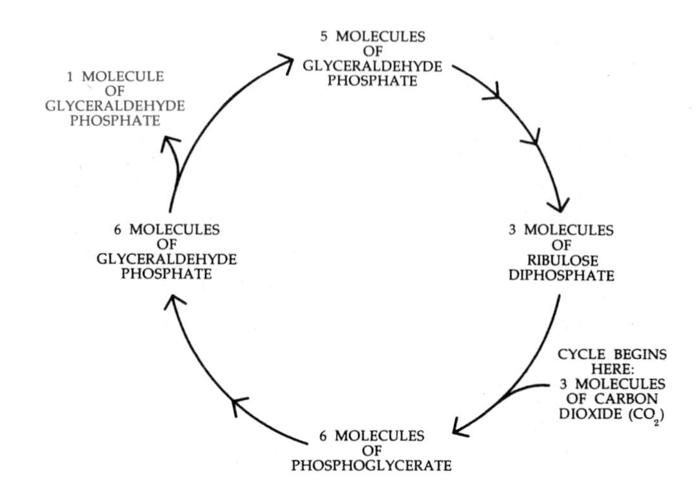

2–59

Summary of the Calvin cycle. Three molecules of ribulose diphosphate (RuDP), a five-carbon compound, are combined with three molecules of carbon dioxide, yielding six molecules of phosphoglycerate, a three-carbon compound. These are converted to six molecules of glyceraldehyde phosphate. Five of these three-carbon molecules are combined and rearranged to form three five-carbon molecules of RuDP. The "extra" molecule of glyceraldehyde phosphate represents the net gain from the Calvin cycle. The energy that "drives" the Calvin cycle is in the form of ATP and $NADP_{red}$, *produced by the light reactions.*

THE PRODUCTS OF PHOTOSYNTHESIS

The net gain from the Calvin cycle is one molecule of glyceraldehyde phosphate. What happens to this molecule? There are many possibilities. One of the most likely is that two molecules of glyceraldehyde phosphate will be put together by the cell to form glucose; if you start with step 4 of glycolysis and run the sequence backward, which is what the cell does, you will end up with glucose. Free glucose, although the most important sugar in vertebrates, is not commonly found in plant cells. Much of the glucose formed in the plant cell is immediately linked to fructose to form sucrose. (Look again at the glycolytic sequence. If you stop before glucose, after step 3, you will have fructose with one phosphate attached. This fructose can be combined with glucose to form sucrose; breaking the phosphate bond supplies the energy for this reaction.) The sucrose is then transported to the other parts of the plant body and taken up by other cells, which, in turn, can break it down for energy or build it up into structural materials.

Much of the ATP of the photosynthetic cell comes directly from the light reactions. However, particularly during periods of dark, the cell may oxidize some of its products of the Calvin cycle to form ATP. In this case, the glyceraldehyde phosphate need not be built up to glucose but could enter the glycolytic sequence at step 5.

Or glucose can be combined with other glucose units to make starch or cellulose. This can take place either in the photosynthetic cell (you will recall the starch grains in the chloroplasts of the photosynthetic cell of the tomato in Chapter 2–3) or in another cell of the plant to which the sugar was transported.

There are, of course, many other chemical modifications which can be made of this single molecular structure, to produce amino acids, oils, flower pigments, and all the other special compounds that the plant body requires for its structures and functions.

We have emphasized the chemical details of photosynthesis in this chapter. To understand photosynthesis in its larger sense, it is necessary to look at the process from the point of view of its role in the biosphere, which we shall do in Part III.

SUMMARY

Photosynthesis is the process by which light energy is converted to chemical energy and so made available for use by living systems.

In the chloroplast, light energy strikes chlorophyll molecules and boosts their electrons to higher energy levels. Two separate light reactions are involved. In the course of these reactions, water is lysed to protons (H^+), electrons, and oxygen; the latter is released as gas. Electrons are pushed "uphill," and as they return "downhill," their energy is harnessed for the production of ATP from ADP and $NADP_{red}$ from $NADP_{ox}$.

In the dark reactions, the ATP and $NADP_{red}$ produced in the light reactions are used in the reduction of carbon dioxide to form a sugar. This process is known as the Calvin cycle. The sugar formed by the Calvin cycle provides both the energy source and the starting material for the functions and structures of living systems.

GLOSSARY

ACID [L. *acidus,* sour]: A substance that on dissociation, releases hydrogen ions (H^+) but not hydroxyl ions (OH^-); having a pH of less than 7; *see* base.

ACTIVE SITE: That part of the surface of an enzyme molecule into which the substrate fits during a catalytic reaction.

ACTIVE TRANSPORT: The energy-expending process by which a cell moves a substance across the cell membrane, often from a point of lower concentration to a point of higher concentration, against the diffusion gradient.

ADENOSINE TRIPHOSPHATE (ATP) (ad-**dih**-n′-seen): The major source of usable chemical energy in metabolism. On hydrolysis, ATP loses one phosphate to become adenosine diphosphate (ADP), with release of usable energy.

ADHESION [L. *adhaerere,* to stick to]: A sticking together of substances.

AMINO ACIDS (am-**ee**-no) [Gk. *Ammon,* referring to the Egyptian sun god, near whose temple ammonium salts were first prepared from camel dung]: Organic molecules containing nitrogen in the form of NH_2; the "building blocks" of protein molecules.

ÅNGSTROM (Å) [after A. J. Ångström, Swedish physicist, 1814–1874]: A unit of length equal to 0.0001 micron.

ATOM [Gk. *atomos,* indivisible]: The smallest unit into which a chemical element can be divided and still retain its characteristic properties.

BASAL BODY: A complex cell structure found at the base of flagella or cilia, which is believed to activate and coordinate their movement.

BASE: A substance that, on dissociation, releases hydroxyl ions (OH^-) but not hydrogen ions (H^+); having a pH of more than 7; *see* acid.

BULK FLOW: The overall movement of a liquid induced by gravity, pressure, or an interplay of both.

CALORIE [L. *calor,* heat]: The amount of energy in the form of heat required to raise the temperature of 1 gram of water 1°C. In making metabolic measurements, the large, or dietary, Calorie (kilocalorie) is generally used; this is defined as the amount of heat required to raise the temperature of 1 kilogram of water 1°C.

CALVIN CYCLE: The process by which carbon dioxide is reduced to carbohydrate during photosynthesis.

CAPILLARY ACTION [L. *capillus,* hair]: The movement of water along a surface, against the action of gravity, resulting from the combined effect of water molecules cohering to each other and adhering to the molecules of the alien surface.

CARBOHYDRATE [L. *carbo,* ember, + *hydor,* water]: An organic compound consisting of a chain of carbon atoms to which hydrogen and oxygen are attached in a 2:1 ratio; includes sugars, starch, glycogen, cellulose, etc.

CAROTENE [L. *carota,* carrot]: A yellow or orange pigment found in plants.

CATALYST [Gk. *katalysis,* dissolution]: A substance that accelerates the rate of a chemical reaction but is not used up in the reaction; enzymes are catalysts.

CELL [L. *cella,* small room]: The structural unit of living organisms, composed of cytoplasm and one or more nuclei and surrounded by a cell membrane. In plants, some protists, and prokaryotes, there is often a cell wall outside the membrane.

CELL MEMBRANE: A phospholipid protein layer which forms the outermost membrane of plant and animal cells.

CELLULOSE: An insoluble complex carbohydrate formed of microfibrils of glucose molecules; the chief component of the cell wall in most plants.

CELL WALL: The rigid outermost layer in cells of plants, some protists, and prokaryotes.

CHLOROPHYLL (**claw**-ro-fill) [Gk. *chloros,* green, + *phyllon,* leaf]: The green light-trapping pigment necessary for photosynthesis; found in the cells of green plants.

CHLOROPLAST: A membrane-bounded chlorophyll-containing organelle in green plant cells; the site of photosynthesis.

CILIUM, *pl.* CILIA (**sil**ly-um) [L. *cilium,* eyelid]: A short hairlike structure present on the surface of some cells, usually in large numbers and arranged in rows. Each cilium has a highly characteristic internal structure of two inner fibrils surrounded by nine pairs of outer fibrils.

COHESION [L. *cohaerere,* to stick together]: The union or holding together of the parts of a substance.

COVALENT BOND: A chemical bond formed between atoms as a result of the sharing of electrons; usually a pair of electrons, one from each atom or both from one atom, is shared.

CYTOPLASM (**sight**-o-plasm) [Gk. *kytos,* container, + *plasma,* form, mold]: The protoplasm of the cell exclusive of the nucleus.

DIFFERENTIATION: The process by which a relatively unspecialized cell or tissue develops more specific functions or comes to produce a specialized substance.

DIFFUSION [L. *diffundere,* to pour out]: The movement of suspended or dissolved particles from a more concentrated to a less concentrated region as a result of the random motion of individual molecules; the process tends to distribute the particles uniformly throughout a medium.

DISACCHARIDE: A sugar composed of two monosaccharide molecules.

DYNAMIC EQUILIBRIUM: With reference to diffusion, the state at which the concentration of the molecules of a substance is equal in all regions and the random movement of the molecules no longer has an overall direction.

ELECTRON: A subatomic particle with a negative electric charge equal in magnitude to the positive charge of the proton but with a mass 1/1,837 that of the proton; normally orbits the atom's positively charged nucleus.

ELECTRON TRANSPORT CHAIN: In cellular respiration, a series of electron-carrier molecules along which the electrons produced by the Krebs cycle are passed to successively lower energy levels, resulting in production of ATP and ultimately of free oxygen and water.

ENDOPLASMIC RETICULUM (end-oh-**plaz**-mick re-**tick**-you-lum): An extensive system of double membranes present in most cells, dividing the cytoplasm into compartments and channels where separate chemical reactions take place.

ENZYME [Gk. *en,* in, + *zymē,* leaven]: A protein of complex chemical constitution, produced in living cells, which even in very low concentration speeds the rate of (catalyzes) a chemical reaction.

EPITHELIAL CELL (epi-**theel**-yull) [Gk. *epi,* upon, + *thēlē,* nipple]: A type of cell that makes up the tissue which covers a body or structure or lines a cavity; epithelial cells form one or more regular layers with practically no intercellular material.

FATS: Organic compounds containing carbon, hydrogen, and oxygen, as in carbohydrates, but with a much smaller proportion of oxygen to carbon. Fats in the liquid state are called oils.

FLAGELLUM, *pl.* FLAGELLA (fla-**jell**-um) [L. *flagellum,* whip]: A fine, long, threadlike structure, composed of protoplasm, which protrudes from a cell body; it is longer than a cilium but has the same internal structure. Flagella are capable of vibratory motion and are used in locomotion and feeding.

GLUCOSE: A six-carbon sugar ($C_6H_{12}O_6$), the chief fuel substance of most organisms; the most common monosaccharide.

GLYCOGEN: A complex carbohydrate (polysaccharide), one of the main stored food substances of most animals and fungi; it is converted into glucose by hydrolysis.

GLYCOLYSIS (gly-**cahl**-y-sis) The process by which the glucose molecule is changed anaerobically to pyruvic acid with the liberation of a small amount of useful energy. This reaction takes place in the cytoplasm.

GOLGI BODY (**goal**-jee): An organelle present in eucaryotic cells consisting of flat, disk-shaped sacs that are often branched into tubules at the margins. The sacs are believed to function as collecting and packaging centers for substances the cell manufactures.

GRANUM, *pl.* GRANA (**grain**-'m): A structure within the chloroplast, seen as a green granule with the light microscope and as a series of stacked thylakoids with the electron microscope. The grana contain chlorophylls and carotenoids and are the actual site of the light reactions of photosynthesis.

HYDROCARBON: An organic compound composed only of hydrogen and carbon atoms.

HYDROGEN BOND: A weak molecular bond linking a hydrogen atom which is covalently bonded to another atom—usually oxygen, nitrogen, or flourine—to the oxygen, nitrogen, or fluorine atom of another molecule.

HYDROLYSIS (high-**drahl**-y-sis) [Gk. *hydor,* water + *lysis,* a loosening]: The splitting of one molecule into two by the addition of H^+ and OH^- ions.

HYDROSTATIC PRESSURE: The pressure exerted by a liquid at rest on an adjacent body.

HYPERTONIC [Gk. *hyper,* above, + *tonos,* tension]: Having a concentration high enough to gain water across a selectively permeable membrane from another solution.

HYPOTONIC [Gk. *hypo,* under]: Having a concentration low enough to lose water across a selectively permeable membrane to another solution.

IMBIBITION [L. *imbibere,* to drink in]: The absorption of water by a substance and its consequent swelling, resulting from the absorption of water molecules onto the internal surfaces of that substance.

ION (**eye**-on): An atom or molecule that has lost or gained one or more electrons. By this process, known as ionization, the atom becomes electrically charged.

IONIZATION: The gaining or losing of electrons by atoms.

ISOTONIC [Gk. *isos,* equal]: Having the same osmotic concentration as that of the substance on the other side of a selectively permeable membrane.

KINETIC ENERGY [Gk. *kinema,* motion]: Energy in the form of motion.

KREBS CYCLE: The series of reactions, occurring in the mitochondrion, that results in the oxidation of pyruvic acid to hydrogen atoms, electrons, and carbon dioxide.

LAMELLA (la-**mel**-la) [L. *lamella,* thin metal plate]: A layer of cellular membranes, particularly photosynthetic, chorophyll-containing membranes.

LIPID [Gk. *lipos,* fat]: One of a large variety of organic fat or fatlike compounds, including fats, oils, steroids, phospholipids, and carotenes.

MITOCHONDRION, *pl.* MITOCHONDRIA (mite-o-**kon**-dri-on): An organelle in eukaryotic cells which contains the enzymes used in energy conversions. It is bounded by a double membrane.

MONOSACCHARIDE [Gk. *monos,* single, + *sakcharon,* sugar]; A single sugar, such as a five- or six-carbon sugar.

NANOMETER (nm): A unit of measure equaling one billionth of a meter (0.000001 millimeter; 0.000025 inch), formerly termed millimicron.

NICOTINAMIDE ADENINE DINUCLEOTIDE (NAD) (nick-oh-**teen**-uh-mid **add**-eh-neen dye-**new**-klee-oh-tid): A coenzyme that functions as an electron acceptor in the chemical processes of respiration.

NICOTINAMIDE ADENINE DINUCLEOTIDE PHOSPHATE (NADP): A coenzyme present in the pigment of the chloroplast that functions as an electron acceptor in photosynthesis.

NUCLEAR ENVELOPE: The double membrane surrounding the nucleus of a cell.

NUCLEOTIDE (**new**-klee-o-tide): A single unit of nucleic acid.

NUCLEUS [L. *nucleus,* a kernel] (1) A specialized body within the eukaryotic cell; it is bounded by a double membrane and contains the chromosomes. (2) The central part of an atom.

ORGAN: A structure composed of different tissues whose functions are coordinated to perform a composite task.

ORGANELLE: Any specialized body in the cytoplasm of a cell.

OSMOSIS [Gk. *ōsmos,* impulse or thrust]: The movement of water between two solutions separated by a membrane that permits the free passage of water and that prevents or retards the passage of the solvent; the water tends to move from the side containing a lesser concentration of solute to the side containing a greater concentration.

OXIDATION: The loss of an electron by an atom. Oxidation and reduction (the gain of an electron) take place simultaneously, since an electron that is lost by one atom is accepted by another. Oxidation-reduction reactions are an important means of energy transfer within living systems.

pH: A symbol denoting the relative concentration of hydrogen ions (H^+) in a solution. In biological systems, pH values normally run from 0 to 14; the lower the value, the more acid a solution is, that is, the more hydrogen ions it contains. pH 7 is neutral, less than 7 is acid, and more than 7 is alkaline.

PHAGOCYTOSIS (fag-o-sigh-**toe**-sis) [Gk. *phagein,* to eat, + *kytos,* container]: The process of ingestion—usually involving isolation and destruction—of particulate substances into cells.

PHOSPHOLIPID: A complex compound consisting of a glycerol molecule linked to two fatty acids by a molecule of phosphoric acid.

PIGMENT: A substance that absorbs light.

PLASMOLYSIS (plaz-**mah**-leh-sis) [Gk. *plasma,* form, + *lysis,* a loosening]: The shrinkage of a plant cell from its cell wall due to the removal of water from the protoplasm by osmosis.

POLARITY: The condition of a substance, such as a magnet, that exhibits opposite or contrasted properties in opposite or contrasted parts.

POLYPEPTIDE: A substance consisting of amino acids linked together by peptide bonds.

POLYSACCHARIDE: A carbohydrate composed of many monosaccharide units joined in a long chain; examples are glycogen, starch, and cellulose.

POTENTIAL ENERGY: The energy present in a particle or body that derives from its position in relation to its surroundings.

PRIMARY STRUCTURE: The organization of a protein molecule in which the amino acid groups are linked in a linear arrangement in a specific sequence.

PROTEIN [Gk. *proteios,* primary]: A complex organic compound composed of many amino acids joined by peptide bonds.

PROTIST: A one-celled organism; one of the Protista, a group that includes the protozoans and some of the true algae but not the bacteria or blue-green algae.

PROTON: A subatomic, or elementary, particle with a single positive charge equal in magnitude to the charge of an electron and with a mass of 1; a component of every atomic nucleus.

PROTOPLASM: The living substance of all cells.

QUATERNARY STRUCTURE: A complex protein structure formed by the association of two or more folded protein chains.

REDUCTION [L. *reducere,* to lead back; originally the "bringing back" of a metal from its oxide, for example, iron from iron rust or ore]: The gain of an electron by an atom; *see* oxidation.

RESPIRATION: [L. *respirare,* to breathe]: (1) In organisms, the intake of oxygen and the liberation of carbon dioxide. (2) In cells, the oxidative breakdown and release of energy from fuel molecules.

RETINENE: An eye pigment, related in structure to carotene, which is believed to activate the sensation of light.

RIBOSOME (**rye**-bo-sohm): A small organelle composed of protein and RNA; the site of protein synthesis.

SECONDARY STRUCTURE: The organization of a protein molecule in which the long polypeptide chains assume a helical configuration maintained by hydrogen bonds.

SELECTIVELY PERMEABLE MEMBRANE: A membrane that is permeable to water but not to solutes.

SOMATIC CELLS [Gk. *sōma,* body]: The differentiated cells making up the body tissues of multicellular plants and animals.

SPECIFIC HEAT: The amount of heat (in calories) required to raise the temperature of 1 gram of a substance 1°C; the specific heat of water is 1 calorie.

STARCH: [M.E. *sterchen,* to stiffen]: A complex insoluble carbohydrate, the chief food-storage substance of plants, which is composed of several hundred glucose units ($C_6H_{12}O_6$); it is readily broken down enzymatically into these glucose units.

STROMA: The granular substance that makes up the interior of the chloroplast, in which the grana are dispersed.

SUBSTRATE [L. *substratus,* strewn under]: The substance acted on by an enzyme.

SUCROSE: Cane sugar, a disaccharide (glucose plus fructose) found in many plants; the primary form in which sugar produced by photosynthesis is transported in the plant body.

SURFACE TENSION: The cohesion of molecules on the free surface of a liquid, caused by the unequal distribution of intermolecular forces on the individual surface molecules.

TERTIARY STRUCTURE: The organization of a protein molecule in which the protein helix is bent at several points (frequently forming a globular structure).

THERMODYNAMICS [Gk. *thermē,* heat, + *dynamis,* power]: The study of the relationships between energy and heat. The first law of thermodynamics states that in all processes, the total energy of the universe remains constant. The second law of thermodynamics states that the entropy, or degree of randomness, tends to increase.

THYLAKOID: A saclike membranous structure in the chloroplast; stacks of thylakoids form the grana.

TISSUE [L. *texere*, to weave]: A group of cells, of one or a few similar types, organized into a structural unit with a specific function.

TURGOR [L. *turgere*, to swell]: The distension of a plant cell by its fluid contents.

VACUOLE [L. *vacuus*, empty]: A space within a cell which is bounded by a membrane and filled with water and various substances in solution or as crystals.

WATER POTENTIAL: The potential energy of water; water potential results from gravity, pressure, or differences in solute concentration, or a combination of any of these factors.

SUGGESTIONS FOR FURTHER READING

DUPRAW, E. J.: *Cell and Molecular Biology*, Academic Press, Inc., New York, 1968.

A stimulating and thoroughly up-to-date account of cell biology which clearly indicates the ongoing nature of this exciting field.

LEDBETTER, M. C., and KEITH PORTER: *Introduction to the Fine Structure of Plant Cells*, Springer Publishing Co., Inc., New York, 1970.

An excellent atlas of electron micrographs of plant cells, with a detailed explanation of each.

LEHNINGER, ALBERT L.: *Biochemistry*, Worth Publishers, Inc., New York, 1970.

This advanced text is outstanding both for its clarity and for its consistent focus on the living cell.

LEHNINGER, ALBERT L.: *Bioenergetics: The Molecular Basis of Biological Energy Transformation*, W. A. Benjamin, Inc., New York, 1965.*

A brief, well-written account of the flow of energy in living systems.

LOEWY, A. G., and P. SIEKEVITZ: *Cell Structure and Function*, 2nd ed., Holt, Rinehart and Winston, Inc., New York, 1969.

An outstanding elementary text on cell structure and formation.

O'BRIEN, T. P., and M. E. MCCULLEY: *Plant Structure and Development*, Collier Books, The Macmillan Company, New York, 1969.

An atlas of plant cells and tissues, emphasizing the structures visible with the light microscope.

PALADE, G.: In Lester Goldstein (ed.), *Cell Biology*, Wm. C. Brown Company, Publishers, Dubuque, Iowa, 1966.

This definitive article by Palade is one of the best contemporary accounts of cellular function and structure.

PORTER, KEITH R., and MARY A. BONNEVILLE: *An Introduction to the Fine Structure of Cells and Tissues*, 3rd ed., Lea & Febiger, Philadelphia, 1968.

An atlas of electron micrographs of animal cells; detailed commentaries accompany each. These are magnificent micrographs, and the commentaries describe not only what the picture shows but also the experimental foundations of knowledge of cell ultrastructures.

* Available in paperback.

SECTION 3

Genetics

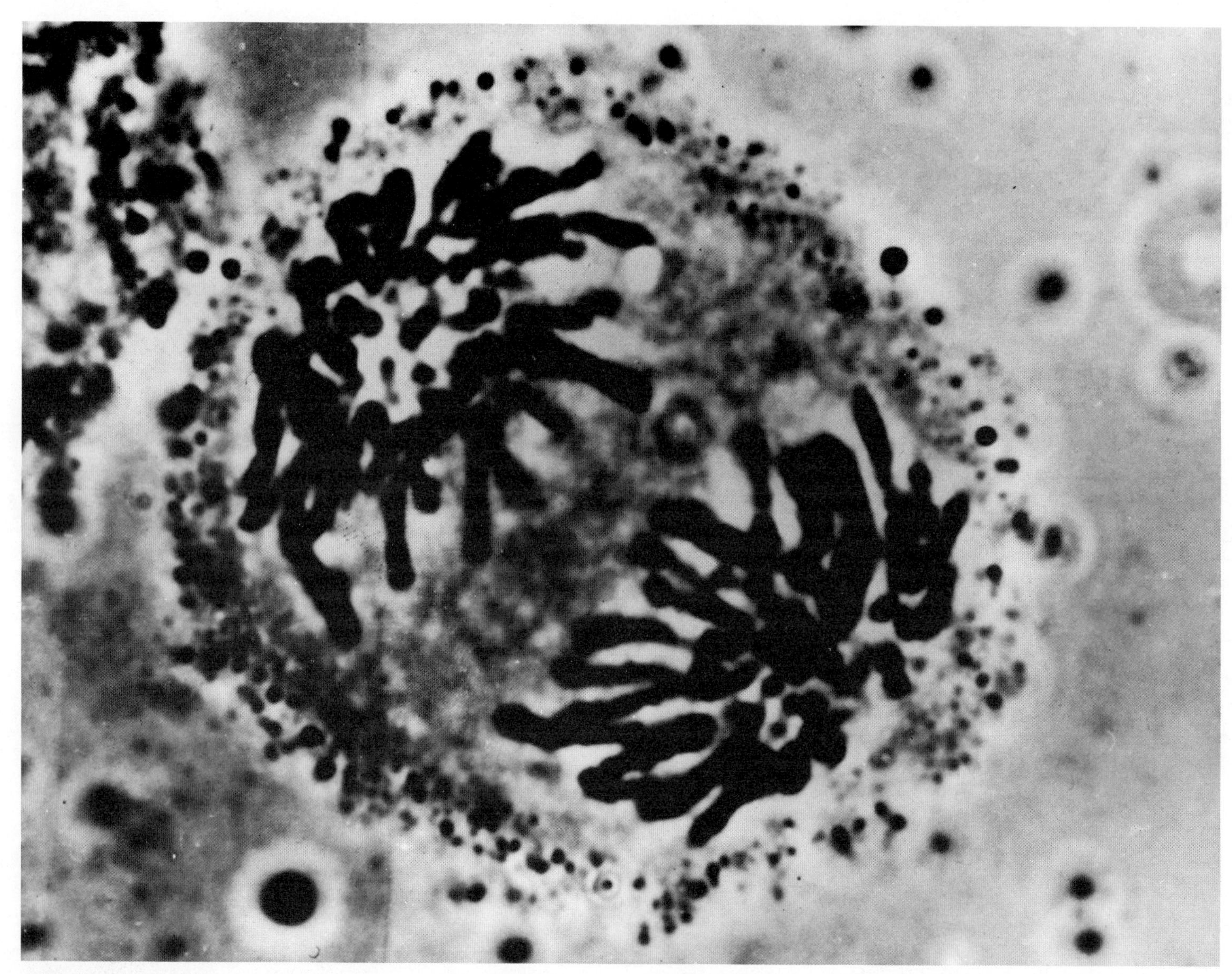

3–1

A cell divides. The dark bodies are chromosomes, the carriers of the genetic information. The chromosomes have duplicated and moved apart. Each set is an exact copy of the other. Thus both of the new, daughter cells contain the same hereditary material.

CHAPTER 3–1

The Nucleus and Cellular Reproduction

Genetics is the branch of biology that deals with heredity, the process by which traits are passed from parents to offspring. As we saw in Section 1, the capacity to reproduce is one of the special characteristics of living matter, and for more than 3 billion years, living things have been reproducing themselves, generation after generation, with each parent passing on to its offspring all the biological instructions necessary for the offspring to develop into the same kind of organism as the parent. A primary question of genetics is how this hereditary information is transmitted by the parent to its young. Or to put it more simply, why do cats have kittens, dogs have puppies, and hens have chicks? Why do you resemble your father or mother? What is the basis of this hereditary information? How is it stored between generations? How is it passed from one generation to another?

Since every organism—whether a paramecium, an oak tree, or a human being—starts as a single cell, the hereditary information must be carried somewhere, somehow in that single cell. More than a century ago, biologists began to suspect that the key to the mystery of heredity was to be found within the cell's nucleus, and as it turned out, they were quite right.

THE CELL'S NUCLEUS

As we saw in the micrographs in Section 2, the nucleus of the cell is a prominent, roughly spherical body. It consists of a nuclear envelope filled with what is known as nuclear sap. Unlike the cytoplasm, which contains many distinct structures, the nucleoplasm contains little that can be seen with any kind of microscope. All that is usually visible is a tangle of threadlike material and one or more dark spots, the *nucleoli* (singular: nucleolus). Yet despite its quiet appearance, the nucleus plays a crucial role in the life of the organism, both as the repository of the hereditary information and as the control center of the individual cell.

One of the most important early investigations of the role of the nucleus in the life of the cell was made about a hundred years ago by a German embryologist, Oscar Hertwig, observing the eggs and sperm of a sea urchin. Sea urchins produce eggs and sperm in great numbers. The eggs are relatively large—about as large as a human ovum, which they closely resemble—and so are easy to observe. They are fertilized in the open water, rather than internally, as is the case with land-dwelling vertebrates such as ourselves. Watching the eggs being fertilized under his microscope, Hertwig observed that only a single sperm cell was required and that when this sperm cell penetrated the egg, its nucleus was released and fused with the nucleus of the egg. This observation, confirmed by other scientists and in other kinds of organisms, was important in establishing the fact that the nucleus is the carrier of heredity: the only link between father and offspring is the nucleus of the sperm.

Since Hertwig's time, a number of experiments have been performed concerning the role of the nucleus in the cell.

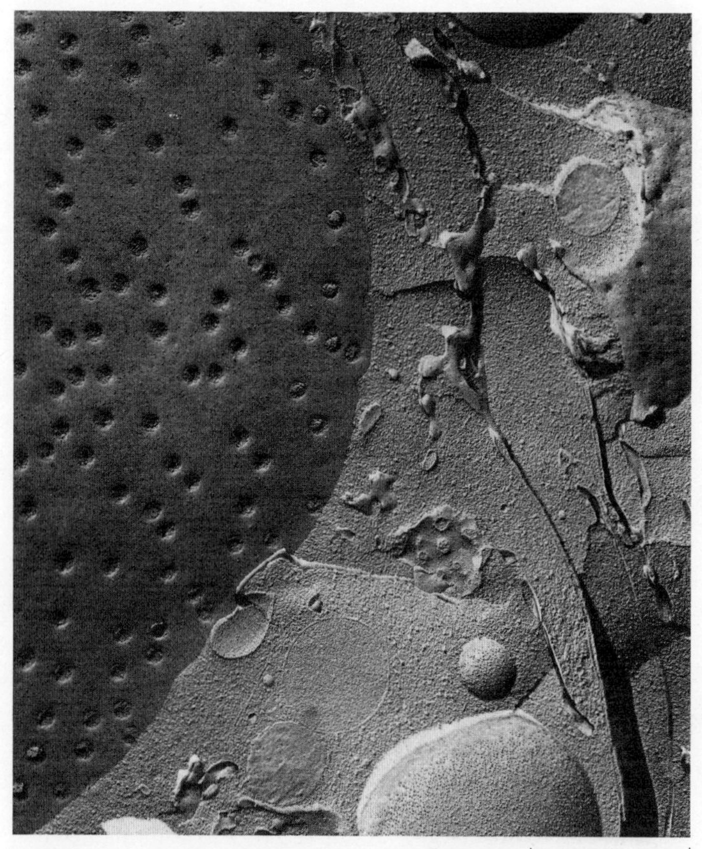

3–2

The cell nucleus fills the left half of this electron micrograph. What you see here is the surface of the nuclear envelope. Clearly visible on this surface are pores, through which, it is believed, the nucleus and the cytoplasm are able to "communicate."

Joachim Hammerling, working in the early 1930s, studied the comparative roles of the nucleus and the cytoplasm by taking advantage of some unusual properties of the marine alga *Acetabularia.* The body of *Acetabularia* consists of a single huge cell 1 to 2 inches in height. Individuals have a cap, a stalk, and a "foot," all of which are differentiated portions of the single cell. If the cap is removed, the cell will rapidly regenerate a new one. Different species of *Acetabularia* have different kinds of caps.

Hammerling took the "foot," which contains the nucleus, from a cell of *A. mediterranea,* a species that has a compact umbrella-shaped cap, and grafted it onto a cell of *A. crenulata,* from which he had first removed the "foot" and the cap. (Normally, *A. crenulata* has a cap of petal-like structures.) The cap that then formed had a shape intermediate between the two. When this cap was removed, the next cap that formed was completely characteristic of *A. mediterranea.*

Hammerling interpreted these results as meaning that certain cap-determining substances are produced under the direction of the nucleus. These cap-determining substances accumulate in the cytoplasm, which is why the first cap that formed after nuclear transplantation was of an intermediate type. By the time the second cap formed, however, the cap-determining substances already in the cytoplasm had been exhausted and the form of the cap was completely under the control of the new nucleus.

In a more recent experiment, performed only a few years ago, J. B. Gurdon at Oxford University, England, removed the nucleus from the intestinal cell of a tadpole and placed it into a frog's egg cell in which the nucleus had been destroyed. The egg cell developed into a normal frog which had the characteristics of the animal that had donated the nucleus. (See Figure 3–4.)

The Gurdon experiment was particularly important because it provided a conclusive answer to a question that biologists had been asking for more than half a century. It showed that *every* nucleus in the body of an organism contains all the hereditary information that is present in

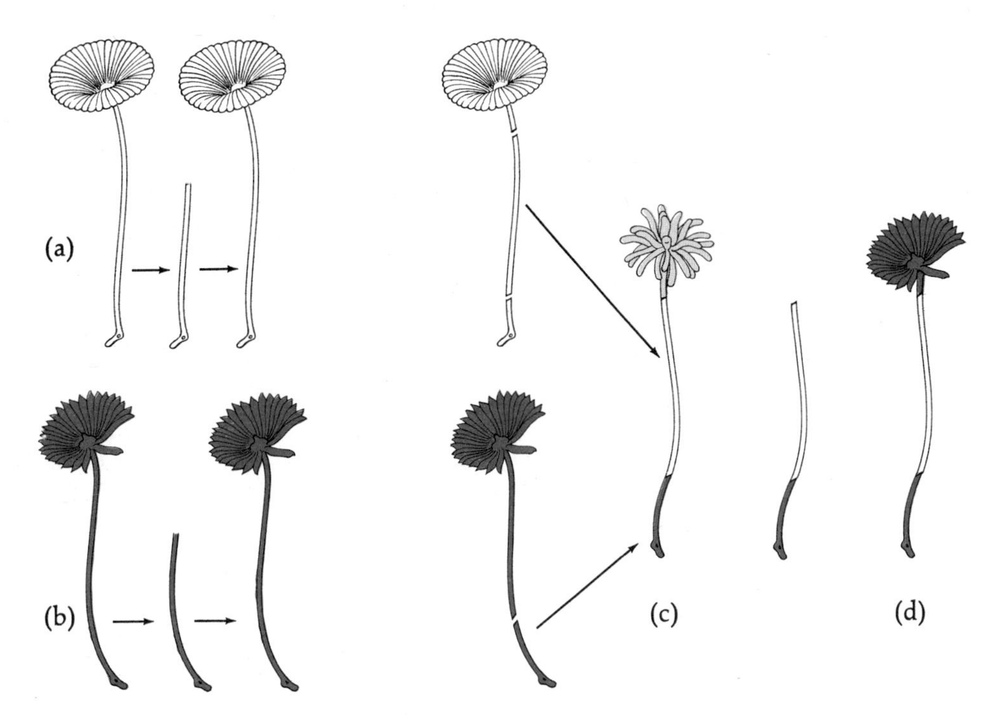

3–3
One species of Acetabularia *has an umbrella-shaped cap, and another has a ragged, petal-like cap. If the caps are removed, others form, similar in appearance to the amputated one* (a *and* b). *However, if the nucleus (which is in the "foot" of the cell) is removed at the same time as the cap and a new nucleus from another species is transplanted, the cap which forms will have a structure with characteristics of both species* (c). *If this cap is removed, the next cap that grows will be characteristic of the cell which donated the nucleus, not of the cell which donated the cytoplasm* (d).

the nucleus of the fertilized egg cell. (We shall be discussing this point further in Chapter 3–8.)

Conclusion: The Functions of the Nucleus

We can see from these experiments that the nucleus performs two crucial functions for the cell. First, it carries the hereditary information for the cell, the instructions that determine whether a particular organism will develop to become a paramecium, an oak tree, or a human being—and not just any paramecium, oak tree, or human being, but one which resembles the parent or parents of that particular, unique organism. Second, as Hammerling's work indicated, the nucleus directs the ongoing activities of the cell, ensuring that the complex molecules that cells require are synthesized in the number and of the kind needed. These molecules, in turn, are involved in carrying out the cell's various activities and also in forming organelles and other structures.

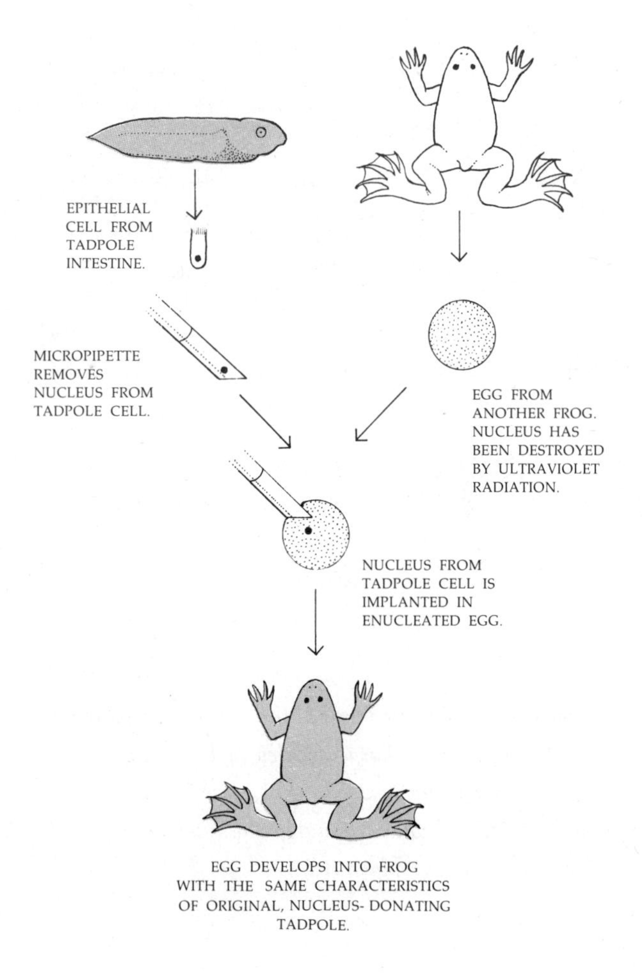

EPITHELIAL CELL FROM TADPOLE INTESTINE.
MICROPIPETTE REMOVES NUCLEUS FROM TADPOLE CELL.
EGG FROM ANOTHER FROG. NUCLEUS HAS BEEN DESTROYED BY ULTRAVIOLET RADIATION.
NUCLEUS FROM TADPOLE CELL IS IMPLANTED IN ENUCLEATED EGG.
EGG DEVELOPS INTO FROG WITH THE SAME CHARACTERISTICS OF ORIGINAL, NUCLEUS- DONATING TADPOLE.

3–4

In experiments by J. B. Gurdon, the nucleus was removed from an intestinal cell of a tadpole and implanted into an egg cell in which the nucleus had been destroyed. In many cases, the egg developed normally, indicating that the intestinal cell nucleus contained all the information required for the whole organism.

CELL DIVISION

At the time of cell division, the nucleus changes its appearance. The threadlike material condenses into a group of structures, the *chromosomes*,* which appear rodlike under the light microscope. Because of the behavior of these bodies at the time of cell division, it was long suspected that they were the carriers of the hereditary instructions, a fact which has now been proved by studies we shall describe later in this section.

Cell division takes place in two closely related events. The nucleus divides first, in a process called *mitosis*, and this is followed by division of the cytoplasm. Mitosis is the process by which the hereditary information stored in the nucleus is transmitted from cell to cell.

Prior to mitosis, the chromosomal material is duplicated so at the beginning of mitosis each chromosome consists of two identical daughter chromosomes called *chromatids*. During mitosis, the chromatids are separated, and each of the two new, "daughter" cells receives one complete set of the chromosomes that were present in the parent cell. The division of the cytoplasm is not as precise as chromosome division. The various cytoplasmic organelles are apportioned more or less equally between the daughter cells. Some organelles, such as the mitochondria, reproduce themselves, and others, such as the ribosomes, are synthesized by the new cell from materials in the nucleus. Usually, at the end of cell division, the two daughter cells will be of about the same size, will have about the same number and kinds of organelles, and will contain *exactly* the same chromosomal material.

* Chromosomes ("colored bodies") were given this name by the early microscopists because they took up certain kinds of stains used in preparing specimens for microscopy and so were first visualized as brightly colored.

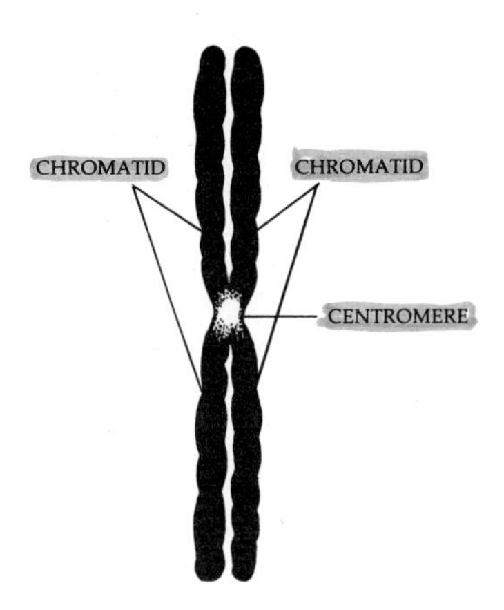

3–5
Diagram of a chromosome at the beginning of mitosis. The chromosomal material has replicated during interphase so that each chromosome now consists of two identical parts, called chromatids.

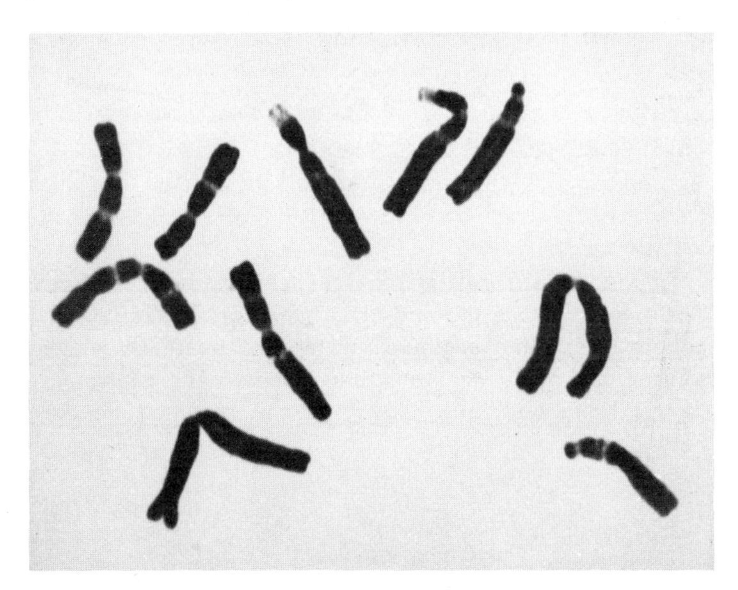

3–6
If you look carefully, you can see that each of these chromosomes actually consists of two longitudinal halves. The centromere, by which they are held together, is the constricted area visible in each. Chromosomes are irregular in shape and also in density, as indicated by the light bands visible in some of them.

Mitosis

The process of mitosis is conventionally divided into four phases: prophase, metaphase, anaphase, and telophase. Of these, prophase is usually much the longest; if a mitotic division takes 10 minutes (which is about the minimum time required), during about 6 of these minutes the cell will be in prophase. Between mitotic divisions, the cell is in interphase. Although little can be seen during interphase, a crucial event takes place then: interphase is the period during which the chromosomal material is duplicated.

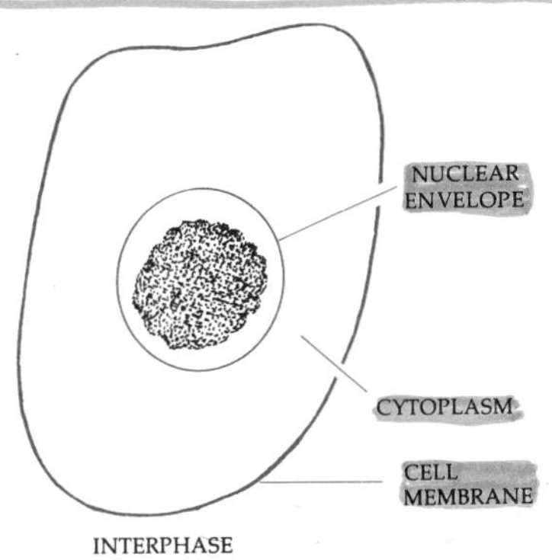

In early prophase, the chromosomal material condenses and the individual chromosomes begin to become visible.

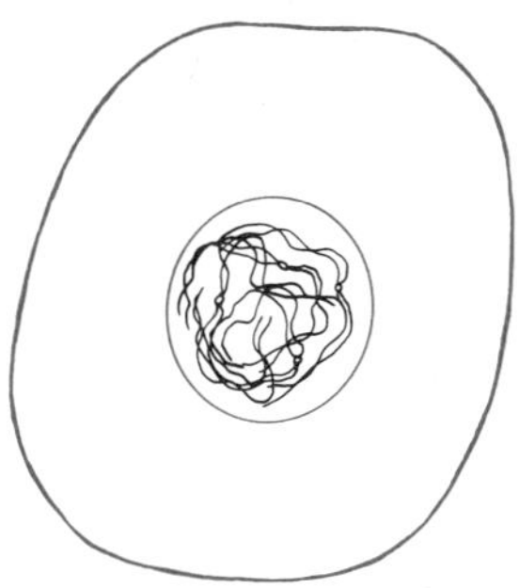

With careful observation, it can be seen that each chromosome consists of two longitudinal parts, held together in a particular area, the *centromere*. These two identical parts are known as *chromatids*.

During prophase, the nuclear envelope breaks down and the nucleolus disappears.

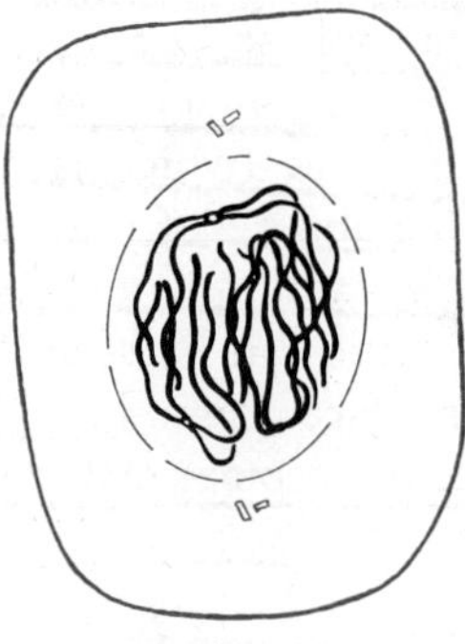

MID-PROPHASE

At the same time, a new structure—the spindle—appears in the cell. The spindle is made up of spindle fibers. Each spindle fiber is a bundle of microtubules (see page 82f). Some of the spindle fibers stretch from one spindle pole to the other in the cell; others are attached to the centromere of the chromosome and appear to play a role in maneuvering the chromosomes during mitosis.

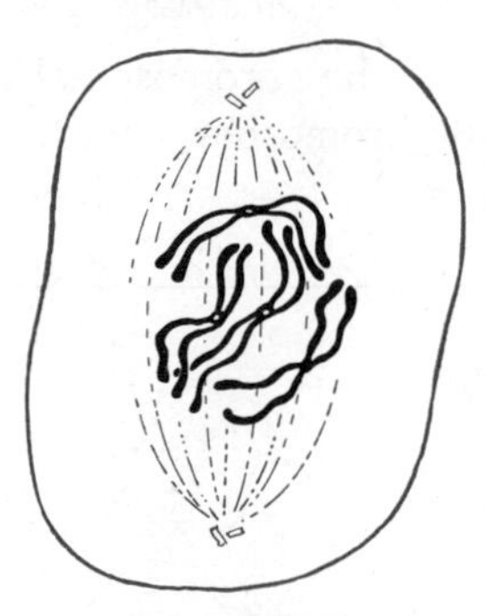

LATE PROPHASE

In animal cells and some plant cells, there is a pair of *centrioles* at each pole of the spindle. Centrioles are essentially basal bodies, the structures from which cilia are formed. In a developing sperm cell, for instance, one of the centrioles involved in the preceding cellular division moves to the cell membrane, where it organizes the formation of the flagellum ("tail") of the sperm. Plant cells—except for flagellated cells, such as fern sperm cells and some algae (*Chlamydomonas*, for example)—have no centrioles, although they do have a spindle apparatus. Prior to metaphase, the chromosomes move within the spindle and arrange themselves on its equatorial plane.

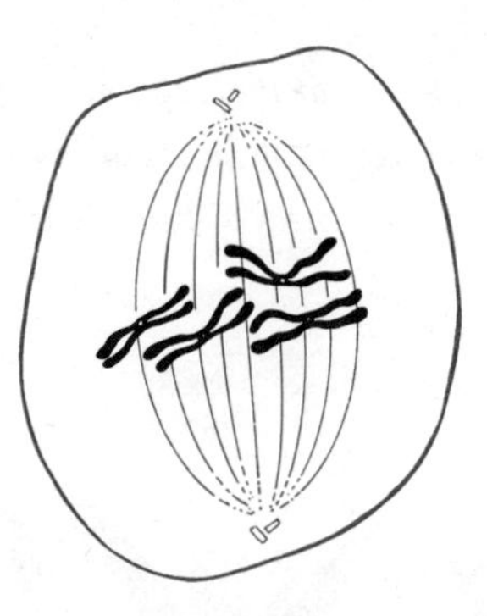

METAPHASE

At the beginning of anaphase, each centromere divides. The two duplicate chromosomes (the former chromatids) adhere briefly and begin to pull apart, apparently tugged toward the poles by the spindle fibers. The centromere moves first, while the "arms" of the chromosome drag behind.

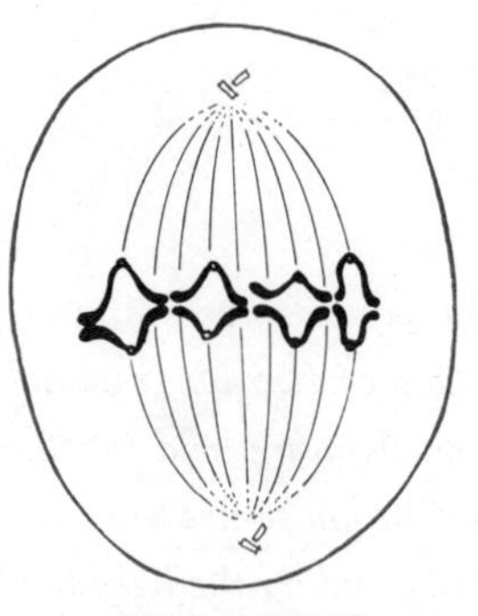

EARLY ANAPHASE

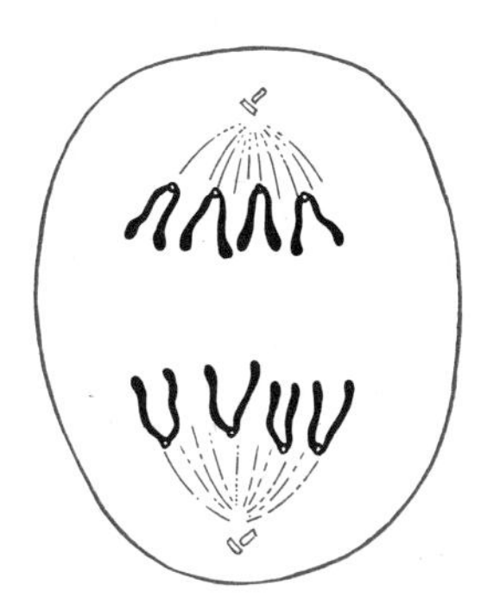

LATE ANAPHASE

Although the fibers lengthen and then shorten during mitosis, they do not appear to get thicker or thinner. This suggests that they do not stretch or contract but that new material is added to the fiber or removed from it as the spindle changes shape. In a recent experiment, a small mark was made at one point on a moving spindle fiber by "branding" it with ultraviolet light. The mark could be seen to move from a point near the equator to the pole and off the end of the fiber, indicating that protein is added at the equator and removed at the pole.

During anaphase, the identical sets of chromosomes move toward the opposite poles of the spindle. By the beginning of telophase, the chromosomes have reached the opposite poles.

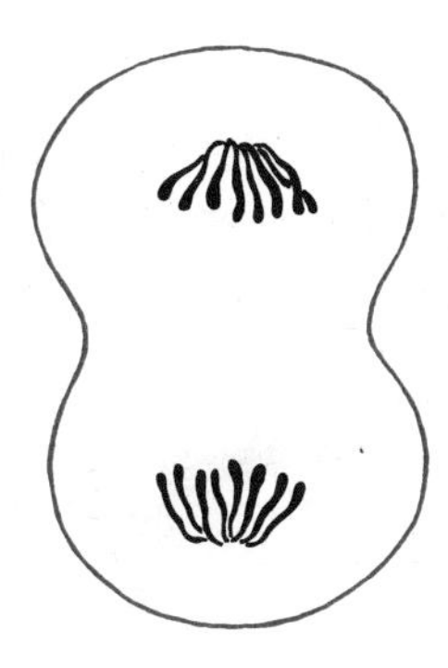

EARLY TELOPHASE

Cytoplasmic Division

Division of the cytoplasm occurs differently in plant and animal cells. In animal cells, during late anaphase or early telophase, the cell membrane begins to pinch in along a circle where the equator of the spindle used to be. At first a furrow appears on the surface, and this gradually deepens into a groove. Eventually the connection between the daughter cells dwindles to a slender thread, which soon parts.

In plant cells, the cytoplasm is divided by the formation of a cell plate, which is a line of vesicles that eventually fuse to form a flat membrane-bound space. As more vesicles fuse, the edges of the growing plate fuse with the membrane of the cell. In this way, space is established between the daughter cells. Wall material is eventually laid down in this space, completing the separation of one cell into two.

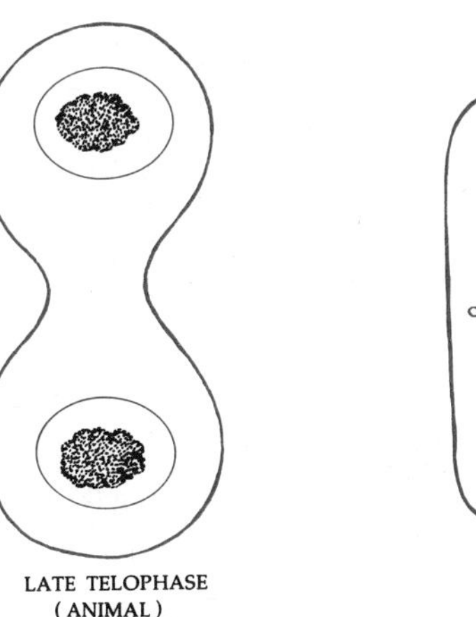

LATE TELOPHASE (ANIMAL)

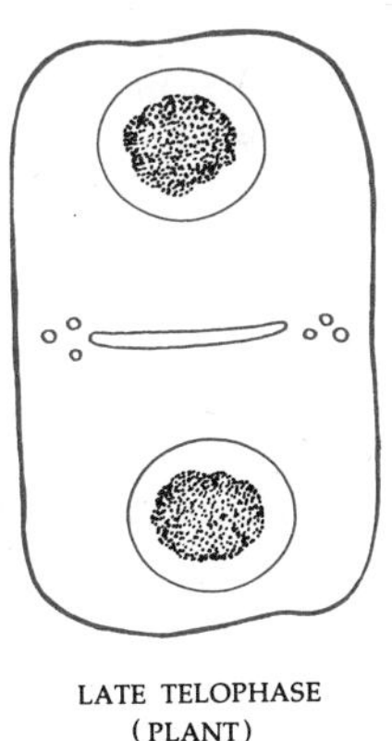

LATE TELOPHASE (PLANT)

During late telophase in both plant and animal cells, the nuclear envelopes re-form and the chromosomes once more become diffuse. Finally, cleavage or the formation of the cell plate separates the two cells.

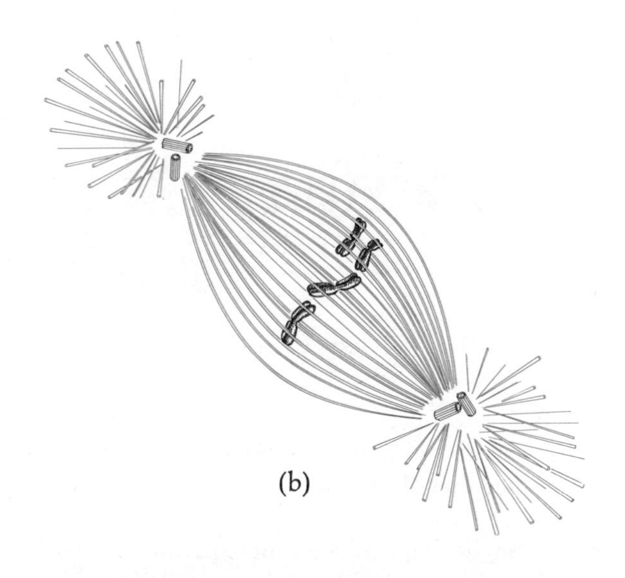

3–7
(a) *This unusual micrograph of a spindle in an egg cell of a marine worm emphasizes the spindle's three-dimensional qualities. The bright streaks are the spindle fibers. The chromosomes appear as indistinct gray bodies near the equator of the spindle. The curved line at the right is part of the cell membrane.* (b) *As this drawing shows, some spindle fibers extend from pole to pole and others are attached to the chromatid pairs at the centromeres. In animal cells, and in some plants and protists as well, a pair of centrioles is found at each pole of the spindle, as shown here.* (c) *Centrioles are structurally identical to basal bodies. The one shown here in cross section is from a mouse embryo tissue culture cell.*

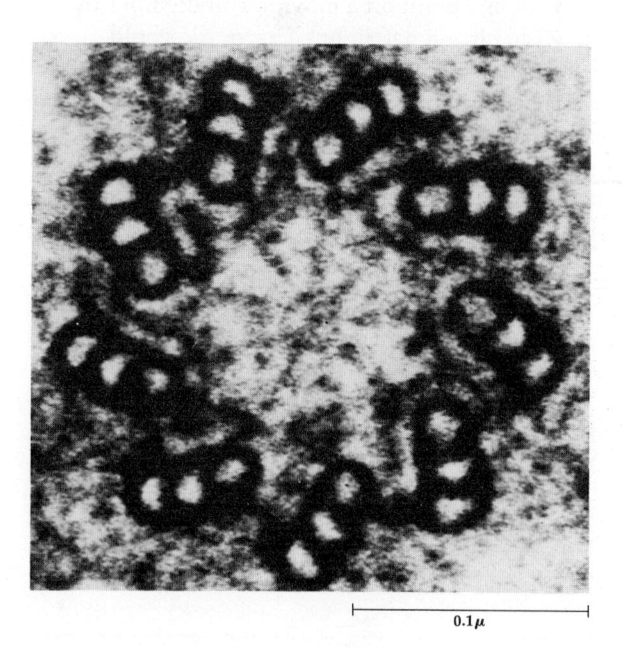

(c)

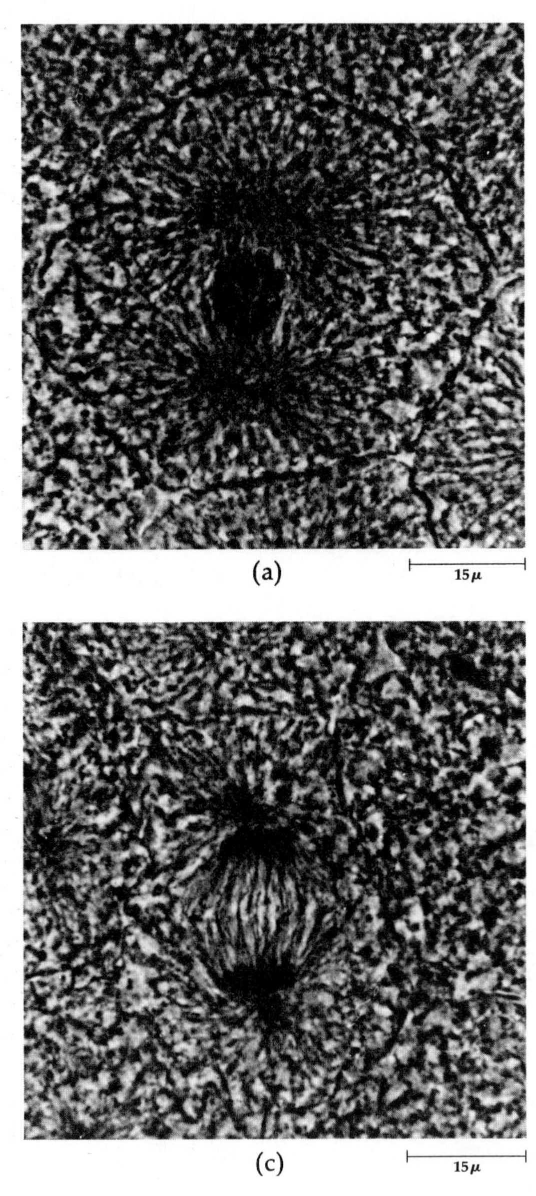

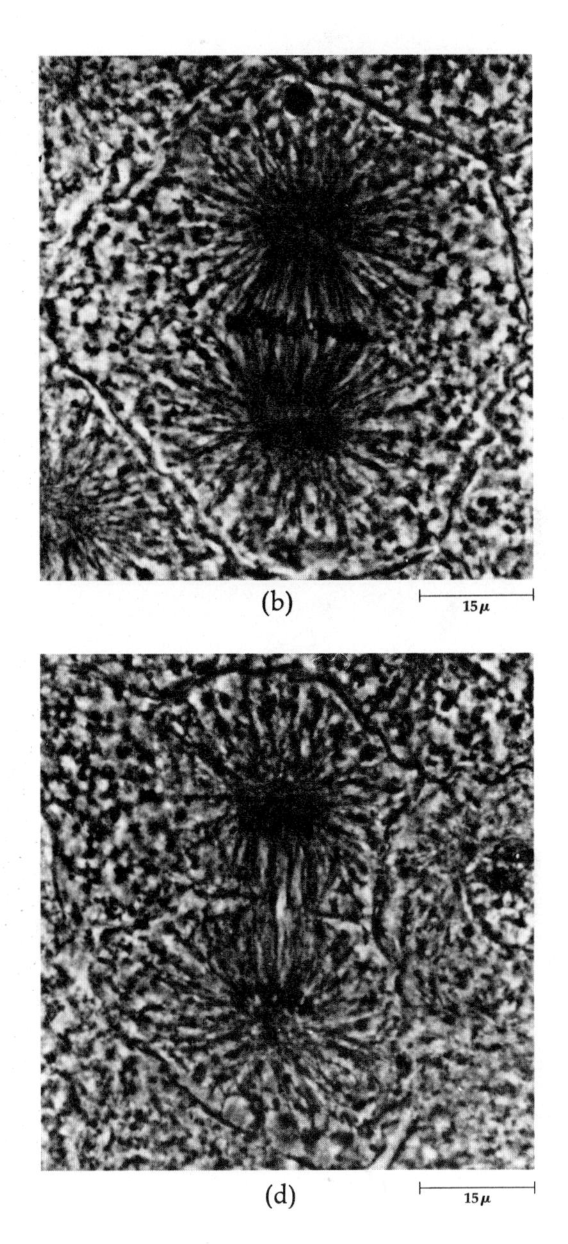

3–8

Mitosis in the embryonic cells of a whitefish. (a) *Prophase. The chromosomes have become visible, the nuclear envelope has broken down, and the spindle apparatus has formed.* (b) *Metaphase. The chromosomes, guided by the spindle fibers, are lined up at the equator of the cell. Some of them appear to have just begun to separate.* (c) *Anaphase. The two sets of chromosomes are moving apart.* (d) *Telophase. The chromosomes are completely separated, the spindle apparatus is disappearing, and a new cell membrane is forming which will complete the separation of the two daughter cells.*

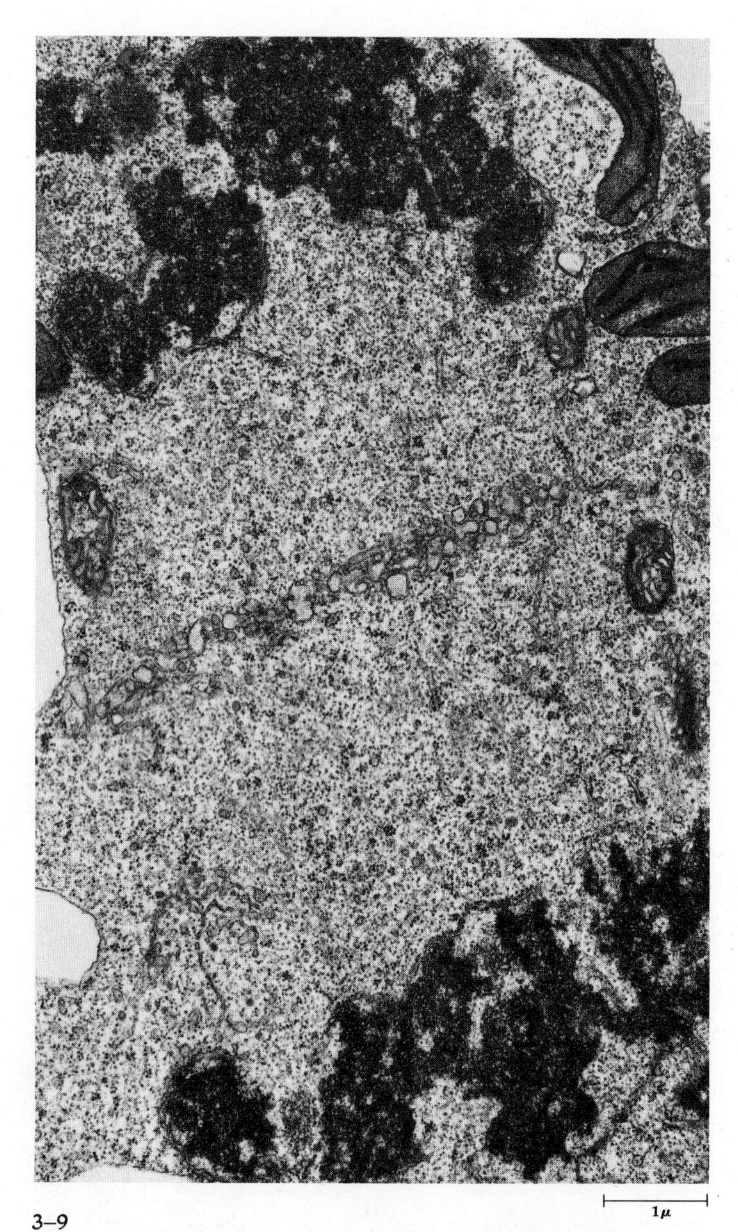

3–9

In plants, the final separation of the two cells takes place by the formation of a structure known as a cell plate. Small droplets appear across the equatorial plate of the cell and gradually fuse, forming a disk. This disk gradually extends outward until it reaches the wall of the dividing cell.

ASEXUAL VERSUS SEXUAL REPRODUCTION

Organisms reproduce either sexually or asexually. In sexual reproduction, two individuals contribute characteristics to the offspring, which therefore have new and unique combinations of traits. When an organism reproduces asexually—as when an amoeba divides—the offspring grow to be exact copies of the single parent. Amoebas and a few other protists reproduce only asexually. Plants and some invertebrates—sea anemones, for instance—can reproduce either sexually or asexually. In these many-celled organisms, asexual reproduction is the result of budding, producing roots or runners, or simply breaking into fragments. In all cases, however, the cell produced from a single parent becomes a new individual. And this individual is just as likely to be successful as its parent was; amoebas have been around for a long, long time.

If sex is not absolutely necessary, why does it occur? In other words, why does any organism invest so much of its resources in the formation of sex cells (eggs or sperm) and the seeking of a mate? The answer that biologists offer to this question is that sexual reproduction provides a source of variations in a population and these variations provide the rich store of material upon which natural selection can operate. The fact that all higher animals—the most complex and advanced organisms, evolutionarily speaking—reproduce sexually appears to confirm this interpretation.

FORMATION OF THE SEX CELLS: MEIOSIS

Asexual reproduction requires only cell division with mitosis. For sexual reproduction, another type of nuclear division, *meiosis*, is necessary. To understand meiosis, we must look again at the chromosomes. Every organism has a chromosome number characteristic of its particular species.* A

* A notable exception is found in higher plants, in which many cells, for reasons which are not known, tend to have more than one set of chromosomes.

mosquito has 6 chromosomes per cell; a cabbage, 18; corn, 20; a frog, 26; a sunflower, 34; a cat, 38; a man, 46; a plum, 48; a dog, 78; and a goldfish, 94. Every set of chromosomes in the organism, as Gurdon's experiment indicates, carries the same hereditary information as every other set.

However, the sex cells—eggs and sperm—have exactly half the number of chromosomes as are characteristic of the somatic cells of the organism. The number of chromosomes in the sex cells is referred to as the haploid ("single") number, and the number in the somatic cells as the diploid ("double") number. For brevity, the haploid number is designated n and the diploid number $2n$.* (Cells that have more than one double set of chromosomes, such as those in higher plants, are known as polyploid.) In man, for example, $n = 23$ and $2n = 46$. When a sperm fertilizes an egg, the two haploid nuclei fuse and the diploid number is restored.

In every diploid cell, half of the chromosomes are derived from the chromosomes that originally came from one parent and half are derived from those of the other. Any particular chromosome from one parent has a partner, originally from the other parent, which resembles it in size and shape and also, as we shall see, in the type of hereditary information it contains. These pairs of chromosomes are known as *homologues*.

In the special kind of nuclear division called meiosis, the diploid number of cells is reduced to the haploid number. If this reduction did not occur, the chromosome number would double each generation. Moreover, as we shall see, meiosis is in itself a source of new genetic combinations.

* In counting, it is often difficult to know whether to count a chromosome that has duplicated but has not divided as 1 or 2. It is customary to count such a chromosome as 1. The trick is to count centromeres.

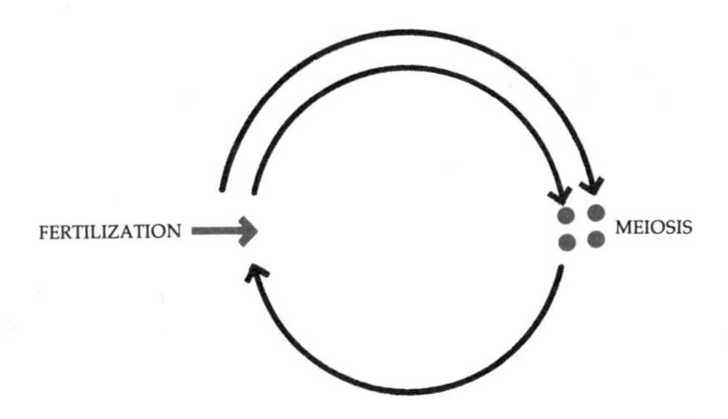

3–10

Sexual reproduction is characterized by two events: the coming together of the sex cells (fertilization) and meiosis. Following meiosis, the number of the chromosomes is single, or haploid (n). *Following fertilization, the number is double, or diploid* (2n).

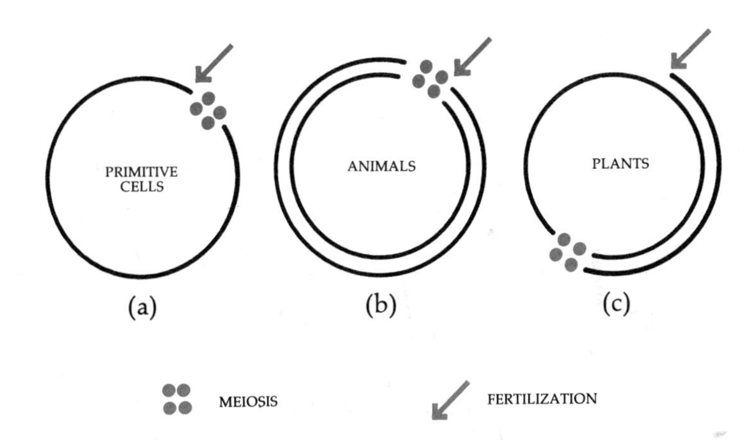

3–11

Fertilization and meiosis occur at different points in the life cycle of different organisms. In primitive cells (a), *meiosis occurs immediately after fertilization and most of the life cycle is spent in the haploid state (signified by the single line). In animals* (b), *meiosis is immediately followed by fertilization. As a consequence, during most of the life cycle the organism is diploid. In plants* (c), *fertilization and meiosis are separated and the organism characteristically has a diploid phase and a haploid phase.*

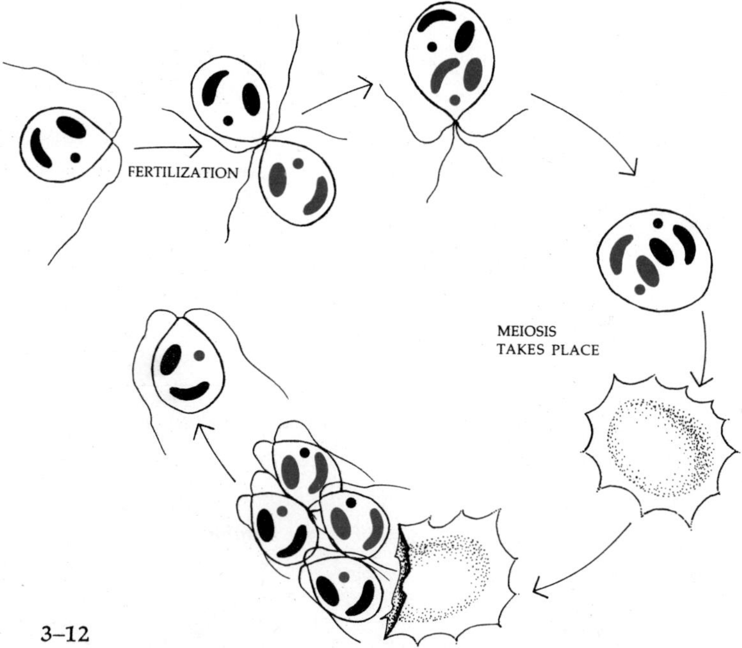

3–12
In Chlamydomonas, *the cell is haploid for most of its life cycle. Fertilization, which restores the diploid state, is followed by meiosis. The chromosomes of one parent cell are in color, and those of the other are in black. Look at the chromosomes of the four offspring.*

3–13 (on facing page)
The life cycle of a fern. Following meiosis, spores, which are haploid, are produced in the sporangia and shed (extreme right). *The spores develop into haploid gametophytes. In many species, the gametophytes are only one layer of cells thick and are somewhat heart-shaped, as shown here* (bottom). *From the lower surface of the gametophyte, filaments, the rhizoids, extend downward into the soil.*

On the lower surface of the gametophyte are borne the flask-shaped archegonia, which enclose the egg cells, and the antheridia, which enclose the sperm. When the sperm are mature and there is an adequate supply of water, the antheridia burst and the sperm cells, which have numerous flagella, swim to the archegonia and fertilize the eggs. From the fertilized (2n) *egg, the* 2n *sporophyte grows out of the archegonium within the gametophyte. After the young sporophyte becomes rooted in the soil, the gametophyte disintegrates. When it becomes mature, the sporophyte develops sporangia, in which meiosis occurs, and the cycle begins again. As in Figure 3–11, the haploid phase of the cycle is indicated by a single arrow, diploid phase by a double arrow.*

Meiosis and the Life Cycle

Meiosis occurs at different times during the life cycle of different organisms. In animals, it occurs at the time the *gametes,* the egg and sperm cells, are formed. In *Chlamydomonas* it occurs following fusion of the mating cells (as shown in Figure 3–12); the cells are ordinarily haploid, and meiosis restores the haploid number. In most higher plants, a haploid phase alternates with a diploid phase in the life of each organism. For example, the fern we see in the woodland is a diploid organism. By meiosis, ferns produce spores, usually on the underside of their leaves. These spores, like the sex cells of animals, have only the haploid number of chromosomes. They germinate to form much smaller plants, often only a few cell layers thick; in these plants, all the cells are haploid. The small, haploid plants produce gametes, which fuse and then develop into a new, diploid fern.

This process, in which the haploid stage is followed by the diploid stage and again by the haploid stage, is known as *alternation of generations.* The spore-producing plant is the *sporophyte,* and the gamete-producing plant is the *gametophyte.* A gametophyte that produces only ova is known as a female gametophyte, and one that produces only sperm is known as a male gametophyte. In the flowering plants, the male gametophyte is the pollen grain. The sex cells produced by the male gametophyte and the female gametophyte—the sperm and the ovum—fuse (fertilization), restoring the diploid number to the egg cell in the seed. A new sporophyte develops from the seed.

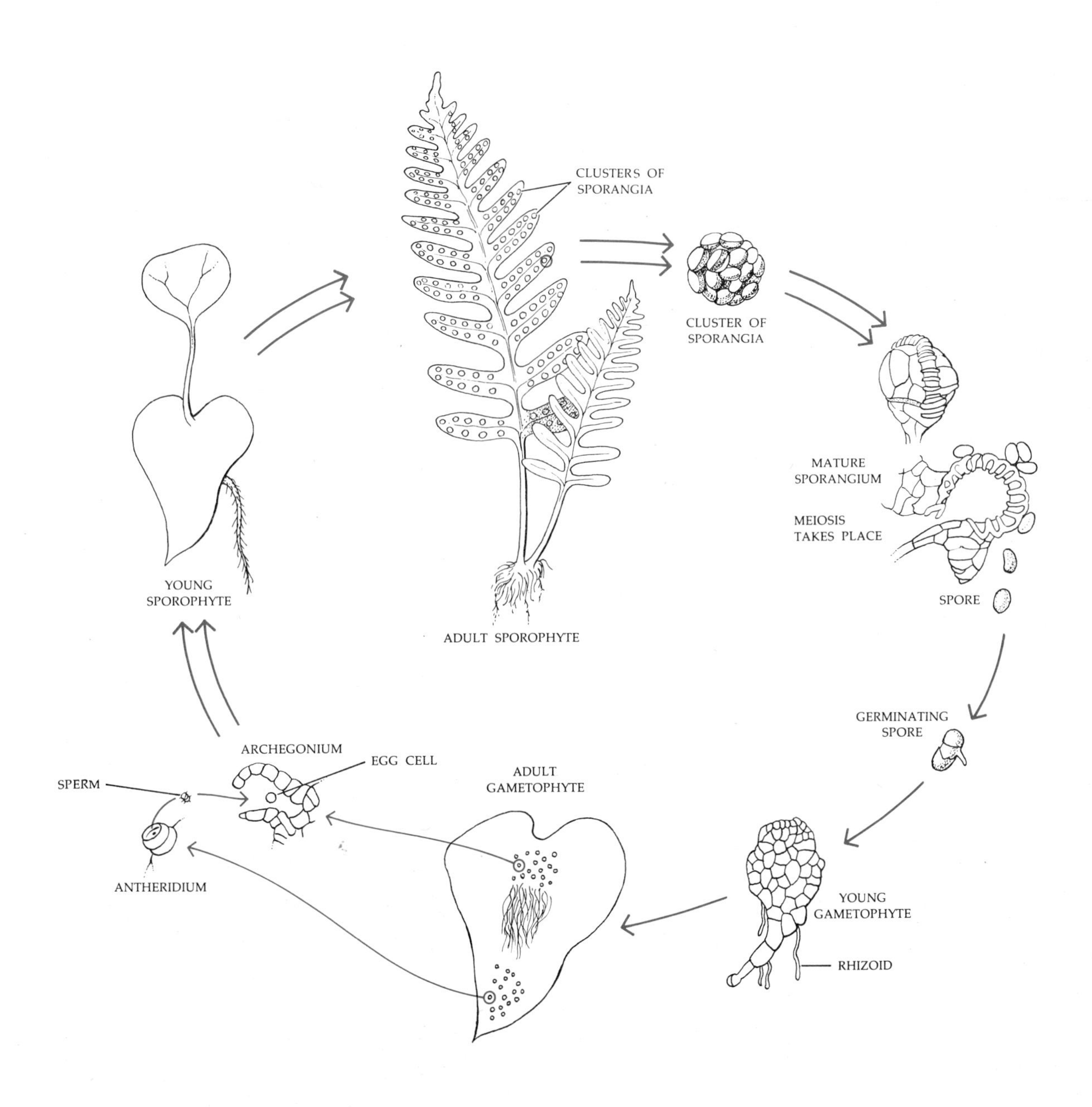
CLUSTERS OF
SPORANGIA
CLUSTER OF
SPORANGIA
MATURE
SPORANGIUM
MEIOSIS
TAKES PLACE
SPORE
YOUNG
SPOROPHYTE
ADULT SPOROPHYTE
GERMINATING
SPORE
ARCHEGONIUM
EGG CELL
ADULT
GAMETOPHYTE
SPERM
ANTHERIDIUM
YOUNG
GAMETOPHYTE
RHIZOID

The Phases of Meiosis

The events that take place during meiosis somewhat resemble those that take place during mitosis, and meiosis, which is believed to have evolved from mitosis, uses much of the same cellular machinery. But the differences are important. First, and most obvious, meiosis takes place in two stages, resulting in four new cells instead of two. Second, mitosis produces two cells which are identical to each other and to the parent cells from which they are formed, but by contrast, meiosis results in four cells which are not identical to each other and which have only half the number of chromosomes present in the parent cell. In the following discussion, we shall describe meiosis in a plant cell which has eight chromosomes. Four of the eight chromosomes were inherited by the plant in which the cell exists from one of its parents, and four from the other parent. For each chromosome from one parent there is an homologous chromosome, or homologue, from the other—a chromosome of similar size and shape which contains the other parent's genetic instructions for the same set of hereditary characteristics; flower color, for instance.

During interphase, the chromosomes are duplicated, so that by the beginning of meiosis each chromosome consists of two identical chromatids held together at the centromere. In the first prophase of meiosis, the chromosomes come into view and the nuclear membrane begins to dissolve.

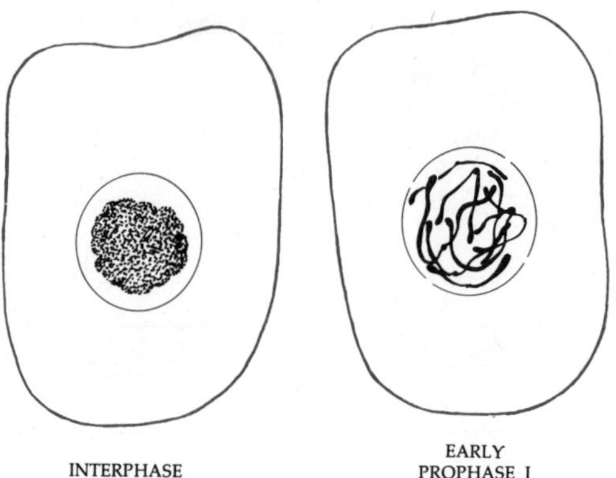

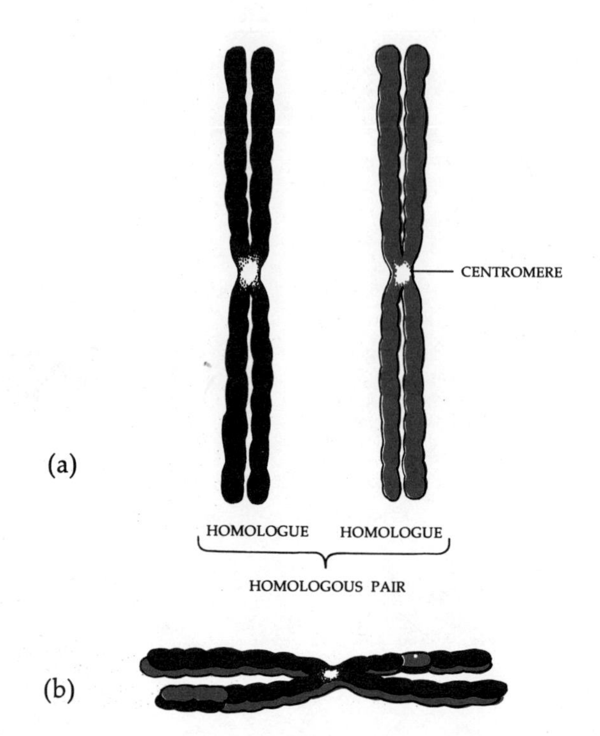

3–14
(a) *Chromosomes in meiosis assort themselves in homologous pairs. One chromosome (homologue) of each pair is a duplicate of a chromosome donated by the maternal parent; the other is a duplicate of a chromosome donated by the paternal parent.* (b) *At the beginning of meiosis, the homologous chromosomes pair and come together. During this process, chromosomal material is exchanged between the chromatids of homologous chromosomes. This exchange is known as crossing over.*

During prophase, the homologous chromosomes (each consisting of two identical chromatids) come together in pairs. Since each homologue consists of two identical chromatids, the pairing process actually involves four chromatids. Crossing-over, the interchange of segments of one chromosome with corresponding segments from its homologue, takes place at this time (see Figure 3–14). Such exchanges of chromosomal material are important sources of variation in the hereditary material. (We shall discuss crossing-over in more detail in Chapter 3–4.)

If you remember the arrangement of the homologous chromosomes at this stage of meiosis, you will be able to remember all the subsequent events with little difficulty.

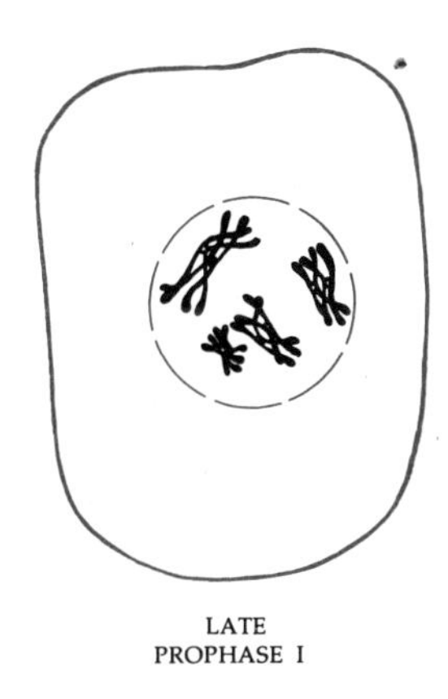

LATE
PROPHASE I

In metaphase I, the four homologous pairs line up along the equatorial plane of the cell.

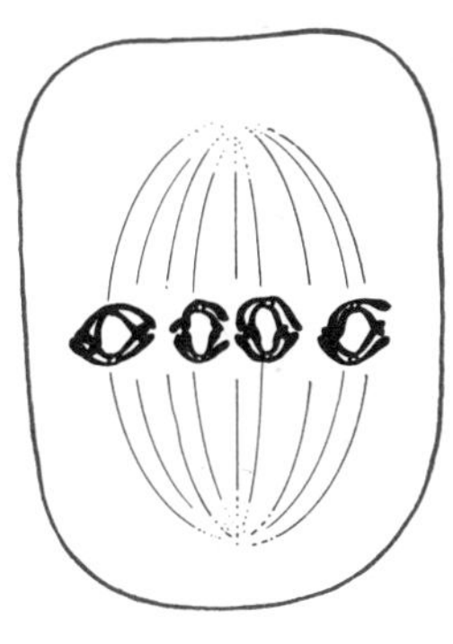

METAPHASE I

The spindle has formed, and spindle fibers attach to each homologue in the area of the centromere. At anaphase I, the homologues, each consisting of two chromatids, separate, apparently pulled apart by the spindle fibers attached to the centromeres. The homologues, which seem to have a strong affinity for one another, pull apart slowly. The centromeres do not divide, as they did in mitosis, and so the two chromatids of each chromosome do not separate.

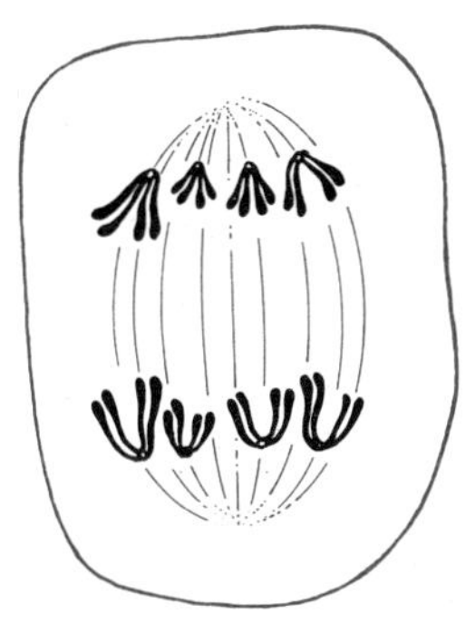

ANAPHASE I

By the end of the first meiotic division, the homologues have separated. The cell enters interphase. The chromosomes disappear from view and nuclear membranes may re-form.

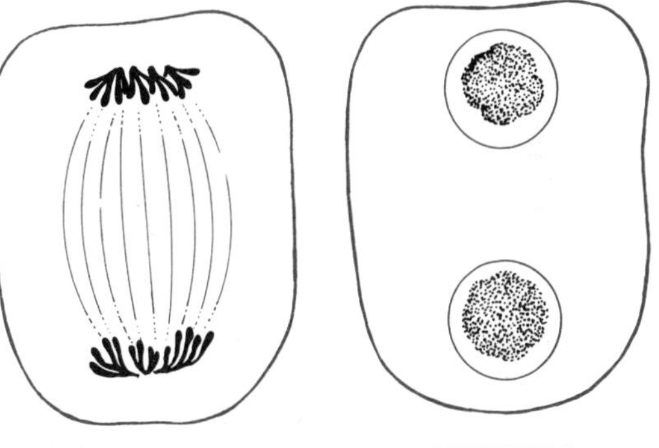

TELOPHASE I INTERPHASE II

Each nucleus now contains only half the number of chromosomes of the original cell: these chromosomes are probably different from any one that was present in the original cell because of changes taking place during crossing over. This interphase differs from the interphase preceding mitosis and from the interphase preceding the first meiotic division in one crucial respect: no duplication of the chromosomal material takes place.

The second meiotic division resembles mitosis. At the beginning of the second meiotic division, the chromosomes condense again. There are four in each nucleus (the haploid number), and they are still in the form of chromatids held together at the centromere.

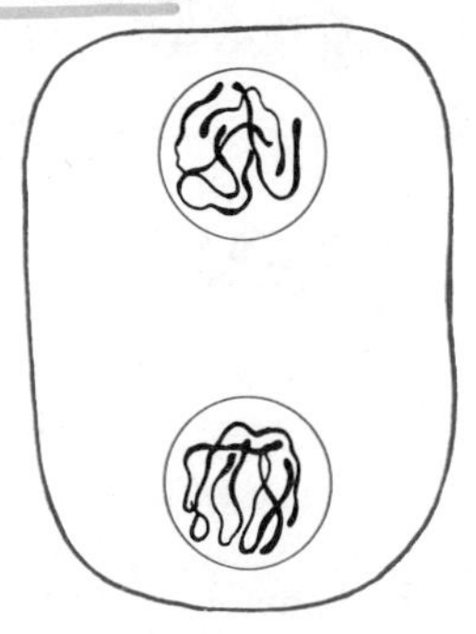

PROPHASE II

During prophase II, the nuclear envelopes, if present, dissolve, and the spindle fibers begin to appear. During the second metaphase, the four chromosomes in each nucleus line up on the equatorial plane. At anaphase II, as in mitosis, the chromosomes split apart at their centromeres and each chromatid (now a single chromosome) moves toward one of the poles.

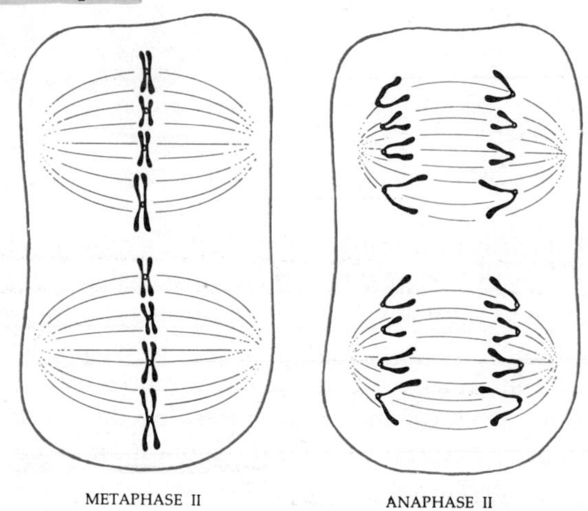

METAPHASE II ANAPHASE II

During telophase, a nuclear membrane forms around each set of chromosomes. There are now four in all, each containing the haploid number of chromosomes.

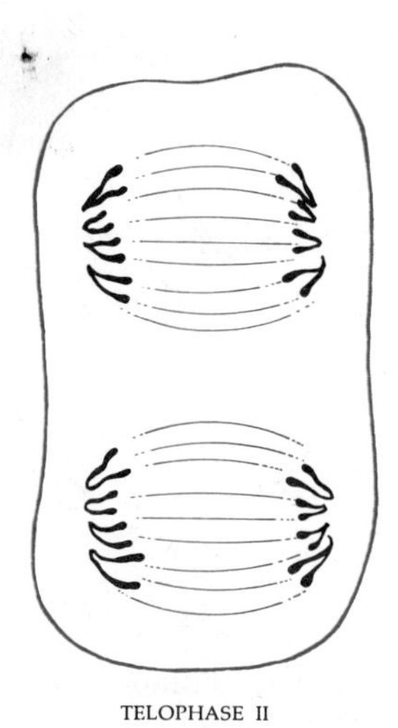

TELOPHASE II

Cell membranes and cell walls form, dividing the cytoplasm, and the cells begin to differentiate into gametes or spores. Each of the cells illustrated here will eventually develop to form a single pollen grain.

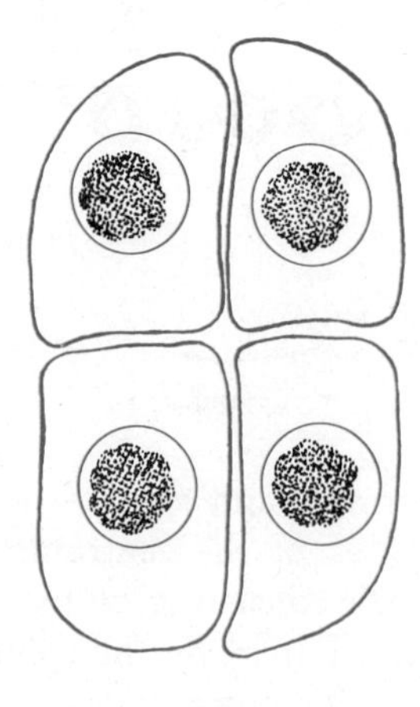

TETRAD

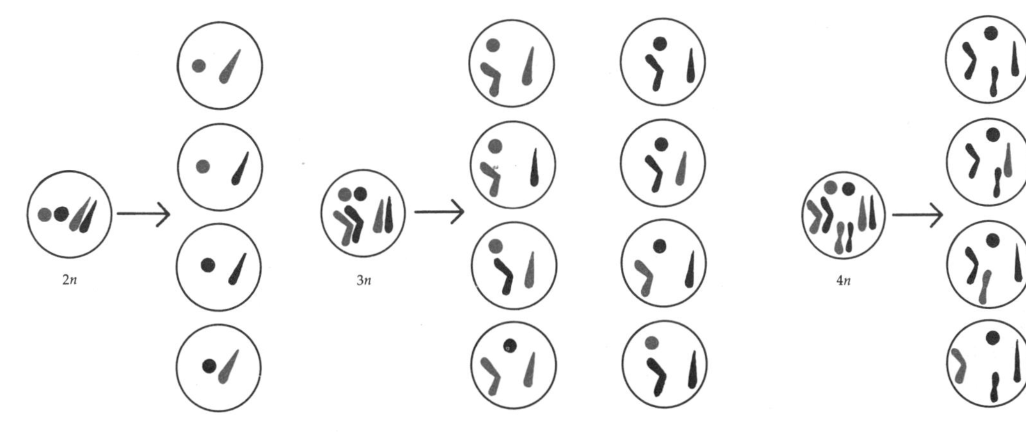

3–15
The distribution of the chromosomes at meiosis in a diploid organism. The black chromosomes were originally of paternal origin, and the chromosomes in color of maternal origin. They were transmitted by replication and mitotic division to the cells from which the eggs and sperm were formed. In the course of meiosis, these chromosomes are sorted out among the haploid cells. As you can see, chromosomes of maternal or paternal origin do not stay together but are sorted independently. If the original number of chromosomes is 4 (n = 2), the number of possible combinations of chromosomes is 2^2, or 4. If the original number is 6, the number of possible combinations is 2^3, or 8. Because maternal and paternal chromosomes differ in some of their genetic material, each of these cells is genetically different. Man with his 46 chromosomes is capable of producing 2^{23} different kinds of sperm cells— 8,388,608 different combinations of chromosomes, as many as there are different individuals in New York City! And this does not take into account the additional variations introduced by crossing over.

SUMMARY

The nucleus is the most prominent structure within most cells. The nucleoplasm is surrounded by a nuclear envelope. Within it are the nucleolus and the chromosomes, the bodies that contain the genetic information of the cell. As indicated by experiments with one-celled algae and frogs, the nucleus controls the structures of the cell and contains the hereditary information.

Cell division includes division of the nucleus (mitosis), during which the chromosomes are apportioned between the two daughter cells, and division of the cytoplasm. When the cell is in interphase (not dividing), the chromosomes are visible only as thin strands of threadlike material. During this period, if mitosis is to take place, the chromosomal material is duplicated. During mitosis, the chromosomal material condenses and each chromosome appears as two identical chromatids, held together at the centromere. At the end of mitosis, the chromatids separate and each chromatid, now a daughter chromosome, moves to one of the daughter cells. In this way, the chromosomes are equally divided between the two new cells. Each daughter cell thus contains the same genetic material as that of the mother cell.

Sexual reproduction involves a special kind of nuclear division called meiosis. Meiosis is the process by which the genetic material is reassorted and cells are produced which have the haploid chromosome number ($1n$). The other principal component of sexual reproduction is fertilization, the coming together of haploid cells, which restores the diploid number ($2n$).

At the start of meiosis, the chromosomes arrange themselves in pairs. Each pair consists of a chromosome which is a double copy of a chromosome originally donated by the father of the organism and a chromosome which is a double copy of one from the mother; the members of the pairs are known as homologues. Each homologue consists of two identical chromatids. Early in meiosis, crossing over occurs between homologues, resulting in exchanges of genetic material.

In the first stage of meiosis, the homologues are separated. Two nuclei are produced, each with a haploid number of double chromosomes. The cell enters interphase, but the chromosomal material is not duplicated. In the second stage of meiosis, the chromatids separate as in mitosis. When the two new nuclei divide, four haploid cells result, each with a single copy of each of its chromosomes.

Meiosis provides a source of variations on which natural selection acts.

QUESTIONS

1. What is happening at each step in the illustrations which appear below?

2. Draw a diagram of a cell with eight chromosomes ($n = 4$) at meiotic prophase I. Label each pair of chromosomes differently (e.g., one pair labeled A^1 and A^2, another B^1 and B^2, etc.)

3. Diagram the possible gametes resulting from a single meiosis in a plant with six chromosomes ($n = 3$). If this were a self-fertilizing plant—a plant such as the garden pea, in which the same plant provides both pollen and ova—how many chromosome combinations would be possible in the fertilized egg? Diagram four of these combinations.

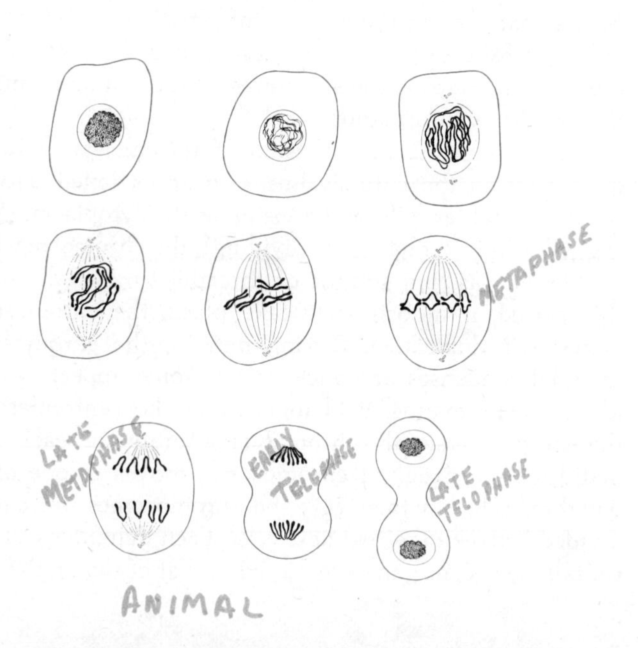

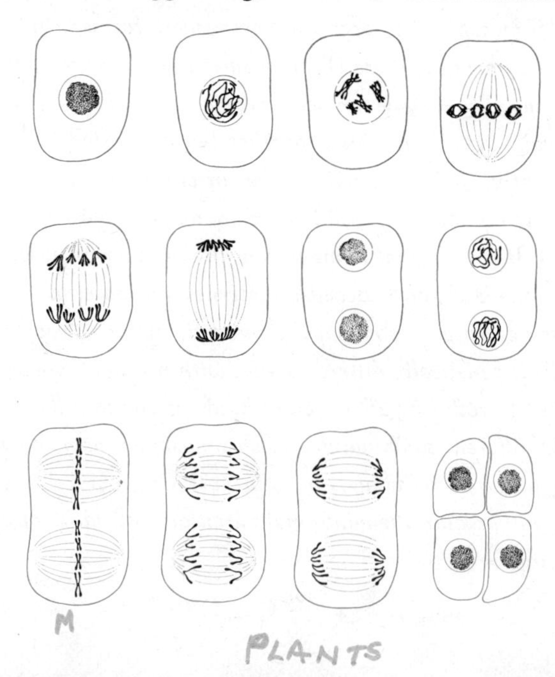

CHAPTER 3–2

From an Abbey Garden: The Beginning of Genetics

In the preceding chapter, we saw that the nucleus carries the information that is passed from parent to offspring and that controls cellular structure and function. Each time a cell divides, this information is passed to daughter cells by complex movements of bodies known as chromosomes, which bear this information. But what do we mean when we speak of "hereditary information"? What is passed from parent to child? What shape and form does it take?

At about the same time that Hertwig was observing the fusion of sperm and egg and Darwin was preparing *The Origin of Species* for publication, Gregor Mendel was conducting the first experiments that provided useful answers to these questions. His work, carried out in a quiet monastery garden and ignored until after his death, marks the beginning of modern genetics.

THE CONCEPT OF THE GENE

By Mendel's time, breeding experiments with domestic plants and animals had shown that both parents contribute to the characteristics of the offspring and that these contributions are carried in the sex cells, the eggs and sperm (or, in the case of plants, the pollen).

Mendel's great contribution was to demonstrate that inherited characters are carried as discrete units, which are parceled out in different ways, or *reassorted,* in each generation. These discrete units eventually came to be known as *genes*.

Mendel was successful where his predecessors had failed largely for two reasons. First, he planned his experiments carefully and imaginatively, choosing for study definite and measurable hereditary differences. Second, probably because of his background in physics and mathematics, he approached the question of inheritance in a thoroughly scientific way; he was one of the first to apply mathematics to the study of biology. Even though his mathematics was simple, the idea that exact rules could be applied to living organisms was startlingly new. He organized his data in such a way that his results could be evaluated simply and objectively and could be checked and duplicated by other scientists, as they eventually were.

A FIT MATERIAL

For his experiments in heredity, Mendel chose the common garden pea. It was a good choice. The plants were commercially available and easy to cultivate. Different varieties had clearly different characteristics, which "bred true," reappearing in crop after crop, and although the plants could be crossbred experimentally, accidental crossbreeding could never occur to confuse the experimental results (see Figure 3–17). As Mendel said in his original paper, "The value and utility of any experiment are determined by the fitness of the material to the purpose for which it is used."

3–16
Mendel and his garden in Brünn, in what is now Czechoslovakia. Mendel's experiments, which showed that hereditary determinants are carried from cell to cell and generation to generation as separate units, showed how such variations could persist, even in a latent form, generation after generation.

Mendel started with 32 different types of pea plants, which he studied for two years before he began his experiments. As a result of these observations, he selected for study seven traits which appeared as conspicuously different characteristics in different types of plants. One variety of plant, for example, always produced yellow peas, or seeds, and another always produced green ones. In one variety, the seeds, when dried, had a wrinkled appearance; in another variety, they were smooth. These seven traits and their alternates are listed in Table 3–1.

Mendel then performed crosses among the different types of pea plants. For instance, he used the pollen from a white

Table 3–1
Mendel's Pea Plant Experiment

			F_2 Generation		
Trait	*Dominant*	*Recessive*	*Dominant*	*Recessive*	*Total*
Seed form	Smooth	Wrinkled	5,474	1,850	7,324
Seed color	Yellow	Green	6,022	2,001	8,023
Flower position	Axial	Terminal	651	207	858
Flower color	Red	White	705	224	929
Pod form	Inflated	Constricted	882	299	1,181
Pod color	Green	Yellow	428	152	580
Stem length	Tall	Dwarf	787	277	1,064

flower to fertilize a red flower and the pollen from a plant which bred true for yellow seeds to fertilize the flower from a green-seeded plant. When he had performed these experimental crosses between the different varieties, he found that in every case in the first generation (now known as F_1 in biological shorthand), one of the alternate traits disappeared completely without a sign. For example, all the progeny of the cross between yellow-seeded plants and green-seeded plants were as yellow-seeded as the yellow-seeded parent. This trait (yellow seeds) and the other such traits of the F_1 generation Mendel called *dominant;* the traits that disappeared in the first generation (such as green seeds) he called *recessive*.

The interesting question was: What had happened to the recessive trait—the wrinkledness of the seed or the greenness of its color—which had been passed on so faithfully for generations by the parent stock? Mendel let the pea plant itself carry out the next stage of the experiment by permitting the F_1 to self-pollinate. The recessive traits reappeared in the second (F_2) generation! In Table 3–1 are the results of Mendel's actual counts. Before you read any further, look at these figures and see if you can detect any relationships among them. Some interpreters of Mendel's work believe that he first formulated his theory and then performed the experiments to test it. Others believe that he examined the data first and then worked out an hypothesis to explain them. In either case, these numbers were the basis for Mendel's first "law."

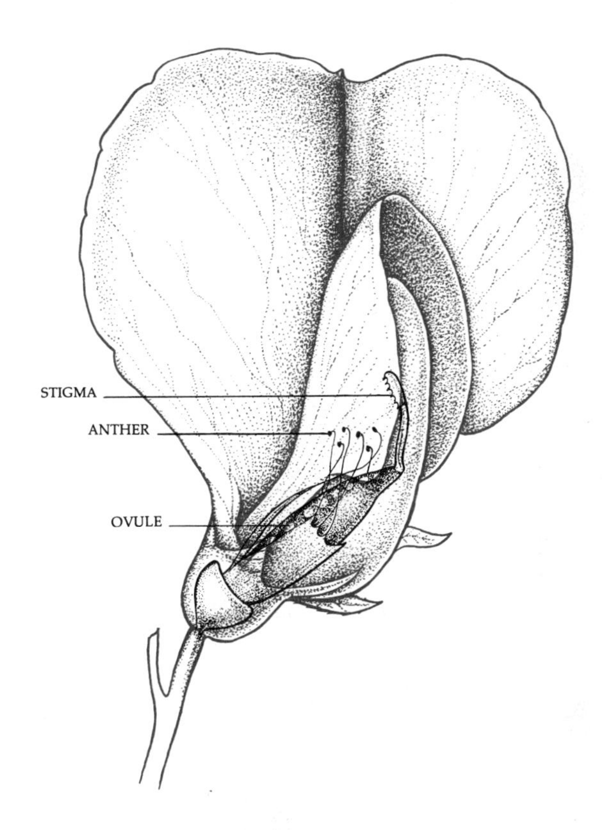

3–17

The sexual organs of the pea blossom are completely enclosed by its petals. The pollen cells develop in the anthers; the egg cells, in the ovules (actually, the immature peas). Since the flower does not open until after fertilization has taken place, the plant normally self-pollinates. However, the anthers can be readily removed before the pollen is mature and the stigma brushed with pollen from another flower, as Mendel did in some of his experiments.

3–18

A pea plant homozygous for red flowers can produce egg cells or pollen grains with only a red-flower (W) gene. The female symbol ♀ (the hand mirror of Venus) indicates that this flower contributed the egg cell.

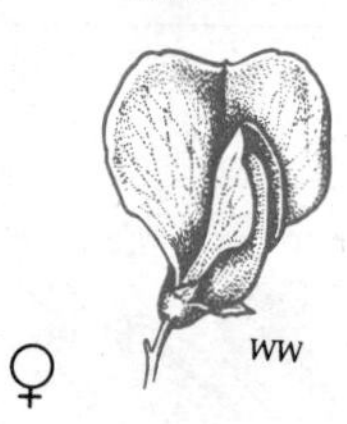

A white pea plant can produce egg cells or pollen grains with only a white-flower (w) gene. The male symbol ♂ (the shield and spear of Mars) indicates that this flower contributed the pollen.

♂ ww

When a w pollen grain fertilizes a W egg cell, the result is a Ww pea plant, which, since the W gene is dominant, will produce red flowers. However, this Ww plant can produce egg cells or pollen cells with either a W or a w allele.

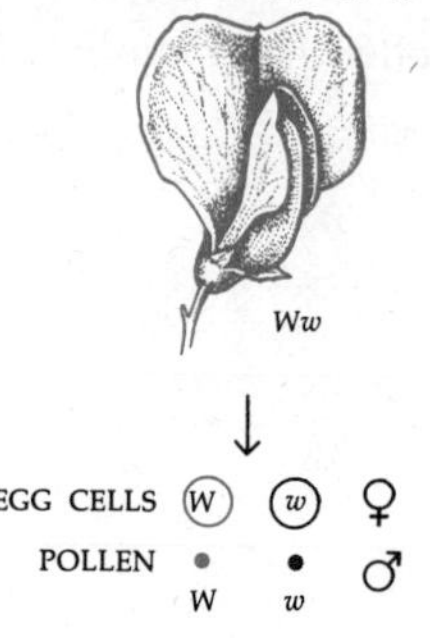

And so, if the plant self-pollinates, four possible crosses can occur:

♀ W × ♂ W ⟶ *red flowers*
♀ W × ♂ w ⟶ *red flowers*
♀ w × ♂ W ⟶ *red flowers*
♀ w × ♂ w ⟶ *white flowers*

These results are summarized in Figure 3–19.

SEGREGATION

Did you notice that the dominant and recessive traits appear in the second, or F_2, generation in ratios of about 3 to 1? How do the recessives disappear so completely and then appear again, and always in such constant proportions? It was in answering this question that Mendel made his greatest contribution. He saw that the appearance and disappearance of traits and their constant proportions could be explained only if hereditary characters are always determined by discrete factors which occur in the offspring as pairs, one inherited from each parent, and which are separated again when the sex cells are formed, producing two kinds of gametes, with one factor of the pair in each.

This hypothesis is known as Mendel's first law, or the *principle of segregation.*

The two factors in a pair might be the same, in which case the self-pollinating plant would breed true. Or the two factors might be different; such different or alternative, forms came to be known as *alleles*. Yellow-seededness and green-seededness, for instance, are determined by alleles, different forms of the gene (factor) for seed color. When the genes of a gene pair are the same, the organism is said to be *homozygous* for that particular trait; when the genes of a gene pair are different, the organism is *heterozygous* for that trait.

3–19
A cross between a pea plant with two dominant genes for red flowers (WW) and one with two recessive genes for white flowers (ww.) The phenotype of the offspring in the F_1 generation is red, but note that the genotype is Ww. The F_2 generation is shown by a Punnett square, named after the English geneticist who first used this sort of checkerboard diagram for the analysis of genetic traits. The F_1 heterozygotes (Ww) produce two kinds of gametes, one W and one w, in equal proportions. When this generation is allowed to self-pollinate, the W and w pollen and eggs combine randomly to form, on the average, 1/4 WW (red), 1/2 Ww (red), and 1/4 ww (white) offspring. It is this underlying 1:2:1 genotypic ratio which accounts for the phenotypic ratio of 3 dominants (red) to 1 recessive (white).

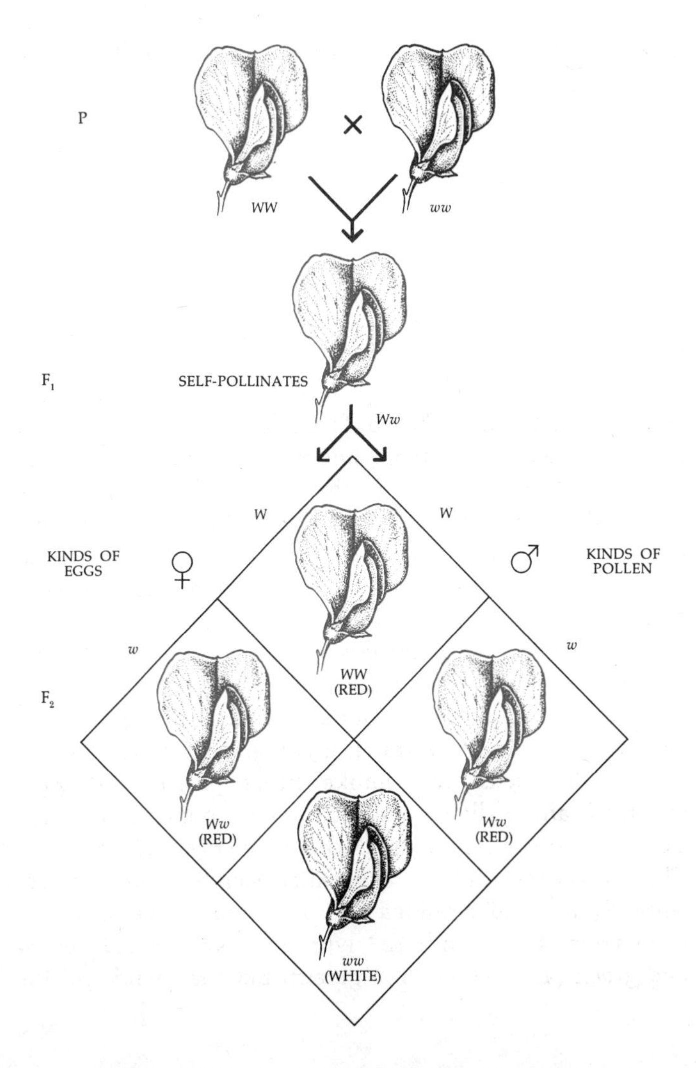

When gametes are formed, genes are passed on to them; but each gamete contains only one of two possible alleles. When two gametes combine in the fertilized egg, the genes occur in matched pairs again. One allele may be dominant over another allele; in this case, the organism will appear as if it had only this gene. This outward appearance is known as its *phenotype*. However, in its genetic makeup, or *genotype*, each allele still exists independently and as a discrete unit even though it is not visible in the phenotype, and the recessive allele will separate from its dominant partner when gametes are again formed. Only if two recessive alleles come together—one from the female gamete and one from the male—will the phenotype then show the recessive trait. When pea plants homozygous for red flowers are crossed with pea plants having white flowers, only pea plants with red flowers are produced, although each plant in the F_1 generation will carry a gene for red and a gene for white. Figure 3–19 shows what happens in the F_2 generation if the F_1 generation self-pollinates. Notice that the result would be the same if the individuals were cross-fertilized with others of the same genotype, which is the way these experiments are performed with plants that are not self-pollinating and with animals.

3–20

A testcross. In order for a pea flower to be white, the plant must be homozygous for the recessive gene (ww). *But a pea flower that is red can come from either a* Ww *plant or a* WW *one. How could you tell its genotype? Mendel solved this problem by breeding such plants with homozygous recessives. As shown here, a phenotypic ratio in the* F_1 *generation of 2 red to 2 white indicates a heterozygous red-flowering parent. What would have been the result if the parent had been homozygous for red flowers?*

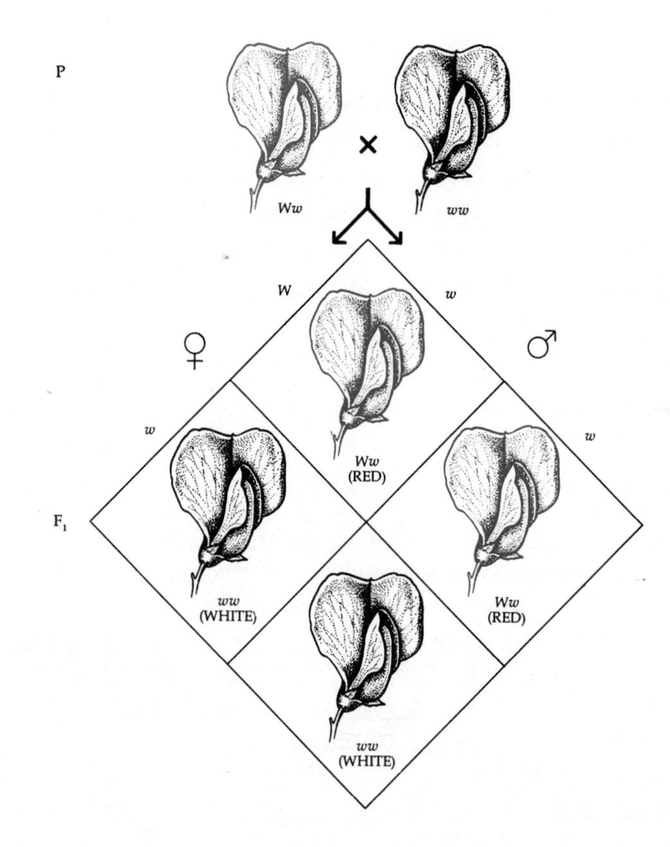

INDEPENDENT ASSORTMENT

In a second series of experiments, Mendel studied crosses between pea plants that differed simultaneously in two characteristics; for example, one parent plant had peas that were round and yellow, and the other had peas that were wrinkled and green. The round and yellow traits, you will recall (see Table 3–1), are both dominant, and the wrinkled and green are recessive. As you would expect, all the seeds of the F_1 generation were round and yellow. When the F_1 seeds were planted and the flowers allowed to self-pollinate, 556 seeds were produced. Of these, 315 showed the two dominant characteristics, round and yellow, but only 32 combined the recessive traits, green and wrinkled. All the rest of the seeds were unlike either parent; 101 were wrinkled and yellow, and 108 were round and green. Totally new combinations of characteristics had appeared. This experiment did not contradict Mendel's previous results. Round and wrinkled still appeared in the same 3:1 proportion (423 round: 133 wrinkled), and so did yellow and green (416 yellow: 140 green). But the round and the

3–21
One of the experiments from which Mendel derived his principle of independent assortment. A plant homozygous for round (RR) *and yellow* (YY) *peas is crossed with a plant having wrinkled* (rr) *and green* (yy) *peas. The* F_1 *generation are all round and yellow, but notice how the traits will, on the average, appear in the* F_2 *generation. Of the 16 offspring, 9 show the two dominant traits* (RY), *3 show one combination of dominant and recessive* (Ry), *3 show the alternate combination* (rY), *and 1 shows the two recessives* (ry). *This 9:3:3:1 distribution is always the expected result from a cross involving two pairs of independent recessive alleles.*

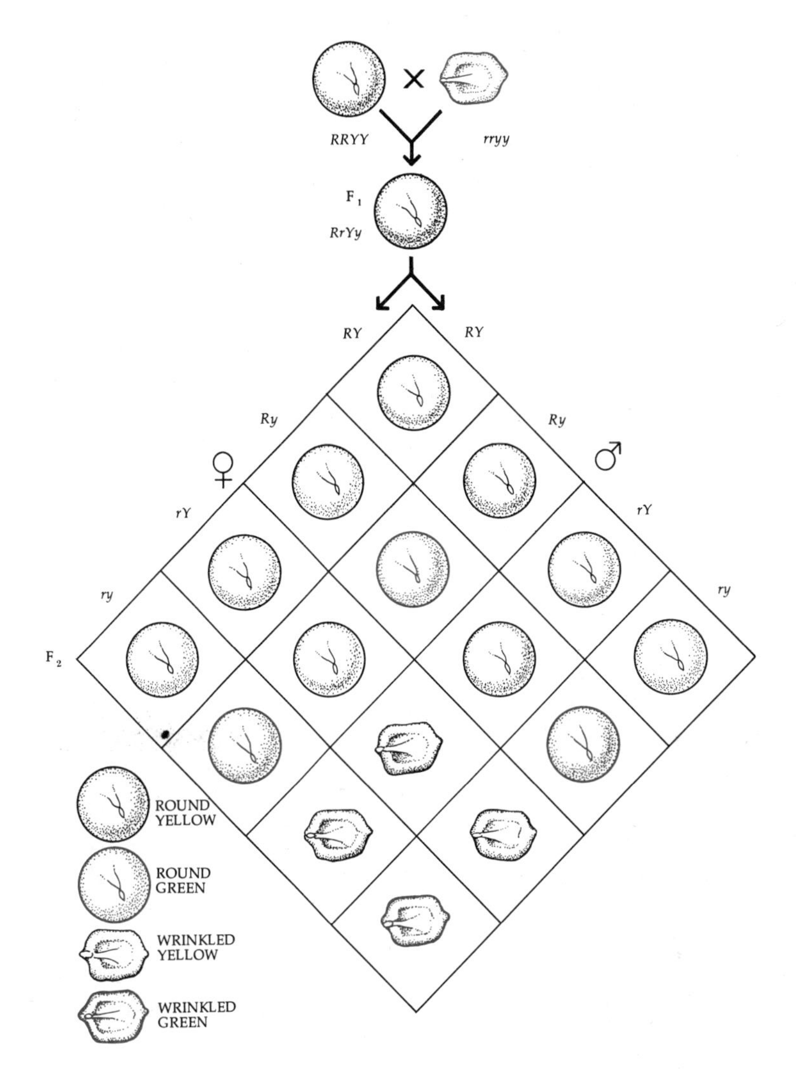

yellow traits and the wrinkled and the green ones, which had originally combined in one plant, behaved as if they were entirely independent of one another. From this, Mendel formulated his second law, the *principle of independent assortment*.

Figure 3–21 diagrams these results and shows why, in a cross involving two pairs of alleles, each pair with one dominant and one recessive allele, the ratio of distribution will be, on the average, 9:3:3:1, with 9 representing the proportion of F_2 progeny that will show the two dominant traits, 1 the proportion that will show the two recessive traits, and 3 and 3 the proportions of the two alternative combinations of dominants and recessives. This is true whether one of the original parents is homozygous for both recessive traits and the other homozygous for both dominant ones, as in the experiment just described (*RRYY* × *rryy*), or each parent is homozygous for one recessive and one dominant trait (*rrYY* × *RRyy*). You can demonstrate this last point by drawing a Punnett square.

INFLUENCE OF MENDEL

Mendel's experiments were first reported in 1865 before a small group of people at a meeting of the Brünn Natural History Society. None of them, apparently, understood what Mendel was talking about. (So do not feel dismayed if you have to read this chapter again.) But his paper was published the following year in the *Proceedings* of the Society, a journal which was circulated to libraries all over Europe. In spite of this, his work was ignored for 35 years, during most of which he devoted himself to the administrative duties of an abbot, and he received no scientific recognition until after his death.

It was not until 1900, 35 years after Mendel first reported his work, that biology was finally prepared to accept his findings. Within a single year, his paper was independently rediscovered by three scientists working in three different European countries. Each of them had done similar experiments and was searching the scientific literature to seek confirmation of his results. And each found, in Mendel's brilliant analysis, that much of their work had been anticipated.

GENES AND CHROMOSOMES

During the 35 years that Mendel's work remained in obscurity, great improvements were made in microscopy and, as a consequence, in the study of the structure of the cell (cytology). It was during this period that chromosomes were first discovered and their movements at meiosis and mitosis, described in the preceding chapter, were observed and recorded.

Two years after the rediscovery of Mendel, in 1902, Walter Sutton, then a graduate student at Columbia University, was observing meiosis in the production of sperm cells in grasshoppers and was struck by the parallels between what he was seeing and the laws of Mendel: Chromosomes come in pairs, so do Mendelian factors (genes). Chromosome pairs (homologues) separate when the gametes are formed; so do genes. And genes and chromosomes come together again in pairs in the offspring. On the basis of these parallels, Sutton proposed that the factors observed by Mendel were carried on the chromosomes.

What about Mendel's law of independent assortment in relation to the movement of the chromosomes at meiosis? This law states that members of pairs of genes assort independently. We can see that this also can be true if—and this is an important point—the genes are on different chromosomes, as shown in Figure 3–22.

Sutton's hypothesis was widely accepted and eventually proved. And genetics was off to a racing start.

3–22 (on facing page)
The chromosome distributions in Mendel's cross of round-yellow *and* wrinkled-green peas, *according to Sutton's hypothesis. Although the pea has 14 chromosomes* ($n = 7$), *only 4 are shown here, the two carrying the genes for round or wrinkled and the two carrying the genes for yellow or green. (This is analogous to what Mendel did when he selected the two traits to study.) As you can see, one parent is homozygous for the recessives, one for the dominants. Therefore, the only gametes they can produce are* RY *and* ry. *(Remember,* R *now stands not just for the trait but for the chromosome carrying the trait, as do the other letters.) The* F_1 *generation, therefore, must be* Rr *and* Yy. *When a mother cell of this generation undergoes meiosis,* R *is separated from* r *and* Y *from* y. *Four different types of* 1n *egg cells are possible, as the diagram reminds us, and also four different types of pollen nuclei. These can combine in* 4 × 4, *or* 16, *different ways. And the ways in which they can combine are illustrated in the Punnett square at the bottom of the figure.*

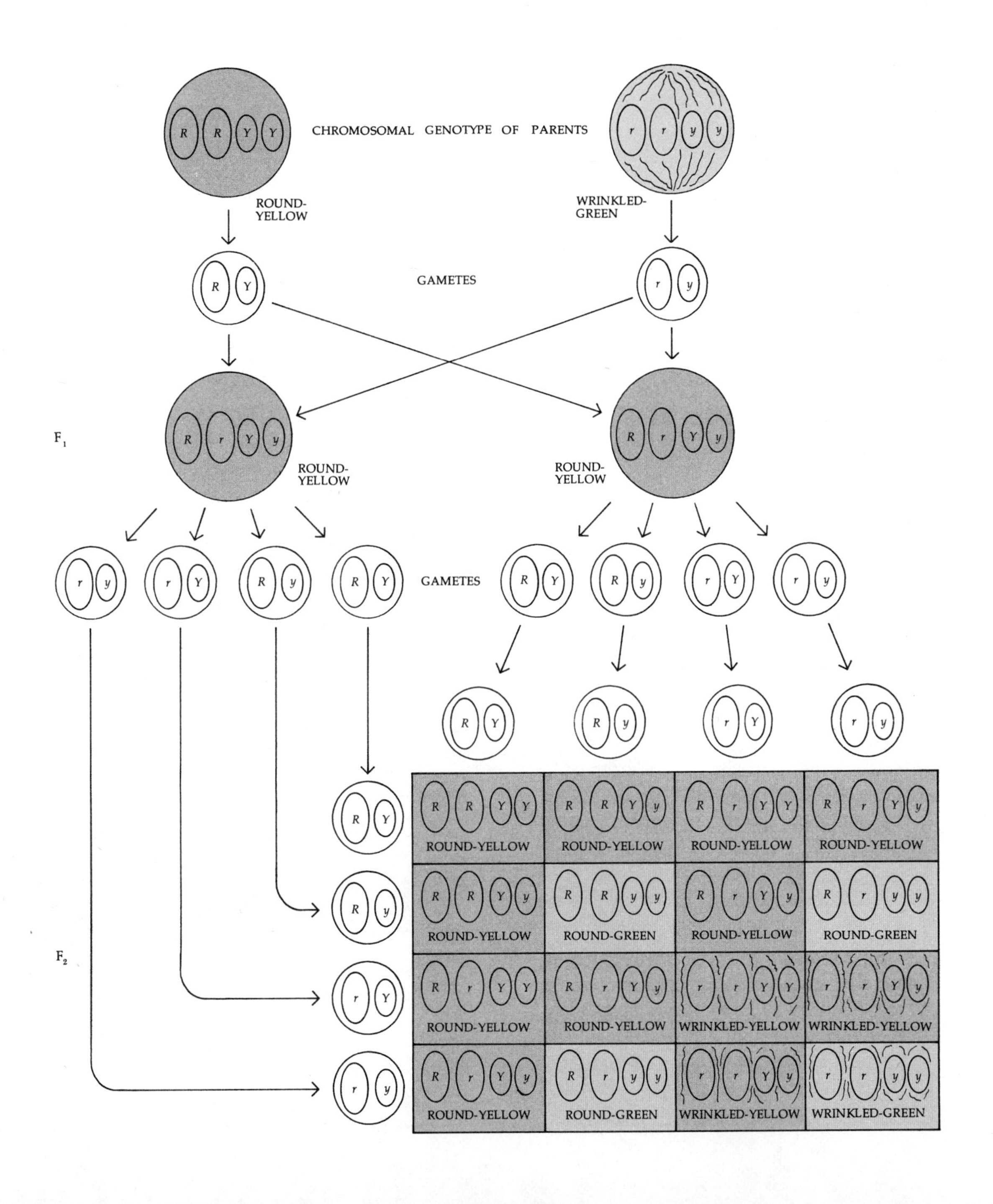
CHROMOSOMAL GENOTYPE OF PARENTS
R R Y Y
r r y y
ROUND-YELLOW
WRINKLED-GREEN
GAMETES
R Y
r y
F1
R r Y y
ROUND-YELLOW
ROUND-YELLOW
GAMETES
F2
ROUND-YELLOW
ROUND-GREEN
WRINKLED-YELLOW
WRINKLED-GREEN

MENDELIAN INHERITANCE IN MAN

Many characteristics in man are governed by simple Mendelian traits. One such characteristic is tongue rolling. The ability to roll one's tongue is a simple dominant. Can you roll your tongue? Can your parents? What is your genotype for tongue rolling? If you cannot roll your tongue, does this mean that neither of your parents can? This may seem to be a very trivial sort of characteristic, one that neither natural selection nor society would favor, but oddly enough, very small differences such as this may have considerable importance over an evolutionary time span.

Of more immediate consequence are a number of congenital diseases which are the result of the coming together of recessive genes. One such disease is sickle-cell anemia, common in the tropics. In persons homozygous for the sickling gene, a large proportion of the red blood cells "sickle"—that is, form a sickle shape—and then clog the small capillaries, causing blood clots and depriving vital organs of their full supply of blood. This produces continuous, painful illness and, usually, death at an early age. About 4 percent of the population in some tropical regions in Africa are born with sickle-cell anemia, and almost half of the members of some African tribes are known to carry the recessive gene.

PKU (phenylketonuria), which produces mental deficiency in infants, is also the result of a "double dose" of a recessive gene.

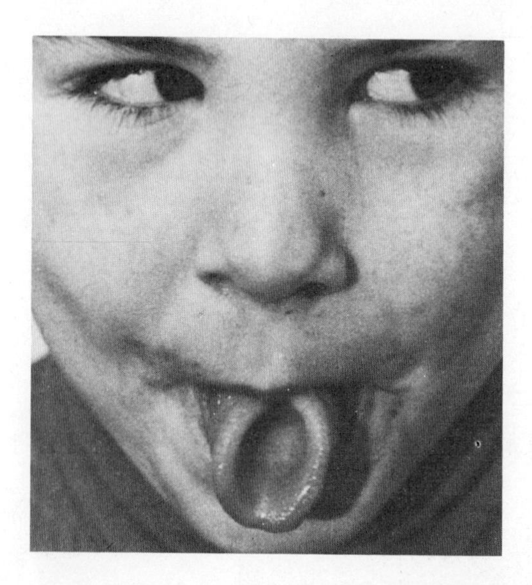

The tongue-rolling ability is transmitted by a dominant gene. Seven out of ten people have this ability. Do you? You can perform a simple experiment in genetics by checking to see if your parents and your brothers and sisters can do it.

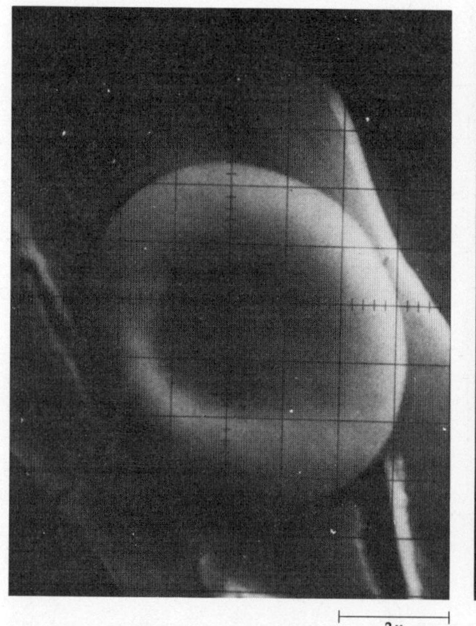

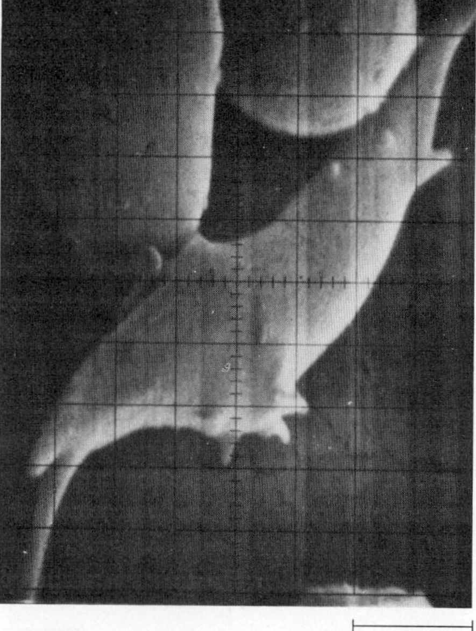

Scanning electron micrograph of normal and sickle red blood cells.

SUMMARY

What is the nature of the hereditary information that is passed from cell to cell and generation to generation? Gregor Mendel provided the first answers to this question when he showed that hereditary traits were passed from parent to offspring as distinct units.

According to Mendel's principle of segregation, these discrete factors (later called genes) appear in pairs in organisms, one of each pair inherited from each parent. When gametes are formed, the factor pairs are separated.

The genetic makeup of an organism is known as its genotype. Its appearance, or outward characteristics, is its phenotype. Both genes in a pair may be alike (a homozygous condition), or they may be different (a heterozygous condition). Two genes forming a heterozygous pair are called alleles. Although both alleles are present in the genotype, only one may be detected in the phenotype. The gene that is expressed in the phenotype is the dominant gene; the one that is concealed in the phenotype is the recessive gene. When two organisms that are heterozygous for different alleles of the same gene are crossed, the ratio of dominant to recessive in the phenotype is 3 to 1.

Mendel's other great principle, the principle of independent assortment, applies to the behavior of two or more genes. This law states that the members of each pair of genes segregate independently (if, as it was shown later, the genes are on different chromosomes). In crosses involving two independent pairs of alleles, the expected phenotypic ratio is 9:3:3:1.

The importance of Mendel's work, first published in 1865, was not recognized until it was rediscovered in 1900. Shortly thereafter, the movement of the chromosomes at meiosis was correlated with Mendel's laws, and an important hypothesis was formed. According to this hypothesis, the Mendelian units, or genes, are located on the chromosomes and are distributed with them.

QUESTIONS

1. Let us assume that brown eyes are dominant over blue eyes (as they usually are). What is the phenotype of a man with two brown-eyed parents, supposing that one parent is heterozygous for the trait and the other parent homozygous? What are his possible genotypes?

2. If such a man marries a woman with blue eyes, what are the possibilities for eye color among their children? Suppose they have one with blue eyes; what would you then know about his genotype? Explain your results by drawing a Punnett square.

3. Suppose this couple had four brown-eyed children. What would you then know about the father's genotype?

4. If two healthy parents have a child with PKU, what are their genotypes with respect to PKU? What are their chances of having another child with the same disease?

CHAPTER 3–3

More about Genes

MUTATIONS AND EVOLUTION

One of Mendel's rediscoverers, a Dutch botanist named Hugo de Vries, was studying genetics in the evening primrose. Heredity in the primrose, he found, was generally orderly and predictable, as in the garden pea, but occasionally a characteristic appeared that was not present in either parent or indeed anywhere in the lineage of that particular plant. De Vries hypothesized that this characteristic came about as the result of a change in a gene and that the trait embodied in the changed gene was then passed along like any other hereditary trait. De Vries spoke of this hereditary change as a *mutation* and of the organism that carried it as a *mutant*.*

During this same period (the early 1900s), biologists were attempting to bring together the new developments in genetics and Darwin's theory of evolution. Mendel's work answered the troublesome question of how genetic traits persist, skipping generations, disappearing and reappearing. It also provided for small variations among populations, with each individual representing a new mixture of traits. But it was hard to see how larger changes could come about—how new species could come into being. This question was made more difficult by the fact that the age of the earth was at that time believed to be much shorter than we know it to be today, and so there was much less time for evolution to have taken place.

* It is one of the ironies of history that only about 2 of some 2,000 changes in the evening primrose observed by de Vries were actually mutations. The rest were due to new combinations of chromosomes rather than to actual changes in any particular gene. However, de Vries' definition of a mutant and his recognition of the importance of the concept of mutation are still valid, although his examples are not.

De Vries and other geneticists seized upon mutations—changes by "leaps and jumps," to quote de Vries—as the principal factor in evolution, and this saltation (leaping) theory of evolution persisted for some time. As one of its supporters, T. H. Morgan, stated, "Evolution is not a war of all against all, but it is largely a creation of new types for unoccupied or poorly occupied places in nature."

Eventually Darwin's original theory was proved right: natural selection acting on small variations is the essential process in evolution. Natural selection acts on the whole organism—the phenotype—and not on individual traits. Mutation, however, is an important source of new variations. These new traits are worked into the genotype, being brought into new combinations and relationships by the process of meiosis.

In 1927, H. J. Muller found that treatment of gametes by x-rays greatly increases the rate of appearance of mutations. It was soon discovered that other radiations, such as ultraviolet light, and some chemicals can also act as *mutagens*, agents that produce mutations. Studies of mutations have shown that whether they are deliberately produced by radiations and chemicals or are "spontaneous,"*

* "Spontaneous," in this sense, means that we do not know why they happen.

3–23
Hugo de Vries standing next to Amorphophallus titanum, *a close relative of the calla lily. This species has the largest flower of any of the flowering plants.*

3–24
An example of the sometimes dramatic effects of mutation. The ewe in the middle is an Ancon, an unusually short-legged strain of sheep which originated in New England in the nineteenth century. The strain is the result of inbreeding by sheep raisers, who saw an advantage in legs too short to permit the sheep to jump the stone walls enclosing the New England sheep pastures.

they are usually detrimental to the organism, and in fact, many of them are lethal. This is not surprising. If one were to change a word at random in a Shakespearean sonnet or a wire at random in a television set, an improvement would be unlikely and the results might well be disastrous. It is, in part, because of the effects on the sex cells—effects whose consequences will be concealed until the next generation—that many authorities are concerned about the increase in radiation exposure caused by atomic testing or by unwarranted or careless use of x-rays. (Radiation exposure can also produce harmful changes in somatic cells which may result in cancer—as shown, for example, by the increased incidence of leukemia among the "survivors" at Hiroshima.)

INCOMPLETE DOMINANCE

During the decade that followed Mendel's discovery, many studies were carried out which, while confirming his work in principle, showed that the action of genes is more complex than it had at first appeared to be. An important example is incomplete dominance.

Dominant and recessive traits are not always so clear-cut as in the pea plant. Some traits do appear to blend. For instance, the cross between a red snapdragon and a white snapdragon produces a first generation that is pink (Figure 3–25). But when this generation is allowed to self-pollinate, the traits begin to sort themselves out once again. As we shall see, what occurs in the snapdragon and in similar crosses is actually a result of the combined effect of gene products.

3–25

A cross between a red snapdragon and a white snapdragon. This looks very much like the cross between a red- and a white-flowering pea plant shown in Figure 3–19, but there is a significant difference. Although the Mendelian genotypic ratio appears in the second generation, the hybrid (Ww), *instead of showing the dominant red, is pink.*

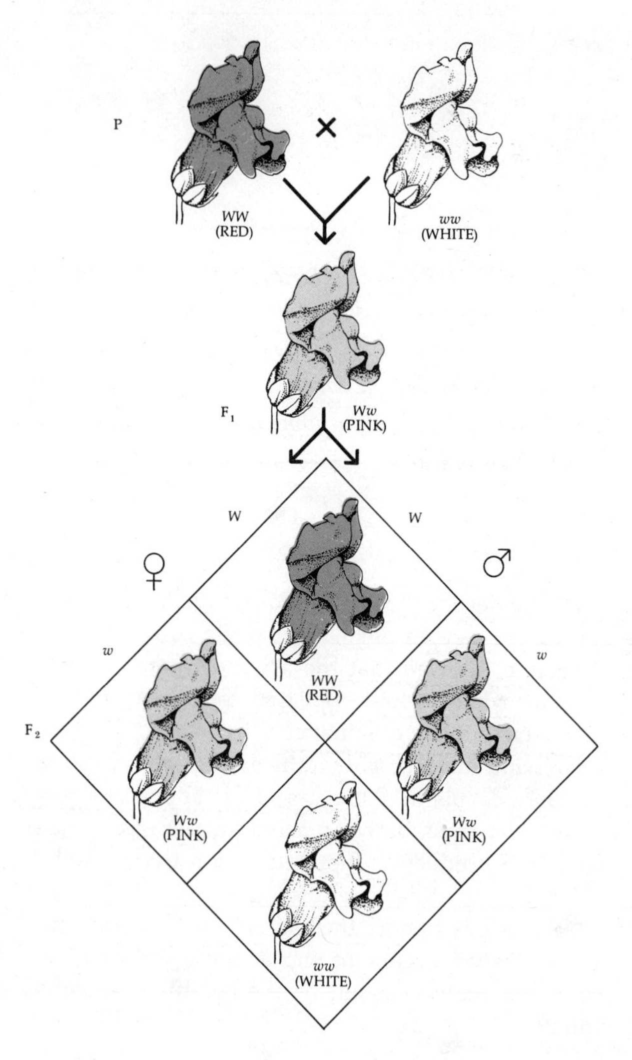

BROADENING THE CONCEPT OF THE GENE

Mendel's experiments seemed to suggest that each gene affects a single characteristic in a one-to-one relationship. It was soon discovered, however, that a single gene sometimes affects many traits in an organism and that, conversely, a single trait is often affected by many genes. The first property is known as *pleiotropy*, the second as *polygenic inheritance*.

Pleiotropy

One of the first investigators to demonstrate pleiotropy was Theodosius Dobzhansky. Dobzhansky arbitrarily selected 12 mutant female fruit flies, each with a single different mutation that changed a specific characteristic, such as eye or body color or wing shape. In each of these flies, he examined the shape of the spermatheca, a special sperm-storing organ associated with the female reproductive tract. Ten out of the twelve mutants showed a variation from normal in the size and shape of this particular organ, although each mutant fly was thought to differ from the norm in only a single obvious characteristic. Thus it was apparent that the genes affected more than one characteristic of the flies.

Sometime later, another investigator, Hans Gruneberg, approaching the problem from the other direction, studied a whole complex of congenital deformities in rats, including thickened ribs, a narrowing of the tracheal passage, a loss of elasticity of the lungs, hypertrophy of the heart, blocked nostrils, a blunt snout, and needless to say, a greatly increased mortality. All these changes, he was able to demonstrate, were caused by a single mutation, that is, a mutation involving only one gene. This particular gene produces a protein involved in the formation of cartilage, and since cartilage is one of the most common structural substances of the body, the widespread effects of such a gene are not difficult to understand. In fact, it is very likely that Mendel's allele for wrinkled, for example, affected other characteristics of the pea.

3–26
"Frizzle" in fowls is the result of a genetic defect leading to defective feather development. Shown here are some of the consequences of this "simple" genetic variation.

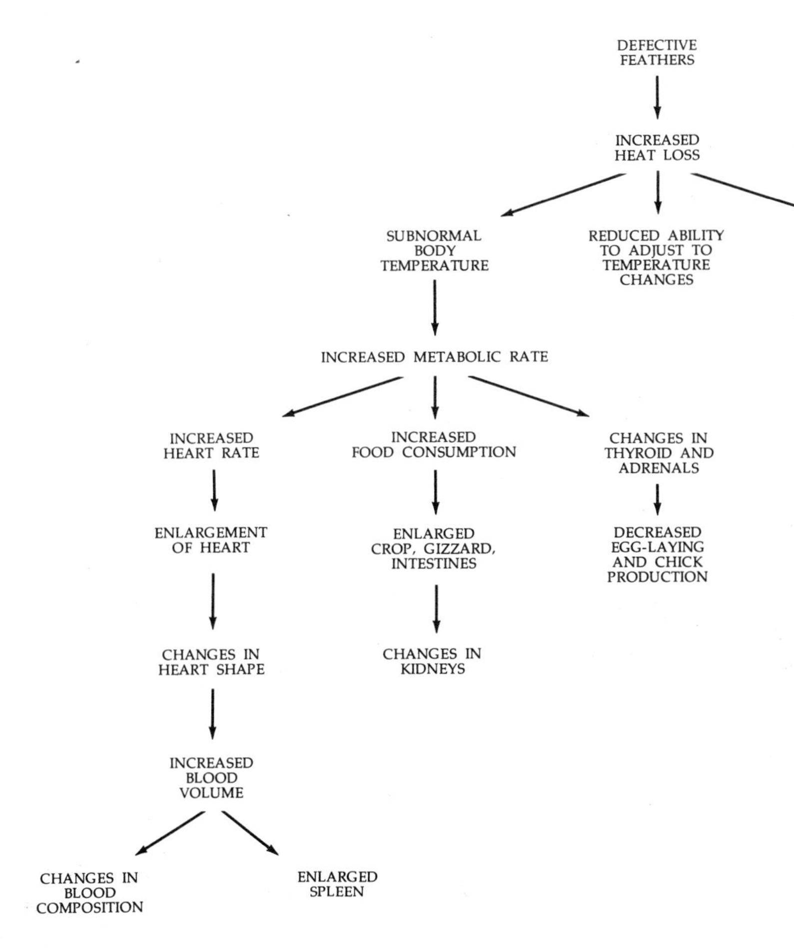

Polygenic Inheritance

A trait affected by a number of genes does not show a distinctive clear-cut difference between groups—such as the differences tabulated by Mendel—but rather shows a gradation of small differences, which is known as continuous variation. If you make a chart of genetic differences among individuals that are affected by a number of genes, you get a curve such as that shown in Figure 3–27.

Fifty years ago, the average height in the United States was less but the shape of the curve was the same; in other words, the great majority fell within the middle range and the extremes in height were represented by only a few individuals. Some of these height variations are produced by environmental factors, such as diet, but even if all the men in a population were maintained from birth on the same type of diet, there would still be a continuous variation in height in the population. This is due to genetic differences in hormone production, bone formation, and numerous other factors. A number of other characteristics in man, notably skin color, are controlled by polygenic inheritance.

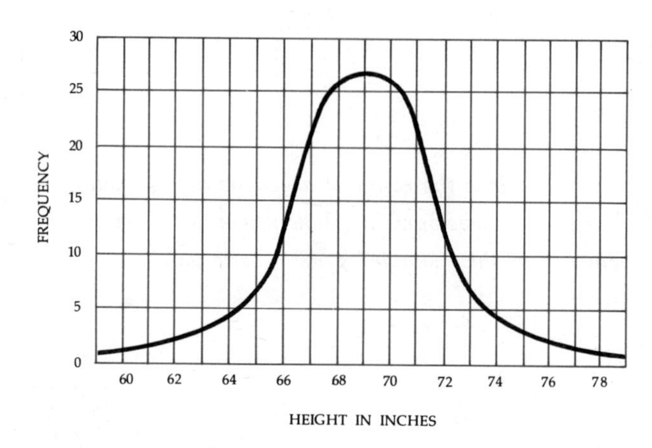

3–27
Height distribution among U.S. males. Height and weight are examples of polygenic inheritance. These genetic traits are characterized by small gradations of difference. A graph of the distribution of such traits always takes the form of a bell-shaped curve, as shown, with the mean, or average, falling in the center of the curve. Can you see why?

GENOTYPE AND PHENOTYPE

There is a crucial difference between genotype and phenotype. From the moment of its conception, every organism is acted upon by the environment, and the expression of any gene is always the result of the interaction of gene and environment. To take a simple, familiar example, a seedling may have the genetic capacity to be green, to flower, and to fruit, but it will never turn green if it is kept in the dark, and it may not flower and fruit unless certain precise environmental requirements are met. Himalayan rabbits are all white if they are raised at high temperatures (above 95°F). However, rabbits of the same genotype, when raised at room temperature, have black ears, forepaws, noses, and tails.

In humans, also, similar genotypes are expressed quite differently in different environments. A simple example is found in height, which, as we have noted, is influenced by a group of genetic factors. This influence is indicated by the fact that tall parents generally have tall children and short parents tend to have short children. On the other hand, adult height is clearly also influenced by environmental factors in childhood, such as diet, sunshine, and incidence of disease. The population of the United States has grown taller with each generation for the past four generations, presumably as a result of general improvements in the standard of living and in the average diet. Similarly, Japanese teen-agers are taller than their parents, and the children raised in the kibbutzim of Israel come to tower over parents who grew up in the ghettos of Central and Eastern Europe.

As René Dubos reminds us, not only are children growing taller, but they are growing faster; a boy now reaches full height at about 19 years, but 50 years ago, maximum stature was not usually attained until age 29. Of more social consequence is the fact that puberty is also being reached earlier. In Norway, for example, the mean age of the onset of menstruation has fallen from 17 in 1850 to 13 in 1960. Historical evidence indicates, however, that also in Imperial Rome and Western Europe in Shakespeare's time, teen-agers reached puberty at an early age (Juliet, remember, was not yet 14). Apparently, the slowing of the growth and maturation rate of the general population was a consequence of the industrial revolution and increasing urbanization, which reduced the standards of nutrition and health. Thus we have a situation in which the genetic potential has remained apparently unaltered over the centuries but the phenotypic expression has undergone fluctuations. A side effect of this change in phenotypic expression is that teen-agers are now reaching physical maturity early in a society in which childhood and dependency have become greatly prolonged.*

* René Dubos, *So Human an Animal*, Charles Scribner's Sons, New York, 1968.

Table 3–2 illustrates a simple example of polygenic inheritance, color in wheat kernels, which is controlled by two pairs of genes.

Table 3–2

The Genetic Control of Color in Wheat Kernels

Parents:	$R_1R_1R_2R_2 \times r_1r_1r_2r_2$ (*dark red*) (*white*)		
F_1:	$R_1r_1R_2r_2$ (*medium red*)		
F_2	*Genotype*	*Phenotype*	
1	$R_1R_1R_2R_2$	Dark red	15 red to 1 white
2	$R_1R_1R_2r_2$	Medium-dark red	
2	$R_1r_1R_2R_2$	Medium-dark red	
4	$R_1r_1R_2r_2$	Medium red	
1	$R_1R_1r_2r_2$	Medium red	
2	$R_1r_1r_2r_2$	Light red	
1	$r_1r_1R_2R_2$	Medium red	
2	$r_1r_1R_2r_2$	Light red	
1	$r_1r_1r_2r_2$	White	

Summary of Phenotypes

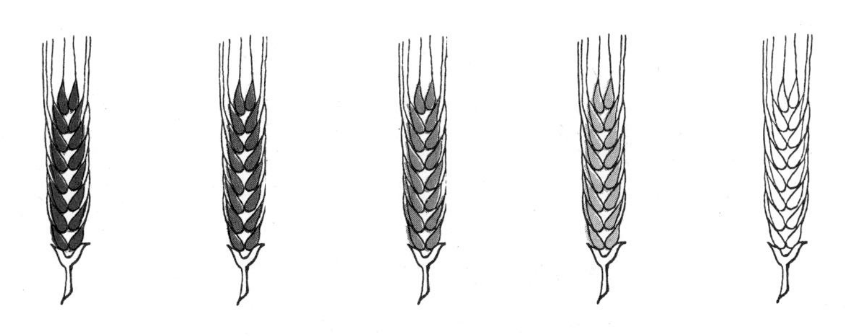

Dark red	*Medium dark red*	*Medium red*	*Light red*	*White*
1	4	6	4	1

SUMMARY

Mutations are abrupt changes in genotype. Together with the genetic recombinations that occur at fertilization and meiosis, they are the source of variations necessary for biological evolution.

Incomplete dominance is a situation in which the effects of the recessive allele are apparent in the heterozygote.

Genes commonly affect more than one characteristic; this property of the gene is known as pleiotropy.

Many characteristics are under the control of a number of separate genes. This phenomenon is polygenic inheritance. Traits under the control of a number of genes typically show continuous variation, as represented by a bell-shaped curve.

Genes determine only potential capacities. Interactions of the genotype and the environment determine the phenotype.

QUESTIONS

1. The so-called "blue" (really gray) Andalusian variety of chicken is produced by a cross between the black and white varieties. What color chickens (and in what proportions) would you expect if you crossed two blues? If you crossed a blue and a black?

2. Skin color in one strain of mice is determined by genes at five different loci. The colors range from almost white to dark brown. Would it be possible for any pair of mice to produce offspring darker or lighter than either parent?

3. Make a diagram, similar to that shown in Figure 3–26, of what you think might be the consequences of the phenotypic expression of the sickling gene in man. Choose as your example a child born in the United States who is homozygous for this recessive.

CHAPTER 3–4

White-eyed and Other Fruit Flies

Early in the 1900s, Thomas Hunt Morgan began a study of genetics at Columbia University, founding what was to be the most important laboratory in the field for several decades. By a remarkable combination of foresight and good fortune, he selected as his experimental material the fruit fly *Drosophila.*

Drosophila means "lover of dew," although actually this useful little fly is not attracted by dew but feeds on the fermenting yeast which it finds in rotting fruit. The fruit fly was a likely choice for a geneticist since it is easy to breed and maintain. These tiny flies, each only an eighth of an inch long, produce a new generation every two weeks. Each female lays hundreds of eggs at a time, and an entire population can be kept in a half-pint bottle, as they were in Morgan's laboratory.

The little fruit fly proved to be a "fit material" for a wide variety of genetic investigations. In the decades that followed, *Drosophila* was to become famous as the biologist's principal tool in studying animal genetics.*

3–28

The fruit fly and its chromosomes. Drosophila *has only four pairs of chromosomes, a fact that simplified Morgan's experiments.*

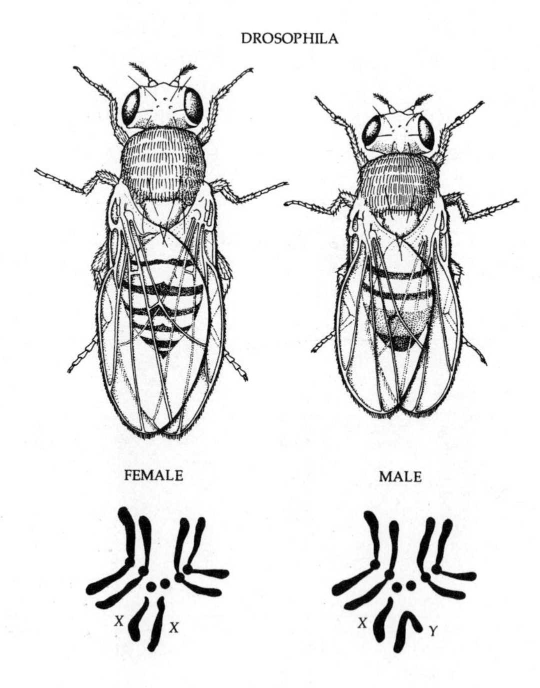

* Geneticists have often used for their experiments such "insignificant" little plants and animals—Mendel's pea plants, for instance, or Hertwig's sea urchins—organisms that seem to occupy very unimportant and out-of-the-way places in the natural order. Underlying this approach is the geneticist's assumption that genetic principles are universal, applying equally to all living things.

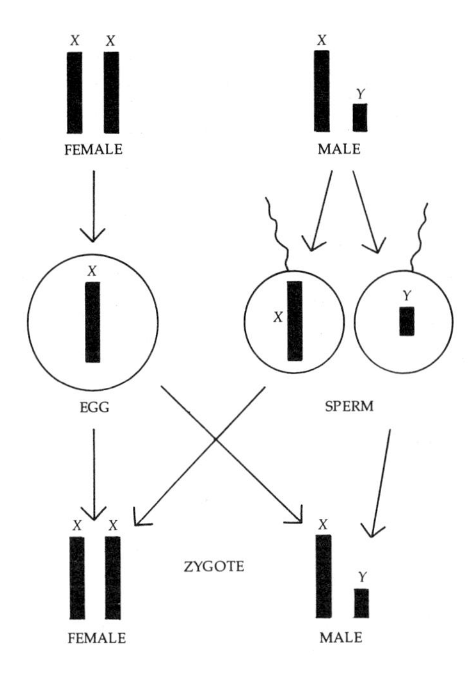

3–29
How the sperm cell determines the sex of human offspring. During meiosis, the egg receives an X *chromosome and four sperm cells are formed, two receiving* X *chromosomes and two receiving* Y *chromosomes. (For simplicity, only two cells are shown in the drawing.) Whether the zygote becomes* XX *(female) or* XY *(male) depends on which of the sex chromosomes is carried in the sperm that fertilizes the egg.*

THE SEX CHROMOSOMES

Fruit flies have only four pairs of chromosomes, a feature that turned out to be particularly useful. Three pairs are structurally the same in both sexes, but the fourth pairs are different; it is this fourth pair that determines the sex of the fly. The two sex chromosomes, as they are called, of the female are structurally the same. These are, by convention, termed *X* chromosomes, and so the female is designated *XX*. The sex chromosomes of the male consist of one *X* chromosome, which is the same as the female *X* chromosome, and one *Y* chromosome, which is smaller in size than the *X*. Thus males in the fruit fly (and also in most species, including man) are designated *XY*. However, there are some species, such as birds, in which the females are *XY* and the males *XX*.

When the sex cells are formed by meiosis, half of the gametes in the male fruit fly (or human) carry an *X* chromosome and half carry a Y chromosome (Figure 3–29). All the gametes formed by a female carry the *X*. The sex of the offspring is determined by whether the female gamete carrying the *X* chromosome is fertilized by a male gamete carrying an *X* chromosome or a male gamete carrying a Y chromosome. Since equal numbers are produced, there is theoretically an even chance of male and female.*

GENES AND THE SEX CHROMOSOMES

One of the first mutants to turn up in Morgan's laboratory was a male fly with white eyes. (Red is the usual color of a fruit fly's eyes.) The white-eyed male was mated to a red-eyed female, and all the offspring had red eyes. Then Mor-

* In actual fact, the ratio of human male to human female births is about 106 to 100. The reason for this is not known, but it has been suggested that because the Y chromosome is slightly lighter than the *X* chromosome, the male-determining sperm may have an advantage in getting to the egg.

gan crossbred the offspring, just as Mendel had in his pea experiments, and this is what he got:

Red-eyed females	2,459
White-eyed females	0
Red-eyed males	1,011
White-eyed males	782

How could he explain this result? Why were there no white-eyed females? Perhaps the gene for white eyes was carried only on the Y chromosome and not on the X. To test this hypothesis, he crossed the original white-eyed male with one of the F_1 females. And the result was:

Red-eyed females	129
White-eyed females	88
Red-eyed males	132
White-eyed males	86

Morgan and his coworkers examined these figures and came to a second conclusion (the right one this time): the gene for eye color is carried only on the *X* chromosome. (In fact, as it was later shown, the *Y* chromosome carries almost no genetic information at all.) The white allele is recessive. Thus a heterozygous female would never have white eyes—which is why there were no white-eyed females in the F_1 generation. However, a male that received an *X* chromosome carrying the allele for white eyes would always be white-eyed since no other allele would be present.

Further experimental crosses proved Morgan's hypothesis to be right. They also showed that white-eyed fruit flies are more likely to die before they hatch than red-eyed fruit flies, which explains the lower-than-expected numbers in the F_1 generation and the testcross.

These experiments introduced the concept of sex-linked traits. As we shall see, such traits are important in human genetics. They also established what Sutton had hypothesized some years before: genes *are* on chromosomes.

3–30

Offspring of a cross between a white-eyed (w) *female fruit fly and a red-eyed* (W) *male fruit fly, illustrating what happens when a recessive gene is carried on an X chromosome. The* F_1 *females, having two X chromosomes, are heterozygous for the recessive and so will be red-eyed. But the* F_1 *males, with their X chromosome carrying the recessive* (w) *trait, will be white-eyed because the Y chromosome carries no gene for eye color.*

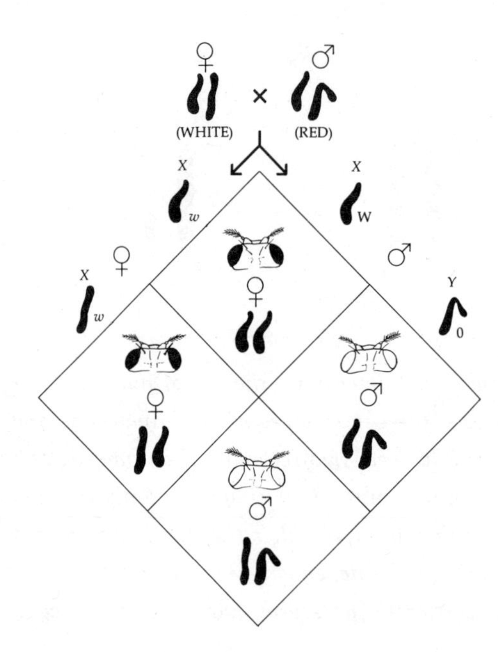

LINKAGE

Mendel, you will remember, showed that certain pairs of alleles, such as round and wrinkled, assort independently of other pairs, such as yellow and green. However, as we noted previously, genes assort independently only if they are on different chromosomes. If two genes are on the same chromosome, they will, of course, be transmitted to the same sex cell at meiosis. The genes on any one chromosome are said to be in *linkage groups*.

As increasing numbers of mutants were found in Columbia University's *Drosophila* population, the mutations began to fall into four linkage groups, in accord with the four pairs of chromosomes visible in the cells. Indeed, in all organisms which have been studied in sufficient genetic detail, the number of linkage groups and the number of pairs of chromosomes have been the same.

CROSSING OVER

Large-scale studies of linkage groups soon revealed some unexpected difficulties. For instance, most fruit flies have gray bodies and long wings. These individuals, which show the common, characteristic features of the population, are known as "wild-type" flies. When wild-type flies were bred with mutant fruit flies having black bodies and short wings (both recessive traits), all the progeny had gray bodies and long wings, as would be expected. Then the F_1 generation was inbred. Two outcomes seemed possible:

1. The two recessives would be assorted independently and would appear in the 9:3:3:1 ratio, indicating that they were on different chromosomes.

2. The two recessives would be linked. In this case, 75 percent of the flies would be gray with long wings and 25 percent, homozygous for both recessives, would be black with short wings.

In the case of these particular traits, the results most closely resembled outcome 2, but they did not conform exactly. In a few of the offspring, the traits seemed to assort independently, not together; that is, some few flies appeared that were gray with short wings, and some that were black with long wings. How could this be? Somehow the genes had moved out of their linkage groups.

To find out what was happening, Morgan tried a testcross, breeding one of the F_1 generation with a homozygous recessive. If black and gray, long and short assorted independently—that is, if they were on different chromosomes—25 percent of the offspring of this cross should be black with long wings, 25 percent gray with long wings, 25 percent black with short wings, and 25 percent gray with short wings. On the other hand, if the two traits (color and wing size) were on the same chromosome and so moved together, half of the offspring of the testcross should be gray with long wings and half should be black with short wings. But actually, as it turned out, over and over, in counts of hundreds of fruit flies resulting from such crosses, 41.5 percent were gray with long wings, 41.5 percent were black with short wings, 8.5 percent were gray with short wings, and another 8.5 percent were black with long wings.

Morgan was convinced by this time that genes are located on chromosomes. It now seemed clear that the two traits were located on the same chromosome since they did not show up in the 25:25:25:25 percentage ratios of separately assorted genes. The only way in which the observed figures could be explained was if one assumed (1) that the two genes were on one chromosome and their two alleles on the homologous chromosome, and (2) that sometimes genes could be exchanged between homologous chromosomes.

How can genes change chromosomes? Recall that at meiosis, the chromosomes pair, or undergo synapsis, before they divide. Consequently four homologous chromatids are lined up alongside each other before the first meiotic division. In 1909, the Belgian cytologist F. A. Janssens had observed that chromatids cross one another and had suggested that during these crosses, or *chiasmata*, as he called them, an exchange of chromosomal material might take place. This exchange is difficult to see and interpret in the living cell, so that the suggestion was not so obvious as it now may seem. Furthermore, there is still no explanation of how homologous chromosomes manage to break in exactly corresponding sites and so exchange equal amounts of chromosomal material following each break. Nevertheless, vast accumulations of data have now confirmed that such exchanges, or crossovers, as they are called, do take place, and at virtually every meiosis.

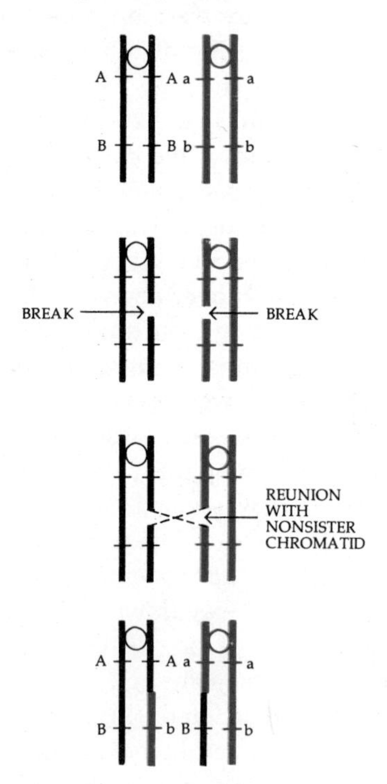

3–31

Crossing over takes place when breaks occur in chromatids at the time of synapsis and the broken end of each chromatid joins with the chromatid of an homologous chromosome. In this way, alleles are exchanged between chromosomes.

"MAPPING" THE CHROMOSOME

With the discovery of crossovers, it began to seem clear not only that the genes are carried on the chromosomes, as Sutton had hypothesized, but that they must be arranged in a definite, fixed linear array along the length of the chromosome. Furthermore, all the alleles must be at corresponding sites on particular chromosomes. If this were not true, exchange of sections of chromosomes could not possibly result in an exact exchange of alleles. In other words, genes are arrayed along the chromosome like beads along a string;

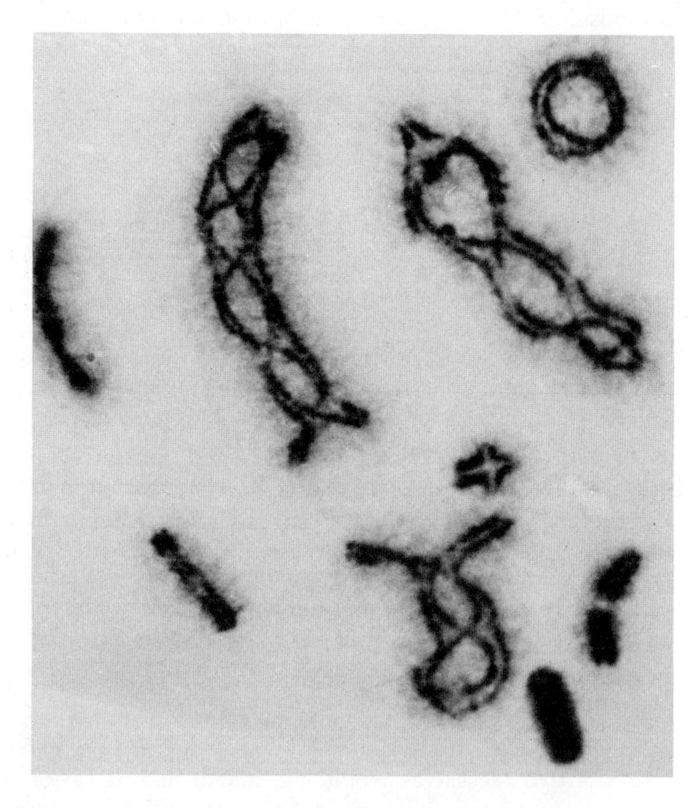

3–32

Chiasmata in the chromosomes of a grasshopper.

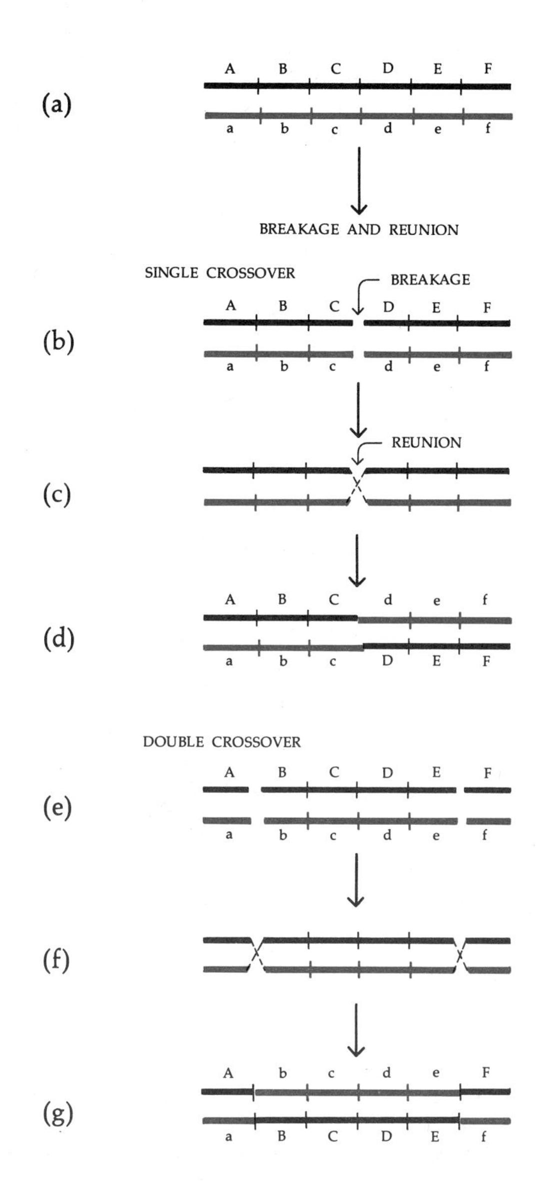

3–33
In (a), *the lines indicate adjacent chromatids at the time of synapsis. The letters* (A, a; B, b; *etc.*) *indicate different alleles of the same gene. If these chromatids break* (b), *and rejoin* (c), *an exchange of alleles takes place* (d). *Breaks* (e) *and rejoining* (f) *may occur at more than one point on adjacent chromatids. The result of a double crossover is shown in* (g).

this simile was widely used in genetics for several decades.

As other traits were studied, it became clear that the percentage of separations, or crossovers, between any two genes, such as gray body and long wing, was different from the percentage of crossovers between two other genes, such as gray body and long leg. In addition, as Morgan's experiments had shown, these percentages were very fixed and predictable. It occurred to A. H. Sturtevant, one of the many brilliant young geneticists attracted to Morgan's laboratory during these golden days of *Drosophila* genetics, that the percentage of crossovers probably had something to do with the distances between the genes, or in other words, with their spacing along the chromosome. (You can see in Figure 3–33, for example, that in a crossover, the chances of a strand breaking and recombining with its homologous strand at the single junction between *B* and *C* is less likely than this happening at any one of the four junctions between *B* and *F*. So it is less likely that *BC* and *bc* will recombine to give *Bc* and *bC* than that *B* and *f* [and *b* and *F*] will end up on the same strand.) This concept opened the way to the "mapping" of chromosomes.

Sturtevant postulated (1) that genes are arranged in a linear series on chromosomes, (2) that genes which are close together will be separated by crossing over less frequently than genes which are further apart, and (3) that it should therefore be possible, by determining the frequencies of crossovers, to plot the sequence of the genes along the chromosome and the relative distances between them. (The distances cannot be absolute because breaking and rejoining of chromosome segments may be more likely to occur at some sites on the chromosome than on others.) In 1913, Sturtevant began a series of crossover studies in fruit flies. As a standard unit of measure, he arbitrarily took the distance that would give (on the average) one crossover per 100 fertilized eggs. Thus genes with 10 percent crossover would be 10 units apart; those with 8 percent crossover would be 8 units apart. By this method, he and other geneticists constructed genetic maps locating a wide variety of genes and their mutants in *Drosophila*.

3–34
A portion of the genetic map of Drosophila melanogaster, *showing some of the genes on chromosome 2 and the distances between them, as calculated on the basis of frequency of crossovers. As you can see, more than one gene may affect a single characteristic, such as eye color.*

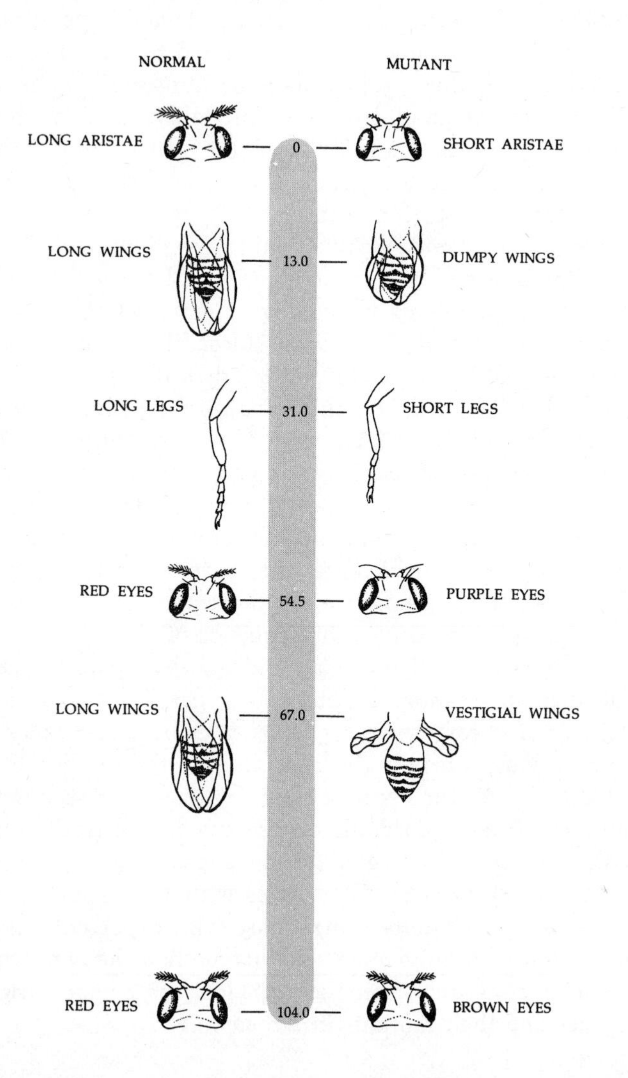

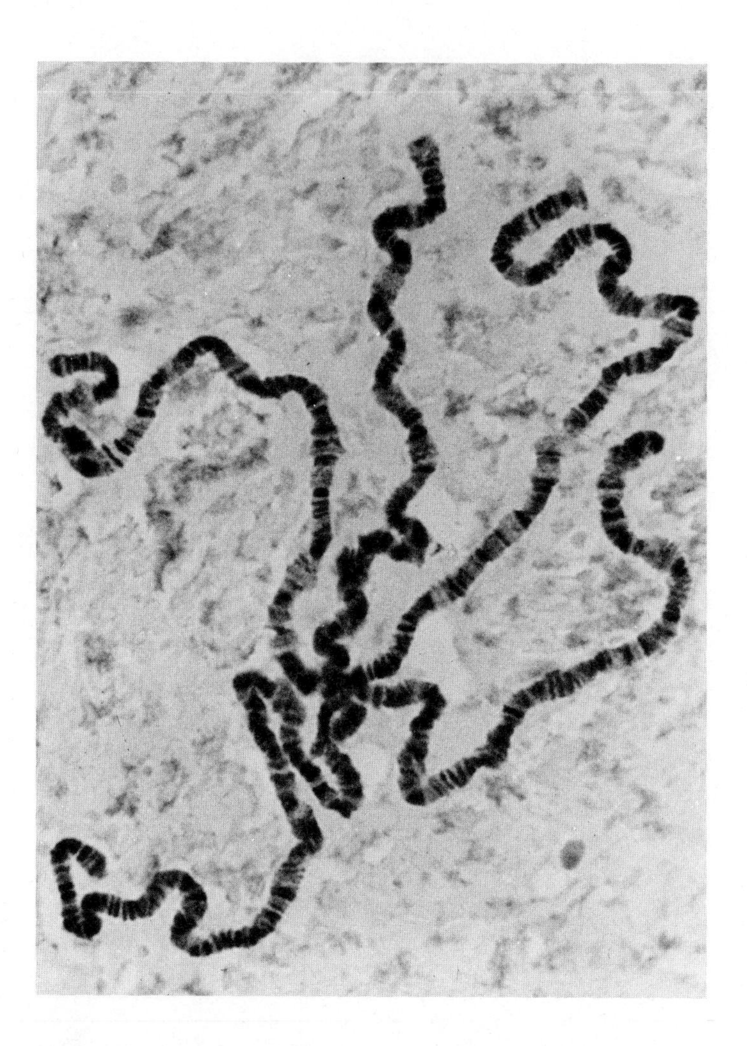

3–35
Chromosomes from the salivary gland of a Drosophila *larva. These chromosomes are* 100 *times larger than the chromosomes in ordinary body cells, and their details are therefore much easier to see. (This is a photomicrograph taken with a light microscope, not an electron microscope.) Experiments showed that given traits were controlled from specific bands on these chromosomes, and the light and dark bandings were found to correspond almost exactly with previously plotted genetic maps.*

Confirmation of Sturtevant's brilliant hypothesis came 20 years later from an unexpected source. In *Drosophila,* as in many other insects, certain cells do not divide after the larval stage of the insect. In such cells, however, the chromosomes continue to replicate, over and over again, but since they do not separate after replication, they simply become larger and larger. In 1933, such chromosomes were observed in the salivary glands of the larvae of *Drosophila.* As you can see in Figure 3–35, these giant chromosomes are marked by very distinctive dark and light bands. Changes in these banding patterns were found to correlate with observed genetic changes in the flies. In addition, crossovers and breaks could actually be *seen.**

SUMMARY

The fruit fly *Drosophila,* because it is easy to breed and maintain, has been used in a wide variety of genetic studies. *Drosophila* has four pairs of chromosomes; three pairs are structurally the same in both sexes, but the fourth pairs, the sex chromosomes, are different. In fruit flies, as in most species, the two sex chromosomes are *XX* in the females and *XY* in the males.

At the time of meiosis, the sex chromosomes are segregated. Each egg cell will receive an *X* chromosome, but half the sperm cells will receive *X* chromosomes and half will receive *Y* chromosomes. Thus it is the sperm cell that determines the sex of the embryo.

In the early 1900s, experiments with mutations in the fruit fly showed that certain characteristics are sex-linked, that is, carried on the sex chromosomes. Recessive genes carried on the *X* chromosome (the *Y* chromosome contains almost no genetic information) appear in the phenotype far more often in males than in females; a female heterozygous for a sex-linked characteristic will show the dominant trait, whereas a single recessive gene in the male, if carried on the *X* chromosome, will result in a recessive phenotype since no allele is present.

Some genes assort independently in breeding experiments, and others tend to remain together. Genes that tend to travel together are said to be in the same linkage groups.

Genes are sometimes exchanged between homologous chromatids at meiosis. Such crossovers could take place only if (1) the genes are arranged in a fixed linear array along the length of the chromosomes and (2) the genes are at corresponding sites on homologous chromosomes. On the basis of these assumptions, chromosome maps, showing the relative locations of gene sites along *Drosophila* chromosomes, were developed from crossover data provided by breeding experiments. The accuracy of these maps was later confirmed by physical observation of giant chromosomes in the salivary glands of *Drosophila.*

QUESTION

In a series of breeding experiments, a linkage group composed of genes *A, B, C, D,* and *E* was found to show the following crossover frequencies:

		Gene					
		A	*B*	*C*	*D*	*E*	
	A	—	8	12	4	1	
	B	8	—	4	12	9	*Crossovers*
Gene	*C*	12	4	—	16	13	*per* 100
	D	4	12	16	—	3	*fertilized eggs*
	E	1	9	13	3	—	

Using Sturtevant's standard unit of measure, "map" the chromosome.

* What is, perhaps, most interesting is that the banding of salivary chromosomes had been observed and recorded by cytologists before the 1900s—and Morgan and his group were not aware of this! While the geneticists were laboriously performing thousands of testcrosses, a simple means of confirming their hypotheses lay buried and forgotten. That they were able to develop such accurate "maps" of the chromosomes without ever having seen the banding was an extraordinary intellectual achievement.

CHAPTER 3–5

Sex Linkage and Human Karyotypes

As in *Drosophila*, the Y chromosome of man carries much less genetic information than the X chromosome. Genes for color vision, for example, are carried on the X chromosome in humans but not on the Y chromosome. Color blindness is produced by a recessive allele of the normal gene. The normal allele is dominant; a woman with one X chromosome with the normal allele and one X chromosome with the allele for color blindness will have normal color vision. If she transmits the X chromosome with the recessive allele to a daughter, the daughter also will have normal color vision if she receives a normal X chromosome from her father (that is, if he is not color-blind). If, however, the X chromosome with recessive allele is transmitted to a son, he will be color-blind since, lacking a second X chromosome, he has only the recessive allele.

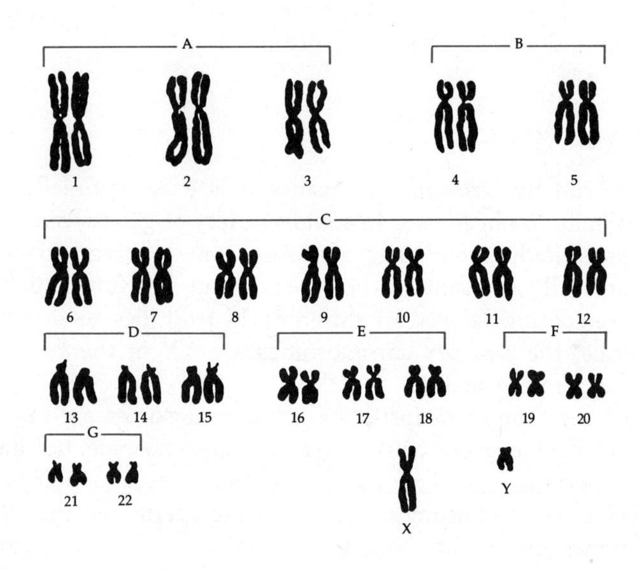

3–36

The normal diploid chromosome number of a human being is 46, 22 pairs of autosomes and the 2 sex chromosomes. In these pictures, the chromosomes of white blood cells of a human male are arranged in what is known as a karyotype. A normal woman has two X chromosomes and a normal man an X and a Y.

Hemophilia

A classic example of a recessive inheritance transmitted on the X chromosome is the hemophilia which has afflicted the royal families of Europe since the nineteenth century. Hemophilia is a disease in which the blood does not clot normally, so that even minor injuries carry the risk of the patient's bleeding to death. Queen Victoria was probably the original carrier in the family. Because none of her forebears or collateral relatives was affected, we conclude that the mutation may have occurred on an X chromosome in one of her parents or in the cell line from which her own

PREPARATION OF A KARYOTYPE

Chromosome typing for the identification of hereditary defects is being carried out at an increasing number of genetic counseling centers throughout the United States. The result of the procedure is known as a karyotype. *The chromosomes shown in a karyotype are actually chromatid pairs, interrupted in the process of meiosis and held together at their centromeres. They do not resemble untreated chromosomes, in which, as we saw in Chapter 3–1, the centromeres divide first and are separated by the spindle fibers, with the arms of the chromatid pairs pulling slowly apart. In some cases, karyotyping can help parents decide whether or not to have children. Karyotypes can also be made of unborn infants and so can reveal the sex of the future child or the presence of genetic defects.*

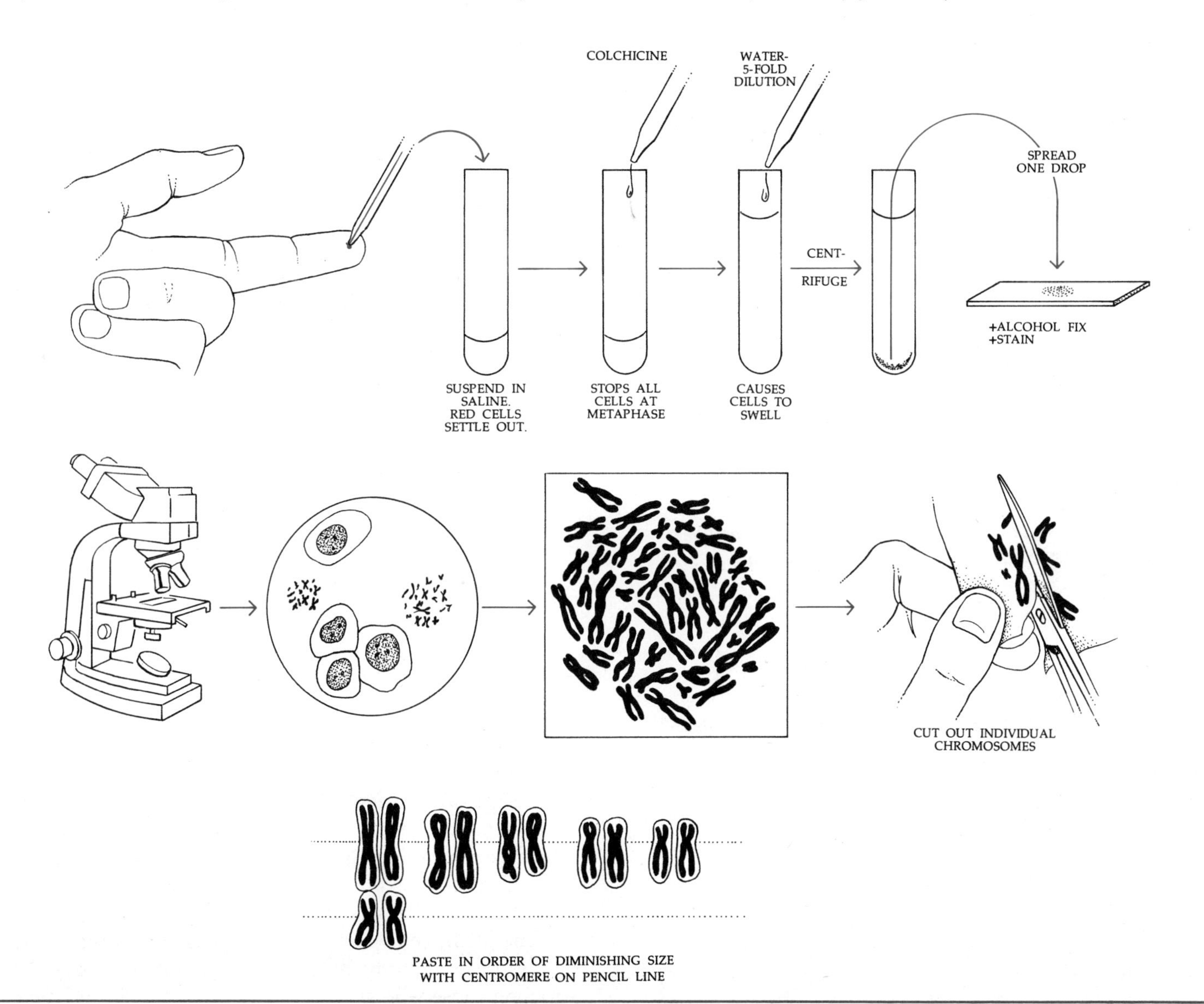

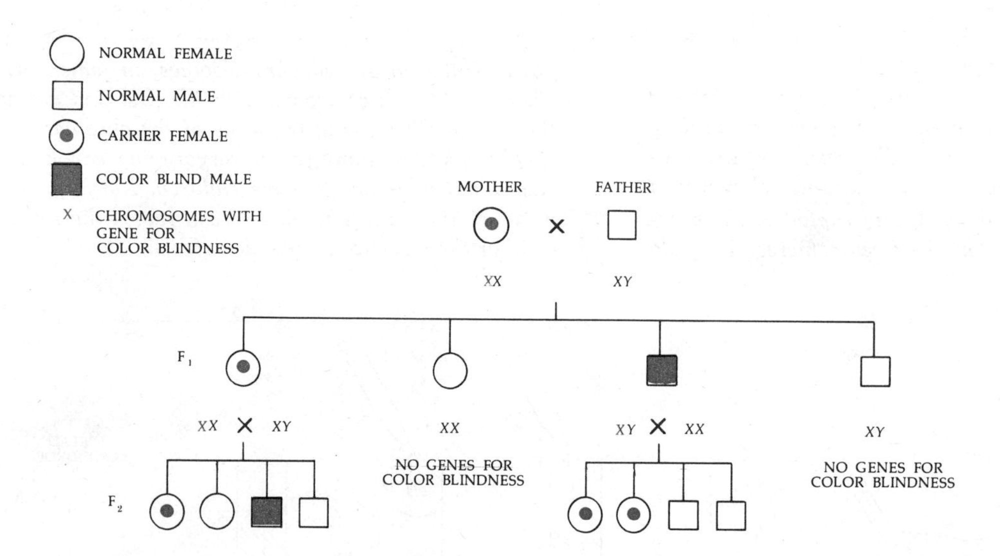

3–37

Color blindness in humans is carried by a recessive allele on the X chromosome. In the chart, the mother has inherited one normal and one defective allele. The normal allele will be dominant, and she will have normal color vision. However, half her eggs (on the average) will carry the defective allele, and half will carry the normal allele—and it is a matter of chance as to which kind is fertilized. Since her husband's Y chromosome, the one that determines a son rather than a daughter, carries no gene for color discrimination, the single gene the wife contributes (even though it is a recessive gene) will determine whether or not the son is color-blind. Therefore, half her sons (on the average) will be color-blind. Assuming that her children marry individuals with normal X chromosomes, the expected distribution of the trait among her grandchildren will be as shown on the chart.

eggs were formed. Prince Albert, Victoria's consort, could not have been responsible; male-to-male inheritance of the disease is impossible. (Why?) One of her sons, Leopold, Duke of Albany, died of hemophilia at the age of 31. At least two of Victoria's daughters were carriers, since a number of *their* descendants were hemophiliacs. And so, through various intermarriages, the disease spread from throne to throne across Europe. In the Czarevitch, son of the last Czar of Russia, and in the princes of Spain, the gene for hemophilia inherited from Victoria had considerable political consequences.

HUMAN CHROMOSOMAL ABNORMALITIES

From time to time, usually because of "mistakes" at the time of meiosis, homologues may not separate. In this case, one of the sex cells has one too many chromosomes and the other one too few. This phenomenon is known as *nondisjunction*. The cell with one too few cannot produce a viable embryo, but the one with one too many sometimes can. The result is an individual with an extra chromosome in every cell of his body.

The presence of additional chromosomes often produces widespread abnormalities. Many such infants are stillborn. Among those who survive, many are mentally deficient. In

3–38

As this chart shows, Queen Victoria was the original carrier of the hemophilia that has afflicted male members of the royal families of Europe since the nineteenth century. This is another case of a recessive trait transmitted on the X chromosome.

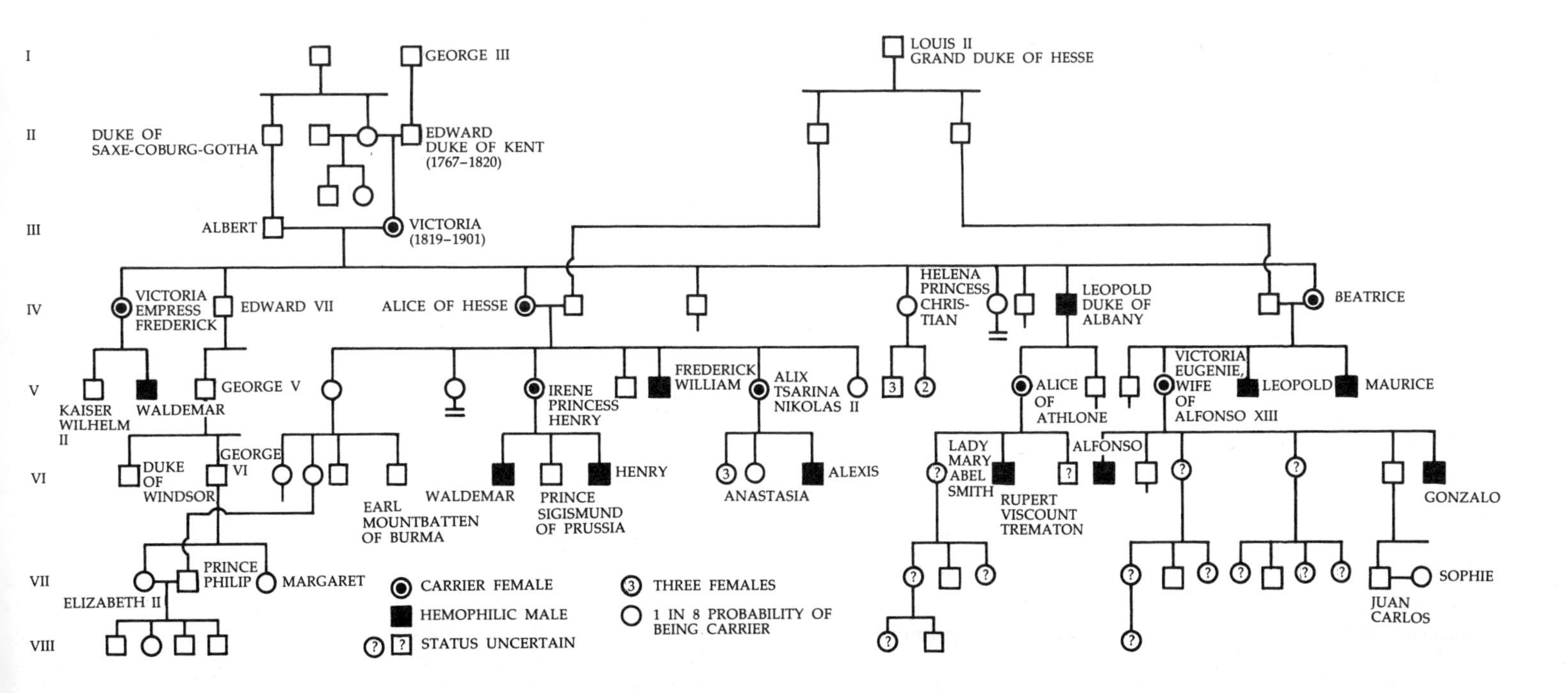

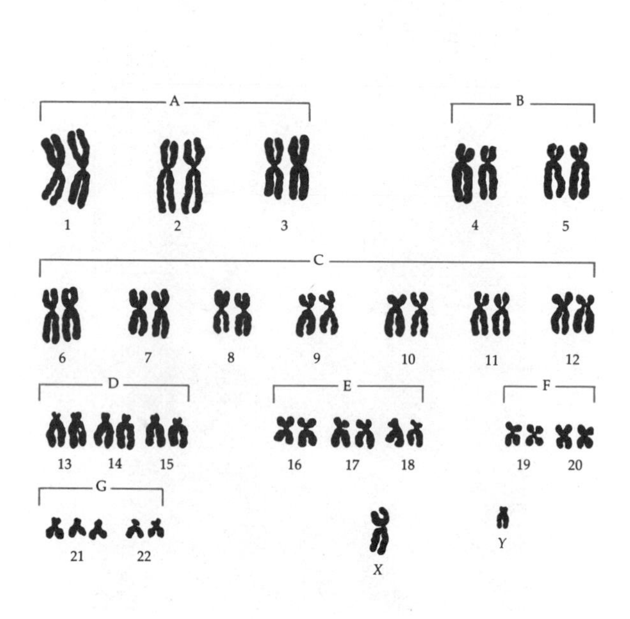

3–39

The karyotype of a patient with Down's syndrome. Note that there are three chromosomes 21.

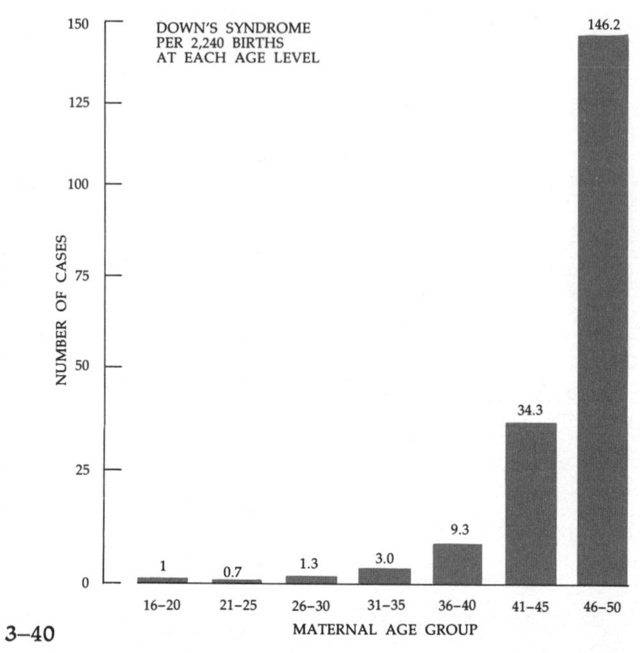

3–40

The frequencies of mongoloid births in relation to the ages of the mothers. The number of cases shown for each age group represents the occurrences of mongolism in every 2,240 births by mothers in that group. As you can see, the risk of having a mongoloid child increases rapidly after the age of 40.

fact, studies of abnormalities in human chromosome number among living subjects are carried out on patients in mental hospitals. These patients often have abnormalities of the heart and other organs as well.

Down's Syndrome

You are probably familiar with the form of mental deficiency known as mongolism. This name derives from a peculiar and typical appearance of the eyefold in these patients, which makes them look "foreign," or mongoloid, to us. Mongolism is actually not a single disease but a syndrome, a group of disorders which occur together, and nowadays it is usually referred to as Down's syndrome, after the physician who first described it. The syndrome includes, in most cases, not only mental deficiency but a short, stocky body type with a thick neck and, often, abnormalities of other organs, especially the heart.

Down's syndrome and a number of other defects involving gross abnormalities of chromosomes are more likely to occur among infants born to older women. The reasons for this are not known, but the formation of the egg cells is well under way in the human female before she is born, so the increasing incidence of abnormalities may be correlated in some way with the aging of the mother's cells.

The most common cause of the genetic abnormality that produces Down's syndrome is nondisjunction involving chromosome 21. This results in an extra chromosome 21 (Figure 3–39) in the cells of the defective child.

Down's syndrome may also be the result of an abnormality in the chromosomes of one of the parents. In the

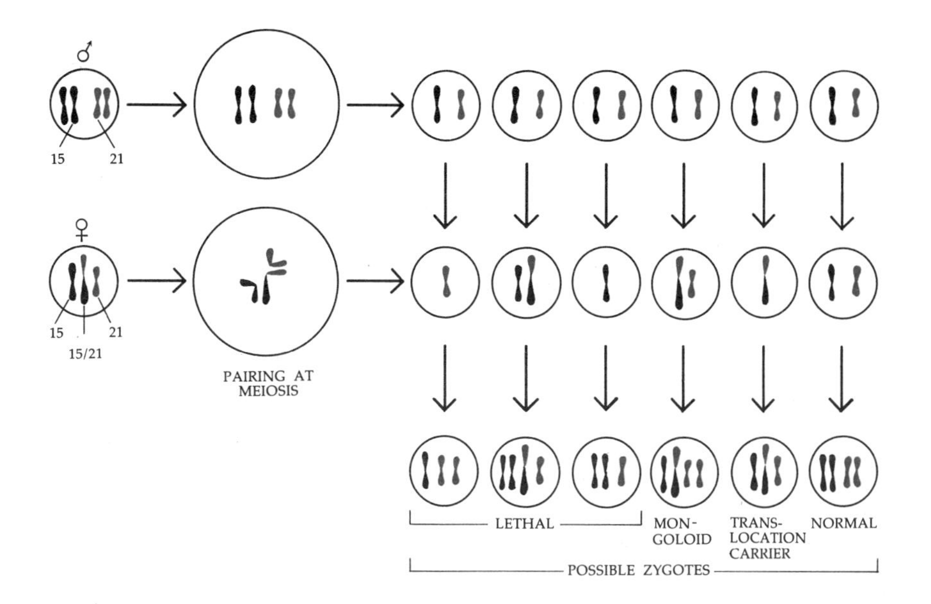

3–41
Transmission of translocation mongolism. The father, top row, has normal pairs of chromosomes 21 and 15, and each of his sperm cells will contain a normal 21 and a normal 15. The mother (more frequently, although not always, the translocation carrier) has one normal 15, one normal 21, and a translocation 15/21. She herself appears normal, but her chromosomes cannot pair normally at meiosis. There are six possibilities for the offspring of these parents: the infant will (1) die before birth, (three of the six possibilities), (2) be mongoloid, (3) be a translocation carrier like the mother, or (4) be normal. Tests for the chromosomal abnormality can be made in prospective parents and in the infant before birth.

case of Down's syndrome, the abnormality involved is *translocation*. Translocation occurs when a portion of a chromosome is broken off and becomes attached to another chromosome. The patient with translocation mongolism usually has a third chromosome 21 (or, at least, most of it) attached to a larger chromosome, such as 15. When cases of translocation mongolism are studied, it is usually found that one parent, although phenotypically normal, has only 45 separate chromosomes—one chromosome being composed of most of chromosomes 15 and 21 joined together. The possible genetic makeups of the offspring of this parent are diagramed in Figure 3–41. Three out of the six possible combinations are lethal. One of the remaining three will produce Down's syndrome, one will be normal, and one will be a carrier. Not a very cheerful prognosis!

Thus parents who have had one abnormal child are faced with the terrible decision of whether or not to risk having another infant. Now there are special clinics throughout the country which can help them in this decision. In these clinics, skin cells from stillborn or abnormal infants are cultivated in a test tube for examination at the time of mitosis, when the chromosomes become visible. If an infant is found to have an abnormal karyotype, the skin cells of the parents can be similarly tested. If the parents show an abnormality, they are warned that they are likely to transmit it to future infants through their gametes.

If, however, the karyotypes of both parents are normal, the parents are advised that the abnormalities in the child were probably the result of nondisjunction—a "mistake" that occurred during meiosis—and that they do not run a greater-than-normal risk for the mother's age group of having another congenitally ill child.

It is now possible to detect the presence of extra chromosomes in the fetus when it is 16 weeks old. This diagnosis can be made from a sample of the amniotic fluid surrounding the embryo. Unfortunately, however, the only treatment is abortion of the unborn infant.

Abnormalities in the Sex Chromosomes

Nondisjunction may also produce individuals with extra sex chromosomes. An *XY* combination in the twenty-third pair, as you know, is associated with maleness, but so is *XXY* and *XXXY* and even *XXXXY*. These males are usually sexually underdeveloped and sterile, however. *XXX* combinations sometimes produce normal females, but many of the *XXX* women and all *XO* women (women with only one *X* chromosome) are sterile.

SUMMARY

Humans have two sex chromosomes and 22 pairs of autosomes. For study, the chromosomes are arranged in a karyotype.

The *Y* chromosome carries fewer genes than the *X* chromosome. As a consequence, if a man receives an *X* chromosome carrying recessive alleles, these alleles (which in the female would be dominated by the normal alleles) will usually be expressed. Characteristics resulting from the expression of such genes are said to be sex-linked or *X*-linked. Color blindness and hemophilia are examples of sex-linked characteristics which are carried by females but seen chiefly among males.

Nondisjunction is the failure of two homologues to separate at the time of meiosis. As a consequence of nondisjunction, children may be born with an extra chromosome. The presence of such an extra chromosome can produce abnormalities such as Down's syndrome.

QUESTIONS

1. Under what conditions would color blindness be found in a woman? If she married a man who was not color-blind, would her sons be color-blind? Her daughters?

2. Mongoloid individuals frequently must be institutionalized for their entire lives, often at state expense. Do you believe that the government would be justified in undertaking measures to reduce the number of mongoloid births? What measures?

CHAPTER 3–6

The Double Helix

The work discussed in the previous chapters of this section is often referred to as "classical genetics." In some ways, this is an unfortunate term because "classical" carries a sense of something that happened a long time ago and that is interesting today only for historical reasons. Actually, classical genetics provides the vital framework for our understanding of all modern genetics and, in particular, as we shall see, is the basis for population genetics, one of the most dynamic of the modern branches of biology.

Classical genetics showed that genes are carried on chromosomes, that they come in alternate forms (alleles), that alleles occupy corresponding positions (loci) on homologous chromosomes, and that chromosomes with their alleles are reassorted at meiosis. Classical geneticists discovered that segments of chromatids are exchanged during meiosis, and they made use of these exchanges to construct chromosome maps. They introduced the concept of mutations and of interactions of genes with other genes, as in polygenic inheritance, to produce the phenotype. All this work—and more—was accomplished in less than half a century!

A turning point in genetics came when scientists began to turn their attention to the question of how it was possible for these little lumps of matter—the chromosomes—to be the bearers of what they had come to realize must be an enormous amount of complex information. The chromosomes, like all the rest of the living cell, are composed of atoms arranged into molecules. Some scientists, a number of them quite eminent in the field of genetics, thought it would be impossible to understand the complexities of genetics in terms of the structure of "lifeless" chemicals. (Although they did not use the term to describe themselves, they were, in fact, the vitalists of the twentieth century.) Others thought that if the chemical structure of the chromosomes was understood, we could then come to understand how chromosomes could function as the bearer of the genetic information. The work of the latter group is now termed "molecular genetics."

The first question was a simple one: What are chromosomes made of? Chemical analysis shows that the chromosomes of higher organisms are made up of protein and, in smaller amounts, a chemical known as deoxyribonucleic acid, now usefully abbreviated as DNA. So the scientists who believed that the key to understanding heredity was to be found in the chemistry of the gene were faced with a second, more difficult question: Are genes protein or DNA?

PROTEIN OR DNA?

In the 1940s, when this question was first being seriously considered, proteins seemed to be the more promising alternative. Biochemists had just begun to work out the structure of proteins. They had discovered that all enzymes are proteins, that protein molecules are large and complex, and most importantly, that the number of structurally different proteins present in the living world—that is, proteins differing in the number and sequence of their component amino acids—is enormous. Therefore, it was possible to see how proteins, with their 20 amino acids arranged in varying sequences, might spell out a "language of life," as our own language is spelled out using only the 26 letters of the alphabet. The various amino acid arrangements might, in other words, be biological "sentences," dictating the directions for the many activities of the cell, with each different protein specific for one particular activity.

Some important research supported this viewpoint. George Beadle and Edward Tatum, who later received the Nobel Prize for their research, demonstrated that a mutation involving a single gene of the bread mold *Neurospora* could result in a change in one of the enzymes produced by the cells. (*Neurospora* is haploid, like *Chlamydomonas*, so a change in a single gene is not masked by the effects of an allele.) On the basis of their experiments, summarized in Figure 3–42, Beadle and Tatum formulated the one-gene–one-enzyme theory. This theory states that genes control cellular structures and functions by controlling enzymes, with each gene responsible for one particular enzyme. This theory was later generalized to "one gene, one protein" and became the basis for much subsequent research.

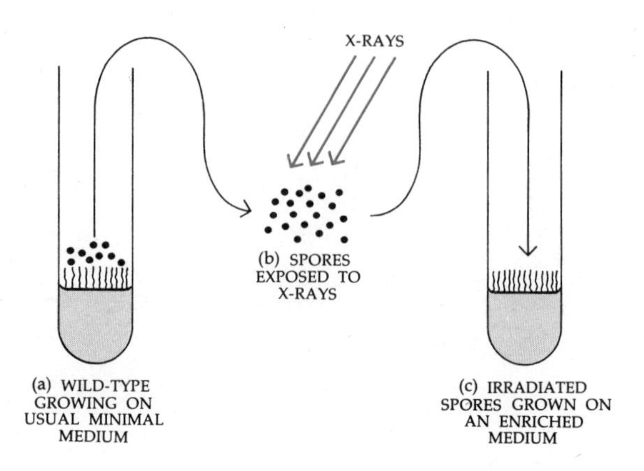

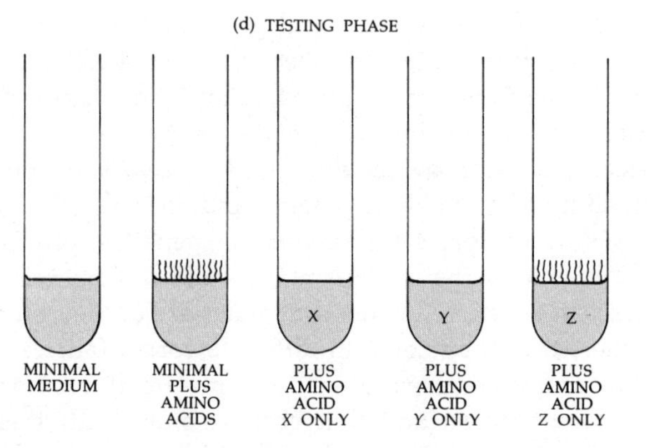

3–42

Beadle and Tatum's experiments on the bread mold Neurospora. *By these experiments, they were able to show that change in a single gene results in change in a single enzyme.* (a) *Wild-type* Neurospora *will grow on a minimal medium, that is, a medium containing only a sugar, some ions, and one vitamin. From these basic materials, it can synthesize, by means of enzymes, all the amino acids and other vitamins it requires.* (b) *Spores of* Neurospora *were x-rayed to increase the mutation rate. (By experimental crosses between these mutants and wild-type* Neurospora, *the investigators were able to show that the mutations they were studying affected only single genes.)* (c) *The mutants, it was found, were unable to survive in the minimal medium, but some of them could be kept alive in an enriched medium, that is, a medium supplemented with amino acids and vitamins.* (d) *The progeny of these mutants were tested for their ability to survive in various other mediums in order to determine which amino acid the mold had lost its ability to synthesize. As in the example shown here, a mold which had lost its capacity to synthesize amino acid Z would be able to survive in a minimal medium plus that amino acid only but would not, under any circumstances, survive in a medium which lacked that amino acid. Further tests were then made to discover, in each case, which enzymatic step had been impaired.*

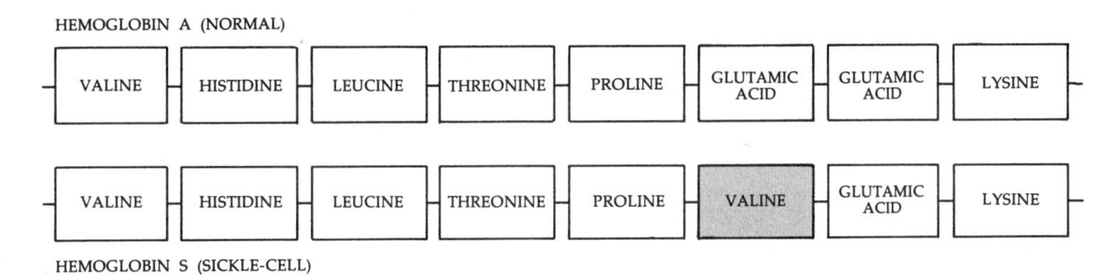

3–43

An example of the remarkable precision of the "language" of proteins. Portions of the beta chains of the hemoglobin A (normal) molecule and the hemoglobin S (sickle-cell) molecule are shown. The hemoglobin molecule is composed of two identical alpha chains and two identical beta chains, each chain consisting of about 150 amino acids, or a total of 600 amino acids in the molecule. The entire structural difference between the normal molecule and the sickle-cell molecule (literally, a life-and-death difference) consists of one change in the sequence of each beta chain: one glutamic acid is replaced by one valine!

The Structure of Hemoglobin

Linus Pauling was one of the first to see some of the implications of these new ideas. Hemoglobin, the substance in vertebrate red blood cells that carries the oxygen, is a protein made up of chains of amino acids. Perhaps, Pauling reasoned, human diseases, such as sickle-cell anemia, involving red blood cells can be traced to a variation from normal in the protein structure of the hemoglobin molecule—a typographical error, so to speak, in one important sentence in the hypothetical language of the cells. He took samples of hemoglobin from normal persons, from persons with sickle-cell anemia, and from persons heterozygous for the trait. To study the differences in these proteins, he used a process known as electrophoresis, in which organic molecules are dissolved in a solution and exposed to a weak electric current. Very small differences, even in very large molecules, are reflected in the electric charges of the molecules, and the molecules will move differently in the electric field. The normal person, he found, makes one sort of hemoglobin, the person with sickle-cell anemia makes a slightly different sort, and the person who carries one copy of the recessive gene for sickling makes both types of hemoglobin.

A few years later, it was found that the actual difference between the normal and the sickle hemoglobin molecules lies in 2 of the molecule's 600 amino acids. The hemoglobin molecule is composed of four polypeptide chains—two identical alpha chains and two identical beta chains—each made up of about 150 amino acids. In a precise location in each beta chain, one glutamic acid is replaced by one valine. (See Figure 3–43.) The language of proteins is fantastically precise, and the consequences of even a small error or small variation may be great.

So it is easy to see why many prominent investigators, particularly those who had been studying proteins, believed that the genes themselves were proteins, that the chromosomes contained master models of all the proteins that would be required by the cell, and that enzymes and other proteins active in cellular life were copied from these master models. This was a logical hypothesis—but, as it turned out, it was wrong.

THE DNA HYPOTHESIS

The Transforming Factor

To trace the beginning of the other hypothesis—the one that proved to be right—it will be necessary to go back to 1928 and pick up an important thread in modern biological history. In that year, an experiment was performed which seemed at the time very remote from either biochemistry or genetics. Frederick Griffith, a public-health bacteriologist, was studying the possibility of developing a vaccine against pneumococci, the bacterial cells that cause one kind of pneumonia. The pneumonia-causing strains, he found, are all surrounded by special capsules, or sheaths, composed mainly of polysaccharides. The existence of the sheath and its composition, it is now known, are both genetically determined, that is, are inherited properties of the bacteria. Griffith thought that injections of heat-killed disease-causing bacteria or of live, closely related but harmless strains—ones without capsules—might immunize mice to live virulent pneumococci. As a result of his experiments, which are outlined in Figure 3–44, Griffith found that something can be passed from dead to live bacteria that changes their hereditary characteristics. This "something" came to be called the *transforming factor*. In 1943, O. T. Avery and his coworkers at the Rockefeller Institute demonstrated that the transforming factor is DNA. Subsequent experiments showed that a variety of genetic traits can be passed from members of one strain of bacterial cells to those of another, similar colony by means of isolated DNA.

DNA had first been isolated by a German chemist named Friedrich Miescher in 1869—in the same remarkable decade in which Darwin published *The Origin of Species* and Mendel presented his results to an audience of 40 at the Natural History Society in Brünn. The substance Miescher isolated was white, sugary, slightly acid, and contained phosphorus. Since he found it only in the nuclei of cells, he called it nucleic acid. This name was later amended to deoxyribonucleic acid, to distinguish it from a similar chemical also found in cells, ribonucleic acid.

By the time of Avery's discovery, it was known that DNA is made up of nucleotides (Figure 3–45). Each nucleotide consists of a nitrogen-containing compound, a deoxyribose sugar, and a phosphate. The nitrogen-containing compounds are of two kinds: purines, which have two rings, and pyrimidines, which have one ring. There are two kinds of purines found in DNA, adenine(A) and guanine(G), and two kinds of pyrimidines, cytosine(C) and thymine(T). So DNA is made up of four types of nucleotides, differing

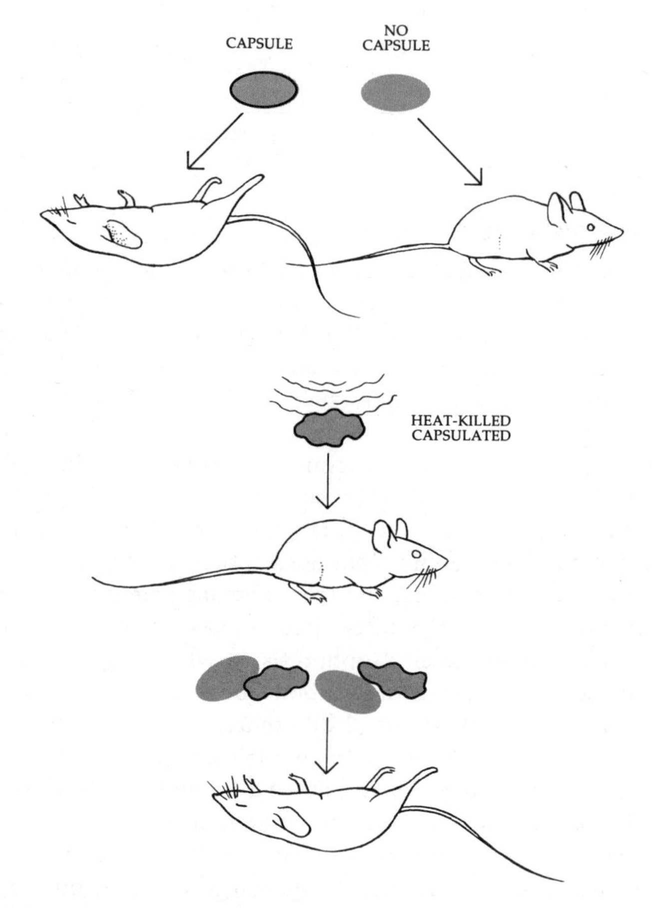

3–44

The studies that led to the discovery of the transforming factor, a substance that can transmit genetic characteristics from one cell to another cell. Pneumococci (pneumonia-causing bacteria) that are surrounded by capsules cause death when injected into mice. The noncapsulated strain is harmless. When the virulent, capsulated bacteria were heat-killed before being injected, they did not cause disease. But death invariably resulted from a mixture of the heat-killed capsulated bacteria and the noncapsulated ones, although both strains are in themselves

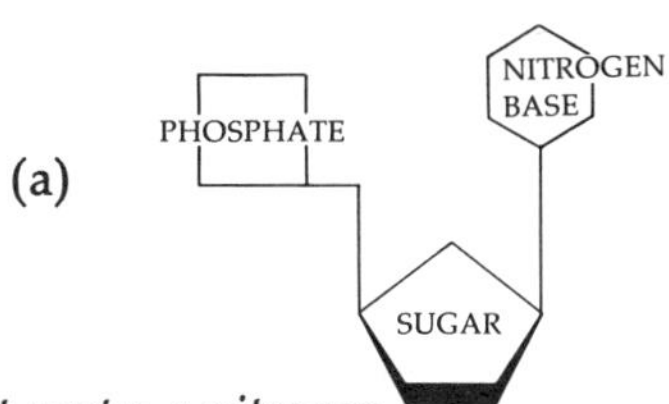

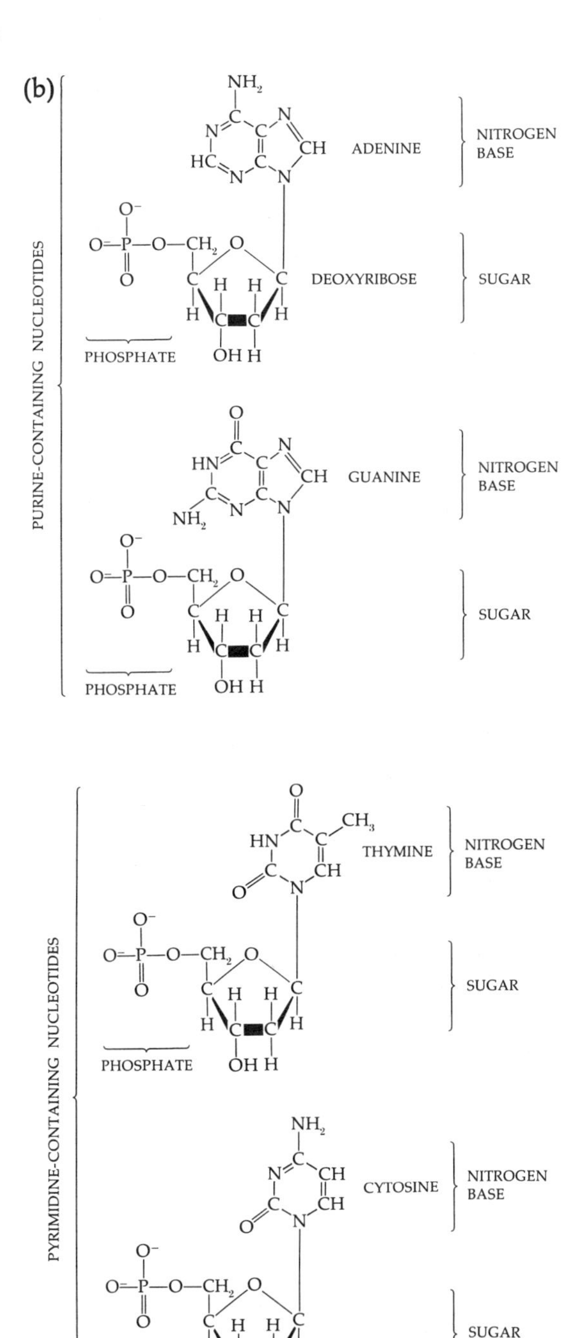

3–45 (at right)
(a) *A nucleotide is made up of three different parts: a nitrogen base, a sugar, and a phosphate.* (b) *Each nucleotide in DNA contains a deoxyribose sugar and one of the four possible nitrogen bases.*

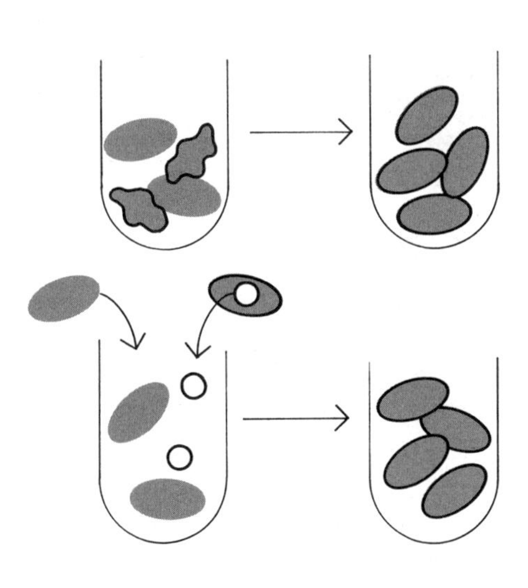

harmless! Autopsies revealed that the dead mice were full of live capsulated *bacteria. The same phenomenon was reproduced in the test tube. When dead, capsulated bacteria were mixed with live, noncapsulated bacteria, bacteria appeared that were live, capsulated and virulent. Apparently, the dead bacteria could transmit to the living harmless bacteria their ability to make capsules and, with it, their virulent character. The transforming factor, the "something" that was transmitted to the non-capsulated bacteria, was later isolated and found to be DNA.*

only in terms of their nitrogen-containing purine or pyrimidine.

Avery's results offered evidence for DNA as the genetic material, but his discovery was slow to gain full recognition. This was partly because bacteria were considered "lower" and "different" and partly because the DNA molecule—made up of only four components—seemed too simple for the enormously complex task of carrying the hereditary information.

The "Bacteria Eaters"

A second, crucial experiment was made with an even "lower" organism, a virus which attacks the bacterial cell *Escherichia coli.* These viruses, known as bacteriophages ("bacteria eaters"), were originally chosen for study because they are so phenomenal at reproducing themselves. They invade bacterial cells, multiply within them, and then escape—usually bursting the cell. The entire infection cycle can take place in as brief a period as 20 minutes, during which time several hundred virus progeny can be produced from a single virus.

Another interesting thing about bacteriophages is that they consist only of DNA and protein, the two contenders for the leading role as carriers of the genetic material. The question of which one carries the viral genes—the hereditary information by which new viral particles are made—was answered in 1952 by Alfred D. Hershey, working with Mary Chase, in the simple and ingenious experiment summarized in Figure 3–46*b*. If you remember that protein contains sulfur and no phosphorus and DNA contains phosphorus and no sulfur, you will see why this experiment showed that only the DNA of the bacteriophages was involved in the replication process and that the protein could not be the hereditary material.

Electron micrographs such as those shown in Figure 3–46 have now confirmed that this type of bacteriophage attaches to the bacterial cell wall by its tail and injects its DNA into the cell, leaving the empty protein coating (the

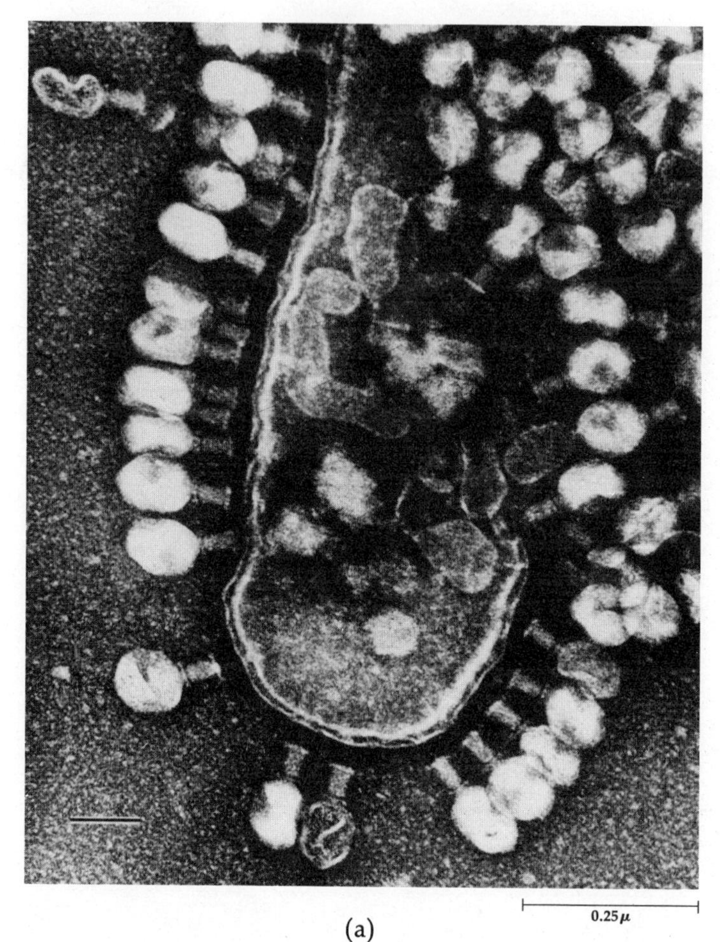

(a)

3–46

(a) *Electron micrograph of bacteriophages. The ones most commonly used in genetic studies are* T-2 *and* T-4. *(The* T *stands simply for "type.")* T-2 *are shown here.* (b) (on facing page.) *An ingenious experiment demonstrating that DNA is the hereditary material. Radioactive sulfur and radioactive phosphorus were substituted for the sulfur and phosphorus in the medium on which an* Escherichia coli *colony was growing. The colony was then infected with a* T-2 *bacteriophage. After a cycle of multiplication, the newly formed viruses all contained radioactive sulfur and radioactive phosphorus. The labeled viruses were used to infect fresh* E. coli *cells growing on a normal medium. Once the*

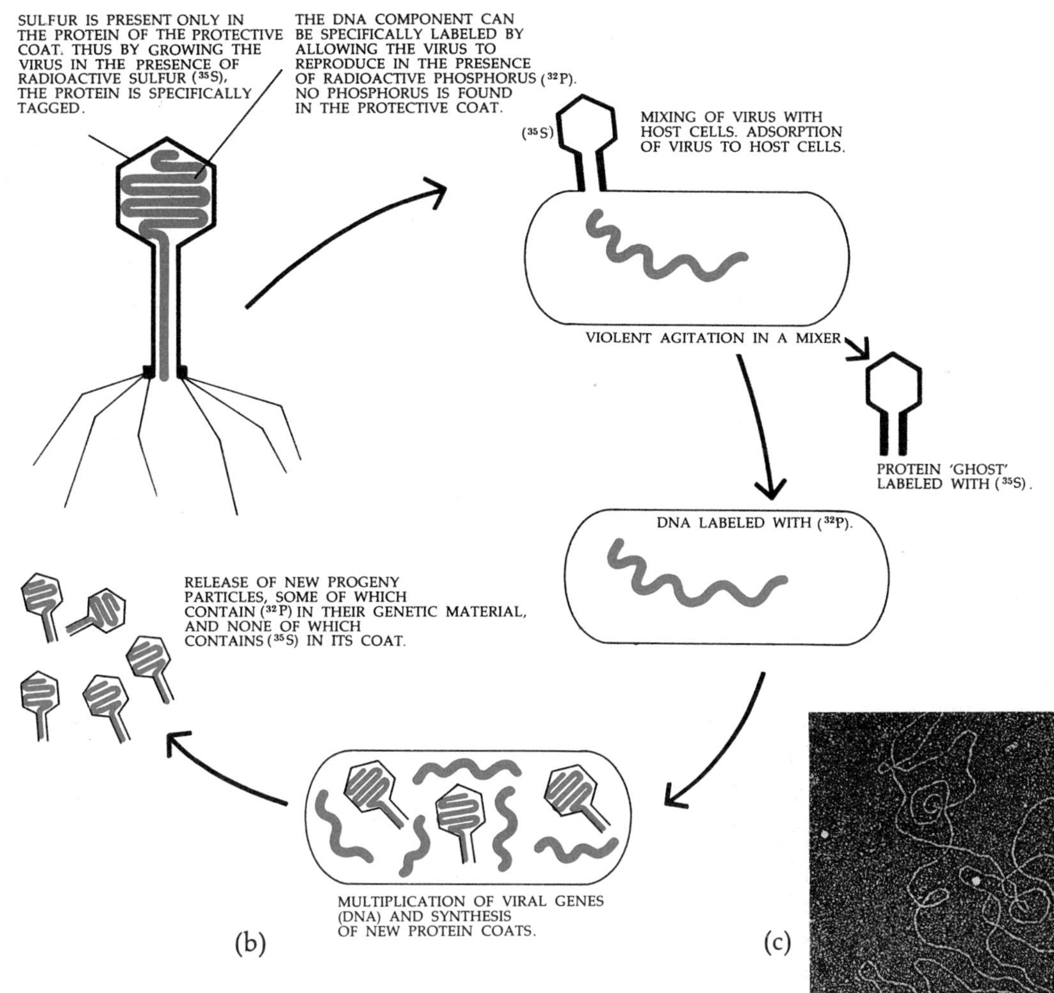
SULFUR IS PRESENT ONLY IN THE PROTEIN OF THE PROTECTIVE COAT. THUS BY GROWING THE VIRUS IN THE PRESENCE OF RADIOACTIVE SULFUR (^{35}S), THE PROTEIN IS SPECIFICALLY TAGGED.
THE DNA COMPONENT CAN BE SPECIFICALLY LABELED BY ALLOWING THE VIRUS TO REPRODUCE IN THE PRESENCE OF RADIOACTIVE PHOSPHORUS (^{32}P). NO PHOSPHORUS IS FOUND IN THE PROTECTIVE COAT.
(^{35}S)
MIXING OF VIRUS WITH HOST CELLS. ADSORPTION OF VIRUS TO HOST CELLS.
VIOLENT AGITATION IN A MIXER
PROTEIN 'GHOST' LABELED WITH (^{35}S).
DNA LABELED WITH (^{32}P).
RELEASE OF NEW PROGENY PARTICLES, SOME OF WHICH CONTAIN (^{32}P) IN THEIR GENETIC MATERIAL, AND NONE OF WHICH CONTAINS (^{35}S) IN ITS COAT.
MULTIPLICATION OF VIRAL GENES (DNA) AND SYNTHESIS OF NEW PROTEIN COATS.
(b)

(c)

0.5μ

infection had begun, the cells were washed free of excess virus and broken apart artificially. When their contents were examined, it was found that almost all the radioactive sulfur was outside *the cells. Now, phosphorus in phosphate form is an important constituent of nucleic acid but is not found in proteins. Sulfur, on the other hand, is found in protein but never in nucleic acid. Thus the experiment clearly demonstrated that the phosphorus-containing DNA was the substance which invaded the cells. The protein, with the radioactive sulfur, could not be the hereditary material.* (c) *Bacteriophage surrounded by its DNA.*

"ghost") on the outside. In short, the protein is just a container for the DNA, which carries all the hereditary information of the virus and directs the synthesis both of new DNA and of new protein for the viral progeny.

Further Evidence for DNA

The role of DNA in transformation and in viral infection formed very convincing evidence for believing that DNA is the genetic material. Two other lines of work also helped to lend weight to the argument. Alfred Mirsky and others, in a long series of careful studies, showed that, in general, the tissue cells of any given species contain equal amounts of DNA. The principal exceptions are the gametes, which regularly contain just half as much DNA as the other cells of the same species.

A second important series of contributions was made by Erwin Chargaff. Chargaff analyzed the purine and pyrimidine content of the DNA of many different kinds of living things and found that the nitrogen bases do not occur in exact proportions. The DNA molecule, he found, can have a great deal of variety in its composition. However, the proportions of the four nitrogen bases are the same in all cells of a given species, even though they vary from one species to another. In other words, the bases, according to Chargaff's analysis, do not occur in a regular sequence —ATCG, ATCG, for instance—because if they did, the proportions would all be the same. Rather, they are arranged irregularly—ATCG, TTAC, for instance. Therefore, these variations could very well provide a "language" in which the instructions controlling cell growth could be written. Some of Chargaff's results are reproduced in Table 3–3. Can you, by examining these figures, notice anything interesting about the proportions of purines and pyrimidines?

Table 3–3

*Proportions of the Nitrogen Bases in the DNA of Several Species**

	Purines, %		*Pyrimidines, %*	
Source	*Adenine*	*Guanine*	*Cytosine*	*Thymine*
Man	30.4	19.6	19.9	30.1
Ox	29.0	21.2	21.2	28.7
Salmon sperm	29.7	20.8	20.4	29.1
Wheat germ	28.1	21.8	22.7	27.4
E. coli	26.0	24.9	25.2	23.9
Sheep liver	29.3	20.7	20.8	29.2

* After Erwin Chargaff, *Essays on Nucleic Acids*, 1963.

THE WATSON–CRICK MODEL

In the early 1950s, a young American scientist, James D. Watson, went to Cambridge, England, on a research fellowship to study problems of molecular structure. There, at the Cavendish Laboratory, he met physicist Francis Crick. Both were interested in DNA, and they soon began to work together to solve the problem of its molecular structure. They did not do experiments in the usual sense but rather undertook to examine all the data about DNA and unify them into a meaningful whole.

The Known Data

By the time Watson and Crick began their studies, quite a lot of information on the subject had already accumulated. It was known that the DNA molecule is very large and also that it is long and thin. It was known that DNA contains nucleotides, each consisting of either one purine or one pyrimidine plus a deoxyribose sugar and a phosphate group. X-ray studies showed that the molecule has spiral-like turns as part of its structure. A spiral of a fixed diameter is called a helix; in 1950, Pauling had shown that proteins sometimes take this form (see page 78) and that the helical structure is maintained by hydrogen bonding between suc-

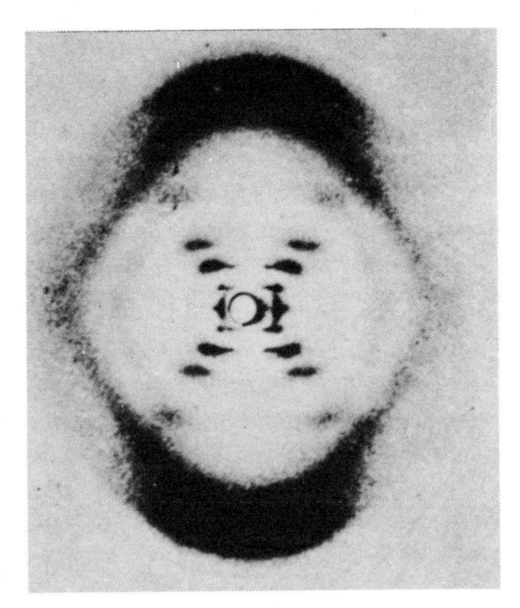

3–47
X-ray diffraction photograph of DNA taken by Rosalind Franklin in the laboratories of Maurice Wilkins, who shared the Nobel Prize with Watson and Crick. The reflections crossing in the middle indicate that the molecule is a helix. The heavy dark regions at the top and bottom are due to the closely stacked bases perpendicular to the axis of the helix.

cessive turns in the helix. Pauling had suggested that the structure of DNA might also be based on a single helical framework. And, finally, there were the Chargaff data (Table 3–3), which indicated that the ratio of DNA nucleotides containing thymine to those containing adenine is approximately 1 to 1 and that the ratio of nucleotides containing guanine to those containing cytosine is also approximately 1 to 1.

Building the Model

From these data, Watson and Crick attempted to construct a model of DNA that would fit the known facts. They were very conscious of the biological role of DNA. In order to carry such a vast amount of information, the molecules would have to be heterogeneous and varied. Also, there had to be some way for them to replicate readily and with great precision in order that faithful copies could be passed from cell to cell and from parent to offspring, generation after generation.

Let us see what the Watson-Crick model looked like. If you take a ladder and twist it into the shape of a helix, keeping the rungs perpendicular, this would form a crude model of the molecule (Figure 3–48). The two longitudinal railings are made up of sugar and phosphate molecules, alternating. The perpendicular rungs of the ladder are formed by the nitrogenous bases—adenine (A), thymine (T), guanine (G), and cytosine (C)—one base for each sugar-phosphate and two bases forming each rung. The paired bases meet across the helix and are joined together by hydrogen bonds, the relatively weak, common, and very important chemical bonds that Pauling had demonstrated in his studies of the structures of proteins. (Hydrogen bonds are described in Chapter 2–2.)

The distance between the two sides, or railings, according to x-ray measurements, is 20 angstroms. Two purines in combination would take up more than 20 angstroms, and two pyrimidines would not reach all the way across. But if a purine paired in each case with a pyrimidine, there would be a perfect fit. The paired bases—the "rungs" of the ladder—would therefore always be purine-pyrimidine combinations.

As Watson and Crick worked their way through these data, they assembled actual tin-and-wire models of the molecule, seeing where each piece would fit into the three-dimensional puzzle. First, they noticed, the nucleotides along any one strand of the double helix could be assembled in any order: ATGCGTACATTGCCA, and so on. (See Figure 3–49.) Since a DNA molecule may be several hundred nucleotides long, there is a possibility for great variety. The meaning of this variety, the molecular heterogeneity of DNA, will be explored more thoroughly in the next chapter.

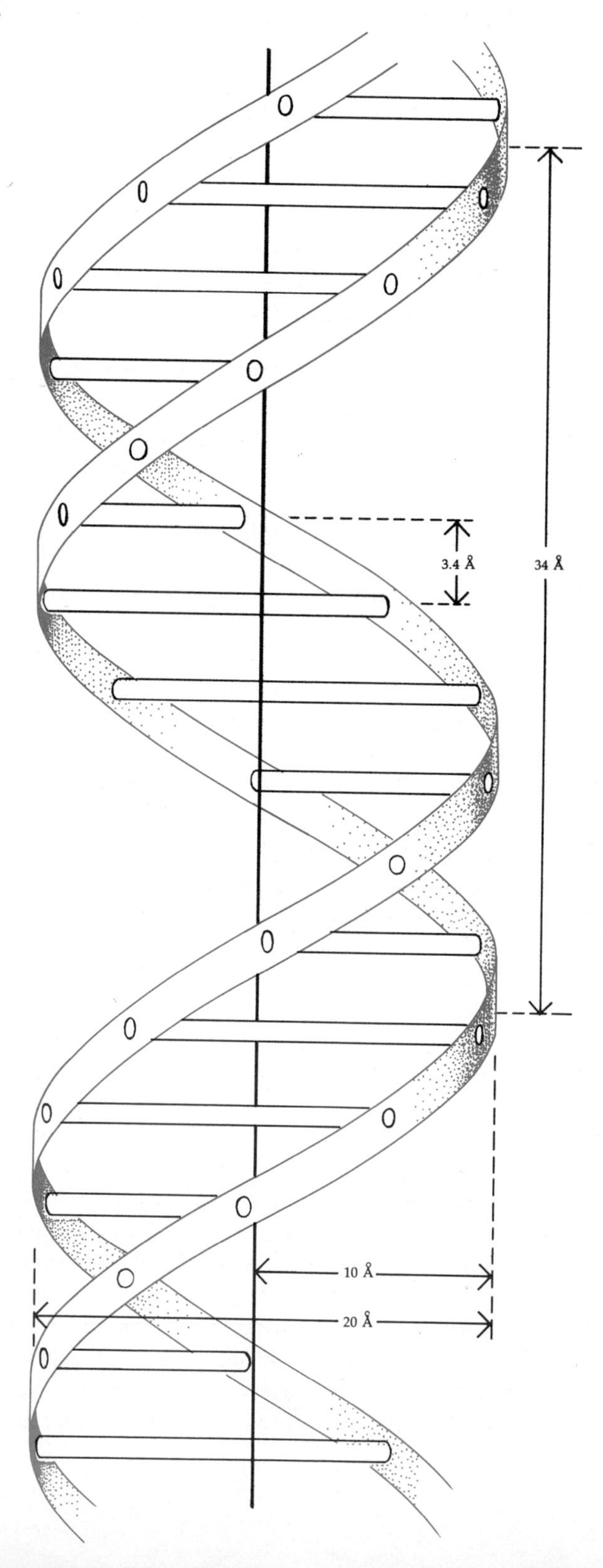

3–48
The double-stranded helix structure of DNA (left) *was first postulated in* 1953 *by Watson and Crick. The framework of the helix is composed of the sugar-phosphate units of the nucleotides. The rungs are formed by the four nitrogen bases —the purines, adenine and guanine, and the pyrimidines, thymine and cytosine. Each rung consists of two bases joined in the center by hydrogen bonds. Knowledge of the distances between the atoms, determined from x-ray diffraction pictures, was crucial in establishing the structure of the molecule. The photograph below shows Watson and Crick with one of their models of DNA.*

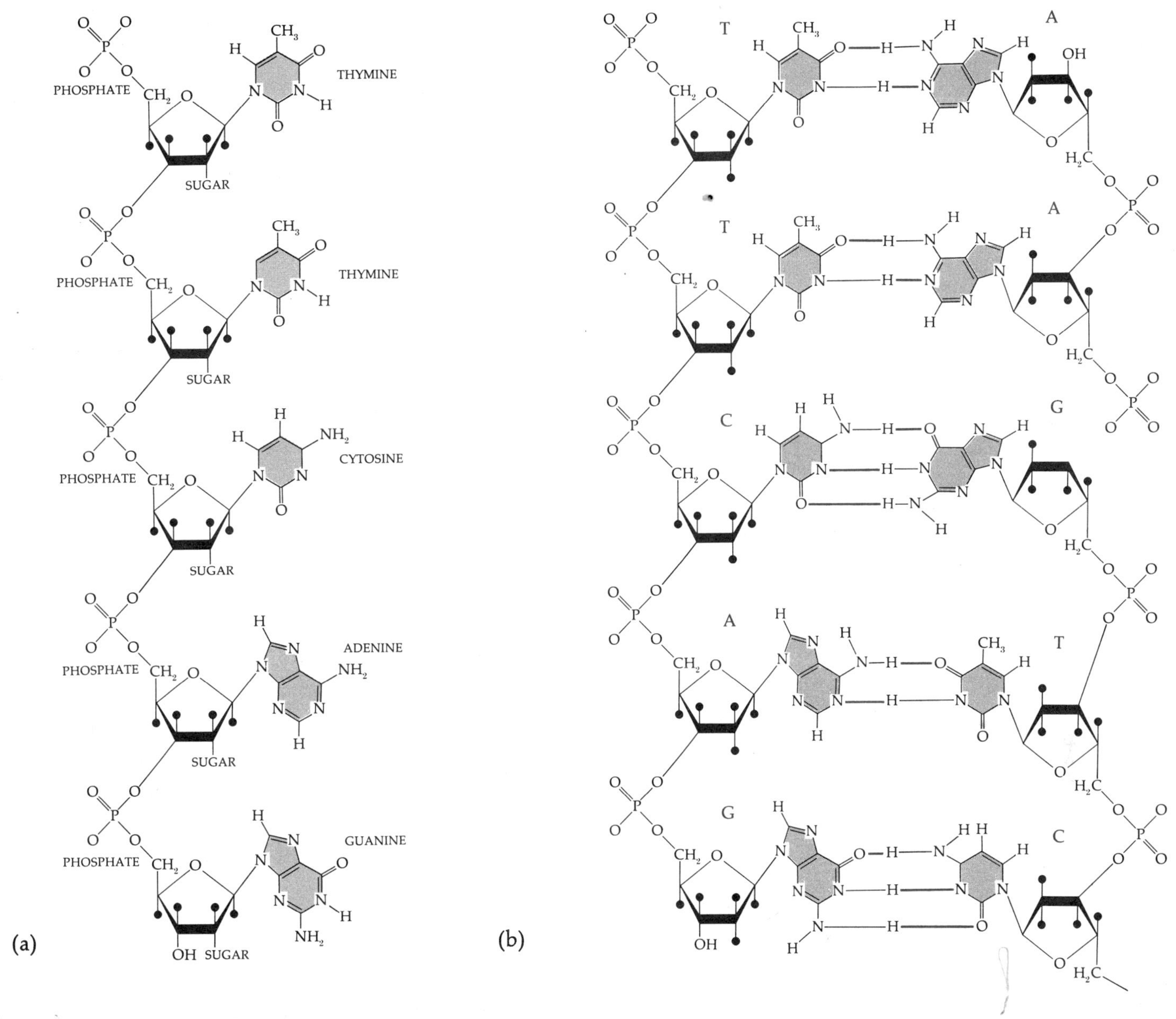

(a) (b)

3–49

(a) *The structure of a portion of one strand of a DNA molecule. Each nucleotide consists of a sugar, a phosphate group, and a purine or pyrimidine base. The sugar of each nucleotide is linked by a phosphate group to the sugar of an adjacent nucleotide. The sequence of nucleotides varies from one molecule to another. In the figure, the order of nucleotides is TTCAG.* (b) ***The double-stranded structure of a portion of the DNA molecule.*** *Do you see why paired combinations of two purines or two pyrimidines would be impossible in this configuration? And why adenine (A) can bond only with thymine (T), and guanine (G) only with cytosine (C)?*

DNA Replication

The most exciting discovery came, however, when they set out to construct the matching strand. They encountered an interesting and important restriction. Not only could purines not pair with purines and pyrimidines not pair with pyrimidines, but because of the configurations of the molecules, adenine could pair only with thymine and guanine only with cytosine. (Look at Table 3–3 again and see how well these physicochemical requirements confirm Chargaff's data.)

This fact about the structure of the DNA molecule immediately suggested the method by which it reproduces itself. At the time of chromosome replication, the molecule opens up, bit by bit, the bases breaking apart at the hydrogen bonds. The two strands separate, and new strands form along each old one, using the raw materials in the cell. If a T is present on the old strand, only an A can fit into place with the new strand; a G will pair only with a C, and so on. In this way, each strand forms a copy of its original partner strand, and two exact replicas of the molecule are produced. The age-old question of how hereditary information is duplicated and passed on for generation after generation had, in principle, been answered.

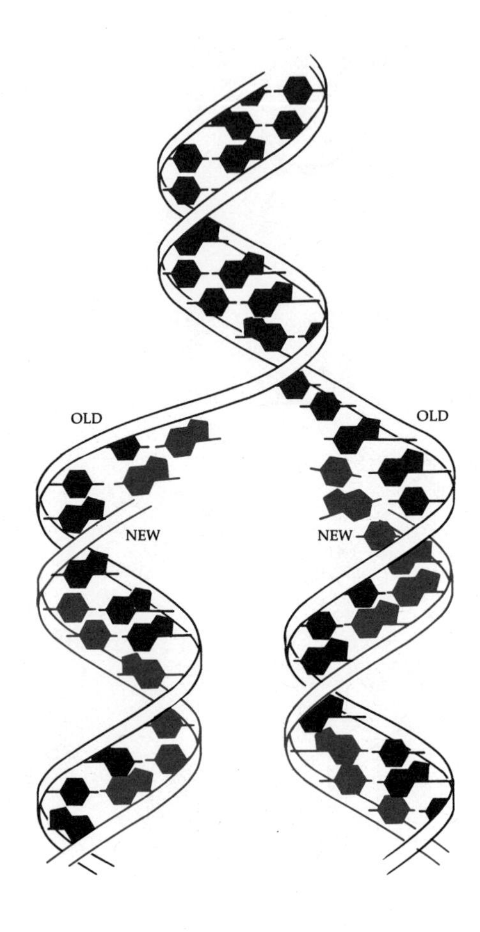

3–50
The DNA molecule shown here is in the process of reproduction, separating down the middle as its base units separate at the hydrogen bonds. (For clarity, the bases are shown out of plane.) Each of the original strands then serves as a template along which a new, complementary strand forms from nucleotides available in the cell.

DNA as the Carrier of Genetic Information

The Watson-Crick model also shows clearly how the DNA molecule is able to carry the genetic information. The information is coded in the sequence of the bases, and *any* sequence of the four pairs (AT, TA, CG, GC) is possible. Since the number of paired bases ranges from about 10,000 for the simplest known virus up to an estimated 10 billion in the 46 chromosomes of man, the possible variations are astronomical! The DNA from a human cell—which, if extended in a single thread would be about 5 feet long—contains genetic instructions that, if spelled out in English, would require some 600,000 printed pages averaging 500 words each, or a library of about a thousand books. Ob-

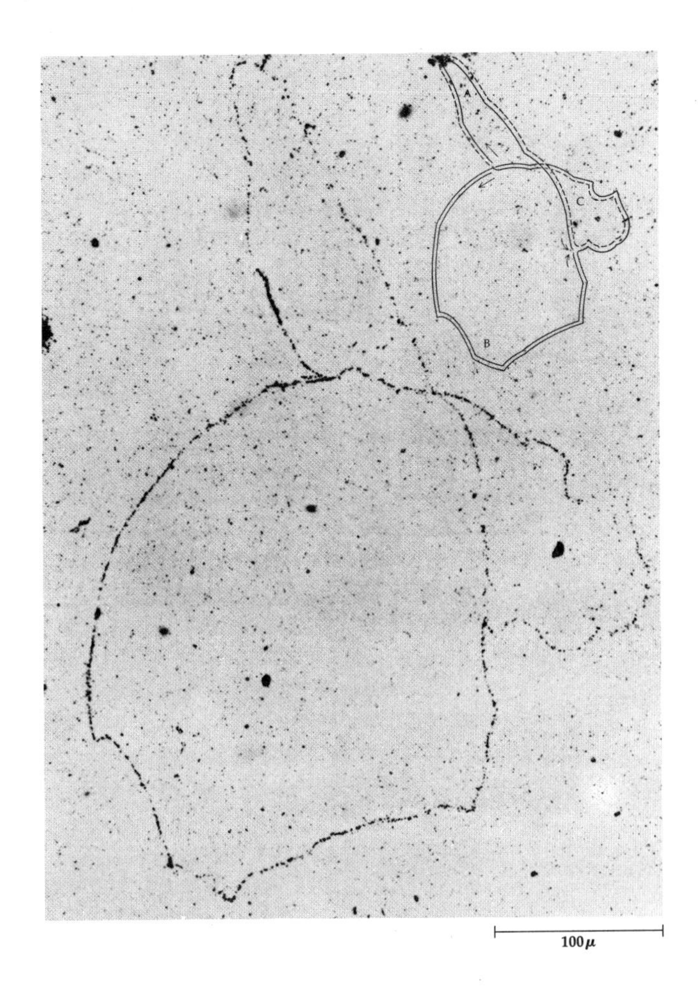

3–51

DNA replicating. The chromosomes of bacteria, unlike the chromosomes of the cells of higher organisms, are composed of only DNA. This DNA exists in the bacterial cell in the form of a circle—that is, of a single molecule with no end. In this electron micrograph, you can see the new DNA (A *and* C) *forming along the strands of the old* (B). *To make this micrograph, the* E. coli *cell was grown in a medium containing thymine which had been labeled with a radioactive isotope. When the cell was placed on a photographic plate, the emissions of radioactivity from the DNA left tracks in the photographic emulsion. This technique is known as radioautography.*

viously, the DNA structure can well explain the endless diversity among living things.

Once the structure of DNA was revealed, there was no longer any serious question as to its genetic role.

SUMMARY

DNA (deoxyribonucleic acid) is the genetic material of the cell. Investigations showing that the transforming factor in bacteria and the carrier of the genetic information in bacteriophages are both DNA provided some of the first evidence for this hypothesis.

Further support for the genetic role of DNA came from two more findings: (1) Almost all tissue cells of any given species contain equal amounts of DNA. (2) The proportions of nitrogen bases are the same in the DNA of all cells of a given species, but they vary in different species.

The structure of DNA was reported in 1953 by Watson and Crick. The DNA molecule, they found, is a double-stranded helix, shaped somewhat like a twisted ladder. The two sides of the ladder are composed of repeating groups of a phosphate and a five-carbon sugar. The "rungs" are made up of paired bases, one purine base pairing with one pyrimidine base. There are four bases—adenine (A), guanine (G), thymine (T), and cytosine (C)—and A can pair only with T, and G only with C. The four bases are the four "letters" used to spell out the genetic message. The paired bases are joined by hydrogen bonds.

The DNA molecule is self-replicating. The two strands come apart down the middle, breaking at the hydrogen bonds, and each strand forms a new complementary strand from nucleotides available in the cell.

On the basis of this structure, as revealed by Watson and Crick, the role of DNA as the carrier and transmitter of the genetic information became widely accepted.

QUESTIONS

1. One of the chief arguments for proteins being the genetic material is that proteins are heterogeneous in structure. Explain why the genetic material must have this property. What feature in the Watson-Crick DNA model was important in this respect?

2. The bread mold with which Beadle and Tatum worked is haploid. Would haploid organisms follow Mendel's rules? Explain. Why would this feature make it easier and faster to conduct genetic experiments?

3. What are the steps by which Griffith demonstrated the existence of the transforming principle? Can you think of any implications of Griffith's discovery for modern medicine?

4. Suppose you are talking to someone who has never heard of DNA. How would you support an argument that DNA is the genetic material? List at least five of the strong points in such an argument.

CHAPTER 3–7

Breaking the Code

Like most important scientific discoveries, the Watson-Crick model raised more questions than it answered.

It was now known that genes are made of DNA and that the products of genes are specific proteins. When it had been thought that genes *were* proteins, scientists had hypothesized that the "gene proteins" formed the models or the molds for the cellular proteins, the basic structural matter of life. During this period, scientists began talking about templates. Templates are patterns or guides, and the word "template" is usually associated with the metalwork patterns used in industry. By extension, the word came to be applied to a biological molecule which, by its shape, directs or molds the structure of another molecule. Each old strand of DNA, for example, serves as a template for the formation of its new partner strand during replication of the DNA molecule.

For a time after the structure of DNA was elucidated, scientists struggled with the problem of how DNA could also be a template for the formation of protein molecules. Several ingenious schemes were proposed, but it was simply not possible to get a satisfactory physicochemical "fit." The relationship between DNA and protein was apparently a more complicated one. If the proteins, with their 20 amino acids, were the "language of life," to extend the metaphor of the 1940s, the DNA molecule, with its four nitrogen bases, could be envisioned as a sort of code for this language. So the term "genetic code" came into being.

THE TRIPLET CODE

As it turned out, the idea of a "code of life" was useful not only as a dramatic metaphor but also as a working analogy. Scientists seeking to understand how the DNA so artfully stored in the nucleus could order the quite dissimilar structures of protein molecules approached the problem by the methods used by cryptographers in deciphering codes. There are 20 biologically important amino acids, and there are four different nucleotides. If each nucleotide "coded" one amino acid, only four could be provided for. If two nucleotides specified one amino acid, there could be a maximum number, using all possible arrangements, of 4^2, or 16—still not quite enough. Therefore, at least three nucleotides must specify each amino acid, following the code analogy. This would provide for 4^3, or 64, possible combinations. This postulate, the *triplet code*, was widely and immediately adopted as a working hypothesis, although it was not actually proved until a decade after the Watson-Crick discovery. Proof depended on answering yet another question: How is the code translated?

THE RNAs

One of the principal clues used in answering this question was that cells which are making large amounts of protein are rich in ribonucleic acid (RNA), a closely related "sister" molecule of DNA. RNA differs from DNA in a few respects. The sugar, or ribose, component of the molecule contains one more atom of oxygen (*deoxy* simply means "minus one oxygen"), and in place of thymine, RNA has another pyrimidine, uracil (U). (See Figure 3–52.) Perhaps more significant, since this part of the story has to do with the function of RNA, the RNA molecule is only rarely found in a fully double-stranded form. As a consequence, its properties and so its activities are quite different from those of DNA.

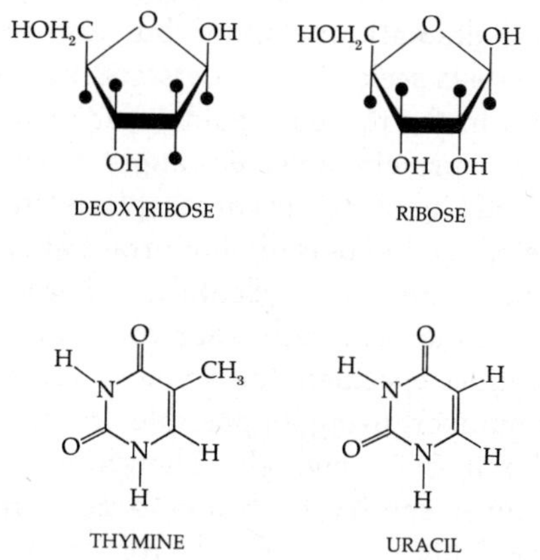

DEOXYRIBOSE RIBOSE

THYMINE URACIL

3–52
Chemically, RNA is very similar to DNA, but there are two differences in its chemical groups. One difference is in the sugar component; instead of deoxyribose, RNA contains ribose, which has an additional oxygen atom. The other difference is that instead of thymine, RNA contains the closely related pyrimidine uracil (U). (A third, and very important, difference between the two is that most RNA does not possess a regular helical structure and is usually single-stranded.)

The role of RNA molecules in "translating" the code was studied by breaking apart *E. coli* cells, separating their contents into various fractions, and seeing which fractions and (finally) which components of which fractions were essential for protein biosynthesis in a test tube. It was found that the machinery is complex, involving, among other things, three forms of RNA.

One chief requirement for protein synthesis, it was found, is the presence of ribosomes, a fact which had already been suspected since electron micrographs had shown that cells making large amounts of protein are rich in ribosomes. Ribosomes are made of protein and RNA, and this form of RNA came to be called *ribosomal RNA*. A second kind of RNA, functionally speaking, found to be involved in protein synthesis occurs in the form of long (from a few hundred to 10,000 nucleotides) single strands. This type, for reasons which will soon become clear, is known as *messenger RNA* (mRNA). The third type of RNA needed for protein synthesis is called *transfer RNA* (tRNA). These molecules are relatively short and partially coiled.

In addition, protein synthesis requires, as you would expect, amino acids, certain enzymes, and ATP. Now, with this rather large cast of characters in mind, let us look at what happens.

THE BIOSYNTHESIS OF PROTEINS

Protein biosynthesis begins when a strand of mRNA, with the help of certain enzymes, forms against a segment of one strand of the DNA helix. (Electron micrographs indicate that the DNA uncoils somewhat to permit the synthesis of RNA.) The mRNA forms along the DNA strand according to the same base-pairing rules as those that govern the formation of a DNA strand, except that in mRNA, uracil substitutes for thymine. Because of the copying mechanism, the mRNA strand when completed carries a faithful transcript of the DNA message.

The strand of mRNA then travels into the cytoplasm, probably passing through the pores in the nuclear enve-

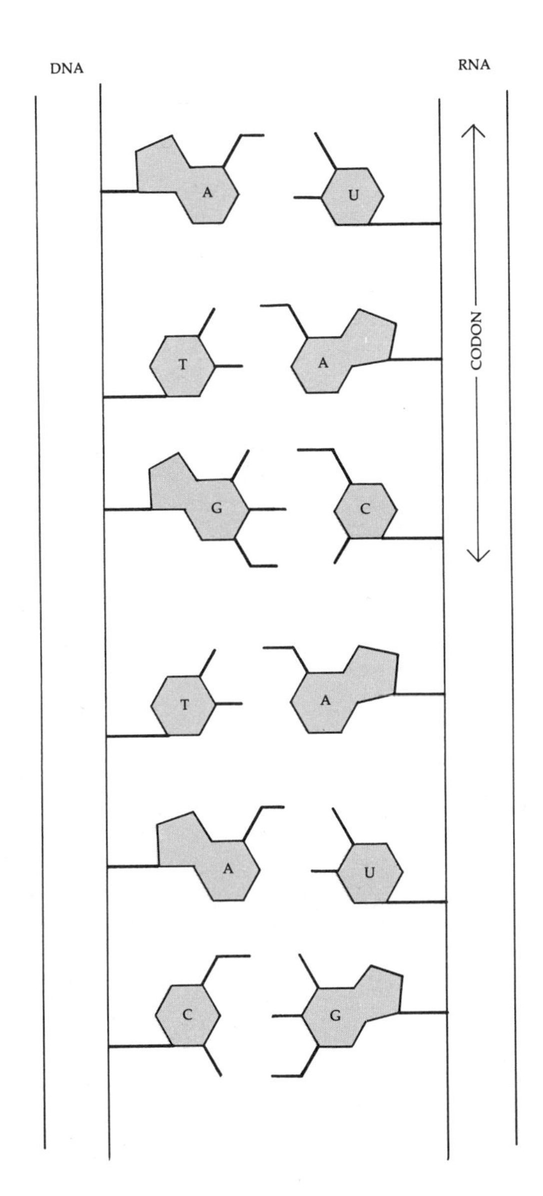

3–53
The beginning of the process of protein biosynthesis is the formation of mRNA on the DNA template. In the cell's nucleus, a strand of DNA which codes one sequence of amino acids for a protein forms a complementary strand of mRNA (instead of a partner DNA strand). The strand of mRNA is now a "negative print" of the sequence of nucleotides in the DNA.

lope. In the cytoplasm are the amino acids, special enzymes, ATP molecules, ribosomes, and molecules of tRNA. There are at least as many kinds of tRNA as there are kinds of amino acids, and all are formed from DNA, just as mRNA is. Each type of tRNA attaches by one end to a particular amino acid; these attachments each involve a special enzyme and a molecule of ATP.

Once in the cytoplasm, the mRNA molecule attaches to a ribosome. The way in which the molecule attaches is not known, but it is very likely that the process involves the ribosomal RNA.

At the point at which the mRNA molecule touches the ribosome, a molecule of tRNA, with its particular amino acid in tow, zeroes into position. Presumably, the tRNA molecule finds its proper place by means of a nucleotide triplet—sometimes called an anticodon—which pairs with the nucleotide triplet (the codon) on the mRNA molecule.

As the mRNA strand moves along the ribosome the next tRNA molecule with its amino acid moves into place. (See the schematic drawing on page 185.) At this point, the first tRNA molecule detaches itself from the mRNA molecule. The energy in the bond that holds the tRNA molecule to the amino acid is now utilized to forge the peptide link between the two amino acids, and the tRNA, released, becomes available once more. These tRNA molecules apparently can be used over and over again.

Messenger RNA appears to have a much briefer life, at least in *E. coli.* It is usually "read" by several ribosomes simultaneously, as shown in Figure 3–55b, thereby producing several polypeptide chains in a matter of seconds. Then it may be read by another group of ribosomes once or twice more, and after that it is destroyed by enzymes. The average lifetime of an mRNA molecule in *E. coli* is two minutes, although its lifetime in other types of cells may be considerably longer. This means that in *E. coli,* continuous production of a protein demands continuous production of the appropriate mRNA molecules. In this way, the bacterial chromosome maintains a very rigid control over the cellular activities.

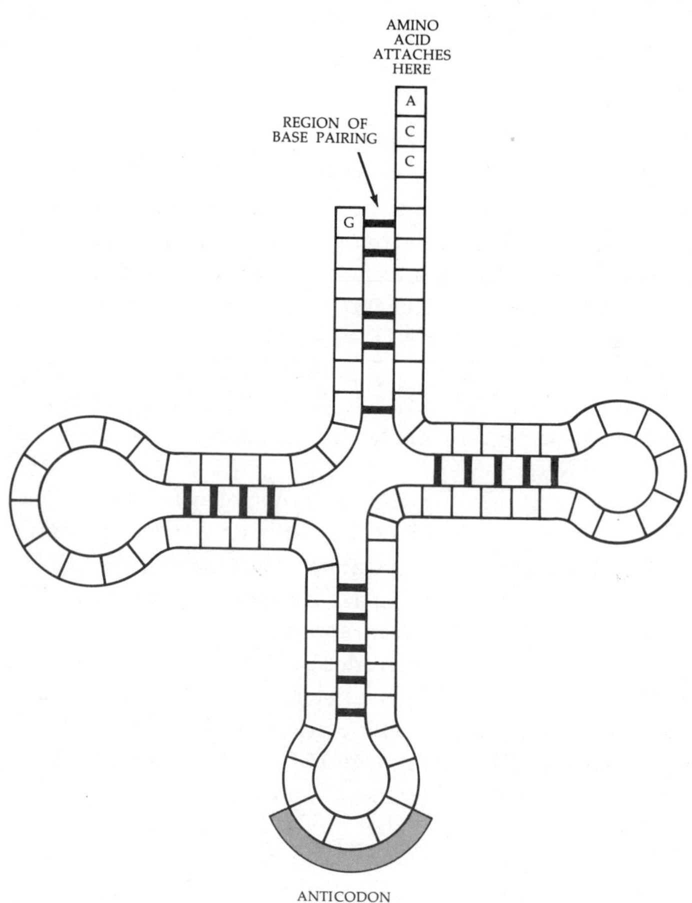

3–54

Structure of a tRNA molecule. These molecules consist of about 80 nucleotides linked together in a single chain. One end of the chain always terminates in a guanine nucleotide, and the other in a CCA sequence. The amino acid is linked to the tRNA at the CCA end. The other nucleotides vary according to the particular RNA. tRNA molecules all appear to have the configuration shown here; in some, however, there is an extra "arm." The "cloverleaf" is, in addition, probably folded in some way. Some of the bases are hydrogen-bonded to one another, following the DNA-type base pairing (A with U, G with C). The unpaired bases at the bottom of the diagram serve as the anticodon and "plug in" the molecule to an mRNA codon.

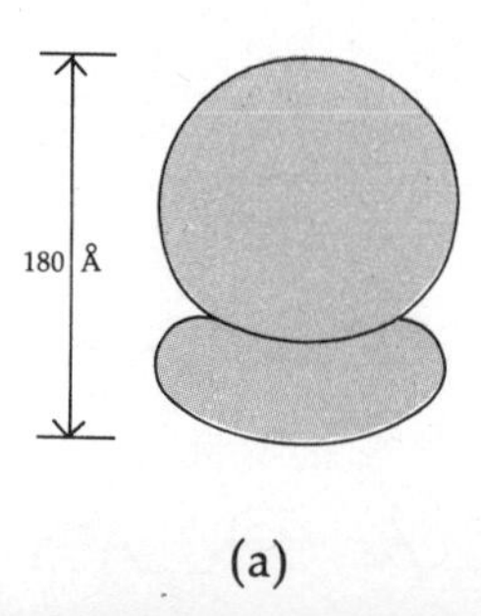

(a)

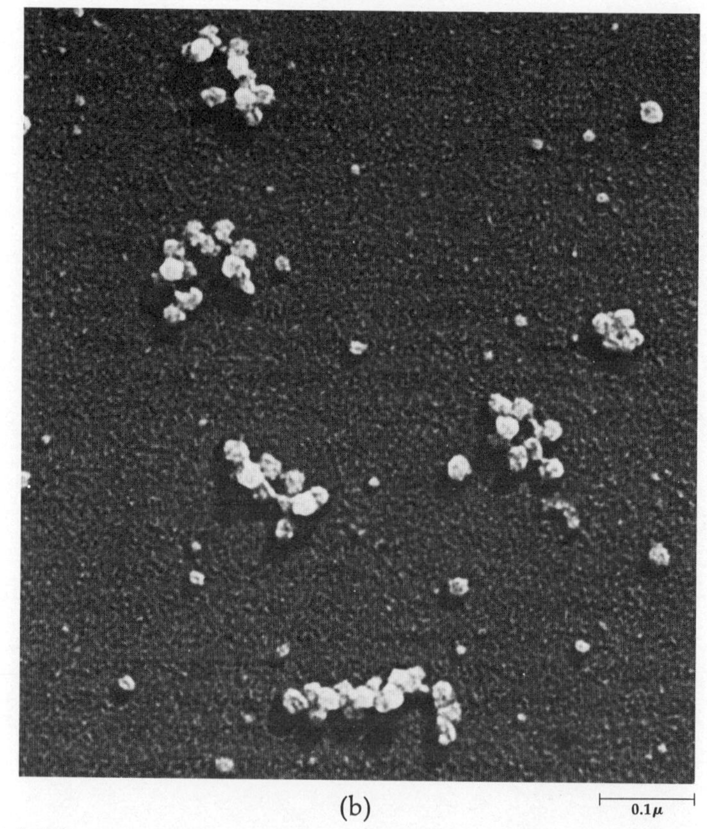

(b)

3–55

(a) *Diagram of a ribosome. As you can see, it consists of two roughly spherical subunits, one slightly larger than the other. Ribosomal RNA is produced on genes of the nucleolus.* (b) *Groups of ribosomes, or polysomes. Each polysome is a group of ribosomes "reading" the same mRNA strand. These polysomes are from yeast cells.*

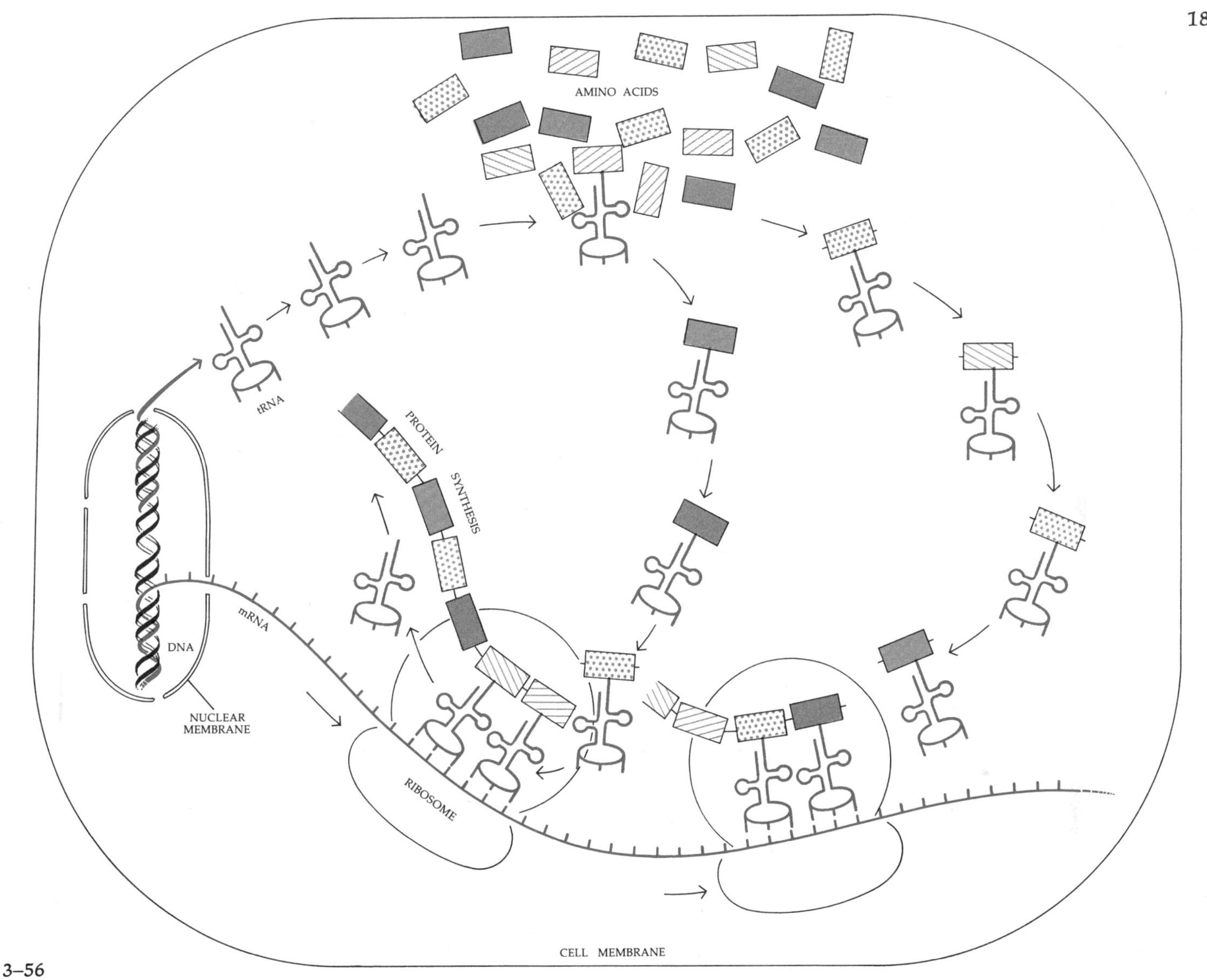

3–56

How a protein is made. At least 20 different kinds of tRNA molecules are formed on the DNA in the nucleus of the cell. These molecules are so structured that each can be attached (by a special enzyme) at one end to a specific amino acid. Each carries somewhere in the molecule an anticodon which fits only the mRNA codon for that particular amino acid. The process of protein biosynthesis begins when an mRNA strand is formed on the DNA template in the nucleus and travels to the cytoplasm. The strand attaches to a ribosome, and at the point of attachment, the matching tRNA molecule, with its amino acid, plugs in momentarily to the codon in the mRNA. As the mRNA moves along the ribosome, another tRNA molecule fits into place against the next codon and the first molecule is released, leaving behind its amino acid. As the process continues, the amino acids are brought into line one by one, following the exact order laid down by the DNA code, and are formed into a protein chain, which may be anywhere from 50 to hundreds of amino acids long.

BREAKING THE CODE

The existence of mRNA was postulated in 1961 by the French scientists François Jacob and Jacques Monod. Almost immediately, Marshall Nirenberg of the United States Public Health Service set out to test the mRNA hypothesis. He added several crude extracts of RNA from a variety of cell sources to extracts of *E. coli*—that is, material which contained amino acids, ribosomes, ATP, and tRNA extracted from *E. coli* cells—and found that they all stimulated protein synthesis in the *E. coli* extract. In other words, the *E. coli* material started producing protein molecules even when the RNA "orders" it received were from a "complete stranger." The "code" seemed to be a universal language.

Perhaps if *E. coli* could read a foreign message and translate it into protein, Nirenberg reasoned, it could read a totally synthetic message, one dictated by the scientists themselves. A man-made RNA was available; Severo Ochoa of New York University had developed a method for linking together ribonucleotides into a long strand of RNA. The trouble with the method, from Nirenberg's point of view, was that there was no way to control the order in which an assortment of ribonucleotides would be assembled. For Nirenberg's purposes, the order was of the utmost importance. He wanted to know the exact contents of any message that he dictated.

A simple solution for this seemingly perplexing problem suddenly presented itself: use an RNA molecule which consisted of only one ribonucleotide repeated over and over again. Nirenberg and his associate, Heinrich Matthaei, selected the ribonucleotide containing uracil. They then prepared 20 different test tubes, each containing cellular extracts of *E. coli* with ribosomes, tRNA, ATP, the necessary enzymes, and the 20 amino acids. In each test tube, one of the amino acids, and only one, carried a radioactive label. The synthetic poly-U, as it was called, was added to each test tube. In 19 of the test tubes, nothing detectable occurred, but in the twentieth one, the one in which the radioactive amino acid was phenylalanine, the investigators

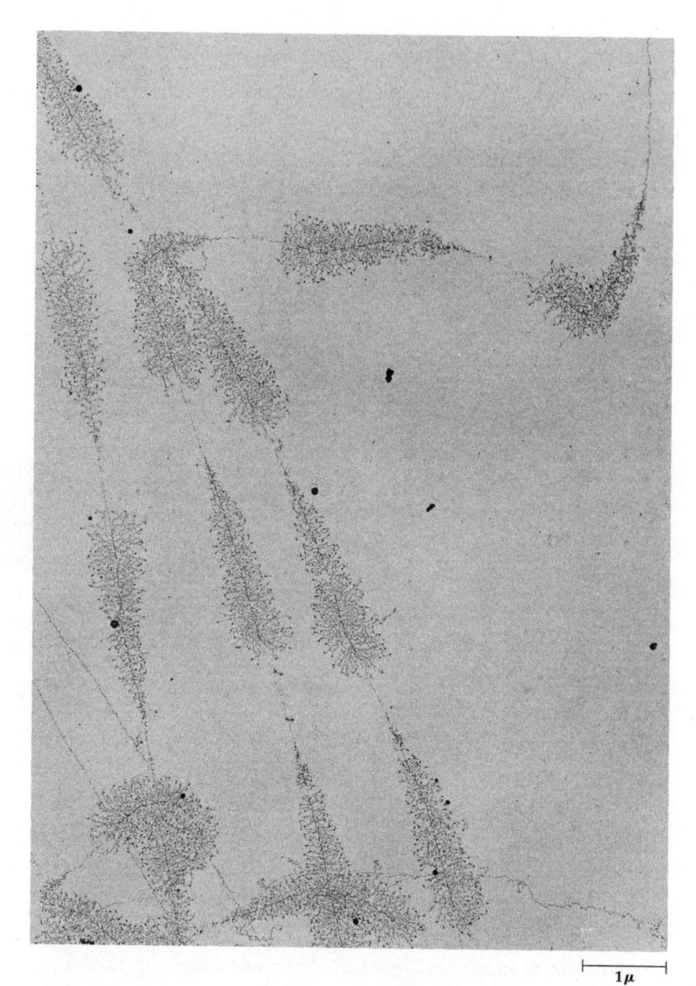

3–57

Genes in action. This electron micrograph shows a section of DNA from the nucleolus of an amphibian egg cell. The fine fibrils are RNA which has formed along the DNA strand. The segment between the active genes is probably made up of genes that were "turned off" at the time the micrograph was taken. By examining the lengths of the RNA molecules, you can tell the direction in which the transcription is proceeding.

were able to detect newly formed, radioactive polypeptide chains. When the polypeptide was analyzed, it was found to consist only of phenylalanines, one after another. Nirenberg and Matthaei had dictated the message "uracil-uracil-uracil-uracil-uracil-uracil-uracil-uracil-uracil . . . ," and a clear answer had come back, "phenylalanine-phenylalanine-phenylalanine. . . ."

Within the year following Nirenberg's discovery, which was first reported in 1961, tentative codes were worked out by Nirenberg and Ochoa and their many coworkers for all the amino acids, using synthetic mRNA. A synthetic polynucleotide made up entirely of adenine (poly-A), for instance, makes a peptide chain composed entirely of lysine. When methods were worked out for controlling the order of the nucleotides in the synthetic RNA, it was possible to determine the rest of the codons. (See Figure 3–58.) All but three trinucleotides have now been identified—61 of the 64 possible combinations. These three are now considered to be punctuation marks, signifying the beginning or end of a particular message.

Since 61 combinations code for 20 amino acids, you can see that there are a number of "synonyms" among the codons. Characteristically, these synonyms almost always differ only in the third nucleotide, leading to the speculation that the first two may be sufficient to hold the tRNA in most instances.

Some of the biological implications of these findings are strikingly clear. Consider mutations, for example. Mutations involve changes in the nucleotides of the DNA. These changes, once they occur, are faithfully replicated and passed on from cell to cell, generation after generation. Let us take another look at sickle-cell anemia in the light of Figure 3–58. Normal hemoglobin contains glutamic acid; sickle-cell hemoglobin contains valine. GAA or GAG specifies glutamic acid; GUA or GUG specifies valine. So the difference between the two hemoglobins lies in two replacements of adenine by uracil in a DNA molecule that, since it dictates a protein which contains more than 600 amino acids, must contain more than 1,800 bases. In other words, the tremendous functional difference between the two hemoglobins can be traced to two "misprints" in over 1,800 nucleotides.

SECOND LETTER

FIRST LETTER	U		C		A		G		THIRD LETTER
U	UUU	PHE	UCU		UAU	TYR	UGU	CYS	U
	UUC		UCC	SER	UAC		UGC		C
	UUA	LEU	UCA		UAA	?	UGA	?	A
	UUG		UCG		UAG	?	UGG	TRP	G
C	CUU		CCU		CAU	HIS	CGU		U
	CUC	LEU	CCC	PRO	CAC		CGC	ARG	C
	CUA		CCA		CAA	GLN	CGA		A
	CUG		CCG		CAG		CGG		G
A	AUU		ACU		AAU	ASN	AGU	SER	U
	AUC	ILE	ACC	THR	AAC		AGC		C
	AUA		ACA		AAA	LYS	AGA	ARG	A
	AUG	MET	ACG		AAG		AGG		G
G	GUU		GCU		GAU	ASP	GGU		U
	GUC	VAL	GCC	ALA	GAC		GGC	GLY	C
	GUA		GCA		GAA	GLU	GGA		A
	GUG		GCG		GAG		GGG		G

3–58

The genetic code, consisting of 64 triplet combinations and their corresponding amino acids. Only three combinations—UAA, UAG, and UGA—have not been found to correspond with an amino acid. These three codons are probably the "punctuation marks" (i.e., stop or start signals) of the code. Since 61 triplets code 20 amino acids, there are "synonyms." Most of the synonyms have a common characteristic; can you see what it is? What do you think might be the significance of this characteristic? ("The Genetic Code: III", F. H. C. Crick.

SUMMARY

Genetic information is coded in the molecules of DNA, and these, in turn, determine the sequence of amino acids in molecules of protein. One gene contains the information needed to specify the complete sequence of one polypeptide chain.

The way in which the gene directs the production of a protein, according to current theory, is as follows: Each series of three nucleotides along a DNA strand is the DNA codon for a particular amino acid. The information is transferred from the DNA by means of a long single strand of RNA (ribonucleic acid). This type of RNA molecule is known as messenger RNA, or mRNA. The mRNA forms along one of the strands of DNA, following the principles of base pairing first suggested by Watson and Crick, and therefore is complementary to it.

The mRNA strand leaves the cell's nucleus and attaches to a ribosome. At the point where the strand of mRNA is in contact with the ribosome, small molecules of RNA, known as transfer RNA (tRNA), which serve as adapters between the mRNA and the amino acids, are bound temporarily to the mRNA strand. This bonding is believed to take place by the same base-pairing principle as that which holds together the two strands of the double helix of DNA. Each tRNA molecule carries the specific amino acid called for by the mRNA codon into which the tRNA plugs. Thus, following the sequence dictated by the DNA, the amino acid units are brought into line one by one and are formed into a polypeptide chain.

Final proof of this hypothesis came when the DNA/RNA code was "broken," that is, when investigators were able to predict what protein would be formed from a given series of nucleotides. Today, 61 of the 64 possible triplet combinations of the four-letter DNA code have each been identified with one of the 20 amino acids that make up protein molecules. The other three triplets, it is believed, serve as "punctuation marks" during protein synthesis.

QUESTIONS

1. Explain the term "genetic code." In what ways is it a useful analogy?

2. What is the minimum number of different types of tRNA molecules needed? Why?

3. In a hypothetical segment of DNA, the sequence of bases is AAGTTTGGTTACTTG. What would be the sequence of bases in an mRNA strand transcribed from this DNA segment? What would be the amino acids coded by the mRNA? Does it matter where the mRNA starts reading? Show why.

4. Certain amino acids are called "essential" because they cannot be made by animal organisms but must be taken in in the diet. A shortage of a single type of essential amino acid can create widespread deficiencies in the organism. On the basis of the information in this chapter, explain why.

CHAPTER 3–8

Work in Progress

HOW GENES ARE REGULATED

One of the most crucial questions remaining to be answered in genetics is how gene function is regulated in the course of development. Human life, for example, begins with one cell, the fertilized egg. This cell divides, and the daughter cells divide, and the process of division continues until eventually there is a baby—composed of some 2 billion cells. But these cells are not all similar to one another or to the original egg cell. They are, in fact, extremely varied in both their form and their function. For example, egg cells do not make hemoglobin, nor do skin or liver cells, but red blood cells hardly produce anything else.

How can an organism's cells be different from the original egg cell and from one another? There are two logical possibilities. One is that the different cells contain different genetic material, and the other is that the genetic information is the same in all the cells but is variously expressed. One of the first definitive experiments to decide between these two possibilities was performed by Hans Spemann in 1914. Using a baby's hair, the finest ligature he could find, Spemann constricted the egg cell of a newt before its first division. The half of the egg that contained the nucleus began to divide normally, whereas the other half, which had no nucleus, did not divide. In organisms, during early cell division, the nuclei condense and get somewhat smaller. For this reason, Spemann was able to slip a single nucleus past the constriction into the undeveloped half of the egg. (See Figure 3–59.) He then tightened the baby-hair noose until the undeveloped half of the egg was completely separated from the part that had been dividing. In many cases, the previously undeveloped half then began to divide and subsequently developed into a separate, whole embryo. What Spemann had succeeded in proving was that, at least early in development, each one of the nuclei in the cells of the developing organism contain all the genetic information present in the original fertilized egg.

More dramatic proof of the hypothesis that every cell of an organism contains all the information needed by the entire organism came in the 1960s, when F. C. Steward demonstrated that it was possible to produce an entire carrot plant from a single adult carrot cell grown in a suitable medium in tissue culture. Shortly thereafter, J. B. Gurdon of the University of Oxford performed a somewhat similar experiment with frogs, which we described in Chapter 3–1. He removed the nuclei from the cells of tadpoles and implanted them in enucleated egg cells. In many cases, the eggs developed into normal adults (Figure 3–4).

If all cells have the same genetic makeup but if different cells produce different proteins and at different times, clearly there must be some way to regulate gene action—to turn the genes off and on.

The Operon

Work with the much studied *E. coli* has provided insight into the way in which at least some genes are regulated. The chromosome of *E. coli*, it is now known, is a long single molecule of DNA, which takes the form of a circle—that is, it has no end. The molecule is large enough to code for between 2,000 and 4,000 different polypeptide chains, yet its

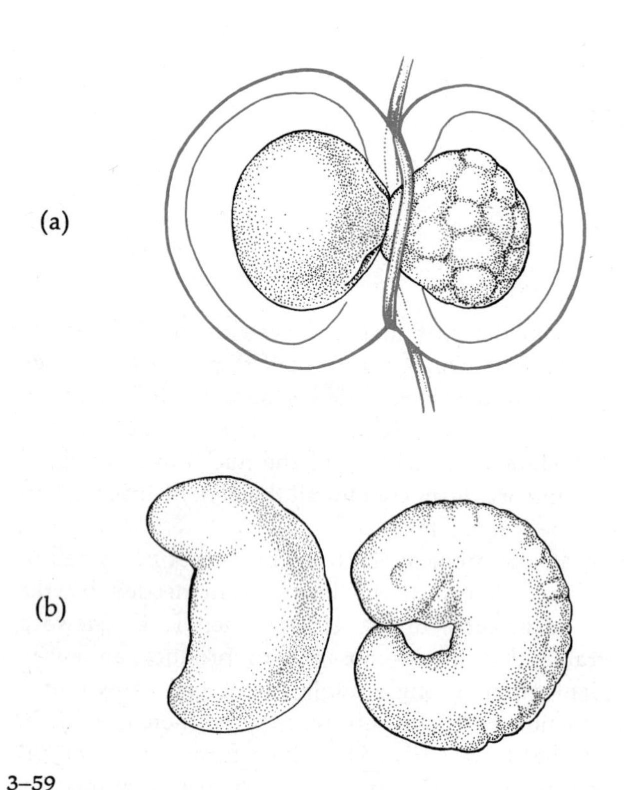

3–59

Spemann's experiment with an amphibian's egg. (a) *Before the first cell division, the egg was constricted across the middle, with the nucleus in the right half. At the stage shown here, the right half had begun to divide normally. One of the nuclei from this half was then allowed to escape through the "bottleneck" into the underdeveloped left half, which presently began to divide.* (b) *Twin embryos developed from the two halves. As you can see, development of the embryo on the left (from the left half of the egg) was delayed, but the embryo was nevertheless normal.*

production of any particular polypeptide seems to be carefully regulated by its needs. For example, *E. coli* that is growing on the disaccharide lactose needs the enzyme beta-galactosidase to split the disaccharide into two monosaccharides, glucose and galactose. When lactose is present, approximately 3,000 molecules of beta-galactosidase are present in every normal *E. coli* cell. This represents about 3 percent of all the protein in the cell. In the absence of lactose, however, it is rare that one detects a single molecule of the enzyme.

Studies of galactosidase have shown that it appears only when needed and that, when needed, it is produced from amino acids, following the production of new mRNA.

Obviously, it is of great advantage to a cell to produce a particular enzyme only when it is needed, and in the required amounts. Mutants of *E. coli* have been found which synthesize almost 15 percent of their protein as beta-galactosidase, even in the absence of lactose, but these are at a selective disadvantage since they are using their energies uneconomically, and they tend to be replaced by the wild-type strains.

How is it possible for a bacterial cell to turn its genes on and off? One way in which gene production is controlled was discovered by François Jacob and Jacques Monod, who introduced the concept of the *operon*. An operon is a group of related genes all aligned along a single segment of DNA. The operon consists of three different types of genes: the *promoter*, which is the site at which formation of the mRNA begins; the *operator*, which is the site of regulation; and one or more *structural genes*, which code for enzymes. In the beta-galactosidase system, the operon includes the gene that codes for beta-galactosidase and two other genes, which also code for enzymes involved in lactose metabolism. These genes are adjacent to one another and are transcribed consecutively, one after the other.

The activity of the operon is controlled by yet another gene, the *regulator*, which is not necessarily adjacent to the operon. The regulator codes for a protein, known as the *repressor*, which apparently binds to the DNA at the site

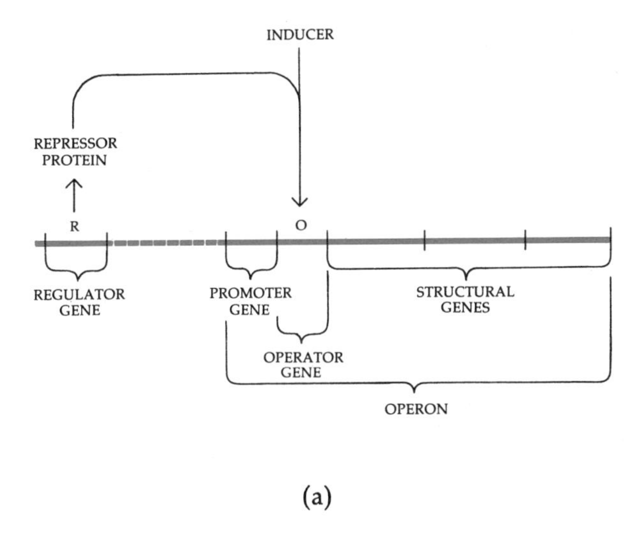

(a)

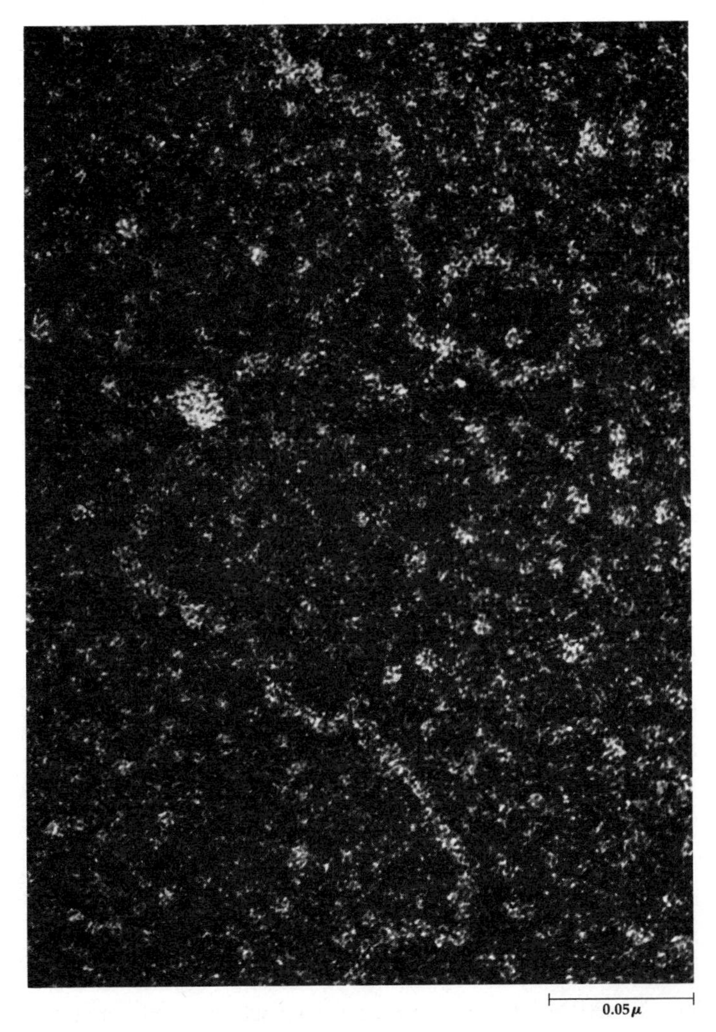

(b)

3–60

(a) *The operon. An operon is a group of genes forming a functional unit. There are usually several structural genes, which code for different enzymes that work sequentially in a particular enzyme sequence. The transcription can begin only at the site marked "promoter." The regulator gene codes for a protein that represses transcription of mRNA from the structural genes. The regulator, which need not be adjacent to the operon on the chromosome, directs production of a repressor protein. This repressor acts upon the operator gene, apparently by binding to the DNA at this site and blocking the formation of mRNA. In order to start transcription again, another compound, the inducer, is needed. The inducer counteracts the effects of the repressor, probably by binding to it and changing its shape. The repressor then vacates the operator site, so that the synthesis of mRNA can proceed.* (b) *A repressor protein (the white spherical form in the upper half of the micrograph) attached to the lactose operon.*

of the operator gene. Since the operator gene is located between the promoter and the structural genes (Figure 3–60), this blocks the production of mRNA.

The regulator is, in turn, controlled by a "signal" compound, the inducer. When this compound is present in the system, it counteracts the effects of the repressor, permitting the production of mRNA to resume. In the beta-galactosidase system, when lactose is present, the inducer (a molecule derived from the lactose) binds with the repressor molecule and changes its shape so that it can no longer attach itself to the DNA. When the supply of lactose is exhausted, the regulator once again assumes control and mRNA production ceases.

Gene Regulation in Higher Cells

The operon, which regulates gene activity in bacterial cells, is not generally believed to be a means of gene regulation in the cells of higher organisms. More than one kind of mechanism is probably involved in the control of gene expression in higher cells, but what these mechanisms are and how they function are still open, and very crucial, questions.

Of particular interest in these investigations is a group of proteins known as histones, which bind tightly to the DNA molecule. (In plant and animal cells, the DNA is always associated with proteins.) Recently, a number of investigators have shown that the rate of RNA synthesis by isolated cell nuclei can be correlated with the amount of histone present. Addition of histones results in a decrease of RNA production, and destruction of histones (by specific enzymes) results in an increase of RNA production. On the basis of these studies, it seems possible that mRNA can be transcribed only if the DNA is not blanketed by histone molecules.

Vincent Allfrey of the Rockefeller University, who is one of the leaders in this research, tells the story, however, about a little boy in a department store who asked his father what made the escalators run. "The people walking

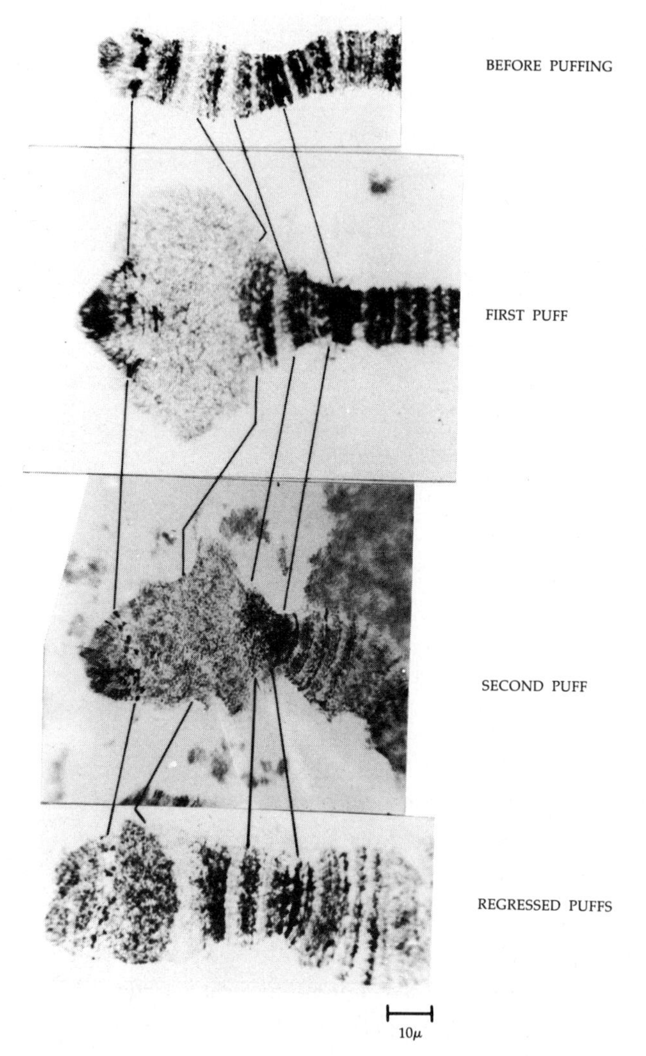

3–61

Observations of chromosomes puffs support the concept that the DNA is somehow unwound to make it available for mRNA transcription. These puffs were observed in chromosomes of the Brazilian gnat which, like the fruit fly, has giant chromosomes in some of its cells. The gnat had previously been treated with a hormone that causes molting and, as the micrographs indicate, the puffs occurred sequentially along one chromosome.

up and down them," his father explained. Then he took his son back that night, and they peered through the glass doors of the locked building. There was the proof: there were no people in the store, and the escalators were not moving.

Despite this word of caution, the histone hypothesis provides an attractive and logical explanation of how some genes—such as the gene that codes for hemoglobin—can be turned off in all except one particular type of cell.

In the many-celled plants and animals, it may well be that genes which are turned off and on are controlled by an inducer-repressor system and those which are turned off for the normal life of the organism—such as a hemoglobin gene in an intestinal cell—are coated with histones or other protein molecules.

NATURE AND NURTURE

The "nature versus nurture" problem, which we touched upon in Chapter 3–3, recurs constantly in new forms. One group of current controversial questions concerns behavior.

In animals, some patterns of behavior are clearly inherited. A spider, for instance, although raised in solitary confinement, can build an extremely complex web on its first try. Among vertebrates, many species of fish and birds have elaborate courtship and territory-defending rituals whose basic patterns are part of their genetic equipment—that is, instinctive rather than learned. Mammals also show inherited behavior. For example, the first time a flying squirrel raised in a bare laboratory cage isolated from other squirrels is given nuts or nutlike objects, it will "bury" them in the bare floor, making scratching movements as if to dig out the earth, pushing the nuts down into the "hole" with its nose, covering them over with imaginary earth, and stamping on them.

Is there such a thing as inherited behavior among human beings? Do we have instinctive likes and dislikes, instinctive fears? It has recently become popular to think of mankind as instinctively aggressive and to use this fact to "explain" the persistence of war and other types of human brutality. Certainly, aggressiveness is inherited in some mammalian species. Wild rats are much more aggressive in their behavior than the white rats used in laboratories, which are generally docile and often affectionate toward their keepers. The offspring of wild rats, even though raised in a laboratory, are also aggressive. (These wild animals have adrenal glands about twice as large as those of their gentle cousins, which appears to provide a simple physiological basis for their behavior.)

There have been a number of recent attempts, by both scientists and popular authors, to elucidate human behavior by comparing it with that of other animals. Depending on which group of animals is chosen for comparison, apparently almost any point can be "proved." Of greater importance, however, is the fact that it is now clear that we never see man's "nature"—if by "nature" we mean the genotype. All the behavioral characteristics of man are the results of a unique interaction involving man's environment and personal experience and his biological makeup.

Another human trait that has become the subject of a nature/nurture controversy in recent years is intelligence. One problem is that it is difficult to generate a scientifically useful definition of intelligence. As a measurable quantity, intelligence can only be defined, essentially, as the ability to do well on intelligence tests, which necessarily reflect our society's conception of intelligence. As a consequence, the concept of intelligence, operationally speaking, has come to involve a rather limited set of socially useful skills, and standard intelligence tests seek primarily those skills alone.

Although there is evidence that intelligence has, to some extent, a genetic basis,* the effects of diet, perception, environmental stimulation, motivation, health, and numerous other factors are so massive that any possible genetic com-

* The genetic basis of intelligence is implied by the results of studies of identical twins. Such twins, developing from the same fertilized egg, have identical genetic material. Identical twins separated at birth and growing up in different homes usually have IQ ratings within 12 points of each other.

ponent is largely obscured. At present, geneticists are reluctant to say more than that intelligence is probably a complex mixture of factors, some inborn, some environmental in nature.

CLONING

A clone is a group of organisms which have been produced from a single parent by a series of mitotic divisions and which are (with the exception of possible mutations) genetically identical. Clones are found in nature only among one-celled organisms and in a few invertebrates and plants which reproduce asexually. However, the recent discovery that it is possible to produce an entire carrot from the single cell of an adult carrot or a frog using the nucleus from the cell of a tadpole points to the possibility of producing strains of identical organisms, say domestic plants and animals, of proven and standard quality. It has even been suggested—although not, so far, by scientists in the field—that in this way a family could "reincarnate" a loved one simply by removing a cell from his body, a society could reproduce its leaders in politics, sciences, and the arts, and a government could produce large numbers of people of a proven useful type.

A way in which the cloning of human cells might be more practically applied is in the production of organs for transplant. There are two main obstacles to the widespread use of transplanted organs to replace diseased ones. One is the lack of suitable replacement organs, which need to be healthy, relatively young, and undamaged. The other, and at present more serious, obstacle is that an animal body reacts more or less strongly against any cells that are not genetically identical to its own cells. For this reason, tissues can be exchanged readily only between identical twins. Currently, transplants are done whenever possible between genetically similar persons, often members of the immediate family. Also, drugs are given to the patient to reduce his reactions to the foreign tissue, but since these drugs reduce his reactions to *all* foreign matter, they greatly increase the dangers of infection. One possible way to overcome these two obstacles might be the production of new organs from the patient's own cells; such organs, clearly, would be acceptable to his body. Needless to say, this possibility is a long way from materializing.

TRANSFERS OF GENETIC INFORMATION

The development of drug resistance in bacteria has long been considered an excellent example of drastic evolutionary changes taking place under man-made pressures. In every large bacterial population, there are a few cells which, as a result of mutation, are resistant to a drug such as streptomycin, for example. If the bacterial population is exposed to streptomycin—as when a patient is being treated with the drug—the susceptible cells are destroyed and only the resistant ones remain. These multiply and produce a population of bacterial cells made up entirely of resistant individuals (which is why most doctors have long frowned upon the indiscriminate use of antibiotics).

In 1959, a group of Japanese scientists made a discovery which revealed an entirely new way in which bacteria can become drug resistant—a way much more rapid and efficient (from the point of view of the bacteria) than that previously known. The process, which is called infectious drug resistance, involves the transfer of genes from resistant to nonresistant cells. These genes, or resistance factors, as they are called, are in the cell in the form of isolated units of genetic material. Since the resistance factors are not actually a part of the bacterial chromosome, they are readily transferable. Under experimental conditions, 100 percent of a population of sensitive cells can become resistant within an hour after being mixed with suitable resistant bacteria. This process, which was, of course, hinted at by Griffith's experiments of nearly half a century ago, was first discovered in a strain of shigella, a bacterium that causes dysentery. Further studies showed that not only can shigella transfer drug resistance to other shigella but also the innocuous *E. coli* can transfer resistance factors to this and other, unrelated groups of bacteria. Infectious resistance is now found among an increasing number of types of bacteria, including those that cause typhoid fever, gastroenteritis, plague, and undulant fever. Fortunately, however, for human populations, the majority of common infectious bacteria do not seem to possess this capacity for infectious drug resistance.

This new knowledge makes it even clearer that antibacterial drugs should be used carefully. (In England, for example, it is no longer legal to give them to healthy livestock with their feed, but almost half of the antibiotics produced in this country are used in this way.) With constant exposure to such antibiotics, the everpresent harmless bacteria will undoubtedly all carry resistance factors, which they may pass on to disease-causing bacteria, which may then infect the livestock—or us.

GENES, VIRUSES, AND CANCER

Cancer is a disease, or group of diseases, in which particular cells in the body cease to respond to whatever controls growth under normal conditions and multiply autonomously—crowding out, invading, and destroying other tissues. There have been many theories of cancer cause. For a number of years, the most widely held theory was that cancer is the result of a mutation in particular cells. Suppose, for example, that there is some sort of regulator gene that controls mitosis. A mutation that rendered such a gene nonfunctional could result in the sort of wild growth characteristic of cancer. This hypothesis was supported by the fact that many agents that cause cancer—x-rays, ultraviolet radiation, and certain chemicals—also cause mutations.

More recently, accumulating information about viruses has led many scientists to believe that viruses are the cause of at least some of the forms of cancer in man. One of the reasons for believing this is the fact that certain types of viruses—the temperate viruses are an example—have been found which enter cells and become part of the cellular chromosome, being replicated with it. Once they are within the cell, they no longer have their characteristic protein coat and so can no longer be detected by usual means. Such viruses can change the genetic characteristic of the cell. They can be carried in the cell indefinitely before they are released to set up a new cycle of infection. Their release can be triggered by outside agents, such as x-ray or ultraviolet light, just as mutations can. More and more biologists are coming to believe that some cancers are caused by such viruses.

An even stronger line of evidence linking viruses and cancer is the proof that many different kinds of cancer in animals are caused by viruses, including cancers in frogs and chickens and in mice, rats, and other mammals. These viruses characteristically have long latent periods between the time of initial infection and the appearance of the disease.

One of the difficulties with the virus theory of cancer has been that many cancer viruses contain RNA instead of DNA. DNA viruses replicate themselves and produce mRNA, from which viral proteins are made. RNA viruses replicate themselves, presumably in a manner analogous to DNA replication, and also serve as mRNA for viral protein production. But RNA cannot serve as a genetic factor within a cell in the way that DNA can. So the theory that cancer viruses act by becoming part of the genetic apparatus of the cell did not seem reconcilable with the existence of cancer-causing RNA viruses. Then, in 1970, a group of virologists discovered that some RNA viruses make DNA! When these viruses infect a cell, they produce, on the cell's ribosomes, a new enzyme, never found in uninfected cells, that makes DNA on the template of the viral RNA. Moreover, most of the cancer-producing RNA viruses tested so far have been found to produce this new enzyme and to make DNA, whereas only one RNA virus not known to cause cancer has been found that makes the enzyme.

So the science of genetics seems to be inching closer to an understanding of one of man's oldest and ugliest enemies.

TRANSDUCTION

When certain types of bacterial viruses infect cells, one of two events may occur. The virus DNA may enter the cell and set up an infection such as we described in Chapter 3–6, or the DNA of the virus may simply lie latent in the cell. Viruses which may follow this second course of action are known as temperate viruses. Temperate viruses were originally detected because from time to time they can become active, set up an infective cycle, multiply, and invade and destroy other bacterial cells. Exposure to x-rays, ultraviolet light, or certain chemicals greatly increases the possibility of a temperate virus becoming active.

When a temperate virus infects a bacterial cell, it may either (a) set up an active infection or (b) remain latent in the cell, in which case, it may become part of the chromosome, replicating with it.

When the DNA of a temperate virus is in the bacterial cell, it has been found, it often becomes a part of the bacterial chromosome. The DNA of a particular temperate virus usually has a particular position on the chromosome of the cell that it infects. When a temperate virus becomes activated, it breaks loose from the chromosome and then sets up its infective cycle. Sometimes, in the course of breaking away, the viruses carry some of the bacterial DNA with them, and when they set up a temperate relationship with a new host cell, they insert the new DNA segment into the bacterial chromosome of this cell. In this way, genes can be transferred from one cell to another. This phenomenon is known as transduction.

The fact that genetic information is carried from cell to cell by viruses under certain natural conditions leads some scientists to speculate that viruses might someday be used to carry specific genes into particular human cells in which the genes were lacking or faulty.

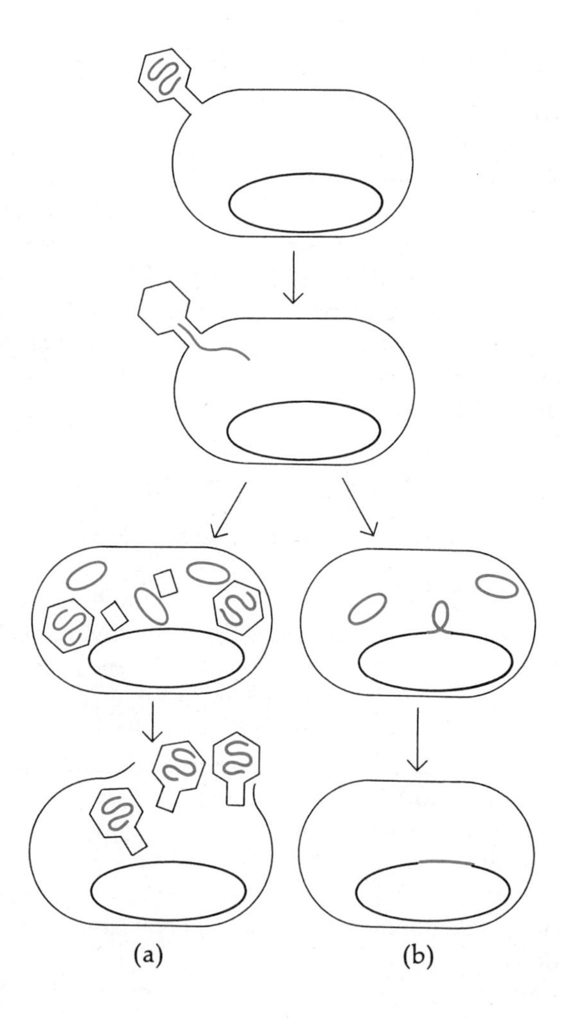

GENETIC ENGINEERING

The discovery that genetic information can be transferred from cell to cell in the form of transforming factors, viruses, or resistance factors has quickened speculations about the possibility of someday inserting selected genes into human cells and so curing genetic diseases such as sickle-cell anemia, PKU, and diabetes.

With biology approaching the day when it may be possible to interfere with human genetic characteristics, almost all scientists would agree that genetics—which began in that quiet monastery garden little more than a century ago—is now no longer the concern of scientists alone but presents problems which are the concern and responsibility of all people.

SUMMARY

Every cell of an organism carries the total genetic information for the entire organism. Since different cells produce different proteins, the use of this genetic information must be regulated in some way.

One way in which genes are "turned off," at least in bacterial cells, involves the production of a repressor substance by a gene known as the regulator. This repressor substance acts on the operator gene, which in turn controls the structural genes, which code for the enzymes. The operator gene, its structural genes, and the promoter (the site at which formation of the mRNA begins) are collectively known as an operon.

Histones are proteins which bind to the DNA molecule. They may be involved in gene regulation in higher organisms.

An organism is the product of an interplay of genetic constitution and environment. The extent to which inherited characteristics contribute to behavioral traits and intelligence is a matter of controversy.

It has been shown that new carrots can be produced from a single cell of an adult carrot and that a frog will develop from the nucleus of a somatic cell inserted into an egg cell in which the original nucleus has been destroyed. This procedure, called cloning, suggests the possibility of producing domestic plants and animals—and perhaps even human organs—of known genetic constitution.

Genetic material can be exchanged between cells. One such form of exchange, infectious drug resistance, involves the transfer of genes involved in the resistance of bacteria to drugs even among unrelated strains of bacteria. This discovery emphasizes the need for control in the use of antibiotics.

Genetic material also may be added to cells in the form of viruses. Some cancers may be caused by the presence of such viruses in cells.

The discovery of ways in which genes can be added to cells quickens hopes—and fears—that the day may be reached when genetic material can be inserted into the cells of human beings.

QUESTIONS

1. List all evidence given in this and earlier chapters that indicates that every cell in the body possesses all the genetic information that was present in the fertilized egg.

2. Why do you think more or different regulatory mechanisms for gene action might be required in multicellular organisms than are required in *E. coli?*

3. Suppose it were possible to produce a genetically identical person from the single cell of one recently dead. To what extent do you think this new person would resemble his "parent"?

GLOSSARY

ADENINE (**add**-eh-neen): A purine base present in DNA, RNA, and nucleotides such as ADP and ATP.

ALLELE *or* ALLELOMORPH, **aleel**; a**leel**-o-morf) [Gk. *allelon*, of one another, + *morphē*, form]: One of the two or more alternative states of a gene that occupy the same position (locus) on homologous chromosomes. Alleles are separated from each other at meiosis.

ALTERNATION OF GENERATIONS: A reproductive cycle in which a haploid ($1n$) phase, the gametophyte, gives rise to gametes; then, after fusion to form a zygote, the gametes germinate to produce a diploid ($2n$) phase, the sporophyte; and finally, spores produced by meiotic division from the sporophyte give rise to new gametophytes, completing the cycle.

ANAPHASE (**anna**-phase) [Gk. *ana*, up, + *phasis*, form]: A stage in mitosis or meiosis in which the chromatids of each chromosome separate and move to opposite poles.

ANTICODON: Three adjacent nucleotides on a tRNA molecule which complement a particular codon on an mRNA strand. The matching of the codon and anticodon ensures the correct positioning of an amino acid in a particular polypeptide.

ASEXUAL REPRODUCTION: Any reproductive process, such as fission or budding, that does not involve the union of gametes.

BACTERIOPHAGE (back-**teer**-ee-o-fage) [Gk. *bakterion*, little rod, + *phagein*, to eat]: A virus that parasitizes a bacterial cell.

BASE-PAIRING RULE: The requirement that adenine must always pair with thymine (or uracil) and guanine with cytosine in the formation of a nucleic acid double helix or in the production of RNA from DNA.

BIVALENT (bye-**vay**-lent) [L. *bis*, twice, + *valens*, having power]: A pair of synapsed chromosomes.

CELL PLATE: A flattened structure that forms at the equator of the spindle in the dividing cells of land plants and a few green algae during early telophase; the predecessor of the middle lamella.

CENTRIOLE (**sen**-tree-ole) [Gk. *kentron*, center]: A cytoplasmic organelle generally found in animal cells and flagellated cells in other groups, usually outside of the nuclear membrane. It doubles before mitosis, and the two centrioles then move apart and organize the spindle apparatus.

CENTROMERE (**sen**-tro-mere) [Gk. *kentron*, center, + *meros*, a part]: That portion of the chromosome to which the spindle fiber is attached; also called the kinetochore.

CHIASMA, *pl.* CHIASMATA (kye-**az**-ma) [Gk. *chiasma*, a cross]: The X-shaped figure formed by the meeting of two nonsister chromatids of homologous chromosomes; the site of crossing over.

CHROMATID (**crow**-ma-tidd): One of the two daughter strands of a duplicated chromosome; the strands are joined by a single centromere.

CHROMATIN (**crow**-ma-tin): The deeply staining nucleoprotein complex of the chromosomes.

CHROMOSOME [Gk. *chrōma*, color, + *soma*, body]: The body in the cell nucleus containing genes in a linear order; threads or rods of chromatin, which appear during mitosis and meiosis and which bear the genes.

CHROMOSOME MAP: A plan showing the relative positions of the genes on the chromosome; determined chiefly by analysis of the relative frequency of crossing over between any two genes.

CLONE (kloan) [Gk. *klon*, twig]: A line of cells all of which have arisen from the same single cell by mitotic division; a population of individuals descended by asexual reproduction from a single ancestor.

CODON (**code**-on): Three adjacent nucleotides on a molecule of mRNA that form the code for a single amino acid.

CROSSOVER: The exchange of corresponding segments of genetic material between chromatids of homologous chromosomes at meiosis.

CYTOSINE (**sight**-o-seen): A pyrimidine base present in DNA and RNA.

DEOXYRIBONUCLEIC ACID (DNA) (dee-**ox**-y-rye-bo-new-**klee**-ick): The carrier of genetic information in cells, composed of two chains of phosphate, sugar molecules (deoxyribose), and purines and pyrimidines wound in a double helix; capable of self-replication as well as of determining RNA synthesis.

DIPLOID: Having two sets of chromosomes; the $2n$ number characteristic of the somatic cells of animals and of sporophyte generation of plants.

DOMINANT GENE: A gene that exerts its full phenotypic effect regardless of its allelic partner; a gene that masks the effect of its allele.

F_1: The first filial generation in a cross between any two parents; F_2 and F_3 are the second and third generations.

GAMETE (gam-**meet**) [Gk. *gametē*, wife, and *gametēs*, husband]: The mature functional haploid reproductive cell. When its nucleus fuses with that of another gamete of opposite sex (fertilization), the resulting diploid cell (zygote) develops into a new individual.

GAMETOPHYTE (gam-**meet**-o-fight): In plants having alternation of generations, the haploid ($1n$), gamete-producing phase.

GENE: A unit of heredity which is transmitted in the chromosome and which by interaction with internal and external environment controls the development of a trait; capable of self-replication. The sequence of nucleotides in a DNA molecule that dictates the amino acid sequence of a particular protein.

GENOTYPE (**jean**-o-type): The genetic constitution, latent or expressed, of an organism, as contrasted with the phenotype; the sum total of all the genes present in an individual.

GUANINE (**gwa**-neen): A purine base present in DNA and RNA.

HAPLOID [Gk. *haploos*, single]: Having only one of each type of chromosome ($1n$), in contrast to diploid ($2n$); characteristic of gametes and of the gametophyte generation in plants.

HETEROZYGOUS: Having different alleles at the same locus on homologous chromosomes.

HOMOLOGUES: Chromosomes that associate in pairs in the first stage of meiosis; each member of the pair is derived from a different parent.

HOMOZYGOUS: Having identical alleles at the same locus on homologous chromosomes.

INCOMPLETE DOMINANCE: A heterozygous condition in which the expression of one allele is not completely masked by the other, resulting in a phenotype which is intermediate between the two possible homozygous phenotypes.

INDEPENDENT ASSORTMENT: *See* Mendel's second law.

INFECTIOUS RESISTANCE: Drug resistance developed in populations of bacteria through the transfer of genes from resistant to nonresistant cells.

INTERPHASE: The stage between two mitotic or meiotic cycles.

KARYOTYPE [Gk. *karyon*, nut]: The general appearance of the chromosomes with regard to number, size, and shape.

LINKAGE: The tendency for certain genes to be inherited together owing to the fact that they are located on the same chromosome.

LOCUS, *pl.* LOCI [L. *locus*, place]: The position of a gene on a chromosome.

LYSOGENIC BACTERIA (lye-so-**jenn**-ick) [Gk. *lysis*, a loosening]: Bacteria carrying viruses (bacteriophages) which eventually break loose from the bacterial chromosome and set up an active cycle of infection, producing lysis in their bacterial hosts.

MEIOSIS (my-**o**-sis) [Gk, *meioun*, to make smaller]: The two successive nuclear divisions in which the chromosome number is reduced from diploid ($2n$) to haploid ($1n$) and segregation and reassortment of the genes occurs; gametes or spores may be produced as a result of meiosis.

MENDEL'S FIRST LAW: The factors for a pair of alternate characters are separate, and only one may be carried in a particular gamete (genetic segregation).

MENDEL'S SECOND LAW: The inheritance of one pair of characteristics is independent of the simultaneous inheritance of other traits, such characters "assorting independently" as though there were no other characters present (later modified by the discovery of linkage).

MESSENGER RNA (mRNA): The RNA that carries genetic information from the gene to the ribosome, where it determines the order of the amino acids in the formation of a polypeptide.

METAPHASE [Gk. *meta,* middle, + *phasis,* form]: The stage of mitosis or meiosis during which the chromosomes lie in the central plane of the spindle.

MITOSIS (my-**toe**-sis) [Gk. *mitos,* thread]: A cellular process during which the chromosomes divide longitudinally and the daughter chromosomes then separate to form two genetically identical daughter nuclei; usually followed by division of the cytoplasm.

MUTAGEN (**mute**-a-jen) [L. *mutare,* to change, + Gk. *genēs,* born]: An agent which increases the mutation rate.

MUTANT: A mutated gene or an organism carrying a gene that has undergone a mutation.

MUTATION: An inheritable change of a gene from one allelic form to another.

NATURAL SELECTION: The differential reproduction of genotypes, a process in nature by which the environment eliminates less well-adapted members of a population, leading to the continuation of one group of organisms or traits and the elimination of another.

NONDISJUNCTION: The failure of homologous chromosomes to separate during meiosis, resulting in one or more extra chromosomes in the cells of some offspring.

NUCLEOLUS (new-**klee**-o-lus) [L. *nucleolus,* a small kernel]: A spherical body, composed chiefly of RNA and protein, present in the nucleus of eukaryotic cells; site of production of ribosomes.

OPERATOR GENE: The gene in an operon which is affected by the repressor protein produced by the regulator gene for that particular operon and which, in turn, inhibits the activity of the other adjacent genes (the structural genes) in the operon.

OPERON: A group of adjacent genes whose functions are related to a particular synthetic activity and which are controlled by the repression or activation of a specific gene (the operator gene) in the group.

PHENOTYPE (**fee**-no-type) [Gk. *phainein,* to show]: The physical appearance of an organism resulting from the interaction of its genetic constitution (genotype) and the environment.

PLEIOTROPISM (plee-**o**-tro-pism): The capacity of a gene to affect a number of different characteristics.

POLYGENIC INHERITANCE: The determination of a given characteristic, such as weight or height, by the complex interaction of many genes.

PROMOTER: A specific site at the beginning of an operon at which a polymerase can start its synthesis.

PROPHASE [Gk. *pro,* before, + *phasis,* form]: An early stage in nuclear division, characterized by the shortening and thickening of the chromosomes and their movement to the equator of the spindle.

PUNNETT SQUARE: The checkerboard diagram used for analysis of gene distributions.

PURINE (**pure**-een): A nitrogenous base, such as adenine or guanine, with a double-ring structure; one of the components of nucleic acids.

PYRIMIDINE: A nitrogenous base, such as cytosine, thymine, or uracil, with a single-ring structure; one of the components of nucleic acids.

RECESSIVE GENE [L. *recedere,* to recede]: A gene whose phenotypic expression is masked by a dominant allele and so is manifest only in the homozygous condition. Heterozygotes involving recessives are phenotypically indistinguishable from dominant homozygotes.

REGULATOR GENE: A gene whose product prevents or represses the activity of the operator gene in an operon, which in turn represses the activity of the structural genes in that operon.

REPRESSOR: The substance produced by a regulator gene that represses protein formation.

RIBONUCLEIC ACID (RNA) (rye-bo-new-**klee**-ick): A nucleic acid formed on chromosomal DNA and involved in protein synthesis; similar in composition to DNA, except that the pyrimidine uracil replaces thymine. RNA is the genetic material of many viruses.

SEGREGATION: The separation of the chromosomes (and genes) from different parents at meiosis.

SEX-LINKED CHARACTERISTIC: A genetic characteristic, such as color blindness, determined by a gene located either on the *X* or *Y* chromosome.

SEXUAL REPRODUCTION: The fusion of gametes, followed by meiosis and recombination at some point in the life cycle.

SPERM [Gk. *sperma*, seed] A mature male sex cell, or gamete, usually motile and smaller than the female gamete.

SPINDLE FIBERS: A group of microtubules that extend from the centromeres of the chromosomes to the poles of the spindle or from pole to pole in a dividing cell.

SPORE: An asexual reproductive cell, usually unicellular, capable of developing into an adult without fusion with another cell, in contrast to a gamete.

SPOROPHYTE: The spore-producing, diploid ($2n$) phase in the life cycle of a plant having alternation of generations.

STRUCTURAL GENE: One of the genes in an operon whose function is controlled by the operator and regulator genes.

SYNAPSIS (sin-**ap**-sis): The pairing of homologous chromosomes that occurs prior to the first meiotic division; crossing over occurs during synapsis.

SYNGAMY (**sin**-gamy): The union of gametes in sexual reproduction.

TELOPHASE [Gk. *telos*, end, + *phasis*, form]: The last stage in mitosis and meiosis, during which the chromosomes become reorganized into two new nuclei.

TEMPERATE PHAGE: A bacterial virus that may remain latent in its host bacterial cell. In this latent (prophage) stage, it is associated with the bacterial chromosome and is replicated with it.

TEMPLATE: A pattern or mold guiding the formation of a negative or complement. DNA replication is explained in terms of a template hypothesis.

THYMINE (**thye**-meen): A pyrimidine present in DNA but not in RNA.

TRANSDUCTION: The transfer of genetic material (DNA) from one bacterium to another by a temperate phage.

TRANSFER RNA (tRNA): The type of RNA that becomes attached to an amino acid and guides it to the correct position on the ribosome-mRNA complex for protein synthesis. There is at least one tRNA molecule for each amino acid. Each tRNA molecule is only about 80 nucleotides in length.

TRANSLOCATION: In genetics, the exchange of chromosome segments between nonhomologous chromosomes.

TRIPLET CODE: The three-symbol system of base-pair sequences in DNA; referred to as a code because it determines the amino acid sequence in the enzymes and other protein components synthesized by the organism.

URACIL (**you**-ra-sill): A pyrimidine present in RNA but not in DNA.

ZYGOTE (**zye**-goat) [Gk. *zygōtos*, joined together]: The cell resulting from the fusion of male and female gametes, usually diploid ($2n$); in higher organisms, the fertilized egg.

SUGGESTIONS FOR FURTHER READING

BEADLE, GEORGE, and MURIEL BEADLE: *The Language of Life*, Doubleday and Company, Inc., Garden City, N.Y., 1966.*

A popular, highly readable book about the chemistry of the gene.

CRICK, F.: *Of Molecules and Men*, University of Washington Press, Seattle, 1967.*

This is a short and extremely interesting account of the DNA-RNA-protein synthesis problem and its unraveling.

DUBOS, RENE: *So Human an Animal*, Charles Scribner's Sons, New York, 1968.*

In this mature, thoughtful book, which won the Pulitzer Prize in 1969, Dr. Dubos, a well-known microbiologist, analyzes the interactions of genetic constitution, environment, and the power to make choices that produce that "unique, unprecedented, unrepeatable" creature, the human being.

DUPRAW, E. J.: *DNA and Chromosomes*, Holt, Rinehart and Winston, Inc., New York, 1970.

An excellent synthesis of information at the cellular and molecular level about chromosomes and the organisms that possess them.

LEVINE, R. P.: *Genetics*, Holt, Rinehart and Winston, Inc., New York, 1969.*

A short, careful exposition of the principles of genetics and the analysis of genetic data, well written and illustrated.

PETERS, JAMES A. (ed.): *Classic Papers in Genetics*, Prentice-Hall, Inc., Englewood Cliffs, N.J.*

Includes papers by most of the scientists responsible for the important developments in genetics: Mendel, Sutton, Morgan, Beadle and Tatum, Watson and Crick, Benzer, etc. You should find this book very interesting; the authors are surprisingly readable, and the papers give a feeling of immediacy that no other account can achieve.

SRB, A., R. OWEN, and R. EDGAR: *General Genetics*, 2nd ed., W. H. Freeman and Company, San Francisco, 1965.

One of the best full-length introductory genetics texts. It is clear, well illustrated, and well balanced between molecular genetics and other aspects of the subject.

STENT, GUNTHER S.: *Molecular Genetics: An Introductory Narrative*, W. H. Freeman and Company, San Francisco, 1971.

A well-written, authoritative account of the genetics of bacteria and their viruses.

WATSON, J. D.: *The Double Helix*, Atheneum Publishers, New York, 1968.

"Making out" in molecular biology. A lively book about how to become a Nobel Laureate.

WATSON, J. D.: *Molecular Biology of the Gene*, 2nd ed., W. A. Benjamin, Inc., Menlo Park, Calif., 1970.*

For the student who wants to go more deeply into the questions of molecular biology, this is a detailed and authoritative account.

* Available in paperback.

PART II

Organisms

SECTION 4

The World of Living Things

4–1
Young black snakes hatching. With the evolution of an egg in which the embryo could develop on land, some groups of vertebrates were able to become fully terrestrial.

SECTION 4

The World of Living Things

In this part of the book we are going to be concerned with organisms, the living things that populate the world around us. The chief focus will be on the higher plants, particularly the flowering plants, and on the higher animals, particularly man. These are the subjects of Sections 5 and 6, respectively. In this first section, we are going to introduce these organisms by examining the modern representatives of some of their predecessors, and, as far as we are able, tracing their evolutionary history.

In Section 8, we shall have more to say about the processes of evolution which have shaped the course of this history. Before we begin, however, we should stress the intimate relationship between an organism and the environment in which it lives. Changes in the environment—in the sources of food or water, in the availability of living space, or in predator-prey relationships—lead, by natural selection, to the production of changes, or adaptations, in populations of organisms. Or, to put it another way, adaptations are solutions to specific environmental problems, and plant and animal physiology, the subject of the next two sections of this book, is the study of these solutions.

It is impossible, therefore, to talk about organisms without reference to the major forces, both past and present, that brought them into being and shaped their existences. So in this introductory section, we shall deal with what Yale ecologist G. E. Hutchinson, who was an ecologist long before ecology became fashionable, aptly termed "the ecological theater and the evolutionary play," introducing some of the leading performers in this drama (in which each one of us, whether he likes it or not, is an actor) and, in broad strokes, sketching in the plot.

Remember, too, that evolution always works with the material at hand; it does not create anything wholly new, in the way that an inventor might devise an entirely new piece of machinery, such as a flying machine, to meet a specific need. Flight for birds was preceded by a long period during which the bone structure that had once served to strengthen the fins of a lumbering lungfish was slowly modified. Furthermore—and this is the hard part, the one that places the most stringent limitations on evolution—every change that took place in the gradual path from fin to wing was, in itself, useful to the animal, or the animal and the change would probably not have survived.

Julian Huxley* put it this way:

> Natural selection has certain obvious limitations. It can only produce results which are of immediate biological utility to the species; and being blind and automatic, it is incapable of purposeful design or foresighted planning. In consequence, *its results will always be relative to the particular environment in which the particular species is living.*

It is because of this need for "immediate biological utility" that evolution is conservative in nature—retaining, for example, the 9-plus-2 cilium through millions of years and millions of changes. All modern organisms bear what Darwin called "the indelible stamp" of their origins.

* In *Evolution as a Process*, G. Allen, London, 1954.

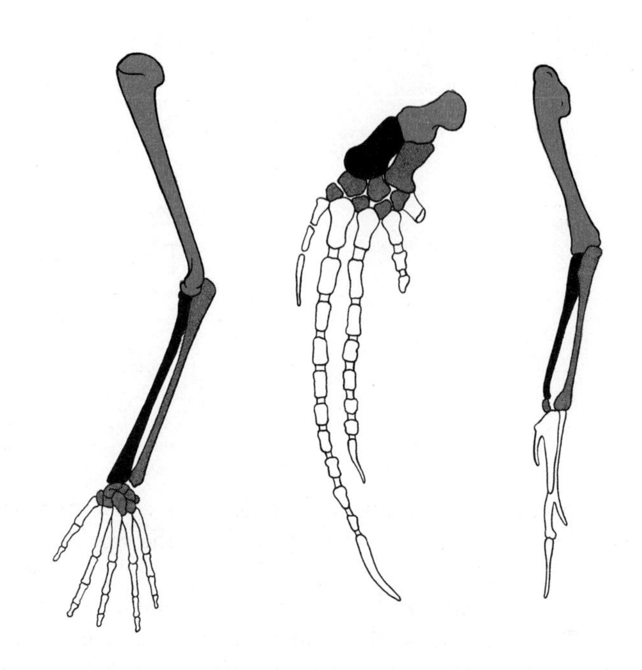

4–2

From left to right, the arm and hand of a man, the flipper of a whale, and the wing of a bird. As you can see, these organs, although providing different types of mobility, are built on the same basic plan. They are therefore homologous. *Organs that have similar functions but are different in basic origin and structure, such as the wings of an insect and those of a bat, are* analogous. *Homology is assumed to reflect a common ancestry, whereas analogy merely notes a common function.*

THE BEGINNING OF LIFE

The story begins more than 3 billion years ago. At that time the ancestors of the modern plants and animals were microscopic cells floating near the surface of the warm shallow seas which covered much of the face of the planet. A few microfossils of such cells have been found; the oldest are of organisms which apparently lived more than 3.2 billion years ago. As these earliest fossils indicate, there were, even by then, well before a third of earth's history had passed, already two forms of life: a rod-shaped bacterialike organism and a spheroid algalike organism.

Modern bacteria and blue-green algae (so called because of their characteristic color) are believed to resemble these primitive cells and to be direct descendants of them. Bacteria and blue-green algae are *prokaryotes* (from *pro*, meaning "before," and *karyos*, "nucleus"). All other cells are *eukaryotes* (from *eu*, meaning "true"). Prokaryotic cells differ so profoundly from the other, nucleated cells that in some systems of classification, including the one followed in this text, they are considered to constitute a different kingdom, the Kingdom Monera. In all, according to the system we follow, which was first proposed by R. H. Whittaker,* there are five kingdoms, large divisions into which all living organisms are grouped: Monera, Protista, Fungi, Plantae, and Animalia.

THE MONERANS

Bacteria and blue-green algae are by far the simplest of all cells. The genetic material, instead of being organized into chromosomes in a nucleus with a surrounding membrane, is dispersed throughout the cytoplasm in the form of a long, continuous coil of DNA. There is no protein associated with the DNA, and since there are no chromosomes, there is no mitosis or meiosis. After the DNA replicates, the cell membrane pinches in, and the cell divides. The cytoplasm contains no complex organelles, such as mitochondria, chloroplasts, and Golgi bodies. The cell is surrounded by a cell membrane and a cell wall. This wall, however, does not contain cellulose, as do the cell walls of plants, but is instead composed of a different type of polysaccharide combined with protein. (It is this unusual cell-wall composition, shared by the bacteria and the blue-green algae, that led to the recognition of the close evolutionary ties between these two groups.)

* R. H. Whittaker, *Science*, **163**:150–160, 1969. For a more complete classification see Appendix C.

THE NAMING OF LIVING ORGANISMS

Most people are concerned with only very limited areas of the natural world. Gauchos, the cowboys of Argentina, who are famous for their horsemanship, have some two hundred names for different colors of horses but divide all the plants known to them into four groups: pasto, *or fodder;* paja, *or bedding;* cardo, *wood; and* yuyos, *everything else.*

Most of us are like the gauchos. Once beyond the range of common plants and animals, and perhaps a few uncommon ones that are of special interest to us, we usually run out of names. In the science of biology, however, all living things are of interest, and biologists face the problem of systematically investigating, identifying, and exchanging information about well over a million different kinds of organisms. In order to do this, they must have a system for naming all these organisms and for grouping them together in orderly and logical ways. The problem of developing such a system has become immensely complicated. Some 700,000 different kinds of insects, for instance, have been named and classified to date!

In the eighteenth century, the binomial ("two-name") system was introduced, and this system is still in use today. The binomial consists of two parts, the name of the genus and that of the species. The wolf, for example, belongs to the genus Canis *(the Latin word for dog), and its official name is* Canis lupus, *which distinguishes it from such near relations as the coyote* (Canis latrans) *and the domestic dog* (Canis familiaris). *Every individual is thus a member of a particular species, which, in turn, belongs to a larger group, a genus (plural: genera).*

According to the modern system of classification, genera are grouped into families, families into orders, orders into classes, classes into phyla (singular: phylum), and phyla into kingdoms.

Despite the fact that monerans have no form of sexual reproduction, they do have a means for exchanging genetic information and so introducing variation into cell lines (which is, of course, the "purpose" of sexual reproduction). In many types of bacteria, it has been discovered, some cells function as males, or donors, and some as females, or recipients. Male cells have a special hollow hairlike structure, a conjugation pilus, through which fragments of DNA are passed into the female cell. This DNA either may re-

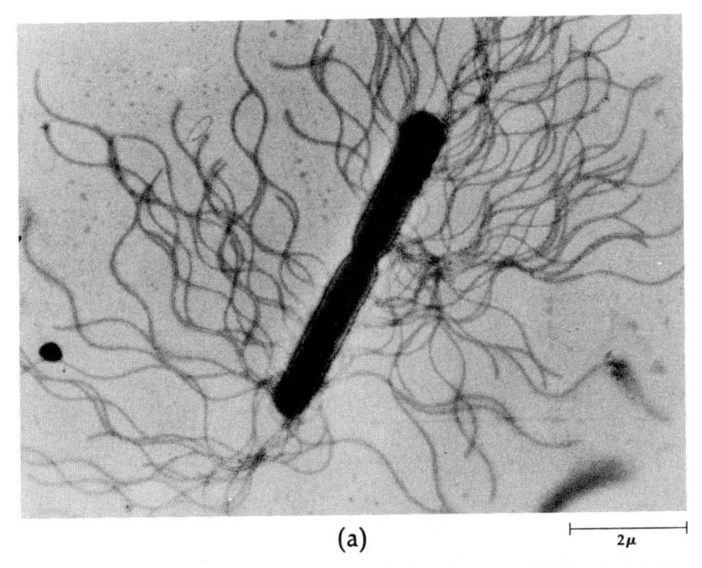

(a)

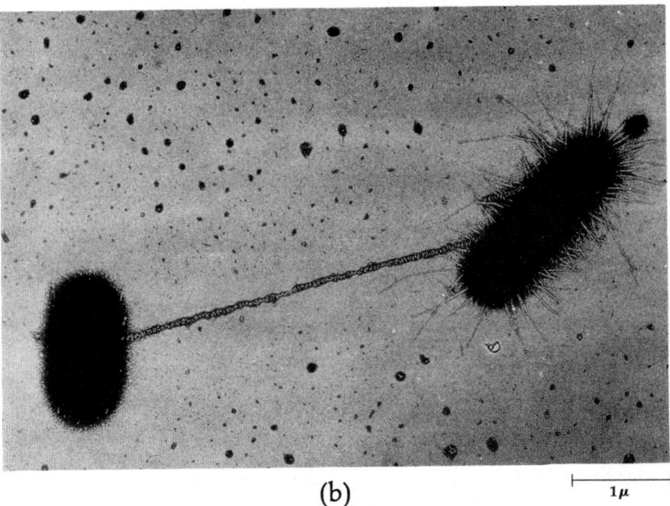

(b)

4–3

(a) *A bacterial cell dividing. Like many bacterial cells, this species* (Proteus mirabilis) *has numerous long flagella. Unlike the flagella of eukaryotes, however, these are simple strands and do not have the 9-plus-2 structure.* (b) *Conjugating cells of the bacterium* Escherichia coli, *an abundant and harmless inhabitant of the human intestine. DNA is passing from one cell to another through a tubular, hairlike structure called a pilus.*

main separate in the cytoplasm or may become part of the female's DNA molecule.

Why do such simple organisms expend relatively large amounts of their energy resources on the exchange of genetic material? These exchanges are known to be a means of creating new combinations of genes and so are an important source of variations. The variations increase the probability of survival under new environmental conditions. (Resistance to drugs such as penicillin, for example, comes about because some few individuals in every large bacterial colony are genetically resistant to penicillin. When the penicillin-sensitive cells are killed, the resistant ones may continue to multiply, forming a new population that is resistant.)

Bacteria

Bacteria are not only the oldest but also the most abundant group of organisms in the world. Despite their small size, they are so numerous that their total weight exceeds that of all other organisms combined. Their great success, evolutionarily speaking, seems to be due to their diversity. They can live in places and under conditions which support no other forms of life. They have been found in the icy wastes of Antarctica, in the near-boiling waters of natural hot springs, and even in the dark depths of the ocean. All else failing, many kinds can take the form of hard, resistant spores, which may lie dormant for years until conditions are more favorable.

Bacteria obtain energy in a great variety of ways. Some are autotrophs. Among the autotrophs are photosynthetic bacteria which, like green plants, capture light energy in forms of chlorophyll but which break apart hydrogen sulfide (H_2S) and other compounds rather than water and do not release oxygen gas. Other autotrophic bacteria are chemoautotrophic; these obtain their energy from the oxidation of inorganic molecules such as certain compounds of nitrogen, sulfur, and iron.

Most bacteria are heterotrophs, and it is this group that is of the greatest importance to man. Some of the heterotrophs live on organic material which is still alive—includ-

---

THE VIRUSES

Viruses, because of their simplicity, are traditionally studied with the bacteria and blue-green algae. Some biologists believe that the earliest forms of life were viruslike; others contend that viruses should not be regarded as living organisms at all but rather as parts of cells that have set up a partially independent existence. S. E. Luria, who recently was awarded the Nobel Prize for his work with bacterial viruses, has called them "bits of heredity looking for a chromosome."

Viruses are parasites that can multiply only within a living cell. All viruses consist of nucleic acid—either DNA or RNA—and protein. The protein forms an outer coat, which protects the nucleic acid and determines what sort of cell the virus can parasitize. Usually particular viruses attack only particular cells: the influenza virus invades the lining of the respiratory tract, the polio virus attacks intestinal cells and sometimes nerve cells, and so forth.

Either before the virus enters the cell or just after (depending on the type of virus), the viral nucleic acid slips out of its protein coat. In the case of the DNA viruses, the DNA of the virus replicates and also codes for messenger RNA. The mRNA, in turn, produces enzymes and coat protein needed by the virus, which uses raw materials (such as amino acids and nucleotides) and cellular machinery (such as the ribosomes and transfer RNA) of the host cell. In the case of the RNA viruses, the nucleic acid both replicates and serves directly as messenger RNA. In either case, the end product is hundreds or often thousands of new viral particles produced and assembled within the infected cell, which often is broken apart as they are released.

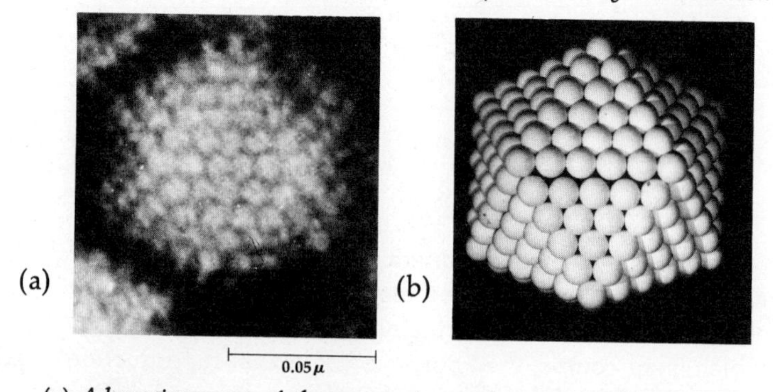

(a) *Adenovirus, one of the many viruses that cause colds in humans. This virus is an icosahedron. Each of its 20 sides is an equilateral triangle composed of identical protein subunits. There are 252 subunits in all.* (b) *A model of the adenovirus, made up of 252 tennis balls.*

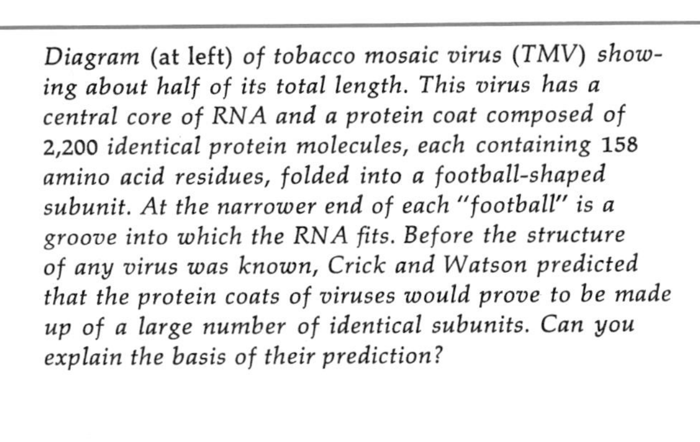

Diagram (at left) *of tobacco mosaic virus (TMV) showing about half of its total length. This virus has a central core of RNA and a protein coat composed of 2,200 identical protein molecules, each containing 158 amino acid residues, folded into a football-shaped subunit. At the narrower end of each "football" is a groove into which the RNA fits. Before the structure of any virus was known, Crick and Watson predicted that the protein coats of viruses would prove to be made up of a large number of identical subunits. Can you explain the basis of their prediction?*

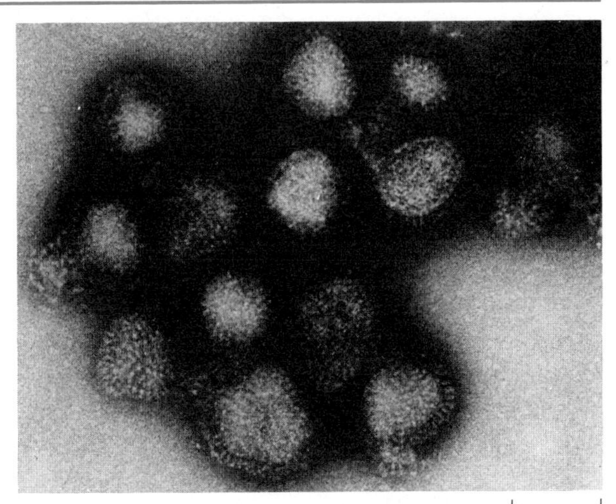

Influenza virus (A_2, Hong Kong 1/68). This virus mutates rapidly. Changes in its genetic material (which is RNA) result in changes in its protein coat. Since immunity to a virus is, in effect, immunity to the specific proteins of its protective coat, these new viral strains are able to infect previously immune populations. The virus is surrounded by a lipoprotein envelope through which protrude stubby protein spikes.

Bacteriophages (below). *Note the tail fibers by which the viruses attach themselves to the bacterial cell walls.*

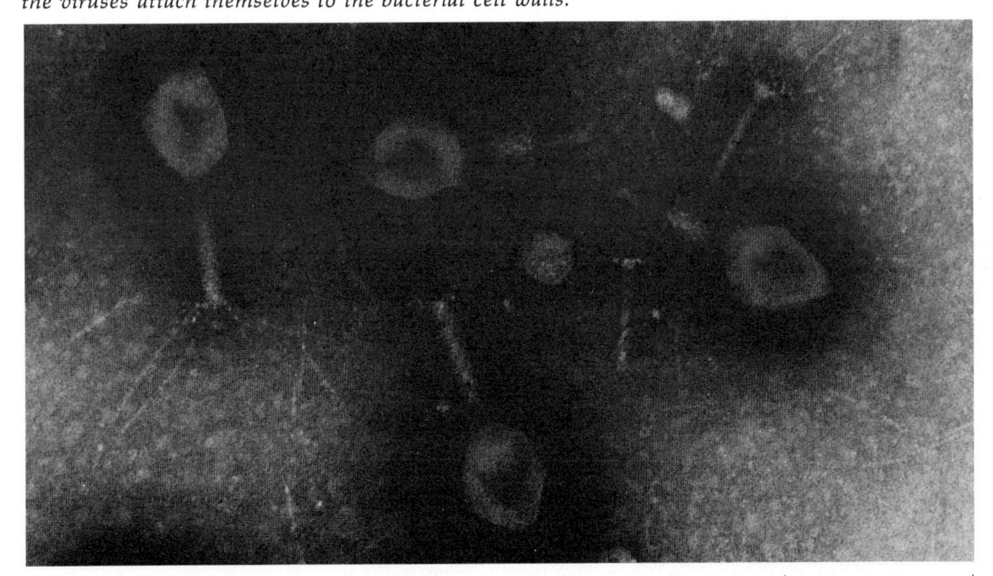

ing us. These are the disease-causing bacteria which have earned the entire group a generally bad reputation. However, by far the largest group of bacteria live on dead organic matter. These bacteria are of the utmost importance to man and all other living things because they are the principal decomposers of the biosphere. Without them, the organic material synthesized by plants and animals would accumulate, and gradually all organisms would be overwhelmed by the products of their own metabolism. Through the action of the bacteria (and the fungi, which are also decomposers), materials incorporated into the bodies of once-living organisms are released and made available for successive generations of living things.

Blue-green Algae

All modern algae are photosynthetic and produce oxygen as a by-product of photosynthesis, but only the blue-greens are prokaryotes and therefore classified as monerans. Strictly speaking, the blue-greens are really a type of photosynthetic bacteria, in which chlorophyll *a* is the chief photosynthetic material.

Blue-green algae grow for the most part in fresh water. They are sometimes single-celled but more often form clusters, threads, or chains. They are among the few microorganisms able to fix nitrogen (that is, combine it with other materials to form organic compounds) and are important contributors of nitrogen to the rice paddies of the Far East. And as we shall see, although they are a small group of organisms, an important evolutionary role has recently been assigned to them.

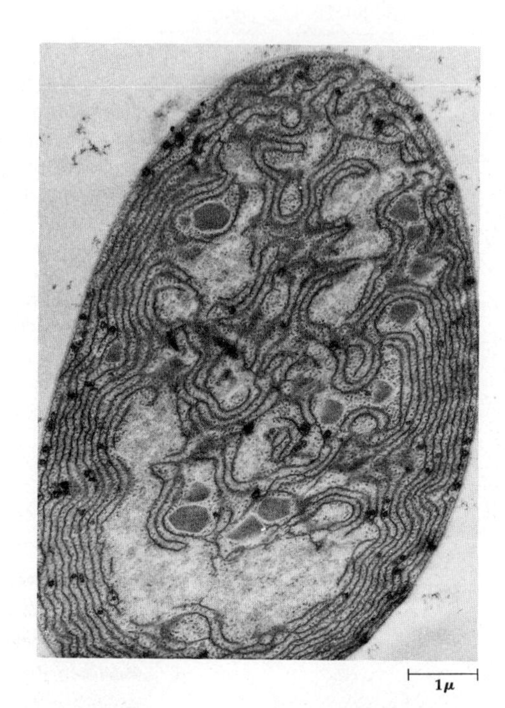

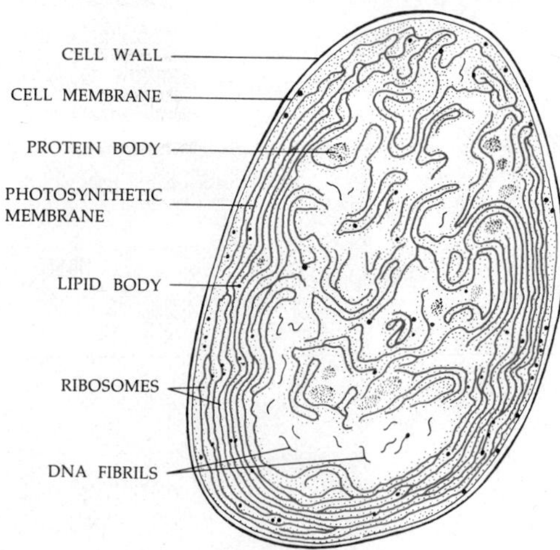

4–4

A blue-green alga. The membranes within the cell, like the chloroplasts of plant cells, contain chlorophyll a *and other pigments and are the sites of photosynthesis. DNA is also visible in the cell. The dark areas are accumulated proteins and lipids.*

THE PROTISTS

All protists are one-celled organisms, and all are eukaryotes. Some are heterotrophs, capturing and eating small organisms or bits of decaying matter or debris or absorbing organic compounds through their membranes, as do somatic cells. Others are photosynthetic organisms—one-

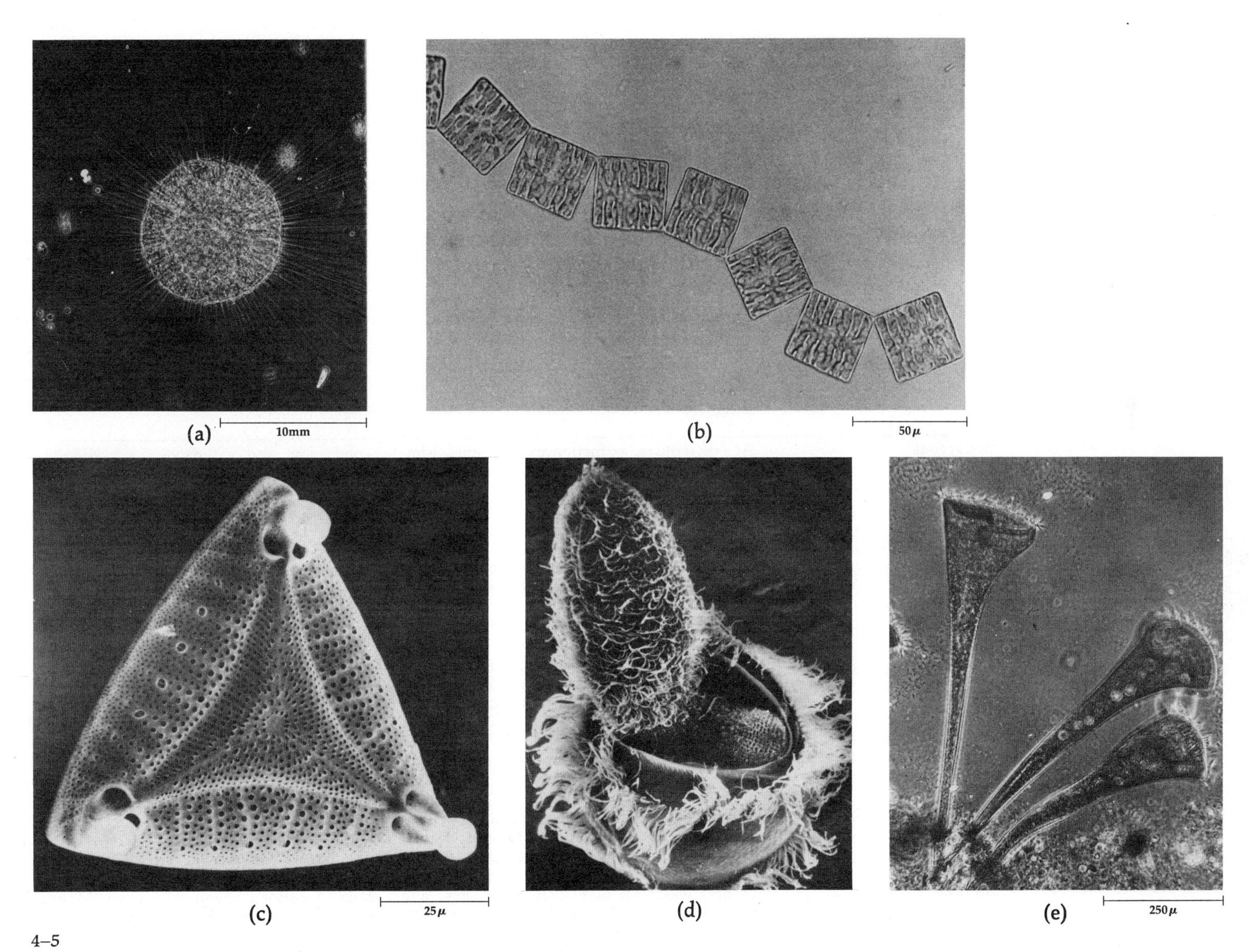

4–5

Members of the Kingdom Protista: (a) Actinosphaerium, *the "sun animal"*; (b) *a chain of* Tabellaria; (c) *the finely carved shell of a diatom*; (d) Didinium, *a barrel-shaped protozoan engulfing a* Paramecium; (e) Stentor coeruleus. *The second and third are one-celled algae, and the other three are free-living protozoans.*

Table 4–1 *Major Physical and Biological Events in Geologic Time*

Millions of Years Ago	*Era*	*Period*	*Epoch*	*Life Forms*	*Climates and Major Physical Events*
	CENOZOIC	Quaternary	Recent Pleistocene	Age of man. Planetary spread of *Homo sapiens;* extinction of many large mammals, including woolly mammoths. Deserts on large scale.	Fluctuating cold to mild. Four glacial advances and retreats (Ice Age); uplift of Sierra Nevada.
1½–7		Tertiary	Pliocene	Large carnivores. First known appearance of man-apes. Herbaceous plants more abundant.	Cooler. Continued uplift and mountain building, with widespread extinction of many species.
7–26			Miocene	Whales, apes, grazing animals. Spread of grasslands as forests contract.	Moderate uplift of Rockies.
26–38			Oligocene	Large, browsing mammals. Apes appear.	Rise of Alps and Himalayas. Lands generally low. Volcanoes in Rockies area.
38–53			Eocene	Primitive horses, tiny camels, modern and giant types of birds.	Mild to very tropical. Many lakes in western North America.
53–65			Paleocene	First known primitive primates and carnivores.	Mild to cool. Wide, shallow continental seas largely disappear.
65–136	MESOZOIC	Cretaceous		Age of reptiles, extinction of dinosaurs. Marsupials, insectivores. Angiosperms become abundant.	Lands low and extensive. Last widespread oceans. Elevation of Rockies cuts off rain.
136–195		Jurassic		Dinosaurs' zenith. Flying reptiles, small mammals. Birds appear. Gymnosperms, especially cycads and ferns.	Mild. Continents low. Large areas in Europe covered by seas. Mountains rise from Alaska to Mexico.
195–225		Triassic		First dinosaurs. Primitive mammals appear. Forests of gymnosperms and ferns.	Continents mountainous. Large areas arid. Eruptions in eastern North America. Appalachians uplifted and broken into basins.
230–280	PALEOZOIC	Permian		Reptiles evolve. Origin of conifers and possible origin of angiosperms; earlier forest types wane.	Extensive glaciation in Southern Hemisphere. Appalachians formed by end of Paleozoic; most of seas drain from continent.
280–345		Carboniferous Pennsylvanian Mississippian		Age of amphibians. First reptiles. Variety of insects. Sharks abundant. Forests, ferns, gymnosperms, and horsetails.	Warm. Lands low, covered by shallow seas or great coal swamps. Mountain building in eastern U.S., Texas, Colorado. Moist, equable climate, conditions like those in temperate or subtropical zones, little seasonal variation, root patterns indicate water plentiful.
345–395		Devonian		Age of fish. Amphibians appear. Shellfish abundant. Lungfish. Rise of land plants. Extinction of primitive vascular plants. Origin of modern subclasses of vascular plants.	Europe mountainous with arid basins. Mountains and volcanoes in eastern U.S. and Canada. Rest of North America low and flat. Sea covers most of land.
395–440		Silurian		Earliest vascular plants. Rise of fish and reef-building corals. Shell-forming sea animals abundant. Modern groups of algae and fungi.	Mild. Continents generally flat. Mountain building in Europe. Again flooded.
440–500		Ordovician		First primitive fish. Shell-forming sea animals. Invasion of land by plants?	Mild. Shallow seas, continents low; sea covers U.S. Limestone deposits; microscopic plant life thriving.
500–600		Cambrian		Age of marine invertebrates. Shell animals.	Mild. Extensive seas. Seas spill over continents.
	PRECAMBRIAN			Earliest known fossils.	Dry and cold to warm and moist. Planet cools. Formation of earth's crust. Extensive mountain building. Shallow seas.

celled algae. (However, all one-celled algae are not protists; the green algae, a group which includes *Chlamydomonas*, are clearly related to many-celled plants and so are classified as plants.) All have their genetic material organized into chromosomes, which divide by mitosis. Many have a sexual cycle similar to that of *Chlamydomonas* (Chapter 3–1), and in those that do not, biologists tend to believe that such a cycle once existed and was lost. Many of the protists have flagella, all with the characteristic 9-plus-2 structure. Some protists move amoeba-fashion, and some change back and forth from amoebalike cells to flagellated cells, depending on whether they are in mud or in water. Others have no means of mobility; these types usually float on the surface of the water or live within other cells as parasites. All protists live in water, although some live out an entire life-span in no more than a droplet. They are much larger and more complex in structure than the prokaryotes.

More than half of the photosynthesis that takes place on this planet is carried out by one-celled algae, both plants and protists, floating on or near the surface of the deep-sea waters. Among these autotrophs, and subsisting on them, are very small animals, mostly tiny crustaceans mingled with the immature forms of larger sea animals. These free-floating populations are known as *plankton*, from *planktos*, the Greek word for "wandering." Small fish and some larger ones and some of the great whales feed directly on the plankton, but the larger fish commonly feed on the smaller ones. In this way, "the great meadow of the sea," as it is sometimes called, is similar to the meadows of the land, which also provide, directly or indirectly, nourishment for animal life.

The step from the prokaryotes (the monerans) to the first eukaryotes (the protists) was one of the big evolutionary transitions. How did it come about? One interesting theory that is gaining more and more support is that larger, more complex cells evolved as a result of certain prokaryotes taking up residence inside of other cells.

As O_2 began slowly to accumulate in the atmosphere, as a result of the photosynthetic activities of the blue-green algae, those bacteria that were able to convert to the use of oxygen in ATP production gained a strong advantage —and so such forms, originally produced as the results of chance mutations, began to prosper and increase. Some of these evolved into modern forms of bacteria. Others, according to the theory, became symbionts within larger cells and evolved into mitochondria! Among the evidence that supports this latter idea is the following: mitochondria contain their own DNA, and this DNA is present in a single, continuous molecule, like the DNA of bacteria; many of the same enzymes contained in the cell membranes of bacteria are found in mitochrondrial membranes; mitochondria appear to be produced only by other mitochondria, which divide within their host cell. We know little about those original cells in which these bacteria first set up housekeeping—or, indeed, if they actually existed. But if they did exist, they probably had no means of their own for using oxygen and so were dependent entirely on some form of fermentation, which, as we have seen, is relatively inefficient. Cells with these respiratory assistants were doubtless more efficient than those which lacked them and so outnumbered and crowded them out.

In an analogous fashion, photosynthetic prokaryotes ingested by larger nonphotosynthetic cells, are believed to be the forerunners of chloroplasts. By this arrangement, the smaller cells gained a position of greater safety and the larger cells were given a new energy source.

Although there is as yet no widely accepted explanation of the origin of the membrane-bounded nucleus, this theory accounts for the presence in eukaryotic cells of the complex organelles not found in the far simpler prokaryotes.

Although the origin of the protists is still in dispute—and mentioned here largely because it is interesting—it is widely agreed among biologists that (1) the protists are a large and varied group of organisms, with little in common beyond their one-celledness, and (2) all the other organisms now occupying the planet had their origins in cells such as these.

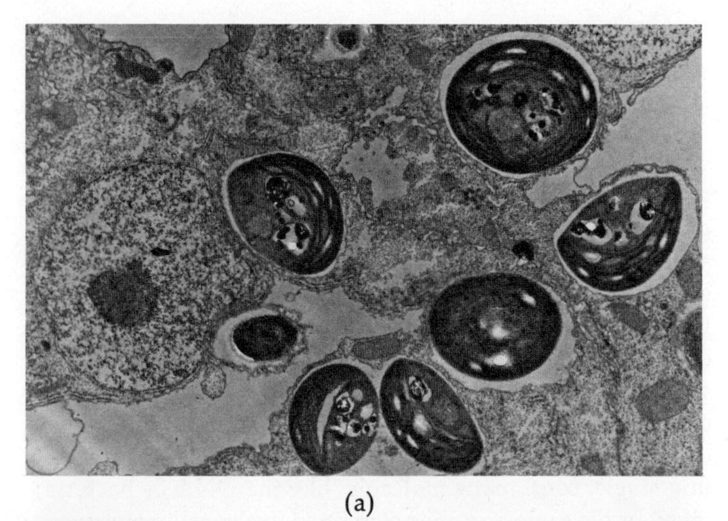

(a)

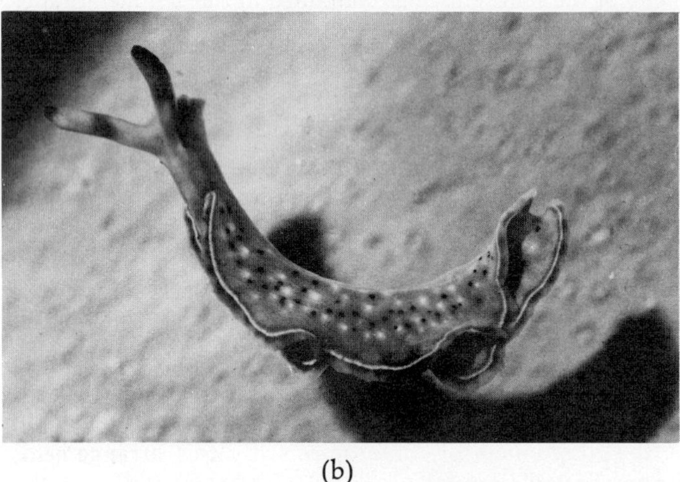

(b)

4–6
Biologists who support the theory that mitochondria and chloroplasts originated as symbiotic microorganisms point out that there are many examples of this sort of symbiosis ("living togetherness"): (a) *These algal cells are in the tissue of a hydra, a common invertebrate; members of this species contain so many intracellular algae that they are bright green and were once thought to be plants.* (b) *This animal is a sea slug, or nudibranch. Its tissues contain thousands of symbiotic photosynthetic green algae.*

THE FUNGI

The fungi are a group of organisms so unlike any others that although they were long classified with the plants, it has come to seem appropriate to assign them to a separate kingdom. Except for some one-celled forms, such as the yeasts, the fungi are basically composed of masses of filaments. A fungal filament is called a *hypha,* and all the hyphae of a single plant are collectively called a *mycelium.* The walls of the hyphae usually contain chitin, a polysaccharide that is never found in plants. (It is, however, the principal component of the exoskeletons—that is, the hard outer coverings—of insects.) Protoplasm containing the fungal nuclei flows within the mycelium. The complex, spore-producing structures of fungi, such as mushrooms, are composed of tightly packed hyphae.

All fungi are heterotrophs. They obtain food passively by absorption of organic compounds, sometimes digesting the compounds first by means of enzymes which they secrete into the food mass. Growth is their only form of mobility (except for sex cells or spores, which may travel through air or water). Consequently, they live in earth, in water, or in some other medium rich in organic substances and are capable of astonishingly rapid growth, as evidenced in the overnight appearance of a lawnful of toadstools.

The fungi, together with the bacteria, are the decomposers of the world, and as we shall see in Section 7, their activities are as vital to the continued survival of higher forms of life as are those of the food producers. Fungi release carbon dioxide into the atmosphere and return nitrogenous compounds and other materials to the soil, where they are used again by green plants and eventually by animals. It has been estimated that the top 6 inches of fertile soil may contain more than 2 tons of fungi and bacteria per acre.

(a)

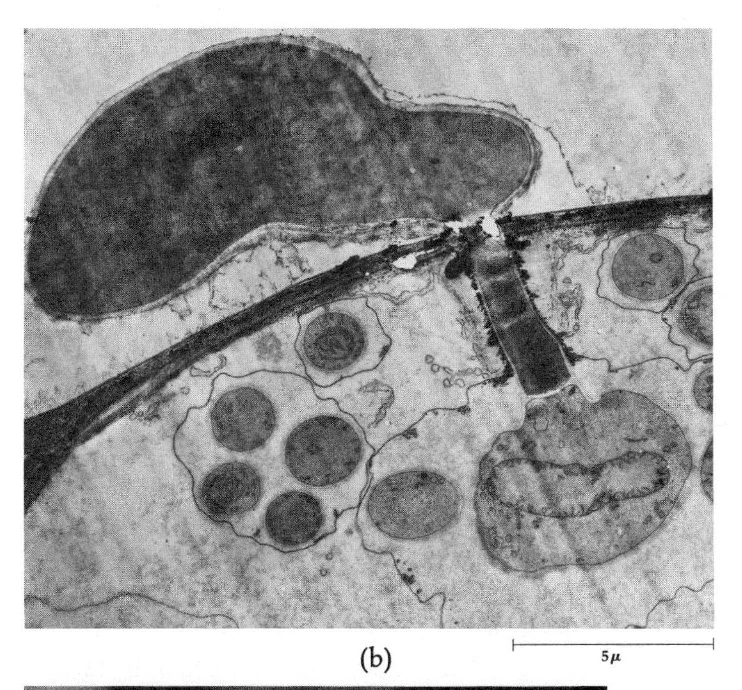

(b)

4–7

Examples of fungi: (a) *Fungal mycelium on a fallen tree trunk.* (b) *A parasitic fungus. The dark body on top is a fungal hypha that is growing through the wall of a barley cell. Some fungi, such as these, have switched from their role as decomposers of dead organic matter to that of attackers of living organisms. Fungi are the most important single cause of plant diseases.* (c) *A morel, one of the several types of fungi prized by gourmets.*

(c)

THE PLANTS

The story of plant evolution begins with one-celled algae floating on or near the surface of the water in the open seas. (Such organisms have left little trace in the fossil record, and so their history must be reconstructed largely on the basis of the observation of modern forms.) Here, sunlight was abundant and the oxygen, hydrogen, and carbon in the air and water were available to every floating cell. Although the cells might form long filaments (as a consequence of not separating after cell division), each remained a separate functional entity and lived an independent existence.

As the cellular colonies multiplied, they probably started to exhaust the supplies of nitrogen, phosphorus, sulfur, and other minerals available in the open ocean. (It is this shortage of essential elements that is the limiting factor in any modern plans to farm the seas.) As a consequence, life began to flourish near the shores, where the waters were rich in nitrates and other minerals washed from the land by rivers and streams and scraped from the coasts by the action of the waves. Here, along the coasts, in a much

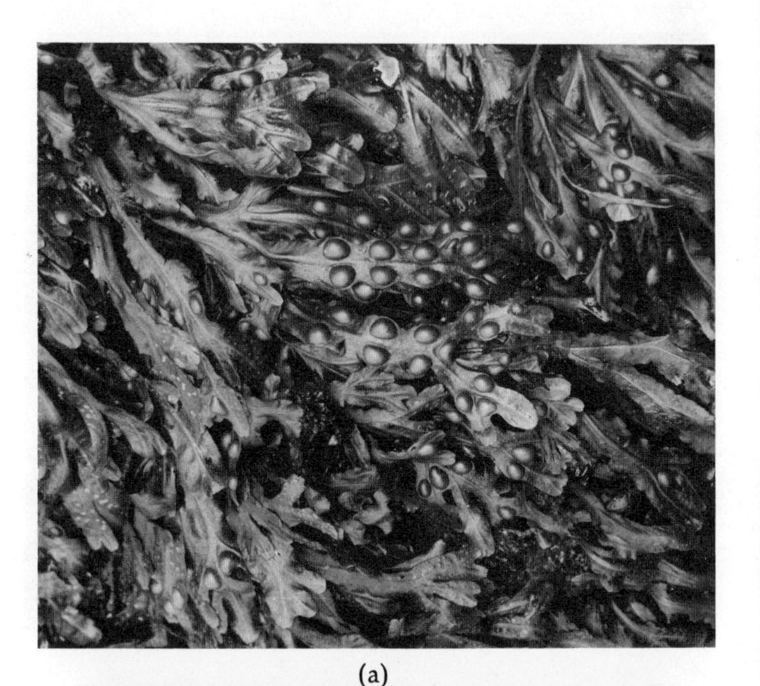

(a)

(b)

4–8

Both the brown algae and the red algae have complex, multicellular bodies and are classified as members of the plant kingdom. (a) *The brown algae, which include the giant kelps, are the most conspicuous seaweeds of the temperate zones. Shown here is a brown alga, rockweed, which covers many rocky shores that are exposed at low tide. The gas bladders float the photosynthetic portions like pontoons.* (b) *The red algae are particularly abundant in warm and tropical waters. Their pigments enable them to survive at greater depths than any of the other algae; some have been found more than 600 feet below the water surface. This red alga was found near Monterey, California.*

more challenging environment than that of the open sea, complex plant life evolved.

Among the challenges presented by life along the shore was the turbulent shore itself. Under evolutionary pressures, groups of plants evolved anchoring structures, *holdfasts*, which adhered to the rocks. These were often below the surface of the water, where less light penetrated. Over millions of years, the upper portions of the plant body, which were near the light, became thinner and more spread out, increasing the surface area. The upper cells became specialists in photosynthesis, producing enough sugars to nourish the overshadowed lower cells which anchored them. Some of these multicellular algae developed specialized conducting tissues that transported the products of photosynthesis downward.

Not only did these marine algae, the seaweeds, take on shapes, but they also developed specialized colors. Water filters out the light, removing first the longer wavelengths, the reds and oranges; at the deeper levels, only a faint blue light penetrates. The red algae, which are red because they contain a special pigment, capture the energy of these blue rays and pass it to chlorophyll. The brown algae (the brown results from a mixture of green and orange pigments) absorb blue-green light, and the green algae, whose principal pigments are the chlorophylls, make fullest use of the longer-waved red light. As a consequence of the evolution of specialized pigments, the various seaweeds came to use the entire spectrum of light as well as the living space along the rocky shore. Although the chronology is not certain, it appears that once the plants had claimed the rocky coasts, animals were able to find favorable living places there and eventually complex communities came into being.

Transition to Land

The true land plants apparently had their origins not in the red algae or the brown algae but among the green algae, of which *Chlamydomonas* is a single-celled representative. Evidence for this statement is based on a comparison with modern plants: like all the land plants, the green algae

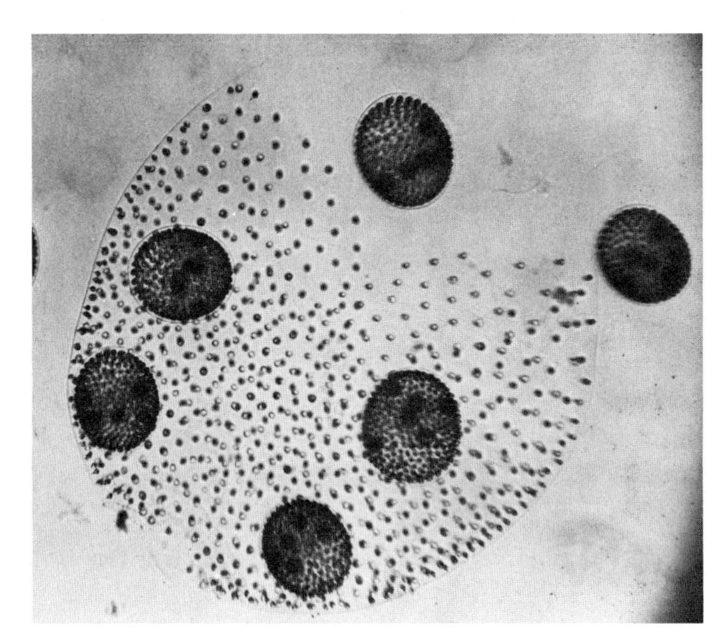

4–9

Volvox, *a multicellular green alga. Flagellated green cells, each much like* Chlamydomonas, *are connected by strands of cytoplasm to form a hollow spherical colony. The number of cells in the sphere ranges from 500 to 50,000, depending on the species. The flagella of each cell beat in such a way that the colony spins slowly as it tumbles through the water. Daughter colonies form within the mother sphere, which finally breaks open to release them, as shown here.*

Table 4–2

The Plant Kingdom

Phylum	*Number of Species*	*Common Name*
Rhodophyta	4,000	Red algae
Phaeophyta	1,100	Brown algae
Chlorophyta	7,000	Green algae
Bryophyta	23,000	Mosses and liverworts
Tracheophyta	510,000	Vascular plants

4–10

Chart of evolution of land plants.

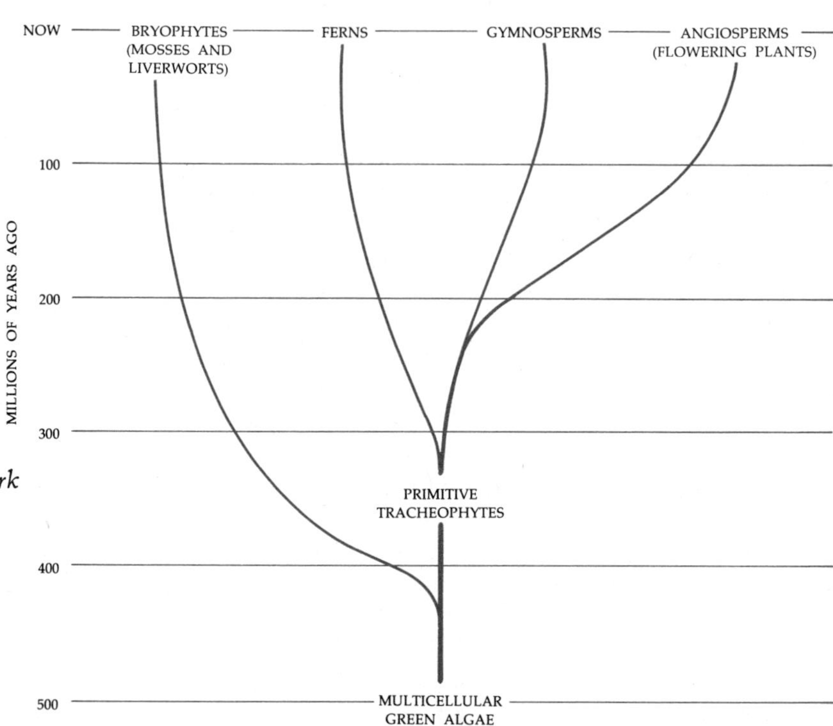

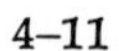

4–11

Reconstruction of a Devonian forest showing western New York as it is believed to have looked 370 million years ago. The tallest plant (just to the right of center) is a relative of the club mosses; although modern club mosses are typically only a few inches high, these primitive trees often grew to heights of 100 feet. To the right and also in a group on the left are tree ferns. In the right foreground is a primitive leafless plant, Psilophyton, *and in the center foreground are several clumps of horsetails, identifiable by their whorled branches.*

contain chlorophylls *a* and *b* and carotenoids as their photosynthetic pigments; they accumulate their food as starch; and their cell walls are composed of cellulose. The common ancestor of all the modern land plants seems to have been a relatively complex multicellular green alga that invaded the land some 500 million years ago. It was probably "prepared" for its invasion of land by its differentiation into a lower rootlike area, or holdfast, and an upper photosynthetic area; such specializations are seen among the modern coastal seaweeds. (This "preparation" is known as *preadaptation*. Preadaptation is the occurrence in an organism of a structure or function that, in the course of evolution, in changing environments takes on a new role.) Once this ancestral plant had made the transition to land, some new evolutionary developments occurred: roots, which anchor the plants and absorb water and minerals from the soil; a waxy coating of cutin, which retards water evaporation; and stomata, specialized openings in the leaves and green stem through which gases are exchanged during photosynthesis and respiration. Presumably all these specializations developed early in the evolutionary history of land plants since all land plants share these features.

Alternation of Generations

Another feature shared by all land plants is alternation of generations described in Chapter 3–1. This reproductive cycle was presumably also present in the ancestral multicellular green alga, as it is found in all modern multicellular green algae. All have a reproductive cycle in which a diploid ($2n$) plant, the sporophyte, produces haploid spores (by meiosis), which give rise to a haploid ($1n$) plant, the gametophyte. The gametophyte produces gametes (by mitosis), which fuse in pairs (fertilization), forming zygotes that develop into new diploid sporophytes. Among the algae, the gametophyte and the sporophyte are always independent organisms. Sometimes they are identical in appearance and size, but in at least one species they are so different that they were long considered separate, quite unrelated plants.

The Major Plants

The two major groups of land plants, the *bryophytes* and the *tracheophytes*, separated early in evolutionary history, probably more than 400 million years ago. The bryophytes, which include the mosses, liverworts, and hornworts, are land plants with simple tissues; they did not develop elaborate conducting systems. These plants are relatively small, usually less than 6 inches in height. They absorb moisture through their leaves as well as from the ground and are most abundant in moist areas. The tracheophytes, the vascular plants that dominate the modern landscape, are characterized by their efficient systems for the transport of water and sugars. Modern tracheophytes include the club mosses, horsetails, and ferns; the *gymnosperms*, a group which includes all the conifers (cone-bearing plants such as pines and spruces); and the *angiosperms*, the flowering plants, which include the trees, grasses, and wild flowers and all the important crop plants on which human survival depends.

Trends among Vascular Plants

Among the vascular plants, there have been three marked evolutionary trends. The first has been the development of increasingly efficient conducting systems. One conducting system, the xylem, transports water from the ground to the leaves, often hundreds of feet. The other conducting system, the phloem, conducts sucrose and other products of photosynthesis from the leaves to the nonphotosynthetic cells of the plants. The structure and function of the vascular system in modern angiosperms will be described in the following chapters.

The second pronounced trend has been the reduction in importance of the gametophyte generation. Among the ferns, as we saw in Chapter 3–1, the gametophyte is separate from the sporophyte but is much smaller. In most species of fern, there is only one type of gametophyte, which is bisexual and produces both male and female gametes. In some few, however, there are male and female gametophytes.

4–12
The major groups of modern land plants include (a) *bryophytes (nonvascular land plants), exemplified by this urn moss. The "urns" enclose spore capsules;* (b) *club mosses, which are vascular plants. This photograph, taken in Fairfax County, Virginia, shows a shining club moss* (Lycopodium lucidulum)*;* (c) *ferns, represented by these "fiddleheads" (young shoots) of cinnamon fern;* (d) *gymnosperms, such as this loblolly pine, a conifer; and* (e) *angiosperms, the flowering plants.*

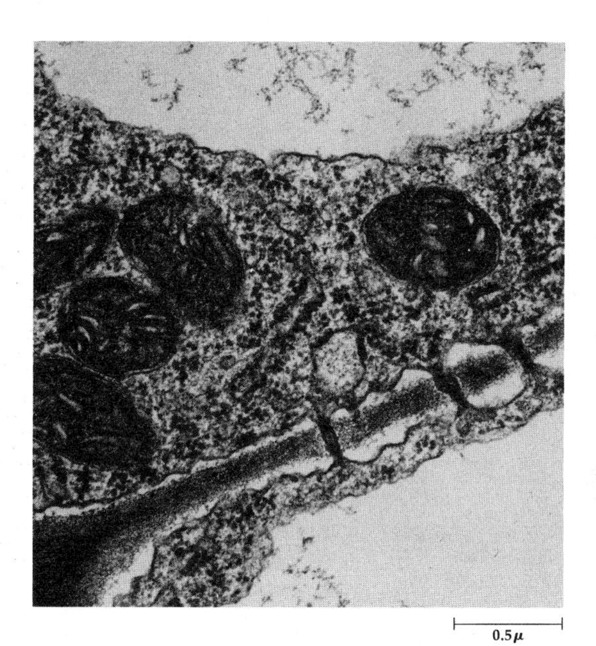

4–13 (at right)

In the higher plants, the individual cells are connected with one another by strands of cytoplasm known as plasmodesmata *(singular: plasmodesma). Three plasmodesmata can be seen in this micrograph running through the cell walls separating two cells of a tomato leaf. Several mitochondria are visible in the upper cell, and both cells have large vacuoles. Plasmodesmata help to unite the individual cells of a complex plant body into an integrated organism.*

4–14 (below at right)

Cones of jackpine shedding pollen. Some of these airborne pollen grains will fall on "female" pinecones. In the female cones, the gametophytes bearing egg cells are enclosed within the ovule on the cone scale. The ovule exudes a sticky fluid to which the pollen grains adhere. As the fluid evaporates, the pollen grain is drawn toward the egg cell. The seed is the mature fertilized ovule, consisting of seed coat (the outer layers of the ovule), female gametophyte, and embryo.

Among the gymnosperms, there are two types of gametophytes, one male and one female, both of which are entirely heterotrophic, being dependent on the parent plant for their nutrition and development. The female gametophyte, in fact, never leaves the protection of the sporophyte but is fertilized there by the male gametophyte, the pollen grain.

In the angiosperms, the most recently evolved of the vascular plants, the gametophytes are microscopic. The female gametophyte, or embryo sac, contains only eight haploid nuclei, and the male gametophyte, the pollen grain, is a single cell with three nuclei. And yet the ancestral pattern persists, and it is impossible to understand reproduction among this most important group of modern plants without knowledge of its legacy from the past.

The Seed

The third major development among the vascular plants, and perhaps the most important to survival on land, is the seed. The seed is a complex structure in which the embryo sporophyte is contained within a protective outer covering, the seed coat. The seed coat, which is derived from tissues of the parent sporophyte, protects the immature plant from drought while it remains dormant, sometimes for years, until conditions are favorable for its germination.

Seeds became important toward the close of the Paleozoic era, which ended some 230 million years ago. Earlier in this era, in the Carboniferous period, the temperature was warm and the lands were low, covered by shallow seas or great swamps. Water was plentiful, and there was little seasonal variation in temperature. This was the period in which our coal deposits were formed from the masses of vegetation in the swamps. At the close of the Carboniferous period, there were worldwide changes of climate, with widespread glaciers and droughts. Amphibians gave way to reptiles, the forerunners of the great dinosaurs. The seed was in existence at this time; according to the fossil record, some of the fernlike plants and even some of the club mosses had seedlike structures. But it was not until the close of the Permian period, when the land became colder and drier, that the seed became a distinct advantage and the seed plants became the dominant plants of the land. The two major groups of modern vascular land plants are the gymnosperms, the naked seed plants, and the angiosperms, flowering plants in which the base of the flower develops into a fruit that encloses and protects the seed and often aids in its dispersal.

The Role of Plants

The only forms of life on land that do not depend on the green plants for their existence are a few kinds of microscopic autotrophs. For all other living things, the chloroplast of the green plant is the "needle's eye" through which the sun's energy is channeled into the biosphere. Even those animals that eat only other animals—the carnivores—could not exist if their prey, or their prey's prey, had not been nourished by green plants.

Moreover, plants are the channels by which many of the simple inorganic substances vital to life enter the biosphere. Carbon is taken from the carbon dioxide of the atmosphere and "fixed" into organic compounds during the process of photosynthesis. Elements such as nitrogen and sulfur are taken from the soil in the form of simple inorganic compounds and incorporated into proteins, vitamins, and other essential organic compounds within green plant cells. Animals cannot make these organic compounds from inorganic materials and so are entirely dependent on plants for these compounds as well as for their sources of energy.

Finally, plants, by the process of photosynthesis, are the producers of oxygen, on which all higher life also depends.

4–15 (on facing page)

Seeds are dispersed in many ways: some are borne on the wind, some are carried from one place to another by animals, some float on water, and some are even forcibly ejected by the parent plant. (a) *Coconut palms grow along sandy seabeaches, by the water's edge. Coconuts, because of their buoyant fibrous husk and gas pockets in the embryo, float and so are dispersed by ocean currents.* (b) *The winged fruit of the maple is carried by the wind. As you can see if you look closely, each "wing" carries a separate seed.* (c) *The seeds of fleshy fruits are usually dispersed by vertebrates that eat the fruits and later excrete the seeds. The tree sparrow shown here is eating the bright red fruits of winterberry, a member of the holly family.* (d) *In milkweed, the fruit is the seed pod which, when ripe, bursts open releasing seeds with tufts of silky hair that aid in their dispersal.* (e) *In the tumbleweed the whole plant breaks off in the wind and is tumbled across open country, scattering the seeds.*

(a) (b) (c)

(d) (e)

THE ANIMALS

Animals are defined as many-celled heterotrophs. They depend directly or indirectly for their nourishment on land plants or algae. Most digest their food in an internal cavity. They move by means of contractile fibers, such as those in muscles or cilia. Reproduction is almost always sexual. The higher animals—both the insects and the vertebrates—are the most complex of all organisms, with many kinds of specialized tissues, including elaborate sensory and neuromotor mechanisms not found in any of the other kingdoms.

For most of us, animal means mammal, and mammals are, in fact, the chief focus of attention in Section 6. However, the mammals represent only a small fraction of the animal kingdom. More than 90 percent of the different kinds, or species, of animals are invertebrates, that is, animals without backbones. We shall discuss briefly only a few members of this large, varied, and fascinating group, which has been the subject of hundreds of volumes far larger than this. In the suggestions for further reading at the end of this section are some books to read for pleasure about these members of the animal kingdom. The section concludes with a survey of the evolution of the vertebrates.

Origins of the Animalia

Animals, like plants, presumably had their origins among the protists, although in the case of animals we have fewer clues about which particular group of protists most closely resemble the ancestral ones. Some students of evolution believe that a colonial form similar to *Volvox* (Figure 4–9) may have been a starting point for animal evolution; this colony of cells forms a hollow sphere, and a hollow sphere of cells is a transient development stage in many species of animals—including, as we shall see, man.

By the Cambrian period, which ended some 500 million years ago, many of the present-day types of invertebrates were already in existence. Although numerous efforts have been made over the past hundred years to trace the evolutionary relationships of these earlier forms, the Precambrian fossil record is inadequate to verify any hypotheses about the chronological order of their appearance. Most of the evidence for charts such as that shown in Figure 4–19 comes from studies of modern forms.

There are about 30 major divisions—phyla—of invertebrates, each distinguished from all the others by a particular type of body plan. Of these, we are going to discuss only a few, concentrating on the most "successful," as judged on the basis of surviving species.

Sponges: Phylum Porifera

Sponges, the most primitive of the modern animals, represent a level of organization somewhere between a colony of cells and a true multicellular organism. As with colonial forms, if the individual cells of the sponge are separated, such as by pressing the tissue through a fine cloth, and then mixed together again, the cells will reassemble to form a whole animal. In contrast to true colonial forms, however, the cells of the sponge are to a certain extent differentiated and specialized. The different cell types include feeding cells (the flagellated collar cells, or choanocytes); epithelial cells, some of which contain contractile fibers; and amoebalike cells (archeocytes), which serve several functions including that of carrying digested food from the collar cells to the epithelial cells and other nonfeeding types. Other cells produce the stiffening structures, either inorganic spicules or organic fibers, which form the skeleton, the only part of the sponge that remains when it is dried and cleaned. Because all the digestive processes of sponges are carried out within single cells, even a giant sponge—and some stand taller than a man—can eat nothing larger than microscopic particles, obtained by filtering the water that is driven through the body of the sponge by the flagella of the collar cells. Sponges are well suited to a slow life on the ocean floor. Sunlight is not necessary to their existence, and they have been found at depths that support few other living creatures. However, they are generally regarded as representing an evolutionary dead end, having given rise to no other forms.

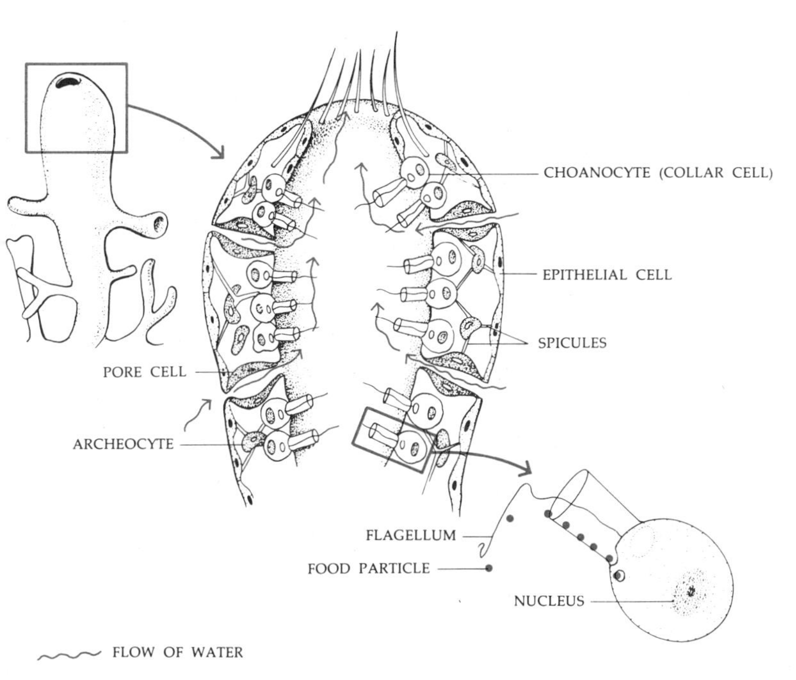

4–16

The body of the sponge is dotted with tiny pores, from which the phylum derives its name (Porifera, or "pore bearers"). Water containing food particles is drawn into the internal cavity of the sponge through these pores and is exhaled out the "mouth." The water is kept in motion by the beating of the flagella within the high collars of the choanocytes. Food particles adhere to the sticky outside of the collar and are drawn down its outer surface to the base, where they are engulfed into the cell body. The digested food is then shared by diffusion with other sponge cells. A sponge filters as much as a ton of water to gain an ounce of body weight.

Coelenterates: Phylum Coelenterata

The coelenterates are a large group of aquatic animals—including jellyfish, hydras, corals, and sea anemones—which are characterized by a hollow, fundamentally two-layered body. The interior cavity of the body is known as the *coelenteron,* from which the coelenterates get their name. (See Figure 4–17.)

Food is captured by means of tentacles and pushed into the coelenteron which extends and engulfs it. Within the coelenteron it is digested to some extent by enzymes secreted by the cells lining the cavity. Digestion is then completed within food vacuoles in the cells lining the coelenteron, very much as in an amoeba. The indigestible remains are then ejected by the same opening through which the food enters. Thus the coelenterates are able to eat almost any prey they can stuff into their remarkably expandable body cavities.

In the simplest types, the body is made up of only two cellular layers: an outer layer, the epidermis, and an inner layer, the gastrodermis. Between these two layers is a jelly-like filling, most conspicuous in the jellyfish. Coelenterates have primitive nervous systems. The medusa form—exemplified by the jellyfish—has two rings of nerve cells that circle the margin of the bell, whereas the polyp form—as in the hydra—has a continuous network of nerve cells just below the surface of the epidermis which links the body into a functional whole and makes possible coordinated movements. Highly specialized stinging cells, cnidoblasts, are found only among this group and are used for defense and to capture their prey.

The life cycle of coelenterates characteristically includes an immature larval form, known as the planula, which is a small free-swimming ciliated organism. In the life cycle of most coelenterates, the planula settles to give rise to a polyp which reproduces asexually and may form extensive colonies. From the polyp, young jellyfish (medusas) may bud off. These are sexually-reproducing adult forms which give rise to planulas again. All adult forms are radially sym-

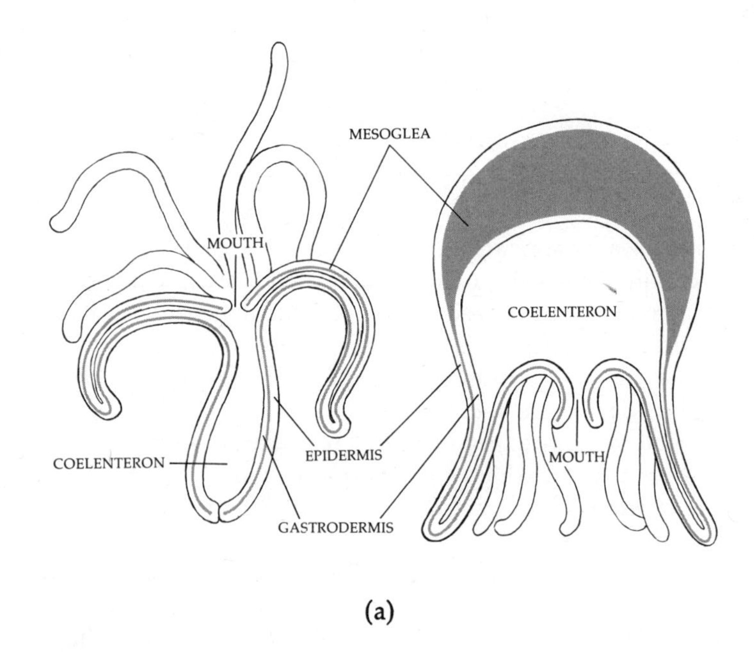

(a)

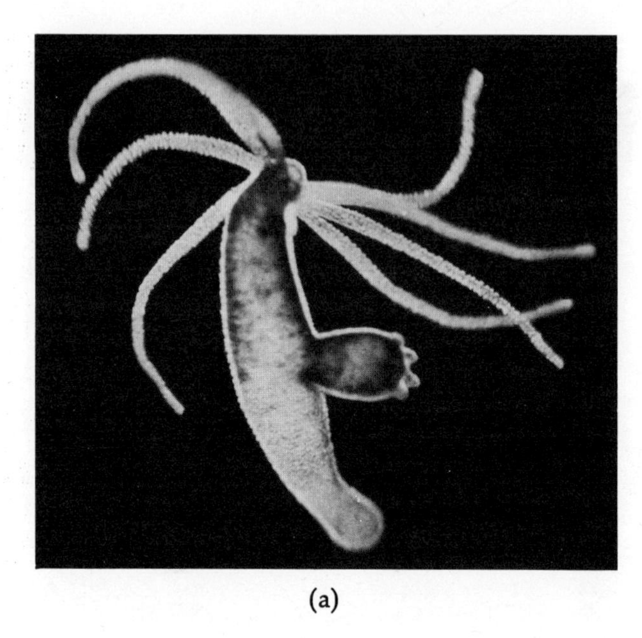

(a)

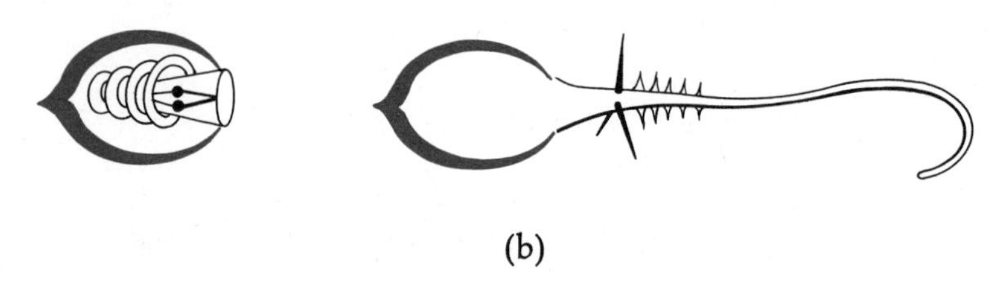

(b)

4–17

(a) *All adult coelenterates are radially symmetrical and conform to one of two basic body plans: the vase-shaped polyp (left) or the bowl-shaped medusa (right).* (b) *A distinguishing feature of coelenterates are their cnidoblasts, specialized stinging cells located in the tentacles and body wall. The interior of the cnidoblast is filled by a nematocyst, which is a capsule containing a coiled tube, as shown on the left. A trigger on the cnidoblast, responding to chemical stimulus and/or touch, causes the nematocyst to discharge, and the tube, turning inside out, explodes to the outside. The tube shown here is armed with barbs and spines, which penetrate the tissues of the prey and inject a paralyzing poison.*

4–18

Some coelenterates. (a) *A hydra, when hungry, extends and waves its tentacles, which surround its mouth. When they come into contact with prey, the nematocysts discharge, paralyzing the victim, and the tentacles contract, conveying the food to the mouth which opens to receive it. The bulge on the side is a bud which will eventually produce a new, independent hydra.* (b) *A jellyfish, a medusa form of coelenterate. The one shown here is swimming vigorously, which it accomplishes by contracting and relaxing a muscular system made up of epithelial cells containing contractile fibers.* (c) *A sea anemone eating a small fish. These animals, which look like flowers, are not mobile and are often found near coral reefs where fish and other prey are abundant.* (d) *A number of different types of corals, each composed of a colony of thousands of individual stinging polyps. Except for the sea fan near the center, these are all soft corals which lack the hard calcium-containing skeleton of the reef corals. Their bodies are stiffened by scattered spicules.*

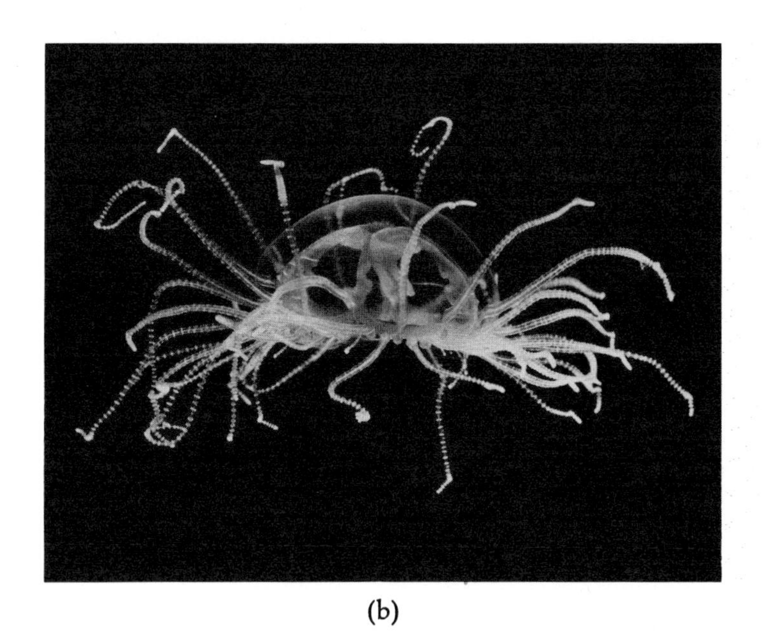

(b)

(c)

(d)

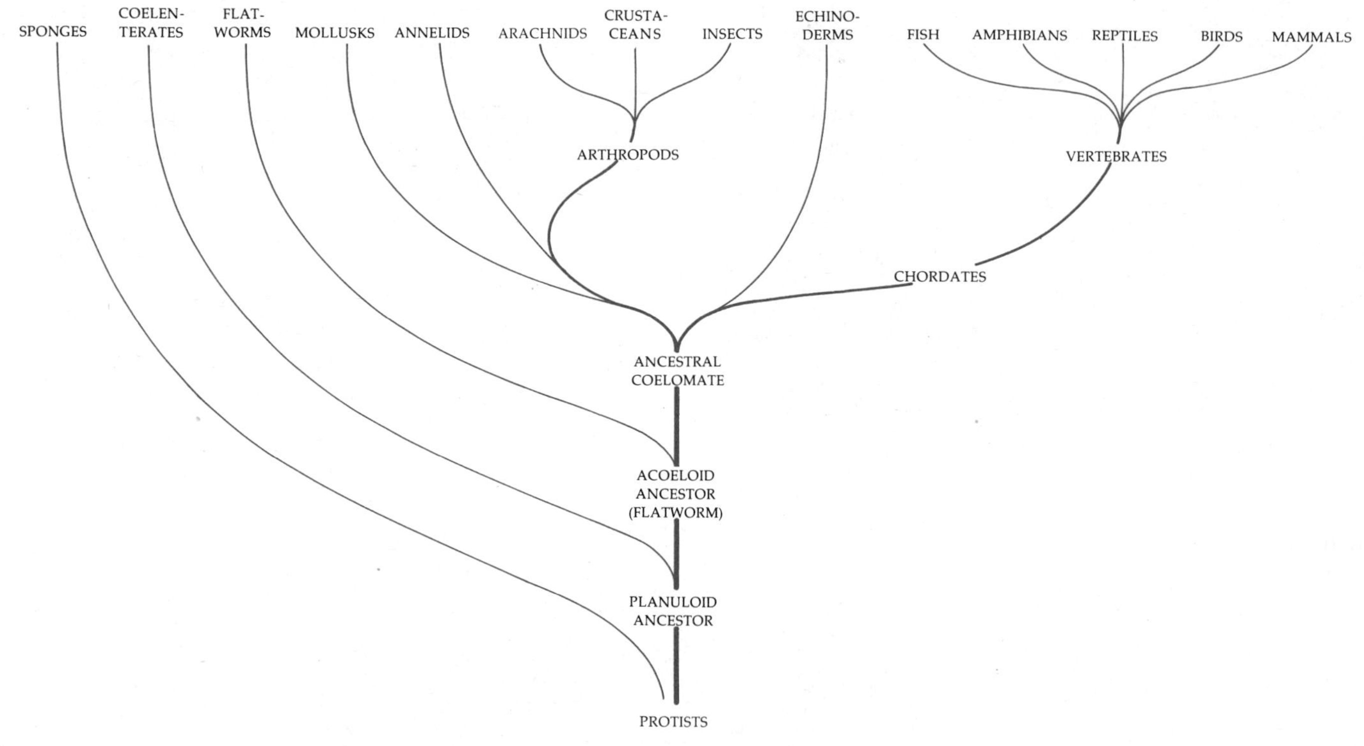

4–19

Evolutionary relationships in the animal kingdom, according to one widely accepted theory.

metrical, which means that their body parts are arranged around an axis, like spokes around the hub of a wheel.

The most ecologically important members of this group are the coral builders, which are responsible for the formation of great land masses in the sea where ordinarily no land would exist. The 1,200-mile-long Great Barrier Reef off the northeast shores of Australia and the Caroline Islands in the Pacific is an example of such a coral-created land mass. A coral reef is composed primarily of the accumulated limestone skeletons of coral coelenterates, covered by a thin crust occupied by the living colonial animals.

Many biologists believe that a primitive coelenterate is to be found on the evolutionary pathway leading from the protists to the higher forms. A principal clue in this evolutionary puzzle, they point out, is the ciliated larva. It is easy to imagine a gradual transition between such a form and the simplest of the flatworms, the next big step in the order of complexity. A form corresponding to the planula of the jellyfish and other modern coelenterates may have been the starting point for wormlike forms in the Kingdom Animalia.

The development of a new group of animals from an immature stage of an ancestral form is known as *paedomorphosis* ("taking shape from a child"). Presumably, in the course of this process, larval forms came to be more and more dominant in successive generations, until eventually they became sexually mature and capable of reproducing themselves. Paedomorphosis, as we shall see, is also believed to have played an important role in the evolution of the vertebrates.

Flatworms: Phylum Platyhelminthes

The flatworms are the simplest animals, in terms of body plan, to show *bilateral symmetry*. In bilaterally symmetrical animals, the body plan is organized along a longitudinal axis, with the right half an approximate mirror image of the left half. A bilaterally symmetrical animal can move more efficiently than a radially symmetrical one. It also has a top and a bottom, or in more precise terms (applicable even when it is turned upside down or, as in the case of humans, standing upright), a dorsal and a ventral surface. Like most bilateral organisms, the flatworm also has a distinct "headness" and "tailness." Apparently when one end goes first, it is advantageous to collect the sensory cells into that end. With the aggregation of sensory cells, there came a concomitant gathering of nerve cells; this gathering is a forerunner of the brain.

Flatworms have three distinct tissue layers—ectoderm, mesoderm, and endoderm—characteristic of all animals above the coelenterate level of organization. Moreover, not only are their tissues specialized for various functions, but two or more types of tissue cells may combine to form organs. Thus while coelenterates are largely limited to the tissue level of organization, flatworms can be said to have gained the organ level of complexity.

The free-living flatworms, of which the freshwater planarians are familiar examples, suck bits of dead animals into their highly branched coelenterons (Figure 4–20), where they are ingested and digested by the cells lining the cavity. The indigestible residue is ejected through the mouth. Two light-sensitive spots at the anterior end resemble a pair of crossed eyes, and two projections on either side of the head resemble ears but are really areas sensitive to chemical stimuli such as emanate from meat or other food. Within the mesoderm is a fairly complex musculature, which enables the animal, when disturbed, to move rapidly

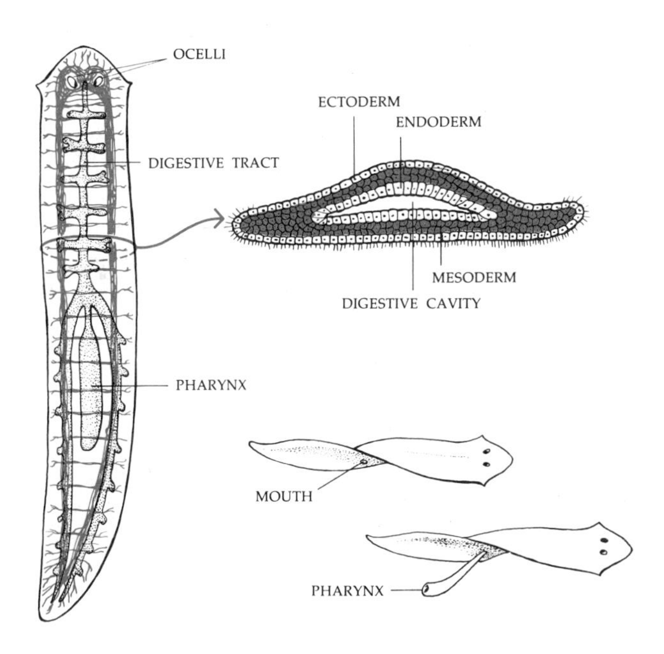

4–20

Example of a flatworm: the freshwater planarian. The planarian, which is carnivorous, feeds by means of its extensible pharynx. The nervous system is indicated in color. Note that some of the fibers have been aggregated into two cords, one on each side of the body, and there is a cluster of nerve cells in the head, the beginnings of a brain. The ocelli are light-sensitive areas.

with a sort of loping motion, resulting from wavelike contractions of muscles. Planarians have an extensive excretory system and complex reproductive organs. The phylum also includes the parasitic flukes and tapeworms, which often have complex life cycles involving several hosts in succession.

Ribbon Worms: Phylum Nemertea

The ribbon worms (nemertines), although they constitute a small phylum, are of special interest to biologists attempting to reconstruct the evolution of the invertebrates because they appear to be closely related to the flatworms with an important difference: they have a one-way digestive tract beginning with a mouth and ending with an anus. This is a far more efficient arrangement than the one-opening digestive system of the coelenterates and flatworms; in the one-way tract, food moves assembly-line fashion, with the consequent possibility of specialization of various segments of the tract for different stages of digestion. The nemertines also have a circulatory system, which aids in the distribution of food molecules.

Roundworms: Phylum Nematoda

The number of species of roundworms has been variously estimated as low as 10,000 and as high as 400,000 to 500,000. Roundworms range from microscopic forms to parasites 6 feet long, but they are all structurally so similar that they are considered to constitute only one class, the nematodes.

Nematodes have a three-layered body plan and a tubular gut with a mouth and an anus. They are unsegmented and are covered by a thick continuous cuticle, which they must shed as they grow, like anthropods. Most are free-living microscopic forms. It has been estimated that a spadeful of good garden soil usually contains about a million nematodes. Some are parasites; all species of plants and animals are parasitized by at least one species of nematodes. Man is host to about fifty species.

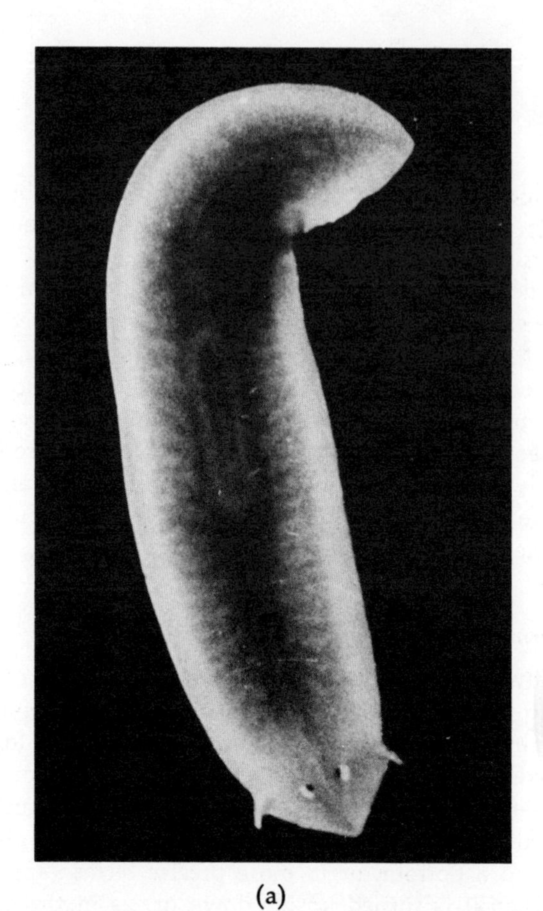

(a)

4–21

Some worms. (a) *A planarian, a freshwater flatworm;* (b) *a marine flatworm which, although different in external appearance, has much the same body plan as the planarians;* (c) *African tapeworm found in a female rhinoceros. Some tapeworms are 15 to 20 feet long. About two-thirds of the species of the phylum of Plathyhelminthes are parasitic forms.* (d) *Annelids, of which earthworms are the most familiar example, have bodies composed of many separate segments. About one-third of the way from the anterior end of each is a thickened band, the clitellum, which during mating secretes mucus that surrounds the fertilized eggs and protects them.* (e) *A more characteristic annelid, a polychaete, or bristle worm. Note the appendages on each segment.*

(b)

(c)

(d)

(e)

Segmented Worms: Phylum Annelida

This phylum includes almost 7,000 different species of marine, freshwater, and soil worms, including the familiar earthworm. The term *annelid* means "ringed" and refers to the most distinctive feature of this group, which is the division of the body into segments, not only by rings on the outside but by partitions on the inside. This segmented pattern is found in a modified form in higher animals, too, such as dragonflies, millipedes, and lobsters, which are thought to have evolved from ancestors that probably gave rise also to modern annelids. Although the earthworm is probably the most familiar of the annelids, it is atypical because of its much-reduced appendages.

The annelids have a three-layered body plan, a tubular gut, and a well-developed, closed circulatory system which transports oxygen (diffused through the skin) and food molecules (from the gut) to all parts of the body. (The circulatory system of the earthworm is shown in Figure 4–22.) The excretory system is made up of specialized paired tubules, *nephridia*, which occur in each section of the body except the head. Annelids have a nervous system and a number of special sense cells, including touch cells, taste receptors, light-sensitive cells, and cells concerned with the detection of moisture.

The annelids also contain *coeloms;* they are the first group of animals discussed so far in which this important evolutionary development is seen. In the flatworms, the mesoderm is packed solid with muscle and other tissues, but in the annelids there is a fluid-filled cavity, the coelom, in this middle layer. (Note that the term "coelom," although it sounds similar to "coelenteron" and comes from the same Greek root, meaning "cavity," refers quite specifically to a cavity *within* the mesoderm, whereas the coelenteron is a digestive cavity lined by endoderm.) The gut is suspended within this cavity by mesenteries, made of double layers of mesoderm. The fluid in the coelom constitutes a sort of hydraulic skeleton for the earthworm, in somewhat the same way water keeps a fire hose distended.

4–22 (at right)
An earthworm, example of an annelid. (a) *The digestive tract and nervous system. The mouth leads into a muscular pharynx which acts as a suction pump, drawing in decaying leaves and other material into the gut. These are stored in the esophagus and ground up in the gizzard, with the help of soil particles. The rest of the tract is a long intestine in which food is digested and absorbed. The nervous tract consists basically of a double nerve cord running down the ventral surface of the animal, with fibers branching from it that innervate each segment. The nerve cord divides at the anterior of the worm, circles the pharynx, and meets again in two clusters of nerve bodies (ganglia) which form the primitive brain.* (b) *The circulatory system comprises longitudinal vessels running the entire length of the animal, one dorsal and several ventral. Smaller vessels in each segment collect the blood from the tissues and feed it into the muscular dorsal vessel through which it is pumped forward. In the anterior segments are five pairs of hearts—muscular pumping areas in the blood vessels—whose irregular contractions force the blood downward to the ventral vessels from which it returns to the posterior segments. Both the hearts and the dorsal vessels have valves that prevent a backflow.* (c) *A segment in cross section. Note the large coelom. The nephridia are primitive kidneys, each of which consists of a long ciliated tubule through which coelomic fluid is excreted through an outer pore. As the fluid masses down the tubule, sugars, salts, and other needed materials are returned to the coelom through the walls of the tubule.*

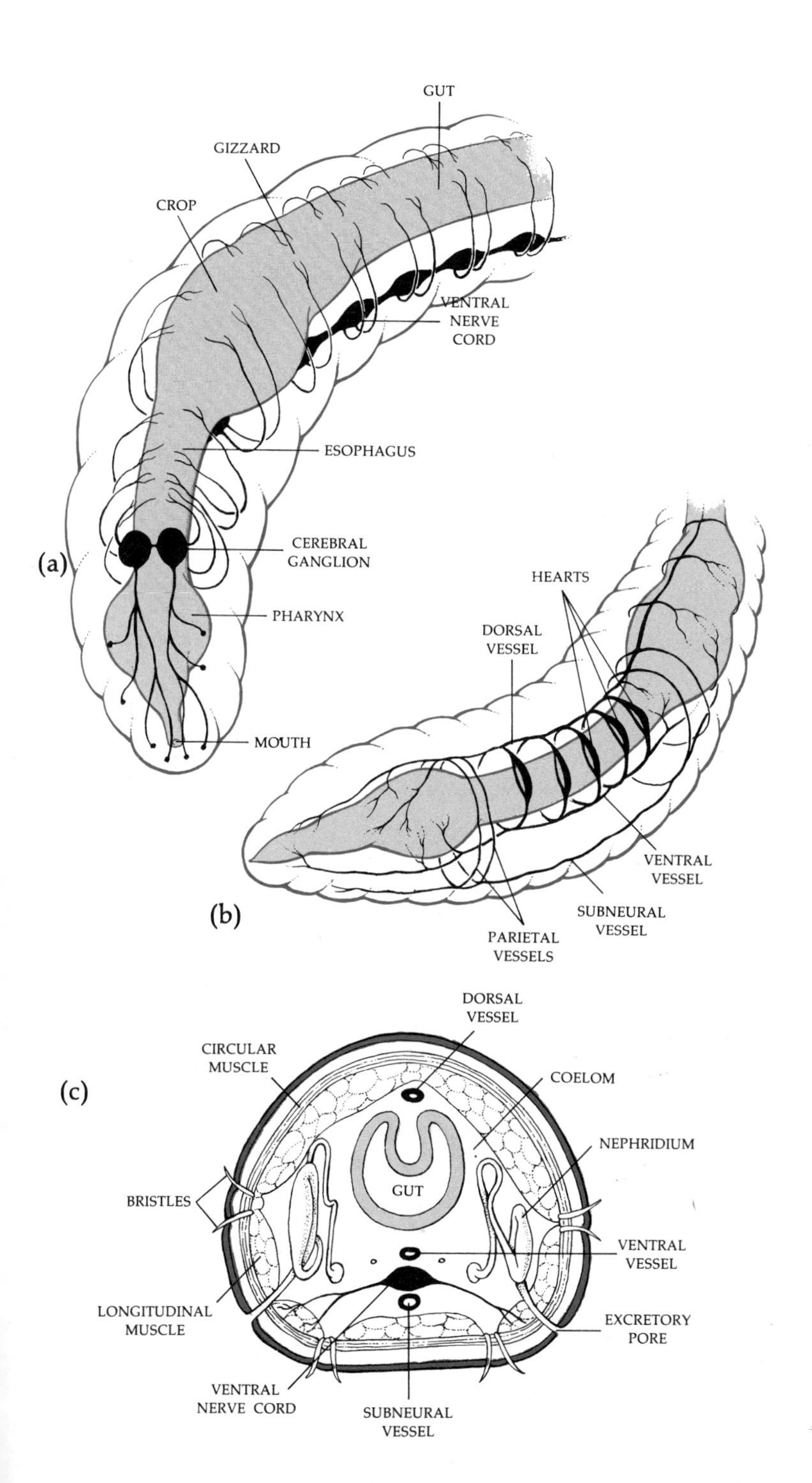

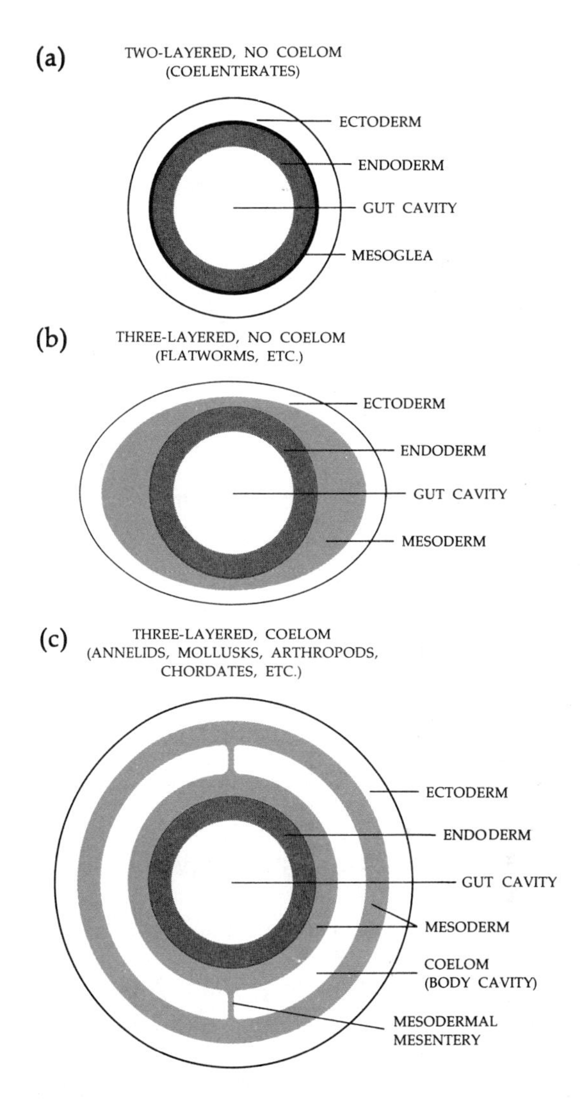

4–23

Basic body plans of the animal phyla, as shown in cross section. A body which consists fundamentally of only two layers (a), *is characteristic of coelenterates. Flatworms and ribbonworms have three-layered bodies, with the layers closely packed on one another* (b). *Annelids and most other animals, including vertebrates, have bodies which are three-layered with a cavity, the coelom, within the middle layer (mesoderm). The mesodermal mesenteries suspend the gut within the body wall* (c).

Most of the higher invertebrates, and all of the vertebrates as well, have a body plan that includes the coelom. Within the coelom, organ systems can bend, twist, and fold back on themselves, increasing their functional surface areas and filling, emptying, and sliding past one another, surrounded by lubricating coelomic fluid. Examples of such organ systems are the human lung, constantly expanding and contracting in the chest cavity, and the 20 or so feet of coiled human intestine, both of which occupy the coelom.

Echinoderms: Phylum Echinodermata

The echinoderms ("prickly skins") include the sea urchins, sand dollars, sea cucumbers, and sea lilies. Adult echinoderms are radially symmetrical, like the coelenterates. The most familiar of the echinoderms are the starfish, which have the spiny skin from which the phylum derives its name. Most starfish have five arms, but some have multiples of five, up to as many as twenty. Other members of the phylum also have bodies arranged in fives or multiples of five. The water vascular system (illustrated in Figure 4–24) provides suction for the clinging and pulling activities of the tube feet, with which many members of the phylum cling, move, and attack prey.

Although the adults are radially symmetrical, the larvae of the members of this phylum are bilaterally organized; it is believed that the radial symmetry is a late, secondary development in the evolution of the group. In the development of these larvae, the first opening of the alimentary canal of the embryo becomes the anus, while the mouth breaks through secondarily at the other end of the larval digestive tract. Animals with such a pattern of early development are known as deuterostomes ("second the mouth") in contrast to the protostomes such as mollusks, annelids and arthropods in which the mouth develops first. The primitive chordates from which the vertebrates arose, as we shall see, also are deuterostomes. For this reason, among others, the vertebrates are believed to have a closer evolutionary relationship with this group than with the other invertebrate phyla, as shown in Figure 4–19.

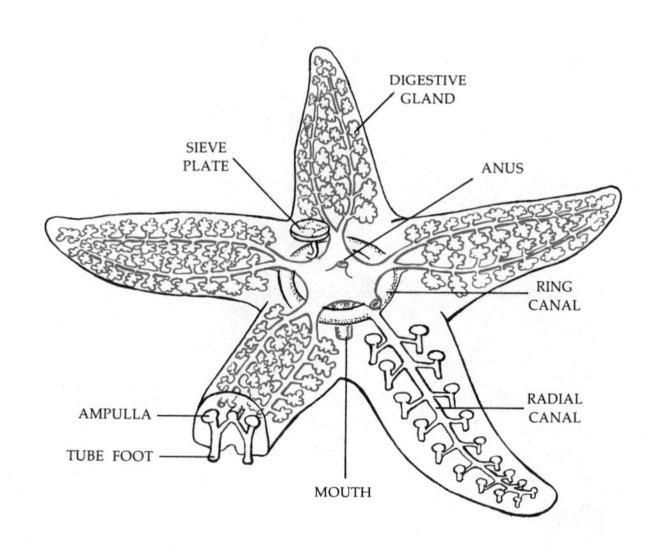

4–24

The water vascular system of the starfish is its means of locomotion. Water enters through minute openings in the sieve plate and is drawn, by ciliary action, down a tube to the ring canal. Five radial canals, one for each arm, connect the ring canal with many pairs of tube feet, which are hollow, thin-walled cylinders ending in suckers. Each tube foot connects with a rounded muscular sac, the ampulla. When the ampulla contracts, the water in it, prevented by a valve from flowing back into the radial canal, is forced under pressure into the tube foot. This stiffens the tube, making it rigid enough to walk on, and extends the foot until it attaches to the substratum by its sucker. The longitudinal muscles of the foot then contract, forcing the water back into the ampulla. If the tube feet are planted on a hard surface, such as a rock or a clamshell, the collection of tubes will exert enough suction to pull the starfish forward or to pull open the clam.

Mollusks: Phylum Mollusca

The mollusks constitute one of the largest phyla of animals, both in numbers of species and in numbers of individuals. They are characterized by soft bodies within a hard, calcium-containing shell, although in some forms the shell has been lost in the course of evolution, as in slugs and octopuses, or greatly reduced in size, as in squids. There are three major classes of mollusks: (1) the gastropods, such as the snails, whose shells are generally in one piece; (2) the bivalves, including the clams, oysters, and mussels, which have two shells joined by a hinge ligament; and (3) the cephalopods, the most active and most intelligent of the mollusks, including the cuttlefish, squids, and octopuses.

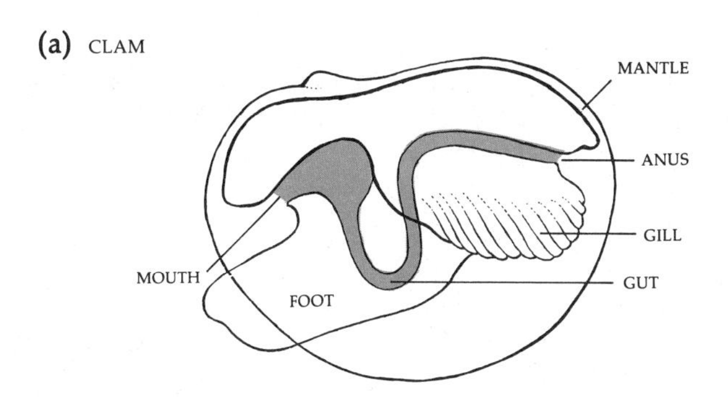

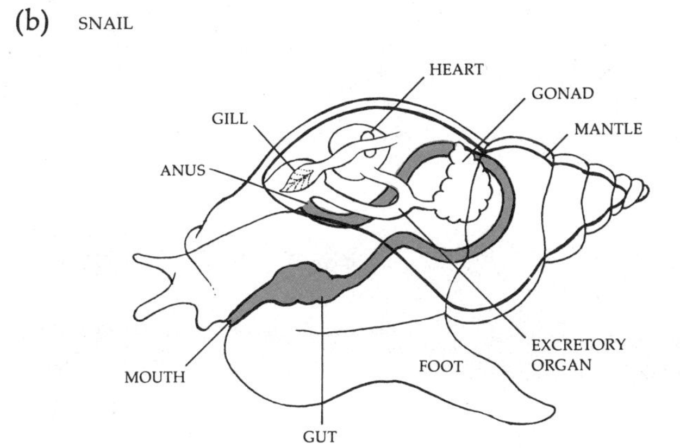

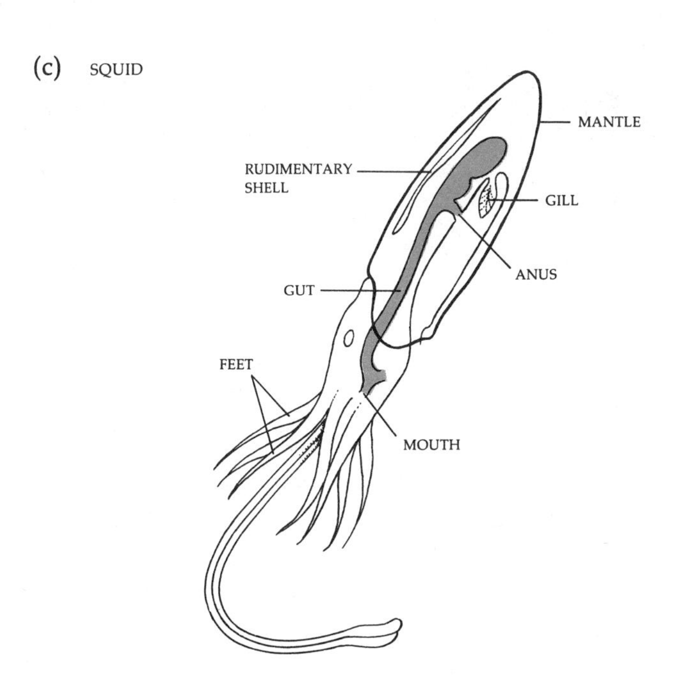

4–25

Mollusks are characterized by soft bodies composed of a head, a foot, and a visceral hump covered by a mantle which secretes a shell. They breathe by gills except for the land snails in which the mantle cavity has been modified for air breathing. The three major groups are the bivalves, such as the clam (a), *which are generally sedentary and feed by filtering water currents, created by beating cilia, through large gills; the gastropods, exemplified by the snail* (b), *in which the visceral hump has become coiled upward and the gut turned back so that mouth, anus, and gills all share the same small aperture in the mantle; and the cephalopods, such as the squid* (c), *in which the head is modified into a circle of tentacles and part of the foot forms a tubelike siphon through which water can be forcibly expelled, providing for locomotion by jet propulsion.*

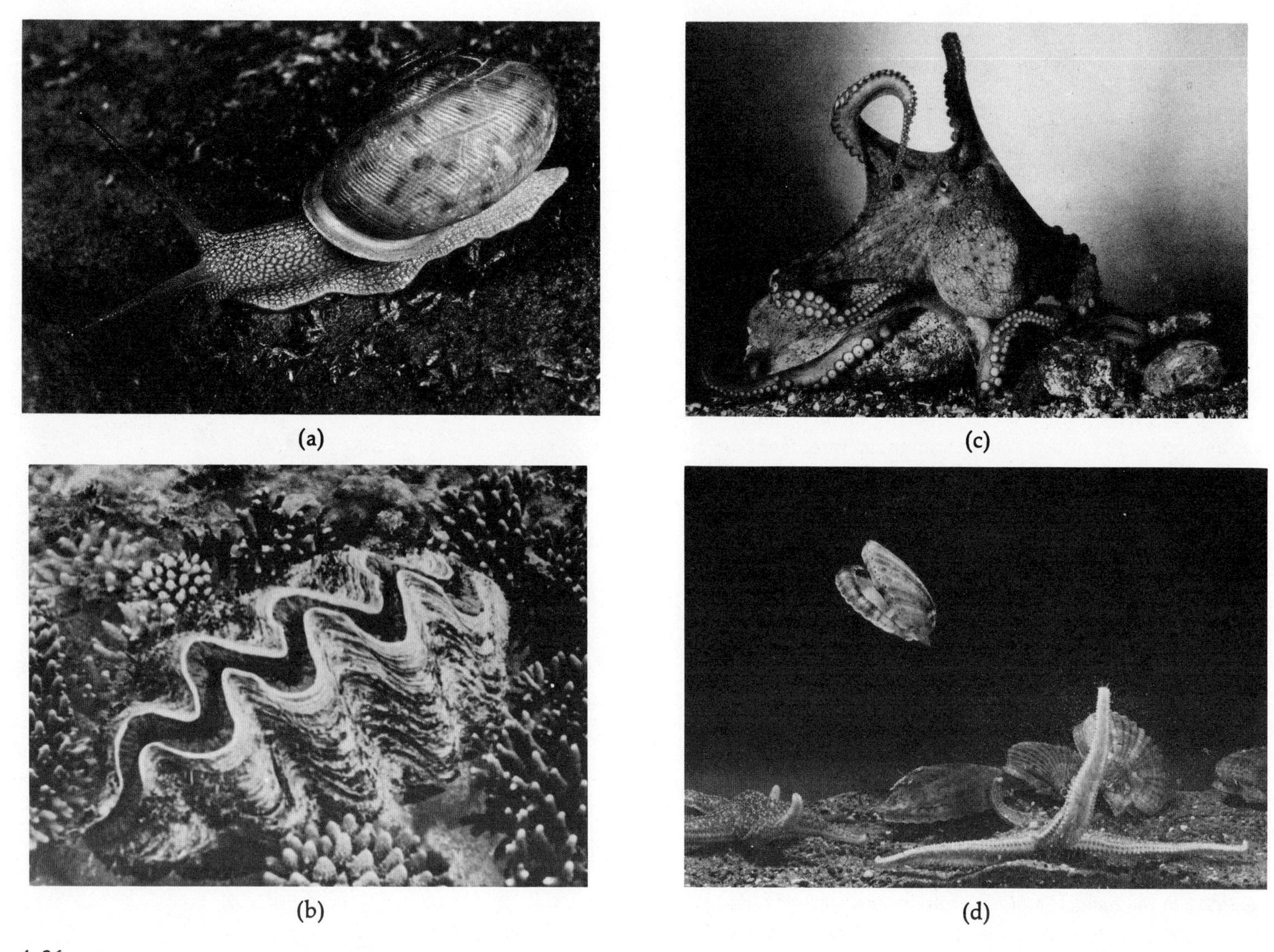

4–26

Four mollusks: (a) *A land snail. At the tip of each long tentacle is a simple eye.* (b) *Giant clam. Although a filter feeder, like other bivalves, these clams, which are often several feet in diameter, are reported to capture divers between the huge "teeth" along the edges of their shells.* (c) *An octopus, advancing.* (d) *An alarmed scallop, jet-propelling itself through the water by rapidly opening and closing its shell, expelling water from its mantle cavity. It can detect an enemy, such as a starfish, by means of chemical receptors. The other nearby scallops are also preparing to take off.*

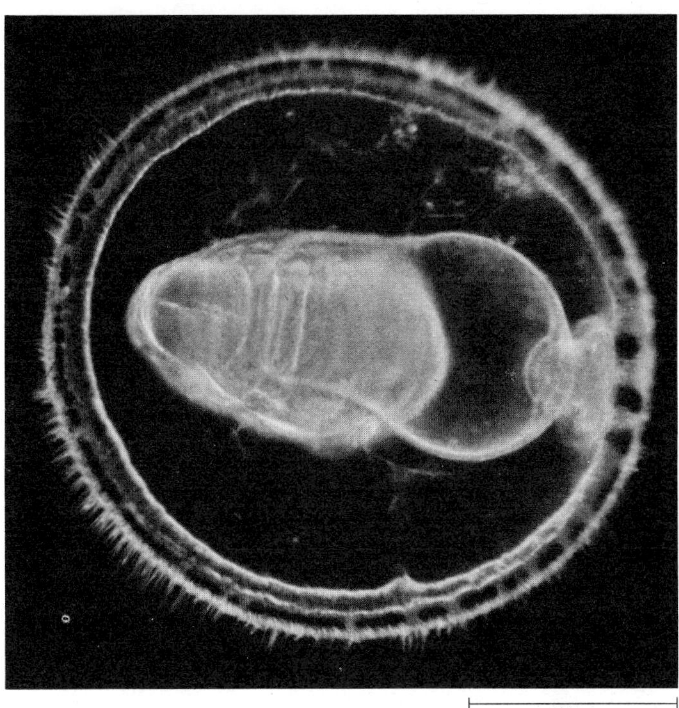

4–27
The trochophore larva (from trochos, *meaning "wheel") is an evolutionary link between annelids, mollusks, and arthropods. Although their adult forms are very different, all have larvae of this type. This particular larva will develop into a polychaete worm.*

Although the mollusks are diverse in size and shape, they all have the same fundamental body plan. There are three distinct body zones: a head-foot, which contains both the sensory and the motor organs; a "hump," or visceral mass, which contains the organs of digestion, excretion, and reproduction; and a mantle, which hangs over and enfolds the visceral mass and which secretes the shell. The mantle cavity, a space between the mantle and the "hump," houses the gills, and the digestive, excretory, and reproductive systems discharge into it.

A characteristic organ, found only in this phylum, is the radula. This tooth-bearing tonguelike organ is variously used by different kinds of mollusks to scrape off algae, to drill holes in the shells of barnacles or other mollusks, or to aid in the ingestion of prey animals. It is present in all mollusks except the bivalves.

The molluscan circulatory system, which includes a chambered heart, is the most efficient circulatory system of the invertebrates. The nervous systems of the gastropods and bivalves are simple, but among the cephalopods, the octopus, in particular, has a highly developed nervous system and a brain and eyes that rival those of the vertebrates in complexity, although different in design. (The light-sensitive cells of the octopod retina point toward the light rather than away from it, as they do in vertebrates.) Octopuses are capable of complex behavior, memory, and learning.

Arthropods: Phylum Arthropoda

The arthropods make up by far the largest of all phyla. Almost 800,000 species have been classified (there are 300,000 different kinds of beetles alone!), and some experts estimate that there may be as many as 10 million species in existence.

The characteristic features of the arthropods are their segmented organization, similar to that of the annnelids, from which they almost certainly evolved, and their hard outer covering, or *exoskeleton,* containing chitin. In the evolution of the arthropods, the various segments of the

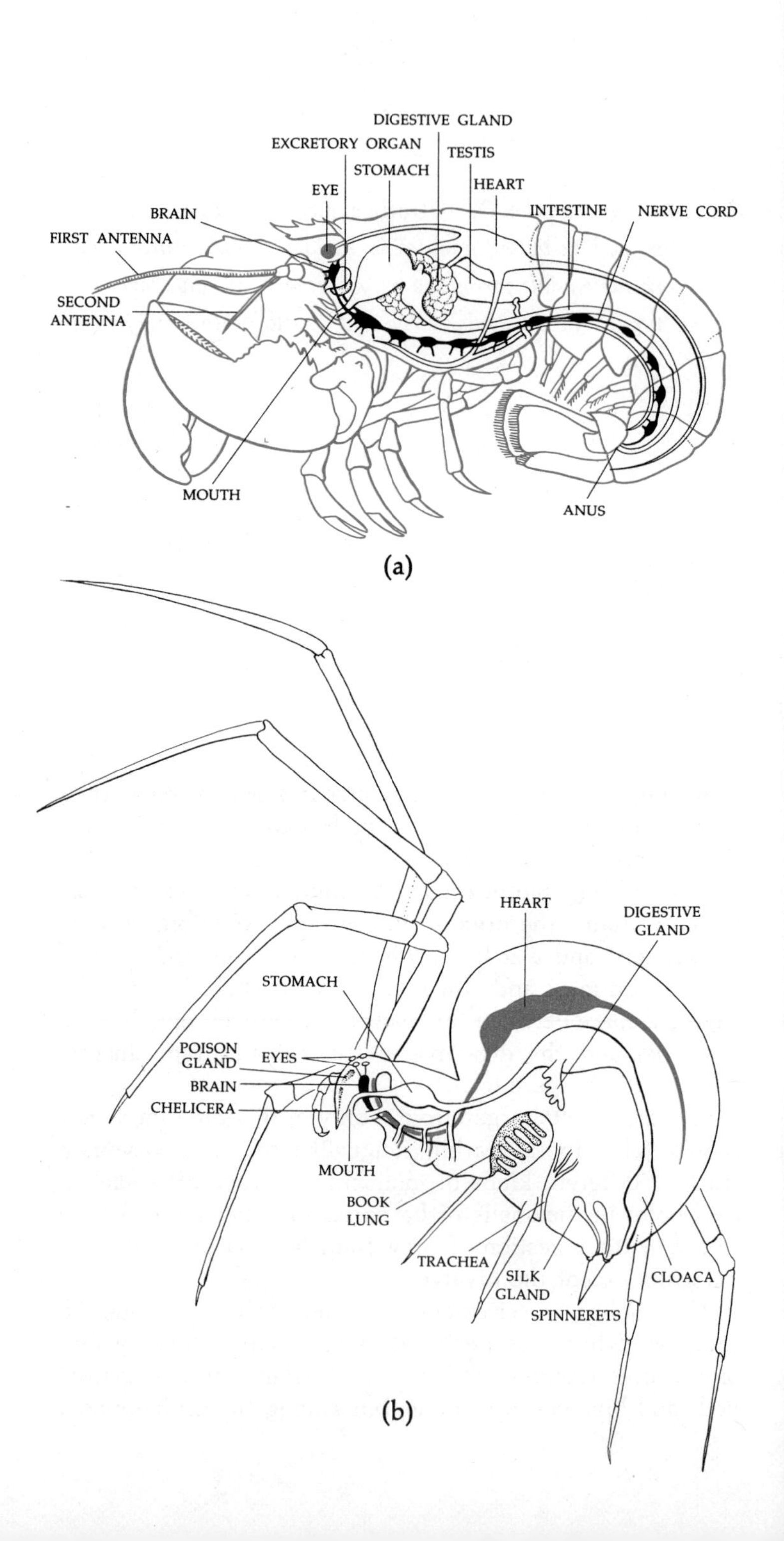

4–28

Three representative arthropods. (a) *In the lobster, a crustacean, the head and thorax are covered by a chitin shield (a carapace). The head, which is made up of the first body segments, has antennae, chewing jaws (mandibles), and feeding appendages (maxillaries), as well as two pairs of antennae, one simple eye and one compound eye. Of the eight segments of the thorax, three have appendages which tear food and pass it toward the mouth (maxillipeds) and five have legs. The principal appendages of the abdominal segment are swimmerets. The lobster, like most other crustaceans, breathes by means of gills, which lie in the body cavity between the thorax and the exoskeleton.* (b) *The spider, an arachnid, has an unsegmented cephalothorax (head and thorax joined) developed from the first eight segments. Unlike insects and arthropods, arachnids have chelicerae, appendages which take the form of pincers or fangs. Spiders capture prey with their chelicerae and inject the prey with poison from poison glands connected with the chelicerae by ducts. Like other arachnids, the spider has four pairs of walking legs.* (c) *In the grasshopper, an insect, the head consists of six fused segments which have appendages specialized for biting and chewing. Each of the three segments of the thorax carries a pair of legs (three pairs in all), and two of them carry wings (in the grasshopper, the forewings are hardened as protective covers). The spiracles in the abdomen open into a network of chitin-lined tubules through which air circulates to various tissues of the body. This sort of tubular breathing system is found only among insects and other land-dwelling arthropods (the centipedes and millipedes).*

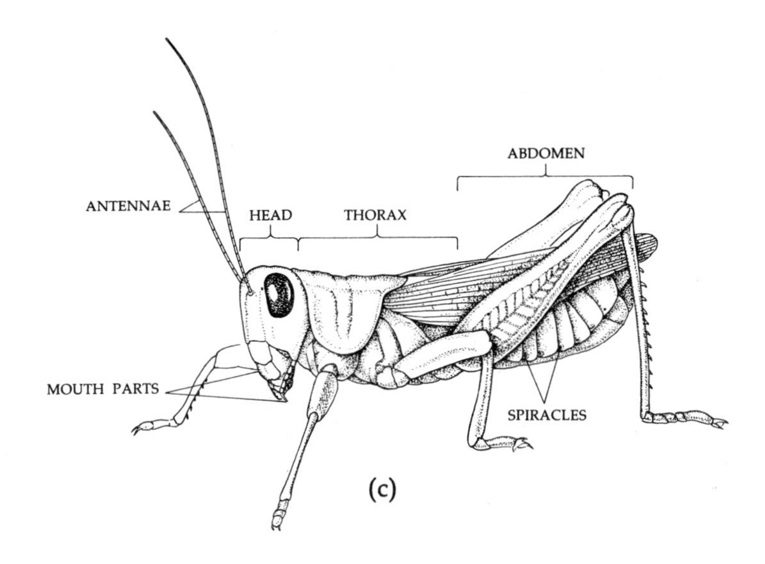

(c)

body became specialized in different ways. The most anterior segments form the head, the next constitute the thorax (head and thorax segments are sometimes fused into a *cephalothorax*), and the final segments are the abdomen. In some of the arthropods, such as the centipedes, which have a pair of legs on almost every segment of the clearly wormlike body, the appendages are uniform in size and structure, but more typically, arthropods have highly differentiated appendages, specialized for walking, swimming, feeling, feeding, chewing, biting, flying, and other such functions, depending on the species.

The exoskeleton deters predators and, in the land forms, protects the animals from desiccation. It is many-jointed and is attached to the musculature so that it moves with the animal, rather than the animal moving inside of it, as do the mollusks within their shells. However, it covers the animal completely and does not grow; as a consequence, arthropods must molt, a process by which the old exoskeleton is discarded and a new and larger one formed. At molting time, the epidermis secretes an enzyme that dissolves the inner layer of the exoskeleton and a new skeleton, not yet hardened, is formed beneath the old one. The animal wriggles out from the old skeleton, which splits open. After emerging, the arthropod expands rapidly by taking up air or water, stretching the new exoskeleton before it hardens.

The anterior section of the arthropod gut is lined with chitin and serves largely to grind food into smaller particles. Large digestive glands secrete enzymes into the midsection, where digestion takes place. Arthropods have a complex endocrine (hormone-producing) system, which plays a major role in molting. Their circulatory system is an open system, in which the blood flows from the blood vessels into blood sinuses, or cavities. The heart is a tube with paired openings suspended within a blood cavity which pumps the blood into the sinuses.

The nervous system of the arthropods is complex, making possible the intricate and finely tuned movements involved in activities such as flight, mating in midair, and building webs and hives. Arthropods have a number of extremely sensitive sensory organs and are capable of complex behavior. Many of them, such as ants, honeybees, and termites, live in highly organized societies. However, they show little capacity to modify their behavior, that is, to learn.

The three major classes of arthropods are (1) the crustaceans, which include lobsters, crabs, barnacles, water fleas (daphnias), and copepods; (2) the arachnids, which are spiders, ticks, mites, scorpions, and daddy longlegs; and (3) the insects, which are by far the largest group. The enormous success of the insect class appears to be related to the high degree of specialization of the individual species. Insects are usually so selective about where they live and how they eat that many different species can live noncompetitively in a very small area—such as on a single plant or in or on one small animal.

Most insects go through definite developmental stages. In some species the infant, although sexually immature, looks like a small copy of the adult; it grows larger by a series of molts until it reaches full size. In others, such as the grasshopper, the newly hatched young is wingless and somewhat different in proportions from the adult, but it is otherwise similar. These immature, nonreproductive forms are known as *nymphs*. Almost 90 percent of all insects, however, undergo a complete metamorphosis, so that the

(a)

(b)

(c)

(d)

4–29

Some arthropods. (a) *Ghost crab. This little crustacean is the same color as the sandy beaches it lives on and emerges from its burrows only at night, hence its name.* (b) *Green spider. Note the unsegmented cephalothorax, the eight walking legs, and the chelicerae.* (c) *Praying mantis. Because its eyes do not move, the insect must turn its entire head to bring a prey animal (usually a small insect) into binocular view. Once its head is in position, it can strike a lightning-swift blow right on target.* (d) *Tiger beetle. The numerous hairs visible on its extremities are mechanoreceptors. A sensory cell at the base of each receptor fires a nerve impulse when a hair is touched or bent.*

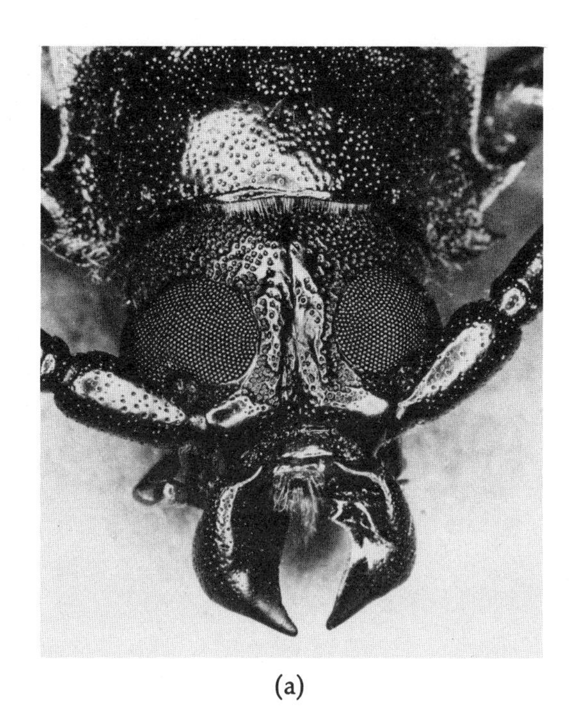

(a)

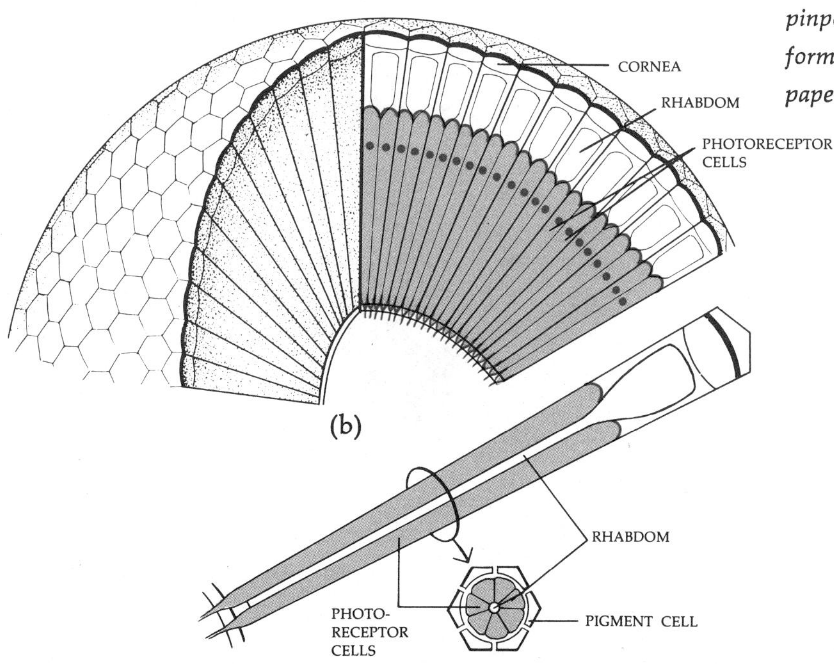

(b)

4–30

The most conspicuous sensory organ of the insects is the compound eye, an evolutionary development shared only by the centipedes, crustaceans, and a few other arthropods. The compound eye is made up of thousands of cone-shaped subunits called ommatidia, *each a small, complete eye with its own nerve endings. The dragonfly has* 30,000 *ommatidia.* (a) *The compound eyes of a long-horned beetle.* (b) *A cutaway view of the compound eye. Each ommatidium is covered by a cornea, usually with a square or hexagonal surface. Underlying the cornea is a crystalline cone, the rhabdom, that conducts the light to the photoreceptor cells. Nerve fibers carry the stimulus from the receptor cells to the brain. The outer surface of the receptor cell is coated with opaque pigment, which prevents light from traveling from one ommatidium to another. Since each ommatidium receives a single pinpoint of light from directly in front of it, the picture formed is a mosaic with very little resolution, like a newspaper picture under high magnification.*

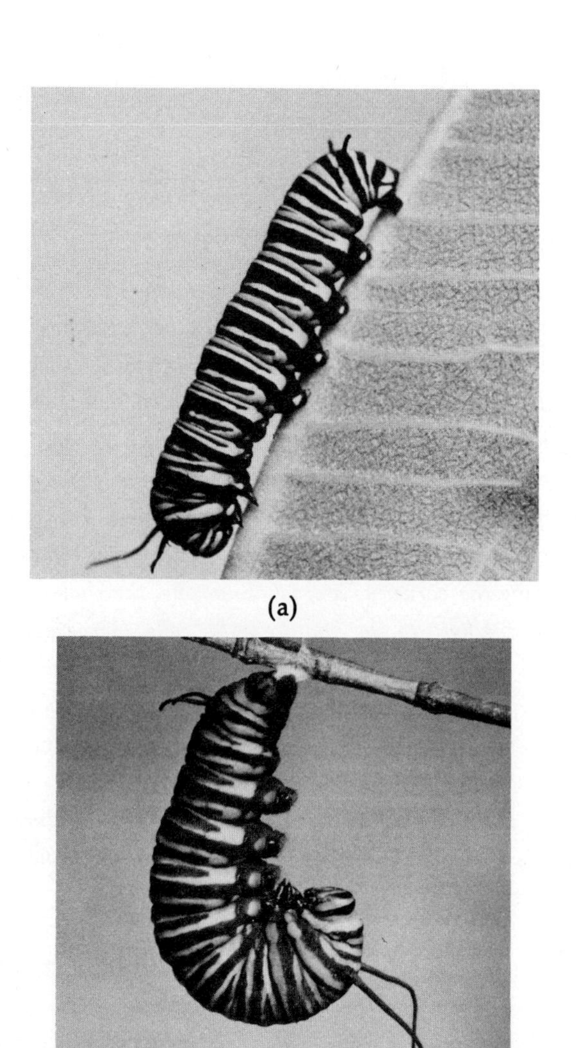

(a)

(b)

(c)

(d)

4–31

Most insects, like the Monarch butterfly shown here, undergo complete metamorphosis. (a) *The caterpillar, the larval form. Note the clearly visible, annelidlike, segmented structure.* (b) *Preparing to spin a cocoon.* (c) *The cocoon completed. Within it is the pupa.* (d) *The adult insect emerges. The adult is the mating, egg-laying form that begins the cycle anew.*

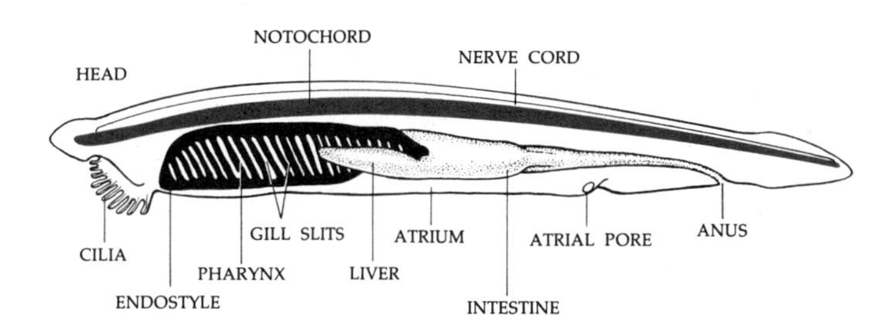

4–32
The amphioxus, a lancelet, exemplifies three distinctive chordate characteristics: (1) *the notochord, the skeletal rod which extends the length of the body and gives support to the soft tissues,* (2) *the dorsal tubular nerve cord, and* (3) *the pharyngeal gill slits.*

adult is entirely different from the immature form. The immature eating and feeding forms may all correctly be referred to as *larvae,* although they are also commonly known as caterpillars, grubs, or maggots, depending on the species. Following the larval period, the insect undergoing complete metamorphosis enters an outwardly quiescent pupal stage, in which extensive remodeling of the organism occurs. The adult insect emerges from the pupa. Thus the insect that undergoes complete metamorphosis exists in four different forms in the course of its life history: the egg, the larva, the pupa, and the adult. The larvae and adults are so different that they do not compete for food or other resources, another example of the extreme specialization found in the insect world.

Chordates: Phylum Chordata

Mammals and other vertebrates belong to a large phylum of animals comprising the chordates. The amphioxus, a lancelet, is a good example of a chordate. The lancelet is a small, sliver-shaped, semitransparent animal which is found in shallow marine waters all over the warmer parts of the world. Although it can swim very efficiently, it spends most of its time buried in the sandy bottom, with only its mouth protruding above the surface. This animal has three features that identify it as a member of the chordate phylum. The first is the *notochord,* a skeletal rod which extends the length of the body and serves as a firm but flexible axis. The notochord is a structural support. Because of it, the amphioxus can swim with strong undulatory motions that move it through the water with a speed unattainable by the flatworms or aquatic annelids. In the amphioxus, the notochord characteristically extends beyond the anus; man is one of the few vertebrates that does not have a tail.

The second chordate characteristic is the nerve cord, a hollow tube that runs beneath the dorsal surface of the animal above the notochord. (The principal nerve cord in the invertebrates, by contrast, is always near the ventral surface.)

The third characteristic is a pharynx with gill slits. The pharyngeal gill slits become highly developed in fishes, in which they serve a respiratory function, and traces of them even remain in the human embryo. In the amphioxus, they serve primarily for collecting food. The cilia around the mouth and at the opening of the pharynx pull in a steady current of water, which passes through the pharyngeal slits into a chamber known as the atrium and then exits through the atrial pore. Food particles are collected in the sievelike pharynx.

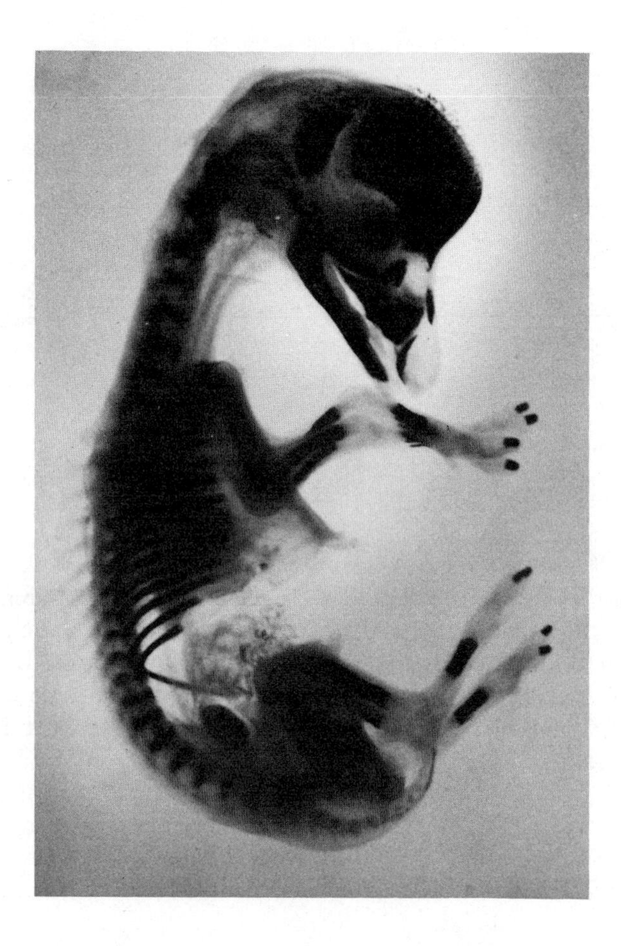

4–33

X-ray of an elk fetus. The bones have been stained to show them more clearly, so that you can see the extent to which the skeleton is still cartilaginous. Notice the legs, for example. Only the dark areas are bone; these will gradually grow and replace the cartilage as the animal matures.

The Principal Chordates: The Vertebrates

The vertebrates constitute the largest and most familiar group of chordates. All vertebrates have a backbone, or vertebral column, as their structural axis; this is a flexible bony support which develops around the notochord, supplanting it entirely in most species. The backbone is made up of bony elements, the vertebrae, which encircle the nerve cord along the length of the spine. The brain is similarly enclosed and is protected by bony skull plates. Between the vertebrae are cartilaginous disks, which give the vertebral column its flexibility. Associated with the vertebrae are a series of muscle segments, the myotomes, by which sections of the vertebral column can be moved separately. This segmented pattern persists in the embryonic forms of higher vertebrates but is largely lost in the course of development.

One of the great advantages of this bony endoskeleton, as compared with the exoskeletons of the invertebrates, is that it is composed of living tissue that can grow with the animal. In the developing vertebrate embryo, the skeleton is largely cartilaginous, the bone gradually replacing cartilage in the course of maturation. The growing tips of the bones remain cartilaginous until the animal reaches its full adulthood. In addition to its powers of growth, bone can also repair itself, unlike the lifeless tissue of a clamshell, for example.

There are seven classes of vertebrates: the fish (comprising three classes), the amphibians, the reptiles, the birds, and the mammals.

The Fish

The first fish were jawless and had a strong notochord running the length of their bodies. Today these jawless fish, once a large and diverse group, are represented only by the hagfish and the lampreys (Class Agnatha), which are the most primitive living vertebrates. They have a notochord throughout their lives, like amphioxus, and a cartilaginous skeleton. The sharks and the skates, the second major class of fish, also have a completely cartilaginous skeleton. Their skin is covered with small pointed teeth (denticles), which resemble vertebrate teeth structurally and give the skin the texture and abrasive capacity of the coarsest grade of sandpaper.

The third major group of fish consists of those that developed bony skeletons, and it is from this group, the osteichthyans, or bony fish, that the land vertebrates developed. This group is represented today by the trout, bass, salmon, perch, and tuna. Some primitive osteichthyans had lungs as well as gills, although the lungs were not efficient enough to serve as more than accessory structures. These lungs were a special adaptation to fresh water, which, unlike ocean water, may stagnate and, because of decay or algal bloom, become depleted of oxygen. A few modern lungfish exist. They surface and gulp air into their lungs and so can live in water that does not have sufficient oxygen to support other kinds of fish life.

Using their supplementary oxygen supply, the primitive osteichthyans could waddle, dragging their bellies on the ground, up the muddy bottom of a drying stream bed to seek deeper water or could perhaps even make their way from one water source to another one nearby. Fins supported by bones helped them to crawl over the land. Lungfish were the most common fish in the later Devonian seas. In most of them, the lung evolved into an air bladder, which is a closed sac that serves as a flotation chamber. The fish is able to raise or lower itself in the water by adding gases to the air bladder or removing oxygen from it. The oxygen is supplied and removed by the bloodstream. Other lungfish retained the lung and increased its efficiency, making way for the first amphibians.

The Amphibians

By the end of the Devonian period, some of the lungfish had begun to develop skeletal supports in their fleshy fins. They looked more like four-legged fish than like the modern amphibians; they were heavy and clumsy, with short sprawling legs, big flattened heads, and stubby tails. Some of them were large, as much as 4 to 5 feet long. Although they could walk on land, they probably spent most of their lives in the water; nor have their descendants, the modern amphibians, freed themselves entirely from the water.

4–34
A modern lungfish. When the dry seasons come, members of this African species wriggle downward into the mud which eventually hardens around them. Mucus glands under the skin secrete a watertight film around the body, preventing evaporation. Only the mouth is left exposed. During this period, they take a breath only about once every two hours.

Amphibians lay their eggs in water, and the males fertilize them externally after they are laid. Consequently, the amphibians, in order to reproduce, must live near water or return to it at regular intervals. Also, if they are not in water or moist air, water evaporates rapidly through their thin, scaleless skin, which serves as an accessory breathing organ.

In their early stages of development, all amphibians are clearly fishlike. The larvae (tadpoles) live entirely in the water and breathe through gills, later changing into lung-breathing forms with limbs. Modern amphibians include toads and frogs (which are tailless as adults) and newts and salamanders (which have tails throughout their lives).

4–35
Alligators and crocodiles have the reptilian characteristic of a skin reinforced with epidermal horny scales. They also lay their eggs on land. Crocodiles have narrow jaws and the fourth pair of teeth on the lower jaw remains exposed when the mouth is closed. Alligator jaws are broader and rounded anteriorly and none of the teeth are exposed when the mouth is closed. The animal shown here is an American alligator.

The Reptiles

The vertebrates freed themselves from dependence on water with the evolution in the reptiles of the *cleidoic* "boxlike") *egg*, an egg which could survive on land without drying up. Another advantage to reptiles of this evolutionary development, it is believed, is that at the time the reptiles emerged onto the land, few land predators existed, in contrast to the numerous aquatic predators. In keeping with their terrestrial existence, the reptiles also evolved a dry skin covered with protective scales (the forerunners of feathers and hair).

From one branch of primitive reptiles all the birds evolved; another branch became the mammals; and several other branches retained mostly reptilian characteristics and are represented today by lizards, snakes, turtles, and crocodiles.

The Birds

In their skeletal structure, birds are essentially reptiles highly specialized for flight. They have feathers, however, which is one of the categorical distinctions, and they maintain a high and constant body temperature, which provides the high energy output required for flight. Their bodies are lightened by air sacs and also by having hollow bones. The oceanic frigate, a seagoing bird with a 7-foot wingspread, has a skeleton that weighs only 4 ounces. The most massive bone in the bird skeleton is the breastbone, or sternum, which provides the keel for the attachment of the huge muscles that operate the wings. Flying birds have jettisoned all extra weight; the female's reproductive system has been trimmed down to a single ovary, and even this becomes large enough to be functional only in the mating season. Most of the common nonflying birds, such as the penguin and the ostrich, are believed to have evolved secondarily from flying types.

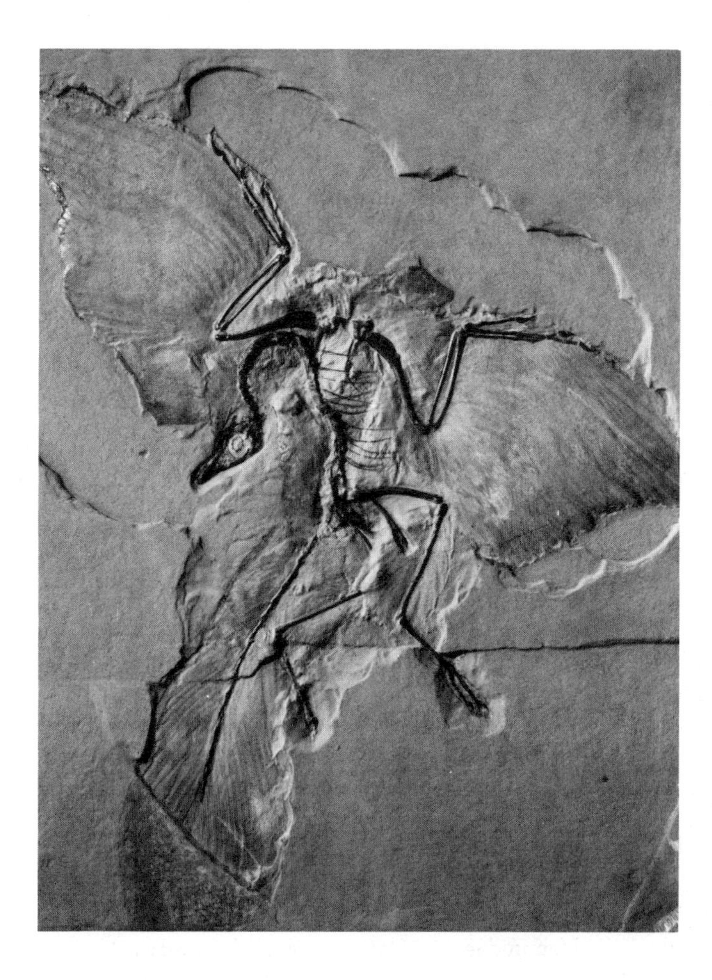

4–36
The oldest known fossil bird, Archaeopteryx, *dates from the middle Jurassic period, about* 150 *million years ago. It still had many reptilian characteristics. The teeth and the long jointed tail are not present in modern birds.*

Table 4–3
Approximate Numbers of Living Species of Animals

Phylum	*Species*	
Coelenterata (Cnidaria)	11,000	Jellyfish, sea anemones, corals, hydroids, etc.
Ctenophora	80	Comb jellies, sea gooseberries
Porifera	4,200	Sponges
Mesozoa	50	......
Platyhelminthes	15,000	Flatworms: planarians, flukes, tape-worms
Nemertea	600	Ribbon worms
Entoprocta	60	Moss animals‡
Rotifera	1,500	Wheel animalcules
Gastrotricha	150	......
Echinorhyncha	100	......
Nematomorpha	250	Horsehair-worms
Acanthocephala	300	Spineheaded-worms
Nematoda	80,000†	Roundworms: free-living and parasitic
Mollusca	110,000	Snails, clams, octopuses, etc.
Annelida	8,800	Segmented worms
Arthropoda	>800,000	Insects, crustaceans, spiders
Onychophora	80	Walking worms
Tardigrada	170	Water bears
Linguatulida	60	Tongue worms
Echiuroidea	80	......
Ectoprocta	4,000	Moss animals‡
Priapulida	5	......
Phoronida	15	......
Brachiopoda	310§	Lampshells
Sipunculoidea	275	......
Chaetognatha	60	Arrow worms
Pogonophora	45	......
Echinodermata	6,000	Starfish, sea urchins, sea cucumbers, etc.
Hemichordata	100	Acorn worms

† This is a compromise figure, between widely differing estimates.
‡ The common name "moss-animals" is applied to two distinct phyla.
§ Plus at least 12,000 described fossil species.

The Mammals

The mammals also descended from the reptiles. Some chief points of difference between mammals and reptiles are (1) mammals are hairy rather than scaly, (2) mammals nurse their young, and (3) mammals maintain a constant body temperature. In nearly all mammalian species, the young are born alive, as they are in some fish and reptiles, which retain the eggs in their bodies until they hatch. Some very primitive mammals, however, such as the duckbill platypus, lay shelled eggs but nurse their young after hatching. The marsupials, which include the opossum and the kangaroo, also bear their young alive, but they differ from the major group of mammals in that the infants are born at a tiny and immature stage and are kept in a special protective pouch in which they suckle and continue their development. Most of the familiar mammals are *placentals,* so called because they have an efficient nutritive connection, the placenta, between the uterus and the embryo. As a result, the young develop to a much more advanced stage before birth. Thus the young are afforded protection during their most vulnerable period without seriously interfering with the mobility of the mother. The earliest placentals were small, shy, and probably nocturnal; they undoubtedly lived mostly on insects, grubs, and worms, devoting much of their energies to avoiding the carnivorous dinosaurs. Shrews, which closely resemble these primitive mammals, have retained their elusive habits.

Mammals have fewer, but larger, skull bones than the fish and reptiles, an example of the fact that "simpler" and "more primitive" may have quite opposite meanings. In the mammals, a bony platform or partition has developed that separates nasal and food passages far back in the throat, making it possible for the animal to breathe while eating. The jawbones have been fused to one, which is far larger and more powerful than the reptilian jaw, although mammals have lost that useful ability of the reptile to unhinge its jaw—an ability which makes it possible for an anaconda, for example, to swallow a pig whole.

Table 4–4
The Biological Classification of Man

Category	*Name*	*Description*
Kingdom	Animalia	Multicellular organisms requiring organic plant and animal substances for food
Phylum	Chordata	Animals with notochord, dorsal hollow nerve cord, gills in pharynx at some stage of life cycle
Subphylum	Vertebrata	Spinal cord enclosed in a vertebral column, body basically segmented; skull enclosing brain
Superclass	Tetrapoda	Land vertebrates; four-legged
Class	Mammalia	Young nourished by milk glands; breathing by lungs; skin with hair or fur; body cavity divided by diaphragm; red corpuscles without nuclei; constant body temperature
Order	Primata	Tree dwellers; usually with fingers and flat nails; sense of smell reduced
Family	Hominidae	Flat face, eyes forward; color vision; upright, two-legged, with hands and feet differently specialized
Genus	*Homo*	Large brain; speech; long childhood
Species	*Homo sapiens*	Prominent chin, high forehead; sparse body hair

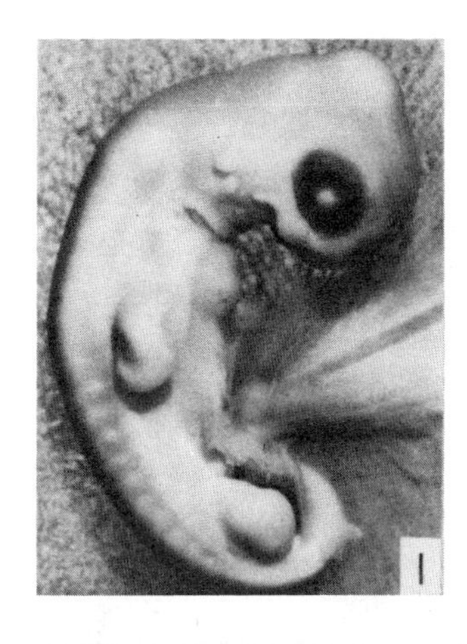

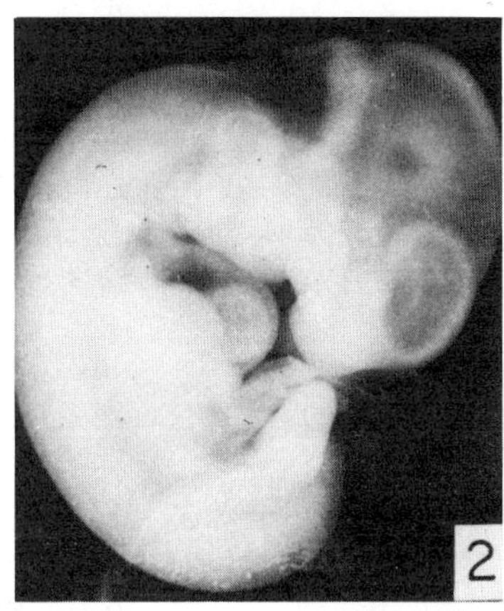

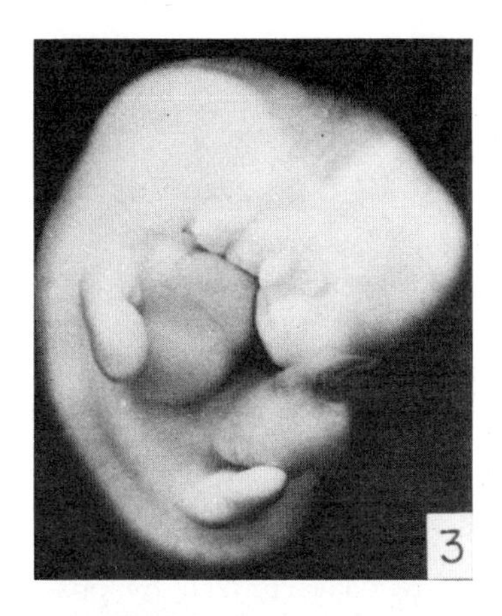

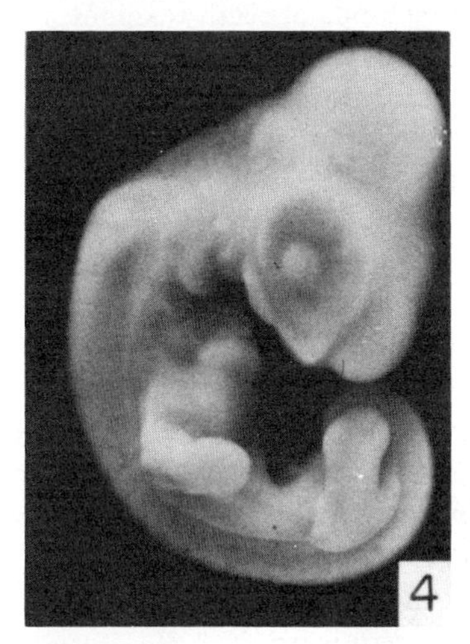

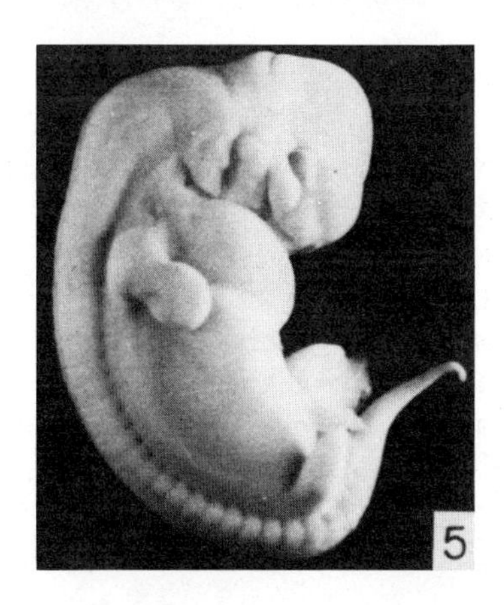

4–37

Which of these is human? As this comparison of the early fetuses of various vertebrates illustrates, the conservative nature of evolution is very evident in embryological development. The human embryo in its earlier stages greatly resembles the embryos of other vertebrates; for instance, at one stage we have gill slits and a tail. At one time, in fact, some students of evolution maintained that every embryo repeats the entire history of its evolution. Actually, what are retained are certain developmental pathways—ways of getting "from here to there." The third embryo is human. The first is a turtle; the second, a mouse; the fourth, a chick; and the fifth, a pig.

Among the placental mammals, there are four major evolutionary lines. One group comprises the rabbits and rodents. A second is made up of the marine mammals, such as whales and seals. The third includes the carnivores, such as dogs and cats, and the hoofed animals—horses, cattle, sheep, pigs, and the like. The fourth group consists of the insect eaters (shrews, moles, and hedgehogs), bats (the only flying mammals), and primates (lemurs, monkeys, apes, and man).

The primates are characterized by possession of (1) a placenta, (2) three kinds of teeth (canines, incisors, molars), (3) apposable first digits (thumbs), (4) two pectoral mammae, (5) expanded cerebral cortex, and (6) a tendency toward single births. Man is distinguished from the other primates by his upright posture and his lack of body hair. Among all the mammals, he is probably the least specialized. He is an omnivore, eating a wide variety of fruits, vegetables, and other animals. His hands closely resemble those of a primitive insectivore, in contrast to the highly specialized forelimbs developed by, for example, the whales, bats, gibbons, and horses. His sensory organs are crude compared with those of insects or of many other mammals. Man has, however, one area of extreme specialization: the brain. Because of his brain, man is unique among all the other animals in his capacity to reason, to speak, to plan, and to learn.

SUMMARY

Living organisms are grouped into five kingdoms: (1) the Kingdom Monera, which consists of the prokaryotes—bacteria and blue-green algae—small cells without nuclei and cytoplasmic organelles; (2) the Kingdom Protista, which includes all single-celled eukaryotes except the green algae, the forms from which the plants are derived; (3) the Kingdom Fungi, made up of heterotrophic cells which feed by absorption and have a filamentous body form and an outer cell wall whose principal component is usually chitin; (4) the Kingdom Plantae, comprising the green plants, that is, multicellular photosynthetic organisms and unicellular green algae; and (5) the Kingdom Animalia, comprising complex multicellular heterotrophs.

The prokaryotes are believed to be the modern representatives of the earliest forms of living organisms. All other living things are eukaryotes and are believed to have evolved from prokaryotic forms, possibly as a result of symbiotic relationships between various groups of these primitive cells. The prokaryotes, it is believed, gave rise to the protists, which are a diverse group of unicellular eukaryotes. Certain protistan lines evolved into the fungi, others into the plants, and one or more into the animals. Both plants and animals evolved in the ocean. Their successful invasion of the land, the result of numerous different adaptations, was achieved only comparatively recently.

GLOSSARY

ADAPTATION [L. *adaptare,* to fit]: **1.** The acquiring of characteristics by an organism (or group of organisms) which make it better suited to live and reproduce in its environment. **2.** A peculiarity of structure, physiology, or behavior of an organism which especially aids in fitting the organism to its particular environment.

ALGA, *pl.* ALGAE (**al**-gah; **al**-jee): A photosynthetic eukaryotic organism lacking multicellular sex organs; the blue-green algae are photosynthetic prokaryotes.

ANALOGOUS [Gk. *analogos,* proportionate]: In biology, applied to structures that are similar in function but different in evolutionary origin, such as the wing of a bird and the wing of an insect.

ANGIOSPERM (**an**-jee-o-sperm) [Gk. *angion,* vessel, + *sperma,* seed]: Literally, a seed borne in a vessel; thus, one of a group of plants whose seeds are borne within a matured ovary (fruit).

ANIMALIA (annie-**male**-ya): The taxonomic kingdom comprising the animals, that is, multicellular heterotrophs.

ANNELID [L. *anellus,* small ring] A member of the phylum Annelida, comprising ringed or segmented worms.

APPENDAGE: Any peripheral extension of an animal body, especially when functioning as a limb.

ARACHNID (a-**rack**-nidd): A member of the Arachnida, a class of arthropods that includes spiders, scorpions, mites, and ticks.

ARTHROPOD [Gk. *arthron,* joint + *pous,* foot]: An invertebrate animal with jointed appendages; a member of the phylum Arthropoda.

BACTERIUM, *pl.* BACTERIA [Gk. dim. of *bactron,* staff]: A small unicellular prokaryotic organism characterized by the absence of a formed nucleus.

BILATERAL SYMMETRY: An anatomical arrangement in which the right and left halves of an organism are approximate mirror images of each other.

BRYOPHYTE (**bry**-o-fight): A member of the phylum Bryophyta; a nonvascular terrestrial green plant, such as a liverwort or moss.

CHITIN (**kite-**'n): A tough, resistant, nitrogen-containing polysaccharide present in the exoskeleton of arthropods, in the epidermal cuticle or other surface structures of many other invertebrates, and in the cell walls of certain fungi.

CHORDATE: A member of the animal phylum Chordata. All chordates possess a notochord, dorsal nerve cord, and pharyngeal gill slits, at least at some stage of the life cycle.

CLASS: The taxonomic category between phylum and order; consists of one or more orders.

CLEIDOIC EGG (cly-**doe**-ick): An egg which is isolated from the environment by a more or less impervious shell during the period of its development and which is completely self-sufficient, requiring only oxygen from the outside.

COELENTERATE (see-**len**-t'-rate [Gk. *enteron,* intestine]: An invertebrate animal possessing a single alimentary opening and tentacles with sting cells, for example, jellyfish, corals, sea anemones, hydroids; a member of the phylum Coelenterata.

COELENTERON (see-**len**-t'-ron): The alimentary cavity within the body of a coelenterate; it has only one opening.

COELOM (**seal**-um) [Gk. *koilos,* a hollow]: A body cavity formed between layers of mesoderm.

COLONY: A group of unicellular or multicellular organisms living together in close association.

COMPOUND EYE: A complex eye composed of a number of separate elements (ommatidia), each with light-sensitive cells and a refractory system which can form an image; common in arthropods.

DORSAL [L. *dorsum,* back]: Pertaining to the upper surface of an animal that creeps or moves on all fours and to the back surface of an animal that holds its body erect. Opposed to ventral.

ECHINODERM (e-**kye**-no-derm) [Gk. *echinos,* spiny, bristly, + *derma,* skin]: A radially symmetrical marine animal; a member of the phylum Echinodermata, which includes sea lilies, starfish, sea urchins, and sand dollars.

ECTODERM [Gk. *ektos,* outside]: The outermost of the three embryonic tissue layers in the gastrula (an early developmental stage of the embryo); gives rise to skin, nervous system, sense organs, etc.

ENDODERM [Gk. *endon,* within]: The innermost of the three embryonic tissue layers in the gastrula; forms the primitive gut.

EPIDERMIS [Gk. *epi,* upon]: In plants and animals, the outermost cells; one cell layer thick in plants, several cell layers thick in vertebrates.

EUKARYOTE (you-**car**ry-oat) [Gk. *eu,* good, + *karyon,* nut, kernel]: An organism having membrane-bound nuclei, Golgi apparatus, and mitochondria; in contrast to prokaryote.

EXOSKELETON: A skeleton covering the outside of the body; common in arthropods.

FAMILY: The taxonomic category between order and genus; consists of one or more genera.

FUNGI (**fun**-jye): The taxonomic kingdom of heterotrophs which live chiefly on decaying organic matter. Some consist of single cells, but most have filaments (hyphae) arranged in a mycelium. Examples are molds, mildews, rusts, smuts, and mushrooms.

GASTRODERMIS: The membrane lining the alimentary tract of an invertebrate.

GENUS, *pl.* GENERA (**jee**-nuss, **jenn**-era): The taxonomic category between family and species; consists of one or more species.

GYMNOSPERM [Gk. *gymnos,* naked, + *sperma,* seed]: A seed plant, such as a conifer, with seeds borne on cone scales rather than in an ovary.

HOLDFAST: The basal part of an algal plant body that attaches it to a solid object; may be unicellular or composed of a mass of tissue.

HOMOLOGOUS [Gk. *homologia,* agreement]: In biology, similar in structure as the result of common ancestry, regardless of function; for example, the wing of a bird and the forelimb of a mammal are homologous.

HYPHA [Gk. *hyphē,* web]: A single tubular filament of a fungus. The hyphae together constitute the mycelium.

KINGDOM: The chief taxonomic division. There are five kingdoms, encompassing all living organisms: Monera, Protista, Fungi, Plantae, and Animalia.

LARVA [L. *larva,* ghost]: An immature animal, such as a caterpillar or tadpole, that is morphologically very different from the adult.

MESODERM: The middle of the three embryonic tissue layers, between the ectoderm and the endoderm; gives rise to skeleton, circulatory system, musculature, excretory system, and most of the reproductive system.

METAMORPHOSIS [Gk. *metamorphoun,* to transform]: In biology, the transformation of a larval form into an adult, such as the transformation of a tadpole into a frog.

MOLLUSK: A member of the phylum Mollusca, a large group of invertebrates, mostly aquatic, that includes chitons, snails, octopuses, and the bivalves.

MONERA (m'-**nir**rah): The taxonomic kingdom comprising all prokaryotic unicellular organisms (bacteria and blue-green algae).

MYCELIUM (my-**seal**y-um) [Gk. *mykēs,* fungus]: The mass of hyphae forming the body of a fungus.

MYRIAPOD [Gk. *myrios,* countless, + *pous,* foot]: An arthropod which has an elongated segmented body; one pair of antennae, and several to many pairs of segmentally arranged legs. Examples are centipedes and millipedes.

NEMERTINE (**nem**-mer-teen) [Gk. *Nēmertēs,* one of the Nereids, sea-nymph attendants on Poseidon]: A member of the phylum Nemertea, comprising the proboscis or ribbon worms, which are soft-bodied, brightly colored, and unsegmented.

NERVE CORD: A pair of closely united ventral longitudinal nerves with their segmented ganglia; a characteristic of many elongate invertebrates, such as earthworms.

NOTOCHORD: A longitudinal, solid, elastic, rodlike structure serving as the internal skeleton in the embryos of all chordates. In most adult chordates, the notochord is replaced by a vertebral column which forms around (but not from) the notochord.

OMMATIDIUM, *pl.* OMMATIDIA (o-ma-**tid**-ee'm) [Gk. *ommat,* eye, + *-idion,* small one]: The single visual unit in the compound eye of arthropods; contains light-sensitive cells and a refractory system able to form an image.

ORDER: The taxonomic category between family and class; consists of one or more families.

OSTEICHTHYAN (oss-tee-**ick**-thi-y'n): A fish that has true bone in its skeleton, in contrast to cartilaginous fish, such as the shark.

PAEDOMORPHOSIS (pee-do-**more**-fo-sis) [Gk. *pais,* child, + *morphē,* form]: The evolution of a new group of animals from an earlier developmental stage of an ancestral form.

PERITONEAL CAVITY (perry-t'-**nee**-'l) [Gk. *peritonos,* stretched over]: The abdominal cavity in a mammal which houses the digestive, excretory, and reproductive organs.

PHARYNX [Gk. *pharynx,* throat]: The part of the alimentary canal between the oral cavity and the esophagus.

PHYLUM [Gk. *phylon,* race, tribe]: The taxonomic category between class and kingdom; the main subdivision of a kingdom.

PLACENTAL [Gk. *plax,* flat surface]: A term applied to those species of mammal in which, during the period of pregnancy, the nutritive exchanges between the blood of the mother and that of the developing embryo are accomplished through the placenta, a structure formed in part from the inner lining of the uterus and in part from the tissues of the embryo.

PLANKTON [Gk. *planktos,* wandering]: Free-floating, mostly microscopic aquatic organisms, both photosynthetic and heterotrophic.

PLANTAE (**plan**-tee): The taxonomic kingdom comprising all multicellular photosynthetic organisms and other, closely related forms.

PLANULA [L. dim. of *planus,* flat]: The ciliated free-swimming larval type occurring in many coelenterates.

PLASMODESMATA (plaz-muh-**dezz**-m'-tuh) [Gk. *plassein,* to mold, + *desmos,* band, bond]: Minute cytoplasmic threads that extend through the pores in cell walls and connect the protoplasts of adjacent cells in higher plants.

PLEURAL CAVITY [Gk. *pleuron,* rib, side]: The chest cavity, housing the lungs and heart. The membranes which line it secrete a lubricating fluid that aids respiratory movements.

PREADAPTATION: The possession by an organism of a structure or function that develops new uses in the course of evolution.

PROKARYOTE (proh-**carry**-oat) [Gk. *pro,* before, + *karyon,* nut, kernel]: An organism which lacks membrane-bound nuclei, mitochondria, chloroplasts, and Golgi apparatus; bacteria and blue-green algae.

PROTISTA: The taxonomic kingdom comprising eukaryotic unicellular organisms such as protozoans, many algae, and slime molds.

PUPA [L. *pupa,* doll]: A developmental stage between the larval and adult phases in insects. Pupae are nonfeeding, immotile, and sometimes encapsulated in a cocoon.

SEED: A complex organ formed by the maturation of the ovule of seed plants following fertilization. In conifers, it consists of seed coat, embryo, and female gametophyte ($1n$) storage tissue. In angiosperms, it consists of seed coat and embryo; some angiosperm seeds also contain endosperm ($3n$), a storage tissue.

SPONGE: One of a large group of chiefly marine animals of the phylum Porifera, consisting basically of two layers of cells surrounding a central cavity. Sponges are permanently attached, either individually or in masses, and vary greatly in size, shape, color, and consistency.

TRACHEOPHYTE (**tray**-key-oh-fight) [Gk. *tracheia,* rough, + *phyton,* plant]: A vascular plant, that is, one possessing xylem and phloem.

VASCULAR: Relating to a tube or system of tubes for conducting fluids.

VENTRAL [L. *venter,* belly]: Pertaining to the undersurface of an animal that creeps or moves on all fours and to the front surface of an animal that holds its body erect; opposed to dorsal.

VERTEBRAL COLUMN: The backbone; in nearly all vertebrates, it forms the supporting axis of the body and a protection for the spinal cord.

VIRUS [L. *virus,* slimy liquid, poison]: A submicroscopic noncellular particle composed of a nucleic acid core and a protein shell. Viruses replicate only within host cells.

SUGGESTIONS FOR FURTHER READING

BUCHSBAUM, RALPH: *Animals without Backbones,* rev. ed., The University of Chicago Press, Chicago, 1948.

A delightful survey of the invertebrates, for the general reader.

BUCHSBAUM, RALPH, AND LOUIS J. MILNE: *The Lower Animals. Living Invertebrates of the World,* Doubleday & Company, Inc., Garden City, N. Y., 1962.

A collection of handsome photos of the invertebrates accompanied by a text prepared by two noted zoologists but directed toward the general reader.

CORNER, E. J. H.: *The Life of Plants,* Mentor Books, New American Library, Inc., New York, 1968.*

A renowned botanist with a flair for poetic prose describes the evolution of plant life, telling how plants modify their structures and functions to meet the challenge of a new environment as they invaded the shore and spread across the land.

CURTIS, HELENA: *The Marvellous Animals,* Natural History Press, Garden City, N. Y., 1968.

An introduction to the protozoa.

KLOTS, ALEXANDER B., AND ELSIE B. KLOTS: *Living Insects of the World,* Doubleday & Company, Inc., Garden City, N. Y., 1962.

A spectacular gallery of insect photos. As with The Lower Animals, *which belongs to the same series, the text is informal but informative, written by experts for laymen.*

RUSSELL-HUNTER, W. D.: *A Biology of Lower Invertebrates,* The Macmillan Company, New York, 1968.*

A short authoritative account.

WELLS, M. J.: *Brain and Behavior in Cephalopods,* Stanford University Press, Stanford, Calif., 1962.

Experimental analyses of behavior in the octopus and squid.

* Available in paperback.

SECTION 5

Plants

5–1
The force that through the green fuse drives the flower
Drives my green age . . . (Dylan Thomas)

CHAPTER 5–1

Plants and the Land

For most of earth's history, the land was bare. A billion years ago, seaweeds may have clung to the shores at low tide and perhaps some gray-green lichens patched a few inland rocks, but had anyone been there to observe it, the earth's surface would generally have appeared as barren and forbidding as the moon's surface is today. According to the fossil records, plants first began to invade the land a mere half billion years ago, and not until then did the earth truly come to life. As a film of green spread from the edges of the waters, other forms of life, the heterotrophs, were able to follow. The shapes of these new forms and the ways in which they lived were determined by the plant life that preceded them because, in freeing themselves from the water, they became increasingly dependent on the land plants not only for their food—their chemical energy—but also for their nesting, hiding, stalking, and breeding places. In all communities except those created by man, the character of the land plants still determines the character of the animals and other forms of life that inhabit that particular area. Even man, who has seemingly freed himself from the life of the land and even, on occasion, from the surface of the earth, is still dependent on the photosynthetic events that take place in the green leaves of the land plants.

The modern land plant can best be understood in terms of its long evolutionary history and, in particular, this transition to land. Like their one-celled ancestors, today's plants have few and relatively simple requirements. They need air, water, light, and certain minerals. From these simple materials they make the sugars, amino acids, and all other organic substances on which plant and animal life depends. But there is an important difference. In the simplest photosynthesizing organism–a single green cell or a filament of cells which floats on the surface of the water—each of the needed materials is immediately available to every cell. In the land plant, however, the single cells can no longer function autonomously but can survive only as part of a specialized structure in the plant body. The reasons for the evolution of these specialized structures are easy to understand. Water and minerals can be obtained by land plants only from below the surface of the earth; the development of a complex and extensive root system can be seen as a response to this selective pressure. Sunlight and air cannot reach these belowground structures, however, so photosynthesis is relegated to another part of the plant body. As the land plants began to crowd the surface and compete for light, selection pressures favored those with more and more extensive and efficient light-collecting areas, and taller plants began to have an advantage over the shorter ones. The stem evolved as the specialized part of the plant which raises the photosynthesizing areas into the sunlight and through which water and minerals travel to the leaves and sugars to the roots.

CONDUCTING SYSTEMS

The separation of the activities of photosynthesis from those of gathering water and minerals led to selection pres-

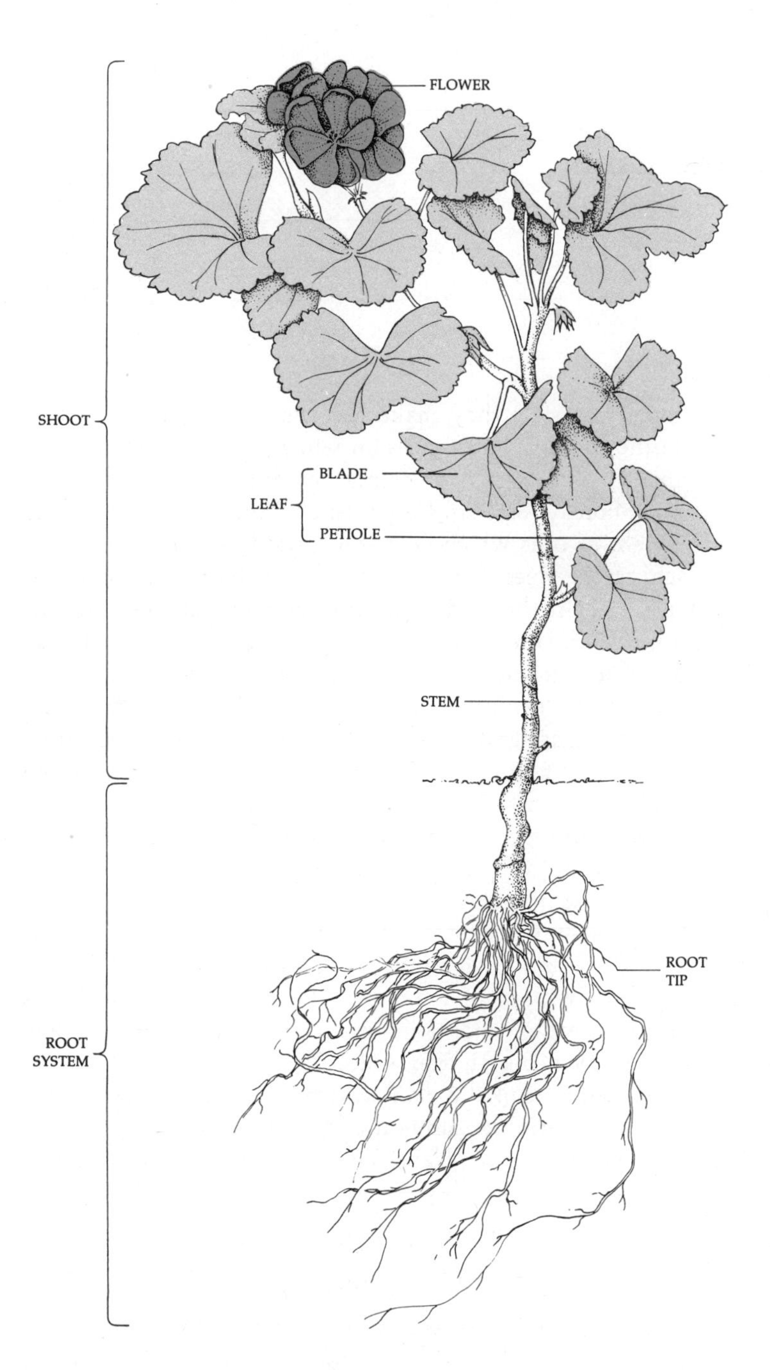

5–2 (at left)

The plant body. The aboveground structures constitute the shoot, consisting of the stem, the leaves, whose primary function is photosynthesis, and the flower, the reproductive organ. The belowground structures, the roots, supply water and minerals (including nitrogen compounds) to the plant body. At least half of a plant is usually underground.

5–3

Root hairs. Most of the uptake of water and solutes takes place through the root hairs, which form just behind the growing tip of the root.

MONOCOTS AND DICOTS

The angiosperms are divided into two broad groups: the dicots (dicotyledons), with about 250,000 species, and the monocots, with about 50,000 species. The names refer to the fact that the plant embryo in the dicots has two cotyledons ("seed leaves") and in the monocots has one.

There are also some other differences, summarized pictorially below. In the dicots, the vascular tissues are arranged around a central core in the stem; in the monocots, they are scattered through the stem. The veins of dicot leaves are usually netlike or fanlike; those of monocot leaves are usually parallel. Dicots characteristically have taproots, and monocot roots are often fibrous. Underground structures for reproduction and food storage, such as bulbs and rhizomes, are found among the monocots.

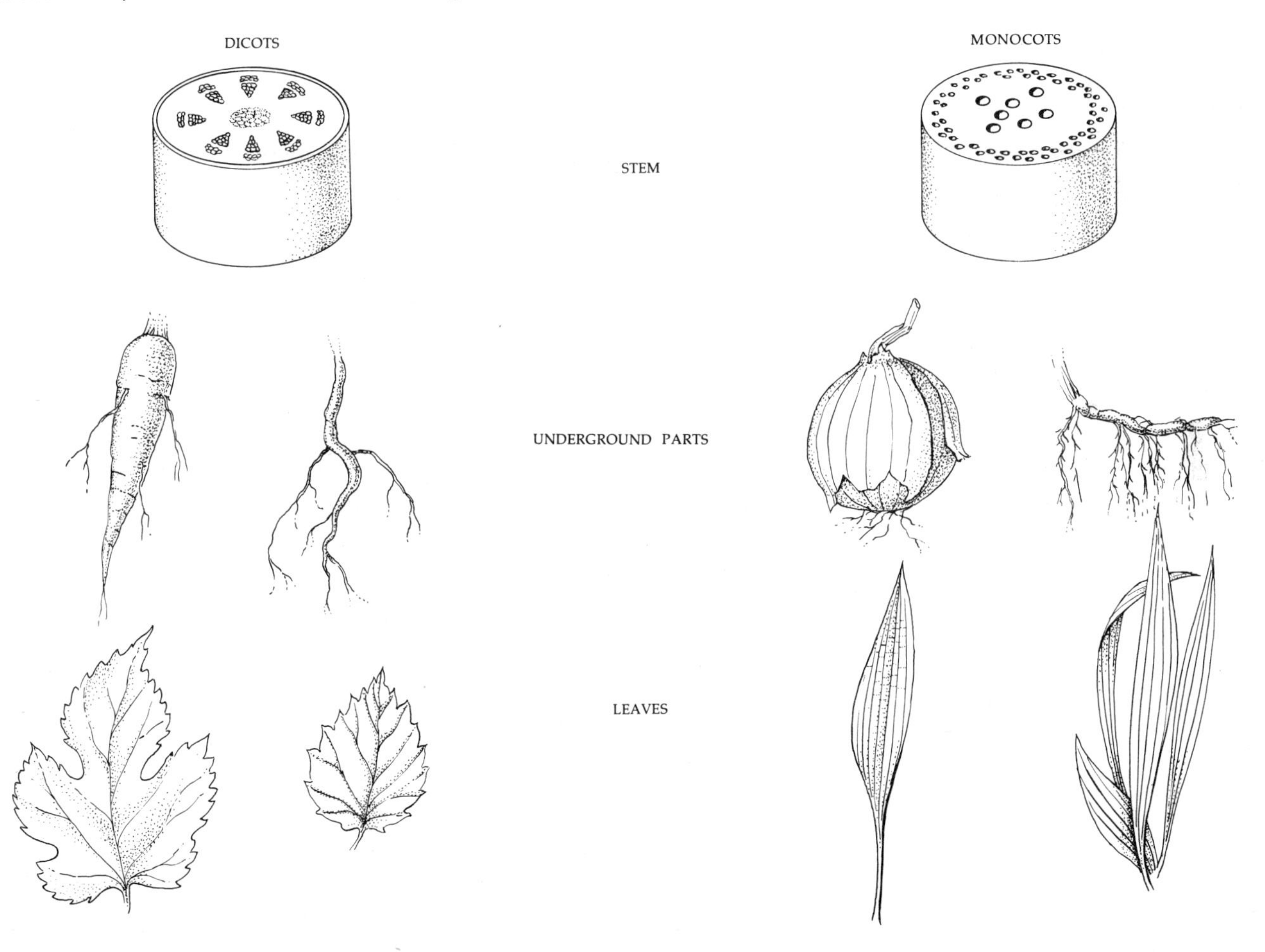

sures for the evolution of conducting and supporting tissues.

All parts of the vascular plant are integrated by conducting tissues. Its vascular system is a complex network of internal plumbing that is continuous from the smallest roots through the main root and stem to the smallest twigs, leaves, and buds. The veins of the leaves are part of this system.

The vascular system has two main components, the *xylem* and the *phloem*. Each of these tissues is made up of several different types of cells, including cells involved with stiffening and support as well as conducting cells. The xylem conducts water and minerals from the roots to other parts of the plant body. The conducting cells of the xylem are either *tracheids* or *vessel elements*. Tracheids are elongated cells with tapering ends. Water passes from tracheid to tracheid laterally through pores in the heavy cell wall. Vessel elements are stacked one on top of another. At maturity, the cell walls at the top and bottom break down, forming a continuous vessel. Many angiosperms have both tracheids and vessel elements in their xylem. Both of these types of cells are dead at maturity.

The phloem conducts the products of photosynthesis, chiefly in the form of sucrose, from the leaves to the stem, roots, reproductive structures, and other nonphotosynthetic parts of the plant. The conducting cells of the phloem are *sieve-tube elements*. These cells are alive at maturity. Sieve cells are largely filled with a watery substance often referred to as slime. The nucleus of a sieve-tube element disintegrates as the cell matures. The cytoplasm forms a film along the longitudinal surface of the cell. Sieve-tube elements are always associated with companion cells, which apparently provide their nuclear functions.

5–4

Portion of an avocado leaf. The clearly visible veins are bundles of conducting tissues, both xylem and phloem. Continuous with the vascular tissue of the stem and root, they branch and divide into finer and finer bundles, reaching close to every photosynthetic cell. Water and dissolved minerals are carried to the leaf through the xylem; sucrose and other organic molecules produced in the leaf are carried away from their sites of synthesis through the phloem.

THE CUTICLE

Along with extensive photosynthetic areas exposed to the air, plants evolved a protective surface which aids in water retention. The surface of the leaves and of green stems is

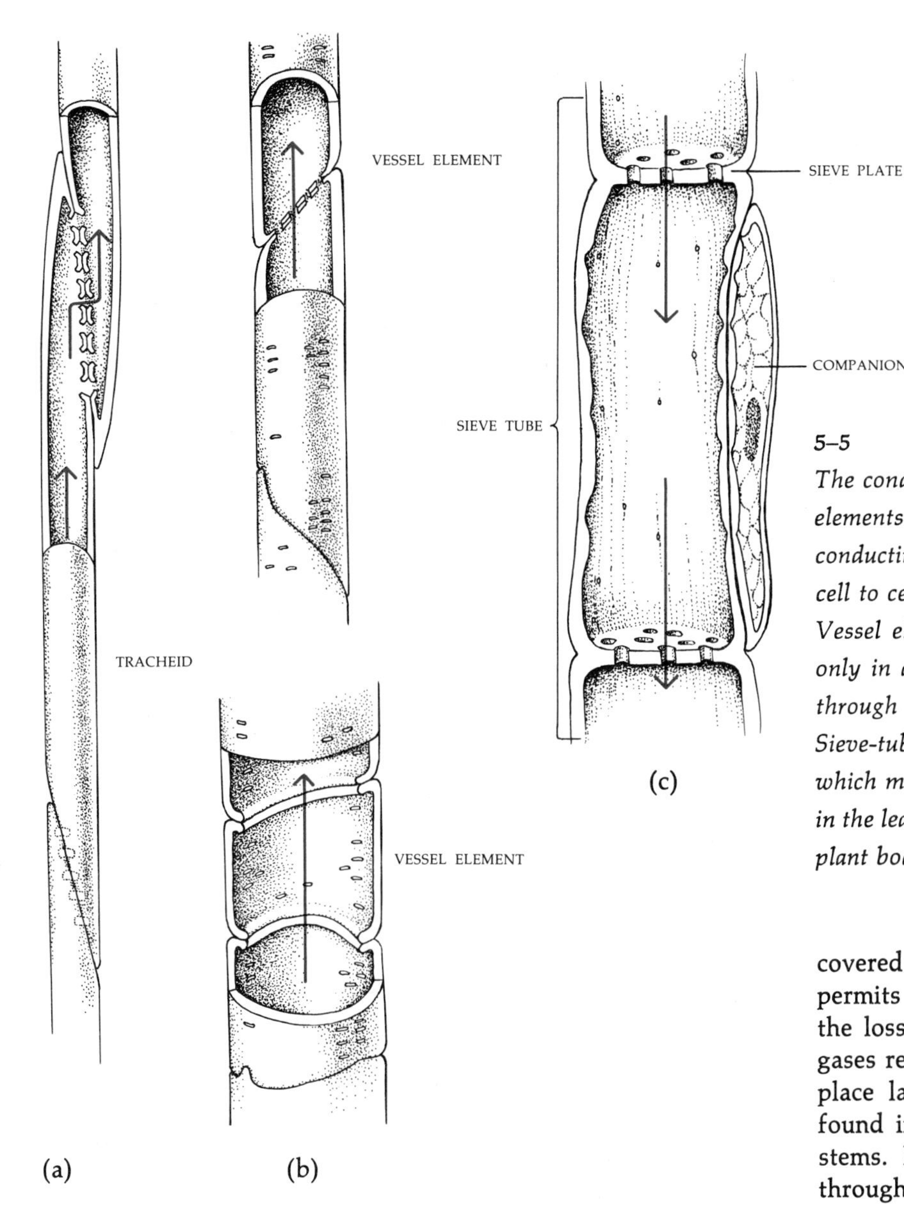

5–5
The conducting cells of xylem are tracheids (a) *and vessel elements* (b). *In tracheids (which are the only type of water-conducting cells found in the conifers), water travels from cell to cell through small holes, or pits, on the lateral surfaces. Vessel elements, a more specialized conducting cell found only in angiosperms, form continuous tubes, or vessels, through which the water travels upward from the ground. Sieve-tube elements* (c) *are the specialized cells of the phloem which make up sieve tubes, through which sugars produced in the leaves by photosynthesis travel to other parts of the plant body.*

covered with a waxy layer of cutin. This layer, the cuticle, permits the passage of light but prevents to a large extent the loss of water from the plant body. The exchange of gases required for photosynthesis and for respiration takes place largely through specialized openings, the stomata, found in the leaves and also, to a lesser extent, in green stems. Most of the water loss of the plant also occurs through these openings. The stomata open and close in response to changes in the external and internal environment of the plant, thus maintaining a finely adjusted compromise between the loss of water and the diffusion of carbon dioxide and oxygen required for photosynthesis and respiration. All the land plants, including the bryophytes, have stomata, indicating that this development occurred early in evolutionary history.

THE FLOWER

The flower is the reproductive organ of the angiosperms, the largest group of land plants and those most important to man. The flower consists of a series of floral parts, which grow in spirals or whorls; each floral part, evolutionarily speaking, is a modified leaf. The outermost parts of the flower are the sepals, which are usually green and obviously leaflike in structure. The sepals, collectively known as the calyx, enclose and protect the flower bud. Next are the petals, collectively called the corolla; these are also usually leaf-shaped but are often brightly colored. Their special function is to advertise the presence of the flower, attracting insects or other animals, which will visit the flowers for the nectar, a sugary liquid, or for other edible substances and so carry pollen from flower to flower.

Within the corolla are the stamens, whose number varies with the species of the plant. Each stamen consists of a single, elongated stalk, the filament, and at the end of the filament, the anther. The pollen grain, which is a male gametophyte, is released when ripe from within the anther, usually through narrow slits or pores.

The innermost parts of the flower are the carpels. Typically, a carpel consists of a hollow expanded base, or ovary, which contains the ovules and later forms the fruit; a stigma, which is a flat, often hairy, sticky surface specialized to receive the pollen; and connecting the stigma and ovary, a slender stalk, the style, down which the pollen tube grows. A flower may have one or several separate carpels.* Often two or more carpels are fused.

Each ovule encloses a female gametophyte with a single egg cell. After fertilization, most of the floral parts usually fall away, leaving the ovary, from which the fruit develops. Within the ovary, each ovule develops into a seed, and inside each seed is a young embryo, the start of the next generation of the flowering plant.

* The carpel or carpels used to be known as the pistil because of their resemblance to an apothecary's pestle. This older term is now being abandoned, however.

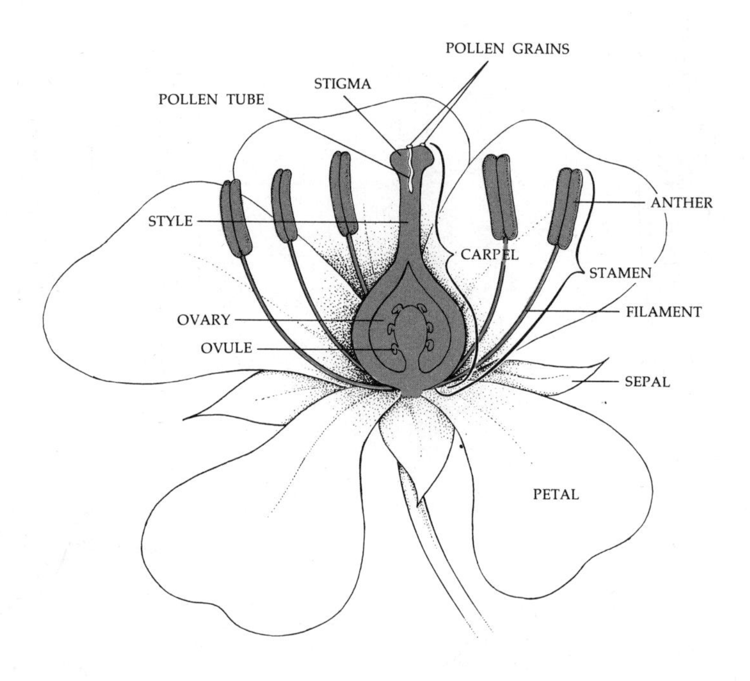

5–6

The generalized structure of a flower. The sex organs are shown in color. In the center is the female sex organ, the carpel, consisting of the ovary, the style, and the pollen-catching stigma. Pollen grains deposited on the stigma grow down through the style to the ovules. Surrounding the carpel are the male organs, the stamens. Each consists of a supporting filament and a pollen-producing anther. The sex organs are surrounded by petals, and at the base of the flower are the sepals, which once enclosed the entire structure in the bud.

Why There Are So Many Kinds of Flowers

As we have seen, sexual reproduction confers certain advantages. However, land plants, unlike most animals, cannot move about seeking other members of their own species with which to mate. The flower is a device by which plants induce animals to make these movements for them.

The early gymnosperms from which the angiosperms evolved were probably wind-pollinated, like modern gymnosperms. And like the modern gymnosperms, the female gametophytes probably exuded droplets of sticky sap in which pollen grains were caught and drawn in to the egg cell. Insects, probably beetles, feeding on plants must have come across the protein-rich pollen grains and the sugary droplet. As they began returning regularly to these newfound food supplies, they inadvertently carried pollen from plant to plant.

Beetle pollination must have been more efficient than wind pollination for some species because, clearly, selection began to favor plants with insect pollinators. The more attractive the plants were to the beetles, the more frequently they would be visited and the more seeds they would produce. Any chance variations which made the visits more frequent or which made pollination more efficient thus offered immediate advantages; more seeds would be formed, and more offspring would survive. Nectaries evolved which lured the pollinators. Plants developed brightly colored flowers which called attention to the nectar and other food supplies. The carpel, originally a leaf-shaped structure, became folded on itself, enclosing and protecting the ovule from the hungry pollinators. By the beginning of the Cenozoic era, some 65 million years ago, the first bees, wasps, butterflies, and moths had appeared. These are long-tongued insects for which flowers are often the only source of nutrition. From this point onward, flowers and insects had a profound influence on one another's evolutionary history, each shaping the other's structures and destinies.

A flower that attracts only a few types of animal visitors and attracts them regularly has an advantage: much less of its pollen is liable to be lost on a plant of another species. Similarly, it is advantageous to the insect to have a private food supply, relatively inaccessible to others. Most of the distinctive features of modern flowers are special adaptations that encourage constancy of particular pollinators. The varied colors and odors are "brand names" for guiding pollinators. The varied shapes represent a tailoring of each type of flower to only one type of pollinator, thus making pollination more efficient.

DORMANCY

Plants in North America and other temperate areas alternate periods of growth with periods of dormancy, following the seasonal variations of heat and cold. In subtropical regions, dormancy occurs among plants in parts of the year when rainfall is scarce or absent. Dormancy enables plants to survive periods when water is scarce, either because of lack of rainfall or because the available water is locked in the ground in the form of ice, and when climatic conditions are unfavorable for delicate growing buds, shoots, new leaves, and root tips.

Annuals, Biennials, and Perennials

There are characteristic differences among plants as to which structures die during dormant periods and which remain alive but dormant. In the annuals, which include many of our weeds, wild flowers, garden flowers, and vegetables, the entire cycle from seed to vegetative plant to flower to seed again takes place within a single growing season. All vegetative organs die, and only the dormant seed bridges the gap between one generation and the next. In annual plants, the stem remains relatively soft. In biennial plants, the period from seed germination to seed formation spans two growing seasons. The first season of growth ends with the formation of a root, a short stem, and a rosette of leaves near the soil surface. In the second growing season, extensive stem elongation, flowering, fruit-

5–7

(a) *Primitive flowers are believed to have resembled the modern hepatica, which is radially symmetrical with numerous separate floral parts arranged in a spiral.* (b) *A more evolutionarily advanced flower, the pale touch-me-not* (Impatiens) *is bilaterally symmetrical. The sepals and petals are fused together, and the flower has a long, slender nectar-filled spur which attracts insects with long enough mouthparts to sip the nectar. The lower "lip" on the flower serves as a landing platform. Pollen is deposited on the front part of the pollinator's body, usually on the head.* (c) *A honey bee foraging in a flower of salvia. She is standing on a landing platform, pushing down into the corolla to reach the nectar. As a consequence, the anthers are pressed down on her head, depositing pollen. On her legs you can see a ball of pollen, scraped off from pollen dusted on her back, a procedure carried out in midflight. Above the anthers is the stigma. As the flower matures, the stigma will droop and so will collect pollen from another pollinator. By this time, the anther will have stopped producing pollen, thus ensuring against the flower fertilizing itself.* (d) *"Honey guides" indicate the position of the nectary in the foxglove* (Digitalis). (e) *Pollinators coevolved with flowers. Note the long sucking tube on this painted lady butterfly* (Vanessa carye). *Some tropical moths have sucking tubes 25 centimeters long.* (f) *Unlike most other angiosperms, grasses are wind-pollinated. They characteristically have enlarged feathery stigmas which catch the air-borne pollen grains.* (g) *Visitor and flower of the daisy* Wyethia, *a composite. Composites, a group which also includes sunflowers, chrysanthemums, marigolds, zinnias, and many others, are the most evolutionarily advanced of dicot flowers. Although it appears to be a single flower, the center is actually a tightly packed mass of hundreds of individual flowers, many of which can be pollinated simultaneously by a single insect such as this beetle.*

(a)

(b)

(c) (d) (e)

(f) (g)

ing, seed formation, and death of vegetative organs occur, completing the life cycle.

Perennials are plants in which vegetative structures survive year after year. Some perennials pass unfavorable seasons as dormant underground roots, rhizomes, bulbs, or tubers. In these, all vegetative structures above ground die. In others, the stem becomes hardened on the inside and develops a heavy protective layer of dead tissue on the outside. These woody perennials, which include vines, shrubs, and trees, survive above ground but usually stop growing or grow much more slowly during the unfavorable seasons.

Many of the woody perennials are deciduous—that is, all their photosynthetic leaves drop in the fall. This is an adaptation which reduces the rate of water loss. The fall of the leaf comes about as a result of the formation of an abscission layer, a special growth of weak, thin-walled parenchyma cells across the petiole. As the leaf ages, the cells separate. Before the leaf falls, a layer of cork develops and seals the leaf scar.

As the leaf dies, new chlorophyll can no longer be formed and the green color gradually disappears, unmasking the yellows and oranges of the carotenoids and producing the blaze of colors characteristic of the deciduous forest in the fall.

SUMMARY

As a consequence of the transition to land, the plant body became differentiated into root and shoot. The root system, which usually accounts for half the volume of the plant body, anchors the plant in the soil and absorbs water and minerals from the soil. These are circulated through the plant body by means of a special conducting system, the xylem. The shoot, which includes the stem and the leaves, is primarily concerned with photosynthesis, which takes place in the chloroplasts of parenchyma cells in the green portions of the plant body, largely in the leaf. The surfaces of the shoot are covered with cutin, a waxy substance that forms an outer layer, the cuticle, which waterproofs the shoot. Openings for the exchanges of carbon dioxide and oxygen that occur during photosynthesis are provided by the stomata, adjustable pores formed of specialized epidermal cells. The products of photosynthesis, largely in the form of sucrose, are transported from the leaves to stem, roots, fruits, and flowers by means of a second specialized conducting system, the phloem.

The flower is the reproductive organ of the angiosperms. The anthers of the flower produce the pollen grain, the male gametophyte; each anther is usually supported by a slender filament. The entire structure, anther plus filament, is known as the stamen. The carpel typically consists of the stigma, an area in which the pollen grains germinate, a long slender style, and at its base, the ovary. The ovary contains one or more ovules. Within each ovule, the female gametophyte, containing the egg cell, develops. The ovule later becomes the seed.

The various shapes and colors of flowers evolved under selection pressures for more efficient pollinating mechanisms.

Plants have evolved mechanisms for remaining dormant during unfavorable periods. In annual plants, only the seeds survive from one season to the next, bridging the gap between generations. In perennials, vegetative structures also survive year after year, but often growth stops as a result of the loss of photosynthetic leaves. The capacity to remain dormant permits plants to survive during periods of drought and cold.

QUESTIONS

1. Sketch a flower. What is the function of each of the floral parts?

2. As fruits, such as apples and pears, ripen, abscission layers form across the stem so that the mature fruits drop to the ground. Why is this phenomenon of survival value to the plant?

CHAPTER 5–2

Growth and Development in Angiosperms

For most flowering plants, a new cycle of life begins when a grain of pollen—brushed from the body of a foraging insect—comes into contact with the stigma of a flower of the same species. By the time this pollen grain is released from its parent flower, it characteristically consists of three haploid nuclei (two sperm nuclei and a so-called "vegetative" nucleus), a small amount of dense cytoplasm, and a tough, protective outer coating.

Once on the stigma, the pollen grain, which is the male gametophyte, germinates, and a pollen tube grows down through the style into an ovule. The ovule holds the female gametophyte, which usually consists of seven cells with eight haploid nuclei, one of which is the nucleus of the egg.

One sperm nucleus moves down the pollen tube and unites with the egg. This fertilized cell, the zygote, will develop into the embryo plant. The second sperm nucleus unites with the two polar nuclei (so called because they move to the center from each end, or pole, of the gametophyte), which are both present in a single large cell of the female gametophyte. From the fertilized $3n$ cell, a specialized tissue called the *endosperm* develops, which nourishes the embryo. This extraordinary phenomenon of "double fertilization" takes place, in all the natural world, only among the flowering plants.

As you will recall (Chapter 3–1), in lower plants, the haploid gamete-producing plants have an autonomous existence. In the angiosperms, however, the female gametophyte completes its existence within the ovule of the mother sporophyte and the new sporophyte also has its beginnings there.

THE EMBRYO

As the fertilized egg cell divides, the cells begin to *differentiate*—become different from one another—and the embryo begins to take on a shape characteristic of the particular species, a process known as *morphogenesis*.

Growth Areas

In the earliest stages of embryonic growth, cell division takes place throughout the body of the infant plant. As the embryo grows older, however, the addition of new cells becomes gradually restricted to certain parts of the plant body: the *apical meristems* of the root and the shoot. During the rest of the life of the plants, all the primary growth —which involves chiefly the elongation of the plant body— will originate in these meristems.

The existence of such meristematic areas, in which growth continues throughout the life of the plant, is one of the principal differences between plants and animals. Higher animals stop growing when they reach maturity, but plants continue to grow during their entire life-span. Growth in plants substitutes in large part for the mobility of animals. For most plants, the only available forms of response to the environment are either extensions of the

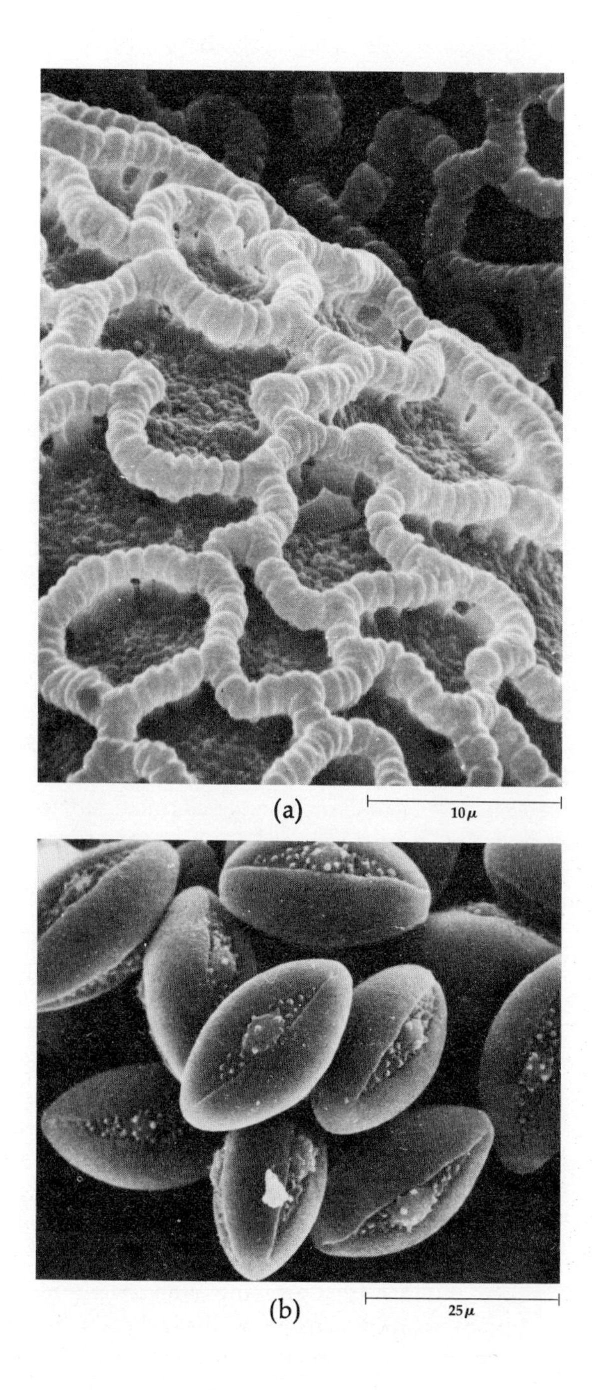

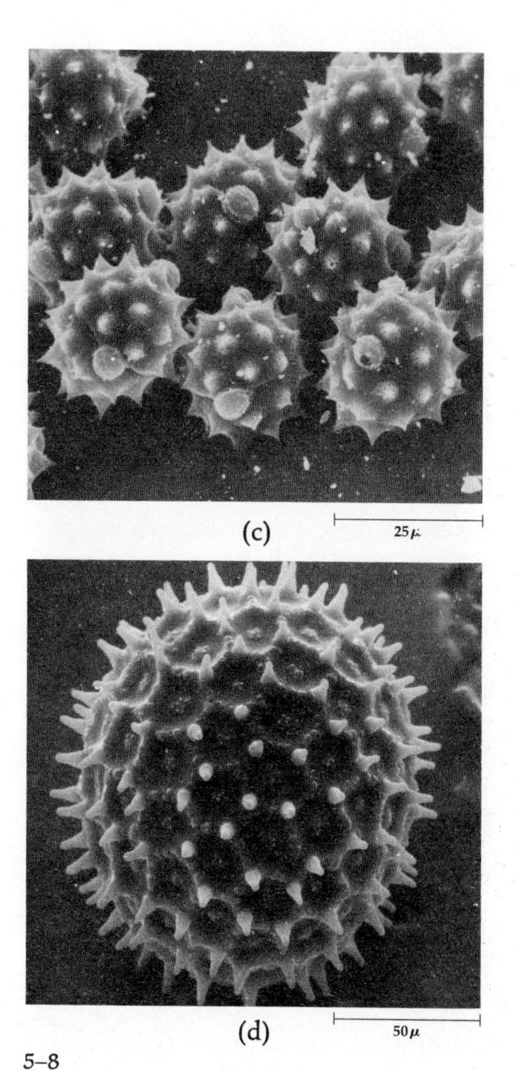

5–8

Pollen grains. The walls of the pollen grain protect the male gametophyte on the journey between the anther and the stigma. These outer surfaces, which are remarkably tough and resistant, are often elaborately sculptured. As you can see, the pollen grains of different species are distinctly different: (a) *a lily,* (b) *a horse chestnut (the three slit-like pores are characteristic of dicots),* (c) *a chrysanthemum (spiny pollen grains such as these are common among composites), and* (d) *a morning glory.*

5–9

Fertilization in angiosperms. The pollen tube of the male gametophyte, or pollen grain, grows down through the style and enters the ovule, in which the female gametophyte has developed to a seven-cell stage. One of the sperm nuclei unites with the egg cell, forming the zygote. The other sperm nucleus fuses with the two polar nuclei that are present in a single large cell (which in the drawing, fills most of the ovule). From the resulting triploid (3n) *cell, the endosperm will develop. The carpel shown here contains a single ovule.*

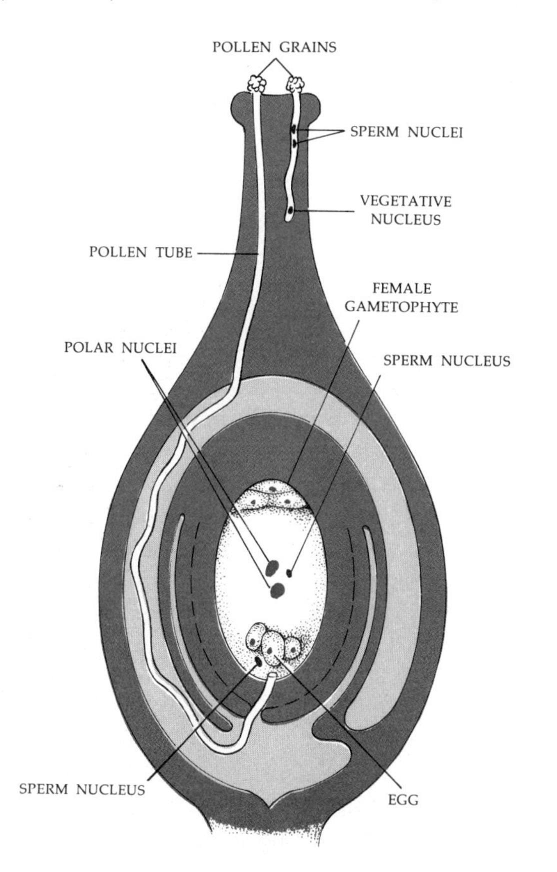

root or shoot or the discarding of parts—leaves, flowers, or seeds—both of which involve changes in size and form. By growth, a plant modifies its relationship with the environment, moving toward the light and extending its roots. The sequence of growth stages in plants thus corresponds to a whole series of motor acts in animals, especially those concerned with the search for food and water. In fact, growth in plants serves many of the functions that we group under the term "behavior" in animals.

The Seed and the Fruit

As the embryo begins to form, the petals and stamens of the parent flower fall away and the wall of the ovary develops into the fruit. The seed of the fruit is the ovule (which comes from the maternal sporophyte), with the $3n$ endosperm (formed from the gametophyte) and the young sporophyte growing within it. In a peach, for instance, which contains only one ovule per ovary, the skin, the fleshy edible portion of the fruit, and the stone are three layers of the matured ovary wall, and the hard almond-shaped structure within the stone is the seed. In a pea, the pod is the mature ovary wall and the peas are the seeds (the ovules and their contents). The raspberry is an aggregate of many fruits from a single flower, each fruit containing a single seed and each formed from a separate carpel.

By the time the embryo is mature, the seed coat has thickened and hardened and the seed, which includes the embryo, falls from the parent plant. The embryo then enters a period of dormancy.

SEED GERMINATION

The seeds of most wild plants require a period of dormancy before they will germinate. This requirement ensures that the seed will "wait" at least until the next favorable growth period. Seeds can remain dormant and yet viable—with the embryo in a state of suspended animation—for hundreds of years. The record for dormancy, as far as is known, has been set by some seeds of Arctic tundra lupine

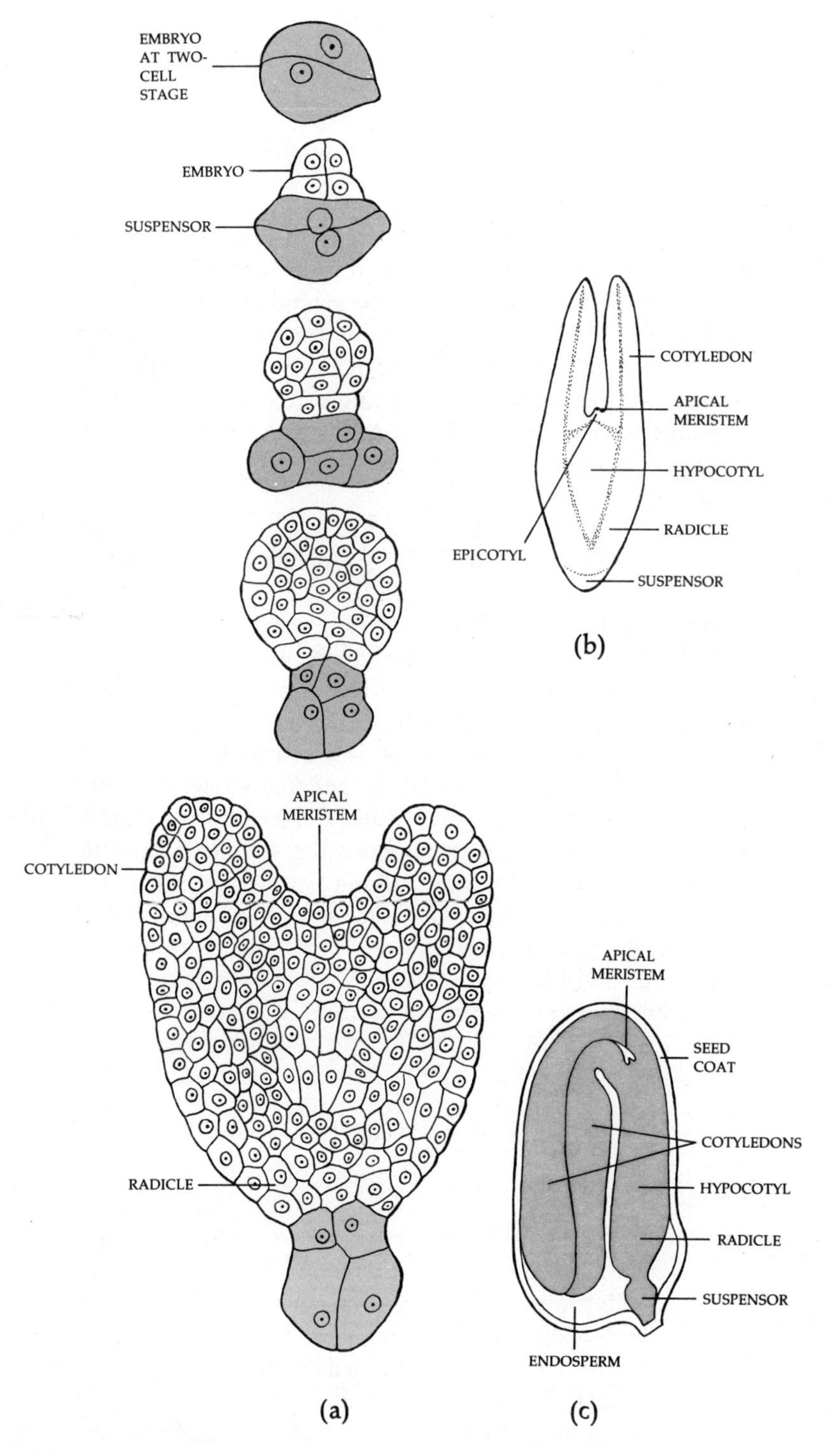

5–10

(a) *Development of a dicot embryo, showing early stages in the differentiation of the parts of the mature plant body. The suspensor, which attaches the embryo to the inner surface of the ovule, disintegrates as the embryo matures.* (b) *Embryo at a later stage.* (c) *The plant embryo at the time the seed is dropped from the mother plant. The hypocotyl is the portion of the stem below the cotyledon. The epicotyl is the portion above the point of attachment of the cotyledons. The radicle is the apical meristem of the root.*

recently found in the Yukon in a lemming burrow. Deeply buried in the permanently frozen silt, they are estimated to be at least 10,000 years old. But when a sample was planted, the seeds germinated within 48 hours!

The seed coat apparently plays the major role in maintaining dormancy. In some species, the seed coat seems to act primarily as a mechanical barrier, preventing the entry of water and gases, without which growth is not possible. In these cases, growth is promoted by the seed coat being worn off in various ways—such as being washed off by rainfall, abraded by sand or soil, burned away by a forest fire, or removal by passing through the digestive tract of a bird or other animal. In other species, dormancy seems to be maintained chiefly by the presence of chemical inhibitors in the seed coat. These inhibitors undergo chemical changes in response to various environmental factors, such as light or prolonged cold or a sudden rise in temperature, which neutralize their effects and so release the seed from dormancy.

The dormancy requirement in seeds apparently evolved only recently, geologically speaking, among groups of plants subjected to the environmental stress of increasing winter cold characteristic of the Ice Age—which explains why even closely related plants have different mechanisms for maintaining and breaking dormancy.

5–11
Seedling of touch-me-not (Impatiens). *In a few months, this plant will pass through an entire growth cycle, leaving only its seeds—from which seedlings like this will emerge the following spring.*

PRIMARY GROWTH OF THE PLANT

The Root

During dormancy, the seed contains almost no moisture (only about 5 to 10 percent). Germination immediately follows a massive uptake (imbibition) of water, which results in the rupture of the seed coat. The first part of the plant body to break through the seed coat, in nearly all species, is the embryonic root.

Figure 5–13 diagrams the growing zone of the root of an angiosperm. At the very tip is the root cap, which protects the apical meristem as the root tip is pushed through the soil. The cells of the root cap wear away and are constantly replaced by new cells from the meristem. The cells in the meristem designated as *apical initials* are those that initiate the growth process; all the other cells in the root are the progeny of these relatively few meristematic cells. The apical initials divide continuously. Some of the daughter cells replace the cells in the root cap, some remain in the tip of the meristem, and some differentiate after they divide. The maximum rate of cell division occurs at a point well above the tip of the meristem. Then, just above the point where cell division ceases, the cells begin to elongate, growing to 10 or more times their previous size, often within the span of a few hours. This elongation process is the immediate cause of root growth, although, of course, growth is ultimately dependent on the production of the new cells which become part of the elongation zone.

At about the place where cell division ceases, some of the cells begin to differentiate into sieve cells, the conducting cells of the phloem, and vessel elements, the conducting cells of xylem. Further up the root, the endodermal ("inside skin") cells take shape; these form a waterproof cylinder around the vascular tissue of the root. At about this same stage, the epidermal ("outside skin") cells differentiate; from these the root hairs grow. Water and soil nutrients are absorbed by the plant principally through the root hairs. As you can see in Figure 5–13, each of these hairs is simply an extension of a single cell. This same basic

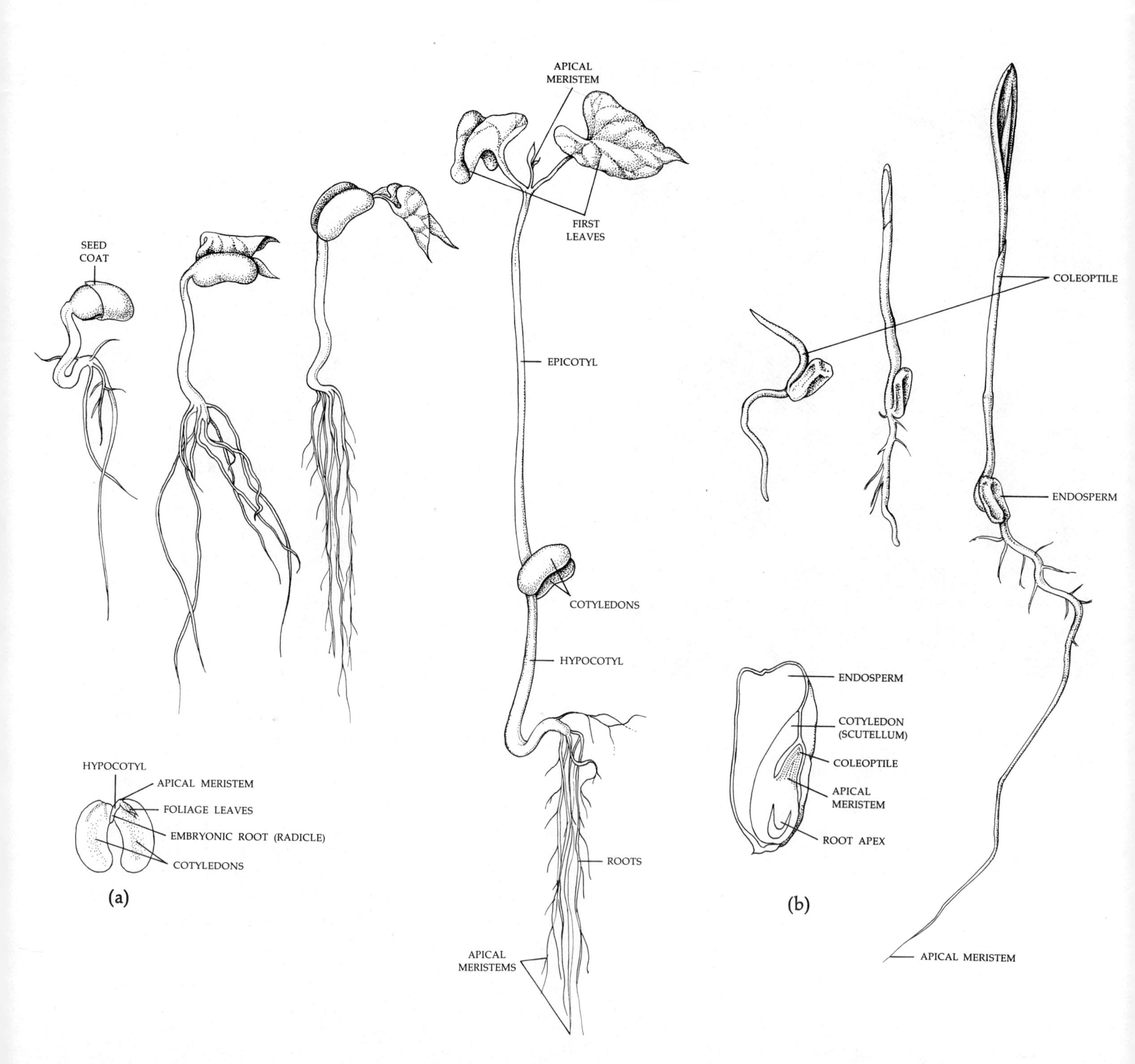
SEED
COAT
APICAL
MERISTEM
FIRST
LEAVES
EPICOTYL
COTYLEDONS
HYPOCOTYL
ROOTS
APICAL
MERISTEMS
HYPOCOTYL
APICAL MERISTEM
FOLIAGE LEAVES
EMBRYONIC ROOT (RADICLE)
COTYLEDONS
(a)
COLEOPTILE
ENDOSPERM
ENDOSPERM
COTYLEDON
(SCUTELLUM)
COLEOPTILE
APICAL
MERISTEM
ROOT APEX
(b)
APICAL MERISTEM

5–12 (at left)
(a) *A bean offers a good example of seedling development in dicots. As the seed matures, the endosperm is digested into and stored by the fleshy cotyledons and the first leaves are formed within the seed. Prior to germination, the seed imbibes water and swells, bursting the seed coat. The young root emerges first, followed by the hypocotyl. The cotyledons eventually will shrivel and fall away.* (b) *In corn plants and other grasses, all of which are monocots, the single cotyledon (known as the* scutellum*) absorbs food from the fleshy endosperm. The first leaf is the coleoptile, which forms a cylinder-like sheath over the growing shoot tip of the plant. Typically, the shriveled endosperm, with the scutellum buried in it, is still present in the young seedling.*

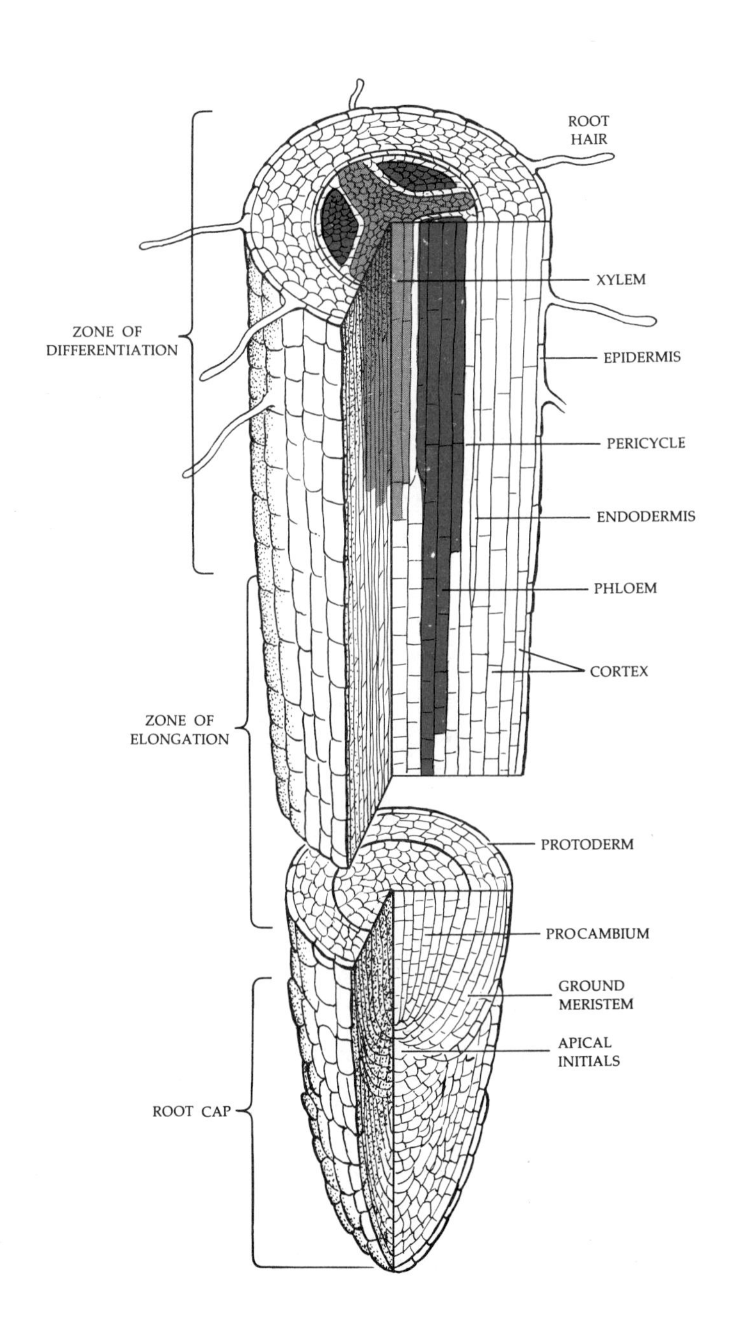

5–13 (at right)
The growing zone of a root. Growth begins in the apical initials, cells within the meristem. As these cells divide and elongate, the root is pushed downward into the soil. The cells above the meristem undergo a characteristic series of changes as the distance increases between them and the root tip. First, there is maximum cell division, followed by cell elongation, which accounts for most of the lengthening of the root. As the cells elongate, they differentiate into various specialized tissues of the plant. The protoderm becomes the epidermis; the ground meristem the cortex; and the procambium becomes the xylem and phloem. Some of the cells produced in the apical meristem form the protective root cap.

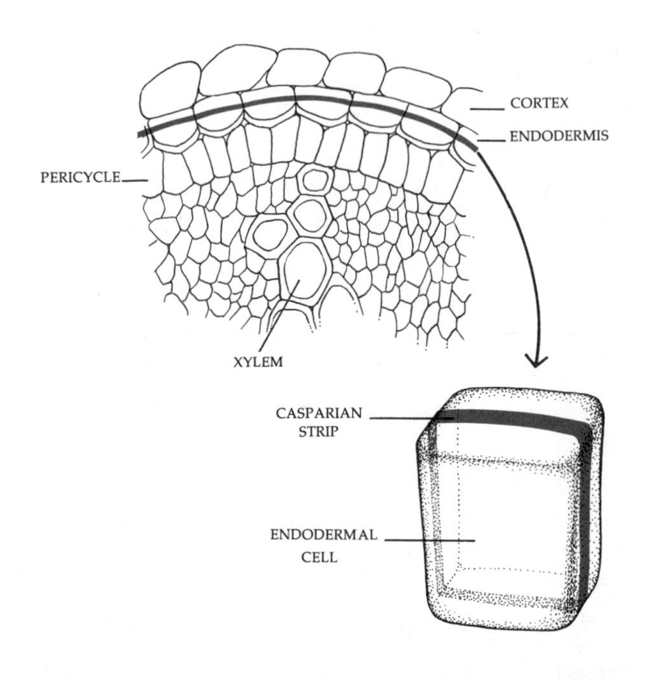

5–14

Endodermal cells from the roots of a morning glory, showing the characteristic casparian strip which extends around the circumference of the cells. This strip, which is actually a thickening of a cell wall, is made up of a waxy waterproof substance that forms a seal between adjacent cells, thus blocking diffusion of water and solutes through the cell walls.

5–15

Types of roots. (a) *Taproot of dandelion.* (b) *Fibrous root of a grass.* (c) *Prop root of corn. Roots such as these prop roots which grow from the stem, are known as* adventitious roots.

(a)

(b)

(c)

pattern of growth is seen in the first tiny root of a seedling and in the growing root tips of a tree 100 feet tall.

The Shoot

The growing region of the shoot is the meristematic tissue of the shoot tip. The organization of this growing region is similar to that seen in the root: first, a zone in which most of the cell division takes place; next, a zone of cell elongation; and finally, a zone of differentiation.

As in the root, the outermost layer of cells develops into the epidermis. In the shoot, these cells have an outer surface or cutin. Meristematic tissue differentiates into the pith, which is in the very center of the stem, and into the cortex, the layer which lies between the epidermis and the vascular tissues. Other cells differentiate into the primary vascular tissues—the primary xylem and the primary phloem. The pattern of development is more complicated, however, than in the root tip since the apical meristem of the shoot is the source of tissues that give rise to new leaves, flowers, and branches.

Figure 5–16 shows the shoot tip of a lilac plant. Here you can see the apical meristem, which is very small (only a few hundred microns in diameter), and the beginnings—primordia—of the forming leaves which make up a terminal bud. As you can see, leaves are formed in an orderly sequence at the shoot tip. The leaf originates by the division of cells in a localized area along the side of the shoot apex. In some species, leaves arise simultaneously in pairs opposite one another, as in our figure. In other species, the leaves occur in circles (whorls) or in spirals. As the stem elongates, the young leaves become separated, so that the leaf pairs clustered so tightly together around the apex in Figure 5–16 will eventually be spaced out along the stem of the plant. The vascular tissue begins to differentiate in the leaf primordia eventually becoming part of the general vascular system that connects the plant from root to leaf tip.

In addition to apical meristems, stems typically develop other bud primordia. These, like the primordia of the apical meristem, may also form leaves, branches, or flowers.

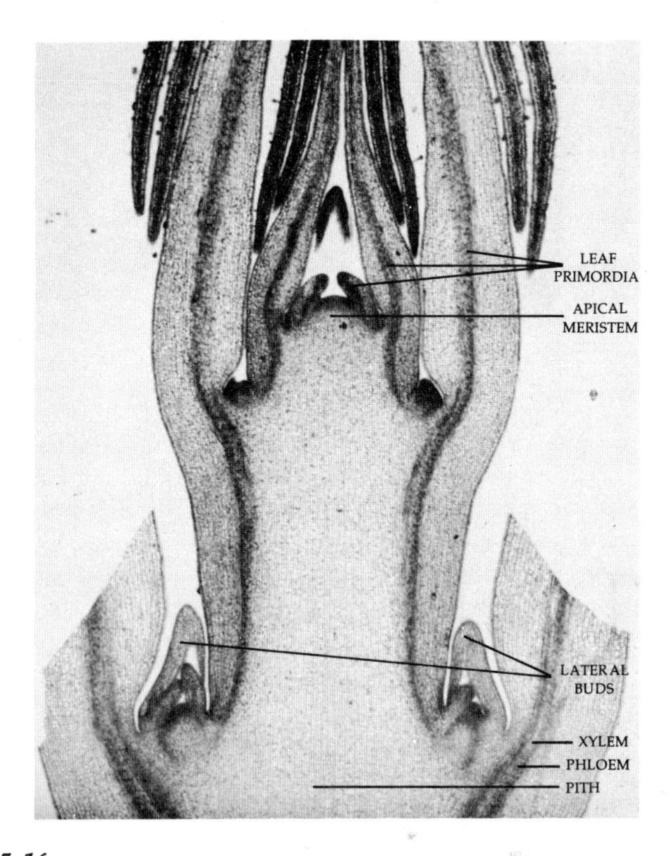

5–16

Longitudinal section of the shoot tip of a lilac (Syringa vulgaris). *At the tip of the stem is the* apical meristem, *the central zone of cell division. The* leaf primordia, *from which new leaves will form, originate along the sides of the shoot apex. In the picture, the bumps on either side of the meristematic dome are two leaf primordia emerging from the apical meristem. Successively older leaf primordia have formed on both sides. The developing stem below the apical meristem and the young leaves on either side are also regions of active cell division.*

SECONDARY GROWTH

Primary growth, then, involves the formation and elongation of stems, roots, and branches and the differentiation of the vascular bundles and other specialized tissues of the young stem or root.

Secondary growth is the process by which woody plants increase the thickness of trunks, stems, branches, and roots. The so-called "secondary tissues" are not derived from the apical meristems; they are the result of the production of new cells by the vascular cambium and cork cambium. The vascular cambium forms a thin sheath of tissue completely surrounding the xylem and completely surrounded by the phloem. The cambium cells divide continually during the growing season, adding new xylem cells, that is, secondary xylem, on the outside of the primary xylem and secondary phloem on the inside of the primary phloem. Some daughter cells remain as a cylinder of undifferentiated cambium. As the tree grows older, the supporting cells of the xylem in the center of the tree, the heartwood, die and the xylem ceases to function.

As the girth of stems and roots increases by secondary growth, the epidermis becomes stretched and torn and is replaced by cork. In response to this tearing process, a new type of cambium forms from the cortex, and from this cork cambium, cork, which is a dead tissue, is produced.

Figure 5–18 shows the cross section of the trunk of a young angiosperm, in which some secondary growth has taken place. In the center of the stem are loosely packed parenchyma cells, the pith. The first cylinders of tissue around the pith are layers of xylem, composed of vessels and tracheids. Around the outermost layer of xylem is a layer of meristematic tissue, the vascular cambium. Each year, during the growing season, the cambium divides mitotically, forming new xylem (secondary xylem) on its inner surface and new phloem (secondary phloem) on its outer surface. This continuous formation of layers of xylem and phloem allows tree trunks to increase their diameter as they increase their height. As they grow, season after season, the new xylem forms visible growth layers, or rings, each

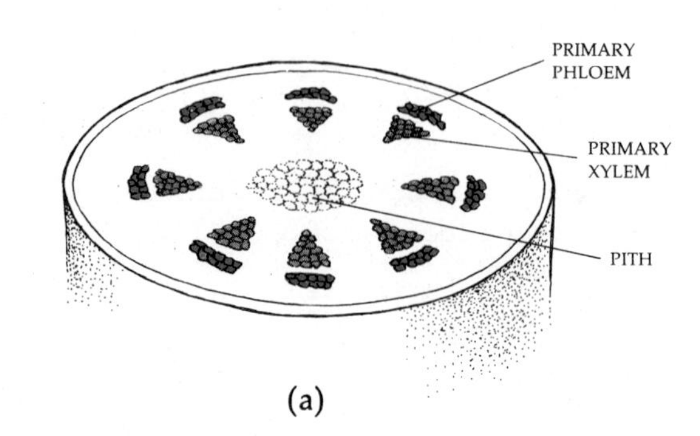

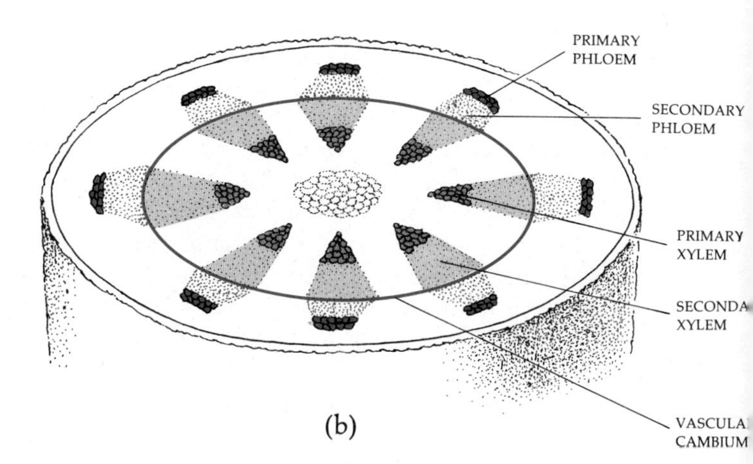

5–17

(a) *Stem of a dicot before the onset of secondary growth.*
(b) *In secondary growth, the secondary xylem and secondary phloem are produced by the vascular cambium, a meristematic tissue formed late in primary growth. The xylem continues to expand, but the cortex, epidermis, primary phloem, and even the outer layers of secondary phloem are eventually replaced by cork.*

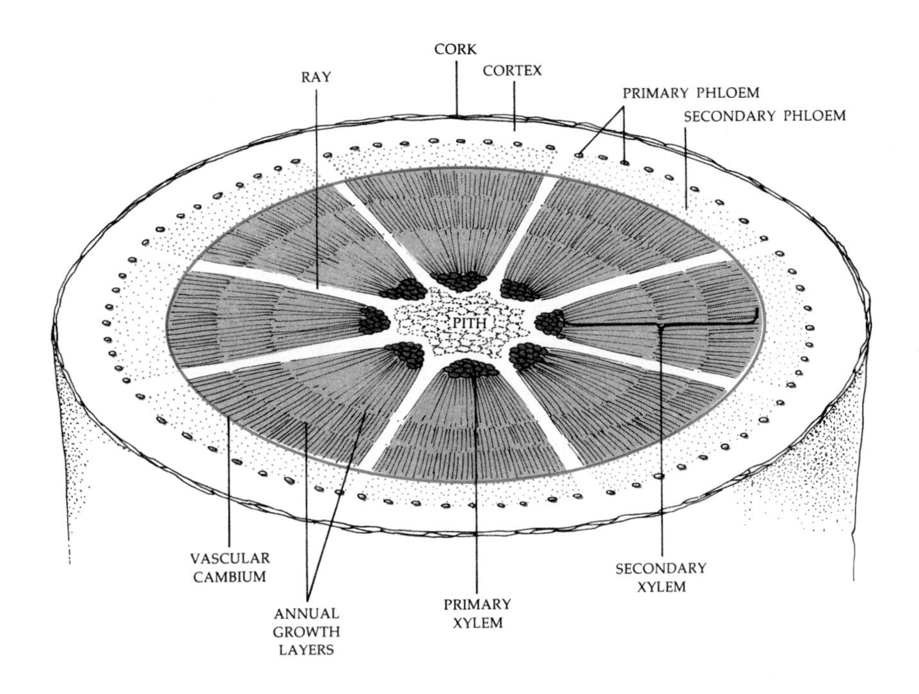

5–18

Cross section of a buckeye stem. The white area in the center is the pith. Outside the pith are several annual growth layers. On the perimeter of the outermost growth ring is the vascular cambium, encircled by a relatively broad band of secondary phloem. Some primary phloem can be seen outside the secondary phloem; the primary phloem will be crushed against the cork as more cork forms and will eventually disappear. The primary phloem is encircled by cortex, cork cambium, and finally, cork. Rays are strands of living cells that transport nutrients laterally (across the trunk).

growth season leaving its trace, so that the age of a tree can be estimated by counting the number of growth rings in a section near its base. Since the rate of growth of the tree depends on climatic conditions, it is possible to determine from the width of the annual growth layers of ancient trees fluctuations in temperature and rainfall that occurred hundreds of years ago.

SUMMARY

In the flowering plants, a new life cycle begins when a pollen grain germinates upon the stigma of a flower of the same species, sending a pollen tube down the style and into an ovule. Within the ovule is the female gametophyte, which usually consists of seven cells with eight haploid nuclei, including the egg nucleus. One of the three haploid nuclei in the pollen grain (a sperm nucleus) unites with the egg cell nucleus, and the other sperm nucleus unites with the two polar nuclei of the female gametophyte. The plant embryo develops from the first union, and the nutritive endosperm, which is a triploid ($3n$) tissue, from the second. This phenomenon of "double fertilization" is found only among the flowering plants.

The immature embryo has two apical meristems—the apical meristem of the shoot and the radicle, or apical meristem of the root—and either one or two cotyledons ("seed leaves"). The cotyledons absorb nutrients from the endosperm which nourish the growing embryo. Early in embryonic life, growth becomes confined to limited areas in the plant body: the apical meristems of the root and the shoot.

The seed consists of the embryo, the endosperm, and the seed coat—that is, the ovule and its contents. The petals, carpels, stamens, and other floral parts of the parent plant fall away as the ovary ripens into fruit and the seed forms. The embryo characteristically then enters a dormant period.

When the seed germinates, growth of the root and shoot proceeds from the apical meristems of the embryo. Certain cells within the meristems divide continuously. Some of these cells remain meristematic, continuing to divide, while others elongate and then differentiate, forming, according to their position, various specialized cells, including those of the xylem and the phloem. In the root, the apical meristem forms a root cap, which protects the root tip as it is pushed through the soil. Root hairs, which are the principal pathways of absorption by the roots, grow from the epidermal cells. Leaves, stems, branches, and flowers form from the apical meristems of shoots.

Secondary growth is the process by which the woody perennials increase their girth. Such growth arises primarily from the vascular cambium, a sheath of tissue completely surrounding the xylem and completely surrounded by phloem. The cambium cells divide continuously during the growing season, adding new xylem cells (secondary xylem) outside the primary xylem and new phloem cells (secondary phloem) on the inside of the primary phloem. As the trunk increases in girth, the epidermis is ruptured and destroyed and is replaced by cork.

QUESTION

Sketch a dicot embryo and label the various areas of the embryo. (Do this first without looking at Figure 5–10; then check your drawing against this diagram.) What structures of the mature plant will arise from each of these parts?

CHAPTER 5–3

Integration of Growth: The Plant Hormones

A plant, in order to grow, needs light, carbon dioxide, water, and minerals, including nitrogen, from the earth. Using the energy from the sun, it makes more of its own substance, turning these simple materials into the complex organic substances of which living things are composed. As we saw in the previous chapter, the plant does far more than simply increase its mass and volume. It differentiates and undergoes morphogenesis, forming a variety of cells, tissues, and organs. How can one single cell, the fertilized egg, be the source of the myriad tissues—shoot, root, flower, fruit, seed—that make up that extraordinary individual known as a "normal plant"? Many of the details of how these processes are regulated are not known, but it has become clear that normal development depends on the interplay of a number of internal and external factors.

Chief among the internal factors are the plant hormones. Hormones, by definition, are substances that are produced in one tissue and transported to another, where they exert highly specific effects. The term "hormone" comes from the Greek word meaning "to excite." It is now clear, however, that many hormones have inhibitory influences. So, rather than thinking of hormones as stimulators, it is perhaps more useful to consider them as chemical messengers. But this term, too, needs qualification. As we shall see, the response to the particular "message" depends not only on its content but upon how it is "read" by its recipient.

Of the many plant hormones that have been identified, the best studied are the auxins, cytokinins, and gibberellins. In this chapter we shall introduce these and a few other hormones by considering some of what is known about their effects in plants. Although we shall discuss each group separately, it should be remembered that the functions of these hormones overlap and that their effects depend not only on the target organ but also on the presence or absence of other hormones. Cytokinins and auxins, for example, interact to produce cell division and cell elongation. In cultures of cells growing in a test tube, an auxin plus a low concentration of cytokinin results in rapid cell expansion, with the production of a relatively few giant cells. On the other hand, a cytokinin plus a low concentration of auxin results in rapid cell division, with the production of large numbers of relatively tiny cells.

THE AUXINS

The effects of the auxins were first observed by Charles Darwin and his son Francis and reported in *The Power of Movement in Plants*, published in 1881. The Darwins were studying the bending toward light (*phototropism*) of grass seedlings. They noted that the bending takes place below the tip, in the lower part of the shoot. Then they showed that if they covered just the terminal portion of the shoot of the seedling with a cylinder of metal foil or a hollow tube of glass blackened with India ink and exposed the plant to a light coming from the side, the characteristic bending of the shoot did not occur. If, however, the tip was enclosed in a transparent glass tube, bending occurred normally. Bending also occurred normally when the lightproof cylinder was placed below the tip:

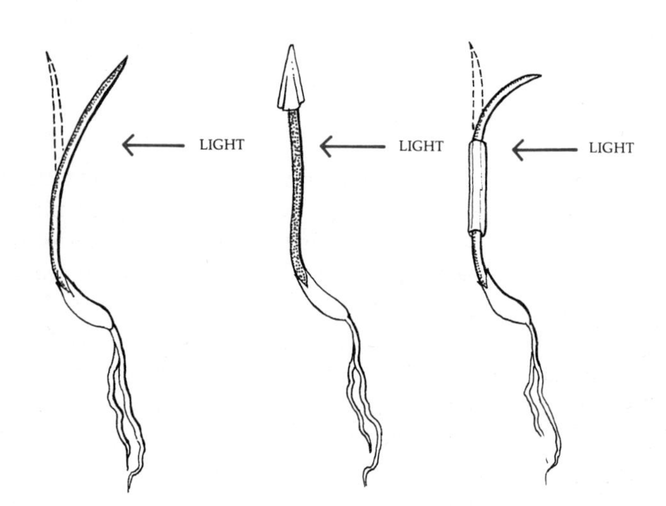

"We must therefore conclude," they stated, "that when seedlings are freely exposed to a lateral light some influence is transmitted from the upper to the lower part, causing the material to bend."

In 1926, the Dutch plant physiologist Frits W. Went succeeded in separating this "influence" from the plants that produced it. Went cut off the coleoptile tips from a number of oat seedlings and placed the tips for about an hour on a slice of agar (a gelatinlike substance), with their cut surfaces in contact with the agar. He then cut the agar into small blocks and placed a block off-center on each stump of the decapitated plants, which were kept in the dark during the entire experiment. Within one hour, he observed a distinct bending *away* from the side on which the agar block was placed. Agar blocks which had not been exposed to a coleoptile tip produced either no bending or a slight bending toward the side on which the block had been placed. Agar blocks which had been exposed to a section of coleoptile lower on the shoot produced no physiological effect:

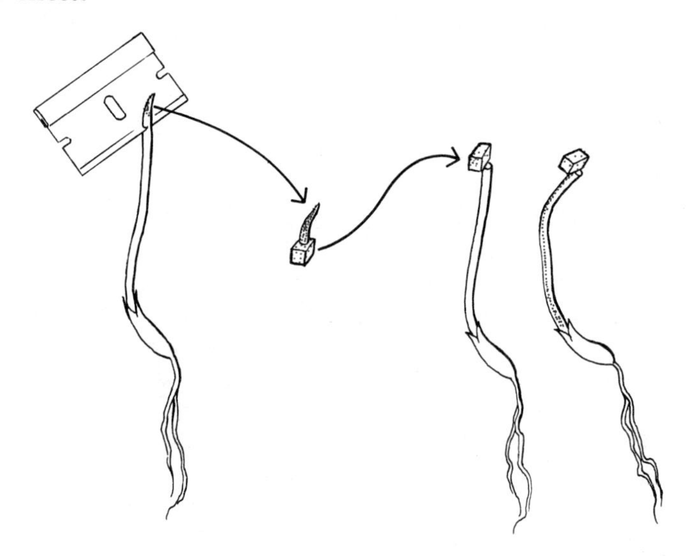

By these experiments, Went showed that the coleoptile tip exerted its effects by means of a chemical stimulus (in short, a hormone) rather than a physical stimulus, such as an electrical charge. This chemical stimulus came to be known as auxin, a term coined by Went from the Greek work *auxein*, "to increase."

Several different substances within auxin activity have now been isolated from plant tissues, and others have been synthesized in the laboratory. One of the most common of the natural auxins is indoleacetic acid, abbreviated IAA. 2, 4-D, one of the synthetic auxins, is commonly used as an herbicide (like many other physiologically active compounds, auxins are toxic in high doses).

Naphthalenacetic acid, also a synthetic auxin, is commonly employed to induce the formation of adventitious roots in cuttings and to reduce fruit drop in orchard crops:

IAA

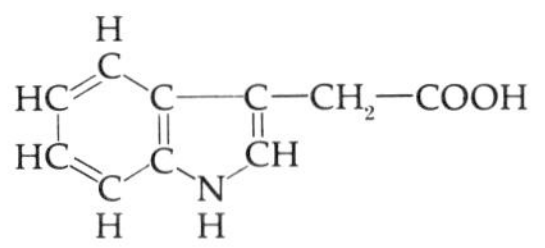

```
        H
        C
HC    C ——— C—CH2—COOH
|     ||    ||
HC    C     CH
   C     N
   H     H
```

2,4-D

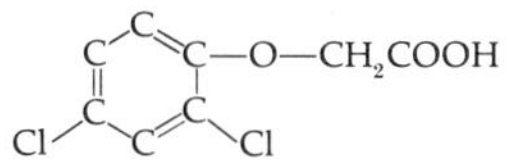

```
       C
   C      C—O—CH2COOH
   ||     |
   C      C
Cl    C      Cl
```

NAPHTHALENACETIC ACID

```
     C        C—CH2—COOH
  C     C       C
  |     ||      |
  C     C       C
     C       C
```

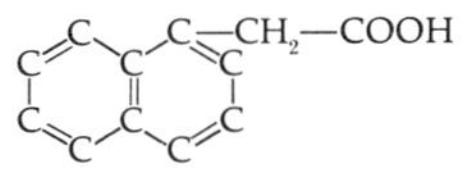

The synthetic auxins, unlike IAA, are not readily broken down by natural plant enzymes and so are better suited for commercial purposes.

Auxins and Phototropism

Auxins promote cell enlargement by increasing the plasticity of the cell wall. When the cell wall softens, the cell enlarges owing to the pressure of water from within its vacuole. As the water potential is reduced, more water enters. The cell thus continues to enlarge until it encounters sufficient resistance from the wall, which ultimately becomes hard again.

The phototropism observed by the Darwins results from the fact that under the influence of light, auxins migrate from the light side to the dark side of the tip. The cells on the dark side, having more auxin, elongate more rapidly than those on the light side, causing the plant to bend toward the light, a response with high survival value for young plants. It is the unequal distribution of auxin, rather than a change in the production of it or in the response of cells to it, that accounts for the difference in growth rate of the two sides of the coleoptile tip.

Auxins and Growth Regulation

In the intact plant, auxins promote growth of the shoot. The hormone is produced in the rapidly dividing cells of the meristem and migrates downward, causing the cells below the apex to elongate. If you cut off the shoot tip, elongation stops. If an auxin—either in an agar block or in a paste—is applied to the cut surface, growth resumes. If the hormone is applied to the apex of an intact plant, there is usually no growth effect. Apparently, under normal circumstances, the tip produces all the auxins that the cells of the stem can respond to.

Auxins, in very small amounts, sometimes stimulate the growth of roots. In somewhat larger amounts, they inhibit the growth of the main roots, although they may promote the initiation of new branch roots or adventitious roots.

Auxins and Bud Growth

Although auxins stimulate growth—mainly at the apex of the plant—in some species they also inhibit growth in the lateral buds.

If you nip off the terminal bud of *Coleus*, a common houseplant, for example, the lateral buds begin to grow vigorously, producing a plant with a bushier, more compact body. If you apply an auxin to the cut surface of the tip, the growth of lateral buds will be inhibited. In a potato plant, the edible portion is actually a thickened modified stem (a tuber) and the "eyes" are lateral buds. Treating the tuber with a synthetic auxin inhibits bud growth, permitting the potatoes to be stored for longer periods.

5–19
Two buds forming on a mass of undifferentiated (callus) tissue following treatment with both an auxin and a cytokinin. Callus from some types of plants—tobacco for instance—will produce undifferentiated tissue, roots, or buds, depending on relative proportions of auxins and cytokinins.

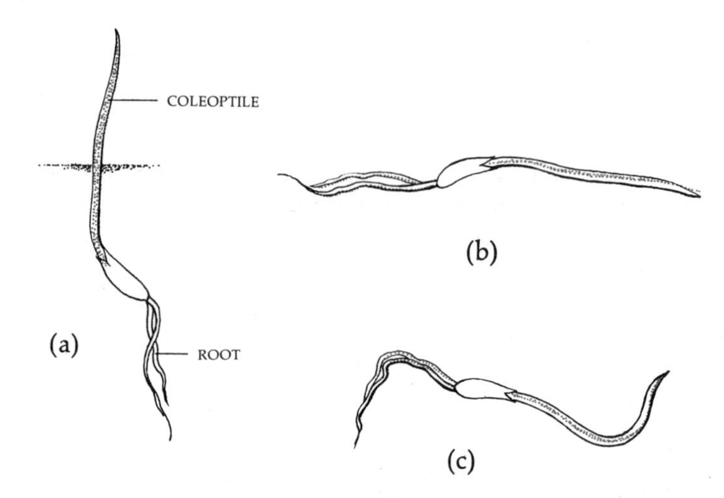

5–20
Why roots always grow downward and shoots upward. When a seedling is perpendicular (a), *the auxin is distributed evenly and the plant grows straight up. However, when a seedling is placed on its side* (b), *auxin accumulates in the lower half of the plant body, inhibiting growth on the underside of the root and promoting growth on the underside of the shoot. This differential effect of auxin on root and shoot causes the seedling to right itself* (c).

In other words, the response of a tissue to auxins may be either promotive or inhibitive, depending on the target organ.

Auxins and Fruit Growth

Growth and ripening in fruit involve changes in color, texture, flavor, and odor. In fleshy fruits, the chlorophyll content drops and other pigments change the fruit color. Simultaneously, the fleshy part of the fruit softens. This is a result of the enzymatic digestion of pectin, the principal component of the middle lamella, which separates adjacent cells. When the middle lamella is weakened, cells are able to slip past one another. During this same period, starches are hydrolyzed to sugars. As a consequence of these changes, fruits become conspicuous, palatable, and fragrant and so are attractive to animals, which eat the fruit and thus scatter the seed.

Ordinarily, if the flower is not pollinated, the fruit will not develop. In some plants, fertilization of one egg cell is sufficient for normal fruit development, but in others, such as apples or melons, which have many seeds, unless several egg cells (in as many ovules) are fertilized, the ovary wall will not mature and become fleshy. By treating the stigma with auxin it is possible to induce seedless fruit such as seedless tomatoes or melons, produced from unfertilized egg cells.

Apparently, the developing embryo is the source of auxin for the fruit. In the strawberry, for example, if the seeds are removed during the fruit's development, the strawberry stops growing altogether. If a narrow ring of

(a)

(b)

5–21
(a) *Normal strawberry (right) and strawberry from which all seeds have been removed (left).* (b) *Strawberry in which three horizontal rows of seeds were left.*

seeds is left, the fruit (actually the fleshy receptacle) forms a bulging girdle of growth in the area of the seeds. If auxin is applied to the denuded receptacle, growth proceeds normally.

Auxins and Ethylene

Another natural substance involved in fruit ripening is ethylene. In the early 1900s, many fruit growers made a practice of improving the color of citrus fruits by "curing" them in a room with a kerosene stove. (Long before this, the Chinese used to ripen fruits in rooms where incense was being burned.) It was long believed that it was the heat that ripened the fruits. Ambitious fruit growers who had installed more modern equipment found to their sorrow that this was not the case. As experiments showed, it is actually the incomplete combustion products of the kerosene that are responsible. The most active gas was identified as ethylene. It has now been found in fruits (in all types tested), flowers, leaves, leafy stems, and roots of many different species and also in certain types of fungi.

In some plants, auxin causes a burst of ethylene production, and it is believed that some of the effects on fruits and flowers once attributed to auxin are related to auxin's role in ethylene production.

THE GIBBERELLINS

Gibberellins, like auxins, are involved in growth, stimulating both cell division and cell elongation in plants.

The most remarkable results are seen when gibberellins are applied to certain plants that, because of a single mutation, have lost the capacity to synthesize gibberellins. Such plants are genetic dwarfs. Under gibberellin treatment, these dwarfs become indistinguishable from normal tall plants. The effect on dwarf plants of gibberellins cannot be duplicated by auxins or any of the other known hormones.

Gibberellins also cause stem elongation in normal (nondwarf) plants. Too much gibberellin typically causes the stems to become long and thin, with few branches and pale stems. In fact, gibberellin was first discovered by a Japanese scientist who was studying a disease of rice plants, "foolish seedling disease." The diseased plants grew rapidly but were spindly and tended to fall over. The cause of the symptoms, the scientist found, was a chemical produced by a fungus, *Gibberella fujikuroi,* which parasitized the seedlings. The substance, which he named gibberellin, was subsequently isolated not only from the fungus but from many species of plants. More than 20 gibberellins have now been identified which vary slightly in structure and in activity.

Gibberellins and Flowering

Some plants, such as mustard and cabbage, form leaf rosettes before flowering. (In a rosette, leaves develop but the stem does not elongate between the developing leaves and separate them.) These plants can be induced to flower by exposing them to long days, to a period of cold (as in the biennials), or to both. Following the appropriate exposures, the stems elongate—a phenomenon known as bolting—and the plants flower. Application of gibberellin causes bolting and flowering without these exposures.

Gibberellins and Seed Germination

The role of gibberellins in the germination of grass seeds has recently been defined. In grass seeds, there is a specialized layer of cells, the aleurone layer, just inside the seed coat. These cells are rich in protein. When the seeds imbibe water, just prior to germination, gibberellin diffuses from the cotyledon to the aleurone layer. In response to the gibberellin, the aleurone cells produce enzymes that hydrolyze the starch in the stored food reserves of the endosperm, converting it to soluble sugars which the embryo can use. In this way, the embryo itself calls forth the substances needed for its survival and growth at the time it requires them.

When the endosperm breaks down under the action of gibberellin, it is believed, it releases other growth-promoting hormones.

5–22
Bolting in cabbage following gibberellin treatment. The plant on the left was untreated.

5–23
Action of gibberellin in barley seeds. Before the picture was taken, each of these three seeds was cut in half and the embryo removed. Forty-eight hours earlier, the seed at the bottom was treated with plain water, the seed in the center was treated with a solution of 1 part per billion of gibberellin, and the seed at the top was treated with 100 parts per billion of gibberellin. As you can see, digestion of the starchy storage tissue has begun to take place in the seeds treated with gibberellin.

THE CYTOKININS

In 1941, Johannes van Overbeek, a Dutch plant physiologist, found that in unripe coconuts, the "milk" (which is a liquid endosperm) contains a potent growth factor—one that was different from anything known at the time. This factor, or factors, it was found, greatly accelerated the enlargement of plant embryos and also promoted the division of cells isolated in test tubes. Because of this latter effect, this group of growth regulators became known as the cytokinins, from *cytokinesis,* a term for cell division.

Cytokinins have now been found in about 40 different species of higher plants, largely in actively dividing tissues, including seeds, fruits, roots, and bleeding sap (the sap that drips out of cracks and pruning cuts).

Cytokinins have been shown in some experiments to have effects that oppose those of the auxins. For example, application of cytokinins to lateral buds will sometimes cause them to develop even in the presence of auxin. And unlike auxin (but like gibberellin), cytokinin stimulates growth in leaves.

The seeds of certain varieties of lettuce which ordinarily require light for germination will germinate even in the dark if treated with cytokinin solutions. Cytokinins are not naturally present in the dry seed, however; they do not appear until the seed starts to grow, indicating that they are a result rather than a cause of germination. According to present theory, cytokinins (and probably auxins also) are released by the endosperm as a consequence of its breakdown by enzymes under the influence of gibberellin.

FLORIGEN

In many species of plant, flowering can be induced by exposing the leaves of the plant to an appropriate period of light. (Such exposure is known as photoinduction; we shall have more to say about this phenomenon in the next chapter.) Apparently something is transmitted from leaf to bud that causes the apical meristem of the bud to form floral primordia rather than leaf primordia. This "something" has been called the flowering hormone, although its existence has not yet been proved.

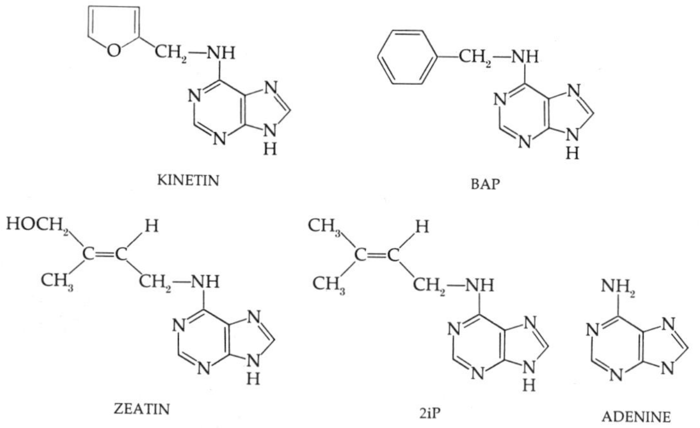

KINETIN BAP

ZEATIN 2iP ADENINE

5–24

Note the resemblances between the purine adenine and these four cytokinins. Kinetin, the first of this group to be discovered, and 6-benzylamino purine (BAP) are commonly used synthetic cytokinins. Zeatin and 2iP have been isolated from plant material. The significance of the resemblance between adenine and the hormones of this group is not known. It may merely be another example of biological economy.

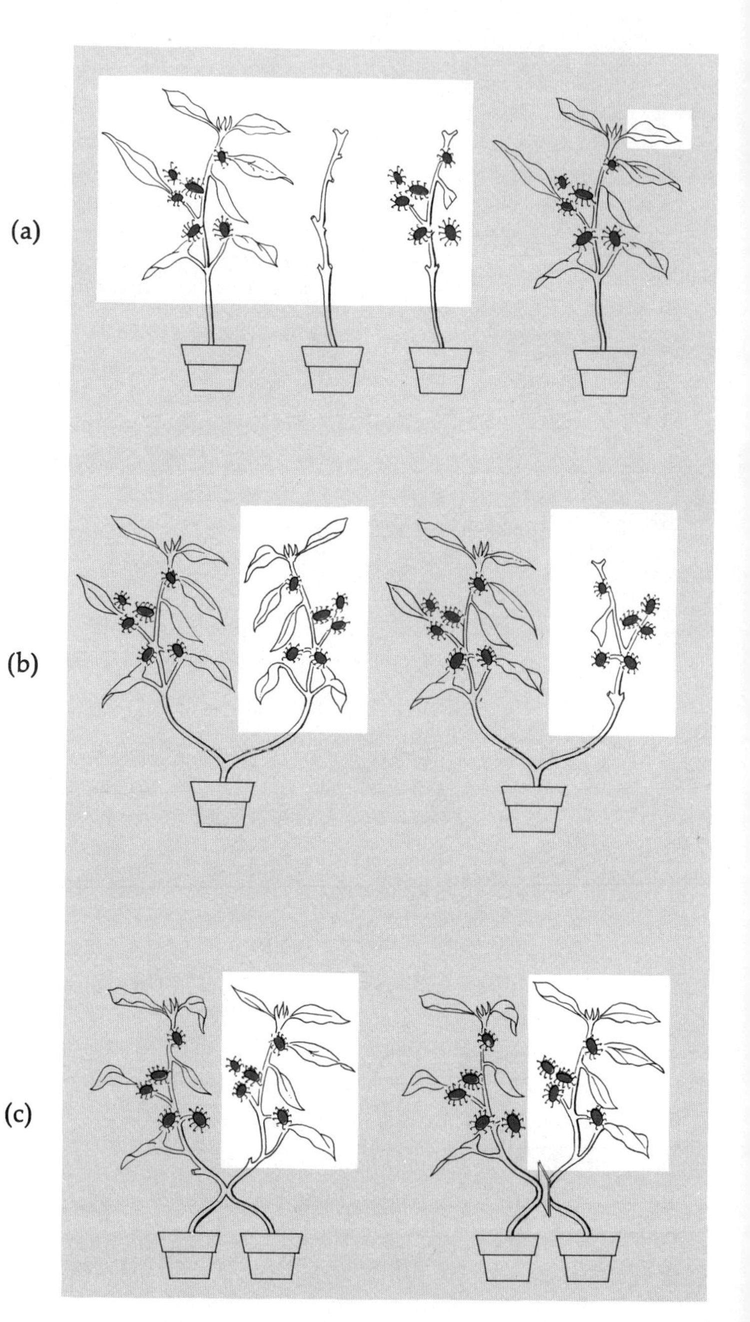

5–25

Experiments which indicate the existence of florigen: (a) *When plants, such as the cocklebur shown here, are exposed to an appropriate light cycle, plants with leaves flower and leafless plants do not. When even one-eighth of a leaf remains on a plant, flowering occurs, and the illumination of a single leaf—not necessarily the whole plant—suffices. These experiments indicate that a chemical originating in the leaves causes the plant to flower.* (b) *This conclusion is supported by experiments on branched plants. Exposure of one branch to the light induces flowering on the other branch as well, even when only a portion of a leaf is present on the lighted branch.* (c) *When two plants are grafted together, exposure of one of the plants to the light cycle induces flowering in both the lighted plant and the grafted one. This effect occurs even when a piece of paper is inserted across the graft.*

The earliest experiments on this hypothetical flowering hormone were carried out by a Russian scientist, M. H. Chailakhyan, in the 1930s. Working with a species of chrysanthemum, Chailakhyan found that if the upper portion of the stem was defoliated and the leaves on the lower part exposed to the appropriate induction period, the plant would flower. If, however, only the upper, leafless stem and its buds were exposed to the appropriate induction period, no flowering occurred. He interpreted these results as indicating that the leaves form a hormone which moves to the apex and initiates flowering. He named this hypothetical hormone florigen, the "flower maker."

Subsequent experiments showed that the flowering response does not take place if the leaf is removed immediately after photoinduction. But if the leaf is left on the plant for a few hours after the induction cycle is complete, it can then be removed without stopping flowering. The flowering hormone can pass through a graft from a photoinduced plant to a noninduced plant. If a branch is girdled, that is, if the "bark" (the cortex and phloem) is removed, florigen movement ceases. This led to the conclusion that florigen moves by way of the phloem system, the means by which most organic substances are transported.

In some plants—the Biloxi soybean is an example—leaves must be removed from the grafted receptor plant or it will not flower. This observation suggests that in noninduced plants, the leaves may produce an inhibitor. In fact, some investigators have concluded on the basis of such evidence that there is no substance that initiates flowering but rather a substance that inhibits flowering unless it is removed by the proper conditions. Strong evidence now suggests that both inhibitors and promoters are involved in the control of flowering.

ABSCISIC ACID

Abscisic acid is a hormone that was first isolated from the dormant buds of ash trees and potatoes. As dormancy in the buds was broken, it was found, the concentration of abscisic acid declined. The hormone was subsequently shown to accelerate abscission in the leaves and fruit of some species.

Abscisic acid and growth-promoting hormones often act as direct antagonists. For instance, if one repeats Went's original experiment and then treats the seedling stump with abscisic acid as well as IAA, the abscisic acid blocks the effect of the auxin. Conversely, auxins can block the effects of abscisic acid in promoting abscission in leaves and fruits.

5–26

Winter bud of a pignut hickory. These buds form in one growing season and develop in the next. The delicate tissues inside the bud are protected from mechanical injury and dessication by bud scales, small tough modified leaves.

Application of abscisic acid to active buds changes them to dormant "winter" buds, altering the differentiation of the leaf primordia from photosynthetic leaves to bud scales. The inhibitory effects of abscisic acid on buds can be cancelled by gibberellin. The appearance of hydrolyzing enzymes, induced by gibberellin in the barley seed, is inhibited by abscisic acid. Abscisic acid has little effect on dwarf plants, but it reduces the growth of normal plants. This inhibition, too, can be counteracted by gibberellin. In short, abscisic acid, in its many varied effects, serves as a direct opponent to the action of the growth-promoting hormones. This again emphasizes the concept that growth is the result of a balance of different factors.

SUMMARY

Hormones are important regulators of growth in both higher animals and higher plants. A hormone is a chemical produced in particular tissues of an organism and carried to other tissues of the organism, where it exerts specific effects. Characteristically, it is active in extremely small amounts.

Auxins are hormones that are produced principally in rapidly dividing tissues, such as coleoptile tips and apical meristems. They cause lengthening of the shoot and the coleoptile, chiefly by promoting cell elongation. They often inhibit growth in lateral buds, thus restricting growth principally to the apex of the plant. The same quantity of auxin that promotes growth in the stem inhibits growth in the main root system. Auxins promote the initiation of branch roots and adventitious roots, however. They retard abscission in leaves and fruits. In fruits, auxin produced by seeds or the pollen tube stimulates growth of the ovary wall, perhaps as a result of increased ethylene production. The capacity of auxins to produce such varied effects is believed to result from the different responses of the various target tissues.

Ethylene is a gas produced by the incomplete combustion of hydrocarbons; it also emanates from fruits during the ripening process. Ethylene promotes the ripening of fruits and is now considered a natural growth regulator.

The gibberellins were first isolated from a parasitic fungus that causes abnormal growth in rice seedlings. They were subsequently found to be natural growth hormones present in higher plants. The most dramatic effects of gibberellins are seen in dwarf plants, in which application of gibberellins restores normal growth. Gibberellins are also involved in seed germination in grasses. They stimulate the production of hydrolyzing enzymes which act on the stored starch of the endosperm, converting it to sugar, which nourishes the embryo.

The cytokinins, a third class of growth hormone, promote cell division.

Florigen is the name given to the hypothetical hormone produced in the leaves which promotes flowering. Although there is evidence for the existence of this hormone (or hormones), it has not yet been isolated.

Abscisic acid is a growth-inhibiting hormone that has been found in dormant buds and in fruits, with a maximum amount present just before the fruit drops. Abscisic acid induces dormancy in vegetative buds and accelerates abscission. It opposes the effects of all three types of growth hormone in various situations.

QUESTION

If you secured several potted seedlings on the rim of a wheel and then mounted the wheel on a turntable so that it revolved continuously at about 78 rpm, in which direction would the roots and shoots grow? Why?

CHAPTER 5–4

Biological Clocks: Circadian Rhythms and Photoperiodism

Living things, in order to survive, must regulate their activities in accordance with the world around them. Birds migrate in the spring and fall, arctic animals change their coats or feathers seasonally, and plants flower at the same time each year. In other words, organisms are characterized by remarkable abilities to adjust to—and moreover to anticipate—changes in their environment. The ways in which plants detect and respond to environmental cues—particularly to the changing cycles of light and dark—are among the most elusive and exciting of research subjects in plant physiology, and offer important clues.

CIRCADIAN RHYTHMS

In some plants, the flowers open in the morning and close at dusk or the leaves spread in the sunlight and fold toward the stem at night. As long ago as 1729, the French scientist Jean-Jacques de Mairan noticed that these diurnal (daily) movements continue even when the plants are kept in dim light. More recent studies have shown that less evident activities, such as photosynthesis, auxin production, and the rate of cell division, also have regular daily rhythms, which continue even when all environmental conditions are kept constant. These regular day-night cycles have come to be called circadian rhythms, from the Latin words *circa*, meaning "about," and *dies*, "day." Circadian rhythms now have been found throughout the plant and animal kingdoms.

Are the Rhythms Endogenous?

Are these rhythms internal—that is, caused by factors within the plant or animal itself—or is the organism keeping itself in tune with some external factor? For a number of years, biologists debated whether it might not be some environmental force, such as cosmic rays, the magnetic field of the earth, or the earth's rotation, that was setting the rhythms. Attempts to settle this recurrent controversy have led to countless experiments under an extraordinary variety of conditions. Organisms have been taken down in salt mines, shipped to the South Pole, flown halfway around the world in airplanes, and most recently, orbited in satellites. Although there is still a vocal minority that believes that circadian rhythms are under the influence of a subtle geophysical factor, most workers now agree that the rhythms are endogenous. Nothing, however, is known about the physical or chemical nature of this internal timing device, which is often referred to as a biological clock.

Entrainment

Although circadian rhythms probably originate within the organisms themselves, they can be modified by external conditions—a fact that is, of course, important to the survival of both individuals and species. For instance, a plant whose natural daily rhythm shows a peak every 26 hours when grown under continuous dim light can adjust its rhythm to 14 hours of light and 10 of darkness. It can also adjust to 11 hours of light and 11 of dark (or 22 hours).

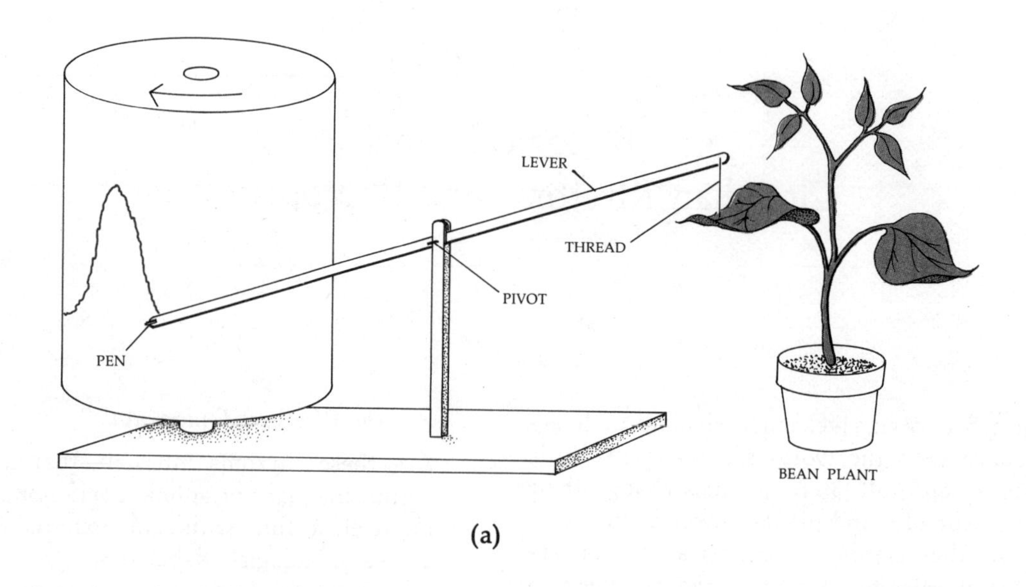

(a)

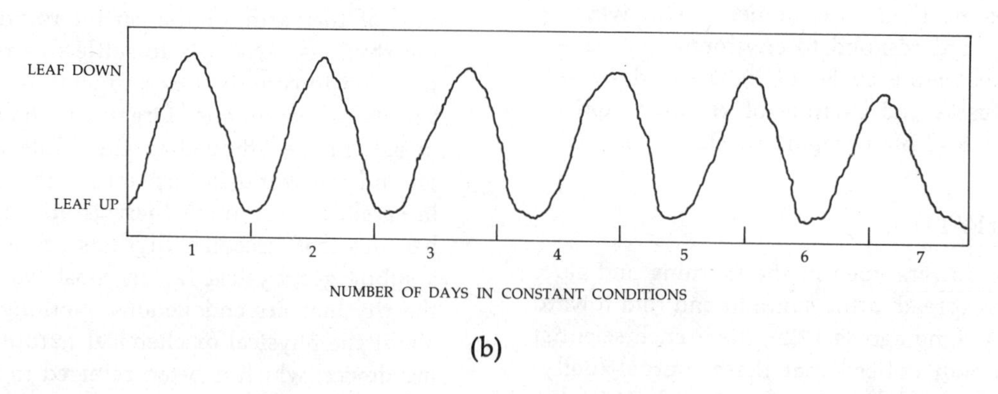

(b)

5–27
"Sleep movement" rhythms in the bean plant. Many legumes, such as the bean, orient their leaves perpendicular to the rays of the sun during the day and fold them up at night. These "sleep movements" can easily be transcribed on a rotating chart by a delicately balanced pen-and-lever system attached to the leaf by a fine thread (a). *The rhythm will persist for several days in continuous dim illumination. A representative recording is seen in* (b).

CIRCADIAN RHYTHMS AND MAN

Man, too, is tied to his own endogenous rhythms. It has long been known that he is more likely to be born between 3 and 4 A.M. and also to die in these same early morning hours. Body temperature fluctuates as much as 2° F during the course of the day, usually reaching a high about 4 P.M. and a low about 4 A.M. Alcohol tolerance is greatest at 5 P.M. Many people have an automatic internal alarm system that wakes them at the same hour every morning—whether they want to or not. Hormone secretion, heart rate, blood pressure, and urinary excretion of potassium, sodium, and calcium in man all vary according to a circadian rhythm. A recent study by the Federal Aviation Agency showed that pilots flying from one time zone to another —from New York to Europe, for instance—exhibit "jet lag," a general decrease in mental alertness and ability to concentrate and an increase in decision time and physiological reaction time. Measurements of circadian rhythms show that the body may be "out of sync" for as much as a week after such a flight. This brings into question present policies of diplomats speeding to foreign capitols at times of international crisis or of troops being air transported into combat.

Such adjustment to an externally imposed rhythm is known as entrainment. If the new rhythm is too far removed from the original one, however, the organism will "escape" the entrained rhythm and revert to its natural one. A plant that has been kept on an artificial or forced rhythm, even for a long period of time, will revert to its normal internal period when returned to continuous dim light.

Clock Functions

Biological clocks are believed to play an essential role in many aspects of plant and animal physiology. For instance, insects are more active in the early evening hours. Bats which feed on insects begin to fly each evening just when the insects are most available. Moreover, caged bats under controlled and constant laboratory conditions continue to show this sort of activity about every 24 hours, indicating that they are following an internal rhythm, not merely responding to environmental cues.

Some plants secrete nectar or perfume at certain specific times of the day. As a result, insects—which have their own biological clocks—become accustomed to visiting these flowers at these times, thereby ensuring maximum rewards for both the insects and the flowers.

The ability to tell time also appears to be involved in a number of complex and fascinating phenomena such as the extraordinary navigatory ability of migrating birds and turtles.

Biological clocks enable organisms to recognize the changing seasons of the year by "comparing" external rhythms of the environment, such as changes in day length, to their own relatively constant internal rhythms. This capacity, as we shall see, is an important factor in regulating the growth cycle of many plants.

PHOTOPERIODISM

Fifty years ago a mutant appeared in a field of tobacco plants growing near Washington, D.C. The new variety had unusually large leaves and stood over 10 feet tall. As the season progressed, the regular plants flowered, but Maryland Mammoth, as it came to be called, merely grew bigger and bigger. Scientists from the Department of Agriculture took cuttings from the Mammoth and put them in the greenhouse, where they would be safe from frost. These cuttings flowered in December, although by then they were only 5 feet tall, half the size of their parent. New Maryland Mammoths grew from their seed, and these, too, did not flower until December.

Coincidentally, these same researchers were carrying out experiments with the Biloxi variety of soybean. Agriculturalists were interested in spacing out the soybean harvest by making successive sowings of seeds at two-week intervals from early May through June. But spacing out the planting had no effect; all the plants, no matter when the

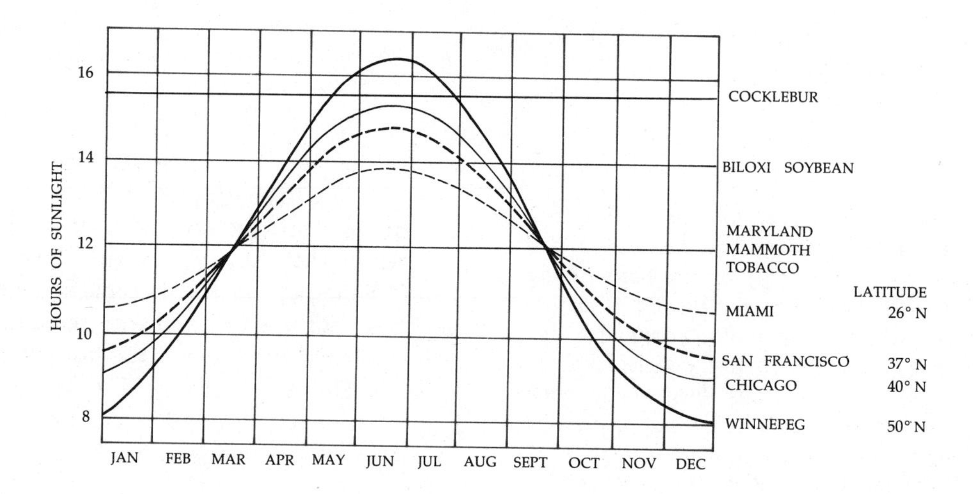

5–28

The length of the day determines when plants flower. The four curves depict the annual change of day length in four North American cities at four different latitudes. The heavy horizontal lines indicate the effective photoperiod of three different short-day plants. The cocklebur, for instance, requires 15½ hours or less *of light. In Miami, it can flower as soon as it matures, but in Winnipeg, the buds do not appear until early in August, so late that the frost will probably kill the plants before the seed is mature.*

seeds were sown, came into flower at the same time—in September.

The investigators started growing these two kinds of plants—Maryland Mammoth tobacco and Biloxi soybeans—under a wide variety of controlled conditions of temperature, moisture, nutrition, and light. They eventually found that the critical factor in both species was the length of day. Neither plant would flower unless the day length was shorter than a critical number of hours. Consequently, Biloxi soybeans, no matter when they were planted, all flowered as soon as the days became short enough, which was in September, and the Maryland Mammoth, no matter how tall it grew, would not flower until December, when the days became even shorter.

This phenomenon is called photoperiodism. It is a biological response to a change in the proportions of light and dark in the 24-hour cycle. Photoperiodism has now been demonstrated in many species of insects, fish, birds, and mammals, influencing such diverse phenomena as the metamorphosis from caterpillar to butterfly, sexual behavior, migration, molting, and seasonal changes in coat or plumage.

Long-day Plants and Short-day Plants

The botanists went on to test and confirm this discovery with many other species of plants. Following this single lead, they were able to answer a host of questions that had long troubled both professional botanists and amateur gardeners. Why, for example, is there no ragweed in northern Maine? The answer, they found, is that ragweed starts making flowers when the day is about 14½ hours long. The long summer days do not shorten to 14½ hours in northern Maine until August, and then there is not enough time for ragweed seed to mature before the frost. Why doesn't spinach grow in the tropics? Because spinach needs 14 hours of light a day for a period of at least two weeks

5–29
Photoperiodic control of flowering serves to bring all the plants of a particular species growing in the same area into flower at a particular time. All the Japanese cherry trees, for example, that grow along the tidal basin in Washington, D.C., always flower on the same day each year.

in order to flower and such long days never occur in the tropics. As you can see, the discovery of the photoperiodic control of flowering not only provided an explanation of plant distribution but was of great practical importance.

The investigators found that plants are of three general types, which they called day-neutral, short-day, and long-day. Day-neutral plants flower without regard to day length. Short-day plants flower in early spring or fall; they must have a light period *shorter* than a critical length. For instance, the cocklebur flowers when exposed to 15½ hours or less of light. Other short-day plants are poinsettias, strawberries, primroses, and some chrysanthemums.

Long-day plants, which flower chiefly in the summer, will flower only if the light periods are longer than a critical length. Spinach, potatoes, some wheat varieties, clover, henbane, and lettuce are examples of long-day plants.

Note that cocklebur and spinach will both bloom if exposed to 14 hours of daylight, yet one is designated as short-day and one as long-day. The important factor is not the absolute length of the photoperiod but rather whether it is longer or shorter than a particular critical interval for that variety. And in some varieties, 5 or 10 minutes difference in exposure can determine whether or not a plant will flower.

Measuring the Dark

In 1938, another pair of investigators, Karl C. Hamner and James Bonner, began a study of photoperiodism, using the cocklebur as their experimental tool. As we mentioned previously, the cocklebur is a short-day plant requiring 15½ hours or *less* of light per 24-hour cycle to flower. It is particularly useful for experimental purposes because a single exposure under laboratory conditions to a short-day cycle will induce flowering two weeks later, even if the plant is immediately returned to long-day conditions. The cocklebur can withstand a great deal of rough treatment, surviving even if its leaves are removed. Hamner and Bonner showed that it is the leaf blade of the cocklebur that responds to the photoperiod. A completely defoliated plant cannot be induced to flower, as we saw in Figure 5–25. But if as little as one-eighth of a fully expanded leaf is left on the stem, the single short-day exposure induces flowering.

In the course of these studies, in which they tested a variety of experimental conditions, Hamner and Bonner made a crucial and totally unexpected discovery. If the period of darkness was interrupted by as little as a one-minute exposure to a 25-watt bulb, flowering did not occur. Interruption of the light period by darkness had no effect on flowering whatsoever. Subsequent experiments with other short-day plants showed that they, too, required periods not of uninterrupted light but of uninterrupted darkness!

What about long-day plants? They also measure darkness. A long-day plant that will flower if it is kept in a laboratory in which there is light for 16 hours and dark for 8 hours will also flower on 8 hours of light and 16 hours of dark if the dark is interrupted by even a brief exposure to light.

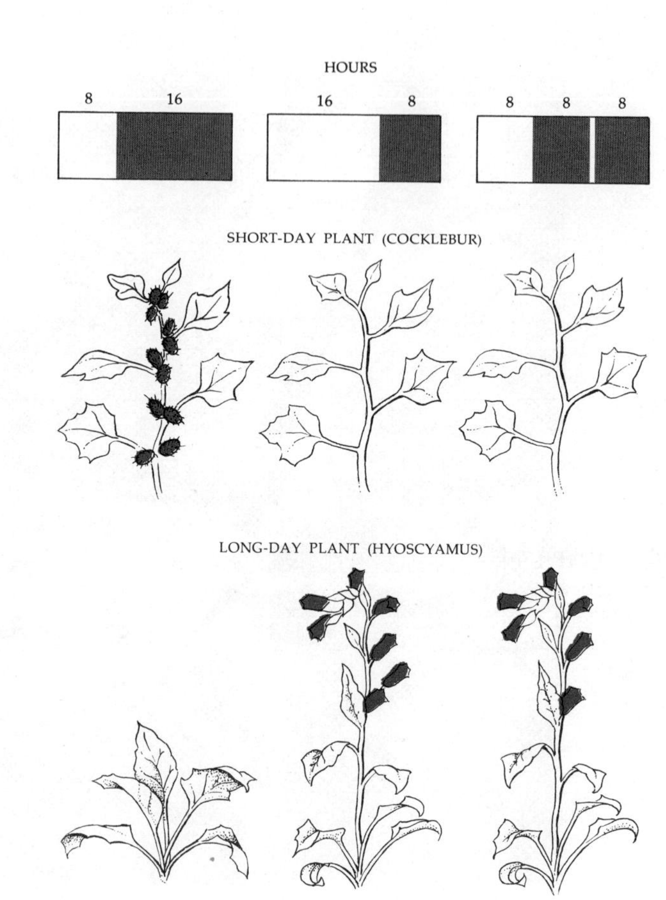

5–30
Photoperiodicity in flowering plants. Short-day plants flower when the day length is below some critical value, and long-day plants flower only when it exceeds some critical value. The short-day plant in the drawing requires a day length of 8 hours or less to bloom; if the 16-hour period of darkness is interrupted even very briefly, as shown on the right, the plant will not flower. The long-day plant, on the other hand, will flower only when the day length exceeds eight hours or if the period of darkness is interrupted.

PHYTOCHROME: THE CHEMICAL BASIS OF PHOTOPERIODISM

Hamner and Bonner had shown that if the dark period is interrupted by a single exposure to light from an ordinary bulb, the cocklebur will not flower. Following this lead, a team of research workers at the U.S. Department of Agriculture Research Station in Beltsville, Maryland, began to experiment with light of different wavelengths, varying the intensity and duration of the flash. They found that red light of about a 660-nanometer wavelength was the most effective in preventing flowering in the cocklebur and other short-day plants. It was also the most effective, they found, in promoting flowering in long-day plants. (Effectiveness is measured in terms of percentage of plants that respond.)

The Beltsville group found their next clue in the report of an earlier study performed with lettuce seeds. Lettuce seeds of the Grand Rapids variety germinate only if they are exposed to light. (This is true of many small seeds, which need to germinate in loose soil and near the surface in order for the seedlings to be able to break through.) The earlier workers, in studying the light requirement of lettuce seeds, had shown that red light stimulated germination and that red light of a slightly longer wavelength (far-red) inhibited germination. In fact, in lettuce seeds exposed to far-red light, the percentage that germinated was even lower than that of the control seeds (those not illuminated at all). The Beltsville group found that when red light was followed by far-red light, the seeds did not germinate. The red light most effective in inducing germination in seeds was light of the same wavelength as that involved in the flowering response—about 660 nanometers. Furthermore, they found that the light most effective in inhibiting the response in seeds to red light was the far-red light, of a wavelength of 730 nanometers. The series of exposures could be repeated over and over; the number of exposures did not matter, but the nature of the final one did. If the series ended with an exposure of red (660) light, the great majority of the seeds germinated. If it ended with a far-red (730) exposure, the great majority did not.

Far-red light was then tried on short-day and long-day plants, with the same on-off effect. Far-red light alone, when given during the dark period, had no effect. But an exposure of far-red light immediately following an exposure of red light canceled the effects of the red light.

Discovery of Phytochrome

Here is how the Beltsville group and others have interpreted these results. The plant contains a pigment that exists in two different forms; these two forms came to be known as P_{660} and P_{730}. P_{660} absorbs red light and is converted to P_{730}, which is the active form. This conversion takes place in daylight or in incandescent light; in both of these lights, red wavelengths predominate over far-red. When P_{730} absorbs far-red light, it is converted back to P_{660}. The P_{730}-to-P_{660} conversion can also take place, although much more slowly, in the dark, which is how it usually occurs in nature. In short-day plants, P_{730} inhibits flowering under conditions in which flowering would otherwise occur. In long-day plants, P_{730} promotes flowering under appropriate conditions.

In 1959, the scientists first isolated this (until then) hypothetical pigment, which they called phytochrome.

Isolation of Phytochrome

Phytochrome is present in plants in very small amounts. To detect it, a spectrophotometer was needed which was sensitive to extremely small changes in light absorbancy. (Large changes in light absorbancy can be detected by the eye, of course, as changes in color.) Such a spectrophotometer was introduced some seven years after the existence of phytochrome was proposed, and it was used to first detect and subsequently isolate the pigment. The pigment proved to be blue in color (why might you expect this color?) and to show the characteristic red–far-red conversion in the test tube by reversibly changing color slightly in response to red or far-red light.

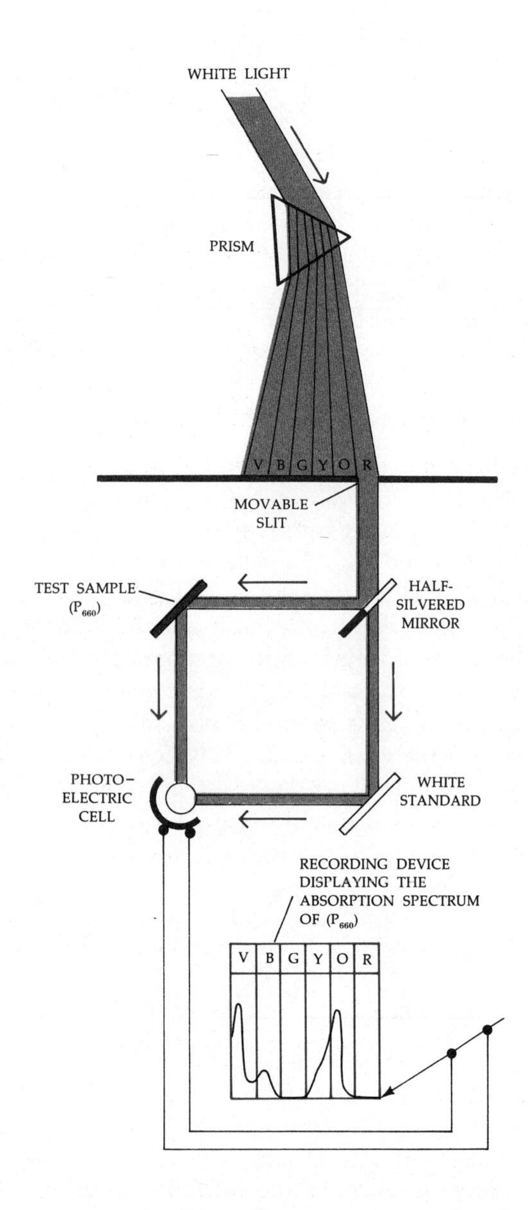

5–31
The absorption spectrum of a pigment is measured by a spectrophotometer. This device directs a beam of light of each wavelength at the object to be analyzed and records what percentage of light of each wavelength is absorbed by the pigment. The development of a highly sensitive spectrophotometer permitted the identification of phytochrome.

There are two portions to the phytochrome molecule, it was found: a light-absorbing portion and a large protein portion. The light-absorbing portion is a phycobilin which is a type of pigment found in blue-green and red algae and which functions in these organisms as an accessory pigment in photosynthesis.

Phytochrome is found in small amounts throughout the plant body, with the largest concentration in the apical meristems of shoots and roots. The red–far-red system has been observed in a number of plant responses other than seed germination and flowering. Dark-grown seedlings, for example, are elongated, have small leaves, and often have stems that curve, forming a hook at the top (all of which are associated with the necessity for the seedling to break its way through the soil if it is to survive). On exposure to light, the elongation process slows down, the leaves expand, and the hook straightens out. This light response is promoted by an exposure to red light; if the red light is followed by exposure to far-red, the effect is neutralized. Sleep movements also apparently involve phytochrome, with red inhibiting closing—an effect that can be observed within a few minutes—and far-red reversing the inhibition.

The way in which phytochrome works is not known. There are three current hypotheses. One is that the molecule acts as an enzyme. According to this theory, when light is absorbed by the pigment portion of the molecule, the energy of the light changes the structure of the pigment and this, in turn, changes the structure of the protein portion, rendering the enzyme active. A second hypothesis, which is based on the study of rapid responses such as those seen in sleep movements, is that the primary action of phytochrome is to alter membrane permeability, thus permitting or prohibiting the passage of hormones or other substances. The third hypothesis is that phytochrome regulates gene activity in accordance with the operon model proposed by Jacob and Monod (page 191). P_{730}, according to this concept, inactivates the repressor substance produced by the regulator gene, permitting RNA synthesis to proceed along the operon.

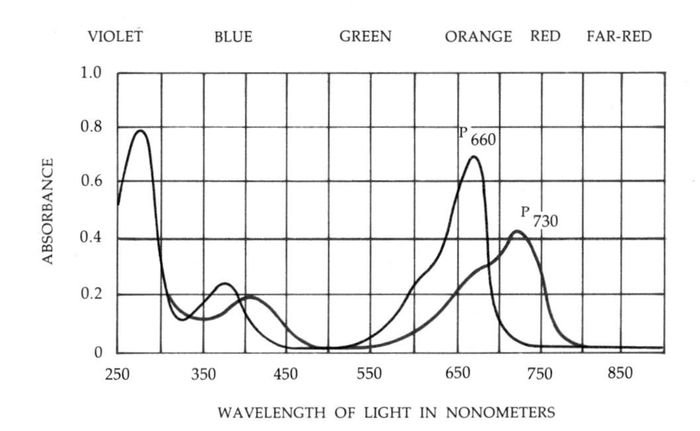

5–32
Absorption spectra of the two forms of phytochrome, P_{660} *and* P_{730}. *This shift in absorption spectra made it possible to isolate the pigment.*

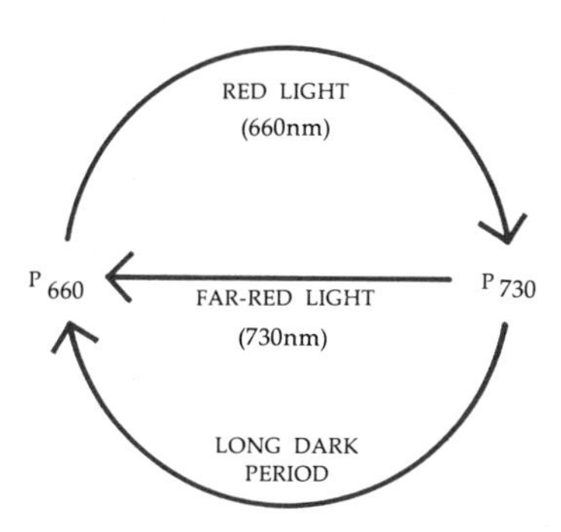

5–33
P_{660} *changes to* P_{730} *when exposed to red light.* P_{730} *reverts to* P_{660} *when exposed to far-red light. In darkness,* P_{730} *reverts to* P_{660}.

Phytochrome and Photoperiodism

Photoperiodism requires, of course, both an environmental change—a change in the duration of light (or dark)—and the existence within the organism of some means of appreciating this change—in other words, a biological clock.

When the existence of phytochrome was first demonstrated, its discoverers hypothesized that the red–far-red conversion might be this time-measuring mechanism, a biological clock functioning rather like an hourglass. According to this hypothesis, in short-day plants, P_{730}, which inhibits flowering in these plants, accumulates in the light and reverts back to the inactive form, P_{660}, during the hours of darkness. When the nights are long enough, all (or a critical amount) of P_{730} is converted to P_{660} and flowering is no longer inhibited. Long-day plants, on the other hand, require short nights, during which the P_{730} is not completely destroyed; if the night is shorter than a critical interval, enough P_{730} remains at the end of it to promote flowering.

How would you test this hypothesis? One way that investigators tested it was by measuring the actual amount of time required for all the P_{730} to revert to P_{660} in the dark. (This is a fairly simple test to perform using the spectrophotometer.) The dark conversion in all the plants studied was found to take only about three hours! What happens during the rest of the dark period? Remember that in some plants, as little as 5 to 10 minutes can make the difference. What happens during that last important 10 minutes, which in some plants is decisive in determining whether flowering occurs?

On the basis of these and other observations, it is now generally agreed that the time-measuring phenomenon of photoperiodism is not controlled solely by the dark reversion of P_{730} to P_{660}. There seem to be two variables involved in the phytochrome response. One is the amount of P_{730}. The second is the amount of some other, not yet identified, substance or substances, perhaps a substrate or a hormone. Both P_{730} and the unknown material have to be available in sufficient quantities for the crucial reactions to occur. The amount of P_{730} is determined by light-dark periods; the P_{730} must be used fairly promptly since it will revert in the dark to the inactive form. The amount of the unknown material must vary according to the circadian rhythm of the plant.

Only when the P_{730} and the other substance or substances are both present in the optimal amounts will flowering occur, according to this hypothesis. Neither accumulates, which may be why even such short intervals of time may make such a difference.

SUMMARY

Plants possess a variety of adaptations for detecting and responding to alterations in their environment. Circadian rhythms are cycles of activity in an organism that recur at intervals of approximately 24 hours. They are probably endogenous, caused not by an external factor such as alternating light and darkness or the earth's rotation but by some internal timing mechanism within the organism. Such a timing mechanism, the chemical and physical nature of which is unknown, is called a biological clock. The possession of a biological clock makes it possible for an organism to perceive changes in daily cycles, such as the lengthening and shortening of days as the seasons progress. In this way, activities such as sexual reproduction, dormancy, and in the case of many animals, seasonal migrations, mating, and coat and plumage changes can be brought into synchrony with the external environment.

Photoperiodism is the response of organisms to changing periods of light and darkness in the 24-hour day. Such a response controls the onset of flowering in many plants. Some plants will flower only when the periods of light exceed a critical length. (The absolute length is not important, only the fact that it must be exceeded for flowering to take place.) Such plants are known as long-day plants. Other plants, short-day plants, flower only when the periods of light are less than some critical period. Day-neutral plants flower regardless of photoperiods. Interruption of the dark phase of the photoperiod, even by a brief exposure to light, can serve to reverse the photoperiodic effects, indicating that it is really the dark period rather than the light period which is critical.

Phytochrome, a pigment commonly present in small amounts in higher-plant tissues, is the receptor molecule in the transitions between light and darkness. The pigment can exist in two forms, P_{660} and P_{730}. P_{660} absorbs red light of a wavelength of 660 nanometers and is thereby converted to P_{730}, which absorbs far-red light of a 730-nanometer wavelength. With ordinary day-night sequences, P_{730} is converted back to P_{660} over a period of hours in the dark; it can also be converted to P_{660} by exposure to far-red light. P_{730} is the active form of the pigment; it promotes flowering in long-day plants and inhibits flowering in short-day plants. Its mechanism of action is under active study but is still unknown.

QUESTIONS

1. Explain the difference between biological clocks and photoperiodism. How are they related?

2. What is the survival value to the plant of the photoperiodic control of flowering?

3. Traveling from north to south, it is possible to find varieties of the same species with different photoperiodic requirements. How would you expect these to differ?

CHAPTER 5-5

Transpiration and Translocation: Water and Sugar

The successful establishment of tall plants on dry land, as we have seen, was dependent primarily on the development of a system for conducting water and minerals from under the ground to the top of the plant. Since plants have openings (stomata) to the air for diffusion of carbon dioxide and oxygen, they lose water by evaporation and their water supply has to be renewed constantly. The quantity of water passing through a plant is enormous—far greater than that used by an animal of comparable weight. An animal requires less water because a great deal of its water recirculates through its body over and over again, taking the form, in vertebrates, of blood plasma and other fluids. In plants, more than 90 percent of the water which enters the roots is given off into the air as water vapor. A single corn plant needs 300 to 400 pounds of water, or 40 to 50 gallons, from bud to harvest, and an acre of corn requires 4 to 5 million pounds, or more than half a million gallons, of water a season.

LOSS OF WATER

In plants, the loss of water by evaporation is known as *transpiration*. To understand the process of transpiration and why the loss of water from the plant body is so large, it will be helpful to look again at the anatomy of the leaf (Figure 2-13). As you will recall, the epidermis of the leaf, although thin and transparent, is nearly waterproof and airproof because of its waxy outer surface of cutin. In the typical dicot leaf, most photosynthesis takes place in the palisade cells. Beneath the palisade layer is a loose, spongy layer of parenchyma, containing extensive air spaces. The veins, bundles of xylem and phloem, pass through this spongy layer. The lower surface, like the upper, is transparent cutinized epidermis.

Within the epidermis are special pores, the stomata, which close in response to a water deficit. In some plants, the stomata are located in the lower surface of the leaf only; in others, they are found in both sides. Stomata are also found in the epidermis of young stems, in floral organs, and in fruits.

Photosynthesis takes place in the sunlight and requires carbon dioxide. This carbon dioxide, which enters the leaf through the stomata, diffuses from the air spaces within the leaf into the photosynthetic cells. During the process, moisture is lost by evaporation from the wet cell walls bounding these intercellular spaces. Unless this water is continuously replaced, the stomatal guard cells will collapse and the stomata will close. Closing of the stomata prevents the further escape of water, but it also cuts off the needed carbon dioxide, thus stopping photosynthesis. Therefore, while photosynthesis is taking place, plants must continuously take in water through their root systems to replace the water lost by transpiration. A corn plant, for example, transpires more than 98 percent of the water it takes up.

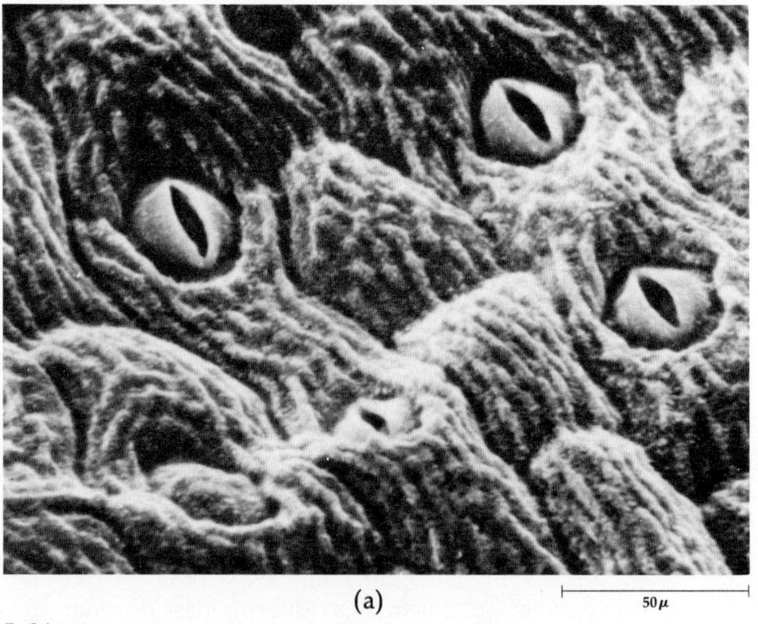

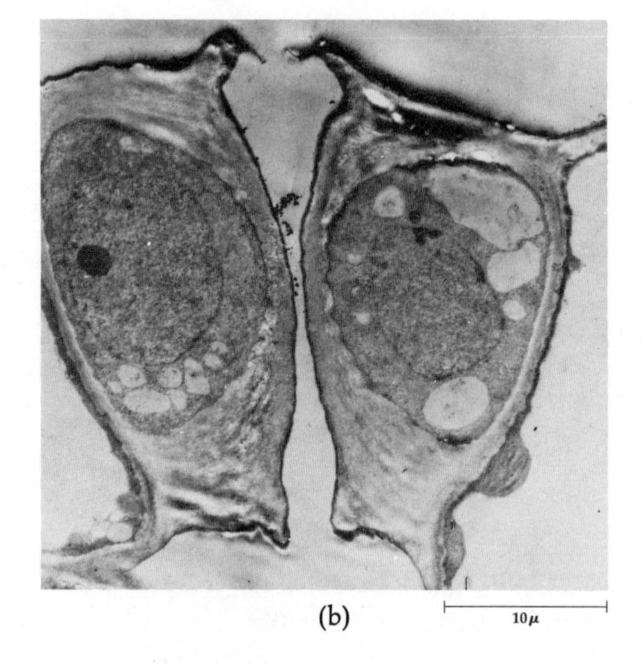

5–34
(a) *Stomata on the lower surface of a leaf of a dianthus, one of the pink family, as recorded by the scanning electron microscope. The paired guard cells, by expanding and contracting, regulate the exchange of oxygen and carbon dioxide and the passage of water out of the leaf.* (b) *Cross section through the guard cells making up a stoma in the leaf of a bean plant. Note the thick walls bordering the stoma and the "lips" at the outer surface of the guard cells.*

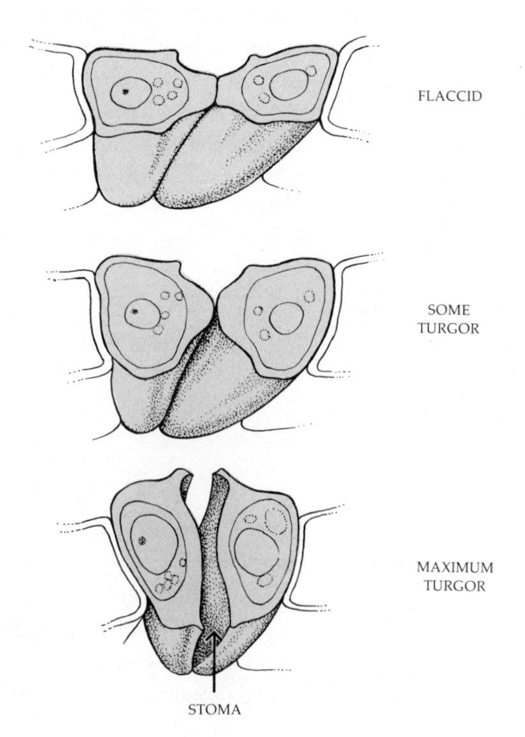

5–35 (at right)
Mechanism of stomatal movements. Each stoma is flanked by two guard cells which open the stoma when they are turgid and close it when they lose turgor. In many species, the guard cells have thickened walls adjacent to the stomatal opening. As turgor pressure increases, the thinner parts of the cell wall are stretched more than the thicker parts, causing the cells to bow out longitudinally and forcing the stoma to open.

MOVEMENT OF WATER

Water enters the body of land plants largely through the root hairs, which are fine tubular extensions of the epidermal cells of the roots. It then moves through the cytoplasm of the cells of the cortex—and also, probably, between the cells, passing through the cell walls and intercellular spaces—to the vascular tissue.

During periods of rapid transpiration, water may be removed from around the root hairs so quickly that the earth becomes depleted; water will then move from some distance away toward the root hairs through the soil. By and large, however, the roots come into contact with water by growing. Under normal conditions, roots of apple trees grow an average of 1/8 to 3/8 inch a day; roots of prairie grasses may grow more than 1/2 inch a day; and the main roots of corn plants average 2 to 2 1/2 inches a day. A four-month-old rye plant has 7,000 miles of roots and many billions of root hairs.

As the roots elongate, they poke their way into the small crevices between particles. The root cells, like the other living parts of the plant, contain a higher concentration of salts and minerals than the soil water. Therefore, water from the soil enters the roots by osmosis, which, you will recall, is the movement of water across a membrane to the point of higher solute concentration.

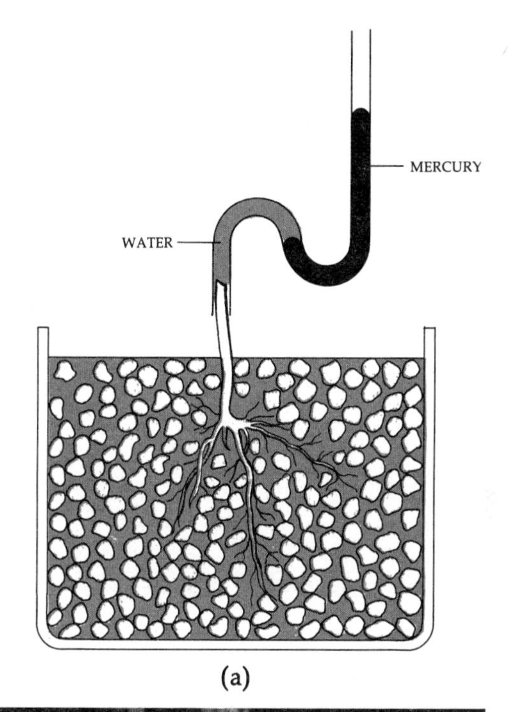

(a)

(b)

5–36 (at right)

(a) *Demonstration of root pressure in the cut stump of a plant. Uptake of water by the plant roots causes the mercury to rise in the column. Pressures of 3 to 5 atmospheres have been demonstrated by this method.* (b) *Guttation droplets on the edge of a wild strawberry leaf. Guttation, the loss of liquid water, is a result of root pressure. The water escapes through specialized stomata located near the ends of the principal veins of the leaf. Guttation usually occurs at night when the air is moist.*

How Water Moves Up in the Plant

The hydrostatic pressure in the roots created by osmosis may be sufficient to push the water some distance up the stem. But how can water reach 60 feet high to the top of an oak tree, travel three stories up the stem of a vine, or move 300 feet up in a tall redwood? As students of physics know, it requires a strong force to raise water more than a few stories off the ground. Yet a vine creeping to the top of a tall building has sufficient water for its topmost leaves, as does a redwood tree 300 feet tall.

One important clue is the observation that during times when the most rapid transpiration is taking place—which is, of course, when the flow of water up the stem must be the greatest—xylem pressures are characteristically negative (below 1 atmosphere of pressure). The existence of negative pressure can be demonstrated readily. If you peel a piece of bark from a transpiring tree and make a cut in the xylem, no sap runs out. In fact, if you place a drop of water on the cut, the drop will be drawn in.

Is the water pulled up from the top? If so, how? We know that atmospheric pressure pushes water up by suction, as occurs when we withdraw the air by sipping through a straw. A sipping straw or a pipette or a suction pump acts by removing air from one part of a system—for example, from the end of a tube. This lowers the pressure at that point. If the fluid in one part of the system is under atmospheric pressure and all the air is removed from the first part, 1 atmosphere of difference in water pressure can be created. But 1 atmosphere of pressure—the amount resulting from the creation of a perfect vacuum—will raise a column of water only about 32 feet at sea level, and no further! How can suction get water to the top of a Douglas fir or a sequoia?

According to the most widely accepted theory, the explanation is to be found not in the properties of the plant but in the remarkable properties of water, to which the plant has become exquisitely adapted. As we pointed out in Chapter 2–2, water has a tremendous power of cohesion, each molecule clinging by hydrogen bonds to several adja-

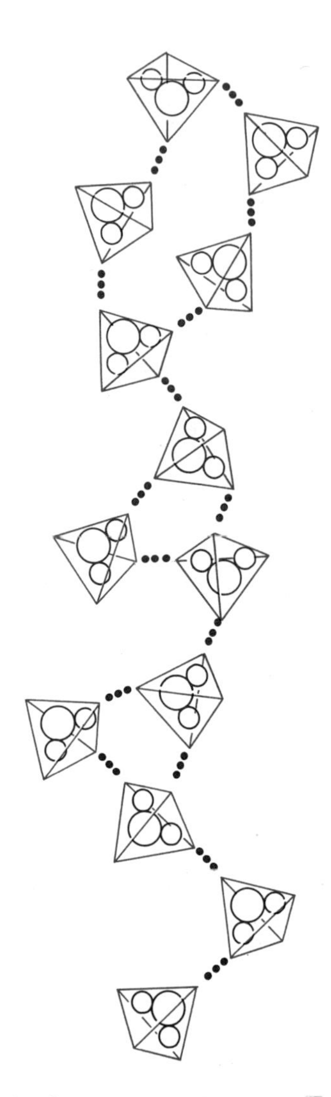

5–37

Water's power of cohesion is important in plant transpiration. In every water molecule, the pair of hydrogen atoms is linked to a single oxygen atom. The hydrogen atoms are also held to the oxygen atom of the nearest water molecule by hydrogen bonds, indicated by the three dots. This secondary attraction can produce a tensile strength of as much as 2,000 pounds per square inch in a thin column of water. Cohesion thus makes it possible for a plant to raise water hundreds of feet in the air.

cent ones. In the leaf, water evaporates, molecule by molecule, from the cell walls as a consequence of the lower water potential of the air cells. The water potential of the leaf cell falls, and water from the vascular tissue moves, molecule by molecule, into the leaf cell. But each molecule in the xylem vessel is linked to other molecules in the vessel, and they, in turn, are linked to others, forming one long, narrow, continuous strand of water reaching right down to a root tip. As the molecule of water moves into the leaf cell, it tugs the next molecule along behind it. This pulling action, molecule by molecule, is the cause of the negative pressures observed in the xylem. The technical term for a negative pressure is *tension*, and this theory of water movement is consequently known as the *cohesion-tension theory*.

MINERALS

In addition to carbon, hydrogen, and oxygen, which can be absorbed from water or from the atmosphere, plants require a number of minerals, which they obtain from the soil. These mineral requirements have been determined by studying the capacity of plants to grow in distilled water to which small amounts of various soil constituents are added. (Sometimes, it has been found, a substance—chlorine, for example—is needed in such small amounts that it is almost impossible to set up experimental conditions that exclude it; in such minute concentrations, it is difficult to prove that a substance is essential.) As a result of such studies, six elements that the plants require in relatively large amounts and seven that are needed in smaller quantities have been identified. As you can see in Table 5–1, nitrogen, which is absorbed by higher plants in the form of nitrogen salts, is used in the largest quantities. It is inaccurate, however, to say that nitrogen is "more essential" than the other elements, since the complete deprivation of *any* material that a plant requires, even in minute quantities, is fatal to the plant.

You might anticipate that organisms make use of what

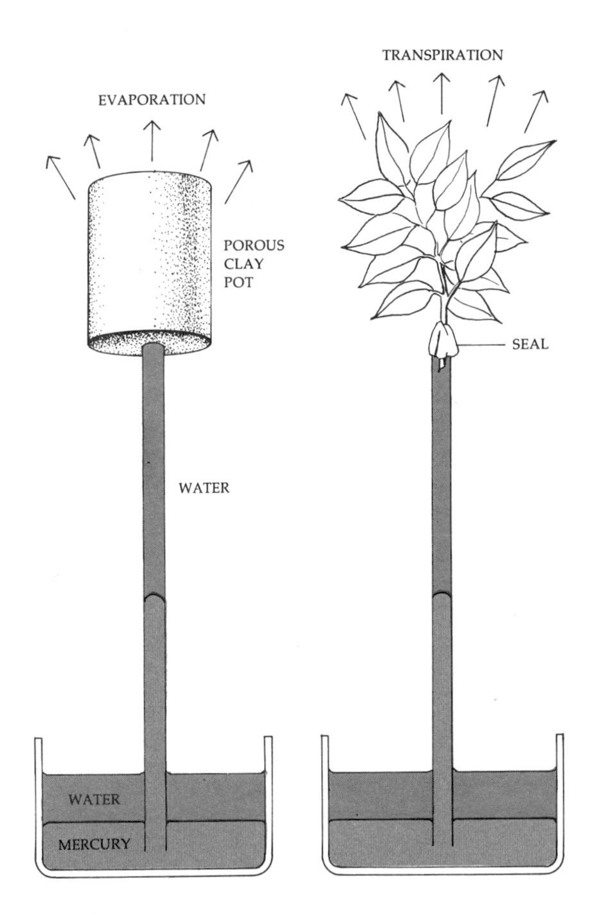

5–38

A simple model that illustrates the cohesion-tension theory. A porous clay pot is filled with water and attached to the end of a long, narrow glass tube also filled with water. The water-filled tube is placed with its lower end below the surface of a volume of mercury contained in a beaker. As water molecules evaporate from the pores in the pot, they are replaced by water "pulled up" through the narrow glass tube in a continuous column. As the water evaporates, mercury rises in the tube to replace it. Transpiration from plant leaves results in sufficient water loss to create a similar negative pressure.

Table 5–1
A Summary of Minerals Essential to Plants

Element	*Form in Which Absorbed*	*Approx. Concentration in Whole Plant (as % of dry weight)*	*Some Functions**
MACRONUTRIENTS, from 5 to 6% of dry weight of plant:			
Nitrogen	NO_3^- (or NH_4^+)	1–3%	Component of amino acids, proteins, nucleotides, nucleic acids, chlorophyll, and coenzymes.
Potassium	K^+	0.3–6%	Involved in enzyme, amino acid, and protein synthesis. Increases permeability of cell membranes.
Calcium	Ca^{2+}	0.1–3.5%	Combines with pectin in cell walls. Decreases cell permeability.
Phosphorus	H_2PO_4 or HPO_4^{2-}	0.05–1.0%	Formation of "high-energy" phosphate compounds (ATP and ADP). Component of nucleic acids and phospholipids.
Magnesium	Mg^{2+}	0.05–0.7%	Component of the chlorophyll molecule. Activator of many enzymes.
Sulfur	SO_4^{2-}	0.05–1.5%	Component of some amino acids and proteins and of coenzyme A.
MICRONUTRIENTS, 1,500 parts per million (ppm) or less:			
Iron	Fe^{2+}, Fe^{3+}	10–1,500 ppm	Required for chlorophyll synthesis. Component of cytochromes and ferredoxin. Activator of some enzymes.
Chlorine	Cl^-	100–300 ppm	Probably essential in photosynthesis in the reactions in which oxygen is produced.
Copper	Cu^{2+}	2–75 ppm	Activator of some enzymes.
Manganese	Mn^{2+}	5–1,500 ppm	Activator of some enzymes.
Zinc	Zn^{2+}	3–150 ppm	Activator of some enzymes.
Molybdenum	MoO_4^{2-}	Trace	Needed for nitrogen metabolism.
Boron	BO^{3-} or $B_4O_7^{2-}$ (borate or tetraborate)	2–75 ppm	Influences Ca^{2+} uptake and utilization.

* All major ions play a role in osmosis and in the distribution of positive and negative charges. All also can affect the configuration of enzymes and other proteins since this is determined, to a large extent, by attractions between differently charged areas of the molecule or molecules.

5–39
Most plants obtain their nitrogen in the form of inorganic salts taken up with the soil water, but for some unusual species, such as the Venus flytrap shown here, insect bodies are the chief nitrogen source. The leaves of the Venus flytrap are hinged in the middle, and each leaf half is equipped with three sensitive hairs. When an insect walks on one of these leaves, attracted by the nectar on the leaf surface, it brushes against the hairs, triggering the traplike closing of the leaf. The toothed edges mesh, the leaf halves gradually squeeze closed, and the insect is pressed against digestive glands on the inner surface of the trap. The plant has evolved a triggering mechanism that enables it to distinguish between living prey and inanimate objects, such as pebbles and small sticks, that fall on its leaves by chance: the leaf will not close unless two of its hairs are touched in succession or one hair is touched twice.

is most readily available, as indeed they seem to have done when life originated from elements in the gases of the primitive atmosphere. But the table reveals some findings that you might not expect. Sodium, for instance, which is one of the most abundant of the elements, is not required at all by plants except for some salt-marsh species. The fact that plants did not find a use for sodium appears even stranger when you consider that sodium is vital to the function of animals. In the seas, where both plants and animals seem to have had their origins, sodium is the most abundant element and is far more readily available than potassium, which it closely resembles in its essential properties. Similarly, although silicon and aluminum are almost always present in large amounts in soils, few plants require silicon and none requires aluminum. On the other hand, all plants need molybdenum, which is relatively rare.

Comparisons of the mineral content of a plant cell with the mineral content of the water in which it grows show that there can be marked differences in the concentrations of various components. Therefore, substances from the soil do not diffuse passively into the root cells but are carried

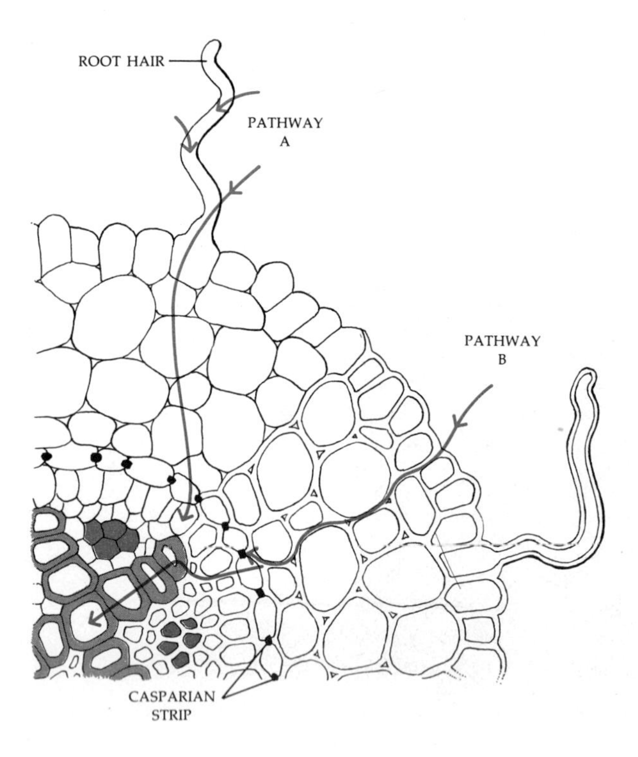

5–40
Diagrammatic cross section of a root, showing the two pathways of uptake of water and minerals. Along pathway A, water moves by osmosis and salts by active transport through the cellular membranes of a series of living cells. Along pathway B, the water flows through the cell walls and intercellular spaces and the solutes flow with the water or by diffusion. Notice how the casparian strip blocks off pathway B all around the vascular cylinder of the root. In order to pass the casparian strip, the solutes must be taken up actively by the cells of the endodermis.

across cell membranes, often against the diffusion gradient, by the energy-requiring processes of active transport.

By growing plants in solutions that contain the required minerals in the form of radioactive isotopes, it has been shown that, once they enter the root cells, the mineral ions are carried up the plant by the flow of water in the xylem.

Minerals are an important factor in determining what plants grow in certain areas and the rates at which they grow. In fact, experienced geologists can often tell the mineral content of the soil in a particular area by observing its characteristic vegetation.

TRANSLOCATION AND THE PHLOEM

In the course of plant evolution, competition for light pushed the photosynthetic areas of the plant higher and higher into the air. As these areas, in which the sugars are formed, were moved further and further from the nonphotosynthetic cells of the stems and roots, plants evolved structures for transporting the products of photosynthesis to other parts of the plant body. The structures, as you know, are the phloem. The transport of sugars from the photosynthetic cells of the plant through the phloem to other parts of the plant body is called *translocation.*

The conducting elements of the phloem system, as we saw in Chapter 5–1, are sieve tubes. Unlike the conducting elements of the xylem, sieve tubes are composed of living cells. The solution they carry is rich in sugar, mostly in the form of sucrose (some 10 to 25 percent by volume), and also contains amino acids and minerals.

The way in which this material moves through the plant body is still not completely understood. It is clearly not a case of simple diffusion. When plants are grown in carbon dioxide in which some of the carbon atoms are radioactive, they form sugars containing radioactive carbon. By following the movement of these radioactive labels, it is possible to show that the sap commonly moves at rates of 100 centimeters per hour. Yet sugar diffusing from a 10 percent solution may take many months to move 100 centimeters.

(a)

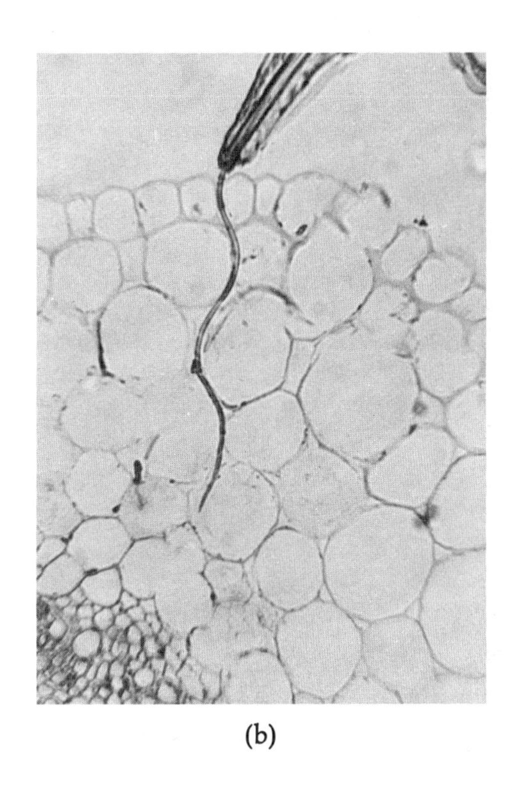

(b)

5–41

Assistance by aphids. (a) *Aphids are small insects that feed on plants, sucking out their juices; you probably have seen them on rosebushes. The aphid drives its sharp mouthpart, or stylet, like a hypodermic needle through the epidermis of a plant and then, as electron micrographs reveal* (b), *taps the contents of a single sieve-tube element. If the aphid is anesthetized, it is possible to sever the stylet, leaving it still in place in the cell. The fluid will continue to exude through the stylet from the sieve cell for several days, and pure samples can be collected for analysis without damaging the sieve tube or interfering with its function.*

The Mass-flow Hypothesis

The theory that was most widely accepted in the past proposes that the solution moves as a result of differences in water pressure. This is known as the *mass-flow theory*. The principle underlying this theory can be illustrated by a simple physical model consisting of cells permeable only to water and connected by glass tubes. The first cell contains a solution in which there is a dissolved material, such as sucrose, and the second cell, to make the example as simple as possible, contains only water. When these interconnected cells are placed in distilled water, water will enter the first cell by osmosis. The entry of water will increase hydrostatic pressure within this cell and cause the solution in it to move along the tube to the second cell, where the pressure again builds up. If the second cell is connected with a third cell, which contains water or a sucrose concentration lower than that which is now in the second cell, the solution will flow from the second to the third by the same process, and so on indefinitely, down a

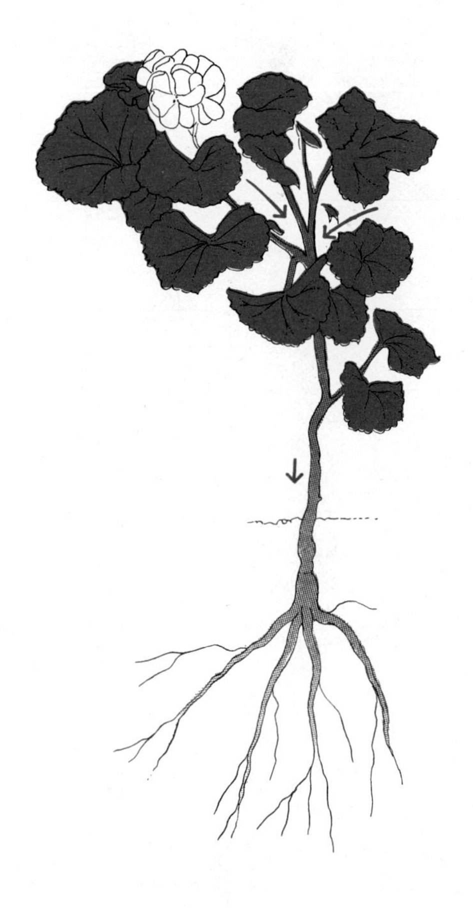

5–42
Diagram of mass flow theory. Sugar (indicated by color), produced in the photosynthetic cells, enters the sieve tubes of the leaf by active transport. As a consequence of the increased concentration of sugar in the sieve tubes, water enters the sieve tube elements by osmosis, causing water pressure to increase. As sugar is utilized in the roots and other nongreen tissues, the sugar concentration in the sieve tube falls. Water leaves the sieve tubes by osmosis and water pressure falls as a result of the decrease in sugar concentration. Because of the differences in water pressure between leaf and root, the solution moves through the sieve tubes toward the root.

line of cells. In support of this hypothesis, it is possible to show that there are distinct gradients in the concentrations of sucrose and other sugars along the phloem tissues during the summer months. Undoubtedly, mass flow plays an important role in the movement of sugar in plants.

Mass flow alone is not sufficient to account for all the data, however. Under some conditions sugar is transported from cells of lesser hydrostatic pressure, or turgor, to cells of greater turgor. Wilted leaves of the sugar beet, for example, continue to export sugar to the roots. Nutrients move from cotyledons and senescing petals into the plant body, against the concentration gradient, presumably by active transport. In a photosynthesizing leaf, a parenchyma cell may contain about 3 percent sugar while the phloem contains 30 percent. To explain these observations, a modified mass-flow theory has been formulated. According to this theory, translocation takes place by mass flow combined with some sort of energy-requiring activities on the part of the phloem tissues. In support of this theory, companion cells and phloem parenchyma have been shown to be sites of extremely high rates of respiration.

SUMMARY

In plants, water, as a component of protoplasm, serves as a solvent, maintains turgor, and participates in many cellular reactions (including photosynthesis, as we saw in Chapter 2–6).

Water is lost from plants by evaporation from plant surfaces (transpiration). Most water loss takes place from the leaf by way of the stomata. Stomatal opening and closing are controlled by the hydrostatic pressure (turgor) of the guard cells surrounding the stomata. Stomata close when the cells lose turgor, as when the plant begins to wilt.

Water enters the plant through the roots and makes its way to the leaves by means of the xylem. The current and widely accepted theory of how water moves through the xylem is the cohesion-tension theory. According to this theory, water within the vessels is under negative pressure because the water molecules cling together in continuous columns pulled by evaporation from above.

A total of 16 elements are known to be required by higher plants for normal growth. Of these, carbon, hydrogen, and oxygen are derived from air and water. The rest are absorbed in soil water by the roots in the form of ions. These 13 elements are sometimes categorized as macronutrients and micronutrients. The macronutrients are nitrogen, potassium, calcium, phosphorus, magnesium, and sulfur.

Translocation, which is the movement of nutrients from the leaves through the plant body, takes place in the phloem. According to the mass-flow hypothesis, the nutrients are carried along by the flow of water from an area of higher solute concentration in the phloem conduits of the leaves to areas of lower solute concentration in the phloem conduits of the roots. The modified mass-flow hypothesis postulates that active, energy-requiring processes are also involved. These processes, which take place in the phloem cells, move sugar into or out of the phloem system against the concentration gradient.

QUESTION

Most plants cannot get enough water in areas in which there is a high salt concentration, such as salt marshes. Why not?

GLOSSARY

ABSCISIC ACID [L. *abscissus*, cut off]: A plant hormone involved in abscission and dormancy.

ABSCISSION: The dropping of leaves, flowers, fruits, or stems following the formation of an abscission layer.

ABSCISSION LAYER: A layer of specialized cells which forms across the stalk of a leaf, flower, or fruit; toward the end of the growing season, the wall of these cells are dissolved by enzyme action.

ADVENTITIOUS [L. *adventicus*, not properly belonging to]: In biology, referring to a structure arising from an unusual place, such as roots growing from stems or leaves.

ALEURONE LAYER (**al**-u-ron) [Gk. *aleuron*, flour]: A layer of cells rich in proteinaceous and fatty material, usually just inside the seed coat of wheat and other grains.

ANDROECIUM (an-**dre**-shi-um) [Gk. *andros*, man, + *oikos*, house]: All the stamens in the flower of a seed plant.

ANNUAL: A plant in which the entire life cycle is completed in a single growing season.

ANTHER [Gk. *anthos*, flower]: A pollen-bearing portion of stamen.

ANTHOCYANIN (an-tho-**sigh**-an-in) [Gk. *anthos*, flower, + *kyanos*, dark blue]: A blue, purple, or red pigment that is water soluble.

APICAL DOMINANCE: The hormonal influence of a terminal bud in suppressing the growth of nearby lateral buds.

APICAL INITIALS: The growth-initiating cells of the apical meristems. These cells divide continuously during periods of growth and are the source of all new cells of the plant.

APICAL MERISTEM: The growing point, composed of meristematic tissue, at the tip of the root or stem in vascular plants.

AUXIN [Gk. *auxein*, to increase]: A plant hormone with a variety of growth-regulating effects, including promotion of cell elongation.

BIENNIAL: A plant that normally requires two growing seasons to complete its life cycle. Only vegetative growth occurs in the first year, often resulting in the formation of an overwintering rosette or an enlarged storage root; flowering and fruiting occur in the second year.

BULB: An underground storage organ that is a specialized bud with enlarged and fleshy leaf bases.

CALYX (**kay**-licks) [Gk. *kalyx*, a husk, cup]: Sepals collectively; the outermost flower parts, typically enclosing the flower in bud.

CAMBIUM [L. *cambiare,* to exchange]: *See* cork cambium, vascular cambium.

CAROTENOID: One of a class of fat-soluble pigments that includes the carotenes (yellows and oranges) and the xanthophylls (yellows).

CARPEL (**kar**-pel) [Gk. *karpos,* fruit]: The organ of the flower which encloses one or more ovules; typically divided into ovary, style, and stigma.

CASPARIAN STRIP [after Robert Caspary, German botanist]: A thickened waxy strip that extends around and seals the walls of endodermal cells, thus restricting the diffusion of solutes across the endodermis into the vascular tissues of the root.

CIRCADIAN RHYTHMS [L. *circa,* about, + *dies,* day]: Regular rhythms of growth and activity which occur on an approximately 24-hour basis.

COHESION-TENSION THEORY: A theory accounting for upward water movement in plants. According to this theory, transpiration of a water molecule results in a negative (below 1 atmosphere) pressure in the leaf cells, inducing the entrance from the vascular tissue of another water molecule, which, because of the cohesive property of water, pulls with it a chain of water molecules extending up from the cells of the root tip.

COLEOPTILE (coal-ee-**op**-tile) [Gk. *koleon,* sheath, + *ptilon,* feather]: A sheathlike, pointed structure covering the shoot of grass seedlings.

COMPANION CELL: A small, specialized parenchyma cell associated with the sieve-tube elements in the phloem of angiosperms.

CORK: A secondary tissue produced by a cork cambium; made up of polygonal cells, nonliving at maturity, with walls infiltrated with suberin, a waxy or fatty material resistant to the passage of gases and water vapor.

CORK CAMBIUM: A lateral meristem producing cork in woody and some herbaceous plants; also called phellogen.

COROLLA (**ko**-role-a) [L. *corolla,* dim. of *corona,* wreath, crown]: Petals, collectively; usually the conspicuously colored flower parts.

CORTEX [L. *cortex,* bark]: In a stem or root, the primary tissue bounded externally by the epidermis and internally by the central cylinder of vascular tissue; consists largely of parenchyma cells.

COTYLEDON (cottle-**ee**-don) [Gk. *kotylēdōn,* cup-shaped hollow]: A leaflike structure of the embryo of seed plants which functions primarily to make stored food in the endosperm available to the developing plant or as a storage of photosynthetic organ.

CUTICLE [L. *cuticula,* dim. of *cutis,* skin]: The waxy layer of cutin on the outer wall of epidermal cells.

CUTIN: A waxy substance on the outer surface of plant cell walls exposed to the air.

CYTOKININS [Gk. *kytos,* vessel, + *kinesis,* motion]: Plant-growth substances that promote cell division, among other effects.

DAY-NEUTRAL PLANT: A plant that develops and matures regardless of the relative lengths of alternating exposures to light and dark.

DECIDUOUS [L. *decidere,* to fall off]: Characterized by having parts that fall off at a certain season, for example, a deciduous tree.

DICOTYLEDON (**dye**-cottle-**lee**-don): A member of one of the two subclasses of angiosperms, often abbreviated as dicot, distinguished essentially by having an embryo with two cotyledons. Other characteristics are a stem with a central pith (sometimes used for food storage), one or more xylem layers, a ring of cambium, and an outer layer of phloem; a taproot, consisting of a large central portion growing vertically and smaller lateral roots; irregularly edged, net-veined leaves; and flowers which usually have four or five petals and sepals.

DIOECIOUS (dye-**ee**-shuss) [Gk. *di,* two, + *oikos,* house]: Unisexual. In a dioecious species, the male and female (or staminate and ovulate) elements are on different individuals.

DORMANCY: A period of inactivity in bulbs, buds, seeds, and other plant organs, during which growth ceases. Dormancy is broken only if certain requirements, as of moisture, temperature, day length, or time, have been fulfilled.

ENDODERMIS [Gk. *endon,* within, + *derma,* skin]: A one-celled layer of specialized cells which lies between the cortex and the vascular tissues in young roots. The casparian strip of the endodermis prevents diffusion of materials across the root.

ENDOSPERM [Gk. *endon,* within, + *sperma,* seed]: A usually triploid ($3n$) tissue containing stored food that surrounds the embryo in early stages of seed development. It develops from the union of a sperm nucleus and polar bodies of the embryo sac. It is found only in angiosperms and is digested by the growing sporophyte either before seed maturation or after germination.

ENTRAINMENT: The modification of a plant's circadian rhythm to a period other than its usual one.

EPICOTYL: The portion of the axis of a plant embryo or seedling above the point of attachment of the cotyledon(s).

EPIDERMIS [Gk. *epi,* upon, + *derma,* skin]: The outermost layer of cells in leaves, young stems, and roots.

ETHYLENE: A simple hydrocarbon which is a plant-growth substance active in the ripening of fruit.

FIBER: A greatly elongated, thick-walled supporting cell, tapering at both ends, found in vascular plants. The protoplasm is usually dead at maturity.

FILAMENT: The stalk of the stamen, supporting the anther.

FLOWER: The reproductive structure of angiosperms. A complete flower includes calyx, corolla, stamens, and carpels.

FRUIT: A matured, ripened ovary (or group of ovaries) which contains the seeds.

GAMETOPHYTE (gam-**meet**-o-fight): In plants having alternation of generations, the haploid ($1n$), gamete-producing phase.

GEOTROPISM [Gk. *gē,* earth, + *tropos,* turn, direction]: Growth that is oriented by gravity, for example, roots growing downward.

GERMINATION [L. *germinare,* to sprout]: The resumption of growth by a spore, seed, bud, or other structure.

GIBBERELLINS (jibb-e-**rell**-in) [fr. *Gibberella,* genus of fungi]: A group of growth hormones, the most characteristic effect of which is to increase the elongation of stems in a number of higher plants.

GROUND MERISTEM: A primary meristematic tissue that gives rise to cortex and pith.

GUARD CELLS: Specialized, paired, crescent-shaped epidermal cells surrounding a stoma. The guard cells control the opening and closing of the stoma by expanding or contracting with changes in turgor.

GUTTATION [L. *gutta,* a drop]: The exudation of liquid water from the margins of leaves.

HERBACEOUS (her-**bay**-shus): A term referring to any nonwoody plant.

HORMONE [Gk. *hormaein,* to excite]: A chemical substance produced, usually in minute amounts, in one part of an organism and transported to another part of that organism, where it has a specific effect.

HYPOCOTYL: The portion of the axis of a plant embryo or seedling between the cotyledon(s) and the radicle.

INDOLEACTIC ACID (IAA): A naturally occurring growth regulator; an auxin. Its principal source is the apical meristem of the stem.

KINETIN [Gk. *kinētikos,* causing motion]: A purine which probably does not occur in nature but which acts as a cytokinin.

LATERAL MERISTEMS: Meristems that give rise to secondary tissue; the vascular cambium and the cork cambium.

LONG-DAY PLANT: A plant that requires exposure to light periods of more than a critical length in order to flower.

MACRONUTRIENT: One of the six minerals required in relatively large quantities for plant growth. The macronutrients are nitrogen, potassium, calcium, phosphorus, magnesium, and sulfur.

MASS-FLOW THEORY: The basic theory accounting for sap flow through the phloem system. According to this theory, the solution containing nutrient sugars is moved by turgor pressure from the sieve tubes to the cells of the rest of the plant.

MERISTEM [Gk. *merizein*, to divide]: The undifferentiated plant tissue from which new cells arise. There are apical meristems, lateral meristems, and meristems in leaves, flowers, and fruits.

MICRONUTRIENT: A mineral required in only minute amounts for plant growth. The micronutrients include iron, chlorine, copper, manganese, zinc, molybdenum, and boron.

MICROPYLE: In the ovules of seed plants, the opening through which the pollen tube usually enters the ovule to effect fertilization.

MONOCOTYLEDON (mono-cottle-**ee**-don): A member of one of the two subclasses of angiosperms, often abbreviated as monocot, distinguished essentially by having one cotyledon. Other characteristics are a stem (usually nonwoody) with vascular bundles scattered throughout, fibrous roots, leaves with parallel main veins and straight edges, and sepals and petals usually in threes. Monocotyledons include the grasses, rushes, lilies, palms, and orchids.

MONOECIOUS (mo-**nee**-shuss) [Gk. *monos*, single, + *oikos*, house]: Having the anthers and carpels on the same individual plant of a species but on different flowers.

MORPHOGENESIS [Gk. *morphe*, form, + *genes*, born]: The development of form.

NECTARY [Gk. *nektar*, the drink of the gods]: An organ in angiosperms that holds the nectar, a sugary fluid which attracts pollinators.

OVARY [L. *ovum*, egg]: The enlarged basal portion of a carpel or of fused carpels. The ovary matures to become the fruit.

OVULE [L. *ovulum*, little egg]: An organ in seed plants containing the female gametophyte with its egg cell and surrounded by a protective coat and a tissue specialized for food storage. When mature, the ovule becomes a seed.

PALISADE CELLS: Columnar, chloroplast-bearing cells composing a layer of leaf tissue; they are oriented with their long axes perpendicular to the leaf surface.

PARENCHYMA (pa-**renk**-ee-ma) [Gk. *para*, beside, + *en*, in, + *chein*, to pour]: Living, thin-walled, unspecialized, more or less isodiametric cells with large vacuoles, common as photosynthetic or storage tissues.

PERENNIAL [L. *per*, through, + *annus*, year]: A plant which persists from year to year and usually produces reproductive structures in two or more different growing seasons.

PERICYCLE [Gk. *peri*, around, + *kyklos*, circle]: A cell layer, generally of the root, bounded externally by the endodermis and internally by the phloem. The origin of branch roots.

PETAL: A flower part, usually conspicuously colored; one of the units of the corolla.

PETIOLE (**pet**-ee-ole): The stalk of a leaf.

PHLOEM (**flow**-em) [Gk. *phloos*, bark]: Food-conducting tissue, consisting of sieve cells (in gymnosperms) or sieve tubes and companion cells (in angiosperms), phloem parenchyma, and fibers.

PHOTOPERIODISM: Response to relative day and night length; a mechanism evolved by organisms for measuring seasonal time.

PHOTOTROPISM [Gk. *photos*, light, + *tropos*, turn, direction] A growth movement in which light is the orienting factor, as the growth of a plant toward or away from a light source; turning or bending response to light.

PHYCOBILIN: A water-soluble accessory pigment which occurs in red and blue-green algae.

PHYTOCHROME: A phycobilinlike pigment found in green plants. It is the photoreceptor for red or far-red light and is involved with a number of developmental processes, such as flowering, dormancy, leaf formation, and seed germination.

PITH: The tissue occupying the center of the vascular cylinder in the stem and in some types of roots. It usually consists of parenchyma.

PLASTID: An organelle in the cells of certain groups of eukaryotes that is the site of such activities as food manufacture and storage. Plastids are bounded by a double membrane.

POLLEN [L. *pollen*, fine dust]: The male gametophytes of seed plants at the stage at which they are shed.

POLLEN TUBE: The tube through which the male gametes move down the stigma into the ovule. It is a tubular extension of the pollen grain.

POLLINATION: The transfer of pollen from the anther, where it was formed, to a receptive surface, usually a stigma.

PRIMARY GROWTH: In plants, growth originating in the apical meristems of shoots and roots, as contrasted with secondary growth, which originates in the vascular and cork cambium. Primary growth results in an increase in length.

PRIMARY TISSUES: Cells derived from the apical meristem and primary meristematic tissue of root and shoot, as contrasted with secondary tissue, which is derived from the lateral cambiums.

PROCAMBIUM: A primary meristematic tissue which gives rise to the primary xylem and phloem and, in woody plants, to the vascular cambium.

PROTODERM: A primary meristematic tissue which gives rise to the epidermis.

PROTOPLAST: A plant cell body, not including the cell wall.

RADICLE [L. *radix*, root]: That portion of the plant embryo which develops into the primary root of the seedling.

RHIZOME [Gk. *rhizoma*, mass of roots]: A more or less horizontal underground stem; distinguished from true root in possessing nodes, buds, and usually scalelike leaves.

ROOT: The descending axis of a plant, normally belowground and serving both to anchor the plant and to take up and conduct water and minerals.

ROOT CAP: A thimblelike mass of cells produced by and covering the growing tip of a root.

ROOT HAIRS: Tubular outgrowths of epidermal cells of the root in the zone of maturation, just above the apical meristem. The surface area of the root is enormously increased by root hairs.

SCUTELLUM [L. *scutella*, small shield]: The single cotyledon of a grass embryo; it absorbs foods from the endosperm during germination.

SECONDARY GROWTH: In plants, growth derived from lateral meristems, that is, the vascular cambium and cork cambium. Secondary growth results in an increase in diameter.

SECONDARY TISSUES: Tissues (such as secondary phloem, secondary xylem, and cork) produced by the vascular cambium and cork cambium.

SEPAL [M.L. *sepalum*, a covering]: One of the outermost flower structures, which usually encloses the other flower parts in the bud; a unit of the calyx.

SHOOT: The aboveground portions, such as the stem and leaves, of a vascular plant.

SHORT-DAY PLANT: A plant that requires exposure to light periods of less than a critical length in order to flower.

SIEVE CELL: A long and slender sugar-conducting cell with relatively unspecialized sieve areas and with tapering end walls that lack sieve plates; found in the phloem of gymnosperms.

SIEVE TUBE: A vertical series of sugar-conducting cells (sieve-tube elements) of the phloem of angiosperms.

SIEVE-TUBE ELEMENT: One of the component cells of the sieve tubes of flowering plants; characterized by sievelike end walls. Sieve-tube elements lack a nucleus at maturity and are always found in association with companion cells.

SPOROPHYTE: The spore-producing, diploid ($2n$) phase in the life cycle of a plant having alternation of generations.

STAMEN [L. *stamen*, thread]: The organ of the flower that produces the pollen; composed (usually) of anther and filament. Collectively the stamens make up the androecium.

STEM: That part of the axis of vascular plants that is aboveground, as well as anatomically similar portions belowground (such as rhizomes).

STIGMA: The region of the carpel which serves as a receptive surface for pollen grains and on which they germinate.

STOMA, *pl.* STOMATA [Gk. *stoma,* mouth]: A minute opening bordered by guard cells in the epidermis of leaves and stems through which water, oxygen, and carbon dioxide diffuse.

STYLE [Gk. *stylos,* column]: A slender column between the stigma and the ovary through which the pollen tube grows.

SUSPENSOR: A filamentous structure in the early embryo of many vascular plants that pushes the terminal part of the embryo into the endosperm.

TRACHEID (**tray**-key-idd) [Gk. *tracheia,* rough]: An elongated, thick-walled conducting and supporting cell of xylem, characterized by tapering ends and pitted walls without true perforations. It is found in nearly all vascular plants and is dead at maturity.

TRANSLOCATION: In plants, the transport of the products of photosynthesis.

TRANSPIRATION [Fr. *transpirer,* to perspire]: The loss of water vapor through stomata.

TUBER [L. *tuber,* bump, swelling]: A much-enlarged, short, fleshy underground stem, such as that of the potato.

VASCULAR CAMBIUM: A cylindrical sheath of meristematic cells which divide, producing secondary phloem and secondary xylem, but always with a cambial cell remaining.

VEIN: In plants, a vascular bundle forming a part of the framework of the conducting and supporting tissue of a leaf or other expanded organ.

VESSEL: A tubelike structure of the xylem composed of cells (vessel elements) placed end to end; generally found in angiosperms. Its function is to conduct water and minerals from the soil.

VESSEL ELEMENT: One of the cells composing a vessel; dead at maturity. In evolutionary terms, a specialized tracheid.

XYLEM (**zye**-lem) [Gk. *xylon,* wood]: A complex vascular tissue through which most of the water and minerals of a plant are conducted; consists of tracheids or vessel elements, parenchyma cells, and fibers. The xylem constitutes the wood of trees and shrubs.

ZEATIN: A plant hormone; a natural cytokinin isolated from corn.

SUGGESTIONS FOR FURTHER READING

GALSTON, A. W.: *The Green Plant,* Prentice-Hall, Inc., Englewood Cliffs, N.J., 1968.*

A convenient and concise summary of plant growth and development.

RAVEN, PETER, and HELENA CURTIS: *Biology of Plants,* Worth Publishers, Inc., New York, 1970.

An up-to-date and handsomely illustrated general botany text, particularly strong in the fields of ecology and evolution.

RAY, PETER M.: *The Living Plant,* 2d ed., Holt, Rinehart and Winston, Inc., New York, 1971.*

An outstanding short text.

RICHARDSON, MICHAEL: *Translocation in Plants,* St. Martin's Press, Inc., New York, 1968.*

An extremely useful, brief review of experimental work on the movement of water in plants.

SALISBURY, FRANK B., and CLEON ROSS: *Plant Physiology,* Wadsworth Publishing Co., Inc., Belmont, Calif., 1969.

A good modern plant physiology text, for more advanced students.

TORREY, JOHN G.: *Development in Flowering Plants,* The Macmillan Company, New York, 1967.*

How flowering plants develop, with emphasis on the underlying physiological processes.

* Available in paperback.

SECTION 6

Animal Physiology

6–1
Genus, Homo; *species,* sapiens. *An immature form.*

CHAPTER 6–1

Sexual Reproduction

The animal kingdom is made up of a vast array of organisms, ranging from microscopic aquatic forms to extremely complicated and highly organized multicellular creatures that are by far the most complex of all living things. In this section, we are going to focus our attention on only one segment of the kingdom, the mammals, a group characterized by hairiness, a high constant temperature (homeothermy), development of the embryo within the body of the female, and nursing of the infant.

We have selected these members of the animal kingdom for two reasons. First, they represent the furthest extension of two evolutionary trends. One trend characteristic of the animals is the tendency toward increasing freedom from the external environment; an important move in this regard is the "warm-bloodedness" just mentioned, which is the subject of Chapter 6–4. The second major evolutionary trend, which is closely linked to the first, is toward increasing complexity of the nervous system. The mammals are by far the most intelligent of all groups of organisms and have the most highly developed systems for receiving and processing information from the environment. (Although in complexity and variety of sense organs, they find close rivals in the insect world.)

Our second reason for selecting the mammals is that they are the group to which our own species belongs and so are of particular personal interest.

However, although we shall concentrate on the mammals, we shall sometimes be looking at other members, both present and ancestral, of the animal kingdom. Just as the structures and functions of land plants begin to "make sense" when seen as special adaptations to life out of the water, the special physiological characteristics of the mammals also are determined in large part by an interplay of past history and present environment—the twin themes of evolution and ecology.

The rest of this chapter will be concerned with the physiology of mammalian reproduction. It is followed by a series of pictures on the human prenatal development, which we hope you will find as exciting as we do.

THE FORMATION OF SPERM AND EGGS

The capacity to reproduce is, as we saw in Section 1, one of the characteristics by which we recognize living things. Reproduction may be asexual, involving the transfer of genes to the offspring from only one parent; asexual reproduction is common among one-celled organisms, fungi, plants, and invertebrates. In higher animals, reproduction of the organism is always sexual and always involves the formation by meiosis of two types of gametes—eggs and sperm—and the fusion of these gametes to produce a cell containing genes from both parents. The resulting embryo carries a unique combination of hereditary material and will therefore give rise to an individual which is different from either parent and from any other individual of its species. It is possible that there is a relationship between the emphasis among the vertebrates on sexual reproduction—which introduces more variety on which evolution can act—and their rapid rate of evolution.

The Male Genital Organs

Among the members of the animal kingdom, the gametes are characteristically the only haploid stage in the life cycle. They are formed in the gonads (the testes in the male and the ovaries in the female). Among the mammals, the sperm-producing areas of the testes are long tubules, the seminiferous ("seed-bearing") tubules. Special cells in the tubules, the spermatocytes, divide meiotically, each producing four haploid cells, which then differentiate into the highly specialized spermatozoa. The spermatozoa are released into the sperm duct—the technical term is the vas deferens—which leads from the testes through sacs in which the sperm are stored and then through the prostate gland. In the prostate gland, fluid is added to the suspension of sperm cells. (The fluid nourishes the sperm and gives it additional motility.) Sperm and fluid together form the seminal fluid, which, in mammals, is released through the urethra of the penis.

The penis (plural: penes) is a special adaptation to life on land. Among most aquatic animals—with the notable exception of the octopus, some live-bearing fish, and of course, the aquatic mammals—the eggs are fertilized externally after they are laid, but the cleidoic egg (see page 323) must be fertilized internally, within the reproductive tract of the female. The primary function of the penis, the male sex organ, is to deposit sperm in the vagina of the female, which is the external opening of the female reproductive tract. The penis, in various forms, has evolved independently in a number of species of insects and in other invertebrates. It is found among some reptiles and birds; all flightless birds have penes, and so do all ducks, flightless or not. In most reptiles and birds, however, one opening, the *cloaca*, serves as the passage for eggs and sperm and also for the elimination of wastes, and these animals mate by juxtaposition of their cloacae. Only among mammalian species is the penis found in all male animals.

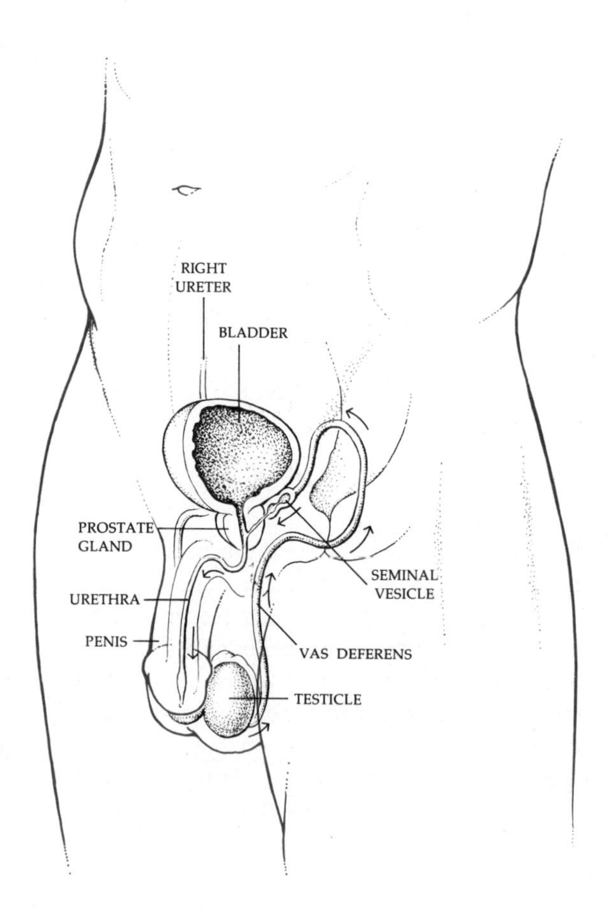

6–2

The human male reproductive tract. Sperm cells formed in the seminiferous ("seed-bearing") tubules of the testicles enter the vas deferens, which leads through a seminal vesicle, where the sperm are stored, and then through the prostate gland. In the prostate gland, the sperm cells are mixed with fluid, and this mixture, the seminal fluid, is released through the urethra of the penis. The urethra is also the passageway for urine, which is stored in the bladder.

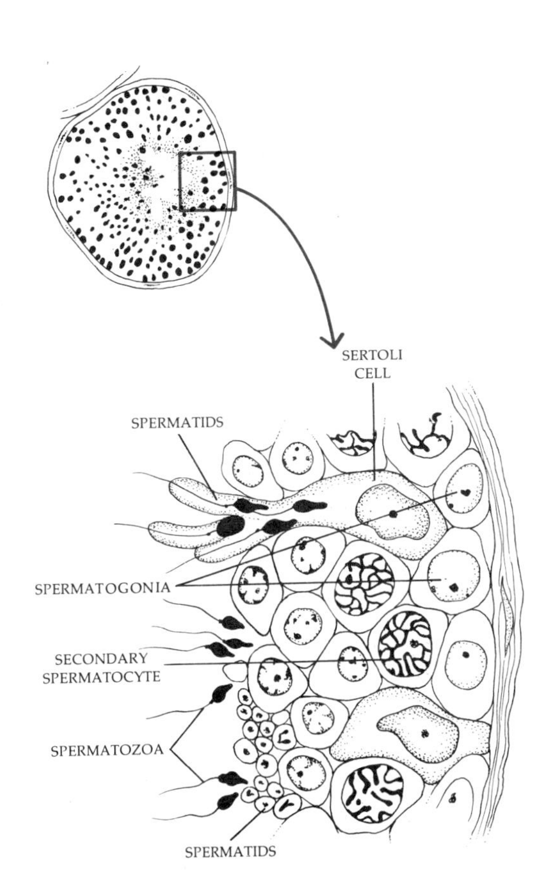

6–3

The seminiferous tubules contain sperm cells in various stages of development. Spermatogonia grow into larger cells known as primary spermatocytes. These divide (first meiotic division) into two equal-sized cells, the secondary spermatocytes. In the second meiotic division, four equal-sized spermatids are formed. These differentiate into functional sperm. The Sertoli cells support and nourish the developing sperm, and the interstitial cells are the source of the male hormone testosterone.

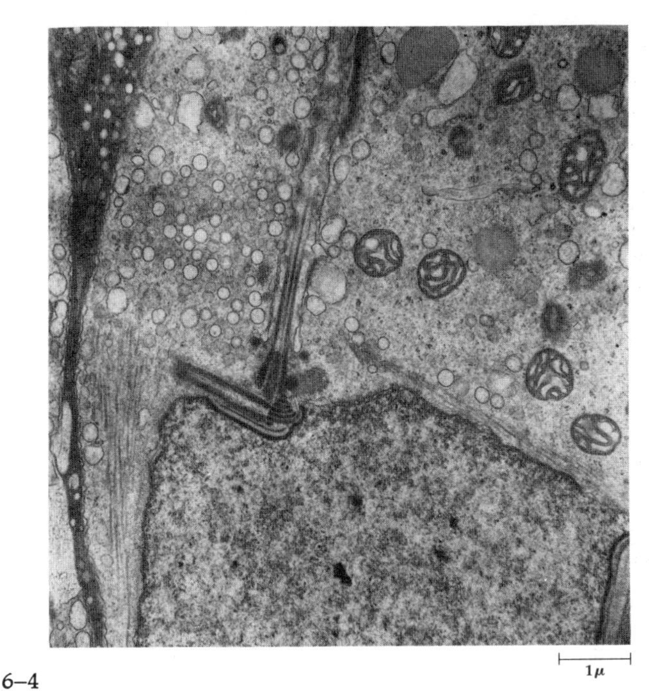

6–4

A mammalian spermatid in the process of differentiation. The dark area in the lower half of the electron micrograph is the nucleus. The two centrioles have moved to a position just behind the nucleus and the tail has begun to form from one of them. To the right are several mitochondria.

The Female Genital Organs

The egg cells, or ova, are formed in the ovaries, which are a solid mass of cells. The oocytes, from which the eggs develop, are in the outer layer of this mass. In human females, the primary oocytes begin to form about the third month of fetal development, and by the time of birth, the two ovaries contain some 400,000 primary oocytes, which have reached prophase of the first meiotic division. These primary oocytes remain in prophase until the female matures sexually. Then, under the influence of hormones, the first meiotic division resumes and is completed at about the time of ovulation.

In the human female after puberty, the primary oocytes usually mature one at a time about every 28 days. Thus as much as 50 years may elapse between the beginning and the end of the first meiotic division in a particular oocyte.

Maturation of the oocyte involves both meiosis and a great increase in size. At meiosis, oocyte does not divide to form four ova. Instead, a single ovum and several polar bodies are formed. When the oocyte is ready to complete meiosis, the nuclear membrane breaks up and the chromosomes move to the surface of the cell. As the nucleus divides, the cytoplasm of the oocyte bulges out. Half of the chromosomal material moves into the bulge, which then pinches off into a small cell, the first polar body. The rest of the cellular material forms the large, secondary oocyte. The second meiotic division is carried out in the same way, producing from the secondary oocyte one haploid cell, which becomes the ovum, and another polar body. In this way, the accumulated food reserves of the oocyte are passed on to a single egg. The first polar body may also divide, although there is no functional reason for it to do so. All the polar bodies eventually die.

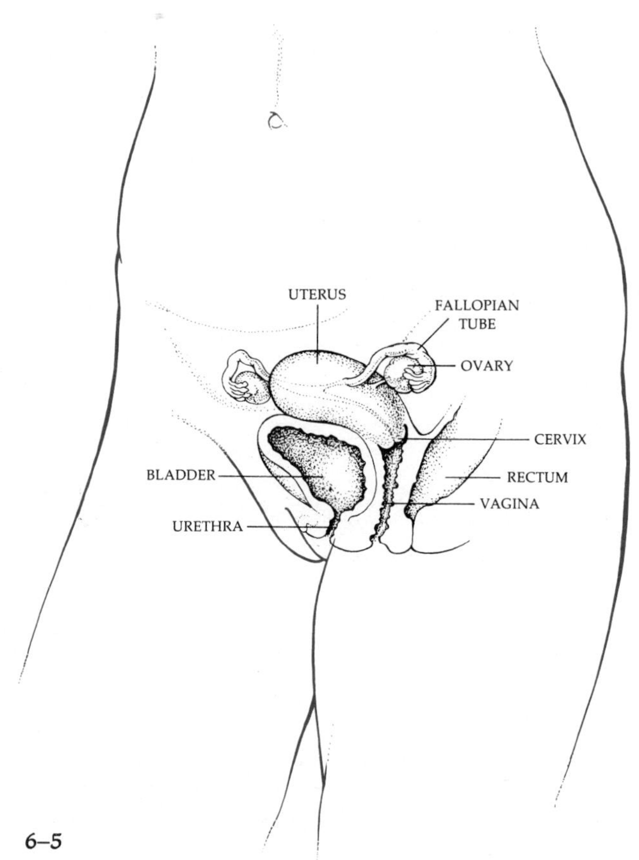

6–5
The human female reproductive tract.

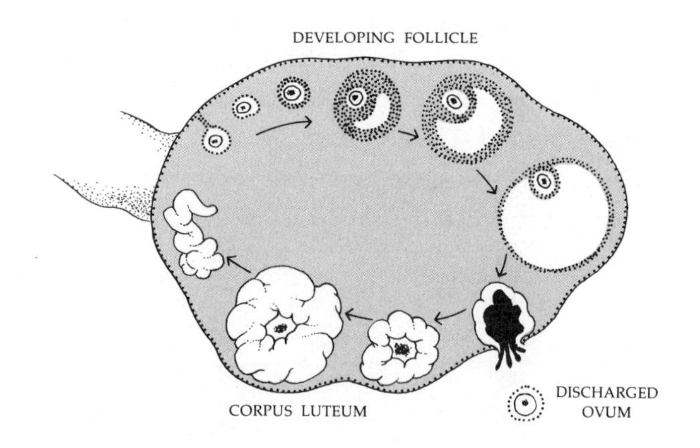

6–6a (at right)
Ova develop near the surface of the ovary within follicles, or cavities, which form in the ovarian tissue. After an egg cell is discharged from the follicle, cells of the ruptured follicle give rise to the corpus luteum ("yellow body"). If the egg is not fertilized, the corpus luteum is reabsorbed in two to three weeks. If the egg is fertilized, the corpus luteum persists, producing progesterone, which prepares the uterus for the embryo.

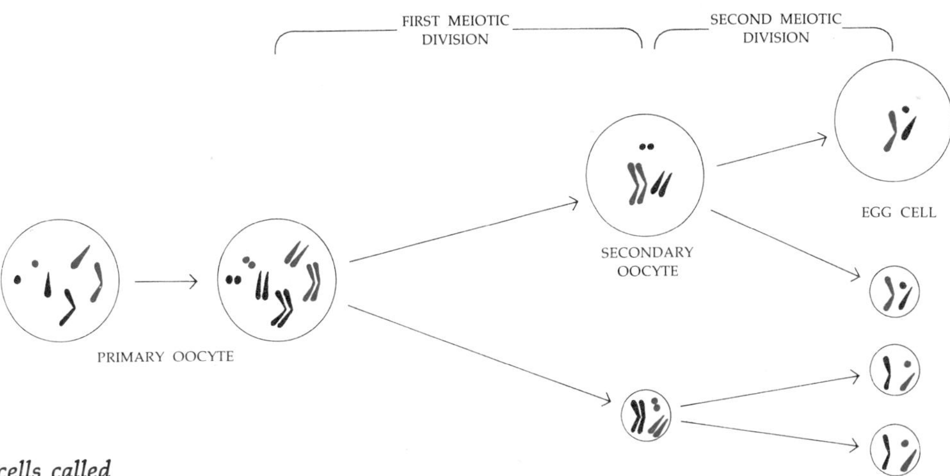

6–6b
Egg cells develop in the ovary from specialized cells called oocytes. The chromosomes are replicated before the first meiotic division, which results in one large cell, the secondary oocyte, and one small cell, the first polar body (so called because it appears as a small speck at one end, or pole, of the egg). Each of these has the same number of chromosomes as was present in the primary oocyte. In the second meiotic division, the secondary oocyte divides into the ootid and a second polar body; the first polar body may also divide. The chromatids are separated in this division. The ootid then becomes an egg cell, or ovum, and the three small polar bodies disintegrate.

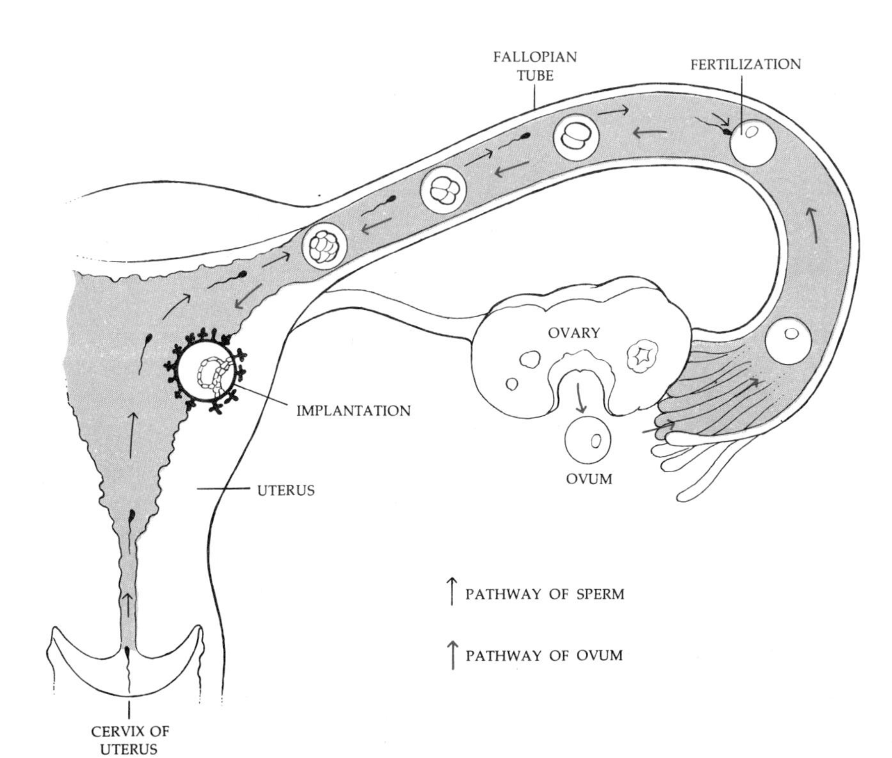

6–7
Fertilization of the egg by sperm. Once a month in the female of reproductive age, an ovum breaks loose from the ovary and enters one of the Fallopian tubes. Drawn by current produced by the beating of cilia lining the tube, the egg travels down the tube toward the uterus. Fertilization, when it occurs, generally takes place within the tube, after which the fertilized egg becomes implanted in the lining of the uterus. If the egg is not fertilized, it degenerates within a few days.

The egg cell develops within a cavity, or follicle, on the surface of the ovary. The cells of the follicle supply food to the growing egg. When the egg is released from the follicle, these cells produce a hormone, progesterone, which helps to prepare the lining of the uterus for the implantation of the egg. From the follicle, the egg drifts into one of the oviducts (called the Fallopian tubes in the human female) as a result of the beating of cilia around the funnel-shaped opening of the oviduct.

A human egg is about 1/10 millimeter, or 100 microns, in diameter, which is very large for a cell. It contains an unusually large supply of ribosomes, enzymes, amino acids, and all the other cellular machinery that will be used in the early, rapid stages of biosynthesis characteristic of embryonic cells.

PROTECTION OF THE EMBRYO

Sperm, eggs, and the developing embryo are all fragile structures whose delicate surfaces are very vulnerable to drying out or other injuries. They must always be kept moist. In the lower animals, as in the lower plants, this moisture is provided by free water—the water found in oceans, lakes, streams, and puddles.

In vertebrate evolution, one of the crucial stages in the transition from water to land—analogous to the development of the seed in plants—was the evolution of a means of reproduction which afforded protection to the gametes and the embryo and yet did not require the presence of free water. The vertebrates became truly terrestrial with the evolution in the reptiles of internal fertilization and, as we noted in Section 4, an egg that can be laid on land. In order to understand the development of the mammalian embryo, it is useful first to take a brief look at the formation and structure of the land egg.

The Cleidoic Egg

The secret of the cleidoic egg is that it carries within its shell its own water supply and all its own life-supporting systems. As the fertilized egg passes down the reproductive tract, a tough calcium-containing shell is deposited around it. Then, when incubation begins, embryonic development starts, and from the rapidly dividing mass of unstructured cells grow both the fetus and a complex membrane system.

Within the egg, the developing embryo becomes completely surrounded by a water-filled membrane, the *amnion*. This water, the amniotic fluid, protects the embryo from drying out and cushions it against injury. The egg also carries a rich food supply, the yolk. The yolk is enclosed within a membrane which is connected directly with the digestive system of the embryo. Because the egg is a closed system—not unlike a spaceship—some method is needed for disposing of wastes. A third membranous sac, the *allantois*, collects liquid wastes from the embryo, sealing them off from the rest of the system. The outer surface of the allantois, which has a rich blood supply, also serves for the exchange of oxygen and carbon dioxide. Both of these gases diffuse in and out of the egg. A fourth membrane, the *chorion*, lies just within the shell, surrounding the embryo and all the other membranes. All these membranes are made of embryonic (not maternal) tissue and develop as the embryo develops.

The Placenta

Although only a few mammals lay eggs (these are the monotremes, such as the duckbill platypus), the mammalian fetus develops within a system of membranes which closely resemble those found in the cleidoic egg. Among most of the mammals, the outermost embryonic membrane, the chorion, has become specialized to form part of an organ known as the *placenta*.

The placenta is a spongy tissue through which oxygen, food molecules, and wastes are exchanged between mother

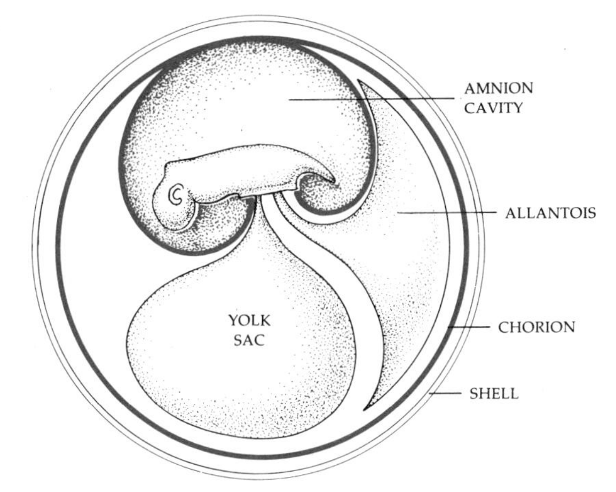

6–8
A diagram of the cleidoic egg.

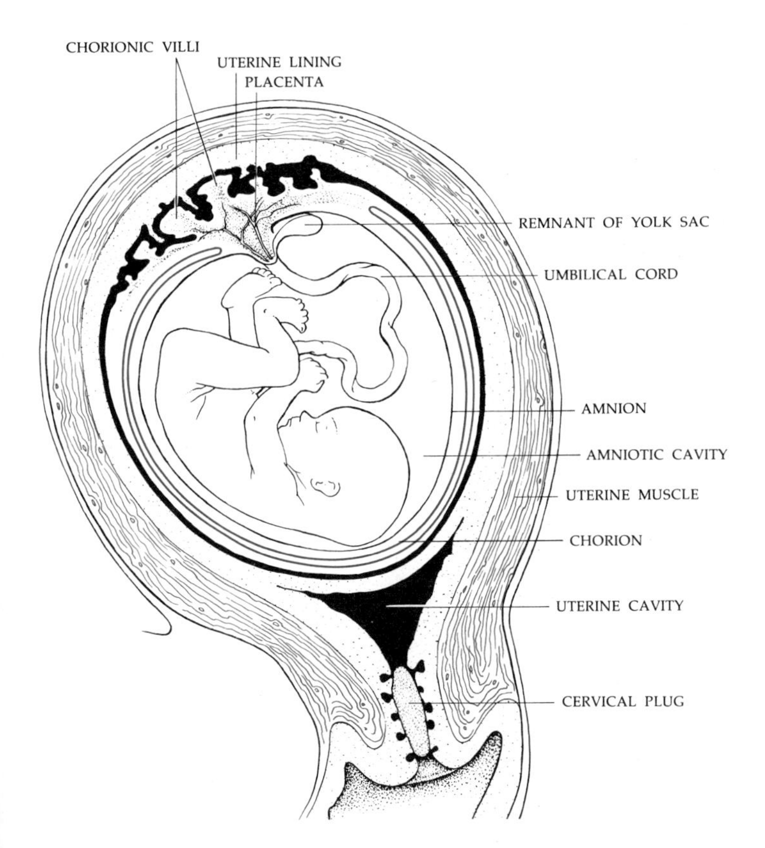

6–9
Human embryo in the uterus.

and embryo. It is formed from a maternal tissue, the inner lining of the uterus, as well as from the chorion, and has a rich blood supply from both.

Within the placenta, the capillaries of the fetal and maternal circulatory systems intertwine but are not directly connected. Molecules, including food and oxygen, enter the placental tissue from the maternal bloodstream, diffuse through the placenta, and are picked up from it by the fetal bloodstream. Similarly, carbon dioxide and other waste products from the fetus are picked up from the placenta by the maternal bloodstream and carried away for disposal through the mother's lungs and kidneys.

As the embryo grows larger, it remains attached to the placenta by the long umbilical cord, which permits it to float freely in its sac of amniotic fluid. The umbilical cord is severed at birth. The embryonic part of the placenta becomes detached from the uterine wall and is expelled along with the remnants of the umbilical cord as the "afterbirth."

HORMONES AND REPRODUCTION

Mammalian reproductive activities are under the control of the sex hormones. These hormones are produced by the testes in the male and the ovaries in the female. In the pregnant female, the placenta also becomes an important source of hormones. The principal male sex hormone is testosterone; the female sex hormones are estrogens (of which there are several with slight chemical differences) and progesterone. The production of the sex hormones is under the influence of a group of hormones produced by the pituitary gland. The pituitary gland, which is about the size and shape of a kidney bean, is located at the base of the brain. Locate, in your imagination, the geometric center of your skull; there is your pituitary. It is lodged just beneath an area of the brain known as the hypothalamus, and the pituitary's production of hormones is under the control of this brain center.

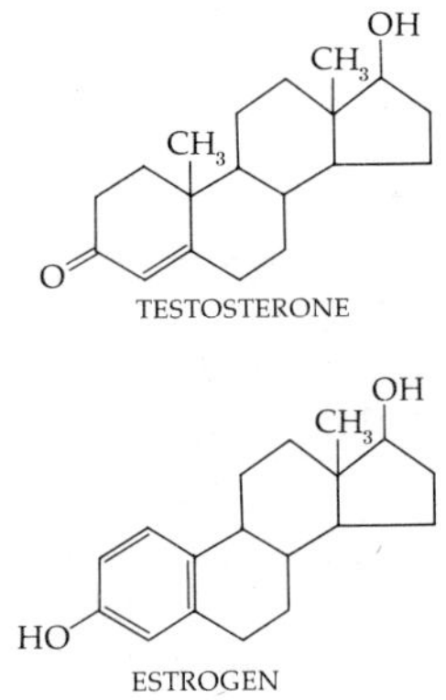

6–10

The chemical structures of testosterone and estrogen. Note that the hormones differ only very slightly chemically, in contrast to their great differences in physiological effects, another example of the extreme specificity of biochemical actions.

The pituitary produces two gonadotropic (gonad-stimulating) hormones: follicle-stimulating hormone (FSH) and luteinizing hormone (LH). The onset of production of these hormones brings about maturation of the reproductive system and all the other many changes associated with puberty.

The gonadotropic hormones and the sex hormones are linked in what is known as a negative feedback system. The simplest example of a negative feedback system is the thermostat on a furnace. When the room temperature drops, the thermostat turns the furnace on. When sufficient heat is produced, the heat acts on the thermostat to turn off the furnace. In the sex hormone system, the hypothalamus is the "instrument" that senses the change. When the concentration of a sex hormone in the blood reaches a certain level, stimulation by the hypothalamus of pituitary output of the related gonadotropic hormone ceases. Secretion of the hormone then ceases, and this, in turn, has the effect of stopping secretion by the sex gland of the sex hormone.

Table 6–1

Mammalian Gonadotropic and Sex Hormones

Hormone	*Source*	*Principal Effects*	*Control*
FSH	Pituitary	Stimulates sperm production; stimulates growth of ovarian follicle; stimulates estrogen production	Hypothalamus
LH	Pituitary	Stimulates testosterone production; stimulates release of egg cell; stimulates progesterone production	Hypothalamus
Estrogen	Ovary (follicle)	Produces and maintains female sex characteristics; thickens lining of uterus	FSH
Progesterone	Ovary (corpus luteum)	Thickens lining of uterus	LH
Testosterone	Testes	Produces and maintains male sex characteristics; stimulates sperm production	LH

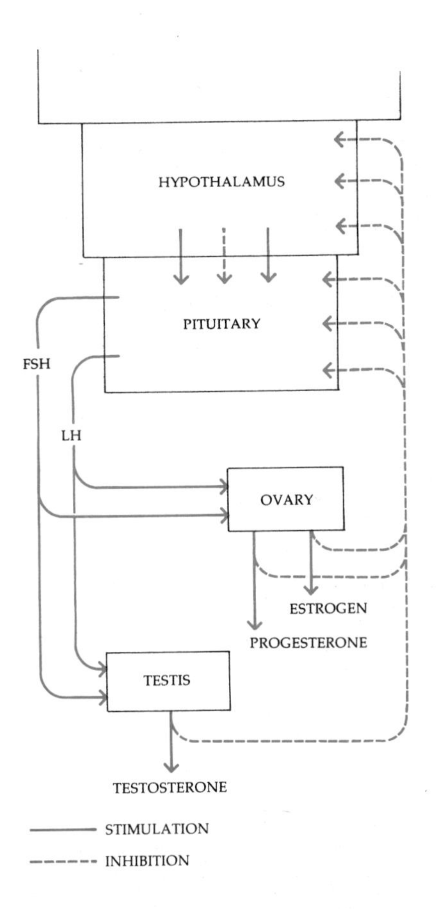

6–11
The production of sex hormone in vertebrates is controlled by a negative feedback system. Production in the pituitary gland of the gonadotropic hormones FSH (follicle-stimulating hormone) and LH (luteinizing hormone) stimulates production of the male sex hormone testosterone or the female sex hormones estrogen and progesterone. Under normal circumstances, the organism never produces more than is needed of any of these hormones because an increase of sex hormones to a certain level of concentration in the blood acts on the hypothalamus, a brain center. The hypothalamus ceases stimulating the pituitary, which, in turn, stops secretion of the related gonadotropic hormone.

Secondary Sex Characteristics

In man, testosterone not only stimulates sperm production but also affects the pitch of the voice, muscle development, skeletal size, and distribution of facial and body hair. In other male animals, it has been shown that many aspects of behavior, including aggressiveness, sex drive, and complex patterns of courtship and mating, are also dependent on male sex hormones. Estrogens stimulate development of the breasts and the female reproductive tract and determine the distribution of body fat. They inhibit growth, particularly of the extremities. Consequently, girls whose estrogen production begins early often have shorter arms and legs and relatively longer trunks than girls whose estrogen production begins at a later age. All these additional changes produced by the sex hormones are known as secondary sex characteristics.

The Menstrual Cycle and the Pill

With the onset of puberty in the human female, the menstrual cycle begins. The primary event of the menstrual cycle is the release of an egg cell, or ovum, which occurs about every 28 days. Under the influence of FSH, the egg cell ripens in a bed of tissue known as the follicle. On about the fourteenth day of the cycle, the egg cell, under the influence of LH, is released from the follicle and begins its journey down the Fallopian tubes. During this same period, the lining of the uterus thickens in preparation for the implantation of the egg. If the egg is fertilized, it becomes implanted. If it is not fertilized, it degenerates and the lining of the uterus is cast off. The casting off of the lining of the uterus, the menstrual period, marks the beginning of the menstrual cycle.

All these events take place as the result of a shifting balance of hormones. LH stimulates the follicle to release the egg cell, which begins its passage to the uterus. Under the continued stimulus of LH, the cells of the emptied follicle grow larger and fill the cavity, producing the corpus luteum ("yellow body"). The cells of the corpus luteum, as they increase in size, begin to synthesize progesterone as

6–12
Secondary sex characteristics among vertebrate males.

6–13
Diagram of the menstrual cycle. An increase in the follicle-stimulating hormone (FSH) of the pituitary promotes the growth of the ovarian follicles and the secretion of estrogen. The estrogen stimulates the lining of the uterus (the endometrium) to thicken. The increased concentration of estrogen turns off FSH production, and under the influence of luteinizing hormone (LH), ovulation occurs. The follicle is converted into the corpus luteum, which begins to secrete progesterone. The presence of progesterone inhibits LH production. By the end of the menstrual cycle, if pregnancy has not occurred, hormone production is low. As a consequence, the endometrium begins to slough off, and the cycle begins again.

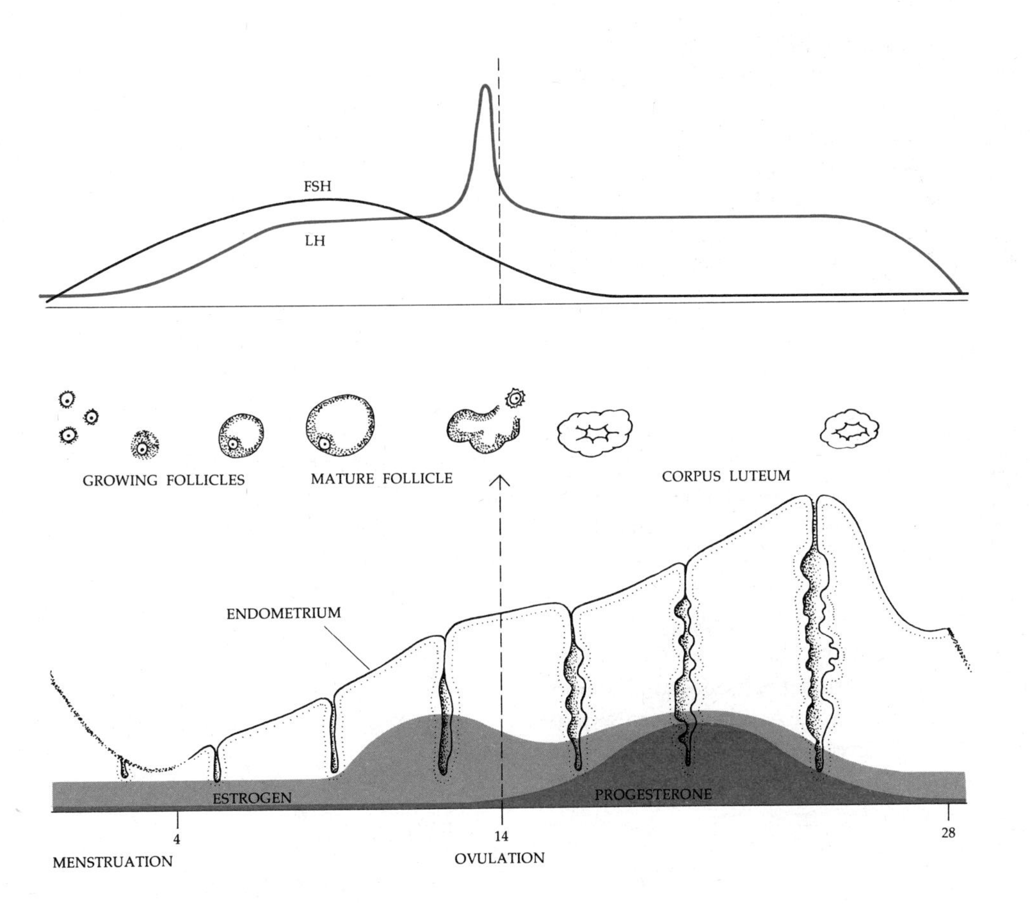

well as estrogen. As the estrogen and progesterone levels increase, they inhibit the production of the gonadotropic hormones from the pituitary. Production of ovarian hormones then drops. The lining of the uterus can no longer sustain itself without hormone support, and a portion of it is sloughed off in the menstrual fluid. Finally, in response to the low level of ovarian hormones, the level of pituitary gonadotropic hormones begins to rise again, followed by a rise in estrogen as the next monthly cycle begins.

"The Pill" consists of a combination of estrogen and progesterone. When taken daily, it keeps the level of ovarian hormones in the blood sufficiently high to shut off production of the pituitary hormones. In the absence of LH, no ovulation occurs. The lining of the uterus thickens under the artificial hormone influence. Then, when a woman stops taking the pill toward the end of the cycle, it sloughs off, producing a menstrual period, although ovulation has never occurred.

Hormones and Human Pregnancy

If the egg cell is fertilized, it becomes implanted in the lining of the uterus and the placenta begins to form. The placenta almost immediately begins to produce gonadotropic hormones of its own. These gonadotropic hormones act upon the corpus luteum so that it continues production of estrogen and progesterone when pituitary gonadotropic production ceases. Pregnancy tests involve the detection in blood or urine of the presence of these gonadotropic hormones from the placenta. Usually their presence is detected by their effects in stimulating the gonadal tissue of a rabbit or other laboratory animal. As the placenta matures, it also begins its own production of estrogen and progesterone, both of which it produces at a relatively high level throughout pregnancy.

FERTILITY AND MATING

Females of almost all species except man will mate only during their fertile period, which is known as *estrus*, or heat. Cows and horses come into heat once a year, usually in the spring; dogs, twice a year; and rats and mice, every few days.

The periods of estrus may last from only a few hours to three or four weeks. In animals, such as dogs, that produce eggs continuously during estrus, the eggs may be fertilized at different times and by different males, which explains in part why a mixed-breed litter can contain such an astonishing variety of siblings and also why there may be a great size variation even among purebred pups, many of which are actually of different ages at birth. In some mammals, such as cats, rabbits, and minks, although the egg is mature and the female receptive during estrus, ovulation occurs only under the stimulus of copulation, obviously a very efficient system, ensuring maximum economy in the utilization of gametes.

The human female is the only female animal which is receptive to mating during unfertile periods. Anthropologists speculate that this receptivity coevolved with the establishment of strong pair-bond relationships between human or prehuman male and females. A consequence of this pair-bond relationship is a society based on a family unit, in contrast to most other primate groups, in which the social and breeding unit is a tribe or band. The establishment of such family units is seen, in turn, as the basis for the division of labor among the sexes, with the female concentrating on childbearing and the home and the male on hunting, defending his territory, and waging war against his enemies. If the anthropologists are right, this behavioral and hormonal adaptation on the part of the human female has profoundly influenced the shape of human civilization.

SUMMARY

In vertebrates, reproduction is always sexual and always involves two parents, one of which produces sperm and the other eggs. Sperm and eggs are formed by meiosis in the gonads (the testes and the ovaries).

The development of the cleidoic egg enabled animals to become fully terrestrial by providing the necessary water and nutrient medium for the embryo within a closed system protected from external hazards. Reptiles and birds lay cleidoic eggs. In the cleidoic egg, the embryo is surrounded by fluid enclosed in a membrane called the amnion. It is nourished by stored food, the yolk, enclosed in another membrane, the yolk sac, which is connected to the embryonic digestive tract. A third membrane, the allantois, collects wastes and exchanges oxygen and carbon dioxide, which diffuse through the porous shell of the egg. Embryo and membranes are surrounded by a fourth membrane, the chorion. Outside the chorion is a tough, waterproof shell containing calcium. The mammalian egg is the evolutionary descendant of the primitive reptilian egg and resembles the modern cleidoic egg in many ways.

Internal fertilization in the mammals is accomplished through specialized structural adaptations—the penis in the male and the vagina in the female. In most mammals, the embryo develops within the female uterus, deriving oxygen and nourishment from the maternal bloodstream by way of the placenta.

Reproductive activity in the vertebrates is under hormonal control. The sex hormones—principally, testosterone in the male and estrogens and progesterone in the female—are produced in the gonads. Production of these hormones is, in turn, regulated by the gonadotropic hormones of the pituitary (FSH and LH).

The sex hormones also stimulate the physical changes associated with puberty, such as voice pitch, height, and muscle development in human males and the development of the breasts and reproductive tract in human females. Puberty in females marks the beginning of the menstrual cycle, the various phases of which are regulated by the changing balance of gonadotropic and ovarian hormones.

QUESTIONS

1. What are the advantages of asexual reproduction?

2. In all species except the humans, females mate only during fertile periods. This system also has advantages from the evolutionary standpoint. Can you name some?

3. A vasectomy is an operation in which the vas deferens is cut. What would be the primary effect of such an operation? (See Figure 6–2.) Some prominent biologists, including Paul Ehrlich of Stanford and Thomas Eisner of Cornell, have urged that all men have such an operation once they have fathered two children. (They themselves have had the operation.) Are you in favor of this proposal?

4. How would you use the information in this chapter to argue against "Women's Liberation"? To argue for it?

5. Explain why doctors emphasize the importance of not skipping a day when taking the Pill.

PICTURE ESSAY

Human Development

When does life begin? At the hour of birth? At the moment of conception? When the baby first moves in its mother's womb? From the biological point of view, the life of each one of us began millions, even billions, of years before any of these events took place. In this sense, there is no "new" life. Every cell arises from a previously existing one, and all cells —including egg and sperm cells—are in themselves as alive as a baby, although their existence may be a more fleeting one. In the words of Nobel laureate Albert Szent-Györgyi, "There is no difference between cabbages and kings; we are all recent leaves on the old tree of life."

Yet each one of us is a unique, unreproducible "leaf" on this ancient tree. Each is a product of one never-to-be-repeated combination of maternal and paternal genes, and these maternal and paternal genotypes are, themselves, the results of a series of unique events that had their beginnings back beyond the reach of human memory. Nor does this combination alone "explain" or "describe" us. We are all also the products of our past experiences, shaped by the interactions of our genetic constitution and the environment in which we live. Some of these experiences are still part of our daily lives, some we remember vividly, but some of the most important events in our existence as individual human beings took place before we drew our first breath and even before our first heartbeat. This part of our lives is described in this section of the book.

The Sex Cells and Conception

Normal males, from adolescence until old age, produce an average of several hundred million sperm cells a day. These sperm cells, as we have seen in Section 2, are composed largely of a condensed, compact nucleus, containing the genetic information and a long "tail," or flagellum. Millions of sperm cells are released into the vagina when intercourse occurs. Of these,

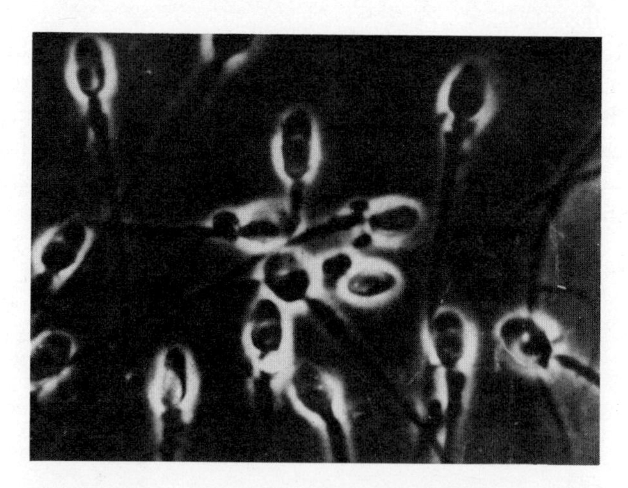

Human sperm cells. About 360 million such cells are present in the testes of one average adult male at any given moment. The glow around the sperm heads is caused by a special lighting technique.

a few hundred thousand, swimming at the rate of about an inch an hour, will make their way up the oviducts, moving in against the beating of the cilia that line the tubes. These sperm cells can survive about 48 hours in the female reproductive tract.

The first sperm cell to reach the ovum is usually the one to fertilize it. The membrane of the sperm cell fuses with the membrane of the egg, and the sperm cell contents are emptied into the cytoplasm of the ovum. The interior of the egg cell begins to vibrate. These movements apparently help the sperm nucleus to travel through the cytoplasm. The two nuclei meet within the cell and merge. This is the most important event in an individual's life. In fact, this event determines whether or not that particular individual—that genetic identity—will ever exist.

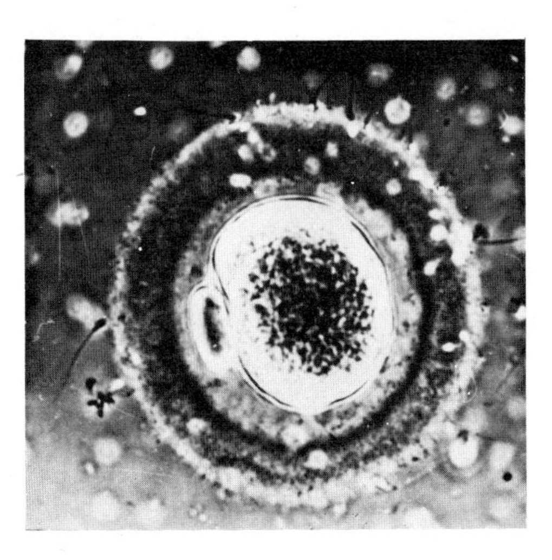

A human egg surrounded by sperm cells. At the left, the first polar body is being extruded. Surrounding the ovum is a layer of mucoprotein, the zona pellucida. The instant a sperm enters the egg, changes take place in the zona pellucida which prevent any additional cells from penetrating.

First Week

After fertilization, the egg continues its passage down the oviduct, propelled by the beating of cilia that line the passageway. Its first cell divisions take place during this journey to the uterus. At 36 hours after fertilization, the fertilized egg divides to form two cells; at 60 hours, the two cells divide to form four cells. At three days, the four cells divide to form eight. The embryo at this early stage is called a blastula. Although the embryo now consists of many cells, it is not significantly larger than the single egg cell from which it originated. As the blastula divides further, it develops an inner space, so that it becomes a hollow ball of cells. One side of the ball of cells, which is thicker than the other side, will develop into the embryo itself, and the remaining cells will develop into the membranes that enclose the embryo.

During this early period of growth, the embryo is nourished by the food reserves that were stored within the egg cell.

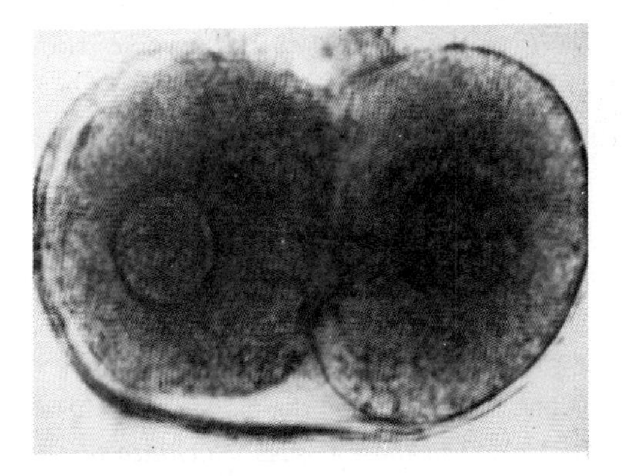

A human egg at the two-cell stage. The nuclei are clearly visible. The cells are surrounded with the remains of the zona pellucida.

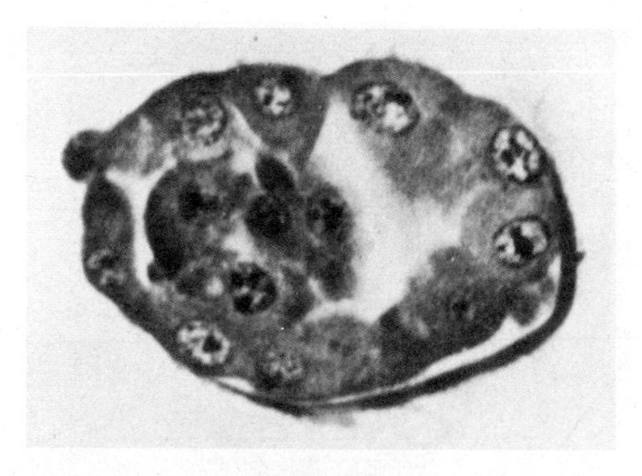

A human blastula, three days after fertilization. At about this time, the blastula completes its journey down the oviduct and enters the uterus.

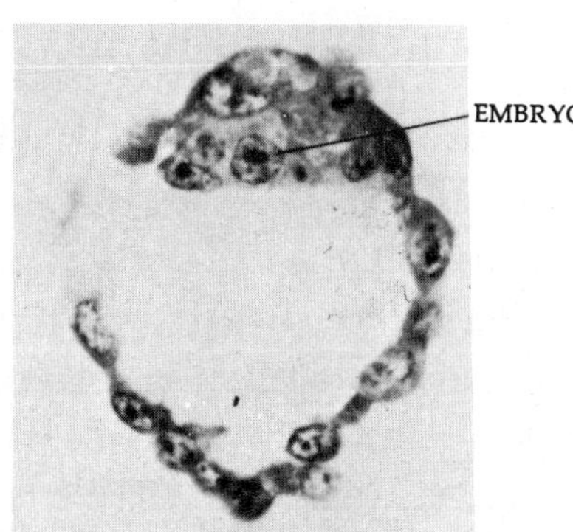

As the tiny embryo continues to develop, a hollow cavity forms within it. The cells at the top, the embroyonic disc, will develop into the embryo itself, and the other cells will form the surrounding membranes.

At about the sixth day, however, the embryo makes contact with the tissues of the uterus. As a result of chemical interactions of the tiny mass of cells and the rich uterine lining, the outer epithelium of the lining breaks down and the embryo becomes embedded in the nourishing tissue. As the embryo descends into the tissues of the uterus, it becomes surrounded by ruptured blood vessels and the nutrient-filled blood escaping from them. By this stage, the embryo is surrounded by membranes analogous to the membranes surrounding the reptilian or chick embryo, which have developed from the extraembryonic ("outside-the-embryo") cells of the blastula (Figure 6–8).

One of these membranes, the chorion, develops fingerlike projections which further invade the mother's uterine tissues. Blood vessels develop in these tissues which connect with the blood vessels of the embryo as they develop. Ultimately, the placenta forms.

Second and Third Weeks

During the second week of its life, the embryo grows to 1.5 millimeters in length and its major body axis begins to develop. During the third week, it grows to 2.3 millimeters long and most of its major organ systems begin to form: the beginnings of the central nervous system (spinal cord and brain), which is the first organ system to develop; the heart and blood vessels; the primitive gut; and the muscle rudiments. This third week of life is probably the most critical week in the entire physical development of a human being. Even a mild virus infection (rubella, or german measles, is a principal and tragic example), the taking of a drug, such as thalidomide, or exposure to low levels of x-ray radiations during this period can produce damage which, if the embryo survives, can leave it with permanent mental and physical abnormalities. Yet at this stage, most women are not yet sure they are pregnant and pregnancy tests may not be decisive.

Fourth Week

By 21 days, the eyes begin to form. Also, by this time, about 100 cells have been set aside (in the yolk sac) as germ cells; from these the ova or sperm cells of the individual will eventually develop. By 24 days, the very rudimentary heart, still only a tube, begins to flutter and then to pulsate; from this time on, it will not stop its 100,000 or more beats per day until the death of the individual.

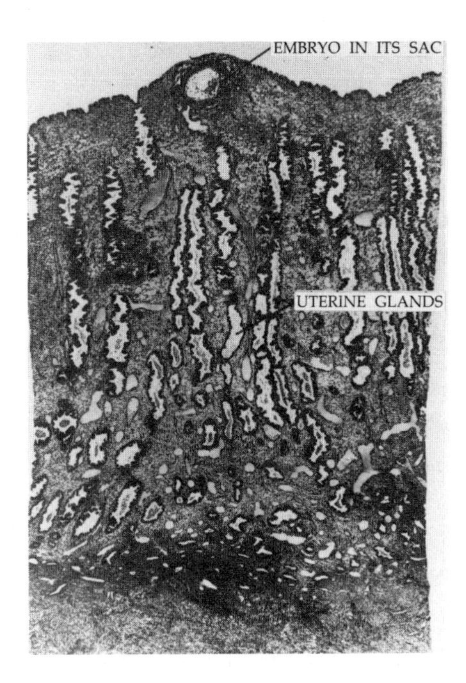

A human embryo invading the maternal tissues. Note the many glands and blood vessels in the uterine lining. The glands secrete a glycogen-rich material, which nourishes the tiny zygote.

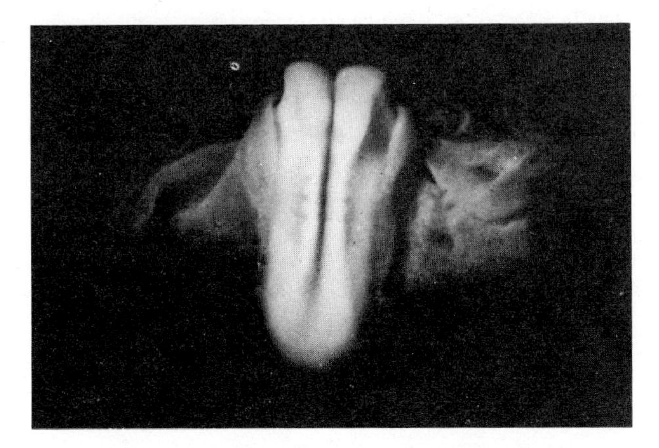

As the embryo develops, two ridges form, the neural folds. The groove between these is called the neural groove. Within the neural groove, the notochord and the nerve cord—the two "trademarks" of the vertebrates—develop. These two ridges soon close. This embryo is 18 days old.

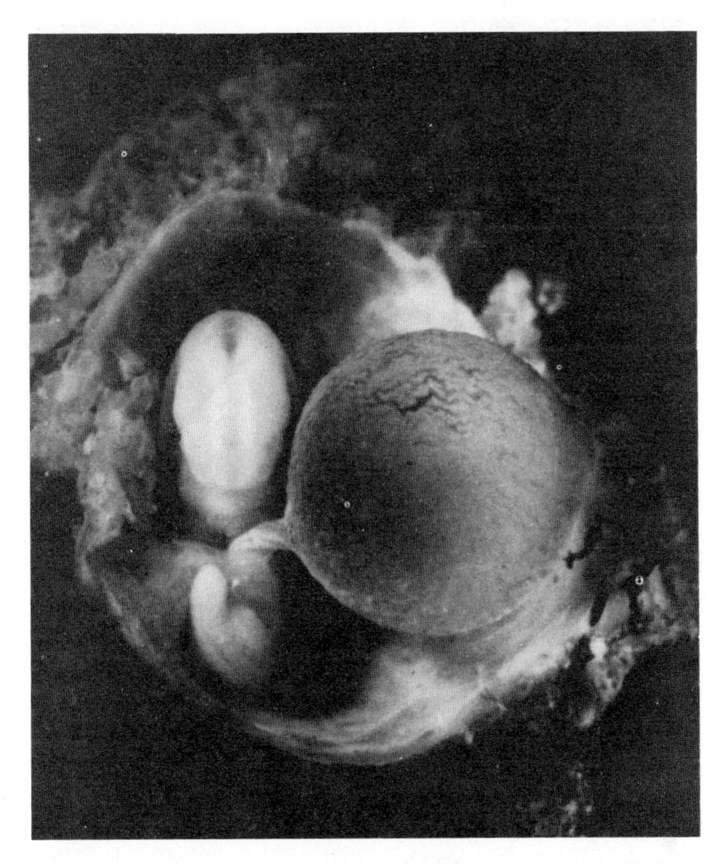

A human embryo at 28 days. The balloonlike structure is called the yolk sac, since it is analogous to the yolk sac in reptilian and avian eggs, although it is yolkless in humans and other placental animals. The embryo is curved toward you. At the top you can see the bulge of the rudimentary brain. The "seam" where the neural ridges closed is still visible. At the posterior of the embryo the somites are visible.

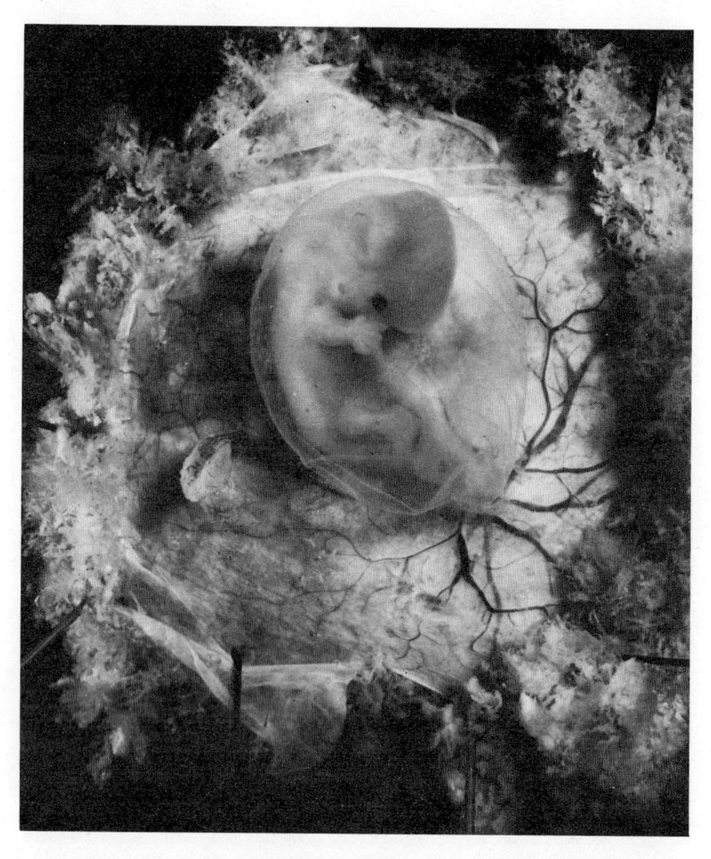

By the end of the fifth week, the amniotic membrane has enveloped the tissues connecting the embryo to the placenta and the yolk sac, forming the umbilical cord, which is the connecting link between the fetus and the placenta.

By the end of the first month, the embryo is 5.2 millimeters (about 1/8 inch) in length and has increased its mass 7,000 times. The neural groove has closed, and the embryo is now C-shaped. At this stage, it can be clearly seen that the tissues lateral to the notochord are arranged in paired segments, or somites. Each baby has 40 pairs of somites. Muscles, bones, and connective tissues will develop from these somites. This segmentation of the muscles persists in the adult forms of lower vertebrates—fish, particularly—but not in the higher, terrestrial vertebrates. The heart, even as it beats, develops from a simple contracting tube to a four-chambered vessel.

Second Month

During the second month, the embryo increases in mass about 500 times. By the end of this period, it weighs about 1/30 ounce, slightly less than the weight of an aspirin tablet, and is about 1 inch long. Despite its small size, it is almost human-looking, and from this time on it is generally referred to as a fetus. Its head is still relatively large, because of the early and rapid development of the brain, but the head size will continue to be reduced in proportion to body size throughout gestation (and through childhood as well). Arms, legs, elbows, knees, fingers, and toes are all forming during this time, and as another reminder of our ancestry, there is a temporary tail. The tail reaches its greatest length in the second month and then gradually begins to disappear; it is entirely gone in 94 percent of all babies by the time they are born. The primitive reproductive organs have begun to form by this time. The liver now constitutes about 10 percent of the body of the fetus and is its main blood-forming organ.

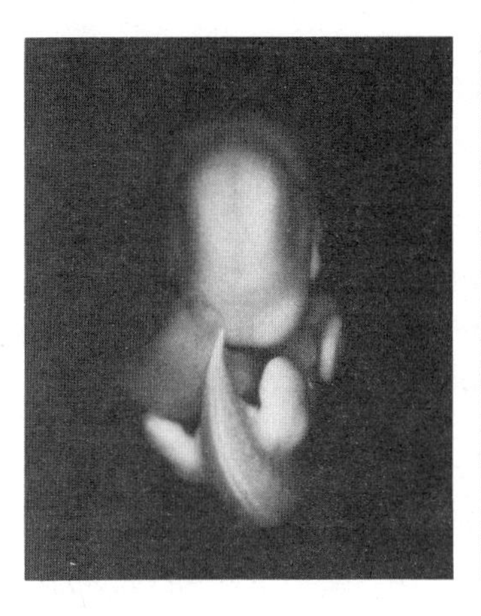
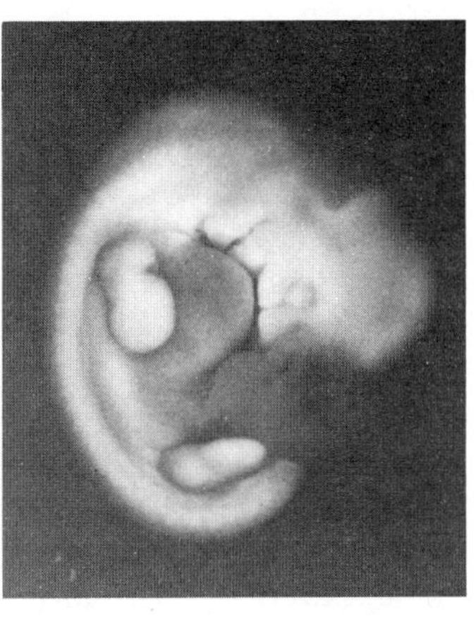

A human fetus at 39 days. The budding limbs, the tail, and the bulge over the heart (now beating) can be clearly seen. The eyes have lenses but are covered by lids, which remain closed for three months. Just below the head structures are a series of branchial ("gill") pouches, another characteristic shared by all vertebrates. In fish and amphibians, perforations develop around which the gills form. In mammals, the pouches develop into the thyroid glands, tonsils, and other tissues of the head and neck.

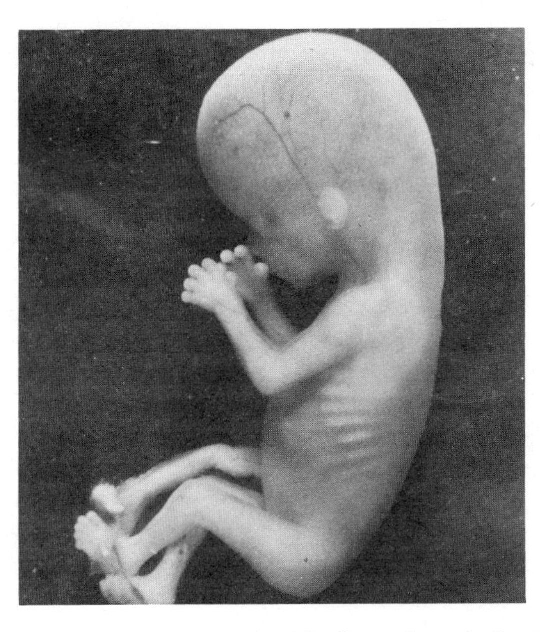

Two-month old fetus. Note the fine blood vessels in the head, the well-formed fingers and toes and facial structures, and the rib cage.

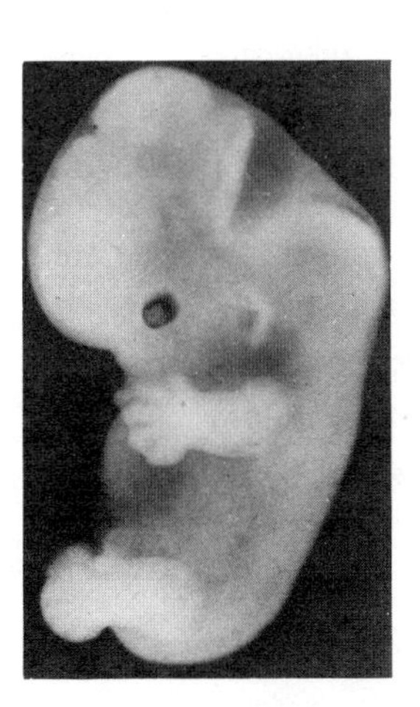
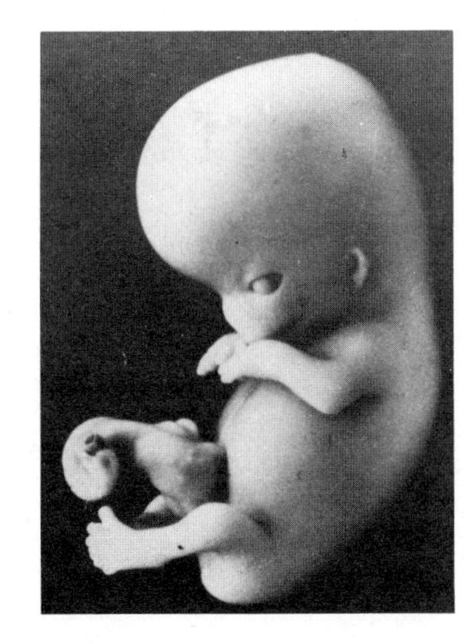

In this 42-day fetus (above left), *the outline of the brain and the developing fingers and toes are clearly visible. The fetus is not quite ½ inch long. By 55 days* (above right), *the fetus is about an inch long in sitting height. Notice how large the head is in relation to the body.*

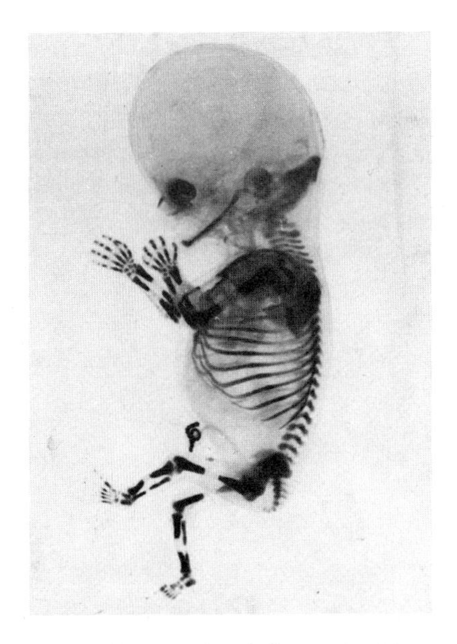

Two-month old fetus showing the skeleton. At this age, most of the bony structures are soft and cartilaginous. The process of bone formation is not complete until after puberty.

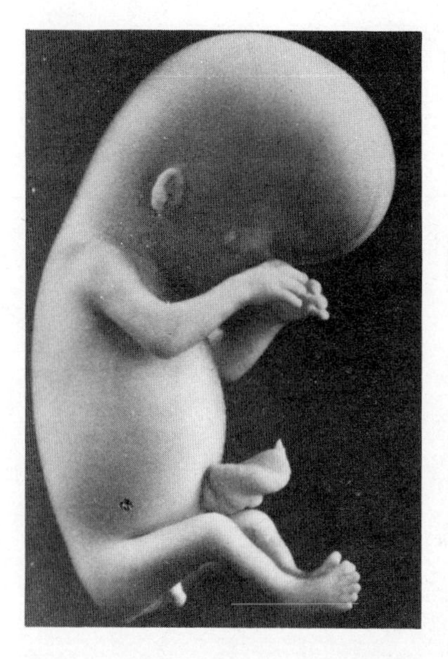

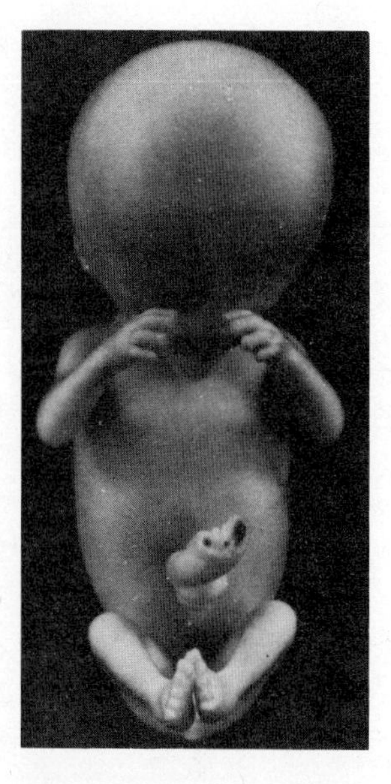

Human fetus at 68 days. It is now almost 2 inches long.

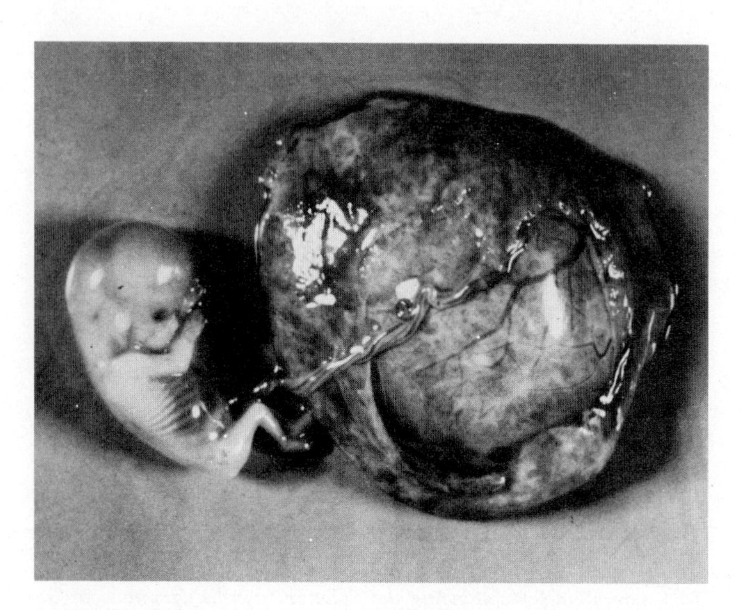

Ten-week old fetus attached to the placenta. You can see the prominent blood vessels in the umbilical cord.

Third Month

During the third month, the fetus begins to move its arms and kick its legs and the mother may become aware of its movements. Reflexes, such as the startle reflex and (by the end of the third month) sucking, first appear at this time. Its face becomes expressive: squinting, frowning, or looking surprised. Its respiratory organs are fairly well formed by this time but, of course, not yet functional. The external sexual organs begin to develop. By the end of this month, the fetus is about 3 inches long from the top of its head to its buttocks and weighs about ½ ounce. It can suck and swallow, and sometimes does swallow some of the fluid which surrounds it in the amniotic sac. The finger, palm, and toe prints are now so well developed that they can be clearly distinguished by ordinary fingerprinting methods. The kidneys and other structures of the excretory system develop rapidly during this period, although waste products are still disposed of through the placenta. By the end of this period—the first trimester of development—all the major organ systems are laid down.

Fourth Month

During the fourth month, the baby's movements become obvious. Its bony skeleton is forming and can be visualized by x-rays. Its body is becoming covered with a protective, cheesy coating. The fetus at this stage is about 5 inches long and weighs about 5 ounces.

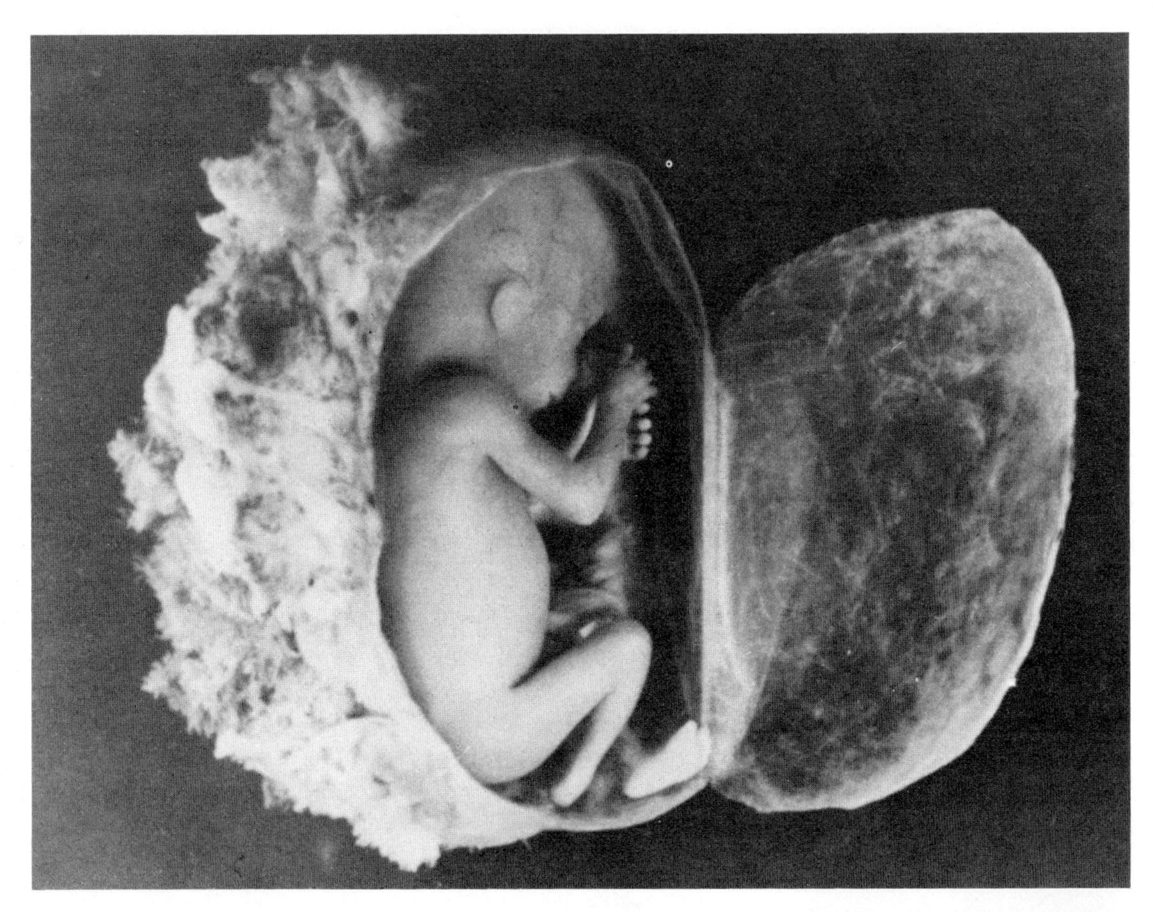

Human fetus at 16 weeks. It is now about 5¼ inches long and fills the uterus, which expands as it grows. You can see the fetal membranes surrounding it.

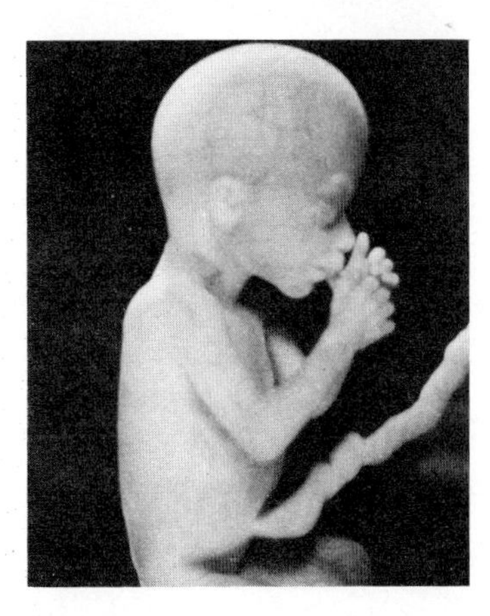

Fetus at 17 weeks, sucking its thumb.

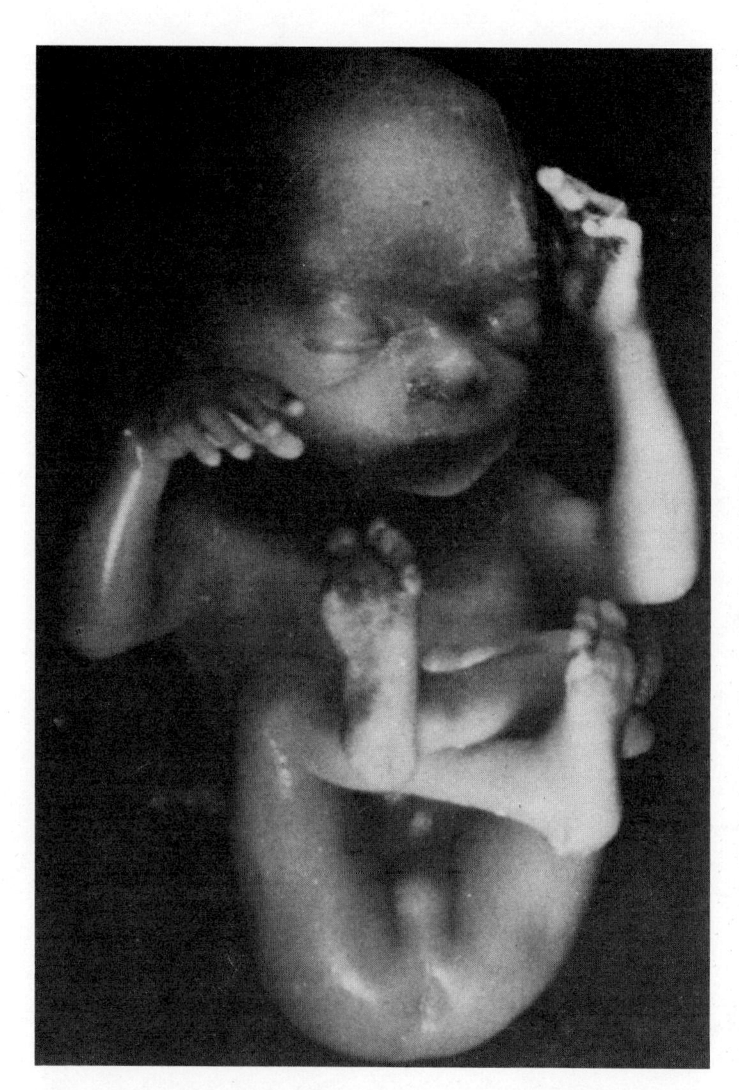

Male fetus at five months. It is now 7¾ inches long and weighs about 10 ounces.

Fifth Month

By the end of the fifth month, the fetus has grown another 2 inches and now weighs half a pound. It has acquired hair on its head, and its body is covered with a fuzzy, soft hair called the lanugo, from the Latin word for "down." Its heart, which beats between 120 and 160 times per minute, can be heard by a stethoscope. The five-month-old fetus is already discarding some of its cells and replacing them with new ones—a process which will continue throughout its lifetime. Nevertheless, a five-month-old fetus cannot yet survive outside of the uterus. The youngest fetus on record which survived was about 23 weeks old and required continuous assistance in vital functions such as breathing, taking food, and maintaining its body temperature.

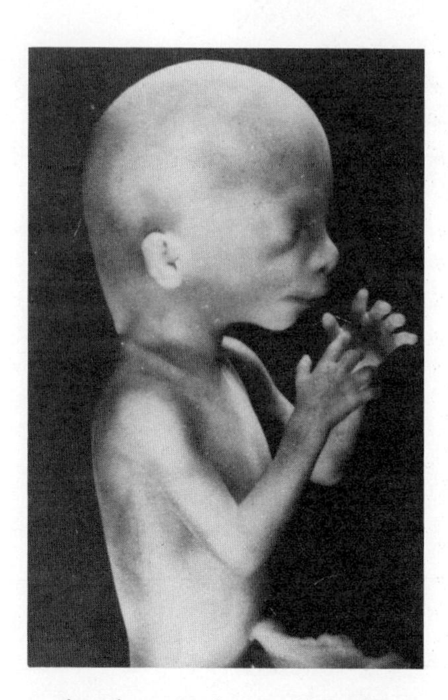

Fetus at five months (almost ready for independence).

Sixth Month

Ninety percent of the fetal weight gain occurs after the fifth month. During the sixth month the fetus has a sitting height of 12 to 14 inches and weighs about 1½ pounds. By the end of the sixth month, it could survive, although probably only in an incubator. Its skin is red and wrinkled, and although teeth are only rarely visible at birth, they are already forming dentine. The cheesy body covering, which helps the fetus to maintain an adequate body temperature as well as protecting it against abrasions, is now abundant. Reflexes are more vigorous. In the intestines is a pasty green mass of dead cells and bile, known as meconium, which will remain there until birth.

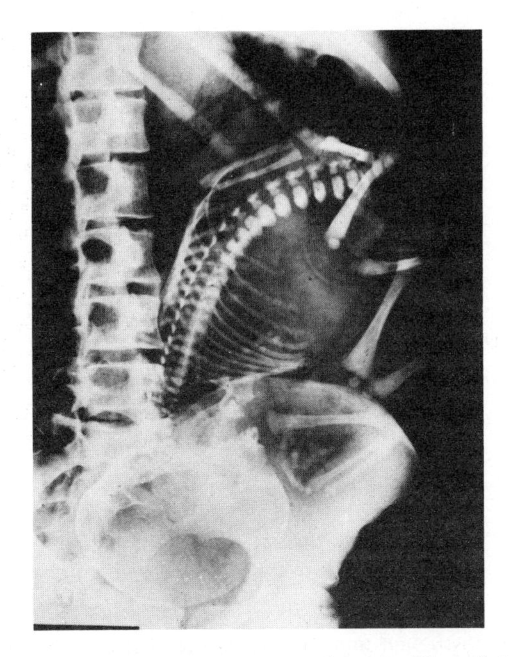

X-ray of a human fetus in position for delivery. The pelvic bones of the mother, clearly visible here, can separate slightly to permit the baby's head and body to pass through.

Final Trimester

During the final trimester, the baby increases greatly in size and weight. In fact, it doubles in size just during the last two months. During this period, many new nerve tracts are forming and new brain cells are being produced at a very rapid rate. Recent research has shown that the protein intake of the mother is extremely important during this period if the child is to have full development of its nervous system and, hence, its intelligence potential.

During the last month in the uterus, the baby usually acquires antibodies from its mother. These are globular proteins —formed against bacteria, viruses, or other foreign invaders— which defend the body against a second attack by such microorganisms. The baby becomes immune to whatever the mother is immune to. This immunity is only temporary. Within one to two months after birth, the maternal antibodies will be gradually replaced by antibodies manufactured by the baby's own immune system.

During the last month of pregnancy, the growth rate of the baby begins to slow down. (If it continued at the same rate, the child would weigh 200 pounds by its first birthday.) The placenta begins to reduce its function during this period.

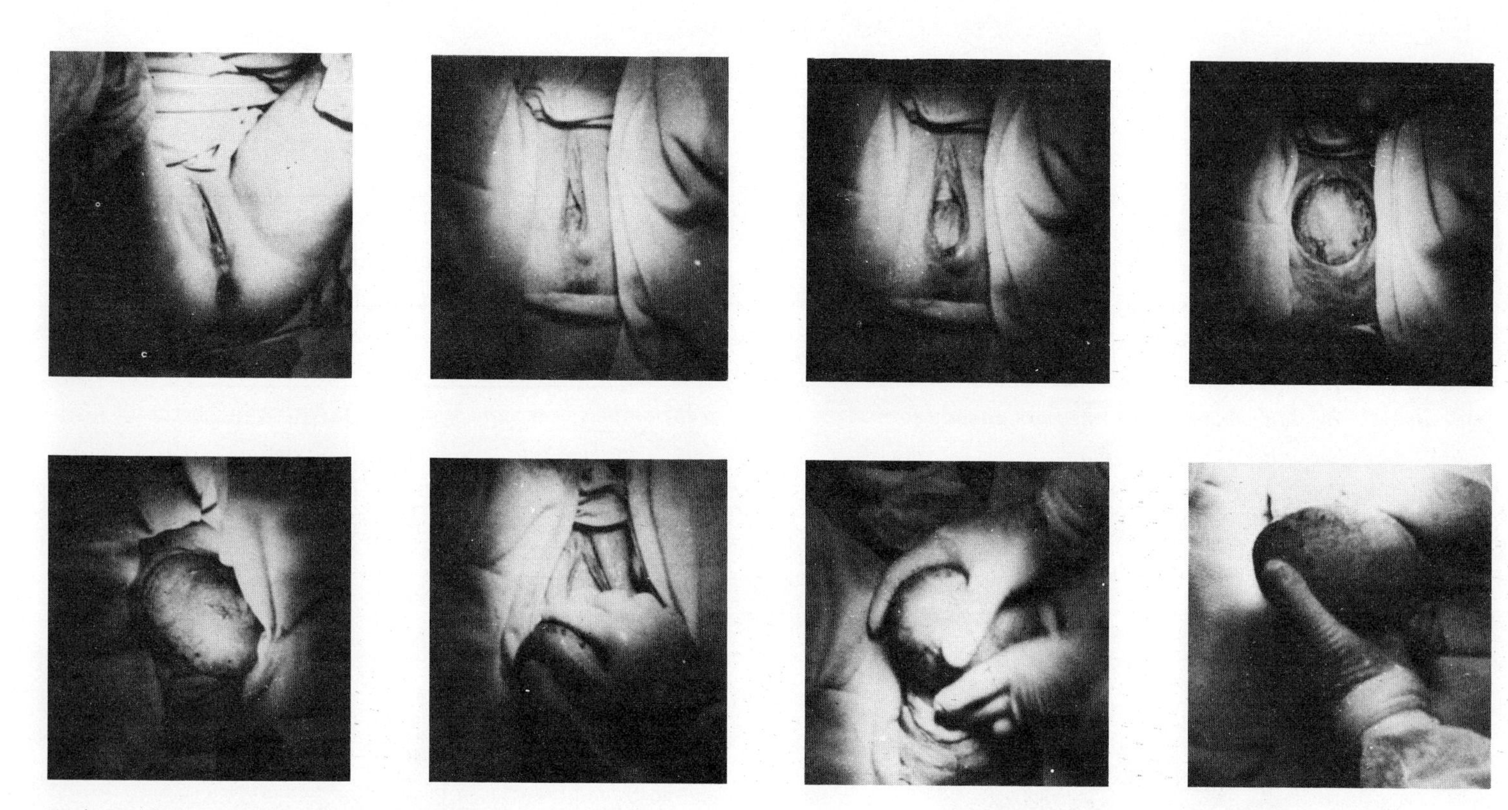

Birth of a baby. Once the head and shoulders emerge, the contracting uterus expels the body quickly.

Birth

Birth takes place about 266 days after conception or 280 days after the beginning of the mother's last menstrual period. Approximately 75 percent of all babies arrive within two weeks of the scheduled date. Labor is divided into three stages: dilatation, expulsion, and the placental stage. Dilatation, which lasts from 2 to 16 hours (it is longer with the first baby), begins with the onset of contractions of the uterus and ends with the full dilation, or opening, of the cervix. At the beginning of the stage, uterine contractions occur at intervals of about 15 to 20 minutes and are relatively mild. By the end of the dilatation stage, contractions are stronger and occur about every one to two minutes. At this point, the cervix is dilated to about 10 centimeters. Rupture of the fetal membranes, with the expulsion of fluids, usually occurs during this stage.

The second, or expulsion, stage lasts 2 to 60 minutes. It begins with the full dilation of cervix and the appearance of the head, called crowning. Contractions at this stage last from 50 to 90 seconds and are one or two minutes apart.

The third, or placental, stage begins immediately after the baby is born. It involves some contractions of the uterus and the expelling from the vagina of fluid, blood, and finally the placenta, with the cut umbilical cord attached. Minor uterine contractions continue, which help to stop the flow of blood and to return the uterus to its prepregnancy size and condition.

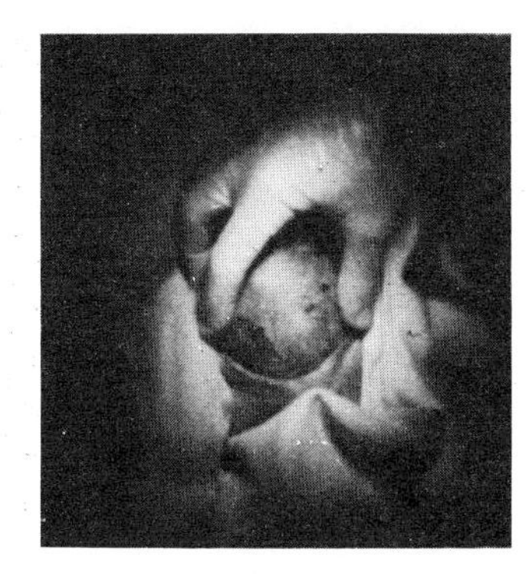

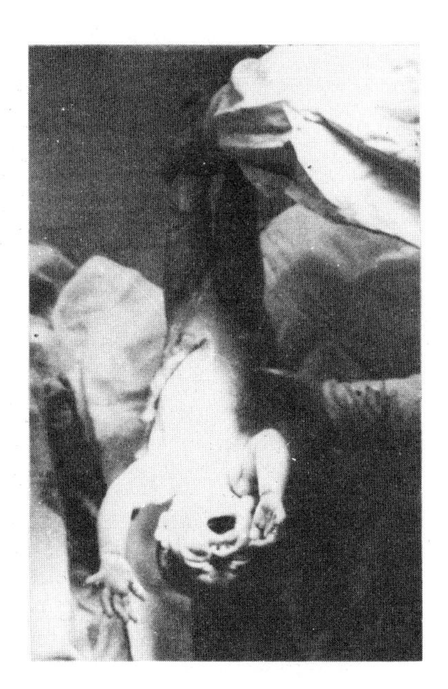

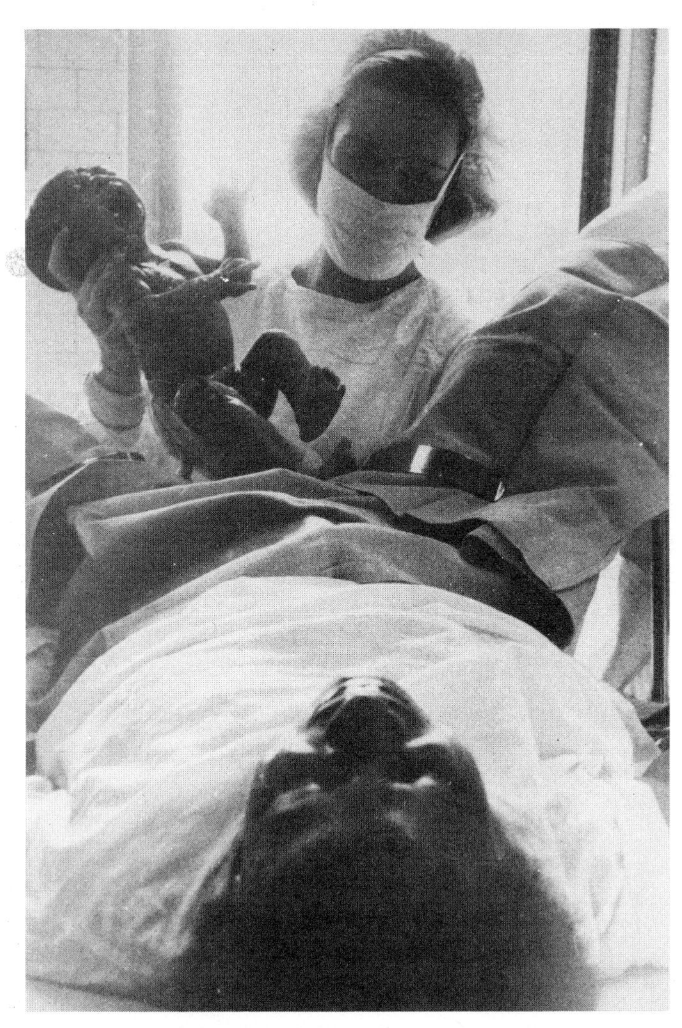

The mother is smiling, but the baby has more serious business to attend to.

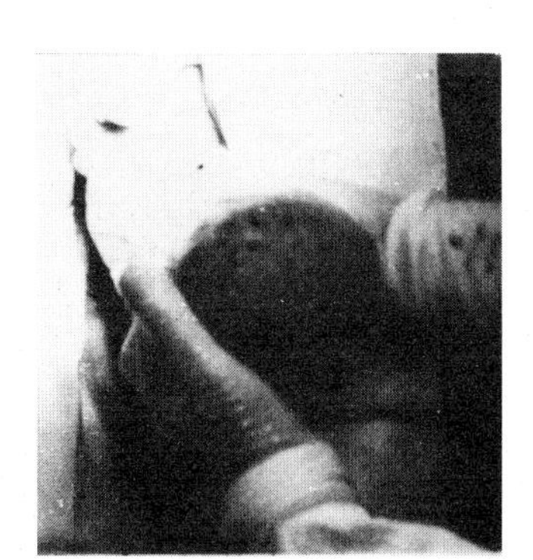

Some people claim to be able to remember the moment of their own birth, which must be a terrifying moment indeed. The baby emerges from the warm, protective enclosure where he has been nourished and permitted to grow for nine months. The umbilical cord—until that moment, his lifeline—is severed. Carbon dioxide, no longer removed by the mother, accumulates in his bloodstream and stimulates the respiratory center of the brain to cause him to take his first breath. The cold air entering his lungs for the first time is a shock. The baby cries (or if he does not, the obstetrician slaps him gently, causing him to cry and hence to breathe) and then starts to breathe regularly. A new human being has entered the world.

CHAPTER 6-2

Integration and Coordination

Do you recall what it feels like to be in an absolute rage? The physical characteristics of a rage result from a simultaneous discharge of many nerve fibers. Certain of these cause the blood vessels in the skin and intestinal tract to contract; this contraction increases the return of blood to the heart, raising the blood pressure and sending more blood to the muscles. The heart beats both faster and stronger, and the respiratory rate increases. The pupils dilate. The muscles underlying the hair follicles in the skin contract; this is probably a legacy from our furry forebears, which looked larger and more ferocious with their hair standing on end. The rhythmic movement of the intestines stops, and sphincters, the muscles that close the intestines and the bladder, contract; these reactions inhibit digestive operations, but the closing of the sphincters may also have the decidedly unuseful consequence of causing involuntary defecation or urination. Chemicals are produced which cause the release of large quantities of sugar from the liver into the bloodstream; this sugar provides an extra energy source for the muscles. The adrenal glands pour out adrenaline. The body is prepared for fight or flight. These changes have come about as a result of an interplay of nerves and hormones.

The nervous and endocrine systems operate continuously —not only on such special occasions as the one just described—to integrate and control the body's activities. In the preceding chapter, we saw several examples of ways in which body functions are regulated by interactions of hormones, and we glimpsed how hormones and the nervous system may interact—in the regulation of estrogen concentration, for example, and in the synchrony of copulation and ovulation. In this chapter, we are going to examine these vast communication networks in order to provide a background for understanding the control of such functions as temperature regulation, circulation, digestion, and excretion, which are the subjects of the chapters that follow. The discussion of the brain and of some of the current research on brain function is reserved for the final chapter.

HORMONES

The endocrine system regulates and coordinates various functions of the body by means of hormones, which are released into the bloodstream by the endocrine glands. Figure 6–14 shows the position of some of the important hormone-producing glands in the human body, and Table 6–2 summarizes the principal actions of the hormones that these glands produce.

Hormones are defined as chemicals produced in specific cells or tissues which act on other cells or tissues that are distant from the production site. In the case of the higher animals, hormones are released into the bloodstream, through which they travel to other tissues of the body. Whether or not a particular cell or group of cells responds to a particular hormone depends on the cells involved and also on their physiological state; cells may be receptive to a hormone at one time and not at another.

Table 6–2

Principal Endocrine Glands of Vertebrates and the Hormones They Produce

Gland	*Hormone*	*Principal Action*	*Mechanism Controlling Secretion*
Pituitary, anterior lobe	Thyrotropic hormone (TSH)	Stimulates thyroid	Thyroid hormone in blood; hypothalamus
	Follicle-stimulating hormone (FSH)	Stimulates follicle of ovary	Estrogen in blood; hypothalamus
	Luteinizing hormone (LH)	Stimulates testes in male, corpus luteum in female	Testosterone or progesterone in blood; hypothalamus
	Adrenocorticotropic hormone (ACTH)	Stimulates adrenal glands	Adrenal cortical hormone in blood; hypothalamus
	Growth hormone	Stimulates bone and muscle growth	
	Prolactin	Stimulates milk secretion, parental behavior (such as nest building in birds)	Hypothalamus
Thyroid	Thyroid hormone	Controls metabolism, some aspects of development	TSH
Parathyroid	Parathyroid hormone	Controls calcium metabolism	Concentration of calcium in blood (not fully known)
Ovary	Estrogens	Develops and maintains sex characteristics in females	FSH
	Progesterone	Promotes growth of uterine tissue	LH
Testes	Testosterone	Develops and maintains sex characteristics in males	FSH
Adrenal cortex	Cortisone, cortisonelike hormones, aldosterone	Controls carbohydrate metabolism, salt and water balance	ACTH
Adrenal medulla	Adrenaline, noradrenaline	Increases blood sugar, dilates blood vessels, increases heartbeat	Nervous system

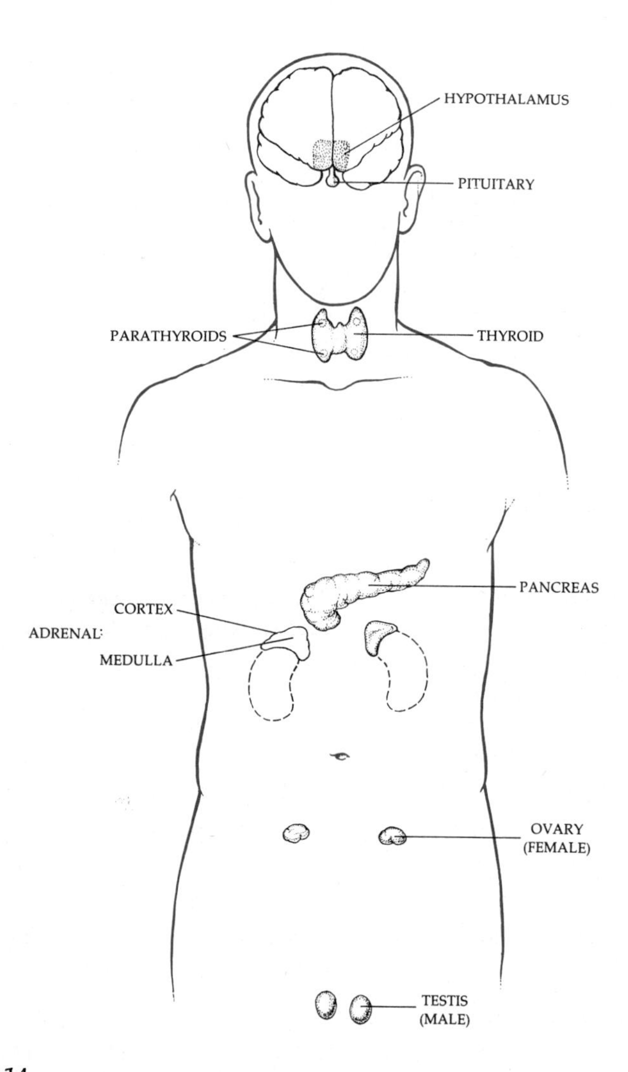

6–14

Some of the hormone-producing organs. The pituitary releases hormones which, in turn, regulate the hormone secretions of the sex glands, the thyroid, and the adrenal cortex (the outer area of the adrenal gland). The pituitary is itself under the regulatory control of an area in the brain known as the hypothalamus, indicated in color.

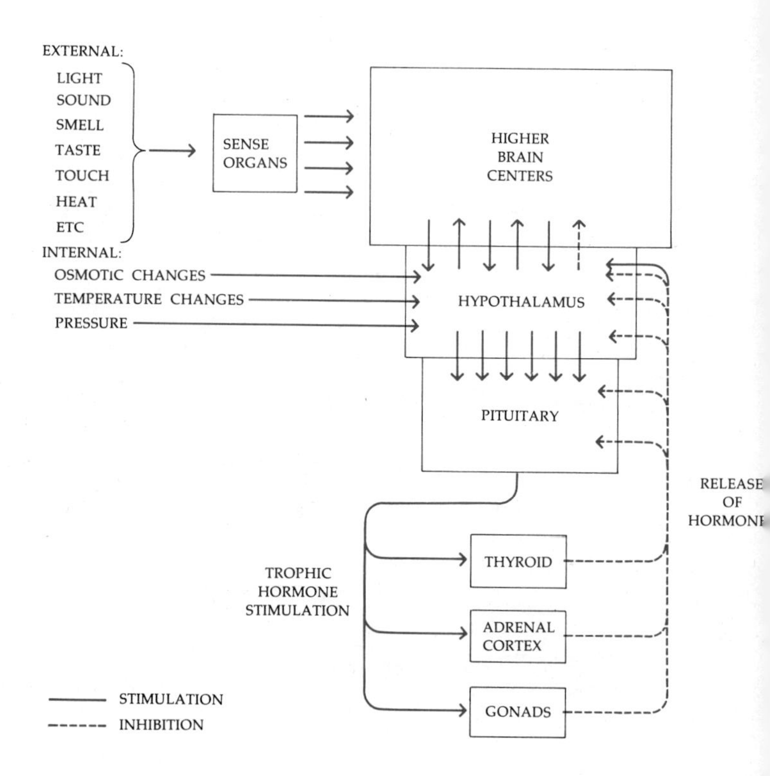

6–15

Hormone production is regulated by a complex negative feedback system. The hypothalamus stimulates the pituitary to secrete hormones, and these, in turn, stimulate the secretion of hormones from the thyroid, adrenal cortex, and gonads (the testes or the ovaries). When the hormones produced by these target glands reach a certain concentration in the blood, the hypothalamus stops stimulating the pituitary, the pituitary stops producing tropic hormones, and production of hormones by the gland also stops. By way of the hypothalamus, which exchanges information with the higher brain centers, hormone production can also be regulated in accord with changes in the external and internal environment.

(a) INSULIN

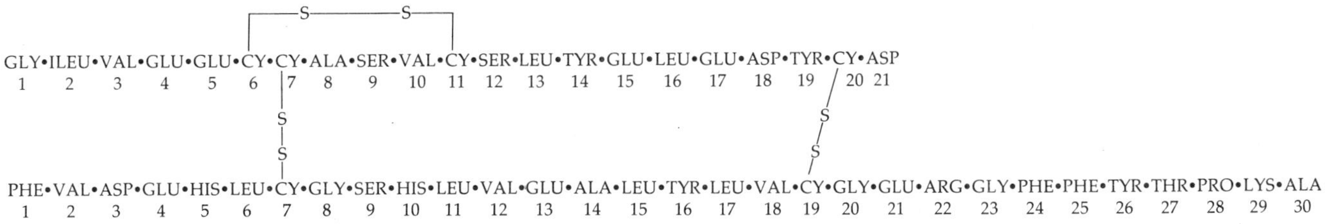

GLY•ILEU•VAL•GLU•GLU•CY•CY•ALA•SER•VAL•CY•SER•LEU•TYR•GLU•LEU•GLU•ASP•TYR•CY•ASP
1 2 3 4 5 6 7 8 9 10 11 12 13 14 15 16 17 18 19 20 21

PHE•VAL•ASP•GLU•HIS•LEU•CY•GLY•SER•HIS•LEU•VAL•GLU•ALA•LEU•TYR•LEU•VAL•CY•GLY•GLU•ARG•GLY•PHE•PHE•TYR•THR•PRO•LYS•ALA
1 2 3 4 5 6 7 8 9 10 11 12 13 14 15 16 17 18 19 20 21 22 23 24 25 26 27 28 29 30

(b) THYROID HORMONE

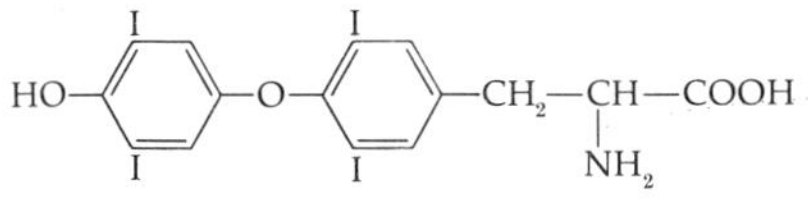

HO—(I,I)—O—(I,I)—CH_2—CH(—NH_2)—COOH

6–16

(a) *The hormone insulin is a protein consisting of only* 51 *amino acids. It was the first protein whose exact primary structure was deciphered. It stimulates the conversion of glucose to glycogen. When insulin is not present in sufficient quantities, as in diabetes, glucose increases in the blood and may be excreted in the urine. The exact way in which insulin works is not clear, but recent studies indicate that it may facilitate the transport of glucose across cell membranes into the cells of liver and muscle, in which it is converted to glycogen and stored. Glucagon, also a small protein, stimulates the breakdown of glycogen to glucose.*
(b) *Thyroid hormone is relatively simple in chemical structure. Each molecule contains four atoms of iodine. Almost all the iodine in the human body is concentrated in the thyroid gland. Thyroid hormone stimulates metabolic activities. Infants with thyroid deficiencies have a greatly reduced growth, particularly in their brains and nervous tissues, leading to a form of mental deficiency known as cretinism.*

Although the general effects on the organism of many of the important hormones are now fairly well known, the way in which hormones act at the cellular level still remains to be clarified. It appears that certain cell types have specific receptor sites for particular hormones; the interaction of hormone and receptor is probably analogous to that of enzyme and substrate (page 80). Some hormones appear to affect the entry of substances into cells. For example, in diabetes, which is caused by a deficiency of the hormone insulin, there is a high concentration of sugar in the blood, so high that sugar appears in the urine. At the same time, the individual cells of the body are starving for lack of glucose. When insulin is administered, sugar enters the cells and leaves the bloodstream. Other hormones, some investigators hypothesize, may act by affecting enzyme systems, either by stimulating synthesis of the enzymes or by promoting their activity.

Recently, cellular physiologists have discovered that a chemical known as cyclic AMP is involved in many hormone actions. The hormone itself, "the first messenger," apparently produces its effects by stimulating the action of an enzyme, bound to the target cell membrane which is responsible for AMP synthesis. The newly synthesized AMP then acts as a "second messenger," activating processes or enzymes within the cell. Almost simultaneously, biologists studying a peculiar group of organisms known as the cellular slime molds isolated a chemical of great importance in this biological system. The cells of the cellular slime mold begin as individual amoebas and then come together to form a single organism (Figure 6–17*b*). The chemical that called them together was named acrasin, after Acrasia, the mythological siren who lured seamen in Homeric legend. Acrasin has now been identified as cyclic AMP, another example of an evolutionary thread reaching far into the past to link organisms now highly diverse.

(a)

CYCLIC AMP

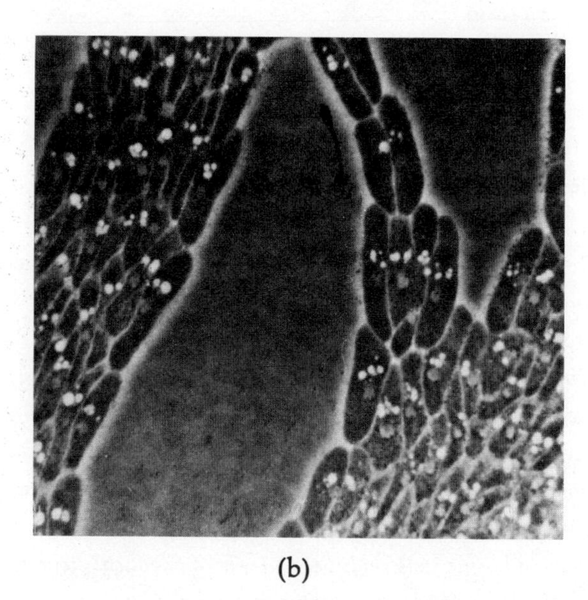

(b)

6–17

(a) *Cyclic AMP (adenosine monophosphate) acts as a "second messenger" within the cells of vertebrates. Following stimulation by various hormones—the "first messengers"—cyclic AMP is formed from ATP. "Cyclic" refers to the fact that the atoms of the phosphate group form a ring.* (b) *Cyclic AMP is also the chemical that attracts the amoebas of the cellular slime molds, causing them to aggregate into a sluglike body, which then behaves like a multicellular organism. This double use of AMP is an example of the conservative nature of evolution, which tends to find new uses for existing materials.*

NEURONS

Functionally, the nervous system differs from the endocrine system chiefly in its capacity for rapid response—a nerve impulse can travel through the body in a matter of milliseconds. Plants, which are very slow-living organisms, have no nervous systems, relying largely on an elaborate interplay of hormones to coordinate their activities. Mammals, birds, octopuses, and many insects have highly developed and highly tuned nervous systems, which monitor internal and external changes and stimulate responses to them.

Neuron Structure

The unit of the nervous system is the neuron, or nerve cell. A neuron has three functional parts: a reception area, consisting of the *dendrites;* the cell body, which contains the nucleus; and the *axon,* or nerve fiber, which transmits signals received by the dendrites to other cells. A single neuron may have many dendrites, but it usually has only one axon, although the axon may be branched.

Vertebrate axons are often enveloped in a myelin sheath formed by Schwann cells. The sheath speeds up the transmission of impulses along the axon. In a large animal, an axon may be many feet in length.

How a Nerve Impulse Is Transmitted

Axons are the communication lines of the nervous system. In order to understand how the nerve impulse travels along an axon, we need to look more closely at the axon itself, particularly at the remarkable properties of its cell membrane. By processes of active transport, the membrane pumps out sodium ions (Na^+) and pumps in potassium ions (K^+). As a consequence, the concentration of K^+ ions is about 30 times higher inside the axon than in the fluid outside, and the concentration of Na^+ ions is about 10 times higher outside of the axons than inside. In its resting state, the membrane is impermeable to Na^+ ions and permeable to K^+; as a consequence, the Na^+ ions are kept on the outside of the axon and the K^+ ions tend to leak out, moving down the steep concentration gradient.

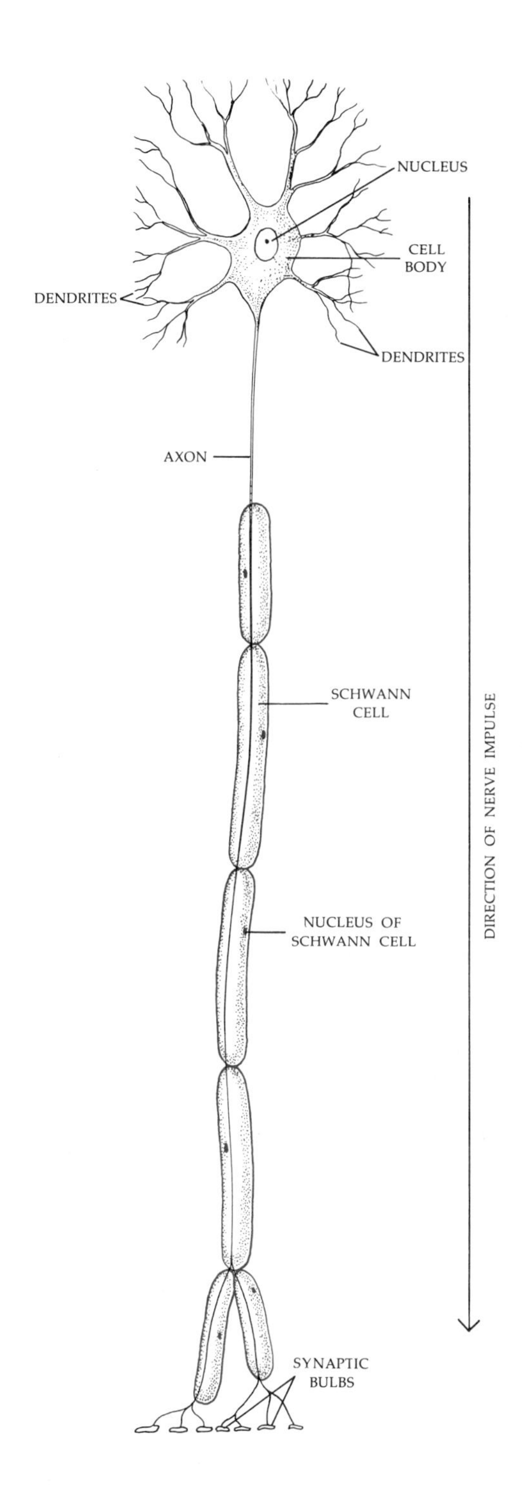

6–18 (at left)

A motor neuron. The stimulus is received by the dendrites, which conduct it to the cell body and to the axon. The axon is a sheathed filament which acts as a self-powered transmission cable. Its power is in the form of a minute voltage difference, derived from a chemical exchange between the axon and its surroundings. The nodes of Ranvier are gaps in the axon sheath which occur at the junctions of adjacent Schwann cells. The nerve endings shown here are tiny unsheathed filaments terminating in bulbous "feet." These "feet" form synapses with other cells. (See Figure 6–21.)

6–19

Cross sections of axons wrapped in myelin sheaths. The sheaths, which are composed of the membranes of Schwann cells, appear here as dark borders enclosing the axons.

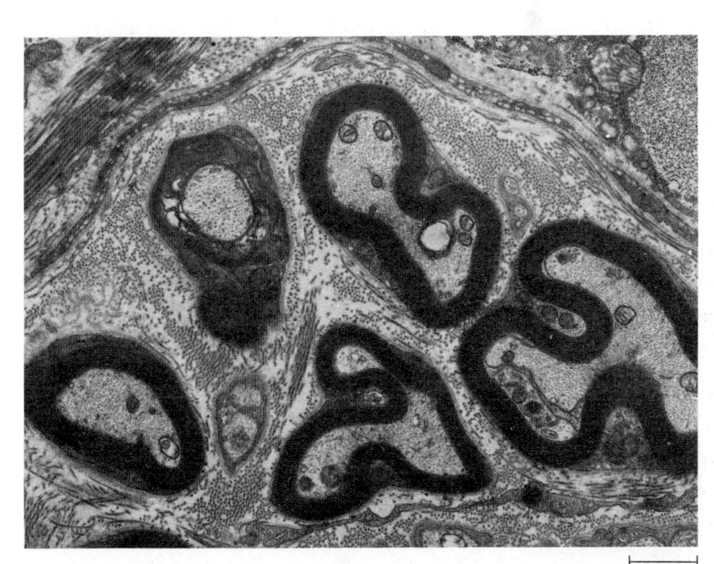

When it is in this resting state, the membrane is electrically polarized; that is, the inside of the axon is negative in relation to the outside:

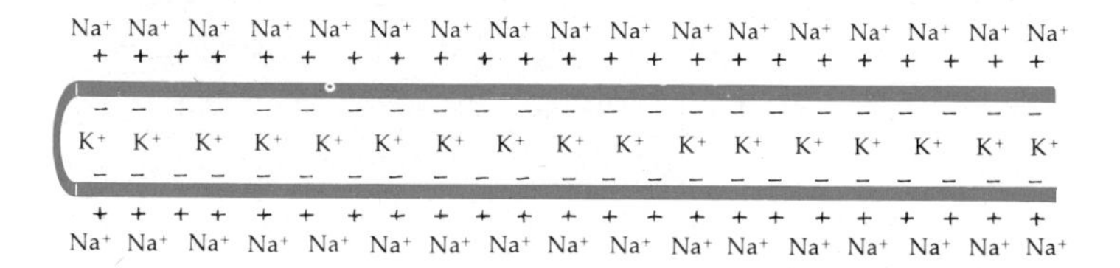

As a consequence of this polarization, an electric potential exists. This electric potential is analogous to the electric potential in the charged battery of a flashlight or the water potential of water behind a dam at the top of a hill. It can be looked upon as a supply of electrons that can be released. When electrons are released, they move in a current; an electric current, as you probably know, is simply the movement of electrons. The force with which the electrons move, analogous to the force of water running downhill, is measured in volts. The electric potential of the membrane at rest is about 70 millivolts (a millivolt is 1/1000 of a volt).

A stimulus affects the axon by depolarizing a small section of its membrane. (In the animal, the stimulus always acts first at the base of the axon—the area nearest the cell body, because, as we saw in Figure 6–18, the nerve impulse moves only in one direction. In the laboratory, any portion of the axon will respond to a stimulus in this way.) The electric potential drops to zero. The depolarized membrane suddenly becomes permeable to Na^+ and impermeable to K^+. The Na^+ rushes in down the concentration gradient. This influx of positively charged ions momentarily reverses the polarity of the membrane so that it becomes positive on the inside and negative on the outside. This change in permeability lasts for only about 1.5 milliseconds. Within a fraction of a second the electric potential between the interior and exterior of the axon switches from a negative potential of about 70 millivolts to a positive potential of about 50 millivolts. The total action potential, as it is called, is therefore about 120 millivolts (see Figure 6–20). Then the membrane regains its previous impermeability to Na^+ and permeability to K^+. The K^+ ions move out and the electric potential is once more restored:

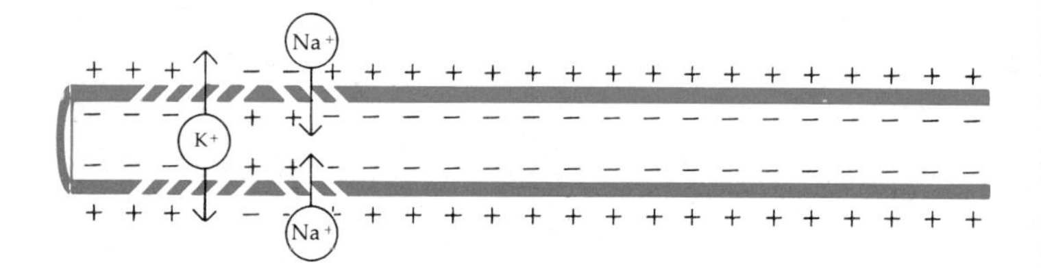

The excited section of the membrane affects the adjacent segment. The positive charges inside the membrane move toward the negative charges inside the membrane, on the adjacent segment, and the negative charges on the outside are attracted toward the positive charges on the outside. As a consequence, the next segment of the membrane is depolarized, its permeability to Na^+ changes, and the entire process is repeated. In this way, the impulse constantly renews itself as it travels along the axon:

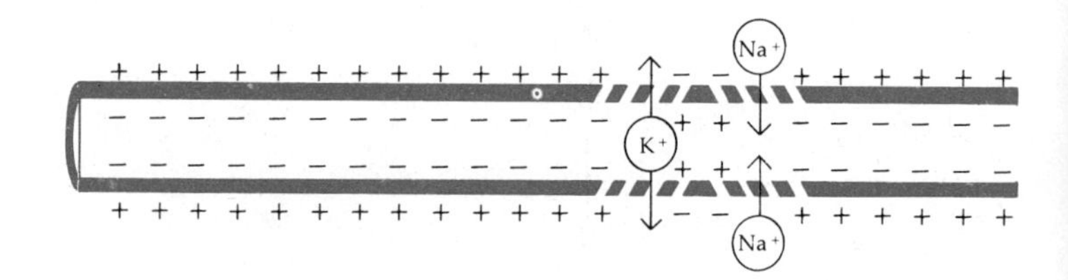

As a consequence of this renewal process, an axon, which would be a very poor conductor of an ordinary electric current, is capable of transmitting a nerve impulse often over a considerable distance with absolutely undiminished velocity.

Once the resting potential is restored, the Na^+ ions are pumped out again and the K^+ ions pumped in. The transmission of the nerve impulse is an all-or-nothing reaction. The size of the action potential is limited by concentration of ions on either side of the membrane, which does not vary to any appreciable extent. As a result, every time the nerve cell is stimulated, the action potential is the same.

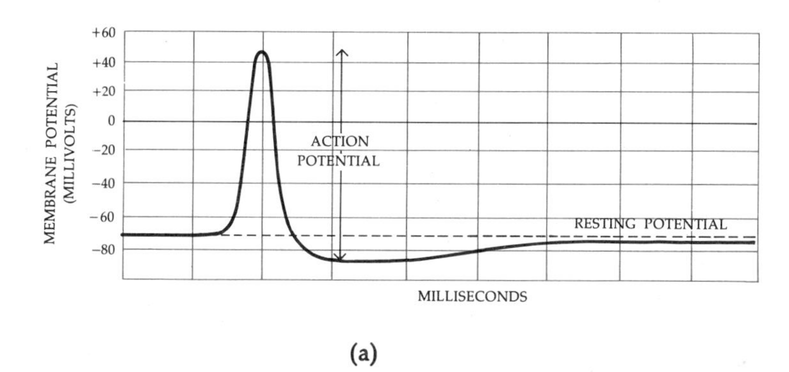

(a)

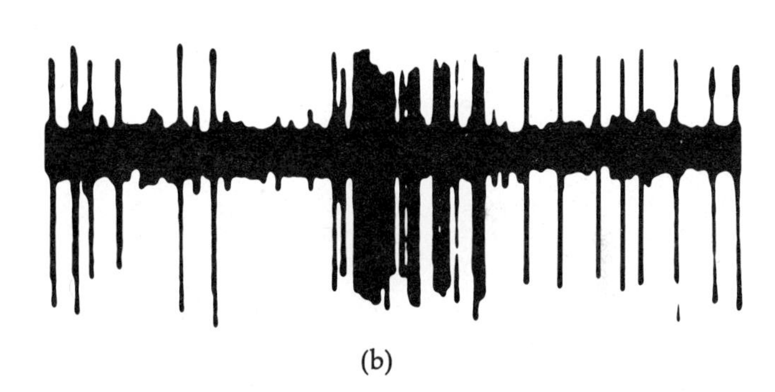

(b)

6–20

Nerve impulses can be monitored by relatively simple electronic recording instruments. The impulses from any one neuron are all the same, that is, each impulse has the same duration and voltage as any other (a). *However, the frequency—the number of impulses per second—and the pattern of impulses vary, depending on the intensity and intensity of the stimulus. This recording from a single neuron of a cat* (b) *shows the change in the nerve code when the nerve is stimulated. In this case, the stimulus was a flash of red light, indicated by the bar.*

The Synapse

Two neurons never quite touch. A nerve impulse travels from one cell to another across a junction known as a *synapse.* Transmission across most synapses in mammals, according to current theory, is by chemical means. In synaptic knobs at the end of the axon are numerous small vesicles, visible in the electron microscope (Figure 6–21), which contain a transmitter substance. Arrival of the nerve impulse causes these vesicles to empty their contents into the synaptic gap. The transmitter substance crosses the gap and combines with receptor molecules on the membrane of the postsynaptic cell, changing the permeability of the membrane.

A number of these chemical transmitters have been tentatively identified, including serotonin, acetylcholine, adrenaline, and noradrenaline (an adrenalinelike compound). The adrenal medulla, a gland which secretes adrenaline, is actually, functionally, a collection of adrenaline-secreting nerve cells, releasing adrenaline and noradrenaline into the bloodstream rather than into a synaptic junction.

The hormones that transmit signals across nerve junctions are rapidly destroyed by specific enzymes after their release. Such destruction is, of course, an essential feature in the control of the activities of the nervous system.

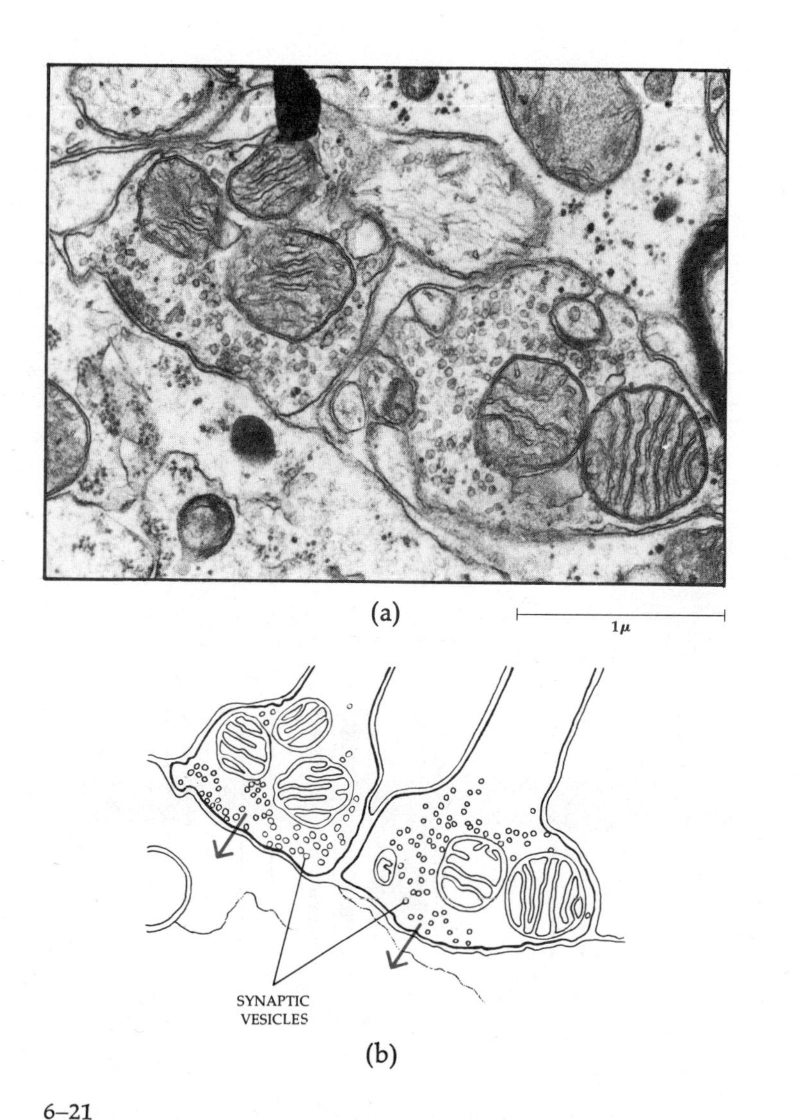

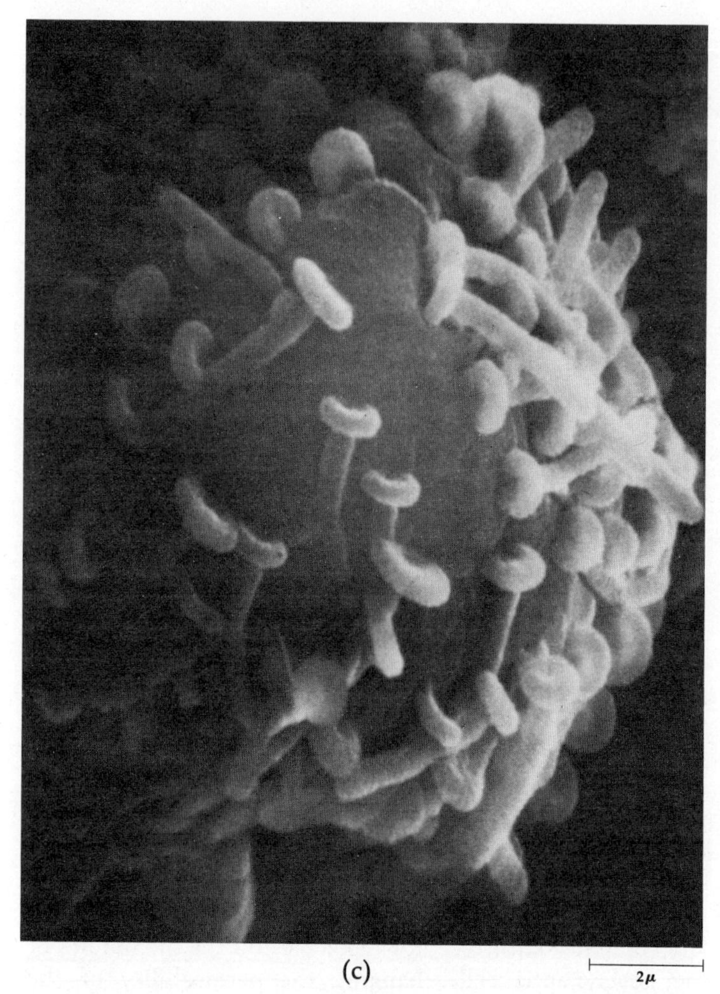

6–21

(a) Electron micrograph of a synaptic area in the spinal cord of a bat, and (b) a diagram of a synapse. In the micrograph, part of the cell body of one motor neuron occupies the left hand corner of the picture. Two of the synaptic knobs, or "feet" (see Figure 6–18), of another motor neuron fill most of the rest of the micrograph. Several nerve terminals are clustered near the cell membranes of the synaptic knobs. These are characterized by numerous synaptic vesicles, which appear in the picture as small, grayish granules. At the synaptic junction, the membranes of the two cells are intact and clearly separated from one another by a gap about 10 millimicrons wide. Apparently, the vesicles, triggered by an electric impulse in the axon, release their chemical contents across this gap. The chemicals stimulate the adjacent neuron and thus relay the impulse from one nerve cell to another, as indicated by the arrows. (c) Synaptic knobs, shown in a scanning electron micrograph.

Acetycholine, which is a transmitter substance in invertebrates as well as vertebrates, is rapidly destroyed by an enzyme called cholinesterase. Many insecticides act by inhibiting the cholinesterase of the insect. The animal's nervous system then runs wild, causing spasms and eventually death.

The basis of the activity of LSD and certain other psychomimetic drugs, such as psilocybin, is thought to be their resemblance to serotonin, which transmits signals within the brain. Mescaline, which produces similar effects, resembles noradrenaline.

Unlike the nerve impulse along the axon—which is an all-or-nothing proposition—signals transmitted by chemicals across a synaptic junction can modulate one another. A single neuron may receive signals from hundreds of synapses, which it processes in order to decide whether or actually, functionally, a collection of adrenaline-secreting not to "fire." Synapses are therefore relay and control points that are extremely important in the functioning of the nervous system.

Types of Neurons

There are three types of neurons in the nervous system: (1) *motor neurons* (sometimes called efferent neurons), which carry impulses outward to the effectors or glands; (2) *sensory neurons* (sometimes called afferent neurons), which receive impulses from sensory receptor cells and transmit these impulses to the central nervous system; and (3) *interneurons,* which interpret and process the messages arriving from the sensory neurons and send out appropriate impulses to other interneurons or to motor neurons.

In the course of evolution, relatively few changes have occurred in the structure of the individual nerve cells, which are remarkably the same throughout the animal kingdom, but vast changes have occurred in the number and deployment of such cells and in the uses made of them. There are about 12 billion nerve cells in the human body, the great majority of which are interneurons.

THE PERIPHERAL AND CENTRAL NERVOUS SYSTEMS

A nerve usually consists of a bundle of hundreds or even thousands of separate nerve fibers (axons), each capable of transmitting separate messages, like the wires in a telephone cable. The nerve has no cell bodies in it, except the cell bodies of the Schwann cells that form the sheaths; the cell bodies of the neurons themselves are centralized in the brain and spinal cord or are aggregated in other parts of the body in clusters known as *ganglia* (singular: *ganglion*). The nervous system lying outside of the brain and spinal cord constitutes the *peripheral nervous system.*

Each nerve contains both motor and sensory fibers, which are sorted out when the nerve makes its connection with the spinal cord. Sensory fibers feed into the spinal column on the dorsal side and ascend toward the brain. Fibers descending from the brain make synapses with motor neurons, the fibers of which emerge from the spinal column on the ventral side.

The brain and the spinal cord constitute the *central nervous system.* The central nervous system of the vertebrates is encased in bony supporting and protecting structures—the vertebral column and the skull. The vertebrate spinal column, the "trademark" of this subphylum, is much the same from fish to man.

Figure 6–23 shows a segment of the human spinal cord, with pairs of spinal nerves entering and emerging from the cord through spaces between the vertebrae. Each of these pairs innervates the skeletal muscles of a different and distinct area of the body. In mammals, there are 31 such pairs. The cell bodies of all these fibers are in the spinal cord.

Within the spinal cord, sensory and motor neurons synapse with interneurons clustered in *nuclei,* which is the term used to describe groups of cell bodies in the central nervous system, and fibers carrying information from various parts of the body (ascending fibers) and those relaying instructions from the brain (descending fibers) are clustered in groups called *fiber tracts.*

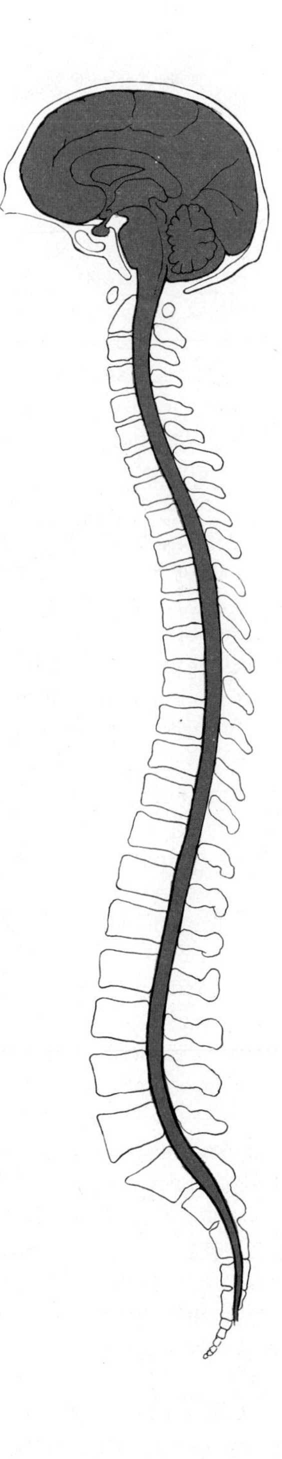

6–22 (at left)

The spinal column and skull form a bony outer covering for the central nervous system of the vertebrate. The central and peripheral nervous systems are connected by fibers carried in the nerves which enter and emerge through the spaces between the vertebrae. There are 31 pairs of spinal nerves, each of which innervates a different segment of the body.

6–23

A segment of the spinal cord. Each spinal nerve divides into two fiber bundles, the sensory root and the motor root, at the vertebral column. The sensory bundle connects with the cord dorsally (toward the back), and the motor bundle ventrally (toward the front). The butterfly-shaped gray matter within the spinal cord is composed of groups of cell bodies and the surrounding white matter consists of ascending and descending tracts of fibers.

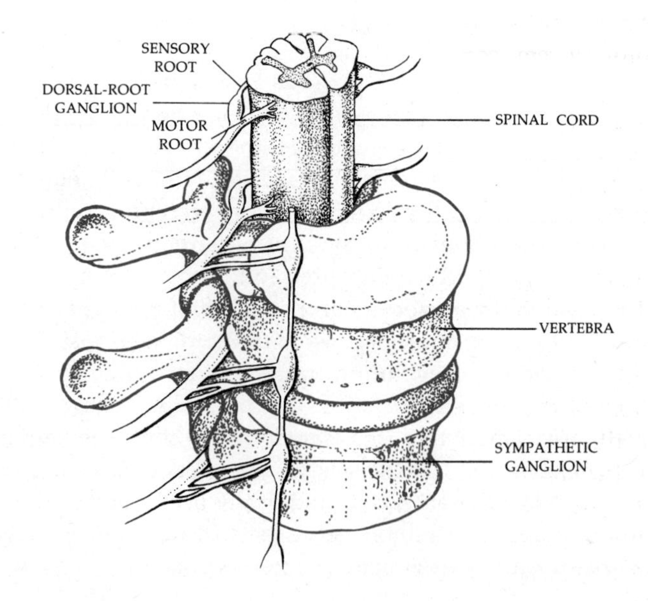

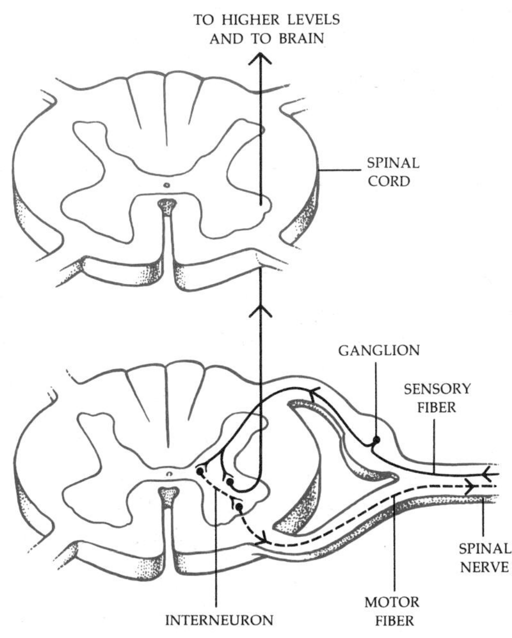

6–24
Diagram of the reflex arc. Impulses from a receptor cell travel along the sensory fiber to the spinal cord. The cell body of the sensory neuron is located in a ganglion lying just outside the spinal cord. The sensory axon enters the cord and synapses with an interneuron in the gray matter of the cord. The interneuron relays the impulse to a motor neuron, which stimulates an effector muscle.

THE SOMATIC AND AUTONOMIC NERVOUS SYSTEMS

There are two important subdivisions of the central and peripheral nervous systems: the *autonomic* and the *somatic.* The somatic nervous system includes both motor and sensory neurons. The autonomic nervous system is entirely a motor system, consisting of the nerves that control cardiac muscle, glands, and smooth muscle (the type of muscle found in the walls of blood vessels and in the digestive, respiratory, and reproductive tracts). The autonomic nervous system is thus generally categorized as an "involuntary" system, in contrast to the somatic system, which controls the muscles that we can move at will, that is, the skeletal muscles.

You will readily recognize that the distinction here between "voluntary" and "involuntary" is not clear-cut. Skeletal muscles often move involuntarily, as in a reflex action, and it is reported that some persons, particularly practitioners of yoga, can control their rate of heartbeat and the contractions of some smooth muscle. Anatomically, the motor neurons of the somatic system are distinct and entirely separate from those of the autonomic nervous system, although fibers of both types may be carried in the same nerve. Also, the cell bodies of the motor neurons of the somatic system are located within the central nervous system, with long nerve fibers running uninterruptedly all the way to the skeletal muscle. The motor fibers of the autonomic nervous system also originate in cell bodies inside the central system; however, they do not travel all the way to their target organ, or effector, but instead form a synapse with a second neuron, which innervates the cell of the muscle or gland. These postganglionic fibers, as they are called, constitute a characteristic difference between the autonomic and somatic systems.

The Somatic Nervous System: The Reflex Arc

Figure 6–24 is a diagram of the reflex arc. In a simple reflex action—your finger touching a hot surface, for example—the stimulus is received by a receptor cell (a pain-sensitive cell, in this case) and is transmitted to the spinal cord. Here this sensory neuron synapses with an interneuron, which, in turn, relays the signal to a motor neuron. The motor neuron causes the appropriate muscle to contract, which moves your arm.

Thus you move your hand away "automatically" before your brain has had a chance to process the information.

6–25
Scientific monitoring of autonomic functions has shown that yogis are able to control activities such as oxygen consumption and the rate of heart beat. In confirmation of the fact that such functions may be brought under voluntary control, Neal Miller of the Rockefeller University has trained rats and, more recently, human volunteers to reduce blood pressure, for example. These techniques hold possibilities for medical treatment which are now being explored.

However, your brain has been informed and can call forth the next appropriate activity—such as swearing or turning off the stove.

In actuality, however, even a simple reaction is the result of interactions among many neurons across many synaptic junctions.

The Autonomic Nervous System

The autonomic nervous system has two divisions, the sympathetic and the parasympathetic, which are anatomically and functionally distinct. The parasympathetic system consists of nerves from the brain and from the lower region of the spinal cord. The sympathetic nervous system originates in the thoracic and lumbar areas of the spinal cord. In the parasympathetic system, the point of synapse is near or in the target organ, whereas in the sympathetic system, the synapses are in a regular chain of ganglia running parallel to the spinal cord. Most postganglionic sympathetic nerve endings release adrenaline or noradrenaline, whereas all parasympathetic endings release acetylcholine.

As you can see in Figure 6–26, most of the major visceral organs of the body are innervated by fibers from both the sympathetic system and the parasympathetic system. The effects of the sympathetic system are often antagonistic to those of the parasympathetic system, and vice versa. As a consequence, whereas the somatic nervous system can only excite or not excite a particular effector, the autonomic nervous system can have both excitory and inhibitory effects.

The changes that occur in a state of rage, described at the beginning of this chapter, are initiated by the simultaneous discharge of nerves of the sympathetic system. The parasympathetic system, on the other hand, is more concerned with the restorative activities of the body, that is, with rest and rumination. Parasympathetic stimulation slows down the heartbeat and increases the muscular movements of the intestine and the secretions of the salivary gland. Most large organs, such as the heart, are under the control of both sympathetic and parasympathetic nerves, and these work in close cooperation for the ultimate regu-

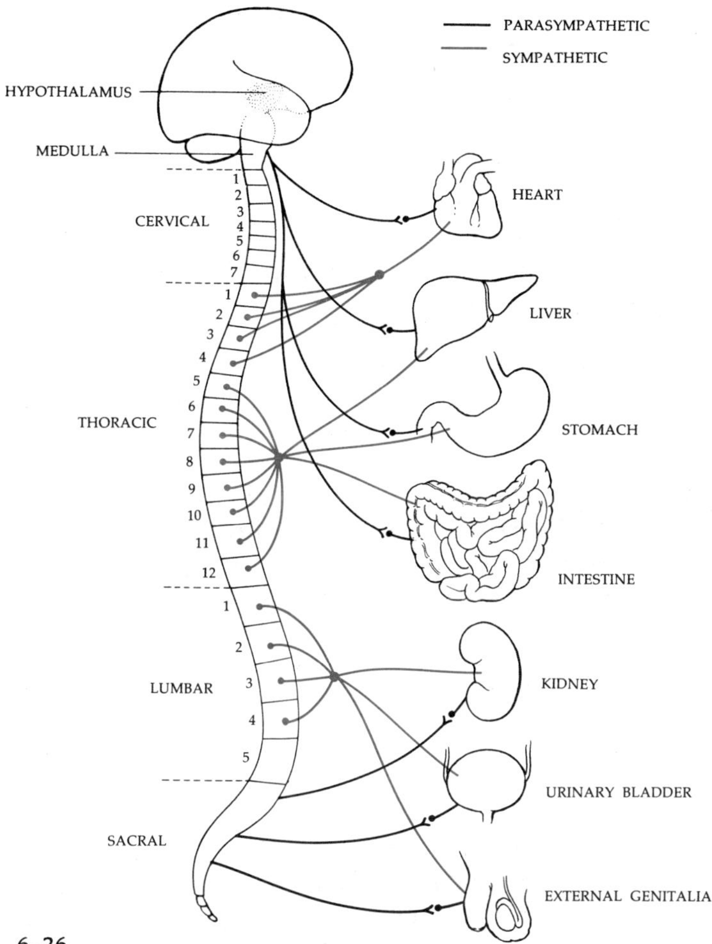

6–26

The autonomic nervous system, consisting of the sympathetic and the parasympathetic systems. The presynaptic neurons of the parasympathetic system exit from the medulla region of the brain and from the sacral region of the spinal cord. The sympathetic system originates in the thoracic and lumbar regions. Most, but not all, internal organs are innervated by both systems, which function in opposition to each other. In general, the sympathetic system produces the effect of exciting an organ and the parasympathetic system produces an opposite effect. Thus, while the one system stimulates activity, the other system inhibits it. Our normal state, somewhere in between absolute excitement and absolute calm, is the result of a constant interplay of these two systems.

lation of these important structures. Some tissues, such as the hair muscles, the small blood vessels, and the sweat glands, are under sympathetic control alone.

In general, of course, we are usually neither in a rage nor in a complete state of vegetation but somewhere in between —a state resulting from a constant interplay of forces, a dialogue between the sympathetic and parasympathetic systems, mediated and reinforced by the hormones and maintaining the state of homeostasis characteristic of animal physiology.

A SPECIAL SENSE ORGAN: THE VERTEBRATE EYE

The central nervous system receives information from the outside world both by means of the sensory neurons of the somatic nervous system and by way of a number of special sense organs, of which the eye is an example.

Vision is a dominant factor in the behavior of many vertebrates, including ourselves. The vertebrate eye shown in Figure 6–27 is a human eye, but a picture of the eye of any other vertebrate species would look much the same. This type of eye is often called a camera eye, and camera fans will be quick to recognize that it has a number of features in common with the ordinary camera and also a number of expensive accessories, such as a built-in cleaning and lubricating system, an exposure meter, and an automatic field finder. As indicated in the figure, light from the object being viewed passes through the transparent cornea and lens, which focus an inverted image of the object on the light-sensitive retina in the back of the eyeball. The iris controls the amount of light entering the eye by regulating the size of the pupil.

Focusing the Eye

In fish and amphibians, the eye is focused in the same way one focuses a camera. Muscles within the eye change the position of the lens, drawing it back toward the retina in order to focus more distant objects. In mammals, the focus

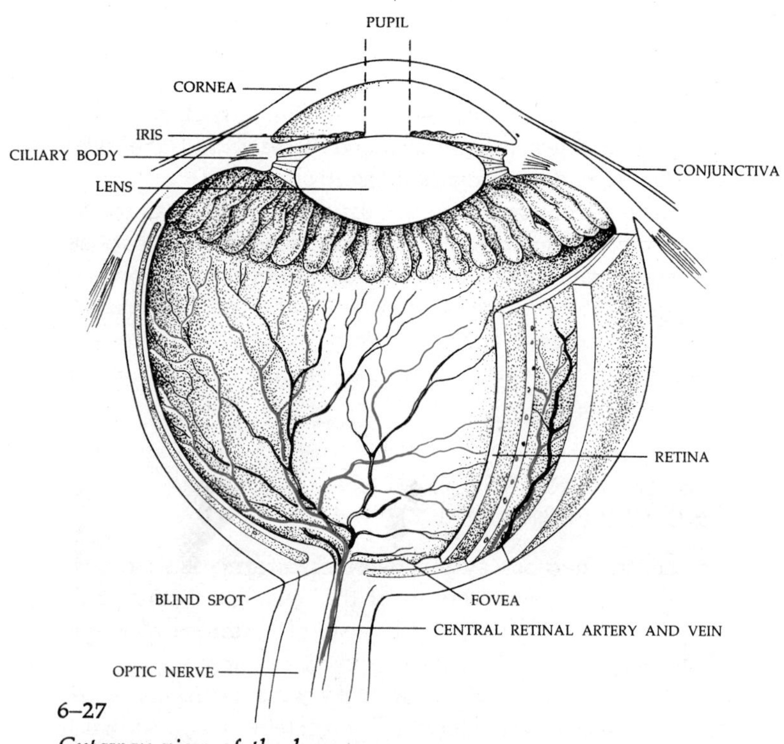

6–27
Cutaway view of the human eye.

THE NATURE OF SENSORY PERCEPTION

We take it for granted that we see with our eyes and hear with our ears and smell with our noses, but actually this is not the case. As you will recall from Figure 6–20, the nature of the impulse received from any type of neuron—whether a touch receptor in the fingertip or a rod in the retina—is the same.

Perception consists of the organization, analysis, and interpretation of stimuli, rather than their reception. In general, the vertebrates perceive more (that is, they receive more information from their surroundings) than the invertebrates, and the primates more than the fishes, not because the special sensory organs are more acute or more efficient but because of the greater capacity of the larger and more complicated brain to make discriminations and associations on the basis of the signals it receives. (In Chapter 6–7, we shall explore the organization of the brain in more detail.)

is adjusted by contrasting or releasing the ciliary muscles and so changing the shape of the lens.

Depth vision, or stereoscopic vision, depends on viewing an object with both eyes simultaneously and seeing one image, not two. When both eyes are trained on an object, the angle at which the two visual axes come together is communicated to the brain by special sensory cells in the ciliary muscles. On the basis of this information, we automatically compute the size and distance of nearby objects. Since our eyes converge only on near objects, we are not able to make such judgments when distances increase. Instead, we estimate these distances on the basis of known size of the objects or on the basis of other visual clues, such as houses, people, or cars. A young child viewing a distant object—an airplane, for example—will think of it as "little" rather than distant.

Tree-dwelling animals, such as the immediate forerunners of *Homo sapiens*, usually have some overlapping of visual fields in the front to provide them with the stereoscopic vision essential for the distance judgments involved in moving from branch to branch. Hunters also tend to have stereoscopic vision, but animals which are more likely to be hunted than to hunt usually have eyes on the side of the head, giving a wider total field. Some birds with large laterally placed eyes—the woodcock, the cuckoo, and certain species of crow—have binocular fields of vision either in front of them or behind them or both.

The Retina

The retina of the vertebrate eye is functionally inside out—that is, the photoreceptors of the eye are pointed toward the back of the eyeball—and light must reach the light-sensitive area by passing through a layer of cell bodies (see Figure 6–28). Light that is not captured by the photoreceptors is absorbed by the pigmented chorion which lines the back of the eyeball.

The photoreceptor cells transmit their impulses to neurons known as bipolar cells, which pass the impulses on to the ganglion cells whose axons form the optic nerves.

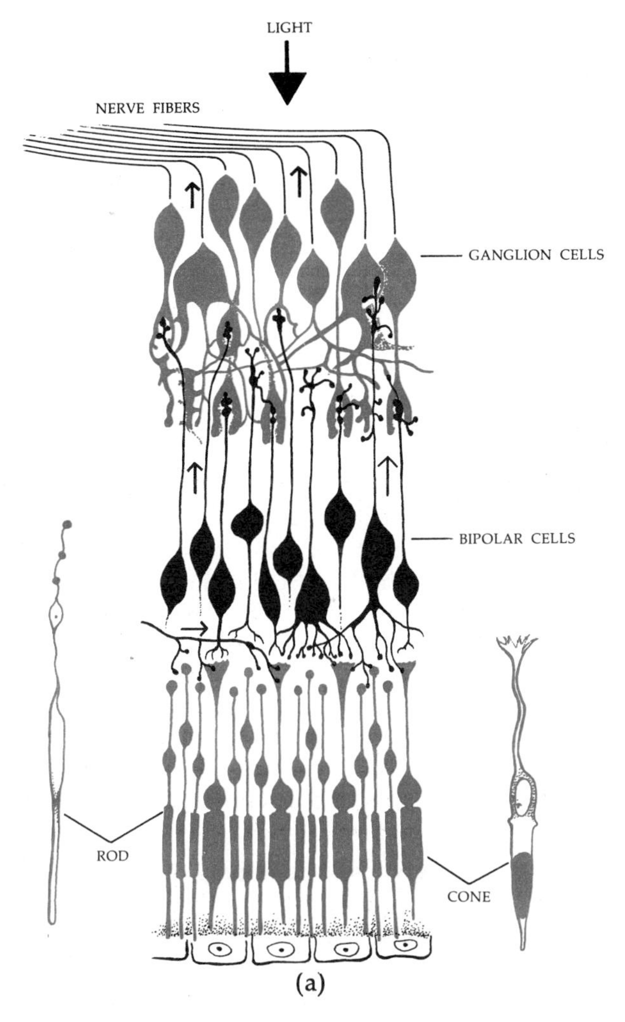

(a)

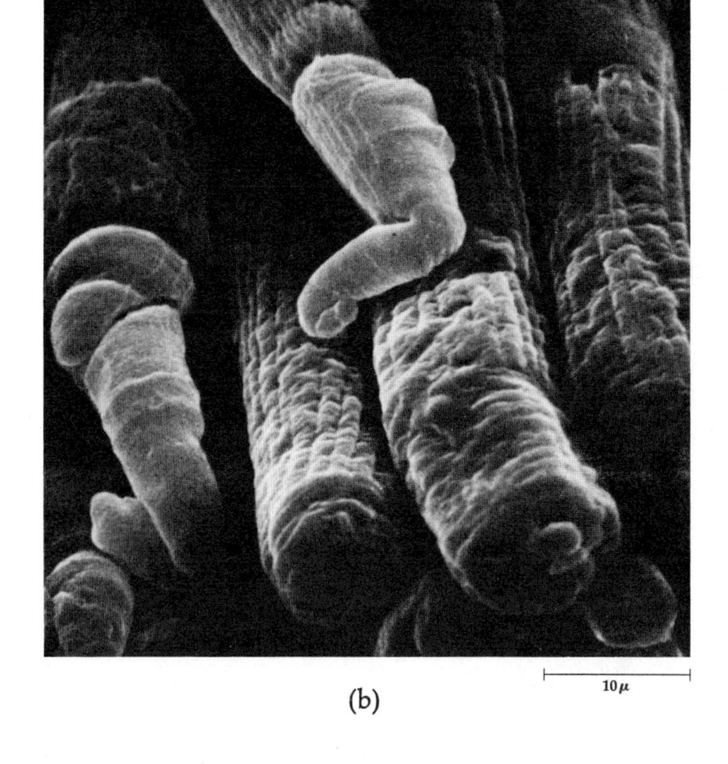

(b)

6–28

(a) *The retina of the vertebrate eye. Light (shown here as entering from the top) must pass through a layer of cells to reach the photoreceptors (the rods and cones) at the back of the eye. Impulses from the photoreceptor cells are transmitted through the bipolar cells to the ganglion cells. The transmission paths are not direct but involve elaborate interconnections.* (b)*Rods and cones, as seen by a scanning electron microscope.*

As you can see in the figure, these pathways of transmission are not direct, but varied and elaborate interconnections are made among the cells even before the impulses leave the retina, indicating that some preliminary analysis of visual information takes place in the retina itself. The human retina contains about 125 million photoreceptors and about 1 million optic nerve fibers.

Ganglion cell fibers from all over the retina converge at the rear of the eyeball and, bundled together like a cable, pass through the retina and out the back. The point at which the fibers pass out of the eyeball is a blind spot in the eye since photoreceptor cells are absent. We generally

X ● / /

6–29

With this diagram, you can prove for yourself that you have a blind spot in each eye. Hold the book about 12 inches from your face, cover your left eye, and gaze steadily at the X while slowly moving the book toward your face. Note that at a certain distance, the image of the dot becomes invisible. Then cover your right eye and gaze steadily at the dot while you move the book toward your face. What happens to the X?

are not aware of the existence of the blind spot since we usually see the same object with both eyes and the "missing piece" is always supplied by the other eye. If you are not aware of your own blind spot, you can demonstrate its existence with the help of Figure 6–29.

As shown in Figure 6–27, there is an area just above the blind spot known as the fovea. This is the only spot in the eye at which we see a really sharp image. In this area, the photoreceptor cells are packed much more closely together and many have a one-to-one connection with the bipolar cells. Birds, which rely on vision above all the other senses, may have two or three foveas. Most vertebrates, however, have no fovea, and their vision is comparable in acuity to what we see out of the corners of our eyes.

Rods and Cones

The human retina contains two types of photoreceptors (Figure 6–28). These are known, because of their shapes, as rods and cones. The cones, of which the human eye contains about 5 million, provide greater resolution of vision, giving a "crisper" picture. The photoreceptors in the fovea are all cones. Man has about 160,000 cones per square millimeter of eye. A hawk, with an eye of about the same size, has some 1 million cones per square millimeter and has a visual acuity about eight times that of man.

Rods do not provide as great a degree of resolution as cones do, but they are more light-sensitive; in dim light, our vision depends entirely on rods. Some nocturnal animals, such as toads, mice, rats, and bats, have retinas made up entirely of rods, and some diurnal animals, such as reptiles, have only cones.

Cones are responsible for color vision, which is why the world becomes gray to us at twilight, when we begin to depend on rods. The physiology of color vision is not entirely understood, but it appears that there are three different types of cones, each one of which contains a pigment sensitive to the wavelength of one of the three "primary" colors—blue, green, or red. Different shades of color stimulate different combinations of these cells, and these impulses, after being processed in the retina, are analyzed by the brain, which translates them into what we perceive as color. Many species of vertebrates do not have color vision. Among the mammals it is found chiefly in the primates.

SUMMARY

The endocrine and nervous systems provide the precise and rapid communication necessary to coordinate the numerous internal functions which enable an animal to regulate its internal environment.

Hormones are the agents of the endocrine system. These are specialized chemicals which are released in response to specific stimuli and which, in turn, produce specific effects on certain organs and tissues. The exact way in which hormones work is still not understood.

In contrast to the endocrine system, the nervous system has a considerably greater capacity for rapid response; mammals and birds have developed highly sensitive nervous systems. The unit of the nervous system is the neuron, or nerve cell, which consists of (1) the dendrites, which receive impulses; (2) the cell body; and (3) the axon, which relays the impulses to other cells. Nerve impulses are transmitted between cells across a junction called the synapse. The signal crosses the synaptic gap by triggering the release of a chemical transmitter, which stimulates adjacent cells. Nerves are bundles of motor and sensory fibers.

The central nervous system consists of the brain and the spinal cord, which are encased, in vertebrates, in the skull and vertebral column. From here, spinal nerves emerge in pairs, each pair innervating the muscles of a different area of the body. Motor neurons convey impulses to the effectors—the muscles and glands. Sensory neurons receive impulses from the sensory cells. The nervous system outside the central nervous system constitutes the peripheral nervous system.

In vertebrates, the overall nervous system has two major subdivisions: the somatic nervous system, which consists of motor and sensory neurons, and the autonomic nervous system, which controls the muscles and glands involved in the digestive, circulatory, respiratory, and reproductive functions. The autonomic system has its own network of effector neurons, separate from that of the somatic system, which are linked indirectly from the central nervous system to the target organs by secondary, or postganglionic, neurons. The system has two subdivisions: (1) the sympathetic system, which is largely responsible for the excitory reactions of the body; (2) the parasympathetic system, which controls the restorative activities, such as digestion and rest.

The central nervous system receives information not only from the somatic neurons but also from special sense organs, of which the eye is an example. In the vertebrate eye, light from the object being viewed passes through the transparent cornea and lens, which focus an inverted image of the object on the light-sensitive retina at the back of the eyeball. The photoreceptors in the human retina are of two types: rods, which are more light-sensitive and are responsible for night vision; and cones, which provide finer resolution, hence more acuity, and color vision. The iris controls the amount of light entering the eye by regulating the size of the pupil. Overlapping of visual fields provides for stereoscopic vision, found in most tree-dwelling and hunting vertebrates and in man.

QUESTIONS

1. What is meant by biochemical specificity? What are the advantages of such specificity to the organism? The disadvantages?
2. Like the other endrocrine glands, the thyroid is regulated by a negative feedback system. The hypothalamus monitors the concentration of thyroid hormone in the blood and stimulates the pituitary to secrete thyroid-stimulatory hormone, which drives the thyroid to produce thyroid hormone. Excess thyroid-stimulatory hormone can cause enlargement of the thyroid. Can you explain why persons with an iodine-deficient diet often develop goiters, that is, overgrowths of the thyroid gland?
3. Would you expect that vision among the early, primitive mammals was dependent chiefly on rods or on cones? Why?

CHAPTER 6–3

Respiration

Respiration has two meanings in biology. At the intracellular level, "respiration" refers to the oxygen-requiring events that take place in the mitochondria and are the chief source of energy for the cell. For multicellular organisms, respiration means acquiring oxygen and getting rid of carbon dioxide. This latter process—which is, of course, essential for the former—is the subject of this chapter.

The means by which organisms extract energy without oxygen are not very efficient, as we saw in Chapter 2–5. Before oxygen came to be present in the atmosphere, the only forms of life were one-celled organisms, and the only present-day forms that can carry on life processes without oxygen are a few types of bacteria and yeasts. The transition to an atmosphere containing free oxygen was responsible for one of the giant steps in evolution.

A second revolution in terms of energy came about with the transition to land. Air is a far better source of oxygen than water; one-fifth of the air of the modern atmosphere is made up of free oxygen, whereas at ordinary temperatures, even when water is saturated with air, only about 1 part in 250 (by weight) is free oxygen. Not only must more water be processed for oxygen, but water weighs a great deal more. A fish spends up to 20 percent of its energy in the muscular work associated with respiration, whereas an air breather expends only 1 or 2 percent of its energy in respiration. Also, oxygen diffuses much more rapidly through air than through water, about 300,000 times more rapidly, and so can be replenished much more quickly as it is used up by respiring organisms. All the higher vertebrates—the birds and the mammals—are air breathers, even those that live in the water.

AIR UNDER PRESSURE

At sea level, the air around us exerts a pressure on our skin of about 15 pounds per square inch. This pressure is enough to raise a column of water 32 feet high or a column of mercury 760 millimeters (29.91 inches) in the air. (Atmospheric pressure is generally measured in terms of mercury simply because mercury is relatively heavy—so the column will not be inconveniently tall—and because air does not dissolve in it.) Of this 760 millimeters, 155 results

Table 6–3
Composition of Dry Air

Oxygen	21%
Nitrogen	77%
Argon	1%
Carbon dioxide	0.03%
Other gases*	0.97%

* Other gases include hydrogen, neon, krypton, helium, ozone, and xenon.

6–30

The French scientist Antoine Lavoisier (1743–1794) *was the first man to recognize that "Respiration is merely a slow combustion of carbon and hydrogen . . . animals that breathe are really combustible bodies which burn and are consumed." As Lavoisier showed, by measuring the amount of oxygen consumed, it is possible to estimate quite accurately the energy requirements of an organism. The oxygen consumption at rest is referred to as the* basal metabolism, *the amount of oxygen required for maintenance only. Metabolic rates increase sharply with exercise; a man exercising consumes* 15 *to* 20 *times the amount of oxygen as a man sitting still. This drawing of Lavoisier conducting an experiment on respiration was made by Mme. Lavoisier, who is seen on the far right.*

6–31

Atmospheric pressure is usually measured by means of a mercury barometer. To make a simple mercury barometer, as Evangelista Torricelli did some 300 *years ago, fill a long glass tube, open at one end, only with mercury* (a). *Closing the tube with your finger, invert it into a dish of mercury. Remove your finger. The mercury level will drop until the pressure of its weight inside the tube is equal to the atmospheric pressure outside* (b). *At sea level, the height of the column will be about* 76 *centimeters. If the tube and dish are enclosed in an airtight covering and the air pressure is lowered by pumping the air out* (c), *the mercury level will fall.*

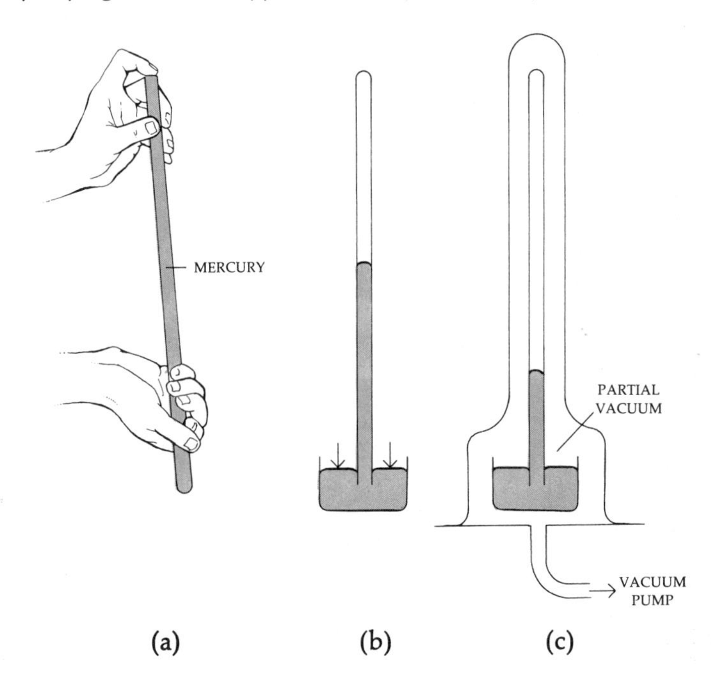

from the pressure of the oxygen in the air. It is as a result of the atmospheric pressure that oxygen enters living organisms.

We are so accustomed to the pressure of the air around us that we are unaware of its presence or of its effects on us. However, if you visit a place—such as Mexico City—which is at a comparatively high altitude and therefore has a lower atmospheric pressure, you will feel lightheaded at first and will tire easily. Regular inhabitants of such places breathe more deeply, have enlarged hearts which circulate oxygen-carrying blood more rapidly, and even have more red blood cells for oxygen transport. The consequences of increased atmospheric pressures are seen in deep-sea divers. Early in the history of deep-sea diving, it was found that if divers come up from the bottom too quickly, they get the "bends," which is always painful and sometimes fatal. The bends develop as a result of breathing compressed air. Nitrogen is relatively insoluble in water, but high pressure forces nitrogen from the air in the lungs into solution in the blood and tissues. If the body is rapidly decompressed, the nitrogen bubbles out of the blood, like the bubbles that appear in a bottle of soda water when you first take off the top. These bubbles lodge in the capillaries, stopping blood flow, or invade nerves or other tissues.

HOW OXYGEN GETS INTO CELLS

Oxygen enters and moves within cells by diffusion. Cell membranes are freely permeable to oxygen, even those that are not permeable to water. This is true of all cells, whether an amoeba, a paramecium, a liver cell, or a brain cell. However, substances can move effectively by diffusion only for very short distances—less than 1 millimeter. Also, the respiratory surface must be exposed to a medium, such as air or water, over which the diffusion process can take place.

You will recall from Chapter 2–2 that diffusion occurs only *down* a concentration gradient. The movement of oxygen into a cell or into an organism takes place because there is a higher concentration of oxygen outside than in-

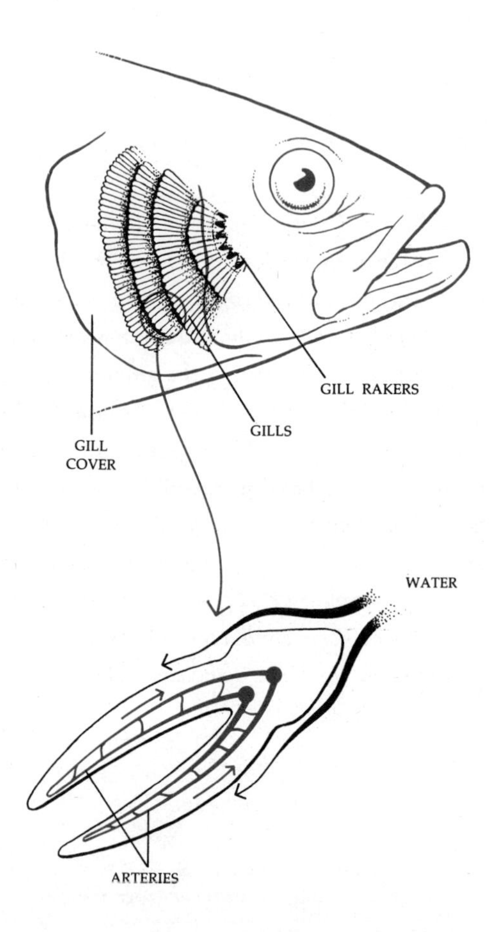

6–32

Rates of diffusion are proportional not only to the surface areas exposed but to differences in concentration. The greater the difference in concentration of a gas or liquid, the more rapid its diffusion. In the gill of the fish, the circulatory vessels are arranged so that the blood is pumped through them in a direction opposite to that of the oxygen-bearing water. As a consequence, the blood with the least oxygen is brought within diffusion distances of the water with the most oxygen. This arrangement, which is an example of what is known as countercurrent exchange (see Figure 6–49) results in a far more efficient transfer of oxygen to the blood than if the blood flowed in the same direction as the water.

KINDS OF BREATHING

Some animals respire through their outer surfaces, just as single-celled organisms do, and the cells receive their oxygen by simple diffusion. This very simple means of respiration is useful only for very small animals, in which each cell is quite close to the surface, or for animals in which much of the tissue is not metabolically active. In jellyfish, for example, the central, jellylike mass is mostly inert tissue. Many eggs and embryos also respire in this simple way, particularly in their early stages of development.

A second major means of respiration also involves diffusion through the general surface area but with the assistance of circulation. Many of the worms respire in this way. Earthworms, for example, have a network of capillaries just one cell layer below the surface. The blood picks up oxygen by diffusion as it travels near the surface of the animal and releases oxygen by diffusion as it travels past the oxygen-poor cells in the interior of the earthworm's body. The tube shape of a worm exposes a proportionally large surface area. Some worms can adjust their surface area in relation to oxygen supply. If you have an aquarium at home, you may be familiar with tubifex worms, which are frequently sold as fish food. If these worms are placed in water low in oxygen, such as a poorly aerated aquarium, they will stretch out as much as 10 times their normal length and thus increase the surface area through which oxygen diffusion occurs.

Gills and lungs are other ways of increasing the respiratory surface. Gills are outgrowths, whereas lungs are ingrowths, or cavities. The respiratory surface of the gill, like that of the earthworm, is a layer of cells, one cell thick, exposed to the environment on one side and to circulatory vessels on the other. The layers of gill tissue may be spread out flat, stacked in layers, or convoluted in various ways. The gill of a clam, for instance, is shaped like a steam-heat radiator, which is also designed to provide a maximum surface-area-to-volume ratio.

Lungs are internal cavities into which the oxygen-laden air is taken. They have certain disadvantages as compared with gills; it is more efficient from the point of view of diffusion to have a continuous flow across the respiratory surface. However, they have one overwhelming advantage in air: the respiratory surface can be kept moist without a large loss of water by evaporation. Lungs are largely a vertebrate "invention"; however, they are found in some lower animals. Land-dwelling snails, for example, have independently evolved lungs which are remarkably similar to the primitive lungs of some amphibians.

side the cell or organism (which, remember, is using up oxygen). Similarly, the passage of carbon dioxide out of the cell or organism takes place because the concentration is higher inside than outside the respiratory surface.

The larger the volume of an animal, the smaller, in proportion, is its exposed surface area (as was demonstrated in Figure 2–4). As a consequence, larger organisms in which the interior cells are metabolically active require a means of increasing the surfaces over which diffusion of oxygen and carbon dioxide can occur and also a means for transporting the oxygen and carbon dioxide to and from this surface area. In vertebrates, these increased surface areas are gills and lungs, and the transport mechanism is the bloodstream.

EVOLUTION OF THE GILL

The vertebrate gill originated primarily as a feeding device. Primitive vertebrates breathed mostly through their skin. They filtered water into their mouths and out of what we now call their gill slits, extracting bits of organic matter from the water as it went through. (Amphioxus, which is believed to resemble closely the ancestral vertebrate, functions in this same way.)

In the course of time, numerous selection pressures, chiefly involved with predation, came into operation. As one consequence, there was a trend toward an increasingly thick skin, even one armored or covered with scales. Such a skin

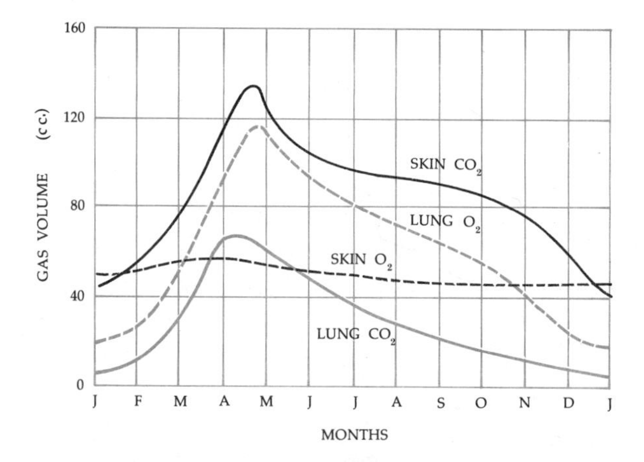

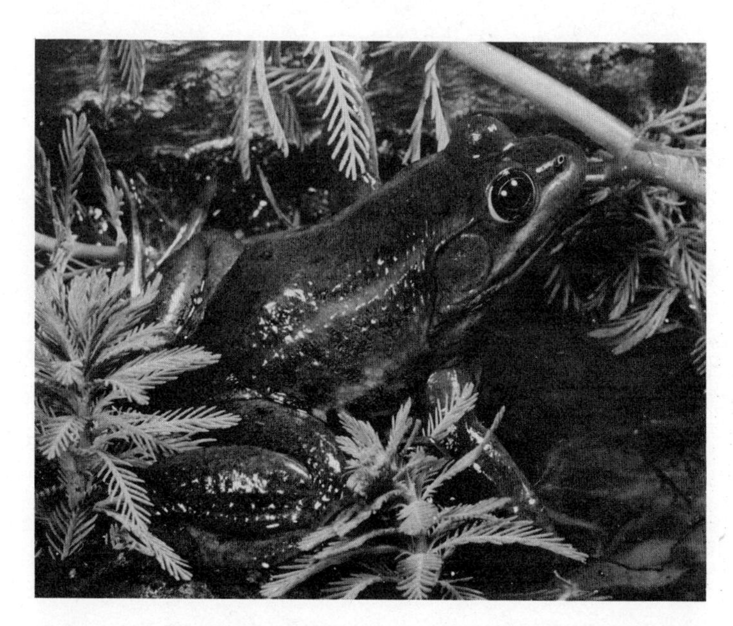

6–33

Frogs have relatively small and simple lungs, and a major part of their external respiration takes place through the skin. The chart shows the results of a study of a frog in which oxygen and carbon dioxide exchange through the skin and lungs were measured simultaneously for one year. Can you explain why the respiratory activity peaks in April–May and is lowest in December–January?

was not, of course, useful for respiratory purposes. At the same time, similar forces were operating to produce animals that were larger and swifter and so more efficient at capturing prey and escaping predators. Such animals also had larger energy requirements—and consequently larger oxygen requirements. These problems were solved by the "capture" of the gill for a new purpose: respiratory exchange. The surface area of the epithelium beneath the gill slowly increased in size and in blood supply over the millenia. The modern gill is the result of this evolutionary process.

In most fish, the water (in which oxygen is dissolved) is pumped in at the mouth by oscillations of the bony gill cover and flows out across the gills. The fish can regulate the rate of flow, and sometimes assist it, by opening and closing its mouth. Fast swimmers, such as mackerel, obtain enough oxygen only by keeping perpetually on the move. Such fish cannot be kept in an aquarium or any other space where their motion is limited because they will die of oxygen lack.

EVOLUTION OF THE LUNG

Some primitive fish had lungs as well as gills, although the lungs were not efficient enough to serve as more than accessory structures. These lungs were a special adaptation to fresh water, which, unlike ocean water, may stagnate and, because of decay or of algal bloom, become depleted of oxygen. A few species of lungfish still exist. These surface and gulp air into their lungs and so can live in water that does not have sufficient oxygen to support other fish life.

Amphibians and reptiles have relatively simple lungs, with small internal surface areas, although their lungs are far larger and more complex than those of the lungfish. The lungs of lungfishes developed directly from the pharynx, the posterior portion of the mouth cavity which leads to the digestive tract. In amphibians, reptiles, and other air-breathing vertebrates, we see the evolution of the windpipe, or *trachea*, guarded by a valve mechanism, the glottis, and

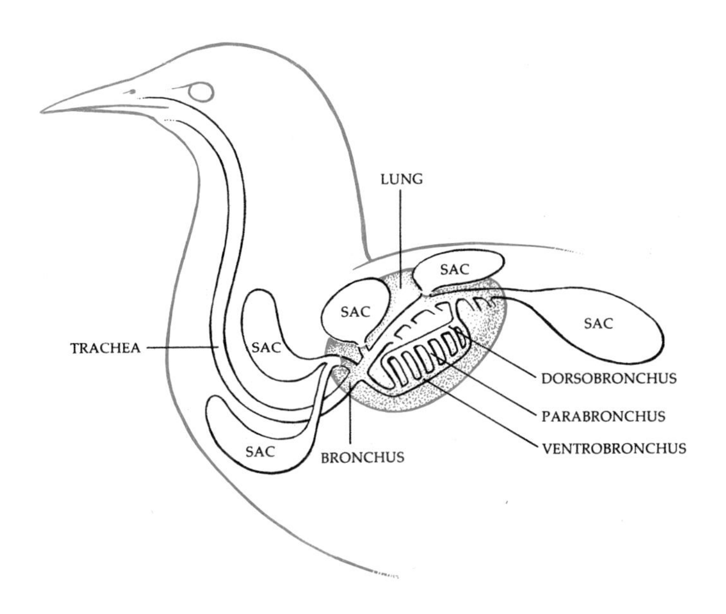

6–34
The lungs of birds are extraordinarily efficient. They are small and are expanded and compressed by movements of the body wall. Also, each lung has several air sacs attached to it which empty and fill like balloons at each breath. The fresh, oxygenated air passes through the main ducts of the lungs into the air sacs and then, at expiration, flows through the smaller lung passages, in which the gas exchange actually takes place. In this way, the lung is almost completely flushed with fresh air at every breath; there is little residual "dead" air left in the lungs, as there is in mammals.

nostrils, which make it possible for the animal to breathe with its mouth closed. Amphibians still rely largely on their skin for exchange of oxygen and carbon dioxide, but reptiles breathe almost entirely through their lungs.

An important feature of all vertebrate lungs is that the exchange of air with the atmosphere takes place as a result of changes in lung volume. Such lungs are known as ventilation lungs. Frogs gulp air and force it into their lungs in a swallowing motion; then they open the glottis and let it out again. In reptiles, birds, and mammals, air is sucked into the lungs as a consequence of changes in the size of the lung cavity, brought about by activity of the chest muscles.

HUMAN LUNGS

The human breathing apparatus, like those of other mammals, is a relatively complicated structure. Air is warmed and cleaned in the nasal passages, after which it passes through the trachea. The trachea is strengthened by rings of cartilage that prevent it from collapsing during inspiration. The trachea leads into the *bronchi* (singular: *bronchus*), which subdivide into smaller and smaller passageways. The actual exchange of gases takes place in small air sacs known as *alveoli*, each only 1 or 2 millimeters in diameter. A pair of human lungs has about 300 million alveoli, providing a respiratory surface area of some 70 square meters, or about 750 square feet—approximately 40 times the surface area of the entire human body.

Mechanics of Respiration

Lung volume is controlled by changes in the size of the thoracic cavity. We inhale by contracting the muscles beneath the ribs, pulling the rib cage up and out, and by contracting the muscular dome-shaped diaphragm, which flattens the diaphragm and increases its diameter. These movements enlarge the thoracic cavity, which encloses the lungs and the heart. This cavity is under negative pressure; that is, the pressure within it is less than atmospheric pressure. Air—which is, of course, at atmospheric pressure—enters the lungs and expands them. The lungs empty as the muscles relax, and the weight of the thorax causes depression of the ribs. Usually, only about 10 percent of the air in the lung cavity is exchanged at every breath, but as much as 80 percent can be exchanged by deliberate deep breathing.

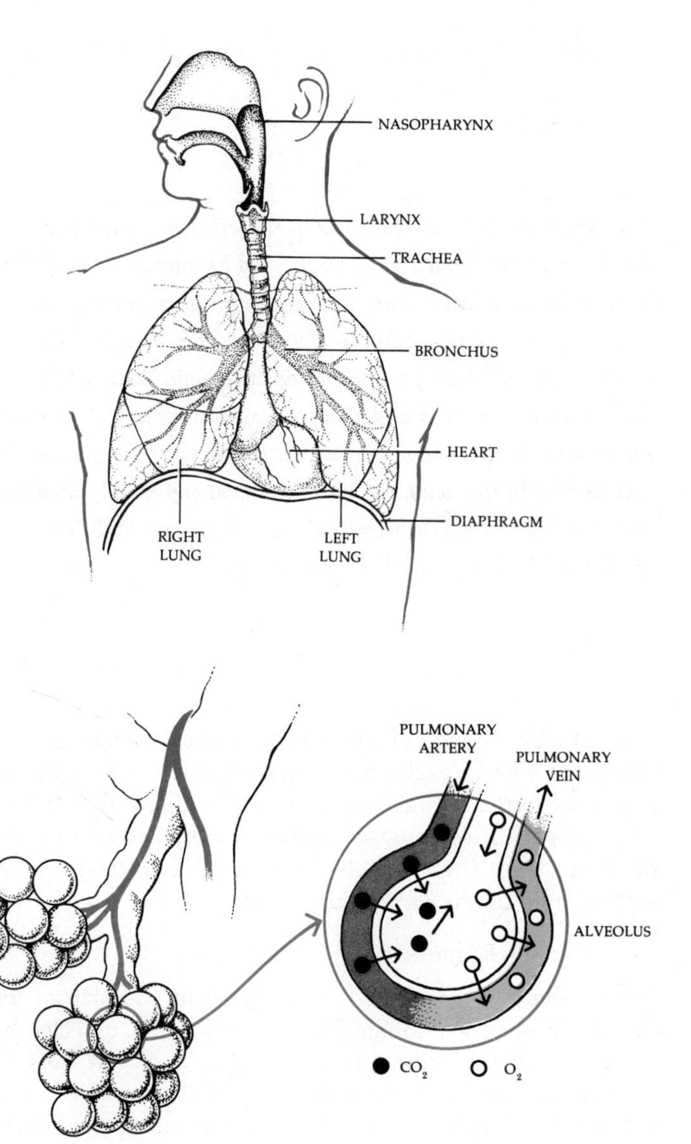

6–35

The human respiratory system. Air enters through the nasopharynx and passes down the trachea and bronchi to the alveoli in the lungs. Within each alveolus, of which there are some 300 million in a pair of human lungs, oxygen and carbon dioxide diffuse into and out of the bloodstream through the pulmonary vein and artery, respectively.

Control of Respiration

The rate of respiration is controlled by a respiratory center at the base of the brain. This center, which is in a portion of the brain known as the medulla oblongata, receives and coordinates information about the carbon dioxide concentration of the blood. Signals are relayed from this center to nerves that originate in the spinal cord and that control the muscles of the chest and the diaphragm. The system is so sensitive to even the smallest change that only the slightest variations in carbon dioxide concentration are permitted to occur. If this concentration increases only slightly, breathing immediately becomes deeper and faster, permitting more carbon dioxide to leave the blood until the carbon dioxide level has returned to normal. You can, of course, deliberately increase your breathing rate by contracting and relaxing your chest muscles, but breathing is normally under involuntary control. It is impossible to commit suicide by deliberately holding your breath; as soon as you lose consciousness, the involuntary controls take over once more.* Deaths due to barbiturates and other drugs with a depressant activity the result are usually of a damping of the activity of this vital brain center.

TRANSPORT OF OXYGEN IN THE BLOOD: HEMOGLOBIN

Oxygen is not normally very soluble in water. The bloodstream's capacity to transport oxygen is greatly increased by the presence of a special carrier molecule, *hemoglobin.* Hemoglobin is composed of a protein (globin) plus an iron-containing porphyrin (heme). Four such globin-heme units are found in each hemoglobin molecule. Each hemoglobin molecule combines with four atoms of oxygen, enabling our bloodstreams to carry about 60 times as much oxygen as

* But in porpoises, for example, which must breathe irregularly in order to stay under water for long periods, breathing is under voluntary control. For this reason, porpoises cannot be anesthetized without special precautions being taken.

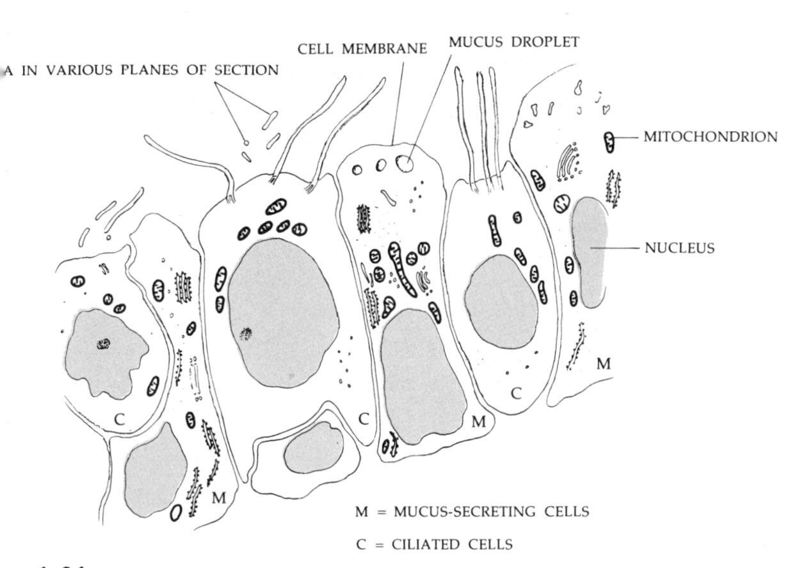

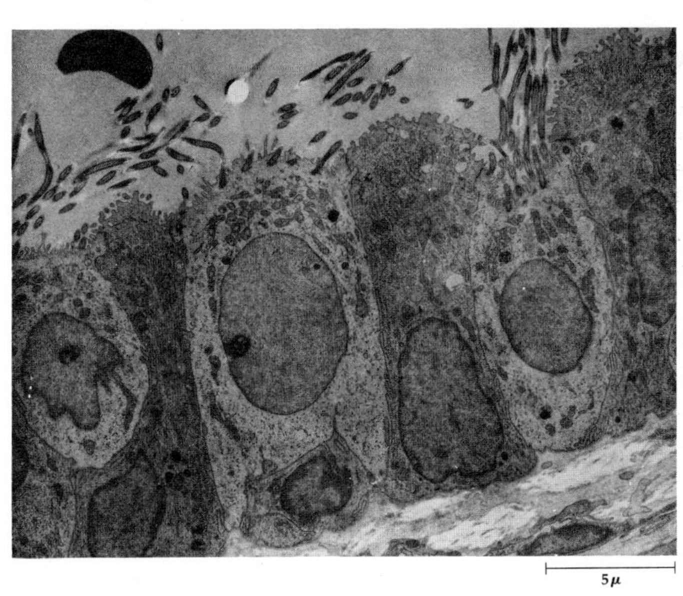

6–36
Electron micrograph and diagram of epithelial cells from the trachea of a bat. As you can see, ciliated cells (marked C in the drawing) and mucus-secreting cells (M) alternate in a regular pattern. The beating of the cilia removes foreign particles from the trachea and distributes a protective coating of mucus. In the micrograph, the white circles in the cytoplasm of the middle M cell are secretion droplets of mucus.

could be transported by an equal volume of water or plasma (the fluid portion of the blood). In invertebrates, hemoglobin and hemoglobinlike molecules often circulate free in the bloodstream, but if the mammalian bloodstream were to carry its vast quantity of hemoglobin molecules in this way, the blood would be more viscous (syrupy) and this would place a great extra load on the heart and blood vessels. In vertebrates, hemoglobin is transported in highly specialized cells: the red blood cells, or erythrocytes. A mature red blood cell carries some 265 million molecules of hemoglobin. Hemoglobin is red, and our blood is red because of its hemoglobin content.

Carbon dioxide, which is highly soluble (about 30 times as soluble as oxygen), is carried largely in the plasma. Carbon monoxide, however, combines readily with hemoglobin, filling the binding sites normally occupied by oxygen. It is for this reason that carbon monoxide is so extremely poisonous.

SUMMARY

Organisms obtain most of their ATP energy from enzymatic processes that take place in the mitochondria. These processes require oxygen and release carbon dioxide. Respiration is the means by which an organism obtains the oxygen required by its cells and rids itself of carbon dioxide.

Oxygen is available in both water and air. It enters into cells and body tissues by diffusion. However, movement of oxygen by diffusion requires a relatively large surface area exposed to the source of oxygen and a short distance over which the oxygen has to diffuse. Selection pressures for increasingly efficient means of obtaining oxygen and getting rid of carbon dioxide led to the evolution in vertebrates of gills and lungs. Both gills and lungs present enormously increased surface areas for the exchange of gases and have a rich blood supply for transporting these gases to and from other parts of the animal's body.

6–37

Lung cancer usually forms in the bronchial epithelium. These ,, indicating stages in cancer development, are all human patients.

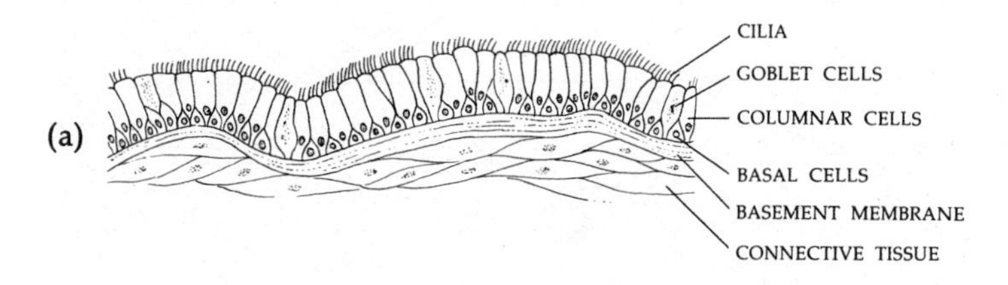

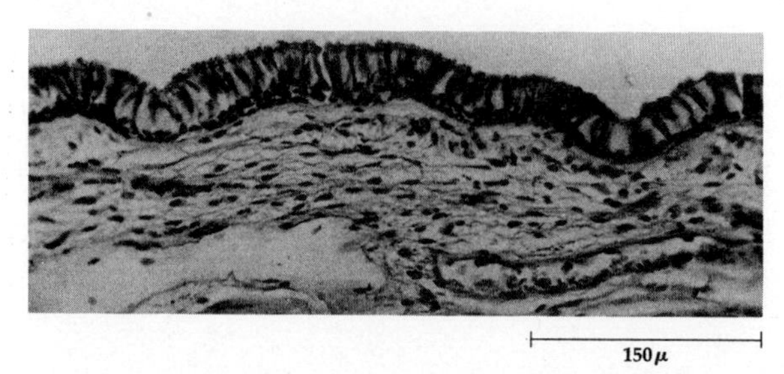

(a) *Normal epithelial tissue, showing ciliated columnar cells similar to those seen in the previous electron micrograph and an underlying layer of basal cells. In healthy tissue, the basal cells divide continuously, replacing the ciliated epithelial cells as these are worn away and slough off. A basement membrane separates the epithelial cells from the underlying tissues.*

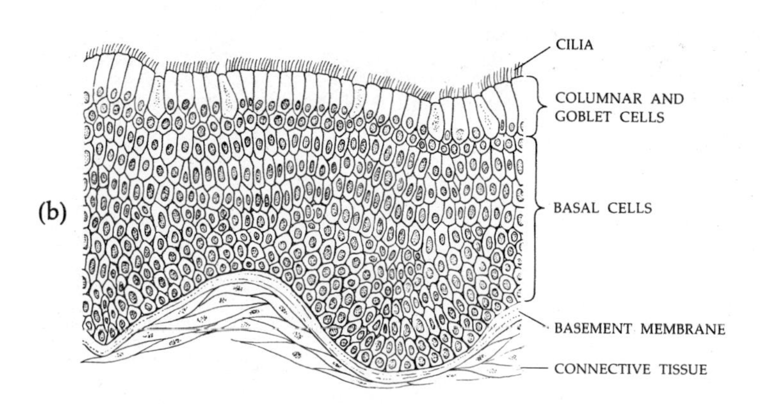

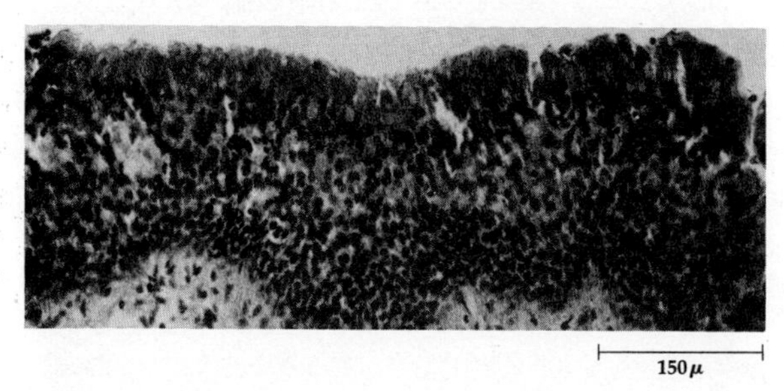

(b) *Tissue showing hyperplasia, an overgrowth of the basal cells. This is typically the first change in the epithelium seen among cigarette smokers.*

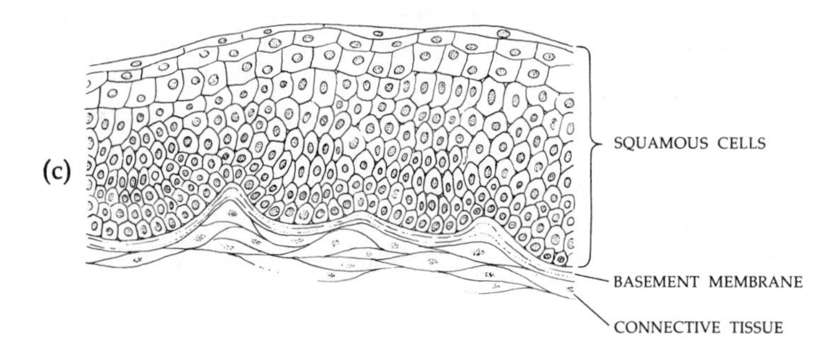

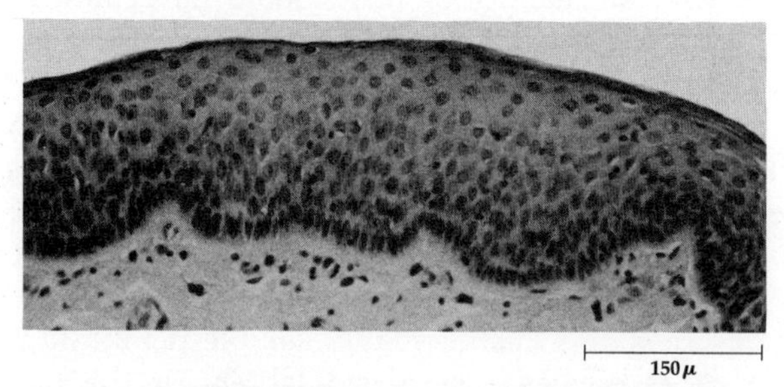

(c) *A later stage. The ciliated cells have been replaced by squamous (flattened) cells. With the disappearance of the cilia, the epithelium has lost a primary defense against tiny particles of soot, tobacco tar, dust, etc.*

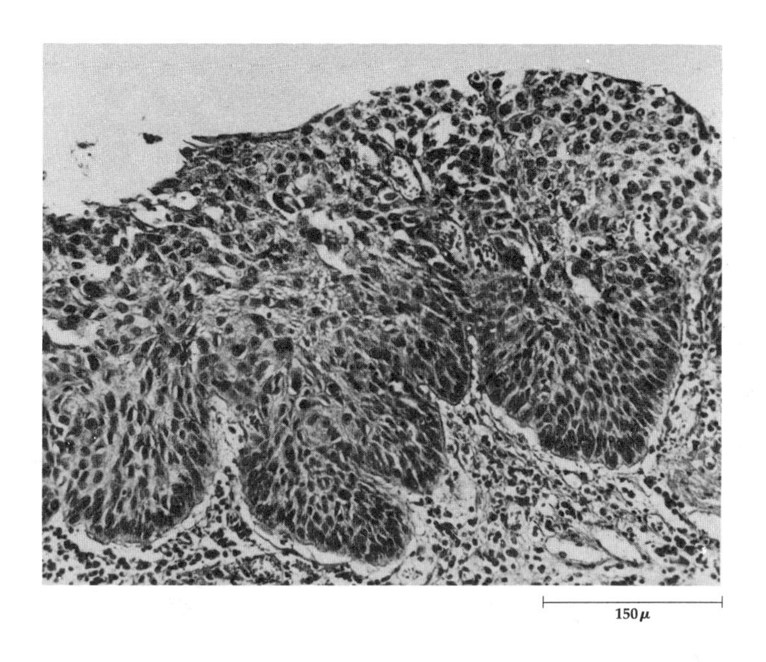

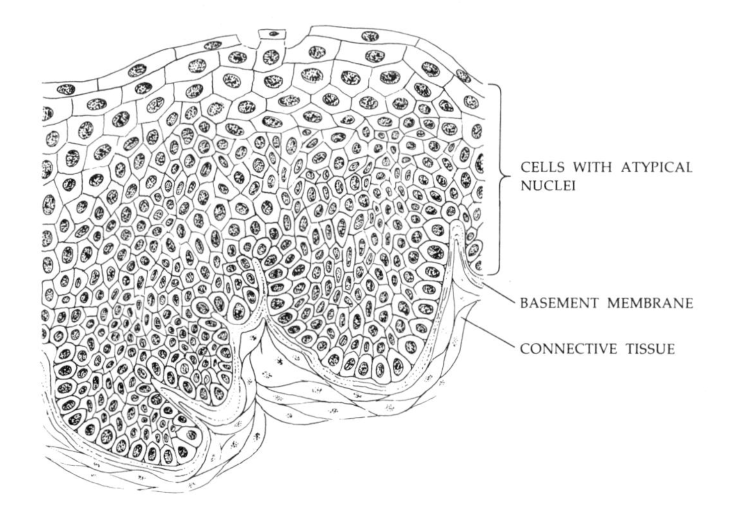

(d) *The cells have developed atypical nuclei and become disordered. This condition is called carcinoma in situ. There are usually no symptoms of disease at this stage.*

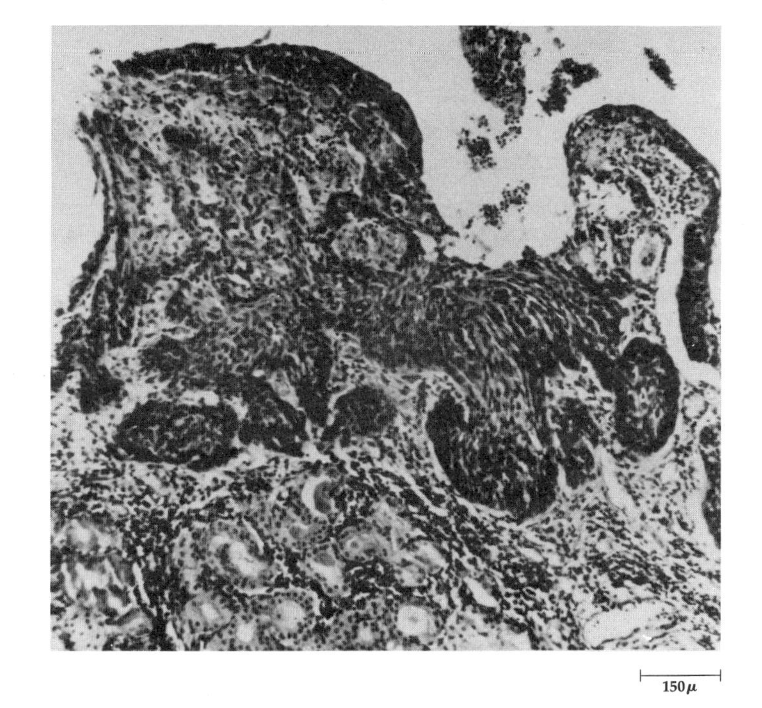

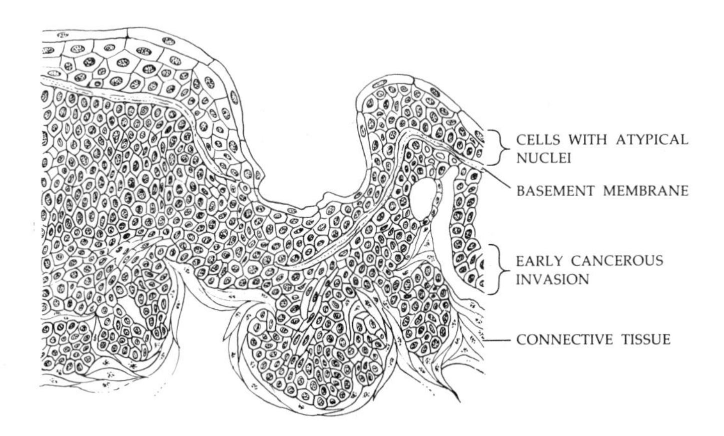

(e) *When these cells break through the basement membrane, as they have here, they may spread throughout the lungs and to the rest of the body. Lung cancer has become the most common form of cancer among males in the United States and is rapidly increasing.*

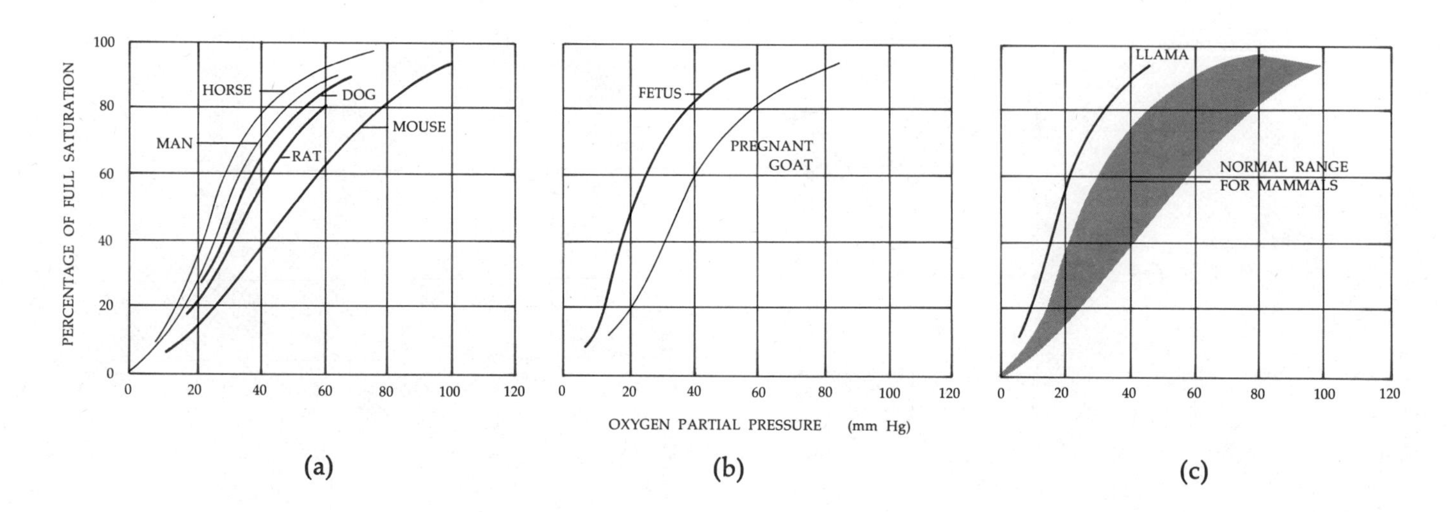

6–38

These curves show how the amount of oxygen carried by the hemoglobin is related to oxygen pressure. When oxygen pressure reaches 100 millimeters of mercury—the pressure usually present in the human lung, for instance—the hemoglobin becomes totally saturated with oxygen. As the pressure drops, the oxygen in the hemoglobin molecule is given up; when the pressure reaches 10 millimeters of mercury, virtually none of the hemoglobin is oxygenated. Therefore, when blood carrying oxygen reaches the capillaries, where pressure is only about 5 to 10 millimeters of mercury, it gives up its oxygen to the tissues. A curve located to the right signifies that the oxygen is given up more readily, that is, at a higher pressure. (a) *Small animals have higher metabolic rates and so need more oxygen per gram of tissue than larger animals. Therefore, they have blood that gives up oxygen more readily.* (b) *The fetus must take up all its oxygen from the maternal blood. Mammalian fetuses have a greater affinity for oxygen than adult mammals, and so the oxygen tends to leave the maternal blood and enter the fetal blood.* (c) *The llama, which lives in the high Andes of South America, has a hemoglobin which enables its blood to take up oxygen more readily at the low atmospheric pressures.*

In humans, air enters the lungs through the trachea, or windpipe, and goes from there into a network of increasingly smaller tubules, the bronchi, which finally end in small air sacs, the alveoli. Gas exchange actually takes place within the alveoli. The lungs empty and fill as a result of changes in the volume of the thoracic cavity. The muscles which control respiration, although they may be moved voluntarily, are under the involuntary control of nerves that originate in the spine. These nerves are stimulated by a respiratory center in the medulla oblongata that responds to very slight changes in the carbon dioxide concentration of the blood.

Carbon dioxide is transported in the blood plasma, the watery fluid of the bloodstream, in which it dissolves readily. Oxygen is carried by special hemoglobin molecules, each of which binds four atoms of oxygen. The hemoglobin is packed within red blood cells. Different types of hemoglobin accept and give up oxygen under slightly different conditions of oxygen pressure.

QUESTION

Why is diffusion much more effective over short distances than over longer ones? If you cannot answer this question, look again at the diagram of diffusion, Figure 2–25.

CHAPTER 6–4

Temperature Regulation

For life forms, the seeming wastelands of the desert and the Arctic represent great extremes of hot and cold. Yet, in fact, measured on a cosmic scale, the temperature differences between them are very slight. Life exists within only a very narrow temperature range. The upper and lower limits of this range are imposed by the nature of biochemical reactions, all of which are extremely sensitive to temperature change. Biochemical reactions take place almost entirely in water, the principal constituent of living things, and the slightly salty water characteristic of living tissues freezes at −1° or −2°C. Molecules that are not immobilized in the ice crystals are left in such a highly concentrated form that their normal interactions are completely disrupted. As temperature increases, the movement of molecules increases and the rate of enzymatic reactions increases rapidly; in fact, the reaction rate doubles for every 10°C increase in temperature. The upper temperature limit for life is apparently set by the point at which the hydrogen bonds that hold the proteins in their functional three-dimensional shape begin to break. (This process is known as *denaturation*.) Once denaturation occurs, enzymes and other proteins that depend on being a specific shape are no longer able to carry out their enzymatic functions. As a consequence of this restriction to a very narrow temperature zone, animals either must find external environments that range from just under freezing to about 40°C or must create internal ones. (An interesting exception are certain algae which, for reasons as yet unknown, are able to thrive at 70° to 80°C.)

The ways in which temperature requirements are met—and, in particular, the internal regulation of temperature—are the subject of this chapter. Before going into the details of these processes, however, it is worth taking a moment to consider what excellent temperature regulators most mammals are. One of the simplest and most dramatic demonstrations of this capacity was given some 200 years ago by Dr. Charles Blagden, then secretary of the Royal Society of London. Dr. Blagden, taking with him a few friends, a small dog in a basket, and a steak, went into a room that had a temperature of 126°C (260°F). The entire group remained there for 45 minutes. Dr. Blagden and his friends emerged unaffected. So did the dog. (The basket had kept its feet from being burned by the floor.) But the steak was cooked.

This regulation of internal temperature is an example of *homeostasis*. Homeostasis is the maintaining of a relatively unchanging internal state under conditions of constant external change. It is one of the principal and most important attributes of living organisms. In fact, evolution may be viewed as a steady increase in the capacity for homeostasis,

6–39

Life processes can take place only within a very narrow range of temperature. The temperature scale shown here is the Kelvin, or absolute, scale. Absolute zero (0°K) is equivalent to −273.1°C, or −459°F, and is the temperature at which all molecular motion ceases.

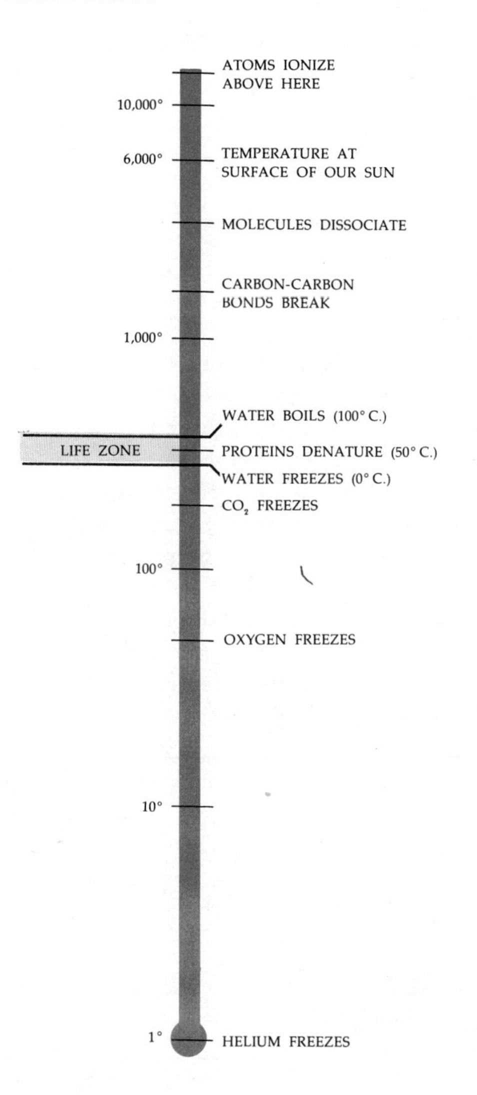

6–40

The rate of biochemical reactions doubles for every 10°C increase in temperature. This type of rapid increase, in which the quantity increases by multiplying (2 × 2 × 2) rather than by adding (2 + 2 + 2), is found often in biology. What other example can you think of?

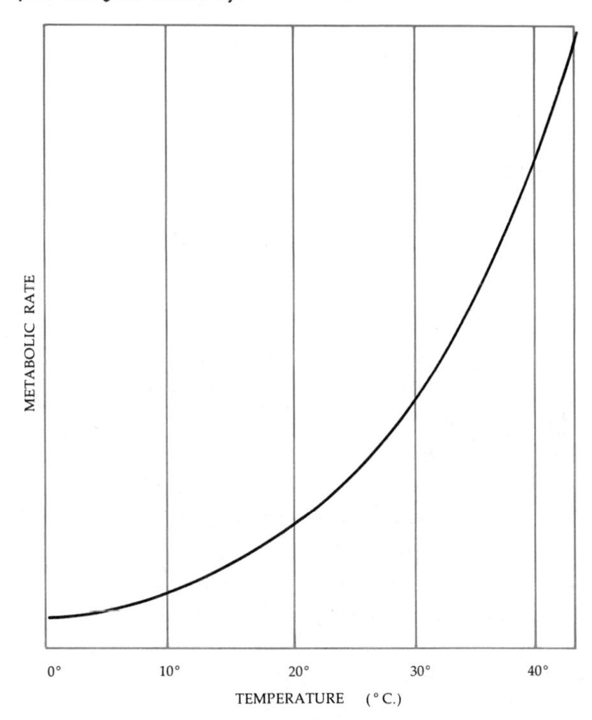

because it is by regulating the internal environment that animals have been able to free themselves to an increasing degree from the external environment.

"COLD-BLOODED" VERTEBRATES

Most animals are what we call "cold-blooded." A more precise term for cold-blooded animals is *poikilotherms*, from *poikilo*, meaning changeable. (Using this terminology, "warm-blooded" animals—the mammals and the birds are the only examples—are *homeotherms*.)

Most fish are poikilotherms. Their metabolic processes produce heat, but this heat is rapidly carried from the core of the body by the bloodstream and lost by conduction into the water. A large proportion of this body heat is lost from the gills. Exposure of a large, well-vascularized surface to the water is necessary in order to acquire enough oxygen, as you will recall from the preceding chapter. This same process rapidly dissipates heat, however. Fish also probably could not maintain a temperature lower than that of the water. Land animals at temperatures higher than that of their bodies unload heat largely by evaporation, a process not available to water dwellers.

In general, large bodies of water, for the reasons we discussed in Chapter 2–2, maintain a very stable temperature, which seldom exceeds 40°F. Because water expands as it freezes, ice floats on the surface of the water, and life continues beneath this surface. In the open ocean, temperatures rarely fluctuate more than a few degrees. In shallower water, where temperature changes do occur, fish seek an optimal temperature, presumably the one to which their metabolic processes are adapted. However, because they can do almost nothing to make their own temperature different from that of the surrounding water, they are quickly victimized by any rapid, drastic changes in water temperature. Although the temperatures of large bodies of water are usually stable—which is an advantage—they are also usually low, and the pace of life in such water is generally a slow one.

With the transition to land, vertebrates gained the first opportunity to maintain a high and constant temperature. Amphibians and reptiles are also poikilotherms, but behavioral adaptations, particularly in the case of the reptiles, leave them less at the mercy of their environment than are the fishes. Moreover, because of their low metabolic rate, they can, when they seek suitable conditions of temperature and humidity during periods of drought or low temperature, go for long intervals without eating. Toads which live in arid parts of Australia have been known to burrow for as long as two years and emerge alive with the advent of a rainy season. The salamanders, which are amphibians, spend most of their lives in water, although they are air breathers, as do aquatic reptiles such as alligators and turtles. These animals, maintaining a body temperature essentially the same as that of the surrounding water, seek water of the preferred temperature.

6–41
In laboratory studies, the internal temperature of reptiles was shown to be almost the same as the temperature of the surrounding air. It was not until observations were made of these animals in their own environment that it was found that they have behavioral means for temperature regulation that give them a surprising degree of temperature independence. By absorbing solar heat, as the desert iguana shown here is doing, reptiles can raise their temperature well above that of the air around them.

Fully terrestrial reptiles, the snakes and the lizards, often maintain remarkably stable body temperatures during their active hours. Unlike mammals, which get warm from the inside out, lizards get warm from the outside in, depending entirely on solar radiation as their source of warmth. By careful selection of suitable sites, such as the slope of a hill facing the sun, and by orienting their body with a maximum surface exposed to sunlight, they can heat themselves rapidly (as rapidly as 1°C per minute) even on mornings when the air temperature, as on deserts or in the high mountains, may be close to 0°C. As soon as their body temperature goes above the preferred level, lizards move to face the sun, presenting less exposed surface, or seek shade.

By staying in motion, lizards are able to keep their temperatures oscillating within a quite narrow range. When temperatures drop, the animals seek the safety of their shelters to avoid being immobilized in an exposed position and thus vulnerable to attack.

"WARM-BLOODED" ANIMALS

The only "warm-blooded" animals—homeotherms—are the birds and the mammals. Early mammals were small, about the size of a rat or mouse. They had sharp teeth, but since they were too small to attack other vertebrates, they are assumed to have lived mostly on insects and worms. They first appeared in the middle of the Mesozoic era, at about the time of the first dinosaurs. For some 80 million years, they coexisted with the huge carnivorous reptiles. Then suddenly, as geologic time is measured, most of the great reptiles disappeared and the mammals began to increase in number and variety, soon becoming by far the most prominent and successful group of large land animals.

Paleontologists are not agreed about why the dinosaurs became extinct. Clearly, however, a major reason for the primitive mammals' being able to survive was their ability to maintain a high and constant body temperature, which enabled them to be both fast and alert regardless of external conditions. Also, and perhaps most important at the beginning, their warm-bloodedness enabled them to be active at twilight and throughout the night, when the reptiles were rendered inactive by the cold. In fact, most modern species of mammals are still largely nocturnal, and even those we regard as the most primitive are intelligent and energetic to a degree seldom attained by any poikilotherm.

Like the mammals, the birds evolved from reptiles. Even today, they share many characteristics, such as certain skeletal and muscular structures and similar eggs, with the reptiles. Paleontologists believe that the first birds began to appear some 150 million years ago, shortly after the first mammals. The principal distinguishing feature of these early birds was the fact that they had feathers, which are important as insulators. This is still a unique avian characteristic; no animals except the birds have feathers, and all birds possess them.

TEMPERATURE REGULATION IN HOMEOTHERMS

In man and other homeotherms, heat is provided for the organism as one of the products of the breakdown of glucose and other energy-containing molecules within cells. A warm-blooded mammal is warmer at the core than at the periphery. (Man's temperature does not reach 37°C or 98.6°F until some distance below the skin surface.) Heat is transported from the core to the periphery by the circulatory system. At the surface of the body, the heat is transferred to the air. Temperature regulation involves increasing or decreasing heat production and increasing or decreasing heat loss at the body surface. The controls are very tight. Normally, a person's internal temperature, as measured by a rectal thermometer, does not vary more than a few tenths of 1°F.

This remarkable constancy of temperature is maintained by an automatic system—a thermostat—in the hypothalamus, which precisely measures the body temperature and triggers the appropriate control mechanisms. Receptor cells in this center monitor the temperature of the blood servicing the cells of the hypothalamus and those of the rest of the brain. Although the surface of the skin is covered with receptors for hot and for cold, these are not directly involved in the regulation of internal temperature, as has been demonstrated by some interesting experiments with human subjects. For example, in a room in which the air is warmer than body temperature, if the blood circulating through a person's hypothalamus is cooled, he will stop perspiring, even though his skin temperature continues to rise. The skin receptors signal changes in external temperature only, and these signals travel to the centers of consciousness in the brain, bypassing the unconscious center in the hypothalamus.

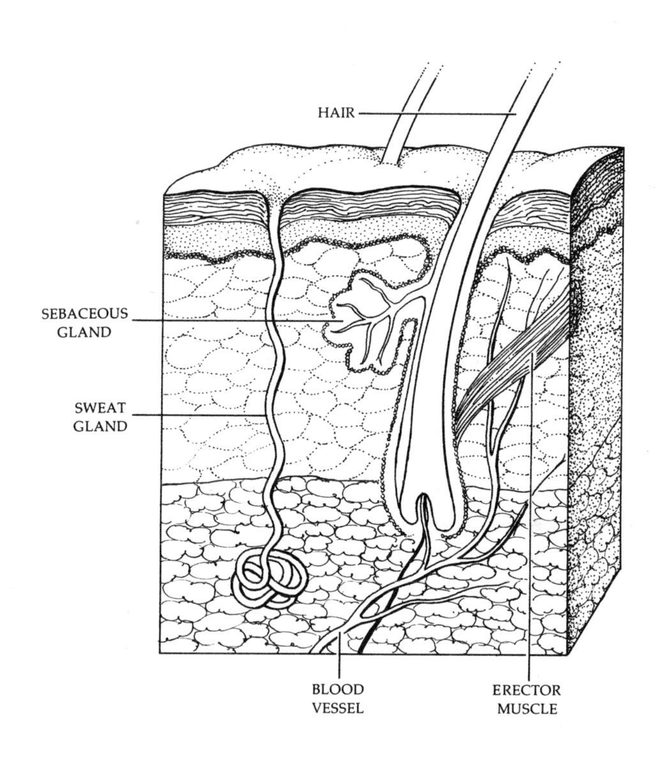

6–42
Cross section of human skin. In the cold, the blood vessels constrict and the hairs, which each have a small muscle at the base under nervous control, stand upright. In animals with body hair (Homo sapiens *is one of the few terrestrial mammals that is virtually hairless), air trapped between the hairs insulates the skin surface, conserving heat. In the heat, the blood vessels dilate and the sweat glands secrete a salty liquid. Evaporation of this liquid cools the skin surface, dissipating heat.*

6–43
Dogs unload heat by panting, which involves short, shallow breaths. When it is hot, a dog breathes at a rate of about 300 *to* 400 *respirations a minute, compared with a respiration rate of from* 10 *to* 40 *a minute in cool surroundings.*

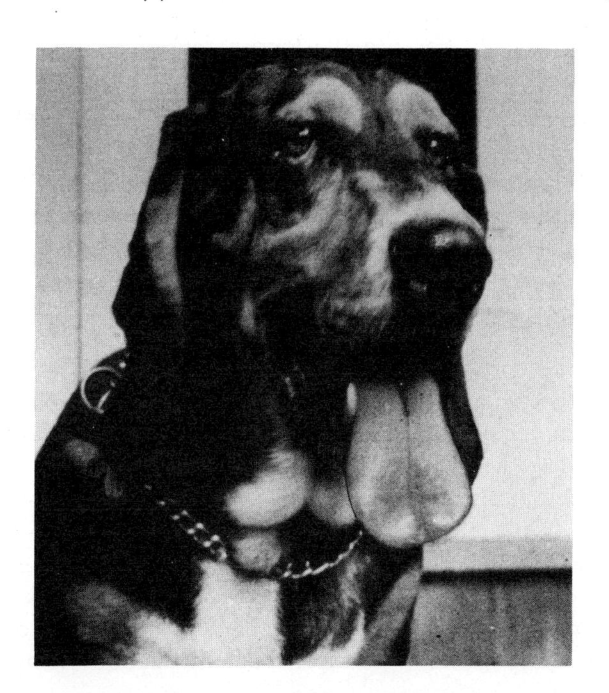

Regulation at High Temperatures

As the external temperature rises, the blood vessels near the skin surface dilate and the supply of blood to the skin surface increases. In man and most other large animals, when the external temperature rises above body temperature, perspiration begins. Evaporation from the body surface requires heat energy (see page 51), which comes from the surface, and the blood just below the skin surface cools as this heat energy is used. Men and horses sweat from all over their body surfaces. Dogs and some other animals pant, so that air passes rapidly over the large moist tongue, evaporating water. Cats lick themselves all over as the temperature rises, and evaporation of this water cools their body surface. For animals that sweat, temperature regulation at high temperatures necessarily involves water loss, a problem which will be discussed in more detail in Chapter 6–6. (Dr. Blagden and his friends were probably very thirsty, illustrating the fact that solution of one evolutionary problem often creates another.)

Regulation at Low Temperatures

As external temperatures go down, cutaneous blood vessels constrict, limiting heat loss across the skin surface. Metabolic processes increase. In response to cold, the thyroid glands, which are two small glands located in the throat, increase their production of thyroid hormone. Thyroid hormone increases the metabolic rate; it apparently exerts its effects directly on the mitochondria, although the exact way in which it works is not known. The skin of a person whose thyroid glands are not functioning at a normal level always feels cold, and such people often complain about never being warm enough. The thyroid, like the sex glands, is under the control of the pituitary, which, in turn, is regulated by the hypothalamus. With a further drop in temperature, the adrenal gland produces adrenaline, which also raises the body's metabolic rate, increasing the supply of heat. A naked man begins shivering when the air temperature reaches 20°C. Shivering is another means of raising heat production; it mobilizes glucose and ATP stored in muscle tissues and releases their energy in the form of motion and heat.

TEMPERATURE, SIZE, AND ENERGY

Heat radiates out from the body surface, and radiation of heat, like diffusion of gases, is proportional to the surface area that is exposed. Animals minimize heat loss by curling up at night with their legs underneath them, reducing the exposed surface area.

It is more difficult for a small animal to maintain a high body temperature than it is for a large one, since the surface-area-to-volume ratio of small animals makes them such efficient radiators. Mice nest, making a communal heat. Rabbits burrow. During the cold season, hamsters and ground squirrels hibernate, permitting their body temperatures to drop very low and their metabolic processes to slow down.

Most birds and mammals maintain an almost constant body temperature. Birds require it for their special activities, including sustained flight. In the mammals, high constant temperature is undoubtedly associated with activity and alertness and, as a consequence, the evolution of intelligence. Maintaining constant temperature adds significantly to an animal's energy requirements, however. Birds maintain a comparatively high body temperature of 40° to 42°C. Unlike other small animals, they spend most of their time exposed, and flying birds have high energy requirements for flight. Flying birds have a particularly difficult problem since they must keep their weight down and therefore cannot store much food for energy; thus they need to refuel constantly to keep warm and to keep flying. A bird eating high-protein foods, such as seeds and insects, commonly consumes as much as 30 percent of its body weight per day. Fall migrations are necessary not so much because of the colder weather, since a well-fed bird can maintain its temperature, but rather because of the reduced food supply and the shortened day, which gives the diurnal bird less time to feed. Similarly, birds return to the temperate zones to breed partly because the longer days give them more time to seek food to meet the high-energy demands of their young. A hummingbird consumes about 85 milliliters of oxygen per gram per hour in hovering and about 10 when awake but not flying. Its reserves are insufficient for this rate of metabolism at night; its body temperature drops every evening, and so it conserves fuel.

HIBERNATION

Hibernation, which comes from *hiber*, the Latin word for "winter," is another means of adjusting energy expenditures to food supplies. Hibernating animals do not stop regulating their temperatures altogether—they turn down their thermostats. Life at a lower temperature conserves energy. Only a few animals truly hibernate; some others, such as the bear, sleep for long periods but do not hibernate in the strict sense of the word. True hibernators include a few insectivores, some rodents, some South American opossums, and some bats. In these animals, the body tempera-

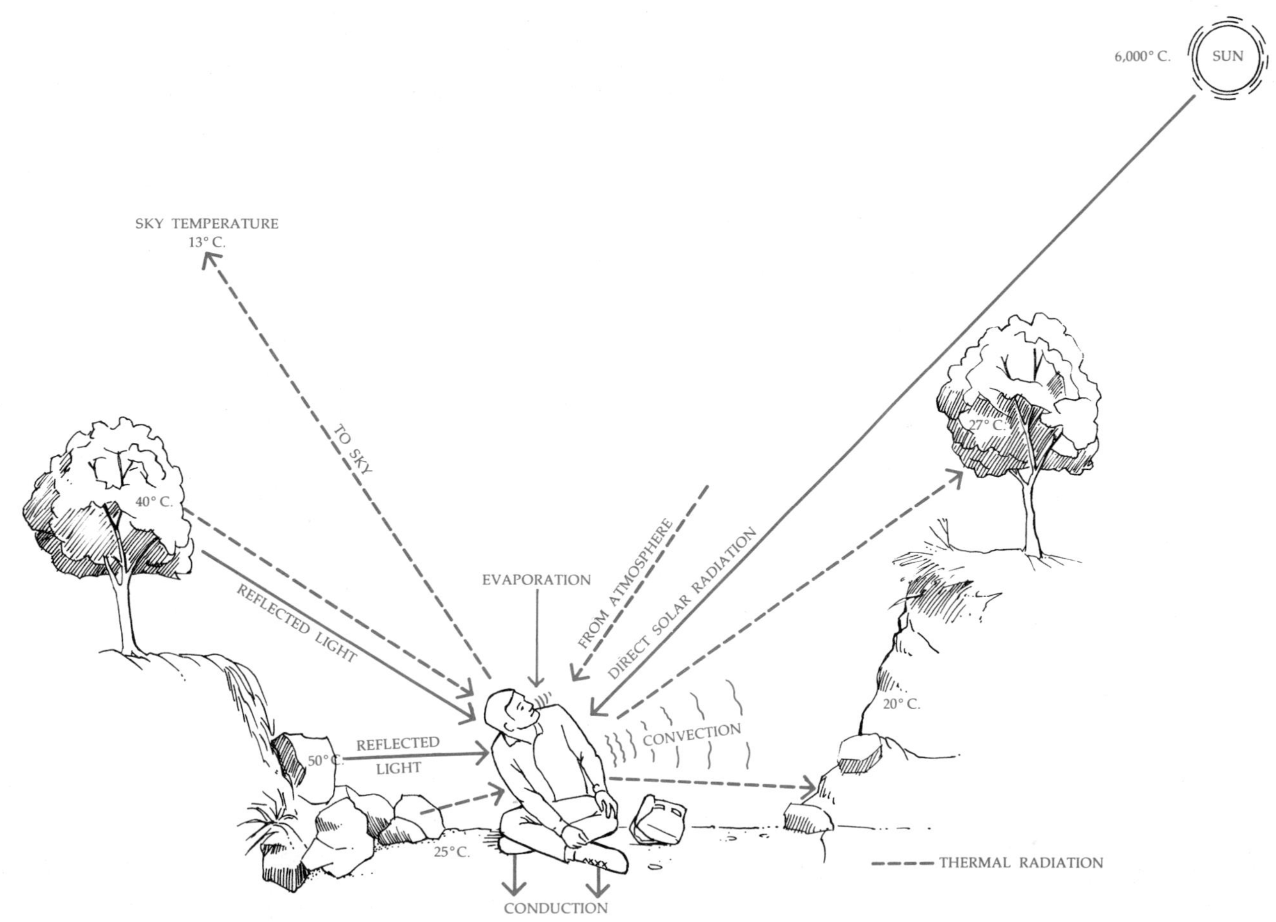

6–44
The energy exchanges between a mammal and its environment under moderately warm conditions.

ture is lowered, often to that of the surrounding air (if the air is above 0°C). The heartbeat slows; the heart of an active ground squirrel, for instance, beats 200 to 400 times a minute whereas that of a hibernating one beats 7 to 10 times a minute. The metabolism as measured by oxygen consumption is reduced 20 to 100 times. However, many of the monitoring processes continue. If the animals are exposed to carbon dioxide, for example, they breathe fast and those of certain species wake up. Similarly, animals of many species wake up if the temperature in their burrows drops to a life-threatening low of 0°C. Hibernators of some species can be awakened by sound or by touch.

Arousal from hibernation can be rapid. In one experiment, bats kept in a refrigerator for 144 days without food were capable of sustained flight after 15 minutes at room temperature. As indicated, however, by the arousals at 0°C, the arousal is a process of self-warming rather than of collecting environmental heat. Breathing becomes more regular and then more rapid, increasing the amount of oxygen available for consumption, and then the animal "burns" its stored food supplies as it returns to its normal temperature and the fast pace of a homeothermic existence.

6–45

(a) *Prior to hibernating, mammals, such as the dormouse shown here, store up food reserves by eating and eating and eating. . . .* (b) *Note the heat-conserving position of this hibernating chipmunk.* (c) *Bats have a number of special adaptations which enable them to conserve heat despite their small body size. Some hibernate. In some, the temperature drops when they are at rest (as in hummingbirds). Members of the species shown here also roll up their ears when they get cold.*

(a)

(b)

(c)

Table 6–4

Oxygen Consumption under Approximately Normal Physiological Conditions

Animals	*Body Weight (gm)*	*Oxygen Consumption (ml of O_2/gm of wet wt/hr)*	*Temperature (°C)*
Mammals:			
Long-tailed shrew	3.6	10.6 (basal); 15.6 (24-hr fed)	
Laboratory mouse	20	1.69	
Laboratory rat	282	0.88	
Guinea pig	460	0.76	
Cat	3,000	0.446	
Dog	20,000	0.360	
Sheep	46,800	0.250	
Woman	57,900	0.204	
Cow	300,000	0.124	
Birds:			
Hummingbird—			
at rest	3.8	10.7	
active		85	
Pigeon—			
at rest		1.5	
active		42	
Poikilothermic vertebrates:			
Trout (*Salvelinus*)		0.349	15
Turtle	3,000–4,000	0.088	16
Alligator	53,000	0.0747	16
Rattlesnake	2,000–3,000	0.068	16
Frog	35	0.056	16

Note the higher energy requirements of the smaller homeotherms.

ADAPTATIONS TO EXTREME TEMPERATURES

Man relies mainly on his technology, rather than his physiology, to permit him to live in extreme climates, but many animals are comfortable in climates that we consider inhospitable or, indeed, uninhabitable.

Adaptations to Extreme Heat

Man is a good regulator of his own temperature at high external temperatures, as Dr. Blagden's experiment showed. His chief limitation in this regard is that he must evaporate a great deal of water in order to unload body heat and so must keep his water consumption high. The camel, the philosophical-looking "ship of the desert," has several advantages over man in the desert. For one thing, the camel excretes a much more concentrated urine; in other words, it does not need to use so much water to dissolve its waste products. In fact, man is very uneconomical with his water supply; even dogs and cats excrete a urine twice as concentrated as man's.

Also, the camel can lose more water proportionally than man and still continue to function. If a man loses 10 percent of his body weight in water, he becomes delirious, deaf, and insensitive to pain. If he loses as much as 12 percent, he is unable to swallow and so cannot recover without assistance. Laboratory rats and many other common animals can tolerate dehydration up to 12 to 14 percent of body weight. Camels can tolerate the loss of more than 25 percent of their body weight! They can go without drinking for one week in the summer months, three weeks in the winter.

Finally, and probably most important, the camel can tolerate a fluctuation in internal temperature of 5° to 6°C. This tolerance means that it can let its temperature rise during the daytime (which the human thermostat would never permit) and cool during the night. The camel begins the next day at below its normal temperature—storing up coolness, in effect. It is estimated that the camel saves as much as 5 liters of water a day as a result of these internal temperature fluctuations. Camels' humps were once thought to be water-storage tanks, but actually they are localized fat deposits. Physiologists have suggested that the camel carries its fat in a hump instead of distributed all over the body because fat acts as an insulator and impedes the heat flow to the skin. An all-over distribution is thus decidedly useful in Arctic animals but not in inhabitants of hot climates.

Small desert animals usually do not unload heat by sweating or panting; because of their relatively large surface area, such mechanisms would be extravagant in terms of water loss. Rather, they regulate their temperature by selecting a diet with a high water yield (discussed further in Chapter 6–6), by excreting a highly concentrated urine, and by avoiding direct heat. Most small desert animals are nocturnal, like the earliest mammals from which they are descended.

6–46

By facing the sun, a camel exposes as small an area of body surface as possible to its radiation. Note the loose-fitting garments worn by the camel driver. When such garments are worn, sweat evaporates off the skin surface. With tight-fitting garments, the sweat evaporates from the surface of the garment instead, and much of the cooling effect is lost. The loose robes, which trap air, also serve to keep desert dwellers warm during the cold nights. Temperatures drop precipitously on the desert—as much as 30°C in a few hours—because the air lacks water, which moderates temperature changes.

Adaptations to Extreme Cold

When you step out of bed barefoot on a cool morning, you probably prefer to step on a wool rug rather than on the bare floor. Although both are at the same temperature, the rug feels warmer. If you touch something metal, such as the door handle, it will feel even cooler than the floor. These apparent differences in temperature are actually differences in the speed at which these different types of materials conduct heat away from your body. The door knob, like all metals, is an excellent conductor, and wood is a better conductor than wool.

Water is a better conductor than air. A man is quite comfortable in air at 70°F but uncomfortable in water at the same temperature. Asbestos is such a poor conductor that it acts as an insulator, interrupting the flow of heat. Fat is also a poor conductor.

Poor conductors, such as asbestos, wool, air, and fat, serve as insulators, materials that interrupt the flow of heat from an object. Animals adapt to extreme cold largely by increased amounts of insulation.

Fur and feathers provide insulation for Arctic land animals. These materials not only are good insulators in themselves but serve to trap air. Fur and feathers are usually shed to some extent in the spring and regrow in the fall.

Whales, walruses, and seals insulate themselves with fat. In general, the rate of heat loss depends both on the amount of surface area and on temperature differences between the body surface and the surroundings. These marine mammals, which survive in extremely cold water, can tolerate a very great drop in their skin surface temperature; measurements of skin temperature have shown that it is only a degree or so above that of the surrounding water. By permitting the skin temperature to drop, these animals, which maintain an internal temperature as warm as that of man, expend very little heat outside of their fat layer and so keep warm—like a skin diver in a wet suit.

Many animals similarly permit temperatures in the extremities to drop. By so doing, they conserve heat. Also, for many animals, particularly those that stand on the ice, this capacity is essential for another reason. If, for example, the feet of a seabird were as warm as its body, they would melt the ice, which would then freeze over them, trapping the animal until the spring thaws set it free. The extremities of such animals are actually adapted to live at a different internal temperature from that of the rest of the animal. For example, the fat in the foot of an Arctic fox has a different freezing point from that of the fat in the rest of its body, so that its footpads are soft and resilient even at temperatures of −50°C.

Countercurrent Exchange

In many Arctic animals, the veins and arteries leading to and from the extremities are juxtaposed in such a way that the chilled blood returning from the legs (or fins or tail) through the veins picks up heat from the blood entering the extremities through the arteries. The veins and arteries are closely intermingled to give maximum surface for heat transfer. This arrangement, which serves to keep heat in the body and away from the extremities, is accomplished,

6–47
It is easy to forget that small animals often do not live in the same environment as ours but live instead in a microclimate which may be entirely different. On our scale, the desert may offer no refuge from the sun, but a prairie dog on the same terrain can readily find a hospitable shelter.

6–48
Animals which live in cold climates, such as this Weddell seal, an inhabitant of Antarctica, have a heavy layer of fat just below the skin which serves as insulation.

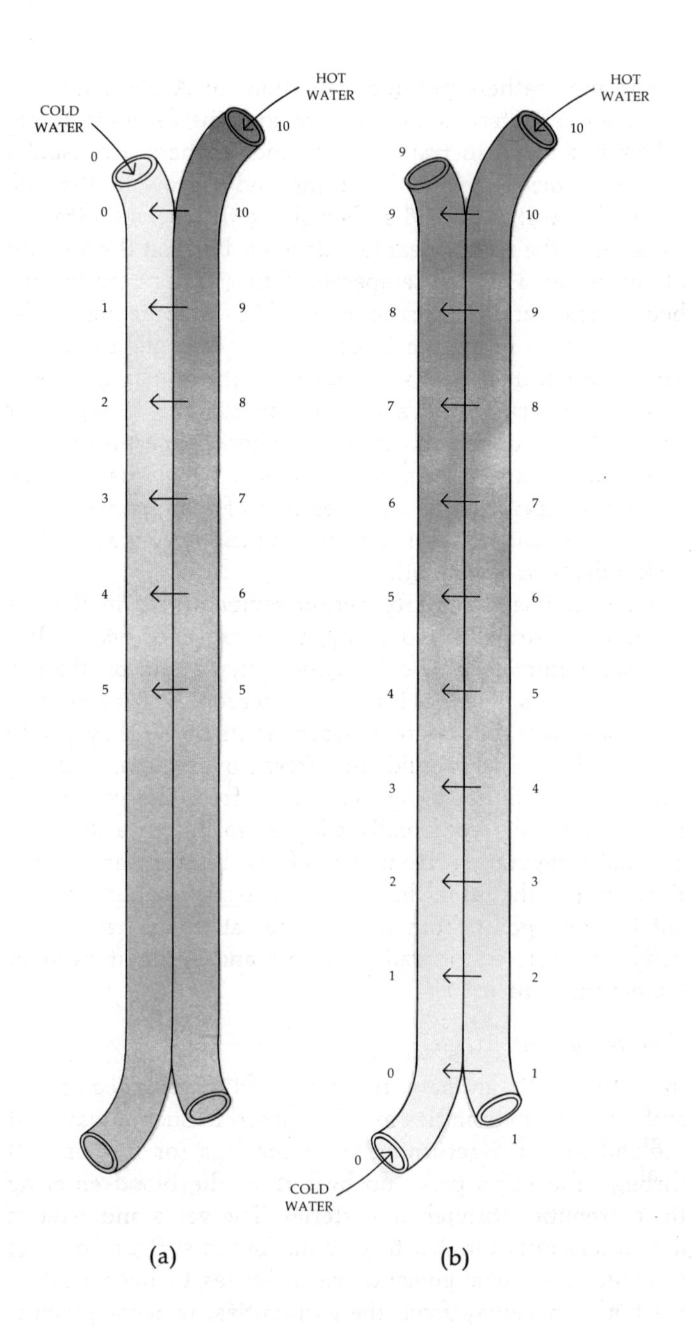

6–49
The principle of countercurrent exchange is illustrated by a hot-water pipe and a cold-water pipe placed side by side to achieve maximum temperature transfer under ideal conditions. In (a), *the hot water and cold water flow in the same direction. Heat from the hot-water pipe warms the cold water until both temperatures equalize, at* 5. *Thereafter, no further exchange takes place and the water in both pipes remains lukewarm. In* (b), *the flow is in opposite directions, so that heat transfer continues for the length of the pipes. The result is that the hot water transfers most of its heat as it travels through the pipe and the cold water is warmed to almost the initial temperature of the hot water.*

largely, by what is known as a countercurrent exchange.

Countercurrent heat-exchange mechanisms have recently been found to operate in two groups of fish—the tunas and the mackerel sharks—to conserve metabolic heat and thus raise body temperature. As we saw earlier in this chapter, the body temperatures of most fish are at or within a degree or two of water temperature, but in these warm-bodied fish, the internal temperatures are anywhere from 5° to 13°C higher than that of the water they inhabit. Both tunas and mackerel sharks are fast and tireless swimmers, and it is believed that their high body temperatures are associated with this ability, since the power available from muscle increases with temperature.

The basic principle of countercurrent exchange is indicated in a simple example borrowed from physiologist P. F. Scholander and illustrated in Figure 6–49. Imagine two pipes lying next to each other so that heat can easily be transmitted from one to the other. Suppose that cold water flows in one and hot water flows in the other. If the flow is in opposite directions and the pipes are long enough, the warm pipe will transfer most of its heat to the cold one.

Countercurrent exchanges are common biological mechanisms; they facilitate the exchange of gases and other substances as well as temperature transfers. The same principle is applied in industry—for example, to use the exhaust gases flowing out from a furnace to preheat air flowing into the furnace thereby making use of machinery "invented" millions of years before the evolution of engineers.

SUMMARY

Life can exist only within a very narrow temperature range, from about 0°C to about 50°C, with few exceptions. Animals must either seek out environments with suitable temperatures or create suitable internal environments. The warmer the environment, the higher the rate of biochemical reactions within the life zone. Poikilotherms are animals whose temperature is variable, that is, dependent on their surroundings. Homeotherms are animals that maintain a constant and comparatively high temperature.

Most poikilotherms live in the ocean or other large bodies of water. In such environments, the temperatures are generally low, however, and the pace of life is slow. Most fish are poikilotherms, as are all invertebrates.

With the emergence to land, animals entered a life zone in which temperatures are much more variable and also in which temperature regulation is possible. Amphibians and, in particular, reptiles modify their temperature by a number of behavioral adaptations. The first homeotherms were the primitive mammals, small nocturnal insectivores.

In homeotherms—the birds and the mammals—temperature is monitored by centers in the hypothalamus. The regulation of temperature requires a means for producing heat and methods of preventing heat dissipation. Heat is produced by the breakdown of glucose and other energy-yielding compounds.

The rates of heat production can be increased to some degree by thyroid hormone and also by adrenaline. Heat is carried by the bloodstream from the core of the body to the surface, where it is dissipated. In hot weather, the cutaneous blood vessels dilate, increasing the flow of blood to the skin; as external temperatures rise above body temperatures, man and most other large animals begin to sweat. Evaporation of sweat from body surfaces requires heat and cools the surface. In the cold, cutaneous blood vessels contract. As temperatures drop, shivering begins; the muscular activity involved in shivering produces heat. Arctic animals tend to have less body surface in proportion to volume than animals of the same species living in more moderate climates. All large Arctic animals are also heavily insulated with fat, fur, or feathers.

Regulation of body temperature is one form of homeostasis, the maintaining of a relatively constant internal state under conditions of external change. The maintaining of a high and constant body temperature makes possible a degree of sustained physical activity and mental alertness generally unattainable by poikilotherms.

QUESTIONS

1. Blood supply to the hypothalamus can be cooled by eating or drinking something very cold. Describe the consequences of ingesting large amounts of ice water or ice cream on a hot day.

2. Compare the surface-area-to-volume ratio of an Eskimo igloo with that of a California ranch house. In what way is the igloo well suited to the environment in which it is found?

3. Compare hibernation in animals with dormancy in plants. In what ways are they alike? In what ways are they different?

4. The gills of fish are countercurrent mechanisms: the blood in the capillaries moves in the direction opposite to that of the oxygen-bearing water. Explain, using a diagram if necessary, why this facilitates the pickup of oxygen.

CHAPTER 6–5

Circulation

In vertebrates, gases (oxygen and carbon dioxide), food molecules, hormones, and wastes are transported in the bloodstream. The blood circulates through the body of higher animals in a closed system of continuous vessels propelled by the contractions of a specialized muscle, the heart. The heart pumps the blood into the large arteries, from which it travels to branching, smaller arteries and then into very small vessels, the *capillaries*. The capillaries are the components of the circulatory system that service the tissues, supplying them with oxygen, food molecules, hormones, and other materials. Substances pass in and out of the capillaries largely by diffusion through the thin (1 micron thick) walls separating the capillaries from the fluids surrounding the individual cells. Passage out of the capillaries is also a result of the hydrostatic pressure of the blood within the vessels.

There is a capillary within rapid diffusion distance of every cell in the body—the total length of the capillaries in a human adult, for example, is more than 50,000 miles. Even the cells in the walls of the veins and arteries depend on this capillary system for their blood supply, as does the heart itself, as well as all the other organs of the body.

From the capillaries, the blood passes back into small veins, then into larger veins, and then back to the heart. In man, the diameter of the opening of the largest artery, the *aorta*, is 2.5 centimeters, that of a capillary only 8 microns, and that of the largest vein, the *vena cava*, 3 centimeters.

By comparing the circulatory system to a complex plumbing system—which is, in effect, what it is—you can see that a great deal of propulsive force is necessary to drive the blood through the small openings of the capillaries. There is a large drop in pressure as the blood goes through the capillaries. The return of blood to the heart through the veins depends in large part on body movements, which squeeze the veins between contracting muscles and force the blood upward. Valves in the veins prevent backflow. Also, the thoracic cavity is under negative pressure (less than atmospheric pressure), and as the thorax expands in respiration, the elastic walls of the veins in the thorax dilate, sucking the blood upward.

EVOLUTION OF THE HEART

The heart, in its simplest form—such as the earthworm heart—is a muscular contractile part of a circulatory vessel. In the course of vertebrate evolution, the heart has undergone some interesting adaptations, as shown in Figure 6–50.

In the fish, there is a single heart divided into the *atrium*, which is the receiving area for the blood, and the *ventricle*, which is the pumping area from which the blood is expelled into the vessels. The ventricle of the fish heart pumps blood directly into the capillaries of the gills, where it picks up oxygen and releases carbon dioxide. From the gills, oxygenated blood is carried to the tissues. By this time, however, most of the propulsive force of the heartbeat has been

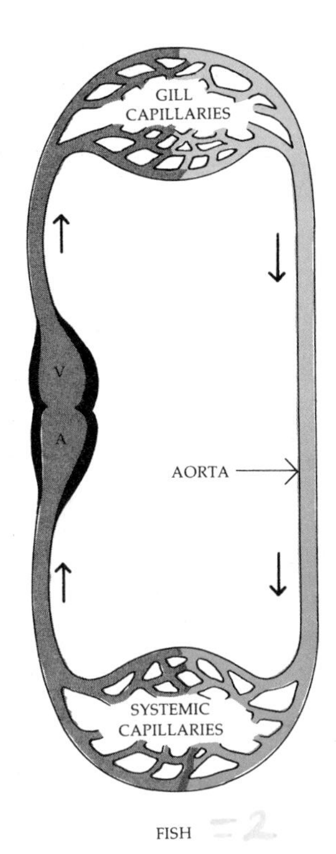

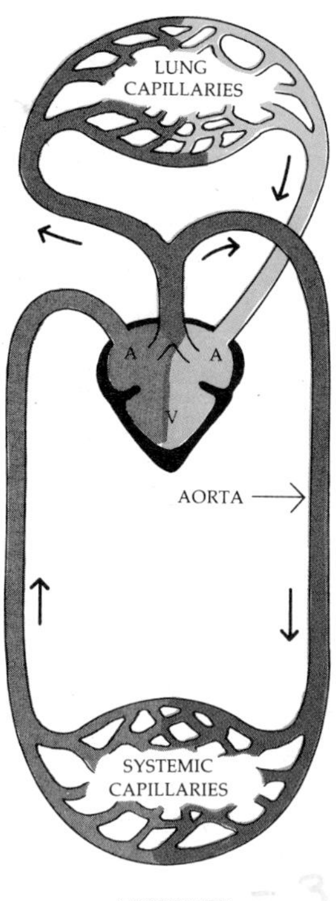

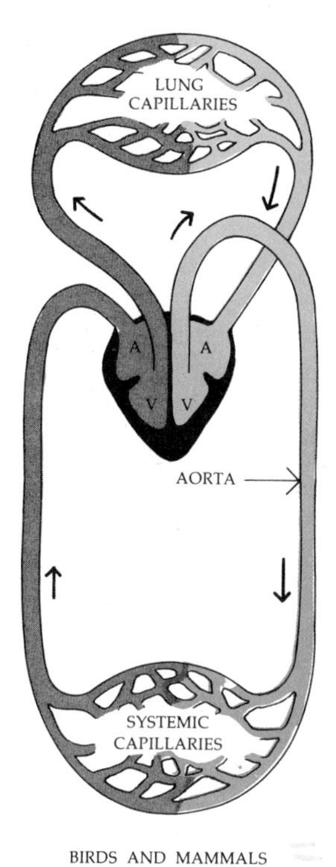

6–50

Vertebrate circulatory systems. In the fish, the heart has only one atrium (A) and one ventricle (V). Blood aerated in the gill capillaries goes straight to the systematic capillaries without first returning to the heart. In amphibians, the single primitive atrium has been divided into two separate chambers. Oxygenated blood (indicated by color) from the lungs enters the left atrium, where it is mixed with the oxygen-poor blood (gray) in the heart; the mixed blood is then passed through the body tissues. In birds and mammals, the ventricle is divided into two separate chambers, so that there are, in effect, two hearts—one for pumping oxygen-poor blood through the lungs and one for pumping oxygen-rich blood through the body tissues.

dissipated by the resistance of the capillaries in the gill, so that the blood flow to and through the systemic circulation is relatively sluggish.

In amphibians, there are two atria; one receives oxygenated blood from the lungs and the other receives deoxygenated blood from the systemic circulation. Both atria empty into the single ventricle, which pumps the mixed blood simultaneously through the lungs and the systemic circulation. By this arrangement, the blood enters the systemic circulation under higher pressure.

In the birds and mammals, the heart is functionally separated longitudinally into two organs. The right heart receives blood from the tissues and pumps it into the lungs, where it becomes oxygenated. From the lungs it returns to the left heart, from which it is pumped into the body tissues. This high-pressure, double circulation system is necessary to maintain the high metabolic rate of both birds and mammals, with their constant body temperature and their generally high level of physical and mental activity.

REGULATION OF THE HEARTBEAT

Figure 6–51 shows a diagram of the human heart. Blood returning from the body tissues enters the right atrium through the superior and inferior venae cavae. Blood returning from the lungs enters the left atrium through the pulmonary veins. The atria, which are thin-walled compared with the ventricles, expand to receive the blood. Both atria then contract simultaneously, assisting the flow of blood through the open valves into the ventricles. Then the ventricles contract simultaneously, closing the valves. The right ventricle propels the blood into the lungs through the pulmonary arteries, and the left ventricle propels it into the aorta, the principal artery leading from the heart, from which it travels to the other body tissues. In a healthy man at rest, this rhythmic process takes place about 70 times a minute. Under strenuous exercise, the rate more than doubles.

What controls the heartbeat? In vertebrates, the heartbeat orginates in the heart itself. A vertebrate heart will continue to contract even after it is removed from the body if it is kept in a nutrient solution. In vertebrate embryos, the heart begins to beat very early in development, before the appearance of any nerve supply. In fact, embryonic heart cells isolated in the test tube will beat.

The beat of the cardiac muscle is synchronized by a special area of the heart known as the pacemaker, which is located in the right atrium. The pacemaker is composed of nodal tissue. Nodal tissue is unique in that it can contract like a muscle and also transmit impulses like a nerve.

From the pacemaker, a wave of contraction spreads to the left atrium. About 100 milliseconds after the pacemaker fires, impulses from both atria stimulate a second area of nodal tissue, the atrioventricular node. The atrioventricular node is the only electrical bridge between the atria and the ventricles. It consists of slow-conducting fibers, which impose a delay between the atrial and ventricular contractions, so that the atrial beat is completed before the beat of the ventricles begins. From the atrioventricular node, stimulation passes to the bundle of His (named after its discoverer), which sets up a simultaneous contraction of the ventricles.

Although the nervous system does not initiate the vertebrate heartbeat, it does control its rate. Fibers from the parasympathetic system travel through the vagus nerve, which is a large nerve that runs through the neck along the windpipe. These fibers secrete acetylcholine, which has an inhibitory effect on the pacemaker and thus slows the rate of heartbeat. Fibers from the sympathetic system secrete adrenaline, which stimulates the pacemaker, increasing the

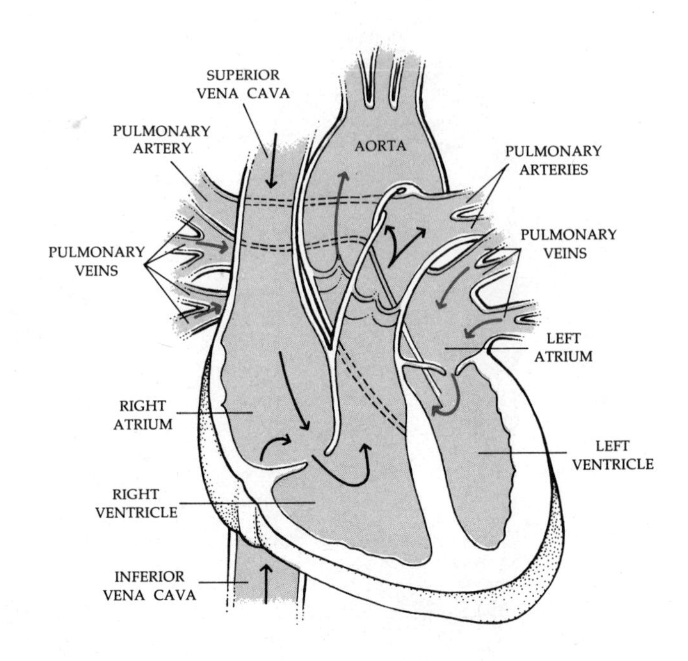

6–51

The human heart. Blood returning from the body tissues enters the right atrium and is pumped to the right ventricle, which propels it throughout the pulmonary arteries to the lungs, where it is oxygenated. Blood from the lungs enters the left atrium through the pulmonary veins and is pumped to the left ventricle and then through the aorta to the body tissues.

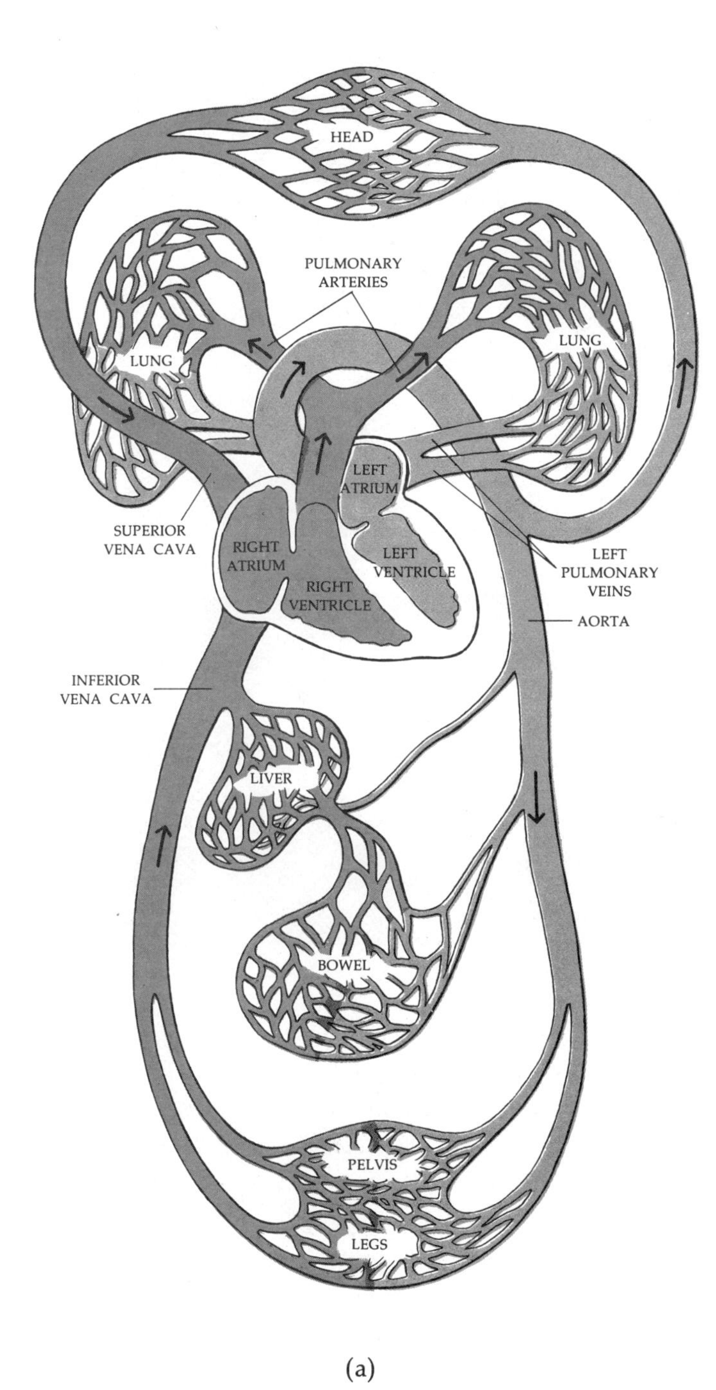

(a)

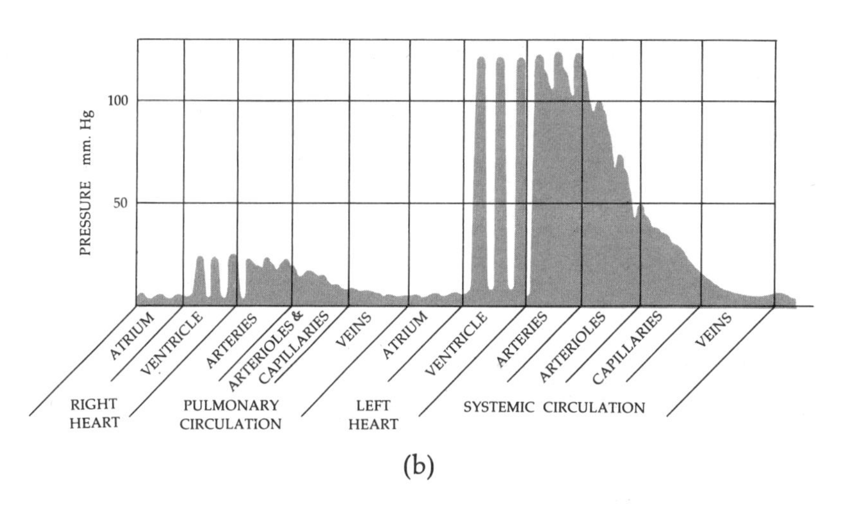

(b)

6–52

(a) *Diagram of the human circulatory system. Oxygenated blood is shown in color.* (b) *Blood pressures in different parts of mammalian circulation. The pressure fluctuations produced by three heartbeats are shown in each section. Note the fall in pressure as blood traverses the arterioles of the systemic circulation.*

rate of heartbeat. Adrenaline is liberated from the adrenal medulla into the circulation in times of stress. Adrenaline also causes peripheral veins to constrict and so increases the amount of blood available to the heart and lungs. Nerve endings that respond to stretching are located in the walls of the aorta and also in the carotid arteries, the arteries carrying blood to the brain. These nerves transmit information to the cardiac brain center, which, in turn, sends out appropriate signals to the controlling nerves, decreasing the heart rate in response to an increase in blood pressure.

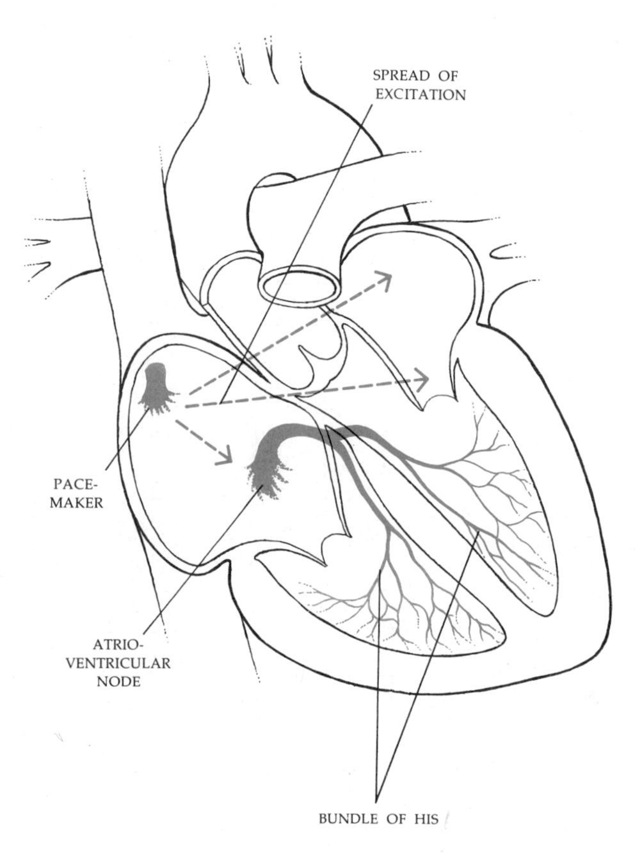

6–53

The beat of the mammalian heart is controlled by a region of specialized muscle tissue in the right atrium appropriately called the pacemaker. The nerves controlling the heart have their endings in this region. Contraction of the pacemaker tissue spreads through the atrial muscles, causing both atria to contract simultaneously. When the wave of excitation reaches the atrioventricular node, its conducting fibers pass the stimulation to the bundle of His, which triggers simultaneous contraction of the ventricles. Because the fibers of the atrioventricular node conduct relatively slowly, the ventricles do not contract until the atrial beat is completed.

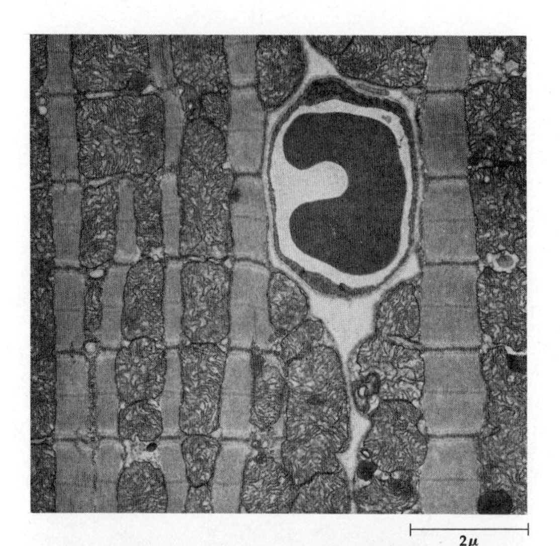

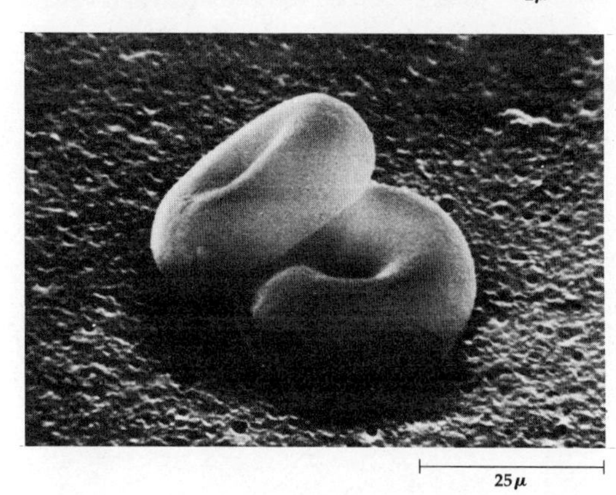

6–54

(a) *Electron micrograph of heart muscle showing a red blood cell lodged in a capillary. The heart, like all the other tissues of the body, is dependent on the fine capillary network for its supply of oxygen and other bloodborne materials. The striated pattern of the muscle is clearly visible. Note also the large and numerous mitochondria.* (b) *Stereoscan of red blood cells. The cells are biconcave enabling them to move easily with the flow of the bloodstream and to bend or twist to slip through tiny capillaries. This shape also provides them with a large surface, which facilitates the exchange of oxygen and carbon dioxide.*

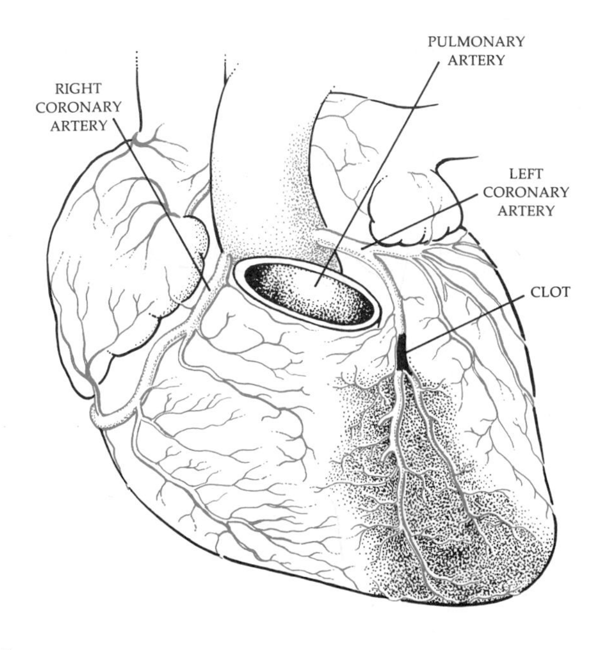

6–55

A heart attack is caused by a clot in a blood vessel supplying the heart tissue. When such a clot occurs, the heart cells normally serviced by this vessel die from lack of oxygen. Recovery from a heart attack depends on how much of the heart tissue is damaged and whether or not other blood vessels in the heart can enlarge their capacity and supply these tissues, which may then recover to some extent.

COMPOSITION OF THE BLOOD

A normal adult male has about 10 pints, or 5 liters, of blood in his body. Fifty-five percent of it is a straw-colored liquid called *plasma*. The plasma, which is more than 90 percent water, contains thousands of different types of compounds in solution or, in the case of some of the larger proteins, in colloidal suspension. The chemicals carried by the bloodstream include fibrinogen, the protein from which clots are formed; nutrients, such as glucose, fats, and amino acids; various ions; antibodies, hormones, and enzymes; and waste materials such as urea and uric acid. A filtrate of plasma, containing no cells and essentially no protein, bathes the cells of the body, seeping out through the capillary walls, flowing into the tissue spaces of the body, and reentering the capillaries. Forty-five percent of the blood is made up of cells: *white cells,* called leukocytes; *red cells,* or erythrocytes; and *platelets.*

Red Cells

Red cells, or erythrocytes, transport oxygen to all the tissues of the body. They are among the most highly specialized of all cells. As a vertebrate red blood cell matures, it extrudes its nucleus and its mitochondria and its other cellular structures dissolve, so that almost the entire volume of the mature cell is filled with hemoglobin. In man, each red blood cell has a life-span of about 127 days; new ones are produced in the bone marrow. There are about 5 billion red cells in a millimeter of blood—25 trillion (25×10^{10}) in the whole body!

White Cells

For every 1,000 red cells, there is one white cell in the bloodstream. The chief function of the white cells is the defense of the body against invaders such as viruses, bacteria, and various foreign particles. White cells accomplish this function in two ways:

1. By taking part in the manufacture of special protein substances known as antibodies, which combine with disease-causing agents and inactivate them.
2. By phagocytosis. Phagocytosis is accomplished in the white blood cells in much the same way as in the amoeba; that is, the cell extends pseudopodia around the foreign substance, incorporates it within a food vacuole, and destroys it with the help of enzymes from the lysosomes.

White cells are often destroyed in the course of fighting infection. Pus is composed largely of dead white blood cells. New white cells are formed constantly in the spleen, in bone marrow, and in certain other tissues to take the place of those cells that are sacrificed.

White blood cells are unusual among somatic cells because of their active locomotion. They travel by amoeboid movement and are attracted to sites of infection, probably by the chemical substances diffused from such sites. They are quite capable of moving against the bloodstream, migrating through the walls of small blood vessels, and making their way into damaged or infected tissues.

Platelets

Platelets, found in the bloodstream of mammals, are colorless round or biconcave disks smaller than erythrocytes (about 3 microns long). They play an important role in plugging breaks in blood vessels and in the formation of clots.

LYMPHATIC SYSTEM

In addition to blood, animals have other body fluids which bathe the tissues and occupy and fill the body cavities. These fluids seep out into the tissues through the thin walls of the capillaries and, in the vertebrates, are collected and returned to the blood by the lymphatic system. In the lymphatic system, this fluid—the *lymph*—makes its way back into the blood system through progressively larger lym-

ANTIBODIES AND IMMUNITY

The invasion of the bloodstream by a foreign organism such as a bacterium or a virus induces the formation in the bloodstream of complex globular proteins known as antibodies. Antibodies, like enzymes, are highly specific, and their specificity, like that of enzymes, is based on their combining in a very precise way with another molecule. The molecule with which an antibody combines is known as an antigen; a single cell, such as a bacterial cell, may carry a number of antigens, all of which can combine with particular different antibodies.

Antibodies are produced in plasma cells—specialized cells that arise from successive divisions of lymphocytes, a kind of white blood cell. Until recently, it was believed that a particular antigen "instructed" the formation of its specific antibody, perhaps by providing a template against which the antibody formed. Present evidence indicates another, much more remarkable mechanism. The body almost from birth possesses capabilities to make antibodies against virtually every kind of antigen it will ever encounter. An antigen, on entering the body, in some way selects the particular lymphocyte that is in charge of the production of that specific antibody. Contact between the antigen and the lymphocyte triggers the latter to enlarge and divide. At each division, the cells become more and more effective at antibody synthesis. Finally a white cell—the plasma cell—is produced that synthesizes antibody at a maximum rate and does not divide further. It takes between four and five days to reach maximum antibody production. If the infection does not proceed too rapidly, antibody production will then catch up with the multiplication of the virus or bacteria and bring the infection under control.

Antibodies can act against invaders in one of three ways: (1) *they may coat the foreign particle in such a way that it can be taken up by the scavenging white cells;* (2) *they may combine with it in such a way that they interfere with some vital activity —such as, for example, covering the site on the protein coat of a virus at which the virus attaches to the cell membrane; or* (3) *they may themselves, in combination with another blood component known as a complement, actually lyse and destroy the foreign cell.*

The first time a particular invader enters the bloodstream, the antibody-producing machinery has to be "tooled up" and, as we have seen, works comparatively slowly. But the second time this same invader is introduced, a host of lymphocytes, sensitized to that particular antibody, is present, and antibody production under these conditions proceeds so rapidly that the infectious agent rarely gains a foothold. It is for this reason that one infection with measles, mumps, or many other infectious diseases immunizes us against a second infection. Vaccines are agents that trigger the immune response of the body to a particular infectious agent without actually producing the disease; thus the invader, when it does appear, is met with a secondary response —the immediate and overwhelming production of antibodies— and so the disease is prevented.

The immune response, *as the antibody-antigen system is called, is a powerful bulwark against disease, but it sometimes goes awry. Hayfever and other allergies are the result of interactions of pollen dust, or other substances which are weak antigens (antigens to which most people do not react), and a special kind of antibody which fixes onto the cells of the skin or mucous membranes where the antigen makes contact.*

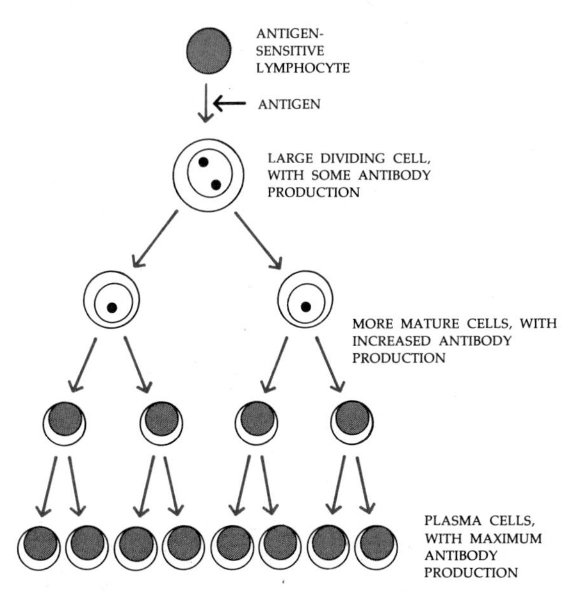

The development of a "clone" of antibody-producing cells from a single antigen-sensitive cell.

Although the body does not normally form antibodies against any substance that has been present since embryonic life, a "mistake" sometimes occurs in this recognition system when the antibody-producing mechanism is first formed, and this can result in serious disease. An example is hemolytic anemia, in which a patient makes antibodies against his own red blood cells, causing their destruction.

You will recall that during the last month in the uterus, the human baby usually acquires antibodies from its mother. Most of these antibodies are beneficial. An important exception, however, is found in the antibodies formed against a blood factor, the Rh factor (named after the rhesus monkeys in which the research leading to its discovery was carried out). The Rh factor is a genetically determined substance which is found on the surface of red blood cells. If a woman who lacks the Rh factor (that is, is Rh negative) has children fathered by a man homozygous for this trait, all the children will be Rh positive. During the course of her first pregnancy she may develop antibodies against this factor. In subsequent pregnancies, these antibodies may be transferred to the fetus, causing erythroblastosis—a dissolving of the red blood cells—which is usually fatal, either shortly before or just after birth. One method of treatment, which is quite an intricate procedure, is to transfuse the fetus completely, replacing all its blood and washing out the mother's antibodies, at about 37 to 40 weeks. Such transfusions save about 35 percent of the babies who would have died of erythroblastosis. Recently, two medical scientists at Columbia have developed a substance, which they have called RhoGAM, that contains antibodies against the antibodies that form in the mother against the Rh factor. RhoGAM, if injected into an Rh negative woman at the birth of her first Rh positive child, destroys her anti-Rh antibodies.

Because it is programmed by evolution to act against foreign materials of all kinds, the immune system works vigorously against tissues—such as skin, kidney, or heart—transplanted from another individual (except an identical twin). Surgeons and medical research workers concerned with extending the use of tissue transplants are seeking ways to suppress or paralyze the immune response in such a way that these foreign but potentially lifesaving tissues can survive.

phatic ducts that empty into the veins. Frogs and many other vertebrates have lymph "hearts," which help to keep the fluid in circulation. Mammals rely chiefly on their muscular movements. Mammals also have lymph nodes, which drain the lymph system. These serve as collecting areas for cellular debris, bacteria, and other infectious organisms picked up by the circulating lymph. In addition, the lymph nodes, along with the spleen, are the primary sites for production of the special white blood cells, the plasma cells, that produce antibodies.

SUMMARY

Transport of oxygen, nutrients, hormones, and wastes is accomplished in vertebrates by closed circulation systems in which the blood is pumped by the muscular contractions of the heart into a vast circuit of arteries, arterioles, capillaries, and veins. This network ultimately services every cell in the body. The essential function of the circulatory system is performed by the capillaries, through which substances diffuse in and out of individual cells.

Heart structure in the vertebrates varies with the demands of their differing metabolic rates. The fish has a two-chambered heart; the atrium receives deoxygenated blood and the ventricle expels blood through the oxygenating gill capillaries into the systemic circulation. In amphibians, both oxygenated and deoxygenated blood are received and mingled in a two-chambered atrium and then are pumped simultaneously through the lungs and through the systematic circulation. Birds and mammals have a double circulation system, made possible by a four-chambered heart that functions as two separate pumping organs, one pumping deoxygenated blood to the lungs and the other pumping oxygenated blood from the lungs into the body tissues.

Synchronization of the heartbeat is controlled by the pacemaker, an area of nodal tissue located in the atrium, and secondarily by the atrioventricular node, which delays the stimulation of ventrical contraction until the atrial contraction is completed. The rate of heartbeat is under neural control. Acetylcholine inhibits the pacemaker; adrenaline stimulates it.

The blood is composed of plasma, white blood cells (leukocytes), red blood cells (erythrocytes), and platelets. The fluid part of the blood is plasma, which is chiefly water in which are dissolved or suspended the nutrients, antibodies, enzymes, waste substances, and other specialized compounds necessary to the life functions. Leukocytes defend the body against invaders by the manufacture of antibodies and by phagocytosis. Erythrocytes are the vehicles of oxygen-bearing hemoglobin. Platelets serve in the formation of clots.

In addition to blood, the body tissues are bathed by lymph, which is transmitted by the blood vessels and returned to the blood by the lymphatic system. The circulating fluids pick up bacteria and other cellular debris, which are subsequently deposited in the lymph nodes. Lymph nodes are also the primary sites for the production of plasma cells, the special white blood cells that produce antibodies.

QUESTION

A heart-lung mechanism is a device through which the blood is shunted during thoracic surgery. It would, for example, remove the blood from the inferior and superior venae cavae and return it to a major artery, thus bypassing both the heart and the lungs. If you were designing such a machine, what features would you have to include in your plan for it?

CHAPTER 6–6

Digestion and Excretion

A principal attribute of living systems—one of the "symptoms of life" that we noted in Section 1—is their capacity to exchange materials with the environment and still maintain their own characteristic chemical composition and structure. At the organism level, this exchange involves the processes of digestion and excretion, the subjects of this chapter.

DIGESTION

Digestion is the process by which food that has been taken in by an organism is broken down into molecules small enough to enter the cells. Such molecules include monosaccharides (such as glucose), amino acids, fatty acids, and iron and other minerals. Once within the cells, these molecules either are processed for entry into the Krebs cycle and the electron transport chain, where they are oxidized and their energy packaged in the high-energy bonds of ATP, or after suitable alterations, are incorporated into cellular structures.

In higher animals, digestion takes place in a long tube, running from mouth to anus, which has different areas that are specialized for particular stages of the digestive process. The food is moved along this tube by successive waves of muscular contractions known as *peristalsis*. The mechanical breakdown of food begins in the mouth. Many vertebrates have teeth especially adapted for tearing or grinding. Some, such as birds and turtles, have horny beaks or bills. The tongue, also a vertebrate development, serves largely to move and manipulate food in mammals, although some vertebrate tongues, such as those of hagfish and lampreys, have horny "teeth" and the tongues of frogs and toads flip out (they are attached at the front, not the back) to catch insects. Mammalian tongues carry the taste buds, and in man the tongue has developed a secondary function of formulating sounds for communication, another example of preadaptation.

Primary Processing: The Mouth

In mammals, enzymatic processes often start in the mouth. The salivary glands of man secrete amylases, which are enzymes that break starch down into sugars. In many human cultures, alcoholic beverages are made by chewing starchy vegetables, usually roots and tubers, long enough to begin the breakdown of the starch and then spitting out the pulp, leaving it for the glycolytic enzymes of airborne yeast cells to convert the sugars into alcohol. The salivary glands, of which there are three pairs, also secrete water, mucus, and sodium bicarbonate, which makes the saliva slightly alkaline. The activity of the glands is under nervous control. The presence of food in the mouth, smelling or tasting food, or even anticipation of food makes the mouth "water."

6–56
The mammalian tongue is covered with sensory receptors—"taste buds"—that are important determinants of whether or not a particular substance will be ingested. In the human mouth, taste buds at the tip of the tongue are more sensitive to sweet tastes, those at the sides to salt (toward the front) and to sour (further back), and those at the back to bitter—which is why saccharin, for instance, changes taste from sweet to bitter as it moves from front to rear. This is a stereoscan of a taste bud.

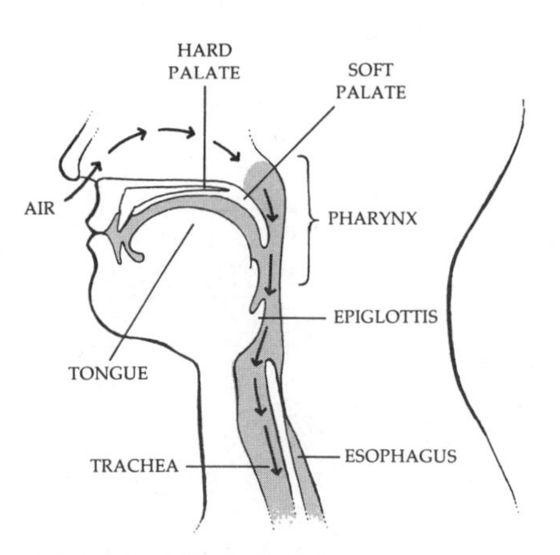

1. FACE AND NECK SHOWING PARTS OF RESPIRATORY AND DIGESTIVE SYSTEMS.

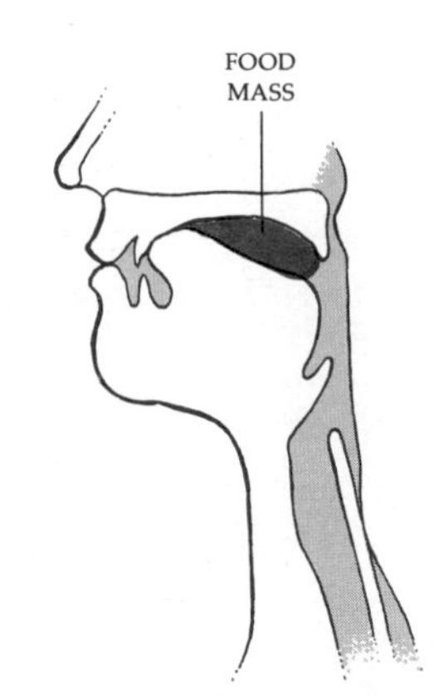

2. FOOD MASS IS IN MOUTH SOFT PALATE IS BEING DRAWN UPWARD.

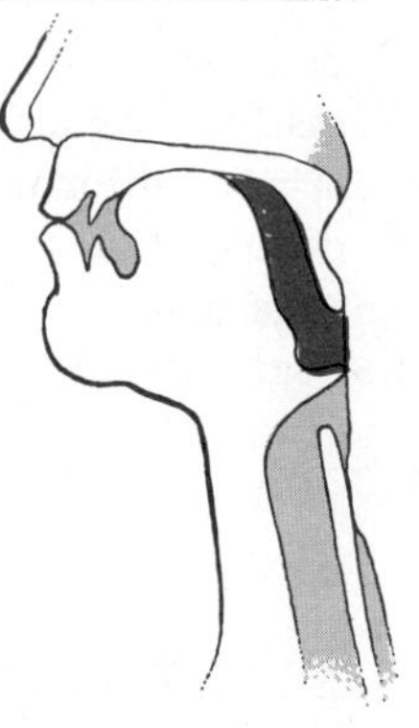

3. TONGUE PUSHES FOOD FURTHER BACK. OPENING BETWEEN SOFT PALATE AND PHARYNGEAL WALL CLOSES.

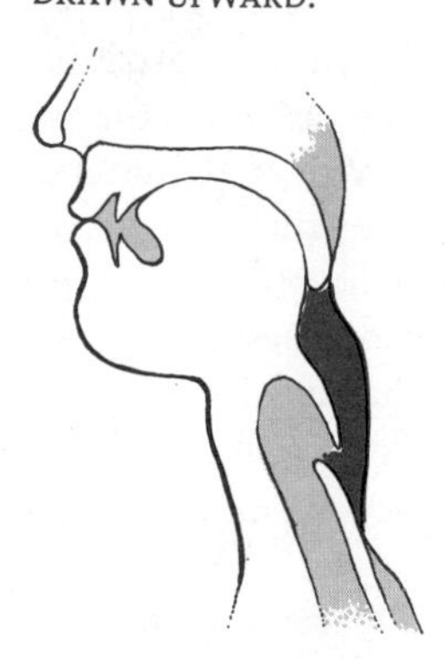

4. AS FOOD MASS DESCENDS, EPIGLOTTIS TIPS DOWNWARD.

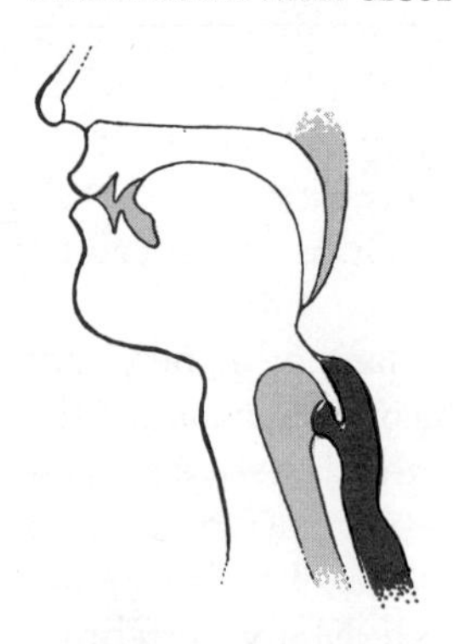

5. FOOD MASS PASSES INTO ESOPHAGUS.

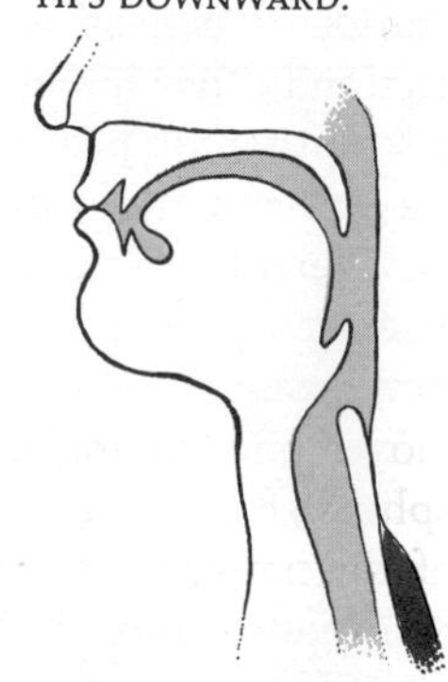

6. AS FOOD PASSES THE TRACHEA, EPIGLOTTIS TURNS UP AGAIN AND PHARYNX IS OPENED ONCE MORE.

6–57 (at right)
Separation of the digestive and respiratory systems in mammals makes it possible for mammals to breathe while eating. The pharynx is the common cavity of the two systems. Notice how an inward bulge of the muscles of the posterior pharyngeal walls moves downward, pushing the food mass ahead of it.

Swallowing begins as a voluntary action, but once the food reaches the pharynx, it is automatically transferred to the esophagus, a muscular tube about 10 inches long which leads to the stomach.

Secondary Processing: The Stomach

The stomach is an elastic muscular bag which acts as a food reservoir. Distended, it can hold 2 to 4 quarts of food. The walls of the stomach are covered by a lining of epithelial cells, which constitute the gastric mucosa. These cells secrete mucus, pepsinogen, and hydrochloric acid (HCl). The HCl kills most bacteria and other living cells in the ingested food and activates pepsinogen, converting it to pepsin, an enzyme which begins the breakdown of protein. [Why is the enzyme released from the cells in an inactive (pepsinogen) form?] The HCl also loosens the tough, fibrous components of tissues and erodes the cementing substances between cells. Because of the HCl, the contents of the stomach are very acid, with a pH of between 1.5 and 2.5 in the normal person. (Pepsin does its enzymatic work best at a pH of 1.9.) Little absorption of food molecules takes place through the stomach, with the exception of glucose molecules and also of alcohol.

The stomach is under both neural and hormonal control. Anticipation of food and the presence of food in the mouth stimulate gastric activity and the production of gastric juices. When food reaches the stomach, its presence causes the release of a hormone, gastrin, from gastric cells into the bloodstream. This hormone acts on the epithelial cells of the stomach to increase their secretion of gastric juices. Food remains in the stomach for about three to four hours. At this time, the pyloric sphincter opens and the food mass moves, by peristalsis, into the small intestine. The pyloric sphincter is a tight ring of muscle surrounding the opening of the stomach to the small intestine. It apparently opens in response to changes in the consistency of the stomach contents, which—owing to the action of the mucus, the pepsin, the hydrochloric acid, and the churning motion of the stomach—are converted into a semiliquid mass.

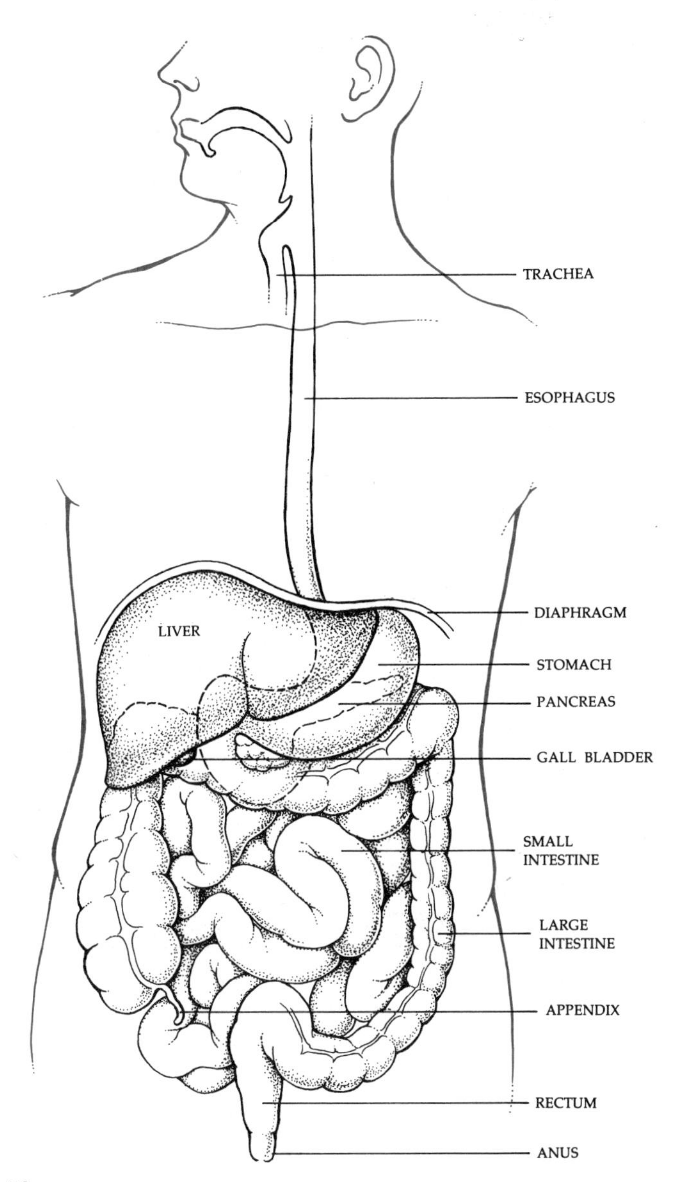

6–58

The human digestive tract. The major stages of digestion occur in the small intestine. In an adult male, the small intestine is about 23 feet long and an inch in diameter.

RUMINANTS

The ruminants—a group that includes cattle, sheep, goats, camels, and giraffes—are the only mammals able to utilize cellulose as a source of energy. Ruminants do not themselves digest this tough polysaccharide; it is broken down by the action of resident microorganisms in their digestive tracts. In the ruminant digestive system is a series of four stomachs, the first two of which (the rumen) contain a rich concentration of bacteria and protozoans. These microorganisms secrete enzymes that break cellulose down into simple fatty acids and gases (carbon dioxide and methane). The fatty acids pass through the walls of the rumen into the bloodstream of the animal and are utilized as energy sources in various parts of the body. The gases are belched forth by the animal. The last two stomachs of the ruminants secrete enzymes which break down the proteins of the bacteria and protozoans that continuously arrive from the first two stomachs. Thus, in effect, the ruminant obtains all its dietary essentials from this rich culture of bacteria and protozoans. The microorganisms, in turn, are provided with an environment rich in carbohydrates and maintained at a constant temperature favorable for growth.

Table 6–5

Gastrointestinal Hormones

Hormone	*Source*	*Stimulus for Production*	*Action*
Gastrin	Stomach	Food in stomach	Stimulates secretion of gastric juices
Enterogastrone	Stomach	Fatty acids in bloodstream	Inhibits secretion of HCl and gastric motility
Secretin	Duodenum	HCl in duodenum	Stimulates secretion of pancreatic fluids and bicarbonate; stimulates production of bile
Pancreozymin	Duodenum	Food in duodenum	Stimulates release of pancreatic enzymes
Cholecystokinin	Duodenum	Fat in food in duodenum	Stimulates release of bile

Digestive Activity: The Duodenum, Liver and Pancreas

The *duodenum* is the upper part of the small intestine. Here, as a result of the action of a variety of enzymes, most of the digestion takes place. Some of these enzymes are secreted by the cells of the intestine, and others come from the pancreas and the liver through ducts opening into the duodenum.

Amylases continue the breakdown of starch and glycogen begun in the mouth. Lipases hydrolyze fats into glycerol and fatty acids. Three types of enzymes break down proteins. One group, which includes pepsin, breaks apart the long protein chains. Each enzyme in this group acts only on the bonds linking particular amino acids, so that several are required to break a single large protein into shorter peptide chains. A second type of enzyme acts only on the end of a chain, some on the amino end and some on the carboxyl end. Finally, when each of these groups has completed its activities, a third group of enzymes comes into action which breaks the remaining dipeptides (pairs of amino acids) into single amino acids, the form in which they are absorbed through the epithelial cells of the intestine and finally enter the bloodstream.

In addition to these many different enzymes, the small intestine receives an alkaline fluid from the pancreas, which neutralizes the stomach acid, and bile, which is produced in the liver and stored in the gallbladder. Bile contains a mixture of salts which emulsify fats, breaking them apart into droplets so that they can be attacked by the lipases.

The digestive activities of the duodenum are almost entirely coordinated and regulated by hormones (Table 6–5). In the presence of food, the duodenum releases secretin, a hormone that stimulates the pancreas to secrete alkaline fluid and the liver to make bile, and pancreozymin, a hormone that acts on the pancreas to stimulate the secretion of enzymes. Fats in the food in the duodenum stimulate production of another hormone, cholecystokinin, which triggers the emptying of the gallbladder. A fourth hormone, enterogastrone, is stimulated by the presence of acid in the duodenum and acts on the stomach to inhibit its release of

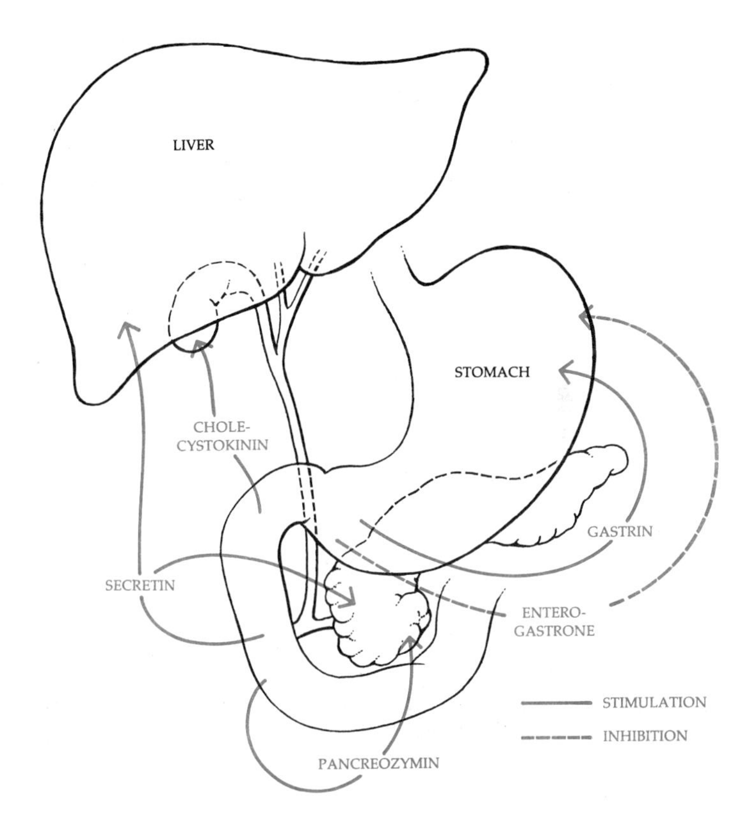

6–59

Gastric hormone interactions in the human digestive system.

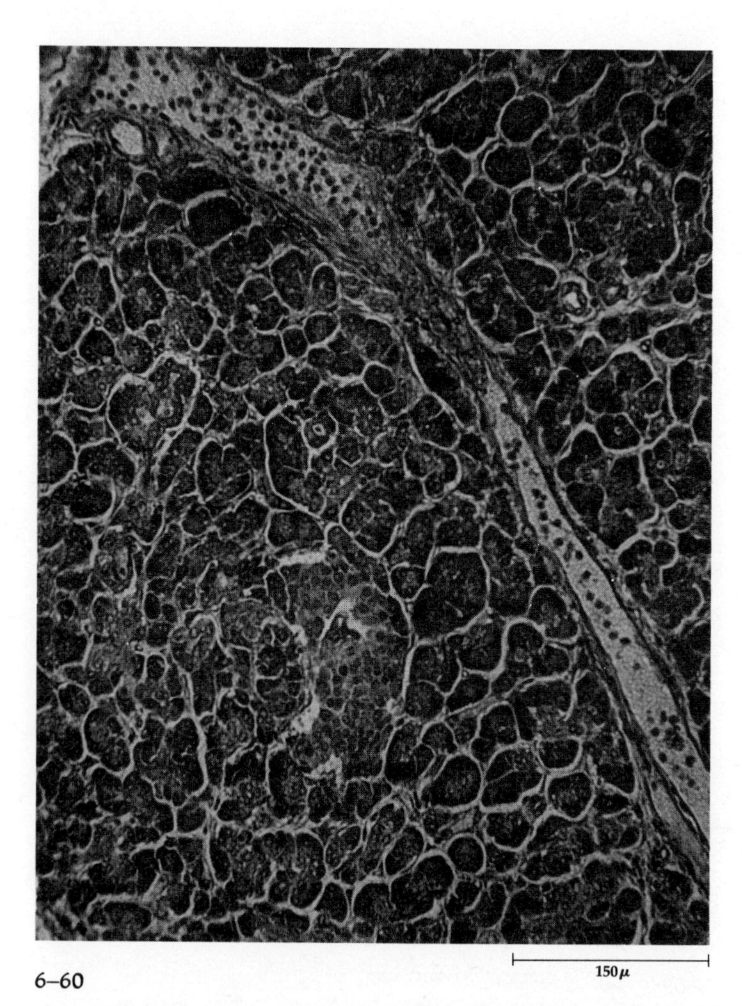

6–60

A cross section of pancreatic tissue. The group of small cells near the center of the micrograph are islet cells which secrete insulin into the bloodstream. The surrounding cells produce digestive enzymes which are carried through the pancreatic duct to the duodenum. The canal-like structure traversing the micrograph is a branch of the duct.

HCl. Do not bother to remember the names or specific actions of these hormones. What is important is to appreciate the complex interplay of stimuli and checks and balances which serve to activate and inactivate digestive enzymes and to adjust the chemical environment.

The pancreas and the liver are specialized digestive organs, developing—both in the embryo and in the course of evolutionary history—from the digestive tract. In addition to its production of digestive fluids, the pancreas also synthesizes insulin and glucagon, hormones which are released directly into the blood and which are concerned with the regulation of glucose concentration in the blood. Glucagon also stimulates the breakdown of glycogen to glucose.

The liver, in addition to its production of bile, serves as a storage place for sugar, picking up glucose from the bloodstream and converting it to glycogen for storage. The glycogen is released as glucose when the concentration of sugar in the blood is low. The liver has numerous other important functions, including synthesis of urea from nitrogenous wastes and breakdown and detoxification of substances in the bloodstream.

Absorption and Excretion: The Lower Intestinal Tract

From the duodenum the food moves by peristalsis into the *ileum*, the lower segment of the small intestine. It is through the walls of the ileum that most of the absorption of food molecules takes place. Here also water is resorbed. This is an important function. In the course of digestion, large amounts of water—some 2 or 3 quarts—enter the stomach and small intestine from the body fluids. When resorption of water is interfered with, as in vomiting or diarrhea, severe dehydration can result.

In the large intestine, most of the rest of the water is removed and the indigestible remains are broken down by the large population of bacteria that live in the gut. As by-products, the bacteria release materials useful to their host. For humans, these bacteria are the chief source of vitamin K, which promotes blood clotting. The residue is stored in the rectum and finally excreted through the anus.

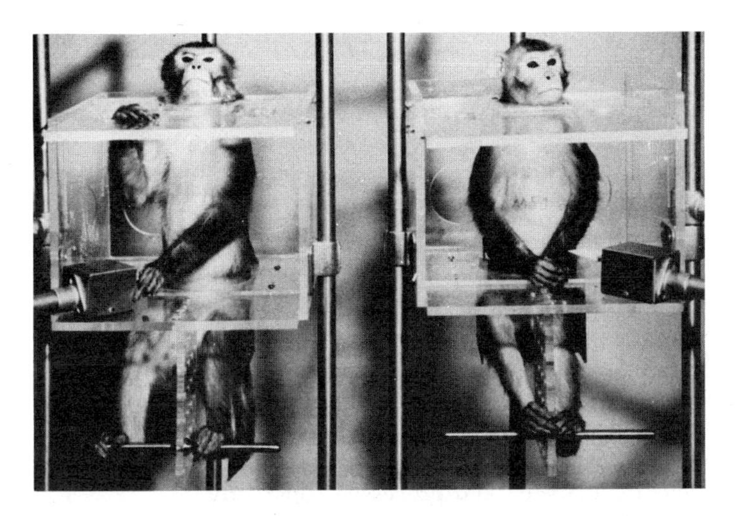

6–61

Nervous influences on gastric secretion are closely tied in with other nervous activities, including those brought on by emotional states. Increased secretion of gastric juices, including acid, has been found to accompany aggressive actions, feelings of hostility, and also emotional stress. In a series of studies carried out by Col. Joseph V. Brady of the Walter Reed Army Institute of Research, pairs of monkeys were placed in restraining chairs and subjected to electric shocks delivered automatically at regular intervals for six-hour periods. Both monkeys were given hand levers to press. If the first, or "executive," monkey pressed his lever at the appropriate time, he could prevent shocks to both monkeys, but the lever for the second monkey had no effect. The executive monkeys quickly learned to press the lever appropriately, and both monkeys actually received very few shocks. The second monkeys soon learned to ignore their levers. In four experiments, the executive monkeys died after 9, 23, 25, and 48 days, respectively, of lever pressing, as a result of extensive gastrointestinal ulcers. No control monkey died or developed ulcers. In these experiments, the decision-making responsibility apparently produced great emotional stress, and the stress, in turn, caused a sufficient increase in gastric secretion to induce ulcers.

EXCRETION, CHEMICAL REGULATION, AND WATER BALANCE

As we saw in Chapter 6–4, one of the primary trends in evolution is toward increasing freedom from the effects of changes in the environment. The excretory system is extremely important in achieving this freedom because it is in large part by means of the excretory system that an animal creates its own chemical environment. In fact, it is possible to trace vertebrate evolution in terms of the evolution of the kidney, as Homer Smith did in his classic book *From Fish to Philosopher.*

Regulation of the chemical environment involves three distinct excretory processes. The first is the elimination of toxic or excess by-products of metabolism. The two chief by-products are carbon dioxide and the nitrogen compounds, such as ammonia, produced by the breakdown of amino acids. Carbon dioxide is eliminated as a gas or diffuses out into water through the skin or the respiratory organs. In simple aquatic organisms, ammonia also diffuses out dissolved in water, but in larger, more complex organisms, it often must be converted to some other form since accumulated ammonia is highly toxic.

Birds and insects eliminate waste nitrogen in the form of *uric acid,* which can be excreted as crystals. In birds, the uric acid is mixed with the undigested wastes in the cloaca and the combination is dropped as a semisolid paste, familiar to frequenters of public parks and admirers of outdoor statuary. This nitrogen-laden substance forms a rich natural fertilizer; guano, the excreta of seabirds, accumulates in such quantities on the small islands where these birds gather in great numbers that it is profitable to harvest it commercially.

Mammals excrete nitrogen by-products largely in the form of *urea,* which must be dissolved in water for excretion. Urea played a crucial role in the history of biology. In 1828, the German chemist Friedrich Wöhler reported that he had been able to prepare the compound in his laboratory. He thereby dealt a blow to the proponents of vitalism,

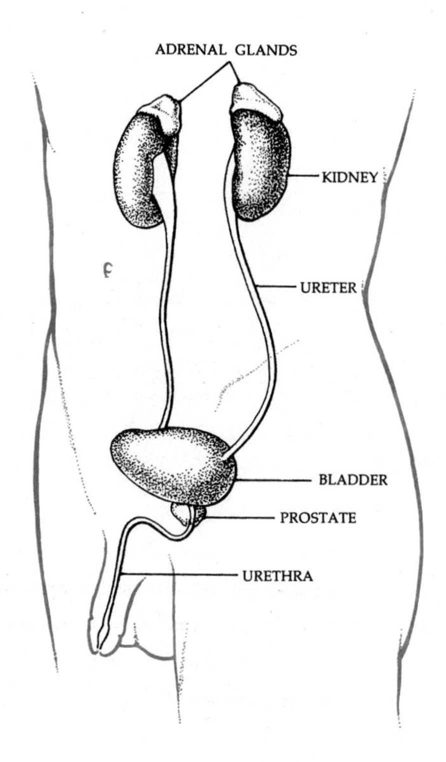

6–62
The urinary system in man.

NH_2—C—NH_2
‖
O

UREA

6–63
Urea, the principal form in which nitrogen is excreted in mammals.

who claimed that the chemistry of living things could not be duplicated in the laboratory or, indeed, even be fathomed by the mind of man.

The second regulatory process is the maintenance of ion concentrations in intracellular and extracellular fluids. Chemical regulation also involves maintaining closely controlled concentrations in the blood and other body fluids of ions such as H^+, Na^+, K^+, Mg^{++}, and Ca^{++} and the important anions (negatively charged ions) Cl^- and HCO_3^-. Such precise regulations demand that the body fluids be analyzed and processed constantly.

The third is the regulation of water content. The concentration of a particular substance in the body depends not only on the absolute amount of the substance but on the amount of water in which it is dissolved. Thus, although the fundamental problem is always the same—the chemical regulation of internal environment—the solution to the problem varies widely, being strongly influenced by the habitat of the animal, that is, by the availability of water.

The earliest organisms probably had a salt and mineral composition much like that of the environment in which they lived. The early organism and its surroundings were probably also isotonic; that is, each had the same total effective concentration of dissolved substances, so water did not tend to move either in or out of the cell. When organisms moved to fresh water (a hypotonic—less concentrated—environment), they had to develop systems for "bailing themselves out," since fresh water tended to move into their bodies by osmosis; the contractile vacuole of *Paramecium* is an example of such a bailing device. With the transition to land, the problem became one of conserving water.

Structure of the Kidney

In the kidney, the blood is filtered and substances are removed from it for elimination from the body. The filtration unit of the kidney is the *nephron*, which consists of a closed bulb, known as the *renal capsule*, and a long tubule. Each of the two kidneys in man contains about a million neph-

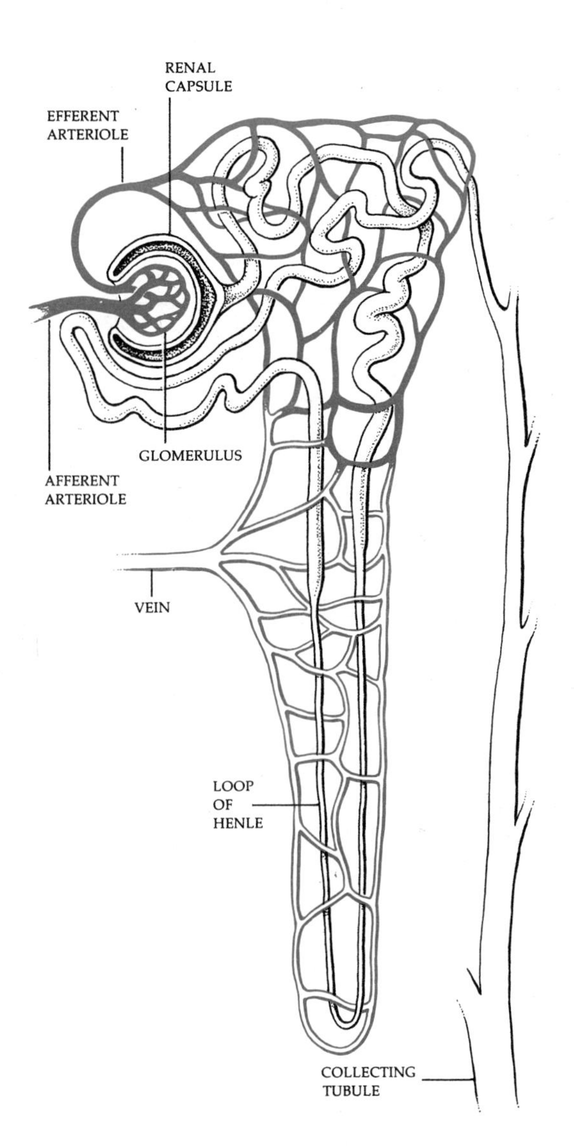

6–64
The human kidney contains about a million filtration units, or nephrons, which filter, analyze, and adjust the blood that flows constantly through them. Blood enters the nephron through the afferent arteriole. In the ball of capillaries known as the glomerulus, blood pressure forces the water of the blood out into the renal capsule, which connects with the long tubule that twists down the length of the nephron and up again. In this water are the small molecules that were contained in the blood but not the large ones, that is, the blood cells and blood proteins. As the fluid travels through the tubule, almost all the water and other useful substances are reabsorbed into the bloodstream through the capillaries surrounding the tubule. Waste materials such as urea pass along the entire length of the tubule into the urine-collecting tubule on the right.

rons, with a total length of some 50 miles in an adult male. The function of the nephron is intimately connected with the circulation of the blood. The kidney receives a rich blood flow from the renal artery. Associated with each renal capsule is a tight, twisted cluster of capillaries known as the *glomerulus*. Blood pressure in the capillaries of the glomerulus forces fluid out into the capsule; blood cells and blood proteins, both of which are too large to pass through the capillary walls, are left behind. The fluid then begins its long passage through the tubule, which is lined with a layer of epithelial cells especially adapted for active transport. The tubules of the nephrons empty into collecting tubules, which lead to a common duct, the ureter. The ureter transmits the urine to the bladder for storage before excretion through the urethra. Each day, some 180 liters (about 48 gallons) of blood passes through the human kidneys (remember, the total blood supply is only 10 pints), where it is filtered, analyzed, and so carefully adjusted that it can be considered to be completely made over.

Most substances and almost all the water contained in the filtrate that enters the tubule are returned to the blood through the blood capillary system that surrounds the long, convoluted tubule. Some of these substances, and some others from the blood, may be taken back into the tubule from the capillary bed surrounding it. For example, glucose, most amino acids, and most vitamins are returned to the bloodstream by a healthy kidney. By-products of amino acid breakdown, such as urea, are excreted. Sodium and other ions may be excreted or may be retained, depending on the physiological status of the body.

Regulation of Water Content

If fish evolved in fresh water, as is generally believed, the first function of the nephrons, phylogenetically speaking, was probably to pump out water and to retain salt. In freshwater fish, the kidney works primarily as a filter and absorber and the urine is hypotonic—that is, it has a lower concentration of solutes than do the body fluids.

Saltwater fish have a different problem. Their body fluids are generally less concentrated than their environment, and so they tend to lose water by osmosis. Their need is to conserve water and thereby keep their body fluids from becoming too salty. This problem has been solved in different ways by different groups of fish. In hagfish, for example, body fluids are about as salty as the salt waters of the surrounding ocean, and so are isotonic with them. Cartilaginous fish such as the shark adapted in a unique way. They developed an unusual tolerance for urea, so instead of constantly pumping it out, as do all other fish, they retain a high concentration of it in the blood. This high concentration of urea makes their body fluids almost isotonic in relation to the seawater. Hence they do not tend to lose water by osmosis; in fact they draw in water through their gills

6–65
Some marine animals, such as the turtle, have special glands in the head which can secrete sodium chloride at a concentration of about twice that of seawater. Since ancient times, turtle watchers have reported that these great armored reptiles come ashore to lay their eggs with tears in their eyes, but it is only recently that biologists have come to know that this is not caused by an excess of sentiment—as in the case of Lewis Carroll's mock turtle—but is, rather, a useful solution to the problem of excess salt.

from the seawater, just enough to meet their requirements. They produce a urine which is about as salty as their body fluids, which are, of course, less salty than seawater.

The bony fish spread to the sea at a much later period than did the cartilaginous ones, and their body fluids are hypotonic in relation to the environment, having an osmotic pressure only about one-third that of seawater. Thus, like terrestrial animals, they have the problem of drying out. In physiological terms, drying out means losing so much water that the solutes in the body fluids become too concentrated and the cells die. In bony fish, this problem has been solved by the evolution of special gland cells in the gills that excrete excess salt. Hence these fish can take in salt water freely and still remain hypotonic.

Since terrestrial animals do not always have automatic access to either fresh or salt water, they must regulate water content in other ways, balancing off gains and expenditures.

Animals gain water by drinking fluids, by eating water-containing foods, and as an end product of the oxidative processes that take place in the mitochondria, as we saw in Chapter 2–5. When 1 gram of glucose is oxidized, 0.6 gram of water is formed. When 1 gram of protein is oxidized, only about 0.3 gram of water is produced. Oxidation of a gram of fat, however, produces 1.1 grams of water because of the high hydrogen concentration in fat (the extra oxygen comes from the air).

Some animals can derive all their water from food and do not require fluids. The kangaroo rat of the American desert, for example, can live its entire existence without drinking water if it eats the right type of food. You should not be surprised to hear that it prefers a diet of fatty seeds. If it is fed high-protein seeds, such as soybeans—which produce a large amount of nitrogen waste and a relatively small amount of water—it will die of thirst unless some other source of water is available.

Animals lose water by respiration, by excretion in the feces, and in sweating. Also, animals need to use water to excrete their soluble wastes as urine.

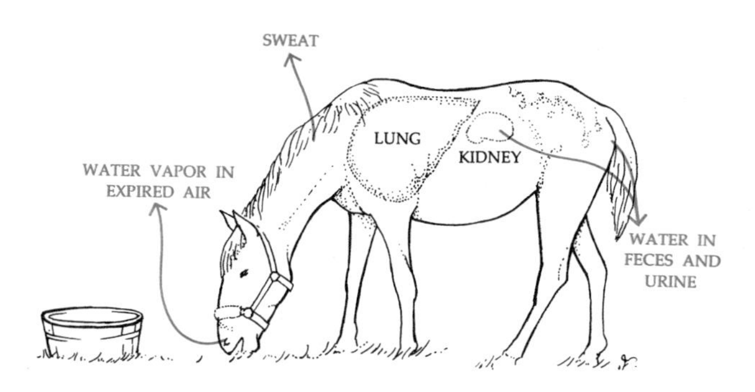

6–66
A mammal is in water balance when total water lost in expired air, sweat, urine, and feces equals the total water gained in food and drinking water and by the oxidation of food molecules.

6–67
The kangaroo rat, a common inhabitant of the American desert, may spend its entire life without a sip of water. It lives on seeds and other dry plant materials, and in the laboratory, it can be kept for months on a diet of only barley or rolled oats. Analysis shows that the rat is highly conservative in its water expenditures. It has no sweat glands, and being nocturnal, it searches for food only when the external temperature is relatively cool. Its feces have a very small water content, and its urine is highly concentrated, far more so than that of any other mammal so far studied. Its major water loss is through respiration.

6–68
Seals and some whales get their water by eating fish, most of which have a high water content and an osmotic concentration only about one-third that of seawater. A man on a life raft cannot survive by eating fresh fish, however, rumor to the contrary notwithstanding. Man cannot form urine with a salt concentration higher than about 2.2 percent. Although a fish's fluids are less concentrated than this, a fish contains a large amount of protein; so our hypothetical shipwreck survivor is placed in the same position as that of the kangaroo rat eating soybeans. He would, of course, be much worse off if he drank seawater, since the salt concentration of seawater is about 3.5 percent and he would have to expend some of his precious body water to dispose of the excess salt. (A kangaroo rat, however, which excretes a highly concentrated urine, can maintain water balance by drinking seawater.)

The Loop of Henle

As a water-conservation measure, birds and mammals have developed the ability to excrete a hypertonic urine, that is, one that is more concentrated than their body fluids. This ability is associated with a hairpin-shaped section of the nephron known as the *loop of Henle* (Figure 6–69). By sampling the fluids in and around the nephron and analyzing the amount of sodium present in each sample, physiologists have been able to determine how this deceptively simple-looking structure makes this function possible.

As shown in the diagram, the fluid entering the renal capsule first enters the proximal tubule, descends the loop of Henle, ascends it, and then passes through the distal tubule into the collecting tubule.

The fluid entering the proximal tubule is isotonic with the blood plasma; that is, it has the same salt concentration as that of the blood plasma. As the urine descends the loop of Henle, it becomes increasingly concentrated. In the ascending branch, it becomes less and less concentrated, and as it enters the distal tubule, it is hypotonic. By the end of the distal tubule, it is once more isotonic, but as the fluid decends to the collecting tubule, it may even be hypotonic in relation to the blood. In animals needing to conserve water, the fluid descending the collecting tubule becomes increasingly concentrated, and the urine that is excreted is distinctly hypertonic in relation to the circulating blood.

This is how these findings are explained: Water diffuses freely through the cells lining the walls of the nephron, while sodium is actively pumped from the tubules back to the fluid surrounding them. The sodium pumped out of the ascending loop passes back into the descending loop by simple diffusion since the fluids surrounding the descending loop are more concentrated than those within it. Thus the salt recirculates to the ascending loop, where it is pumped out again. This recirculation of the sodium has two consequences: (1) As the urine passes through the loop of Henle, much salt but little water is removed; and (2) the lower part of the loop of Henle and also the lower part of the collecting tubule are bathed in a fluid containing nearly 10

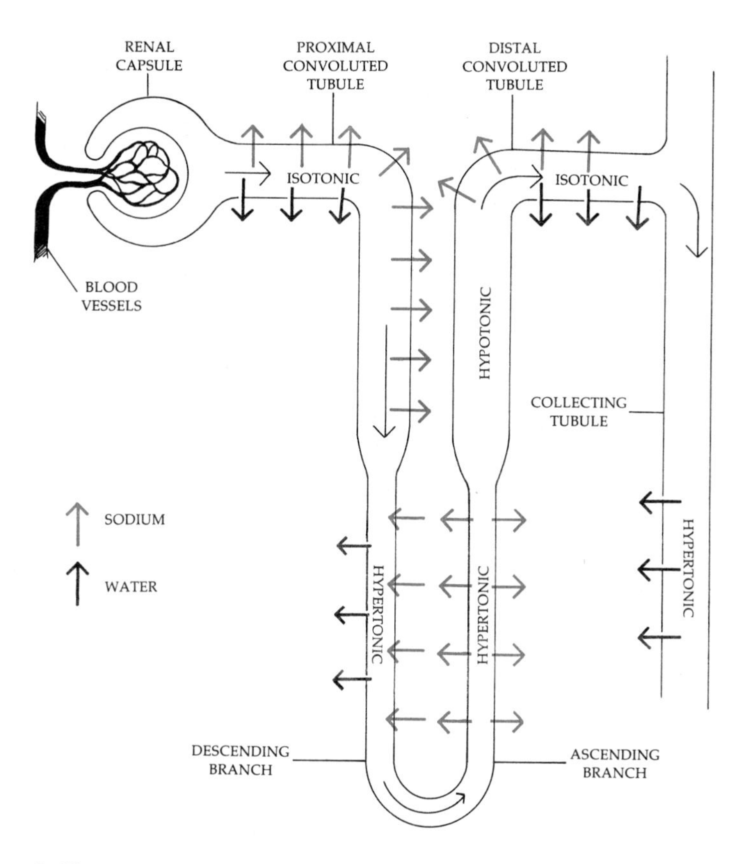

6–69

The formation of hypertonic urine in the human nephron. As the urine passes up the ascending branch of the loop of Henle, the cells lining the tubule pump out sodium into the surrounding fluid. The sodium ions diffuse passively into the descending branch and are recirculated to the ascending branch, where they are pumped out again. As a consequence of this recirculation of sodium, the loop of Henle and the lower section of the collecting tubule are constantly bathed in a salty fluid. The cells of the ascending loop are apparently impermeable to water, since it does not diffuse out when the sodium is pumped out, but in the lower segment of the collecting tubule the membrane becomes permeable again, the water pours out into the surrounding salty fluid, and a hypertonic urine is passed down the tubule to the bladder.

times the level of salt normally found in tissue or blood. The urine—which has become isotonic by the time it leaves the distal tubule—then descends the collecting tubule, which traverses the zone of high salt concentration. Water pours out of the collecting tubule by osmosis, leaving within the tubule a urine that is isotonic with the surrounding briny fluid but hypertonic in relation to the body fluids as a whole. As you can see, this is another circumstance in which structure, the actual physical shape of an organ, is indispensable to function.

Regulation of Water and Salt Excretion

In a normal adult, the rate of water excretion in urine averages 1,200 milliliters a day. Although the actual amount of urine produced may vary from a few hundred milliliters a day to several thousand, there will be a variation of less than 1 percent in the fluid content of the body. However a *minimum* output of 500 milliliters of water is necessary for health since this much water is needed to remove potentially toxic waste products, such as urea.

The physiological adjustments that regulate water balance are under the control of a hormone known as ADH (antidiuretic hormone). ADH acts on the membranes of the collecting ducts of the nephrons and increases their permeability to water. If no ADH is present, the collecting ducts are impermeable to water and so no water passes back into the blood from the urine before it enters the ureters. In short, ADH stimulates the kidney to conserve water.

ADH is a neurohormone secreted by nerve endings of the hypothalamus. It is stored in the posterior region of the pituitary gland. Whether or not ADH is released depends on the osmotic concentration of the blood and also on blood pressure. Osmotic receptors, which monitor the solute content of the blood, are located in the brain, possibly within the hypothalamus itself. Pressure receptors, which detect changes in blood volume, are found in the walls of the heart, in the aorta, and in the carotid arteries. Stimuli received by these receptors are transmitted to the hypothalamus. Factors that increase blood concentration or decrease blood pressure or both stimulate the production of ADH and conservation of water in the body. Such factors include hemorrhage and dehydration. Factors that decrease blood concentration, such as the ingestion of large amounts of water, or that increase blood pressure—adrenaline, for instance—signal the hypothalamus to stop producing ADH, and so water is excreted. Alcohol suppresses ADH secretion and thus increases the amount of water produced; pain and emotional stress trigger ADH secretion and thus decrease urinary flow.

The kidney is also affected by the adrenal cortical hormone aldosterone. Aldosterone stimulates reabsorption of sodium from the kidney. When the adrenal glands are removed or when they function poorly (as in a disease called Addison's disease), sodium chloride is lost in the urine and the tissues of the body become depleted of sodium and chloride ions. Generalized weakness results, and the loss can eventually become fatal, although most of the ill effects can be overcome by feeding the patient enough sodium chloride to make up for the salt lost by excretion.

SUMMARY

Digestion, the process by which food is chemically simplified for absorption by cells, occurs in successive stages of an increasingly complex interplay of chemical stimuli, as food substances are reduced to the specific compounds utilized by cells. In mammals, food is processed initially in the mouth, where the breakdown of starch begins, and then in the stomach, where gastric juices destroy bacteria and begin to break down proteins. Most of the digestion occurs in the duodenum; here, digestive activity is almost completely under hormonal regulation. The breakdown of starch by amylases continues; fats are hydrolyzed by lipases; and proteins are reduced to single amino acids. Digestive enzymes are secreted by the intestinal cells, and these, in turn, stimulate the functions of the pancreas and the liver. The pancreas releases an alkalizing fluid; the liver produces bile. Besides their specialized digestive functions, both organs are involved in the regulation of blood sugar. Most absorption of food molecules occurs through the ileum. Undigested remains and excess water are removed through the large intestine.

The excretory system enables an animal to regulate its internal chemical environment. Regulation is accomplished by (1) the elimination of toxic by-products of metabolism, especially carbon dioxide and the nitrogen compounds produced in the breakdown of amino acids; (2) control of the ionic content of the body fluids; and (3) the maintenance of water balance. The excretory unit of the vertebrate is the nephron; each nephron consists of a long tubule and a closed bulb (renal capsule). The renal capsule is associated with a twisted cluster of capillaries, the glomerulus, from which fluids are forced into the nephron. By a combination of filtration, reabsorption, and secretion, the circulating body fluids are processed during their passage along the tubules.

Problems of water balance are different for animals living in saltwater (hypertonic), freshwater (hypotonic), and terrestrial environments. Terrestrial animals generally need to conserve water. An important means of water conservation is the capacity for excreting a urine that is hypertonic in relation to the blood. The loop of Henle is the portion of the mammalian nephron that makes possible the production of a hypertonic urine.

Water and salt excretion is also subject to hormonal regulation. ADH acts on the collecting ducts of the nephrons, increasing their permeability to water and thus decreasing water loss. Aldosterone stimulates reabsorption of sodium from the kidney.

QUESTION

Uric acid is an insoluble compound and so can be excreted as crystals. Urea, on the other hand, must be excreted dissolved in water. Explain why, in terrestrial animals, uric acid excretion is correlated with egg laying and urea excretion with giving birth to living young.

CHAPTER 6–7

The Human Brain

Philosophers since the time of Aristotle have wondered about the relationship between mind and body. Even René Descartes—"I think therefore I am"—who believed that the body is essentially a complex machine, conceived of mind and body as two separate entities, which come together in a tiny gland, the pineal, located within the skull. The persistence of this dualistic concept is reflected in such phrases, still in use today, as "mind over matter." Many years after the cell doctrine was accepted for all other parts of the body, the brain was considered to be an exception, partly because special staining techniques are required to reveal the brain neurons but undoubtedly, in large part, because of the feeling that mental processes are somehow special, different from other physiological functions. There is, of course, no longer any scientific doubt that mind and matter are one.

However, even today, many of us who accept without difficulty the fact that the tissues and cells of the digestive tract are responsible for digestion or that respiratory gases are exchanged across the surface of the lungs find it less easy to apprehend—to *really* believe—that the ideas, ideals, dreams, fantasies, thoughts, hypotheses, loves, hates, fears, and aspirations which make up the content of the mind somehow can be explained by the interactions of the cells within our heads. Nor, indeed, can scientists yet fully explain "mind" in terms of brain, despite the many intriguing discoveries that have been made in this field.

HOW THE BRAIN IS STUDIED

Scientists use a variety of techniques to probe the machinery of the mind. The electroencephalogram measures the sum of the electrical activity of large groups of cells. Microelectrodes inserted within the brain record the activity of smaller groups of cells and even of single cells. Particular areas of the brain are stimulated by pulses of electric current or by chemicals, and the results are recorded and analyzed. Portions of the brain are removed or connections severed either by accidental injury, in the course of treatment for epilepsy or other diseases, or experimentally in animals. The experimental animals most frequently used are primates, because of their resemblance to man, and cats, because they are readily available to research laboratories and because the skull of any cat has very much the same size and shape as the skull of every other cat, making it easier to locate specific points within it. Fortunately for both experimenter and subject, there are no pain receptors within brain tissue, so that procedures can be carried out painlessly and with a minimal disturbance of function using only local anesthesia.

Perhaps as a consequence of the techniques presently available, researchers have been more successful in "mapping" the brain—that is, correlating certain functions with certain areas—than in discovering the cellular basis of thought, memory, or learning.

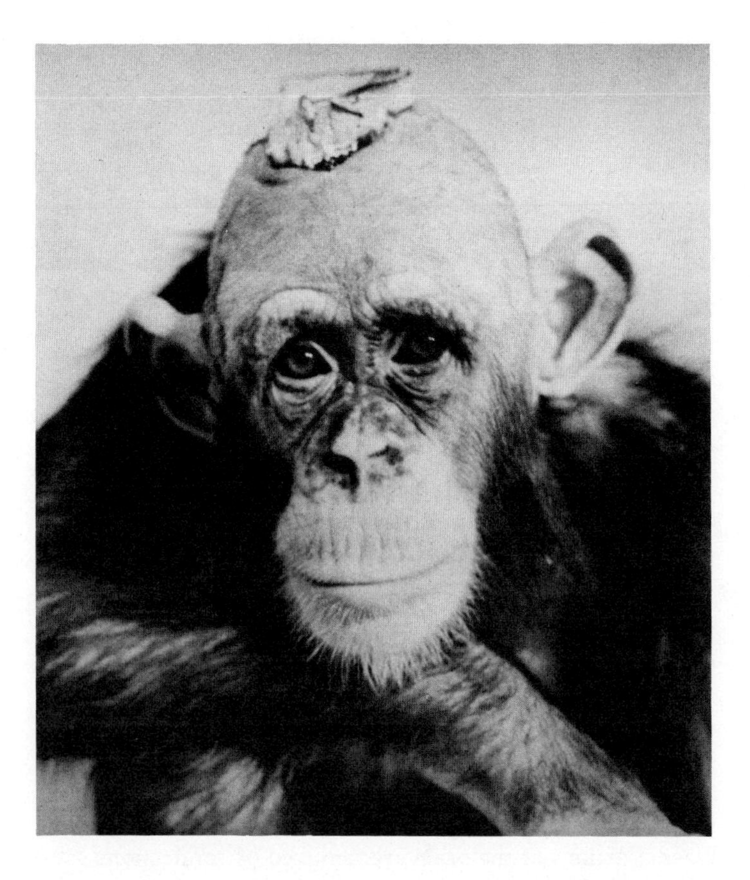

6–70

Paddy, a chimpanzee, is one of a group of experimental animals used by José Delgado of Yale to establish effects of electrical stimulation on various areas of the brain. A hundred fine electrodes planted in various areas of Paddy's brain send signals to a remote computer, which analyzes them and sends back responses to other parts of Paddy's brain. For example, when the computer receives signals from Paddy associated with aggressiveness, it sends back counteracting signals, which have the effect of making Paddy docile.

ARCHITECTURE OF THE BRAIN

The human brain, which weighs about 3 pounds, has the consistency of semisoft cheese. Like the spinal cord, the brain is made up of both "white matter"—the fiber tracts—and gray matter. The gray matter consists of neurons and glial ("glue") cells, which support and nourish the neurons. In some areas of the brain, these cells are so densely packed that a single cubic inch of gray matter contains some 100 million cells, with each cell connected to as many as 60,000 others.

Three subdivisions of the brain are evident:

1. The *brainstem,* which is an enlarged, knobby extension of the spinal cord. The lower part of the brainstem (medulla, pons, and midbrain) retains the tubular shape of the spinal cord, with the knobby protrusions appearing in pairs, one on each side.
2. The *cerebellum* ("little brain"), a bulbous mass of tissue which, in man, lies at the back of the head.
3. The *cerebrum* and its outer surface, a greatly convoluted layer of cells called the *cerebral cortex.* The cerebrum is divided into two hemispheres by a clearly visible deep furrow running from front to back. The cerebral hemispheres overlie the brainstem and the cerebellum.

THE BRAINSTEM

The brainstem contains all the fiber tracts connecting the higher brain structures with the spinal cord. Sensory fibers run dorsally along the brainstem, and motor fibers ventrally. Within the brainstem, these fibers cross from left to right, so that the right side of the brain controls the motor and sensory functions of the left side of the body, and vice versa. The brainstem also contains nuclei for most of the sensory and motor fibers serving the facial muscles and the sensory organs. Nuclei concerned with heart action, respiration, and gastrointestinal function are located at the base of the brainstem, in the medulla. The pons ("bridge") contains, in addition to ascending and descending tracts

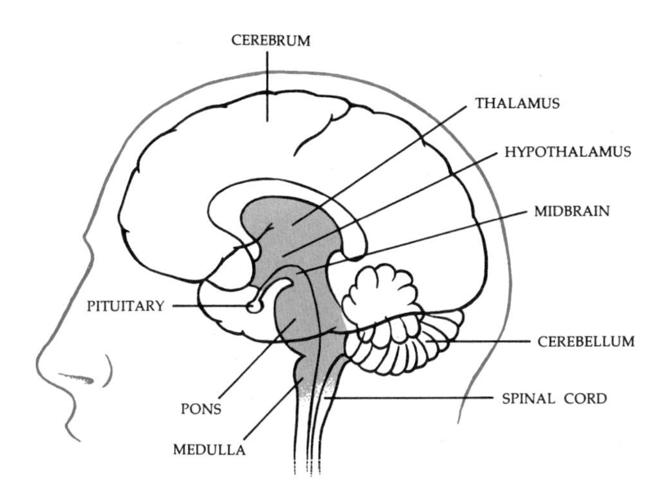

6–71
The principal areas of the human brain. The brain stem is shown in color.

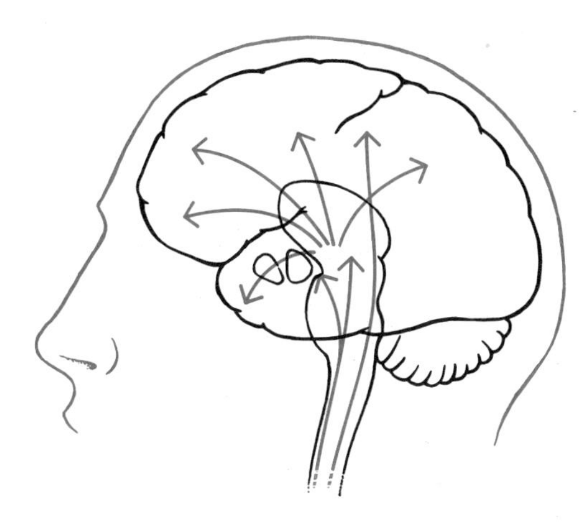

6–72
The reticular formation is a core of tissue running through the brainstem. It monitors incoming stimuli, analyzes them, and sends appropriate arousal signals to other areas of the brain.

and nuclei, a very large bundle of transverse fibers which interconnect the brainstem and cerebellum.

The midbrain evolved originally as the receiving center for fibers coming from the eye. In modern fish and amphibians, it is still the primary visual center but it also receives information from all other parts of the brain and serves as the highest level for overall integration of behavior. In reptiles and birds, the visual impulses travel first to the midbrain but are relayed from there to the more forward centers of the brain, where they are coordinated with other incoming information. In mammals, too, the midbrain is concerned with vision and also with hearing. Conspicuous on its surface are four paired knobs, two large and two small, the optic and auditory lobes, which serve complex sensory reflex functions.

The medulla, pons, and midbrain are much the same in structure and function from fish to man. An animal such as a cat can live for long periods and continue to carry out many basic functions—eating, meowing, purring, walking, etc.—even after all the brain above the midbrain has been removed, although it will lack individual personality and its behavior will become automatonlike.

Reticular Formation

As we mentioned previously, the dorsal part of the brainstem is concerned mainly with sensory input and the ventral part with motor and response processes. Between these two zones is a core of tissue which runs centrally through the entire brainstem. It is made up of a weblike network of fibers and neurons; hence its name: the reticular ("netlike") formation.

The reticular formation is of particular interest to physiological psychologists because it is involved with arousal and attention—that hard-to-define state we know as consciousness. All the sensory systems have fibers that feed into the reticular formation, which apparently filters incoming stimuli and "decides" whether or not they are important. Electrical stimulation of this system, either artificially or by these incoming impulses, results in increased electrical activity in other areas of the brain. The existence of such a filtering system is well verified by ordinary experience. A person will sleep through the familiar blare of a subway train or a loud radio or TV program but will wake instantly at the cry of the baby or the stealthy turn of a doorknob.

Similarly, we may be unaware of the contents of a dimly overheard conversation until something important—our own name, for instance—is mentioned, and then our degree of attention increases.

The most recent research indicates that the reticular formation is not, as was first thought, *the* arousal zone but rather works in conjunction with other areas in the brain to produce wakefulness and consciousness. In fact, modern research is tending to undermine previous concepts of particular "centers" for the control of specific functions and to put more stress on complex interrelationships of various parts of the brain. Nevertheless, the reticular formation continues to be recognized as having important functions concerned with consciousness and sensory filtration.

Thalamus and Basal Ganglia

The specific functions of the other brainstem tissues, the thalamus and the basal ganglia, are not well understood. They are primarily groups of nuclei, and both appear to be involved with relaying information among different areas of the brain. The major known function of the thalamus is to relay sensory information—particularly auditory, visual, and tactile information—to the cerebral cortex.

THE CEREBELLUM

The cerebellum is primarily concerned with the regulation and coordination of muscular activities. There are connections between the cerebellum and nearly all the other parts of the brain, including spinal sensory fibers and areas concerned with hearing and vision as well as the thalamus and other parts of the brainstem. Because of the coordinating capacities of the cerebellum, we are able to carry out complex functions such as walking, riding a bicycle, driving a car, swallowing (Figure 6–57), and playing a piano. Like the brainstem, the cerebellum is an old structure, evolutionarily speaking. It reaches its largest size, proportionately, in the brains of birds, where it is concerned with the activities involved in flying.

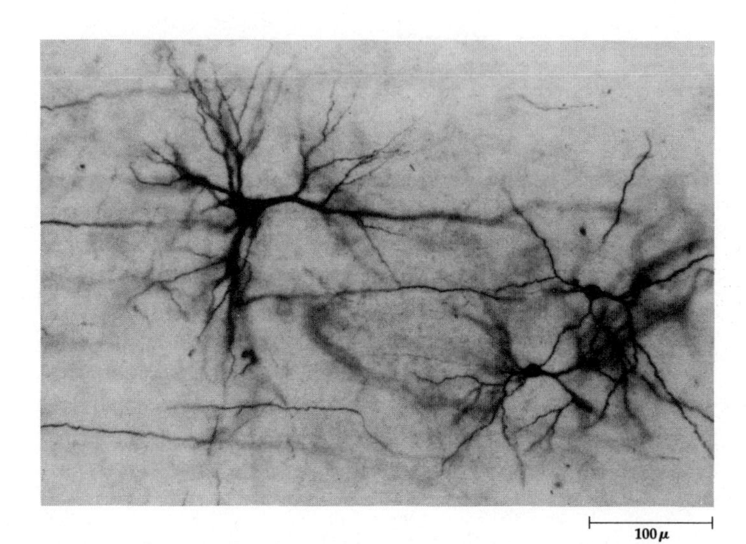

6–73

Photomicrograph of several brain cells from the cortex of the cat.

THE CEREBRUM AND THE CEREBRAL CORTEX

The cerebrum makes up the bulk of the human brain—about 80 percent, in fact. It consists of two large lobes, the cerebral hemispheres, the interior of which are composed of fiber tracts ("white matter") and relay nuclei. The two lobes are connected by a large bundle of fibers, the corpus callosum. The surface of the cerebral hemispheres is a 2-millimeter-thick layer of gray matter, the cerebral cortex. The cortex of the human brain is wrinkled and folded, which greatly increases its surface area.

Evolution of the Cerebrum

The cerebrum has undergone striking evolutionary changes in the vertebrates. In fish, amphibians, and reptiles, the most imposing parts of the cerebrum are the olfactory bulbs, which directly underlie the special organs for smell and which are attached to the rest of the brain by the olfactory stalks. Behind the olfactory bulbs in these primitive

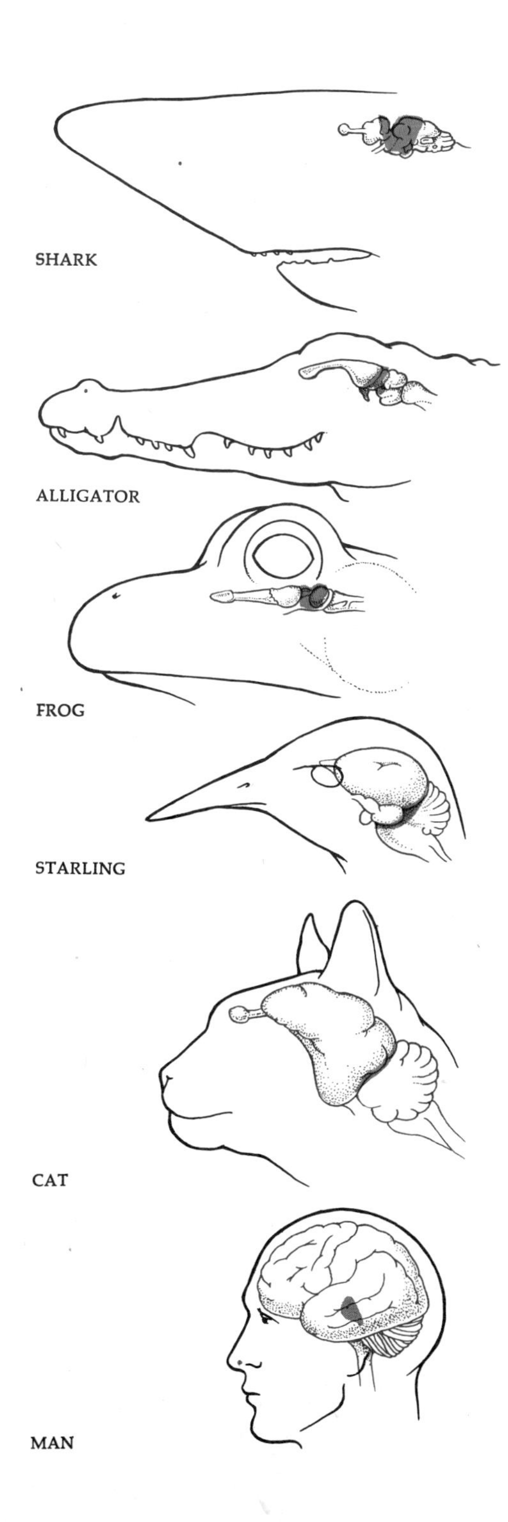

6–74 (at left)
The brains of various vertebrates. Note that the brainstems (indicated in color) are approximately the same in the various species but that the cerebrum is much larger in the homeotherms than in any of the "cold-blooded" animals. The cerebral cortex, the upper surface of the cerebral hemispheres, is relatively smooth and small in area in lower mammals, is larger in more intelligent ones, such as dogs and cats, and reaches its greatest development in the primates, particularly man.

vertebrates are two round swellings that make up the cerebrum. In fish and amphibians, the cerebrum is largely concerned with olfactory sensations, but in reptiles we can find the trace of new evolutionary developments that have proved to be of major importance.

Among the reptiles, there is a small patch of nervous tissue, the neocortex, at the forward upper part of the cerebral lobes. In mammals, this neocortex is greatly expanded, folding back to cover the entire surface of the cerebral hemispheres. In more primitive mammals, such as the rat, the cortex is relatively small and smooth, but in primates and man the cortex is large and convoluted. Generally speaking, there is a correlation between the extent of cortical development in a species, the position of the species on the evolutionary scale, and both the complexity and the modifiability of its behavior.

As the cortex increased in size during mammalian evolution, it took over more and more functions from the older, underlying, and more posterior parts of the brain. As a result of these progressive changes, in which the patterns of control shifted as the brain evolved, many different areas in the brain may now be concerned with the same function, such as vision or hearing, and the higher the mammal is on the evolutionary scale, the more important has become the relative role of the cortex. For instance, if the area of the cortex concerned with vision is removed in a rat, the animal cannot discriminate patterns. If it is removed in a monkey, the animal can only distinguish light from dark. A man whose visual cortex has been damaged is totally blind.

An important difference between the cortices of the higher and lower mammals is that in the latter almost the whole surface of the cortex is concerned with definite sensory or motor activities, whereas in man, about three-fourths of the cortex is made up of so-called association cortex, which does not have definite sensory or motor functions. Scientists agree that learning ability and intelligence in man and the other primates are generally correlated with the size and degree of development of these large areas of association cortex, although at present we know very little about the anatomical organization of these abilities.

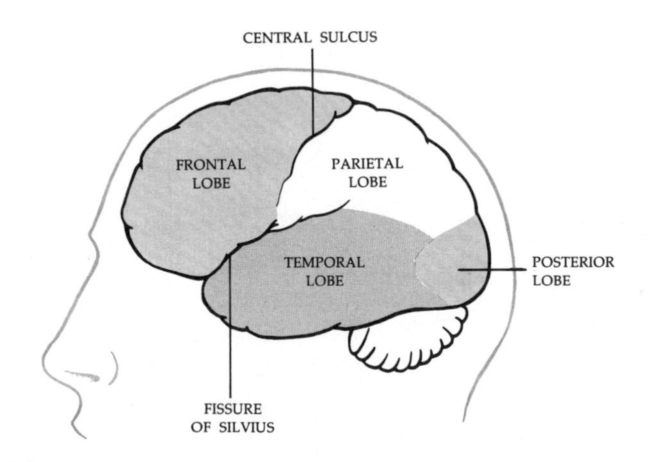

6–75
The principal fissures and lobes of the human cerebral cortex.

Mapping the Cortex

Because of its relative accessibility, just below the surface of the skull, the cerebral cortex is the most thoroughly studied area of the human brain, and some parts of it have been mapped in exquisite detail. In primates, each of the hemispheres is divided into lobes by two deep fissures, or grooves, in the surface: the central sulcus, which runs laterally—from left to right—across the top of each hemisphere, and the fissure of Silvius (Figure 6–75).

The area just anterior to (in front of) the central sulcus is the somatic motor cortex, and the area posterior to it is the somatic sensory cortex. Because of the crossing of fibers in the brainstem, the motor and sensory areas for the left side of the body are on the right hemisphere, and vice versa.

The temporal lobe of the brain lies below and posterior to these sensory and motor areas, separated by the other deep fissure, the fissure of Silvius. On the temporal lobe, buried within this fissure, is the auditory cortex. By measuring electrical discharges in this area of the brain in dogs and cats exposed to sounds of varying frequencies, investigators have been able to show that different regions of the auditory cortex respond to different pitches of sound.

The area behind the sensory cortex is divided into two main lobes, the most posterior of which is concerned with vision. Stimulation of an animal's retina with light results in electrical discharges by cells in this area of the cortex. By using a tiny point of light to stimulate very small regions of the retina, one after the other, investigators have been able to show that each region of the retina is represented by a corresponding but larger region on the visual cortex. This cortical region contains a variety of cells, different groups of which respond to different types of visual stimuli.

The somatic sensory, auditory, and visual zones of the cortex appear to function as terminals for messages from somatic sensory neurons and from the special organs of the eye and the ear. The motor cortex is apparently one of the areas of origin for outgoing signals to the somatic muscles. The function of the rest of the cortex is largely unknown.

Association Cortex

The unmapped area of the cortex is sometimes known as the association, or "silent," cortex. The term "association"

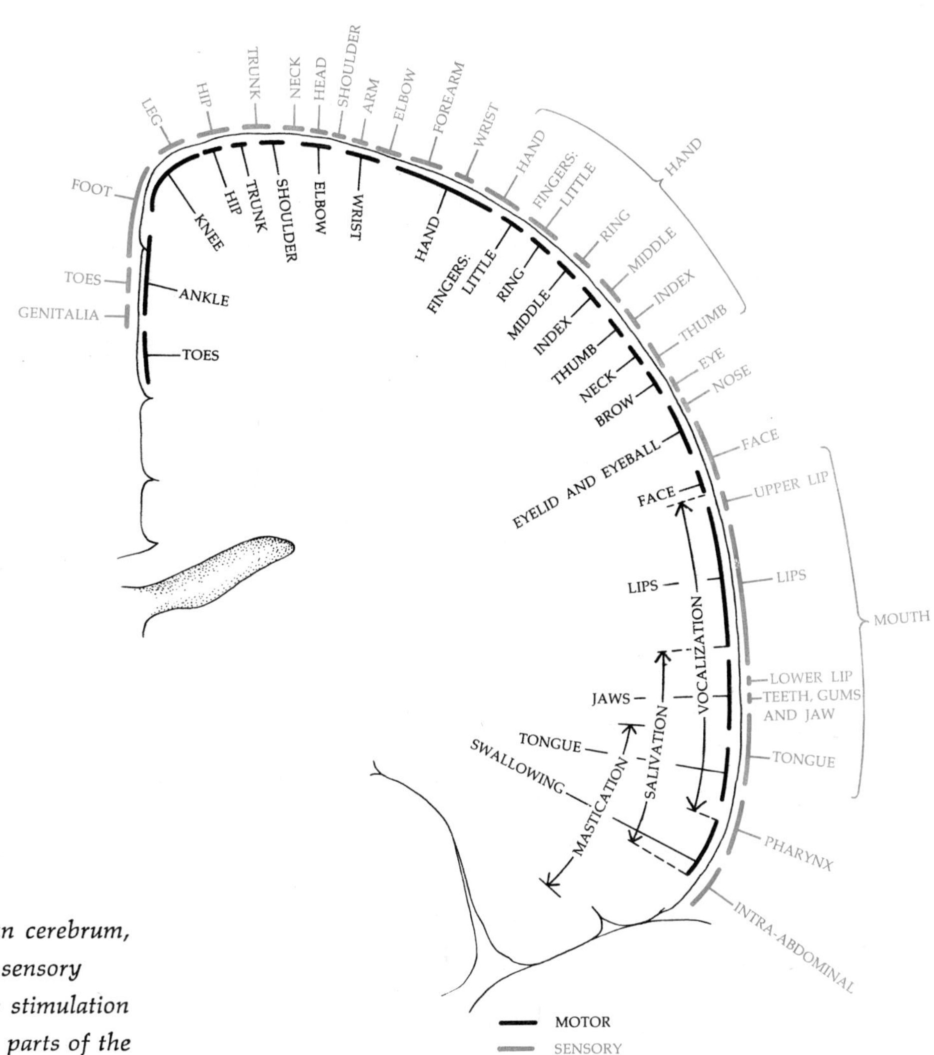

6–76

A cross section of one hemisphere of the human cerebrum, indicating the functional areas of the motor and sensory cortices. The motor cortex is indicated in black; stimulation of these areas causes responses in corresponding parts of the body. The sensory cortex is in color; stimulation of various parts of the body produces electrical activity in corresponding parts of this cortex. Notice the relatively huge motor and sensory areas associated with the hand and the mouth. This map is based largely on studies done by neurosurgeon Wilder Penfield on patients undergoing surgical treatment for epilepsy.

6–77
Pathway of a sensory impulse. A signal received in a receptor cell and transmitted to a sensory neuron travels along the axon of the neuron to the spinal cord where it crosses a synaptic junction to an ascending fiber within the spinal column. It travels upward and passes through the brainstem, producing an arousal pattern in the reticular formation. From the thalamus, it is relayed to the sensory cortex.

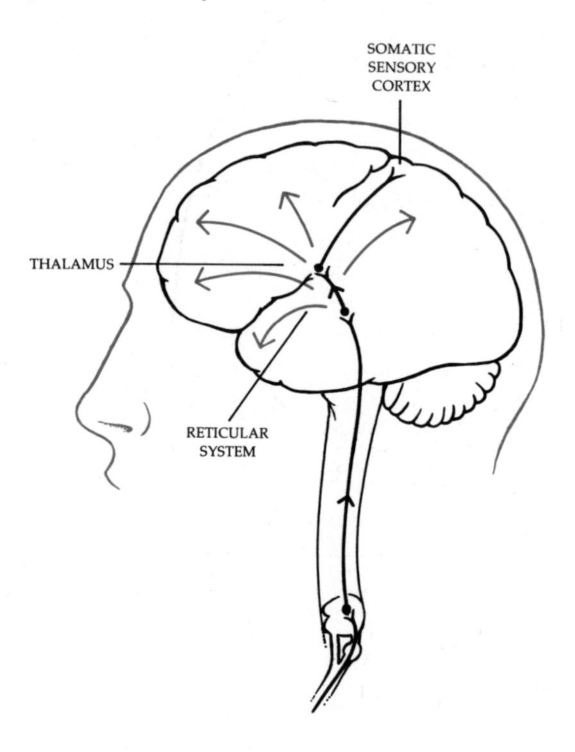

was introduced when this part of the cortex was thought to function as a sort of giant switchboard, interconnecting the motor and sensory zones. More recent studies suggest, however, that organization of the brain is more vertical than horizontal; for example, severing the motor cortex from the sensory areas by deep vertical cuts appears to have no effect at all on an animal's behavior. Anatomical studies support the interpretation that communication between the sensory areas and motor areas of the cortex takes place primarily in lower brain centers, particularly the thalamus.

Nor are the association areas actually "silent." Although sensory stimuli do not evoke local responses in these areas nor does stimulation of these areas elicit muscle movements (as in the somatic sensory and motor areas), the association cortex shows steady spontaneous activity, apparent (but nonspecific) responses to stimuli, and increased activity following electrical stimulation of the thalamus or the reticular formation. Clearly, something is taking place in these "silent" areas. Since the proportion of association area to motor and sensory area is much larger in primates than in other mammals, even higher mammals such as cats, and is very large in man, we assume that these areas have something to do with what is special about the mind of man.

Frontal Lobes

About half of the association area of the cortex is in the frontal lobes, the part of the brain anterior to the motor cortex. This part of the brain developed most rapidly during the recent evolution of man. It is responsible for the high forehead of modern man, as compared with the beetle brow of our most immediate ancestors, and public appraisal of its function is reflected in the terms "highbrow" and "lowbrow."

During a particularly bleak period in the recent history of treatment of the mentally disturbed, a large number of mental patients had their frontal lobes removed or the connections severed between these lobes and the rest of the brain in order to make the patients more docile and hence

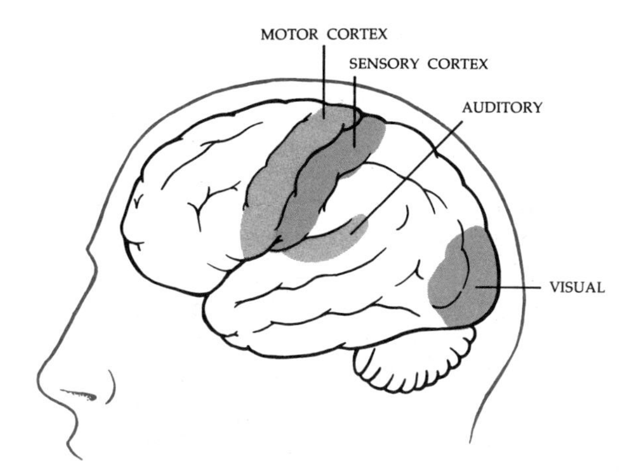

6–78
The human cerebral cortex, showing the location of the motor and sensory areas, on either side of the central sulcus, and of the auditory and visual zones.

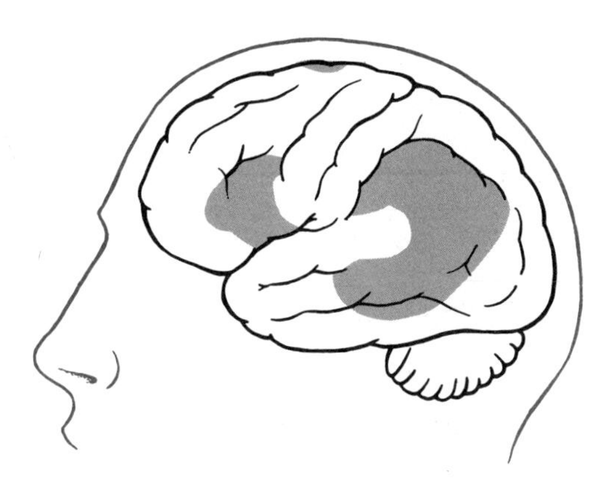

6–79
The human cerebral cortex, showing the areas associated with speech.

less inconvenient to their families and society (a goal now attained by the long-term use of massive doses of tranquilizers). Despite the large number of persons "treated" in this way, there is little agreement on the results of frontal lobotomy—or, more precisely, on the normal role of the frontal lobes. Most, but not all, psychologists agree that the frontal lobes are somehow involved with intention, will, planning, and the marshalling of efforts to carry out long-range programs—all of which are particularly human characteristics.

Posterior Lobes

The functions of the posterior lobes, which have largely been studied by experiments in which parts of the cortex have been damaged, removed, or otherwise interfered with, are not quite so elusive. In both experimental animals and humans, damage to specific areas of the posterior lobes seems to result in problems of discriminating specific stimuli. For example, monkeys can no longer tell the difference between patterns, objects, or sounds that they were previously able to distinguish if particular cortical areas of the posterior lobes are damaged. In humans, damage to specific areas of the cortex of the left posterior lobe (in left-handed as well as most right-handed people) results in impairment of speech. It is probable, although far from proved, that the functions of memory and learning and the organization of ideas take place in the "silent" areas of the posterior lobes.

Split-brain Studies

Additional interesting implications concerning storage of information in the cortex has come from split-brain studies carried out over the past 10 years by Robert Sperry and coworkers at the California Institute of Technology. Sperry severed the connections between the two cerebral hemispheres in cats and monkeys by cutting through the corpus callosum and the optic chiasma (see Figure 6–80). The animals' everyday behavior was normal, but by using specially contrived testing techniques, the investigators were able to show some remarkable and unexpected effects. The animals responded to test procedures as if they had two brains! When the right eye of an animal was covered, for instance, it could learn tasks and discriminations using only

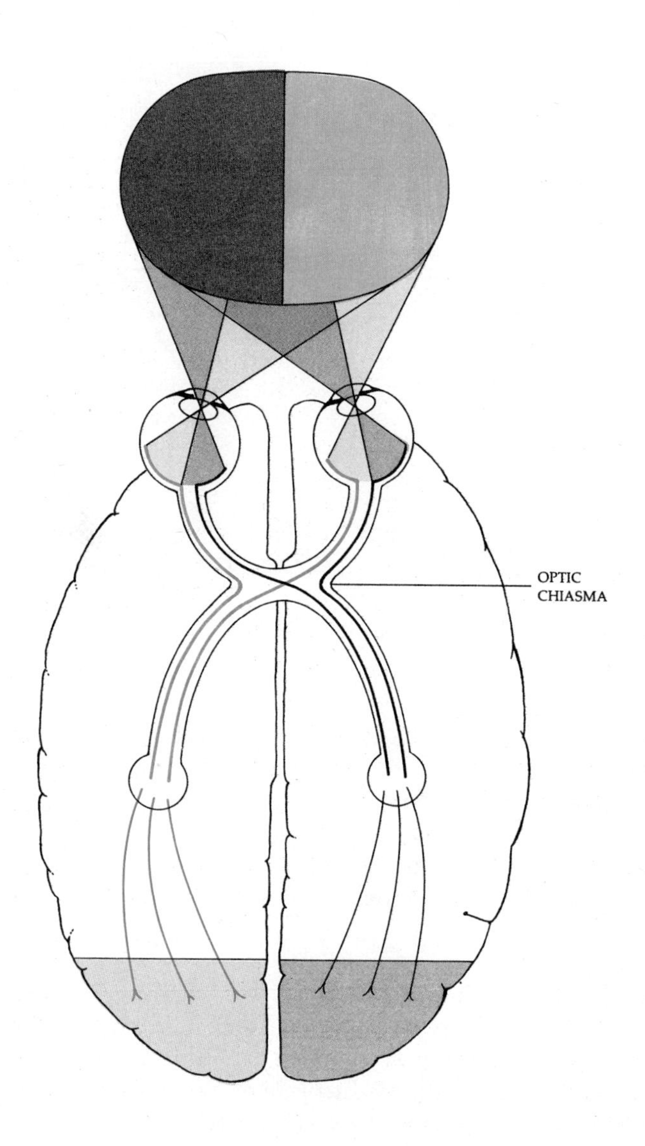

6–80

The optic chiasma viewed from below. Because of a crossing-over of fibers traveling from the retina to the visual lobes of the brain, the right half of each eye transmits signals to the right cerebral hemisphere and the left half to the left cerebral hemisphere.

the left eye. But then if the left eye was covered and the same tests were presented to the right eye, the animal had to learn all over again. In fact, it was possible, without creating any confusion at all, to have both sides of the brain learn opposite solutions to the same problem.

More recently, similar observations have been made in patients in whom the corpus callosum has been severed for the treatment of epilepsy. The patients are able to carry out their normal activities, but specially devised tests indicate the presence of what Sperry calls "two realms of consciousness." If such a patient is asked to identify by touch objects that he cannot see, he can name those he can feel with his right hand but not those he can feel only with his left hand; apparently, because the speech centers of the brain are located in the left hemisphere, the right brain is mute. If a picture is flashed before his eyes as his head is held in a fixed position (in human patients with the optic chiasma intact, the right brain sees only the left side of a picture briefly presented and the left brain sees only the right side) and he is asked what he saw, he reports only what he saw on the right side of the picture; however, the left side of the picture can have an emotional effect on him. The left brain reports, for example, that it is amused, but it does not know why.

One conclusion that can be drawn from these experiments is that there is a great deal of redundancy in the human cortex, perhaps as a protection device against injury. This redundancy has long been suspected by surgeons who have observed patients functioning normally after removal of or damage to extensive areas of the right temporal lobes. In a number of cases, children have, after an interval, learned to speak again after left temperal lobe damage, as a result of the right lobe taking over the speech functions. (In later life, apparently some of this plasticity is lost and the right speech centers cannot take over.)

Sperry's experiments also demonstrate quite clearly that the site of complex learned behaviors in humans *is* the cerebral cortex, rather than the brain stem, since this structure was still intact in all the subjects tested.

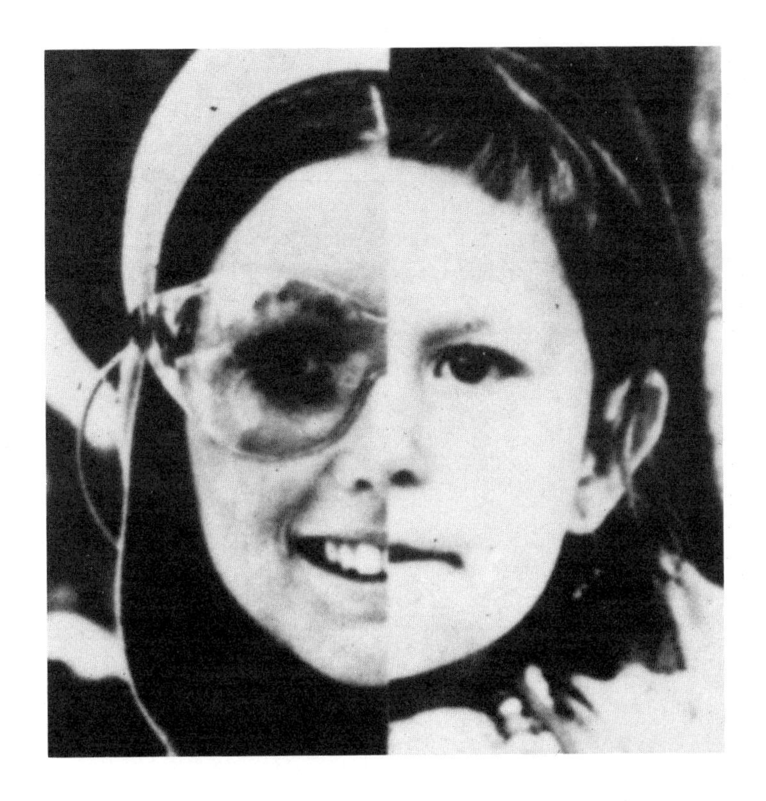

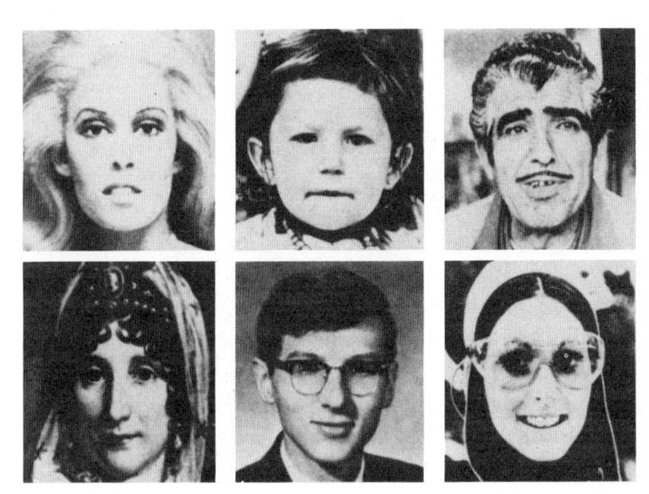

6–81

A split face used in tests of patients whose brains have been surgically divided is made up from two of the faces shown in the selection above. The patient, wearing a headgear that restrains eye movements, sees the picture projected briefly on a screen. The left side of the brain recognizes the pictures as being the one of the child; the right side sees the woman.

There is some indication that there are differences in mental capacity between the two brains, with the left brain excelling in verbal skills, obviously, and the right one in discriminations involving shape, form, and texture. The implication of this finding for the normal person is not clear, however.

LEARNING AND MEMORY

For scientists interested in brain research, one of the biggest present challenges is to understand the mechanisms of memory and learning. If the brain is a physicochemical system—and all scientists are agreed that it is—learning, which involves a change in function, must involve a physicochemical change. But what is this change, and where does it take place?

Almost 50 years ago, the late Karl Lashley of Harvard set out to locate this physical change, the "scar" of memory, which he called the engram. He taught rats and other animals to solve particular problems and then performed operations to see whether he could remove the portion of the brain containing that particular engram. But he never found the engram. As long as he left enough brain tissue to

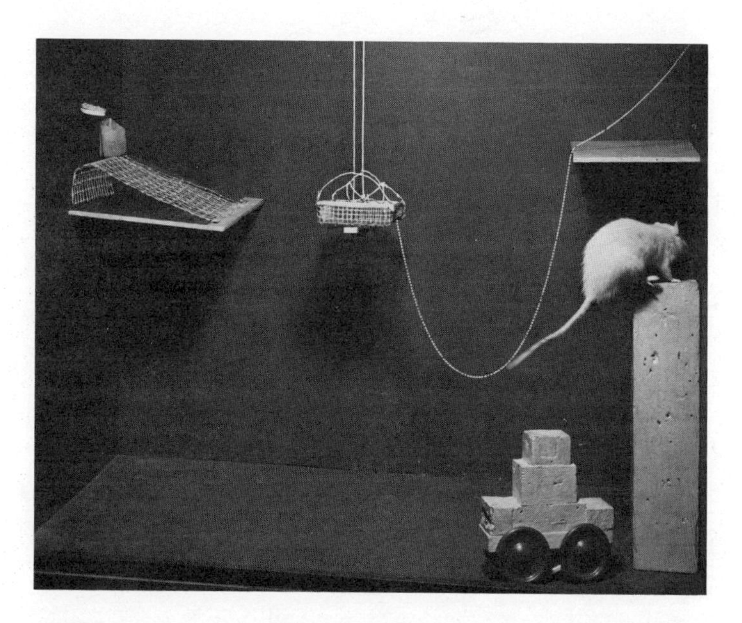

6–82
This rat has learned to push the cart to the tower, climb the tower to the platform, pull the basket to the platform, and swing across in the basket to the opposite platform, where a food reward has been placed. Following the learning of tasks such as this, there is an increase in the RNA content of the brain cells.

enable the animal to respond to the test procedures at all, he left memory as well. And the amount of memory that remained was generally proportional to the amount of remaining brain tissue. Lashley concluded that memory is "nowhere and everywhere present."

There are currently two main theories of memory storage. One school of investigators maintains that memory involves chemical changes in the brain cells, particularly in the RNA. A line of supporting evidence for this hypothesis is the fact that there is a significant increase in the RNA of the brain cells of rats trained in a complex performance, as compared with that of untrained rats. Rats injected with RNA learn more rapidly than uninjected rats. Aged men and women show memory improvement after RNA treatments.

More direct but much more controversial evidence on the role of RNA comes from the studies of James V. McConnell of the University of Michigan. McConnell exposed planarians first to a three-second flash of light and then to an electric shock, which made them contract. Eventually the planarians "learned" to contract when exposed to the light alone. When trained planarians were ground up and fed to other, untrained planarians, the untrained planarians which had eaten the trained ones learned faster than planarians which had simply eaten other, equally ignorant planarians. As might be expected, McConnell's report on this subject, which he referred to as "the transfer of learning by cannibalism," led to a number of jokes about the possibilities for more meaningful student-teacher relationships. It also led to a large number of studies, some of which questioned whether or not planarians actually did learn anything in the first place (planarians do not like light, so light alone can make them contract), whether or not it was possible to transfer this learning by means of the RNA (which McConnell's group held responsible for the transfer effects), and whether or not this sort of transfer could be shown to take place in higher animals, such as rats. None of these questions has been answered conclusively, and investigations of the RNA coding of memory seem to be less intensive than they were previously. However, the subject is still of interest not only in its own right but also as illustrative of the problems involved in trying to design experiments for the study of questions of learning and memory using invertebrates as model systems.

A second major hypothesis concerning the nature of the physicochemical change involved in learning proposes that the learning process sets up new connections between cells. These connections may be only temporary, accounting for short-term memory—such as looking up a number in the phone book and dialing it a few seconds later. Or they may be long-term—as, for example, when frequent use of a telephone number results in the establishment of more permanent memory traces. These more permanent traces presumably take the form of new, enlarged, or more effective synaptic connections. The role of RNA, according to proponents of this theory, is in the production of the protein needed for establishment of the new or altered synapses. At present, this seems to be the most popular hypothesis, but here again, there is no direct proof.

It is not customary for an introductory text to deal with a subject as speculative as the physical basis of learning, but clearly much new work will be done in this field in the next few years and we thought you might enjoy knowing some of the background of this research.

SUMMARY

The brain is made up of "white matter" (nerve fiber tracts) and gray matter (neurons and glia cells). It has three principal subdivisions: (1) the brainstem, which includes the medulla, pons, midbrain, thalamus, hypothalamus, and basal ganglia; (2) the cerebellum; and (3) the cerebrum, which is divided into two cerebral hemispheres and, in the higher mammals, has a greatly convoluted outer layer of cells, the cerebral cortex.

The brainstem, which connects the spinal tract with the brain centers, contains many ascending and descending fibers and relay nuclei. Motor and sensory fibers cross over in the brain stem, which is why the right side of the brain controls the left side of the body, and vice versa. Nuclei concerned with heart action, respiration, and gastrointestinal function are located in the medulla. The optic and auditory lobes, which in mammals serve complex sensory reflex functions, are located in the midbrain. The reticular formation of the brainstem is involved in arousal and sensory filtration. The thalamus relays auditory, visual, and tactile information to the cerebral cortex.

The cerebellum integrates and coordinates motor and sensory information and so makes possible the carrying out of complex muscular activities.

The cerebral cortex, which is the outer layer of the cerebral hemispheres, reaches its greatest degree of development, evolutionarily speaking, in man. Mapping studies of the human cortex have succeeded in identifying a motor cortex, in which stimulation of particular areas results in motor response in particular parts of the body; a sensory cortex, which appears to be a terminal for incoming somatic sensory impulses; and auditory and visual cortices, where impulses from the ear and eye, respectively, are received and analyzed. On the left cerebral hemisphere are also the so-called speech areas; damage to these areas interferes with the complex processes involved in the formulation and organization of speech. The rest of the cortex—which constitutes about three-fourths of the total area—is known as the association cortex. The relative amount of association cortex is much larger in man than it is in other mammals, and these areas are believed to be involved in planning, learning, memory, and other abstract mental processes.

GLOSSARY

ACETYLCHOLINE (a-**sea**-tull-**co**-leen): One of the known chemical transmitters of nervous impulses across synaptic junctions.

ADRENAL GLAND: A vertebrate endocrine gland. The cortex (outer surface) is the source of aldosterone and cortisonelike hormones; the adrenal medulla (inner core) secretes adrenaline.

ADRENALINE: A hormone produced by the medulla of the adrenal gland which increases the concentration of sugar in the blood, raises blood pressure and heartbeat rate, and increases muscular power and resistance to fatigue; also a chemical transmitter across synaptic junctions.

AFFERENT [L. *ad,* to + *ferre,* to bear]: Bringing inward to a central part, applied to nerves and blood vessels.

ALDOSTERONE: A hormone produced by the adrenal cortex which stimulates the reabsorption of sodium from the kidney.

ALLANTOIS (al-**lan**-toe-iss) [Gk. *allantoeides,* sausage-shaped]: A saclike outgrowth from the ventral surface of the posterior part of the embryonic gut. It functions as an embryonic urinary bladder or as a respiratory extension of the hindgut in reptiles and birds and is modified to carry blood vessels to and from the placenta in mammals.

ALVEOLUS, *pl.* ALVEOLI (al-**vee**-o-luss, al-**vee**-o-lie) [L. dim. of *alveus,* tub, cavity, hollow]: An air sac of the lungs; site of actual gas exchange.

AMNION (**am**-neon) [Gk. dim. of *amnos,* lamb]: Inner, fluid-filled sac composed of a thin double membrane that surrounds the embryo in reptiles, birds, and mammals.

ANTIDIURETIC HORMONE (ADH): A hormone secreted by the hypothalamus which inhibits urine excretion by inducing the resorption of water from the nephrons of the kidneys.

AORTA (a-**ore**-ta) [Gk. *airein,* to lift, heave]: The major artery in blood-circulating systems.

ARTERY: A vessel carrying oxygenated blood from the heart to the tissues; it is usually thick-walled, elastic, and muscular.

ATRIOVENTRICULAR NODE: A small mass of special tissue in the hearts of higher vertebrates which transmits the heartbeat stimulus from the atria to the ventricles through the bundle of His.

ATRIUM, *pl.* ATRIA (**a**-tree-um, **a**-tree-a): A chamber of the heart which receives blood and passes it on to a ventricle. In birds and mammals, there are two atria; the left atrium receives deoxygenated blood returning from the body tissues, and the right atrium receives oxygenated blood from the lungs.

AUTONOMIC NERVOUS SYSTEM: A special system of motor nerves and ganglia in vertebrates which is not under voluntary control and which innervates the heart, glands, visceral organs, and smooth muscle. It is subdivided into the sympathetic and parasympathetic nervous systems.

AXON: That part of a neuron which carries impulses away from the cell body; the nerve fiber.

AXOPLASM: The contents of the interior of an axon, plays a role in nerve impulse transmission.

BASAL METABOLISM: The metabolism of an organism when it is using just enough energy to maintain vital processes; measured as the amount of heat (in terms of quantity of oxygen consumed or carbon dioxide given off) produced by an animal at rest (but not asleep) determined at least 14 hours after eating and expressed as calories per square meter of body surface (or per unit weight) per hour.

BILE: A yellow secretion of the vertebrate liver, temporarily stored in the gallbladder and composed of organic salts which emulsify fats in the small intestine.

BLASTULA: The animal embryo during the stage at which it is a solid or hollow sphere of cells.

BOWMAN'S CAPSULE: *See* renal capsule.

BRAINSTEM: The lowest portion of the brain; deals with reflexes controlling homeostasis and connects the spinal cord to the forebrain and cerebrum.

BRONCHUS, *pl.* BRONCHI (**bronk**-us, **bronk**-eye) [Gk. *bronchos,* windpipe]: One of a pair of respiratory tubes branching into either lung at the lower end of the trachea; it subdivides into progressively finer passageways, culminating in the alveoli.

BUDDING: A form of asexual reproduction in animals such as *Hydra;* protuberances (buds) are formed on the parent body, and these eventually develop into new individuals.

BUNDLE OF HIS (hiss) [after Wilhelm His, German anatomist]: A muscular band in the heart which contains nerve fibers connecting the atria with the ventricles. It conveys the stimuli from the atrioventricular node and causes the ventricles to beat in coordinated fashion with the atria.

CAPILLARY [L. *capillus,* hair]: A small, thin-walled blood vessel connecting arteries and veins, through which diffusion and filtration into the tissues can occur.

CEREBELLUM [L. dim of *cerebrum,* brain]: An enlarged part of the dorsal side of the vertebrate brain; chief muscle-coordinating center.

CEREBRAL CORTEX: A layer of nerve cells (gray matter) forming the upper surface of the cerebral hemispheres in the brain; well developed only in mammals; the seat of conscious sensations and voluntary muscular activity.

CEREBRUM [L. *cerebrum,* brain]: An enlarged part of the dorsal and lateral sides of the brain at its front end; controls many voluntary functions and is the source of conscious sensations, learning, and memory.

CHOLINESTERASE: The enzyme which destroys acetylcholine shortly after the latter has been secreted by nerve fibers.

CHORION (**core**-ee-on): The outermost embryonic membrane of reptiles, birds, and mammals; in placental mammals it contributes to the structure of the placenta.

CLEIDOIC (cly-**doe**-ick): A type of egg which is isolated from the environment by a more or less impervious shell during the period of its development and is completely self-sufficient, requiring only oxygen from the outside.

CLOACA (klo-**ay**-ka) [L. *cloaca,* sewer]: The exit chamber from the digestive system; also may serve as exit for the reproductive system and urinary system.

CONE: A type of light-sensitive nerve cell in the vertebrate retina; concerned in the perception of color and in the most acute discrimination of detail.

CORNEA [L. *corneus,* horny]: The transparent epidermis and connective tissue covering the outer surface of the vertebrate eye, overlying the iris and pupil.

CORPUS CALLOSUM [L. *corpus,* body; *callosum,* callous]: Transverse band of fibers connecting the two cerebral hemispheres in mammals.

CORPUS LUTEUM (**core**-puss **lu**-tyum) [L. *corpus,* body; *luteum,* yellow]: The mass of yellow tissue in the reproductive apparatus of female mammals which forms from the empty follicle after an egg has been released. It secretes progesterone, which stimulates the uterine lining to thicken in preparation for the embryo.

DENDRITE [Gk. *dendron,* tree]: The area of a neuron that transmits impulses toward the nerve cell body.

DIAPHRAGM [Gk. *diaphrassein,* to barricade]: A sheetlike muscle forming the partition between the abdominal and thoracic cavities.

DIGESTION: The process by which foods are chemically simplified and made soluble so that they can be used by cells.

DUODENUM (duo-**dee**-num) [L. *duodeni,* twelve each—fr. its length, about 12 fingers' breadth]: The upper portion of the small intestine in vertebrates, where food particles are broken down into molecules that can be absorbed by cells.

EFFERENT [L. *ex,* out + *ferre,* to bear]; Carrying away from a center, applied to nerves and blood vessels.

EMBRYO [Gk. *en,* in, + *bryein,* to swell]: The early developmental stage of an organism produced from a fertilized egg; a young organism before it emerges from the seed, egg, or body of its mother.

ENDOCRINE GLANDS [Gk. *krinein,* to separate]: Ductless glands whose secretions (hormones) are released into the circulatory system. In vertebrates, these glands include, among others, the pituitary, sex glands, adrenal gland, and thyroid.

ENDOMETRIUM: The glandular lining of the uterus in female mammals; undergoes thickening in response to progesterone secretion during ovulation and is sloughed off in menstruation.

ERYTHROCYTE (eh-**rith**-ro-sight) [Gk. *erythros*, red, + *kytos*, vessel]: Red blood cell, which serves as a carrier of hemoglobin.

ESOPHAGUS: The tubular portion of the digestive tract immediately behind the pharynx and leading into the stomach.

ESTROGEN (**ess**-tro-jen): One of a group of hormones produced by the ovary and follicle and responsible for the development and maintenance of female sex characteristics.

ESTRUS [Gk. *oistros*, frenzy]: The mating period in female mammals, except the higher primates; characterized by intensified sexual urge.

FALLOPIAN TUBES: In female mammals, the two reproductive tubes which carry the egg from the ovary to the uterus, where it can be fertilized.

FETUS [L. *fetus*, pregnant]: An unborn or unhatched vertebrate that has passed through the earliest developmental stages; a developing human from about the third month after conception until birth.

FIBRINOGEN: A protein carried in the blood plasma which is active in the formation of blood clots.

FISSION [L. *fissus*, split]: Asexual reproduction in organisms by division of the body into two or more equal, or nearly equal, parts.

FOLLICLE: A vesicle in the ovary of female mammals containing the developing egg. After release of the egg cell into the Fallopian tube, the follicle converts into the corpus luteum.

FOLLICLE-STIMULATING HORMONE (FSH): A hormone secreted by the pituitary which stimulates sperm production in males and ovulation and estrogen production in females.

FOVEA (**foh**-vee-a): The point of acutest vision in the vertebrate retina. Here the photoreceptor cells are packed closest together.

GALLBLADDER: A small sac associated with the vertebrate liver which is used to store bile and which empties into the digestive tract.

GAMETE [Gk. *gametē*, wife, and *gametēs*, husband]: The mature functional haploid reproductive cell. When its nucleus fuses with that of another gamete of opposite sex (fertilization), the resulting cell (zygote) develops into a new individual.

GANGLION, *pl.* GANGLIA (**gang**-lee-on): A mass of nerve tissue containing nerve cell bodies.

GASTRIC MUCOSA: The layer of epithelial cells lining the stomach wall that secretes mucus, hydrochloric acid, and pepsinogen, which effect the initial breakdown of food into its chemical components.

GILL: The respiratory organ of aquatic animals; usually a thin-walled projection from some part of the external body surface or, in vertebrates, from some part of the digestive tract.

GLOMERULUS: A cluster of capillaries enclosed by the capsule of the nephron; passes protein-free fluid from the blood into the nephron.

GONAD (**go**-nad) [Gk. *gonē*, seed]: The gamete-producing organ of multicellular animals; the ovary or testis.

GONADOTROPIC HORMONES: Hormones secreted by the pituitary and stimulating the development and activity of the gonads.

HEMOGLOBIN: The iron-containing protein in the blood capable of carrying oxygen.

HIBERNATION: A period of dormancy and inactivity, varying in length depending on the organism and occurring in dry or cold seasons. During hibernation, metabolic processes are greatly slowed and, in mammals, body temperature may drop to just above freezing.

HOMEOSTASIS (home-e-o-**stay**-sis) [Gk. *homos*, same, + *stasis*, standing]: The maintaining of a relatively stable internal physiological environment or equilibrium in an organism, population, or ecosystem.

HOMEOTHERM: An animal capable of maintaining a uniform body temperature independent of the environment; warm-blooded. Birds and mammals are homeotherms.

HYPOTHALAMUS [Gk. *thalamos*, inner room]: The floor and sides of the vertebrate brain just below the cerebral hemispheres; contains centers of coordination and body temperature control.

ILEUM: The lower portion of the small intestine; here most of the absorption of food molecules takes place through the intestinal walls.

IRIS: The pigmented muscular portion of the vertebrate eye which controls the size of the pupil.

KIDNEY: Chiefly in vertebrates, the organ which regulates water and salt balance and the excretion of waste in the form of urine.

LEARNING: The process that produces change in individual behavior as the result of experience.

LENS: A biconvex transparent structure in the eye which focuses a sharp image on the retina.

LIVER: A large digestive gland in vertebrates which secretes bile, converts sugars in the blood to glycogen, forms urea, and detoxifies and/or inactivates substances in the blood.

LOOP OF HENLE [after F. G. J. Henle, German pathologist]: A hairpin-shaped portion of the kidney tubule system where water conversion and salt excretion occur by osmotic and diffusion processes.

LUTEINIZING HORMONE (LH): A hormone secreted by the pituitary which stimulates ovulation and the development of the corpus luteum.

LYMPH: Colorless fluid occurring in intercellular space cavities and special lymph ducts; derived from the blood by filtration through capillary walls.

MEDULLA (med-**dull**-a) [L. *medulla,* the innermost part]: **1.** The inner as opposed to the outer part of an organ, as in the adrenal gland. **2.** The most posterior region of the vertebrate hindbrain that connects with the spinal cord.

MENSTRUAL CYCLE [L. *mensis,* month]: In certain primates, the periodic discharge of blood and disintegrated uterine lining through the vagina.

METABOLISM [Gk. *metabolē,* change]: The sum of all chemical reactions occurring within a living system.

MIDBRAIN: A section of the brainstem which evolved originally as the receiving center for fibers coming into the eye; concerned with vision and hearing in mammals.

MYELIN SHEATH: A fatty material surrounding the axons of nerve cells in the central nervous system of vertebrates; made up of the membranes of Schwann cells.

NEPHRON [Gk. *nephros,* kidney]: A unit of the kidney structure. A human kidney is composed of about a million nephrons.

NEUROHORMONE: A hormone secreted by a nerve ending; neurohormones include adrenaline, noradrenaline, and cetylcholine.

NEURON: A nerve cell, including the cell body, dendrites, and axons.

NODAL TISSUE: Special muscle tissue in the heart which initiates and transmits the heartbeat stimulus and which is concentrated in the pacemaker and the atrioventricular node.

OOCYTE (**o**-uh-sight) [Gk. *ōion,* egg, + *kytos,* vessel]: One of the cells derived from the oogonia, which eventually gives rise by meiosis to an ovum. *See* primary oocyte, secondary oocyte.

OOGONIUM, *pl.* OOGONIA (o-uh-**go**-nee-um, o-uh-**go**-nee-a): A cell in the outer layer of cells of the ovary which is the precursor of a mature oocyte and which multiplies by mitosis.

OOTID (**o**-uh-tidd): The large haploid cell which results from the meiotic division of the secondary oocyte and which develops into the functional ovum.

OPTIC CHIASMA [Gk. *khiasma,* cross]: Crossing of the fibers of the optic nerves on the ventral side of the brain.

OVARY: The egg-producing organ of female animals.

OVIDUCT [L. *ovum,* egg, + *duccre,* to lead]: The tube serving to transport the eggs to the uterus or to the outside.

OVUM, *pl.* OVA: The egg cell; the female gamete.

PACEMAKER: A mass of nodal tissue which initiates the heartbeat; located where the superior vena cava enters the right atrium.

PANCREAS (**pang**-kree-us) [Gk. *pan,* all, + *kreas,* meat, flesh]: A small complex gland in vertebrates, located between the stomach and the duodenum, which produces digestive fluids and the hormones insulin and glucagon.

PARASYMPATHETIC NERVOUS SYSTEM: A subdivision of the autonomic nervous system of vertebrates, with centers located in the brain and in the most anterior and most posterior parts of the spinal cord; stimulates digestion and has a general inhibitory effect on other functions. Neurohormone: acetylcholine.

PENIS: The male copulatory organ, through which sperm is deposited in the female reproductive tract. In vertebrates, it also serves as the channel for the excretion of urine from the body.

PERISTALSIS: In vertebrates and higher invertebrates, wavelike muscular contractions which move progressively along the digestive tract and serve to mix food matter and move it down the tract.

PITUITARY (pit-**too**-i-terry) [L. *pituita,* phlegm]: An endocrine gland in vertebrates, composed of anterior, intermediate, and posterior lobes, each representing a functionally separate gland. It is stimulated by neurohormones produced by the hypothalamus.

PLACENTA (plass-**sent**-a) [Gk. *plax,* flat surface]: A structure formed in part from the inner lining of the uterus and in part from the tissues of the embryo. The placenta develops in most species of mammals and serves as the connection between the mother and the embryo during pregnancy; through it exchanges occur between the blood of the mother and that of the embryo.

PLASMA [Gk. *plasma,* form]: The clear, colorless fluid component of blood, containing dissolved salts and proteins; in it, the corpuscles (various cell bodies) are suspended.

PLATELET (**plate**-let): In mammals, a minute granular body suspended in the blood and involved in the formation of blood clots.

POIKILOTHERM [Gk. *poikilos,* various, + *thermē,* heat]: An animal whose body temperature remains close to that of its environment; cold-blooded.

POLAR BODY: Minute nonfunctioning cell produced during meiotic divisions in egg cells; contains a nucleus but very little cytoplasm.

PONS [L. *pons,* bridge]: An area of the brainstem; contains a large bundle of transverse fibers which interconnect the brainstem and cerebellum.

PROGESTERONE: A steroid hormone produced by the corpus luteum which helps prepare the uterus for reception of the ovum.

PROSTATE GLAND: A mass of muscle and glandular tissue surrounding the base of the urethra in male mammals, through which the seminal vesicle passes and which secretes an alkaline fluid that has a stimulating effect on the sperm as they are released.

PULMONARY ARTERY [L. *pulmonis,* lung]: In vertebrates, the artery carrying blood to the lungs.

PULMONARY VEIN: In vertebrates, a vein carrying oxygenated blood from the lungs to the left atrium, from which the blood is pumped into the left ventricle and from there to the body tissues.

PYLORIC SPHINCTER: A ring of muscle at the junction of the stomach and duodenum which controls the entrance of the stomach contents into the duodenum.

REFLEX ARC [L. *reflectere,* to bend back]: A unit of action of the nervous system involving a sensory neuron, one or more interneurons, and a motor neuron or neurons.

RENAL CAPSULE: The bulbous unit of the nephron that encloses the glomerulus. It is the site of filtration of the renal fluid from the blood, the initial process in urine formation. Also called Bowman's capsule.

RETICULAR FORMATION: A core of tissue which runs centrally through the entire brainstem; a weblike network of fibers and neurons. Involved with consciousness.

RETINA [L. *rete,* a net]: The innermost nervous-tissue layer of the eyeball, containing several layers of neurons and light-receptor cells (rods and cones); receives the image formed by the lens and is connected to the brain by the optic nerve.

ROD: Photoreceptor cell found in the vertebrate retina; sensitive to very dim light. More light-sensitive than cones, rods are responsible for "night vision."

SCHWANN CELL: A cell which enwraps one or more axons; several membranous layers of Schwann cells constitute the myelin sheath.

SEBACEOUS GLAND [L. *sebum,* grease]: A skin gland near the hair base in mammals that secretes sebum, an oil.

SEMINAL VESICLE [L. *semen,* seed]: The portion of the male reproductive duct in which sperm are stored.

SEMINIFEROUS TUBULE [L. *semen,* seed, + *ferre,* to carry]: One of the minute coiled tubules which constitute the sperm-producing areas of the testes.

SEROTONIN: A chemical transmitter of nervous impulses across synaptic junctions in the brain.

SOMATIC SYSTEM: The nervous system in vertebrates; composed of the motor, sensory, and interneurons.

SPERMATIDS [Gk. *sperma,* seed]: The four haploid cells that result from the meiotic divisions of a spermatocyte. Each spermatid differentiates into a sperm cell.

SPERMATOCYTE: The diploid cell formed by the enlargement of a spermatogonium; gives rise by meiotic division to four spermatids.

SPERMATOGONIUM, *pl.* SPERMATOGONIA: The unspecialized diploid germ cell on the walls of the testes which by meiotic division becomes a spermatocyte, then four spermatids, then four spermatozoons or sperm.

SPINAL CORD: The part of the vertebrate central nervous system that consists of a thick longitudinal bundle of nerve fibers extending from the brain posteriorly along the dorsal side.

SYMPATHETIC NERVOUS SYSTEM: A subdivision of the autonomic nervous system with centers located in the midportion of the spinal cord; slows digestion and has a general excitatory effect on other functions. Neurohormone: adrenaline or noradrenaline.

SYNAPSE (**si**-naps) [Gk. *synapsis,* union]: The region of nerve-impulse transfer between two neurons.

TESTIS, *pl.* TESTES [L. *testis,* witness]: The male gamete-producing organ, which is also the source of male sex hormone.

TESTOSTERONE: A hormone secreted by the testes in higher vertebrates and stimulating the development and maintenance of male sex characteristics and the production of sperm.

THALAMUS [Gk. *thalamos,* chamber]: A part of the vertebrate forebrain just posterior to the cerebrum; an important intermediary between all other parts of the nervous system and the cerebrum.

THORACIC: In vertebrates, pertaining to that portion of the trunk containing the heart and lungs.

THYROID: An endocrine gland of vertebrates, located in the neck region, that secretes an iodine-containing hormone (thyroxine) which increases the oxidative processes (metabolic rate).

TRACHEA (**trake**-ee-a) [Gk. *tracheia,* rough]: An air-conducting tube, such as the windpipe of mammals and the breathing systems of insects.

UREA [Gk. *ouron,* urine]: An organic compound formed in the vertebrate liver from ammonia and carbon dioxide and excreted by the kidneys; the principal form of ammonia disposal in mammals and some other animal groups.

URETER [Gk. *ourein,* to urinate]: The tube carrying urine from the kidney to the cloaca (in reptiles and birds) or the bladder (in amphibians and mammals).

URETHRA: The tube carrying urine from the bladder to the exterior in mammals.

URIC ACID: An insoluble nitrogenous waste product that is the principal excretory product of birds, reptiles, and insects.

UTERUS [L. *uterus*, womb]: The muscular, expanded portion of the female reproductive tract modified for the storage of eggs or for housing and nourishing the developing embryo.

VAGINA [L. *vagina*, sheath]: The part of the female reproductive duct in mammals which receives the male penis during copulation.

VAGUS NERVE [L. *vagus*, wandering]: One of a pair of nerves arising from the medulla of the vertebrate brain which innervate the visceral organs and supply them with autonomic fibers.

VAS DEFERENS (vass **deaf**-er-ens): In mammals, the tube carrying sperm from the testes to the urethra.

VEIN: A blood vessel carrying blood from the tissues and organs to the heart.

VENA CAVA (**vee**-na **cah**-va): A large vein which brings blood from the tissues to the right atrium of the heart. The superior vena cava collects blood from the forelimbs, head, and anterior or upper trunk; the inferior vena cava collects blood from the kidneys, the liver or lower gonads, and the general posterior body region.

VENTRICLE [L. *ventriculus*, the stomach]: A chamber of the heart which receives blood from an atrium and pumps blood out from the heart.

VERTEBRA [L. *vertebra*, joint]: A segment of the spinal column.

VERTEBRAL COLUMN: In vertebrates, the series of bones or cartilages which constitute the vertebrate backbone and which surround the spinal cord and support the body.

VILLI [L. *villus*, a tuft of hair]: In vertebrates, minute fingerlike projections lining the small intestine which serve to increase the absorptive surface area of the intestine.

YOLK: The stored food material formed in the egg cell which nourishes the embryo.

SUGGESTIONS FOR FURTHER READING

CALDER, NIGEL: *The Mind of Man*, The Viking Press, Inc., New York, 1970.

A well-written, fast-moving, and comprehensive report on the "drama of brain research."

CLEGG, P. CATHERINE, and ARTHUR C. CLEGG: *Hormones, Cells, and Organisms: The Role of Hormones in Mammals*, Stanford University Press, Stanford, Calif., 1969.

A short, up-to-date, supplementary text on hormones. The authors stress the role of hormones in homeostasis and in adaptation to the environment.

GORDON, MALCOLM S., et al.: *Animal Functions: Principles and Adaptations*, The Macmillan Company, New York, 1968.

A good animal physiology textbook, intended for an advanced course. The authors emphasize function as it relates to the survival of organisms in their natural environments.

HANDLER, PHILIP (ed.): *Biology and the Future of Man*, Oxford University Press, New York, 1970.*

A widely acclaimed survey of the state of biology today, written by a committee of experts in the various fields. Its purpose is to inform the general public both of present knowledge in major fields of biology and of current problems and probable areas of future investigation. It deals with a wide variety of subjects, including most of the ones covered in this book. There are outstanding sections on animal physiology, development, and behavior.

MACEY, ROBERT I.: *Human Physiology*, Prentice-Hall, Inc., Englewood Cliffs, N.J.*

A clear, thorough text.

MICHELMORE, SUSAN: *Sexual Reproduction,* Natural History Press, Garden City, N.Y., 1965.

A clear, intelligent, and often fascinating book about the many forms of sexual behavior. The subject ranges from conjugation in paramecia, through the tribulations of the male sea horse, to Indian customs of wife lending.

NOSSAL, G. J. V.: *Antibodies and Immunity,* Basic Books, Inc., Publishers, New York, 1969.

An account for the general public of recent research in this important field by a young scientist who has himself made many significant contributions.

ROMER, ALFRED: *The Vertebrate Story,* 4th ed., The University of Chicago Press, Chicago, 1959.

The history of vertebrate evolution, written by an expert but as readable as a novel.

RUGH, ROBERTS, and LANDRUM B. SHETTLES: *From Conception to Birth: The Drama of Life's Beginnings,* Harper & Row, Publishers, Inc., New York, 1971.

This is an account of the history of life before birth. The book describes in detail the development of the unborn child from the moment of fertilization and also the changes in the mother during pregnancy. It discusses such related topics as birth control, congenital malformation, labor, and delivery of the baby. There are a large number of illustrations, including a group of magnificent color photographs of the developing fetus.

SCHMIDT-NIELSON, KNUT: *Animal Physiology,* 2d ed., Prentice-Hall, Inc., Englewood Cliffs, N.J., 1964.*

A useful short paperback text.

SCHMIDT-NIELSON, KNUT: *Desert Animals,* Oxford University Press, New York, 1964.

Although considered the definitive work on the physiological problems relating to heat and water, this readable book also contains numerous anecdotes—such as that about Dr. Blagden—and many fascinating personal observations.

SMITH, HOMER W.: *From Fish to Philosopher,* Doubleday & Company, Inc., Garden City, N.Y., 1959.*

Dr. Smith is an eminent specialist in the physiology of the kidney. Writing for the general public, he explains the role of this remarkable organ in the story of how, in the course of evolution, organisms have increasingly freed themselves from their environments.

THOMPSON, R. E.: *Foundations of Physiological Psychology,* Harper & Row, Publishers, Inc., New York, 1967.

This book is so clearly written that although its subject is specialized, almost all of it can be read easily by anyone who has made his way through Section 2 and 6 of Invitation to Biology.

WOOLDRIDGE, D. E.: *The Machinery of the Brain,* McGraw-Hill Book Company, New York, 1963.*

A fine, thoughtful account of the workings of the central nervous system.

* Available in paperback.

PART III

Ecology and Evolution

SECTION 7

Life on a Small Planet

7–1
Light striking corn plants.

CHAPTER 7-1

The Biosphere

The biosphere is the part of the earth within which life exists. It forms a thin film at the surface of the planet, extending about 5 or 6 miles above the surface of the earth and about equally far into the depths of the sea. This film of life is not uniform in either its depth or its density. In fact, in some areas, such as portions of the deserts or the ice sheets of the Antarctic and Greenland, the only evidence of life to be found is in the form of dormant bacterial spores. The two requirements for the presence of living organisms on our planet are an energy source—the sun—and a temperature which permits the presence of liquid water. Presumably there are many other biospheres in the universe, and it is likely that they also are characterized by the presence of a similar energy source and liquid water.

A number of factors are involved in producing the changing patterns of life on the surface of the earth. The major one is the planet's relationship to the sun. The earth is about 93 million miles from this star, which is the source of its heat and light energy and, indirectly, of its chemical energy as well. At the equator, the sun's rays are almost perpendicular to the earth's surface, and this sector receives more energy from the sun than the areas to the north and south, with the polar regions receiving the least. Moreover, because the earth, which is tilted on its axis, rotates once every 24 hours and completes an orbit around the sun once about every 365 days, the amount of energy reaching different parts of the surface may vary hour by hour and season by season.

The variations in temperature over the surface of the earth and also the earth's rotation establish major patterns of air circulation and rainfall on the earth's surface. These patterns depend, to a large extent, on the fact that cold air is denser than warm air. As a consequence, hot air rises and cold air falls. Because hot, less dense air holds more water vapor than cold air, rising air tends to lose its moisture as it cools in the form of rain or snow. The air is hottest along the equator, the region heated most by the sun. This air rises, creating a low-pressure area (the doldrums), moves away from the equator, cools, loses its moisture, and falls again at latitudes of about 30° north and south, the regions where most of the great deserts of the world are found. This air warms, picks up moisture, and rises again at about 60° latitude (north and south); this is the polar front, another low-pressure area. A third, weaker belt rising at the polar front descends again at the pole, producing a region in which again, as in other areas of descending air, there is virtually no rainfall. These belts of air circulation, combined with rotational movements of the earth, are responsible for the major wind currents across the earth's surface.

The worldwide patterns are modified locally by a variety of factors. For example, along our own West Coast, where the winds are prevailing westerlies, the western slopes of the Sierra Nevadas have abundant rainfall while the eastern slopes are dry and desertlike. As the air from the ocean hits the western slope, it rises, is cooled, and releases its water. Then, after it passes the crest of the mountain range, the

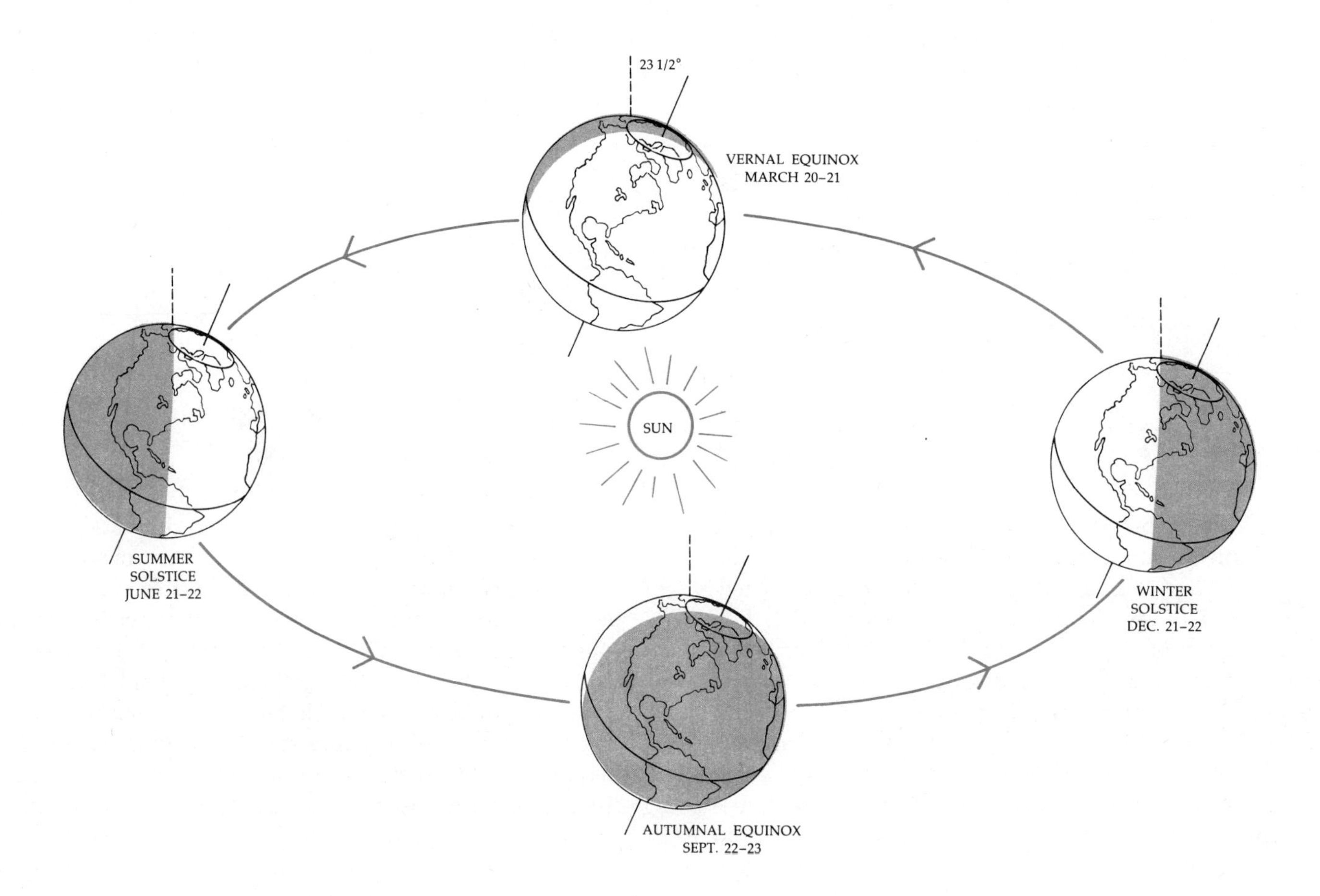

7–2

The earth receives the most heat from the sun when the sun's rays are perpendicular to the earth's surface, as they are at the equator. In the northern and southern hemispheres, temperatures change in an annual cycle because the earth is slightly tilted on its axis in relation to its pathway around the sun. In winter, the northern hemisphere tilts away from the sun, which decreases the angle at which the sun's rays strike the surface and also decreases the duration of daylight, both of which result in lower temperatures. In the summer, the northern hemisphere tilts toward the sun. Note that the polar region of the northern hemisphere is continuously dark during the winter and continuously light during the summer.

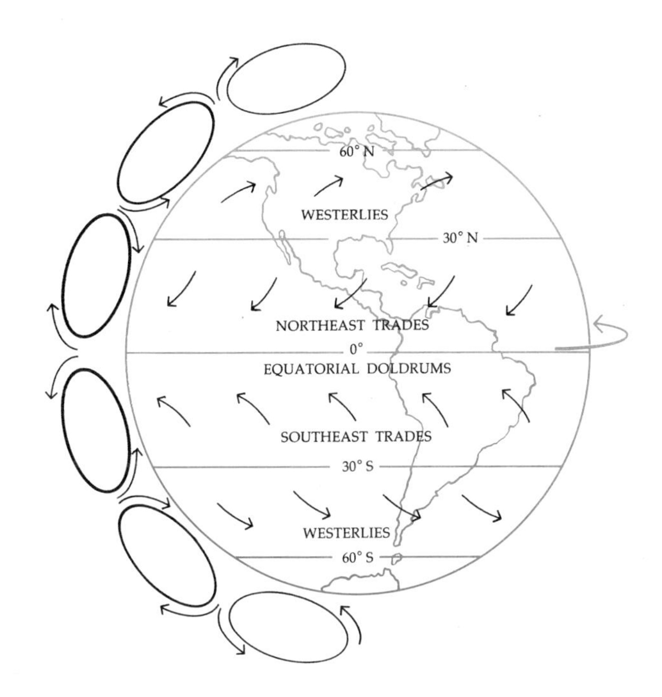

7–3
The earth's surface is covered by many belts of air currents, which determine the major patterns of rainfall and wind. Air rising at the equator loses moisture in the form of rain, and falling air at latitudes of 30° north and south is responsible for the great deserts found at these latitudes.

7–4
The mean annual rainfall (vertical columns) in relation to altitude at a series of stations from Palo Alto on the Pacific Coast across the Coast Range and the Sierra Nevada. Notice the "rain shadows" on the landward side of the two mountain ranges.

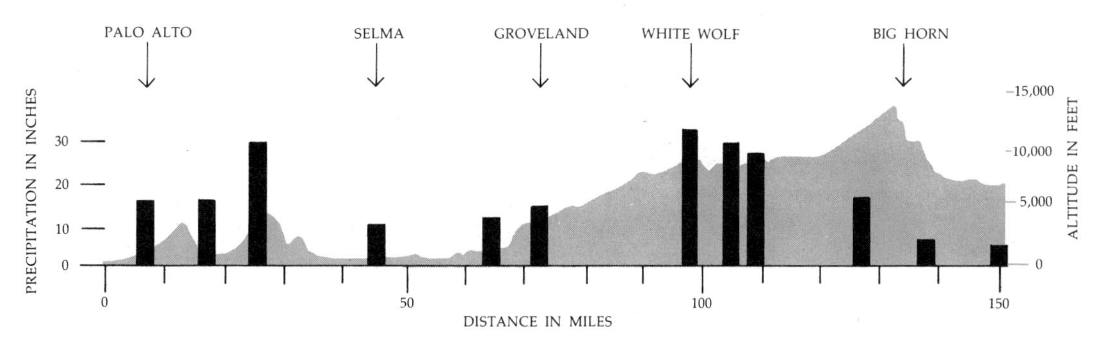

THE WATER CYCLE

The earth's supply of water is constant and is used over and over again. Most of the water (98 percent) is in liquid form—in the oceans, lakes, and streams. Of the remaining 2 percent, some is frozen in polar ice and glaciers, some is in the soil, some is in the atmosphere in the form of vapor, and some is in the bodies of living organisms.

The water cycle is driven by solar energy. Energy from the sun evaporates water from the oceans, lakes, and streams, from the moist soil surfaces, from the leaves of plants (transpiration), and from the bodies of other living organisms. The water molecules are carried up into the atmosphere by air currents and eventually fall again as rain.

Some of the water that falls on the land percolates down through the soil until it reaches a zone of saturation. In the zone of saturation, all pores and cracks in the rock are filled with water. The upper surface of this zone of saturation is known as the water table. Below the zone of saturation is solid rock, through which the water cannot penetrate. The deep ground water, moving extremely slowly, eventually emerges in lakes and streams and even in the ocean floor as seepages and springs, thereby completing the water cycle.

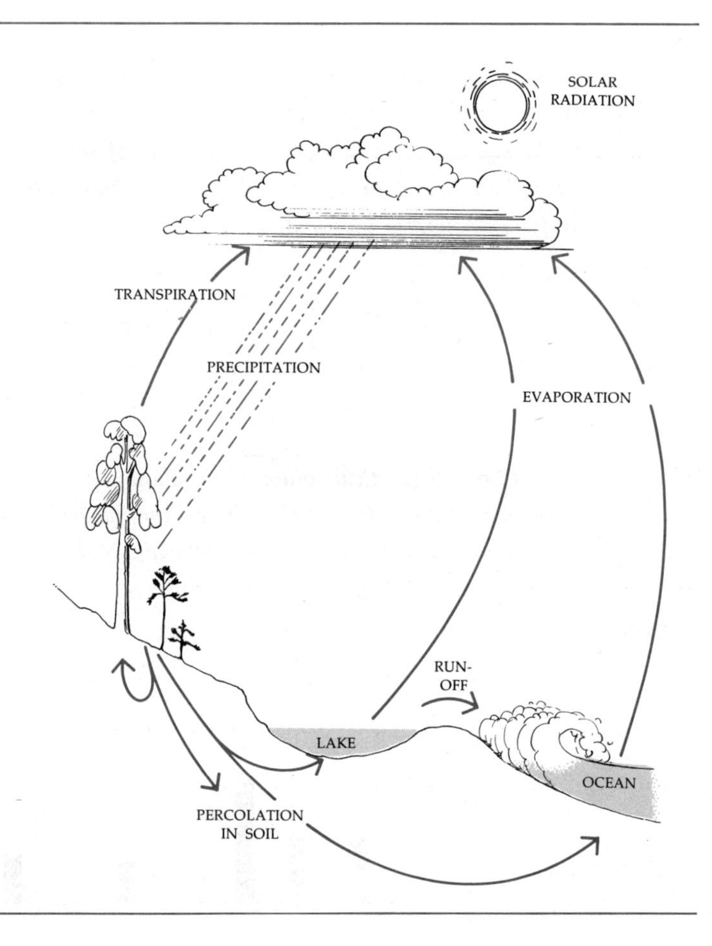

air descends again, becoming warmer, and its water-holding capacity increases.

The patterns of water circulation—clockwise in the Northern Hemisphere and counterclockwise in the Southern Hemisphere—move currents of warm water north and south from the equator. One such current, the Gulf Stream, warms a portion of our East Coast and the western shores of Europe, and another warms the eastern coast of South America. The same patterns of circulation bring cold waters to the western coasts of North and South America. Where the winds move the water continuously away from the shores, as off the coasts of Portugal and Peru, cold water rich in nutrients is brought to the surface in upwellings. Such areas have highly profitable fishing industries.

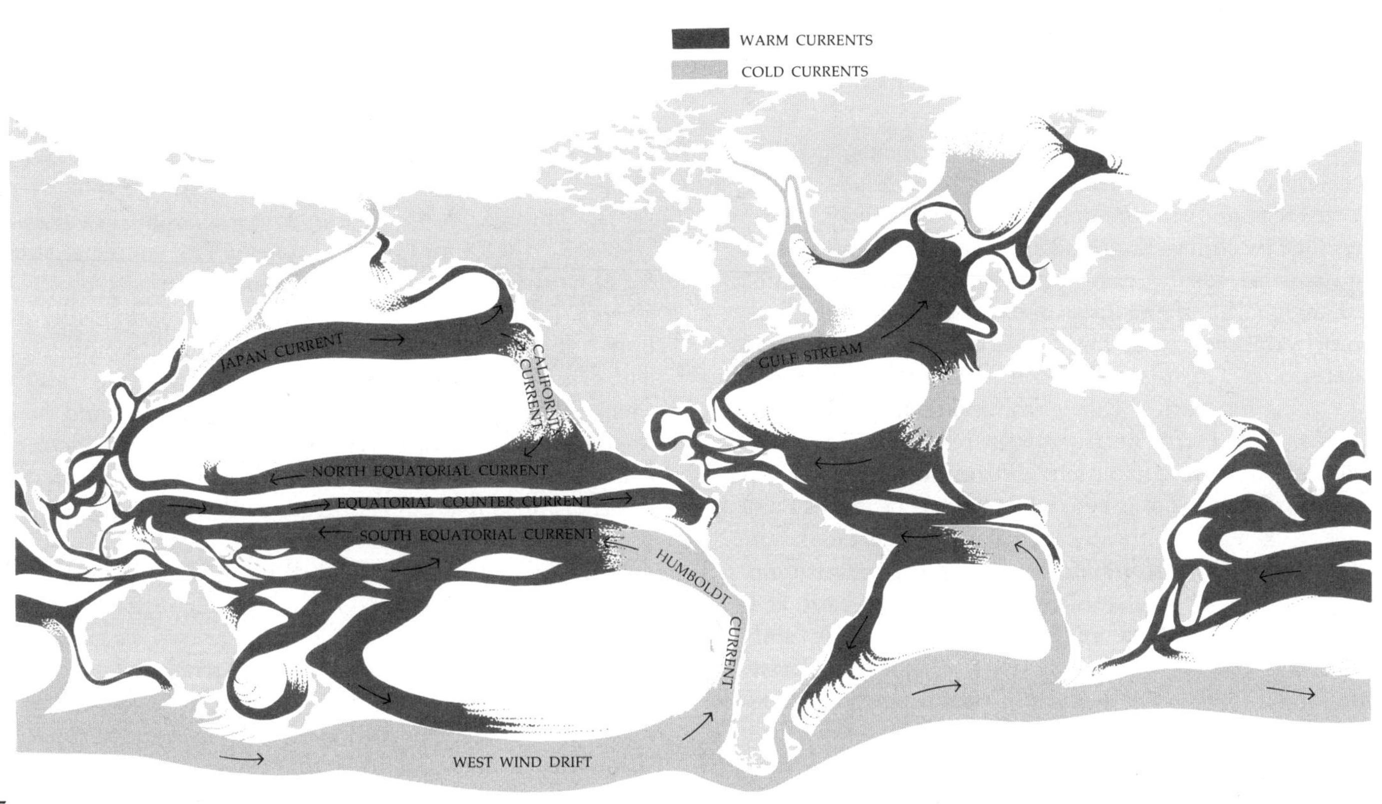

7–5
The major currents of the ocean have profound effects on climate. Because of the warming effects of the Gulf Stream, Europe is milder in temperature than is North America at similar latitudes. The eastern coast of South America is warmed by water from the equator, and the Humboldt Current brings cooler weather to the western coast of South America. Where winds constantly move surface water away from coastal slopes, cold water rich in nutrients is brought to the surface in upwellings, such as those responsible for the rich fishing industry off the coast of Peru.

THE EARTH'S CRUST

The heavier materials of which the earth is composed are collected in a dense core in the center. This core, whose diameter is about half the diameter of the earth, consists chiefly of iron, with a small amount of nickel and cobalt. Surrounding the core is a layer of lighter material, basalt. Both the core and the basalt are hot; below the surface of the earth, temperatures increase rapidly, and at a depth of only 30 to 40 miles, the temperature is as high as 1500°C. Although the temperature of the basalt is higher than its melting point, it is not fluid because it is under pressure. However, it is plastic—that is, it gives a little. This plastic layer lies beneath a relatively brittle layer, the crust, which underlies both the oceans and the continents.

The land surfaces of the earth are composed of granite, which is an igneous rock (from the Latin word *ignis*, for "fire"); igneous rock is rock formed directly from molten material. These masses of granite, the continents, rest on plates of crust which float on the molten basalt.

The surfaces of the continents change constantly. They are crumpled by contractions and shiftings as the continents rise and sink because of the motion of the basalt beneath them. Earthquakes shake the surfaces, cracks form, and molten material forces its way up under pressure from the hot interior. As a consequence, the earth's surface is not at all uniform but varies widely from place to place in its composition and in its height above seawater. Both the mineral content of the earth's surface and the altitude affect the growth of plants and other living organisms, and as we saw in the example of the Sierra Nevadas, the mountain ranges of the continents do much to determine the patterns of rainfall.

The interplay of all these factors makes various parts of the earth's surface very different from one another, and these differences—in heat, light, rainfall, altitude, and composition—are the strongest factors (with the possible exception, as we shall see, of man) in determining the pattern of growth of living things.

LIFE ON LAND: THE BIOMES

The major different life formations on land are called *biomes.* These biomes, the big landscapes, are generally characterized and identified by their most important, or dominant, plants. The dominant plants, which are usually the tallest plants in the area, provide both protection and food for much of the animal life, and the way in which these plants grow strongly determines what can grow around their roots and in their shadows.

Tropical Rain Forest

The richest of all the biomes is the tropical rain forest, where neither shortage of water nor extreme temperatures limit growth during any part of the year. More species of plants and animals live in tropical rain forests than in all the rest of the biomes of the world combined.

The most abundant plants are trees; as many as 3,000 different species of trees may grow in a square mile of rain forest. The trees are tall—the average height is about 150 feet—and the canopy is very dense. The characteristically smooth-barked trees generally branch only near the crown. Their leaves are large, leathery, and dark green with shiny upper surfaces. Many leaves have extended points known as "drip tips"; their smooth surfaces and drip tips help them to shed water. The roots are shallow, and so the trunks are often buttressed near the base to provide a firm, broad anchorage. Woody vines and epiphytes (plants which use the trunks and limbs of trees as their sole physical support) abound. Epiphytes obtain their water and minerals directly from the humid air of the canopy; they often have spongy roots or cupshaped leaves in which they catch moisture and debris. Many classes of plants, including ferns, orchids, mosses, and lichens, have exploited this life-style. An extraordinary variety of insects, birds, and other animals have moved into the treetops along with the vines and epiphytes to make it the area of the tropical rain forest that is most abundantly and diversely populated.

Little light reaches the forest floor. Often the only plants growing beneath the canopy are ferns, some of which reach 20 feet or more in height. There is almost no accumulation of leaf litter, such as we find in our northern forests; decomposition is too rapid. Everything that touches the ground disappears almost immediately—carried off, consumed, or decomposed. In many places the ground is bare.

The soils of tropical rain forests, as a consequence of this rapid decomposition and recycling of the materials, are relatively infertile. They are chiefly composed of a red clay into which the tree roots never penetrate deeply. These red soils are known as laterites, from the latin word for "brick." When laterite soils are cleared, in many cases they either erode rapidly or form thick, impenetrable crusts on which no cultivation is possible after a season or two. As a consequence, the tropical forests are being cut and burned and abandoned as useless at an ever-increasing rate in an attempt to meet the food demands of expanding populations.

(a)

(b)

7–6

(a) *Tropical rain forest of the Orinoco, looking upward toward the canopy. Note the slim, tall trunks of the trees, branching at or near the top, and the many vines and epiphytes. Little sun reaches the forest floor and only ferns and other plants physiologically adapted to filtered light can survive there.*
(b) *An epiphyte, a type of plant that grows on the trunks and branches of other plants. Epiphytes, which have no direct contact with the forest floor, obtain their water and minerals from the humid air of the canopy. This epiphyte is a bromeliad, a relative of the pineapple. Some of its leaves have been cut at the base to show the many water pockets. Small animals, such as the snail at right, live their entire lives in and around such aerial water sources, never leaving the rich canopy.*

In past geologic eras, the tropical rain forests may have extended over great areas of the world, and it is believed that they represent the point of origin for many groups of plants and animals. Although there is little tropical forest in the United States, such forests still account for about half of the forested areas of the earth.

The Deciduous Forest

Throughout most of the eastern half of the United States, the characteristic natural biome is a deciduous forest, although only scattered patches of the original "virgin" forest still remain. A deciduous forest is made up primarily of trees that shed their leaves for the winter and so conserve water during the period when most of the water is locked in the soil in the form of ice.

The dominant trees of the deciduous forests vary from region to region, depending largely on the local rainfall. In the northern and upland regions, oak, birch, beech, and maple are the most prominent trees. Before the chestnut blight struck North America, an oak-chestnut forest ran from Cape Ann, Massachusetts, and the Mohawk River Valley of New York to the southern end of the Appalachian highland. Maple and basswood predominate in Wisconsin and Minnesota, and maple and beech in southern Michigan, becoming mixed with hemlock and white pine as the forest moves southward. The southern and lowland regions have forests of oak and hickory. Along the southeastern coast, the wet warm climate supports an evergreen forest of oak and magnolia.

The Great Smoky Mountains of North Carolina and Tennessee (named for the wisps of fog and haze that hover over them constantly) are the home of 130 species of trees. Each time the glacier moved down the face of the continent, trees were forced southward. The climate just in advance of the glacier was probably near-arctic. The regions just to the south—in Kentucky, for example—probably had a climate much like that of most of modern Canada. Only further south—in what is now Tennessee, for instance—

7–7

A deciduous forest in Wisconsin. The trees along the road are mainly sugar maple, aspen, red maples, and birches. The shrub layer is clearly visible. If you walked further into the forest, you would find hemlock and white pine.

was it temperate enough for most of the deciduous trees to survive. As a consequence, new species were added to the ancient forest of the Great Smokies, and their representatives can still be found there. In Western Europe, where there was no haven for the trees retreating before the glacier, there is a much smaller number of tree species than in the United States.

In deciduous woodlands there are up to four layers of plant growth.

1. The tree layer, in which the crowns form a continuous canopy. The canopy is usually between 25 and 100 feet high.
2. The shrub layer, which grows to a height of about 15 feet. Shrubs and bushes resemble trees in that they are woody and deciduous, but they branch at or close to the ground.
3. The field layer, made up of grasses and other herbaceous (nonwoody) plants, including the wild flowers, which typically bloom in the spring before the trees regain their leaves. Bracken and other ferns, whose large leaf areas make them efficient interceptors of light, also often are important members of the field layer.
4. The ground layer, which consists of mosses and liverworts. It is also often covered with a layer of leaf litter, which breaks down into a rich gray-brown soil. The land where deciduous forests have stood makes good farmland.

Each type of dominant tree is likely to be associated with particular types of undergrowth, and these, in turn, with particular types of animals, fungi, and bacteria.

The Coniferous Forest: The Taiga

Conifers are gymnosperms, "naked-seed" plants. With the exception of a few groups, such as the larch and the tamarack, conifers are evergreens. They possess several adaptations which enable them to live with a persistent cover of snow in regions with long, cold winters and a short grow-

7–8
Ice sheets, or glaciers, form when summer heat is not sufficient to melt as much water as freezes in the winter. In the most recent Ice Age—which is just now ending (or perhaps only subsiding)—four successive sheets of ice pushed down over the face of North America. At their furthest advance, shown on the map above, glaciers covered most of Canada and much of the northern United States and the western mountains. At times, the wall of ice pushed forward as much as a foot a day.

(a) (b)

7–9

Wet and swampy taiga in the interior of Alaska. The northern coniferous forest in this winter scene includes a scattering of spruce, willow, aspen and alder. Summer in the arctic tundra is a brief eight weeks of above-freezing temperatures. In this picture, the small pond in the foreground has melted, enabling the marsh marigold around its edge to blossom. Note, however, that the lake in the background is still covered with ice.

ing season. One of these adaptations is their needle-shaped leaves which have thick cuticles; both the needle shape and the cuticle serve to reduce the rate of transpiration. Conifers also have a thick bark, which protects the inner woody tissues, and a root system that enables them to grow in shallow soils. The northern coniferous forest, composed chiefly of evergreen needle-leaf trees such as pine, fir, spruce, and hemlock, stretches across northern North America. It also covers vast areas of the Soviet Union, where the dominant tree is the deciduous needle-leaf larch. Ecologists call this biome, the northern coniferous forest, by its Russian name, the *taiga*.

Along the Pacific Coast of the United States, where temperatures are moderated by the prevailing westerlies, which also bring in abundant rainfall from the Pacific, is a relatively luxuriant coniferous forest. In Alaska, it is composed largely of hemlock and Sitka spruce. Between British Columbia and Oregon, the Douglas fir is the dominant tree. In southern Oregon and northern California are found the coastal redwoods, *Sequoia sempervirens*, which are also conifers. These trees have a life-span of more than 2,000 years. The giant sequoia, *Sequoiadendron giganteum*, occurs further inland, in scattered groves on the western slopes of the Sierra Nevadas, which is the only place where it is found.

Nearly all these redwoods grow within 20 miles of the ocean. The trees of the Pacific forests are very tall, much taller than those of the deciduous forests. The tallest living specimen of redwood measures 359 feet, 4 inches! Douglas firs range up to 260 feet tall and 50 feet in circumference, and the shorter giant sequoias reach more than 65 feet in girth.

Although the climate of part of the Pacific Coast is now temperate enough to maintain a deciduous forest, all such trees were destroyed by cold and drought during the Ice Age and never regained their position on the West Coast, as they did in the East.

Coniferous forests are also found in warmer areas where drought rather than cold restricts the growing season for deciduous trees or where the soil is so sandy that it holds little moisture. Pine and pine-oak forests of this sort are found along the coastal plains from New Jersey to the border of Texas.

The Tundra

Where the climate is too cold and dry even for the hardy conifers, the taiga grades into the tundra. The tundra is a form of grassland which occupies one-tenth of the earth's land surface. Its most characteristic feature is permafrost —a layer of ground that is permanently frozen below the surface. During the summer the ground thaws to a depth of a few inches and becomes wet and soggy; in winter it freezes over again. This freeze-thaw process, which tears and crushes the roots, keeps the plants small and stunted. Drying winter winds and abrasive driven snow further reduce the growth of the arctic tundra.

The virtually treeless vegetation of the tundra is dominated by herbaceous plants, such as grasses, sedges, and rushes, and heather, beneath which is a well-developed ground layer of mosses and lichens, particularly the lichen known as reindeer "moss." The flowering plants are mainly perennials because the growth season is so short there is often not sufficient time for plants to flower and set seed.

Vegetation which resembles that of the tundra in many respects occurs on mountaintops between the tree line (made up usually of conifers) and the snow line.

7–10
Thompson's gazelles on savanna in Africa.

Grasslands

Grasslands, areas in which grasses are the dominant plants, are found in the interior areas of continents where there is not enough rainfall to support the growth of trees. Grazing animals often help to maintain the tall-grass prairies (*pampas* in South America) by eating young tree shoots which otherwise might invade their borders. Repeated burning, as a result either of man-made fires or of fires started by lightning, also destroys young trees but leaves the tough roots of the grasses unharmed. Where the climate is drier, short-grass prairie, or *steppe*, predominates.

Savannas are tropical grasslands characterized by extensive rainfall during one part of the year and a period of drought during another. The savanna often has occasional trees or bushes, which are typically tough, thorny (which prevents their being eaten by the grazing animals), and deciduous, losing their leaves during the dry periods. Tropical savannas are particularly susceptible to burning as their grasses form a highly flammable straw mat during the dry season.

7–11
Chapparral which is characterized by long, hot, dry summers and moist cool winters is found in many parts of the world, including central Chile, South Africa, western Australia, southern California, and many areas bordering on the Mediterranean. Plants of these areas, although usually very different in their ancestral origin, are almost always leathery-leaved shrubs, such as these shown here. The scene is Greece.

Chaparral

Regions with abundant winter rainfall but dry summers, such as the southern coast of California, are dominated by small trees or often by spiny shrubs with hard, thick evergreen leaves. In the United States, such areas are known as *chaparral.* Similar communities are found in areas of the Mediterranean (where they are called the *maquis,* which became the name of the French underground in World War II), in Chile (where they are the *matorral*), in southern Africa, and along portions of the coast of Australia. Although the plants of these various areas are essentially unrelated, they closely resemble one another in their characteristics.

The Desert

The great deserts of the world are all located at latitudes of about 30°, both north and south. These are areas of falling, warming air and, consequently, little rainfall. The Sahara Desert, which extends all the way from the Atlantic coast of Africa to Arabia, is the largest in the world. Australia's desert, covering 44 percent of that continent, is next in size. About 5 percent of North America is desert.

Desert regions are characterized by less than 10 centimeters of rain a year during the daytime. Because there is little water vapor in the air to prevent the escape of heat, the nights are often extremely cold. The temperature may drop as much as 30°C (54°F) at night, in comparison with the humid tropics, where day and night temperatures vary by only a few degrees.

Most of the plants in the desert are annuals which race from seed to flower to seed during periods when water is available. Because few perennials can survive with such a low supply of water, there is no dense covering to inhibit the growth of annuals—as there is in the grasslands, for example—and during the brief growing seasons, the desert may be carpeted with flowers. The relatively few taller perennials either are succulents, such as the cacti, which store water, or are plants with small leaves that are very leathery or are shed during the long droughts.

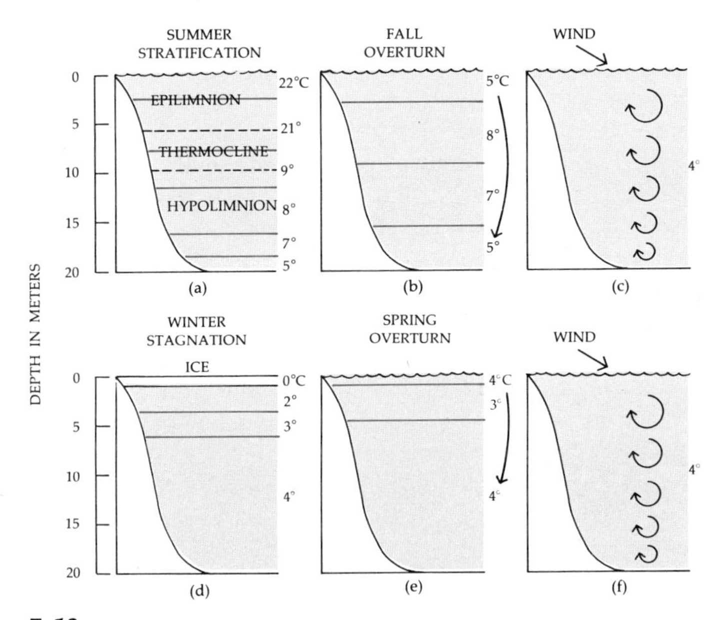

7–12

Seasonal cycle of temperature changes in a temperate-zone lake. Water, like air, increases in density as it cools, reaching a maximum density at 4°C. In the summer, the top layer of water, called the epilimnion, becomes warmer than the lower layers and therefore remains on the surface. Only the water in this warm, oxygen-rich layer circulates. In the middle layer, which is called the thermocline, there is an abrupt drop in temperature. Since the thermocline water does not mix, it cuts off oxygen from the third layer, the hypolimnion, producing summer stagnation (a). *In the fall, the temperature of the epilimnion drops until it is the same as that of the hypolimnion. The warmer water of the thermocline then rises to the surface, producing the full overturn* (b). *Aided by the fall winds, all the water of the lake begins to circulate* (c), *and oxygen is returned to the depths. As the surface water cools below 4°C, it expands, becoming lighter, and remains on the surface; in many areas, it freezes. The result is winter stratification* (d). *In spring, as ice melts and the water on the surface warms to 4°C, it sinks to the bottom, producing the spring overturn* (e), *after which the water again circulates freely* (f).

THE INLAND WATERS

Lakes and large ponds contain three distinct ecological zones:

1. The zone along the shore. This area, which is the lake's most richly inhabited zone, is normally dominated by angiosperms—cattails, rushes, water lilies, and pondweeds—which are rooted to the bottom.
2. The zone of open water, in which small floating algae and other planktonic forms are found. This zone extends down to the limits of light penetration.
3. The deepwater profundal zone, below the limits of light. This area has no photosynthetic organisms but is rich in animal life, bacteria, and fungi.

In large lakes of the temperate region, temperature changes cause two annual turnovers in the lake's waters. (Figure 7–12). As a consequence of these disturbances, these lakes are characterized by a succession of organisms replacing one another seasonally. In the spring and fall, when oxygen and dissolved minerals are redistributed by the overturns, there is often a spectacular "bloom" of plankton.

COASTAL WATERS

The richest of the coastal environments are the salt marshes, which often are found where rivers open to the sea and on the landward side of sandspits and sandbars. The principal vegetation here is grasses, such as *Spartina*, which are specially adapted to survive in salt water. The salt marshes are rich in nutrients brought down from the land and so are the spawning places and nurseries for many forms of marine life. Certain states, such as Massachusetts, have established conservation laws protecting their marshlands, not only because of their ecological richness and diversity but also because of their importance to offshore fishing and shellfish industries.

Algae cling to the bottom along the cliffs and rocky beaches, which they share with barnacles and other forms

(a)

(b)

7–13
(a) *A rocky coast off Fort Fisher, North Carolina.* (b) *Some characteristic inhabitants of the rocky shore include brown algae, whelks, barnacles and predatory starfish, all especially adapted to cling to the rocky surface despite the battering waves.*

of sea life, such as crabs, that can hide in the tidal pools or rock crevices or can cling to the wave-battered coasts—glued like barnacles, tethered like mussels, or spread-eagled like the suction-cupped starfish.

Sandy beaches have fewer bottom dwellers because of the constantly shifting sands. Clams, shrimp, and other small invertebrates live below the surface of the sand, sometimes emerging at low tide to feed on the debris washed in and out by the tide. Along the sand beaches, beach grasses are important for stabilizing the shifting dunes. These grasses spread by means of underground roots, which eventually form a network that knits the dunes together and prepares them for other types of plant life.

THE OCEANS

The oceans cover almost three-quarters of the surface of the earth and reach down to an average depth of 3 miles. There are three zones in the open ocean. In a thin layer on the surface floats the plankton, which is made up of photosynthetic algal cells intermingled with small shrimp and other crustaceans and the eggs and larval forms of many fish and invertebrates. Twelve million single-celled algae have been found in a cubic foot of seawater from this sunlit surface. These planktonic forms provide food for fish that live in the second zone of the ocean, the *pelagic zone*, which is much colder than the surface zone. Much of the pelagic zone is dark; even in clear water, less than 40 percent of the sunlight reaches a depth of 1 meter and less than 1 percent penetrates below 50 meters. On the very bottom, in the *benthic zone*, sessile animals, such as sponges and sea anemones, and a variety of fungi and bacteria live on the sparse but steady accumulation of debris drifting from the more populated levels of the ocean.

The Coral Reef

The coral reef is the richest and most diverse of all the ocean communities, providing food and shelter for a great

7–14
Coral reefs, the richest and most diverse of all marine communities, support a variety of sea animals, such as the brightly colored Moorish idols shown here. If you look carefully, you can see several sea urchins (the type of echinoderm that supplied the gametes for Hertwig's studies) in the center foreground.

variety of fish and marine invertebrates. In fact, the coral reef rivals the tropical rain forest in the number of different kinds of organisms that inhabit it.

Coral reefs are constructed by tiny polyploid coelenterates. Each secretes its own calcium-containing skeleton, and when it dies, the next generation builds on top of it. Symbiotic algae occur on and in the calcium skeletons as well as inside the living tissues of the invertebrates. In fact, the entire colony contains about three times as much plant as animal substance. The algae contribute oxygen and organic compounds to the coelenterates, which also feed by means of their stinging tentacles.

The coelenterates that form the reef can grow only in warm, well-lighted surface water, where the temperature seldom falls below 21°C (70°F). The oceans are rising slowly, 4 to 6 inches every 100 years as the glaciers melt, but the upward growth of the coral keeps the reef tops in the well-lighted zone. The longest reef in the world is the Great Barrier Reef of Australia, which extends some 1,200 miles. Other reefs are found throughout the tropical waters and as far north as Bermuda, which is warmed by the Gulf Stream. The Florida Keys, which begin just south of Miami, were formed by coral animals during a warm period before the last glaciation. When the glacial ice built up in the north, the water level fell, exposing the upper surface of the Keys and killing the coral coelenterates.

ECOLOGY AND EVOLUTION

In the following chapters, we shall examine these groups of living organisms in greater detail and attempt to answer some questions about them. Why is there such an immense variation in size, form, and function among living things? Why are there so many different species? What determines the boundaries of where each lives, and what regulates their numbers?

Two branches of biology deal with these questions: the first is ecology, which is the study of the relationships of living things with one another and with their environment, and the second is population biology, the study of the processes of evolution. Population biology is concerned with the evolutionary effects of ecological relationships, that is, with the population changes, both in number and in genetic composition, that accompany changes in the living and nonliving environment. Although the ecologist and the population biologist traditionally use different tools, these two areas of investigation overlap and, as you will see in the pages that follow, illuminate each other in important ways.

SUMMARY

The biosphere is the part of the earth which contains living organisms. It is a thin film on the surface of the planet, irregular in its thickness and its density. The biosphere is affected by the position and movements of the earth in relation to the sun, the movements of air and water, and the irregularities in the composition of the earth's crust. These conditions cause wide differences in temperature and rainfall from place to place and season to season on the earth's surface. These differences are reflected in differences in the kinds of plant and animal life.

The major life formations on land are called biomes. The richest terrestrial biome is the tropical rain forest, where neither water shortage nor extreme temperatures limit plant growth. Here the trees are broad-leaved evergreens, which are characteristically covered with vines and epiphytes. There is almost no collection of decomposing material or humus. Tropical soils are red clays (called laterites) which erode or solidify when the forest is cleared.

The deciduous forest is an important biome of North America, where temperatures are warm in the summer and cold in the winter. In the deciduous forest, the trees lose their leaves in the fall, which reduces water loss during the winter months, when the water is locked in ice.

North of the deciduous forest is the taiga, the subarctic coniferous forest. The trees of this forest, the conifers, have a number of special adaptations for conservation of water and protection against extreme cold. Other needle-leafed evergreen forests are found along the Pacific Coast and in the southeast.

Between the taiga and the northern polar region is the tundra, on which there are few trees and the dominant vegetation consists of low-growing perennials.

The grasslands lie between the deciduous forests and the deserts and have a rainfall intermediate between the two. The growth of trees is prevented not only by a shortage of rain but also by grazing animals, which eat the young shoots, and by recurring prairie fires. Savannas are tropical grasslands.

Chaparral, a shrubland biome found on the southern California coast and in the Mediterranean area, is characterized by dry summers but abundant winter rainfall.

Desert regions have very little rainfall and high daytime temperatures. The chief vegetation is made up of annual plants with extremely short growing seasons; cacti and other perennial desert plants are highly adapted for the conservation of water.

In the lakes and ponds of inland waters there are three life zones: the zone along the shore, characterized by rooted vegetation; the surface water, on which plankton grows; and the zone beneath the surface, which consists entirely of animal life. Large bodies of inland water in temperate zones characteristically turn over in spring and fall, redistributing oxygen and nutrients and keeping life in a constant flux.

Along the coasts of the continents are found a variety of environments: cliffs, sand dunes, rocky beaches, and salt marshes, each with its characteristic vegetation and animal life. Of these, the salt marshes are the richest, and a great diversity of marine life is found here, including immature forms of many fish and shellfish.

The open ocean has three distinct life zones: the sunlit surface, the pelagic (swimming) zone, and the benthic (bottom) zone. The photosynthetic vegetation of the open ocean is almost entirely plankton—floating, single-celled algae. The plant plankton, mixed with animal plankton, provides the basic food supply for fish and deepwater mammals and also for the sparse but diverse animal, fungal, and bacterial life of the ocean bottoms.

The richest zone of ocean life is the coral reef, composed of the bodies and skeletons of coelenterates combined with symbiotic algae.

QUESTIONS

1. Why are the various biomes characterized by plant life rather than by animal life?

2. What is the position of the earth in relation to the sun at the vernal equinox? At the winter solstice? What is the day-night cycle at the South Pole on these dates?

3. What were the dominant plants in the area in which you now live? How far do you have to go to find examples of this vegetation? What kinds of animals are characteristically associated with it? What other areas of the world are part of this same biome?

CHAPTER 7–2

Ecosystems

In the preceding chapter, we described the major biomes and aquatic communities of the world, some of the forces that determine their characteristics, and some of the organisms that inhabit each of them. If we look more carefully at any one part of any of these communities—at a patch of woodland, a pasture, a pond, or one coral reef—we begin to see that none of the organisms that lives in this particular area exists in isolation; rather, each is involved in a number of relationships both with the other organisms and with factors in the nonliving environment. The details of these relationships vary from place to place. In all cases, however, the interactions of the various organisms in a natural setting have two consequences: (1) a flow of energy through photosynthetic autotrophs—green plants or algae—to heterotrophs, which eat either the green plants or other heterotrophs; and (2) a cycling of inorganic materials, which move from the abiotic (nonliving) environment through the bodies of living organisms and back to the abiotic environment.

Such a combination of living and nonliving elements through which energy flows and minerals recycle is known as an *ecosystem*. Taking a large, astronautical view, the entire biosphere can be seen as a single ecosystem. This view is useful when studying materials that are circulated on a worldwide basis, such as carbon dioxide, oxygen, and water. A suitably stocked aquarium or terrarium is also an ecosystem, and such man-made models may be useful in studying certain ecological problems, such as the details of transfer of a particular mineral element. Most studies of ecosystems have been made, however, on more or less self-contained natural units—on a pond, for example, or a swamp or meadow.

The role of each species of organism within an ecosystem is known as an *ecological niche*. The ecological niche, which is an abstract concept, involves such factors as where an organism lives, what kind and size of prey it eats, and the season in which it appears. We shall have more to say about this concept in Chapter 7–3 and Section 8.

THE SOURCE OF ENERGY

The ultimate source of energy for all natural ecosystems is the sun. The earth receives from the sun an average of 2 calories* of radiant energy per minute for every square centimeter of the earth's surface—a total of 13×10^{23} (13 followed by 23 zeros) every year. About half of this energy never reaches the surface, however, but is reflected from the clouds and dust of the atmosphere. (Because of these reflections, earth, from outer space, is a shining planet, as bright as Venus.)

Of the solar energy that reaches the ground, some is dissipated in the evaporation of water. Most of it is ab-

* A calorie is the amount of energy required to raise the temperature of one gram of water one degree Celsius. As we noted in Chapter 2–4, energy can be converted from one form to another, and heat energy is one of the most convenient forms to measure. When we use calories to measure the energy of light or the energy in an organic compound, we are talking about energy equivalents. One thousand calories equals 1 kilocalorie, the unit usually used in measuring human dietary requirements.

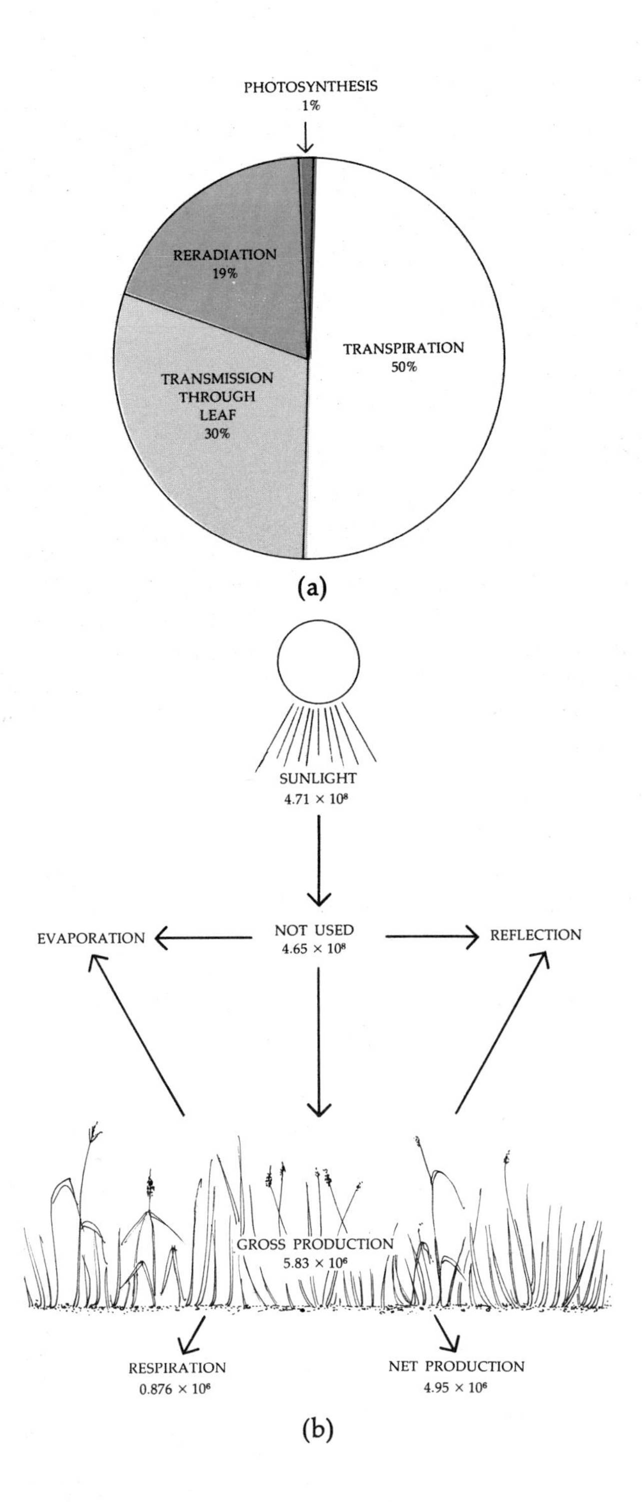

7–15
(a) *What happens to the light energy that falls on a green leaf. Note that only about 1 percent is used in photosynthesis.* (b) *Calculation of the productivity of a field in Michigan in which the vegetation was mostly perennial grasses and herbs. Measurements are in terms of calories per square meter per year. In this field, the net production—the amount of chemical energy stored in plant material—was 4,950,000 calories per square meter per year. Thus, slightly more than 1 percent of the 471,000,000 calories per square meter per year of sunlight reaching the field was converted to chemical energy and stored in the plant bodies.*

sorbed into the ground and radiated back in the form of heat. Only a small fraction of the total available energy from the sun enters the food chain of living organisms. Even if the light falls where vegetation is abundant, as in a forest, a cornfield, or a marsh, only 1 or 2 percent of that light (calculated on an annual basis) is used in photosynthesis. Yet this fraction, as small as it is, may result in the production—from carbon, oxygen, water, and a few minerals—of several thousand pounds (dry weight) of organic matter per year in a single square meter of field or forest, a total of 150 to 200 billion tons of such organic matter per year on a worldwide basis.

TROPHIC LEVELS

The passage of energy from one organism to another takes place along a particular food chain, which is made up of trophic (feeding) levels. The circumstances of this passage and each organism's relative position on the chain constitute one of the principal distinguishing features of an ecological niche.

The first step in the chain is always a primary producer, which on land is a green plant and in aquatic ecosystems is usually a photosynthetic alga. These photosynthetic organisms use light energy to make carbohydrates and other compounds, which then become sources of chemical energy.

Ecologists speak of the *productivity* of ecosystems; productivity is measured by the amount of energy fixed in chemical compounds or by increase in biomass in a particular length of time. (*Biomass* is a convenient shorthand term meaning organic matter; it includes woody parts of trees, stored food, arthropod shells, bones, etc.) *Net productivity* represents the amount of light energy converted to chemical energy less the amount dissipated (that is, converted to heat energy) by the plants in the course of respiration. The term "net productivity" is one you will be hearing more often with increasing discussions of how to improve agricultural yields.

The Consumers

Energy enters the animal world through the activities of the herbivores, animals which eat plants (including fruits and other plant parts). An herbivore may be a blue whale, a caterpillar, an elephant, or a field mouse; each type of ecosystem has its characteristic complement of herbivores. Of the plant material consumed by herbivores, much is excreted undigested. Some of the chemical energy is transformed to other types of energy—heat energy or kinetic energy—or used in the digestive process itself. A fraction of the material is converted to animal biomass.

The next level in the food chain, the second consumer level, involves a carnivore, a meat-eating animal that devours the herbivore. The carnivore may be a lion, a minnow, a robin, or a spider, but in each of these cases, only a small part of the organic substance present in the body of the herbivore becomes incorporated into the body of the consumer. Some chains have third and fourth consumer levels, but five links are usually the absolute limit, largely because of the waste involved in movement from one trophic level to another.

In a stable ecosystem, all the chemical energy that is fixed by the primary producers is stored in plant or animal biomass or is used in respiration (either by the producers and consumers or by the decomposers, a group of which we shall have more to say in the following pages).

7–16

Trophic levels: (a) *a harvest mouse eating the winter seed capsules of a trumpet-creeper;* (b) *a corn snake and a white-footed mouse;* (c) *an osprey catching a bream;* (d) *primary producers, herbivores, and carnivores on African savanna.*

(a)

(b)

(c)

(d)

7–17

The feeding relationships of the adult herring in the North Atlantic, illustrating the complex interconnections of a food web.

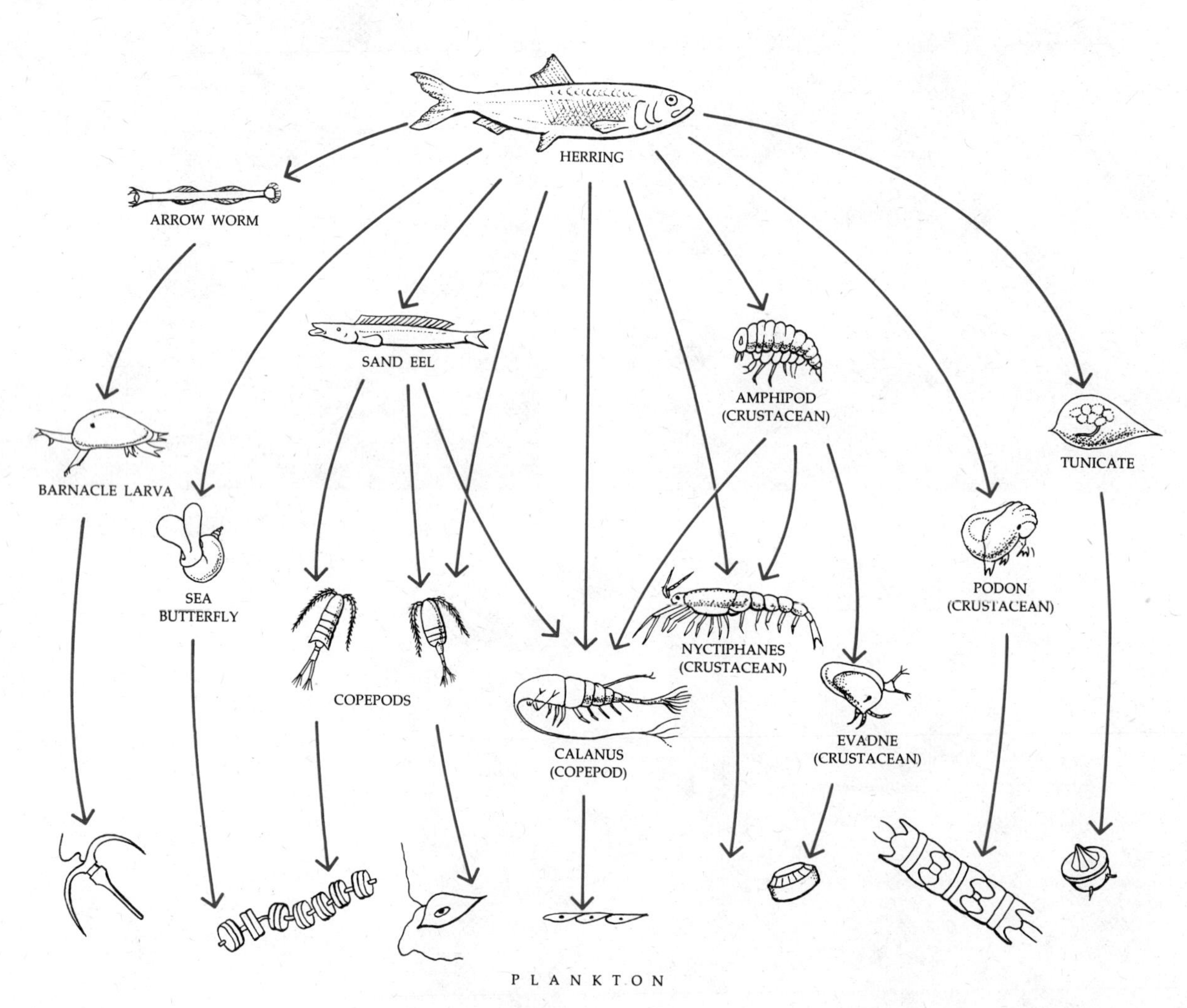

Ecological Pyramids

The flow of energy through a food chain is often represented by a graph of quantitative relationships among the various trophic levels. Because large amounts of energy and biomass are dissipated at every trophic level, so that each level retains a much smaller amount than the preceding level, these diagrams nearly always take the form of pyramids. An *ecological pyramid*, as such a diagram is called, may be a pyramid of numbers, showing the numbers of individual organisms at each level; a pyramid of biomass, based either on the total dry weight of the organisms at each level or on the number of calories at each level; or a pyramid of energy, showing the productivity of the different trophic levels.

The shape of any particular pyramid tells a great deal about the eocsystem it represents. For example, this is a pyramid of numbers for a grassland ecosystem:

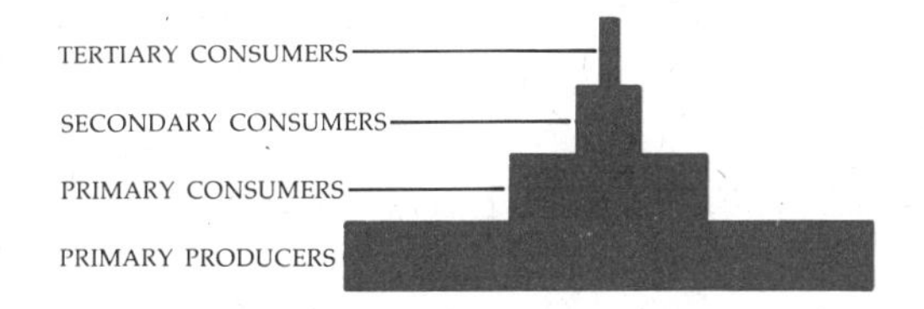

In this type of food chain, the primary producers (in this case, the grasses) are small, and so a large quantity of them is required to support the primary consumers (the herbivores). In a food chain in which the primary producers are large (for instance, trees), one primary producer may support many herbivores, as indicated in this pyramid of numbers:

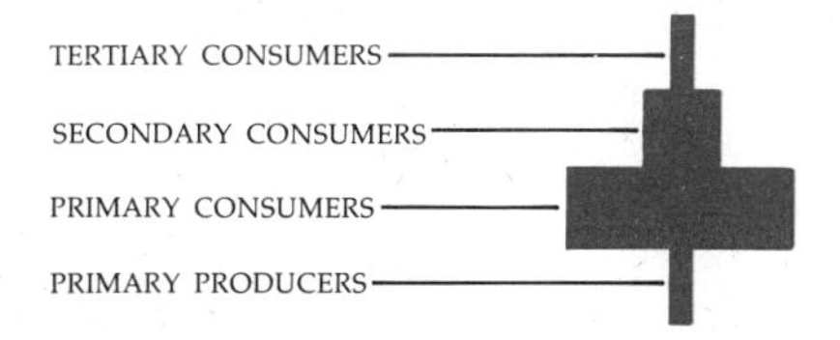

A pyramid of biomass for a grasslands ecosystem, like the pyramid of numbers for that system, takes the form of an upright pyramid:

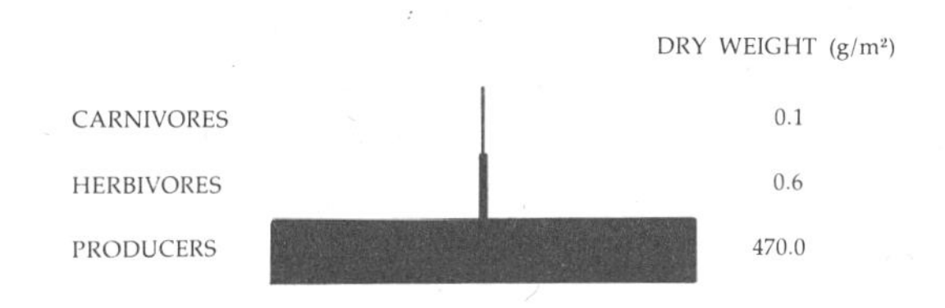

Inverted pyramids of biomass occur only when the producers and primary consumers are small. For example, in the ocean, the biomass of phytoplankton (plant plankton), as measured by the "standing crop" at any particular moment, may be smaller than the biomass of the animal plankton (zoo plankton) that feeds upon it:

Because the growth rate of the phytoplankton is much more rapid than that of the zooplankton (animal plankton) is much smaller, a small biomass of phytoplankton can supply food for a larger biomass of zooplankton. Like pyramids of numbers, pyramids of biomass indicate only the quantity of organic material present at any one time; they do not give the total amount of material produced or, as do pyramids of energy, the rate at which it is produced.

Table 7–1
Trophic Levels

	Arctic Tundra	*Antarctic Ocean*	*Taiga*	*Deciduous Forest*
Carnivores		Killer whales, leopard seals, skuas	Man	Man
Carnivores	Man	Penguins, petrels, etc.; Ross seals, Weddell seals, fish	Wolves, lynx, great horned owls, martins, minks	Foxes, wolves, owls, hawks, skunks, raccoons, opossums
Carnivores	Arctic wolves, polar bears, snowy owls, ravens, etc.; wolf spiders, warble flies	Crabeater seals, blue whales, some seabirds, squids, small fish	Black bears, woodpeckers, nuthatches, thrushes, predatory insects	Tufted titmice, jays, etc.; squirrels, chipmunks, turtles, snakes, woodland mice, spiders, salamanders, toads
Herbivores	Ptarmigans, muskoxen, caribou, lemmings, arctic hares, springtails, bumblebees, etc.	Krill	Moose, crossbills, spruce grouse, mice, hares, squirrels, porcupines, budworms, other plant-eating insects	Woodchucks, woodland deer, slugs, snails, earthworms, mites, millipedes, insects
Producers	Lichens, mosses, arctic willows, etc.	Phytoplankton	Spruces, firs, birches, aspens, etc.; algae, mosses, lichens	Maples, beeches, oaks, hickories, etc.; algae, mosses, ferns, wildflowers, shrubs

Southern Swamp	*Prairie*	*American Desert*	*African Forest*	*Savanna*
Man	Man	Man		Man
Raccoons, egrets, herons, water moccasins, alligators, snapping turtles, soft-shelled turtles	Coyotes, wolves, grizzly bears, ferrets, snakes, badgers, owls, hawks, kit foxes	Desert foxes, roadrunners, owls	Man	Lions, leopards, cheetahs, hyenas, vultures, maribou, storks
Waterfowl, warblers, predatory insects, fish, frogs	Ground squirrels, some birds, praying mantids, prairie dogs	Poorwills, cactus wrens, toads, lizards, snakes, woodpeckers	Leopards, crowned eagles, snakes, frogs, lizards	Herons, flamingos, baboons
Insects, insect larvae, snails, shrimps, etc.	Pronghorns, bison, mice, horned larks, prairie chickens, grasshoppers, jack rabbits, pocket gophers	Peccaries, quail, kangaroo rats, snails, grasshoppers, moths, ants, etc.	Monkeys, lemurs, bongos, okapis, mice, hornbills, parrots, etc.; ants, bees, termites, butterflies, moths, etc.	Ostriches, elephants, hippos, rhinos, giraffes, zebras, impalas, gnus, topis, fish, grasshoppers, other insects, crustaceans
Phytoplankton, water grasses	Grasses	Joshua trees, ocotillos, saguaros, etc.; bunch grasses, creosote bushes, chollas, wildflowers	Woody vines, etc.; broad-leaved trees, shrubs	Palms, acacias, euphorbias, etc.; phytoplankton, grasses, shrubs

Productivity and biomass, although related, are not the same. As we saw, a small biomass of phytoplankton can support a larger biomass of zooplankton because of the phytoplankton's greater productivity. Energy pyramids, which show the productivity relationships of the trophic levels, have therefore the same general shape for every ecosystem. This particular pyramid is based upon calculations for a river system (Silver Springs) in Florida:

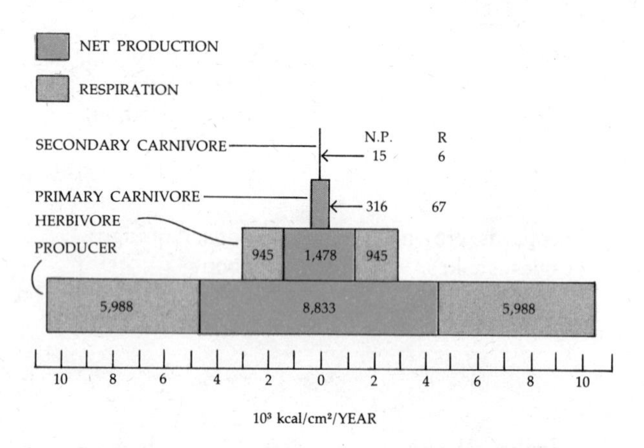

Also, productivity and yield to man are not the same thing; a coral reef is highly productive in terms of the life of the reef, but man profits little from its products.

In general, of the total amount of energy in the biomass that passes from one trophic level to another in a food chain, only about 10 percent is stored in body tissue—whether the plants are single-celled algae and the consumers are tiny crustaceans or the plants are grasses and the consumers are cows. About 90 percent of the total calories is either unassimilated or "burned" in respiration by the eating animals at that level. Hence the tertiary consumers, the carnivores which eat the carnivores which eat the herbivores, are reduced to roughly $1/10 \times 1/10 \times 1/10 = 1/1{,}000$ of the energy stored in the plants that are eaten. Supercarnivores, which eat these secondary carnivores, are reduced to one-tenth of this, or 1/10,000 of the energy in the plant material.

Individuals at the top of the ecological pyramid, such as man, may have to be quite large, as individuals, to capture and eat other animals, but even if they are larger as individuals, they must always be smaller in total number, total biomass, and total captured energy than the animals they eat.

MINERAL CYCLES

Energy takes a one-way course through an ecosystem, but many inorganic substances cycle through the system. Such substances include water, nitrogen, carbon, phosphorus, potassium, sulfur, magnesium, sodium, and calcium and also a number of minerals, such as cobalt, which are required by some living systems in only very small amounts. These movements of inorganic substances are referred to as *biogeochemical cycles*, because they involve geological as well as biological components of the ecosystem. The geological components are the atmosphere, which is made up largely of gases, including water vapor, but also contains dust and other particles; the lithosphere, the solid crust of the earth; and the hydrosphere, comprising the oceans, lakes, and rivers, which cover three-fourths of the earth's surface. The biological components include the producers and consumers, which we have already discussed, and a third and extremely important group, the decomposers. The decomposers, which are primarily bacteria and fungi, break down dead and discarded organic matter, completing the oxidation of the energy-rich compounds formed by photosynthesis. Without their activities, all the living world would soon be inundated with its own waste products. As a result of the metabolic work of the decomposers, these waste products—dead leaves and branches, the roots of annual plants, feces, carcasses, even the discarded exoskeletons of insects—are broken down and the chemicals of which they are composed are returned to the soil or the water. From the soil or water, they once more enter the bodies of plants and begin their cycle again.

Water and gases circulate on a worldwide basis, although, in the case of water in particular, the amount available in any one area may depend on local conditions. Nitrogen and minerals, however, circulate through the ecosystem, and a stable ecosystem is self-sufficient in its supply of these essential components.

The Nitrogen Cycle

Nitrogen makes up 78 percent of our atmosphere. Since most living things, however, cannot use elemental, atmospheric nitrogen to make amino acids and other nitrogen-containing compounds, they must depend on nitrogen contained in soil minerals. So despite the abundance of nitrogen in the biosphere, a shortage of nitrogen in the soil is often the major limiting factor in plant growth. The process by which this limited amount of nitrogen is circulated and recirculated throughout the world of living organisms is known as the *nitrogen cycle.* The three principal stages of this cycle are (1) ammonification, (2) nitrification, and (3) assimilation.

Much of the nitrogen found in the soil reaches it as a result of the decomposition of organic materials and is in the form of complex organic compounds, such as proteins, amino acids, nucleic acids, and nucleotides. However, these nitrogenous compounds are usually rapidly decomposed into simple compounds by soil-dwelling organisms. Certain soil bacteria and fungi are mainly responsible for the decomposition of dead organic materials. These microorganisms use the protein and amino acids as a source of their own needed proteins and release the excess nitrogen in the form of ammonia (NH_3) or ammonium (NH_4^+). This process is known as *ammonification.*

Several species of bacteria common in soils are able to oxidize ammonia or ammonium. The oxidation of ammonia or ammonium, known as *nitrification,* is an energy-yielding process, and the energy released in the process is used by these bacteria as their primary energy source. One group of bacteria oxidize ammonia (or ammonium) to nitrite (NO_2^-):

$$2NH_3 + 3O_2 \longrightarrow 2NO_2^- + 2H^+ + 2H_2O$$

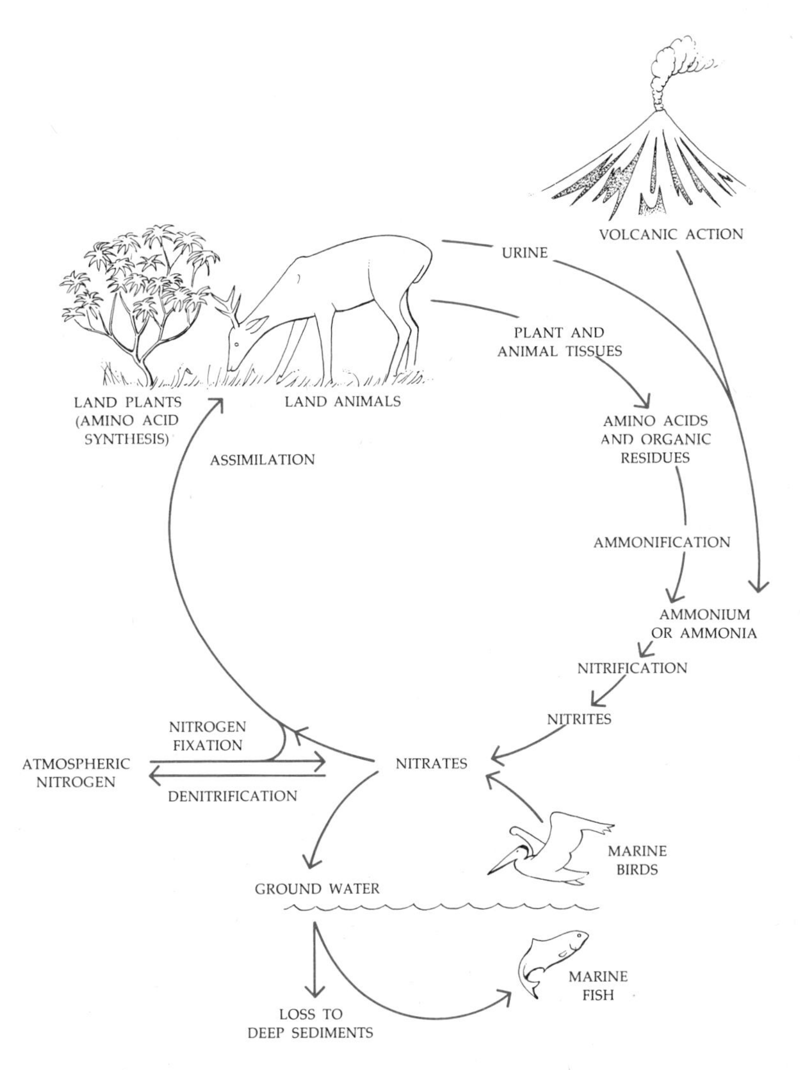

7–18
The nitrogen cycle.

Nitrite is toxic to higher plants, but it rarely accumulates. Members of another genus of bacteria oxidize the nitrite to nitrate, again with a release of energy:

$$2NO_2^- + O_2 \longrightarrow 2NO_3^-$$

Nitrate is the form in which almost all nitrogen moves from the soil into the roots.

$$\underset{\text{KETOGLUTARIC ACID}}{COOH-C{=}O-CH_2-CH_2-COOH} + NH_3 + NAD_{red} \xrightarrow[\text{GLUTAMIC DEHYDROGENASE}]{} \underset{\text{GLUTAMIC ACID}}{COOH-HCNH_2-CH_2-CH_2-COOH} + NAD_{ox}$$

(a)

$$\underset{\text{GLUTAMIC ACID}}{COOH-HCNH_2-CH_2-CH_2-COOH} + \underset{\text{PYRUVIC ACID}}{CH_3-C{=}O-COOH} \xrightarrow{\text{TRANS-AMINASE}} \underset{\text{KETOGLUTARIC ACID}}{COOH-C{=}O-CH_2-CH_2-COOH} + \underset{\text{ALANINE}}{NH_2-CH(CH_3)-COOH}$$

(b)

7–19

(a) *Amination is the process by which amino acids are formed by the combination of ammonia with a carbon-containing compound. It is carried out principally by the root cells of green plants. In this reaction, ammonia is combined with the five-carbon compound ketoglutaric acid to form glutamic acid. Energy for the reaction is provided by the oxidation of* NAD_{red}. (b) *Transamination is the process by which the amino* (NH_2) *group of one amino acid is transferred to another carbon-containing compound to form a second type of amino acid. Animal cells are able to produce some of the amino acids they require by transamination. Others, the so-called essential amino acids, must be taken in already formed in the diet. In this reaction, the amino group is transferred from the amino acid—glutamic acid—to pyruvic acid to form the amino acid alanine. Pyruvic acid is the end-product of glycolysis and ketoglutaric acid is one of the compounds involved in the Krebs cycle, further examples of cellular economy.*

Once the nitrate is within the cell, it is reduced back to ammonium. In contrast to nitrification, this assimilation process requires energy. The ammonium ions thus formed are transferred to carbon-containing compounds to produce amino acids and other nitrogen-containing organic compounds.

Nitrogen Fixation

The nitrogen-containing compounds of green plants are returned to the soil with the death of the plants (or of the animals which have eaten the plants) and are reprocessed by soil organisms and microorganisms, taken up by the plant roots in the form of nitrate dissolved in the soil water, and reconverted to organic compounds. In the course of this cycle, a certain amount of nitrogen is always "lost," in the sense that it becomes unavailable to the land plants.

The main source of nitrogen loss is the removal of plants from the soil. Soils under cultivation often show a steady decline of nitrogen content. Nitrogen may also be lost when topsoil is carried off by soil erosion or when ground cover is destroyed by fire. Nitrogen is also leached away by water percolating down through the soil to the groundwater. In addition, numerous types of bacteria are present in the soil that, when oxygen is not present, can break down nitrates, releasing nitrogen into the air and using the oxygen for the oxidation of carbon compounds (respiration). This process, known as denitrification, takes place in poorly drained (hence poorly aerated) soils.

As you can see, if the nitrogen that is lost from the soil were not steadily replaced, virtually all life on this planet would finally flicker out. The "lost" nitrogen is returned to the soil by *nitrogen fixation.*

Nitrogen fixation is the process by which gaseous nitrogen from the air is incorporated into organic nitrogen-containing compounds and, thereby, brought into the nitrogen cycle. This process is carried out to a significant extent only by green algae, a few free-living microorganisms—chief among which are the blue-green algae and some free-liv-

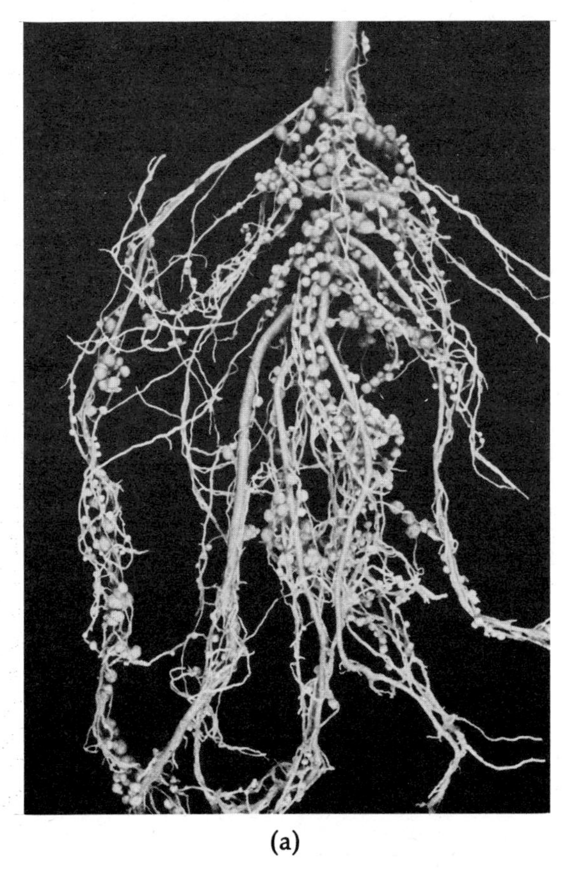

(a)

7–20
(a) *Nitrogen-fixing nodules of the roots of a bird foot trefoil, a legume. These nodules are the result of a symbiotic relationship between a soil bacterium* (Rhizobium) *and root cells.* (b) *Tip of a root hair of a clover seedling, with several rhizobia and some soil particles.* (c) *Cross section of an infected nodule.*

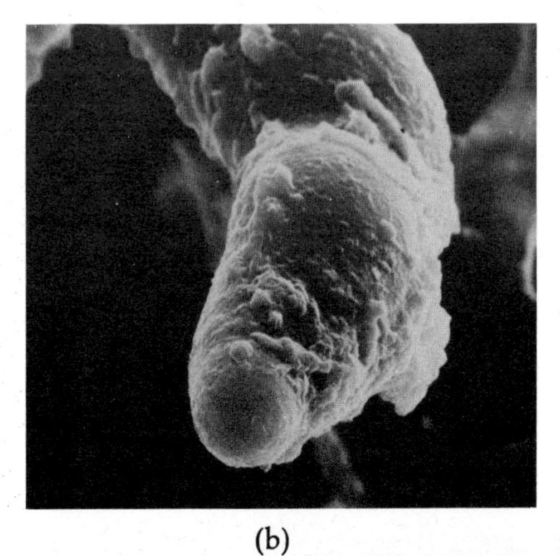

(b)

ing bacteria—and symbiotic combinations of bacteria and higher plants. Nitrogen fixation is the process by which nitrogen must originally have become available to living organisms and on which all living organisms are now dependent, just as all organisms are ultimately dependent on photosynthesis for energy.

One hundred million tons of nitrogen is added to the earth's surface each year, of which 90 million tons is biological in origin. (The other 10 percent is largely in the form of chemical fertilizers.) Of the various classes of nitrogen-fixing organisms, the symbiotic bacteria are by far the most important in terms of total amounts of nitrogen fixed. The most common of the nitrogen-fixing symbiotic bacteria is *Rhizobium*, a type of bacterium that invades the roots of leguminous plants such as clover, peas, beans, vetches, and alfalfa.

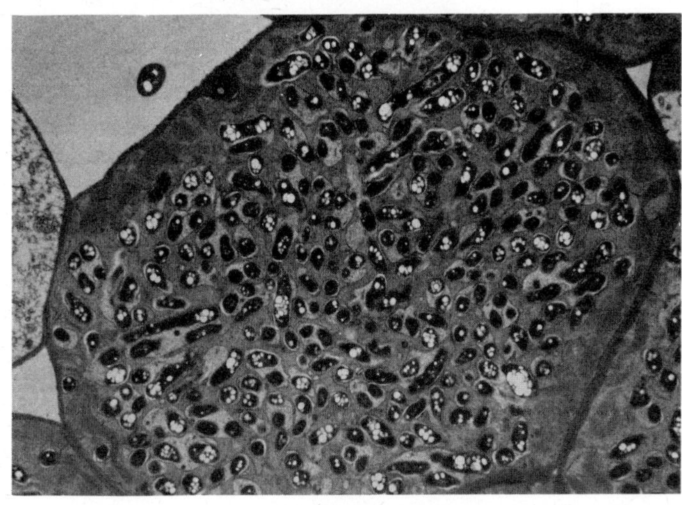

(c)

Phosphorus and Other Minerals

The chief reserve supply of nitrogen is in the air. The other elements that cycle through ecosystems are found in the largest amounts in the soil. Such elements are said to undergo sedimentary cycles, as distinguished from the atmospheric cycles of gases. An example of a sedimentary cycle is the phosphorus cycle, diagramed in Figure 7–21. As we saw in Section 2, phosphorus is directly involved in the transfer of energy in biological systems. It is also a component of nucleic acids, both DNA and RNA. In these capacities, it is required by every living system. Many kinds of rocks contain phosphorus, and when such rocks are eroded by water, minute amounts of phosphorus dissolve and become available to plants and so enter the biogeochemical cycle. A certain amount of phosphate runs off each year, eventually finding its way to the oceans, where most of it is lost to terrestrial ecosystems. This "lost" phosphorus, in natural systems, is replaced by the further release of phosphorus from the rocks beneath the soil.

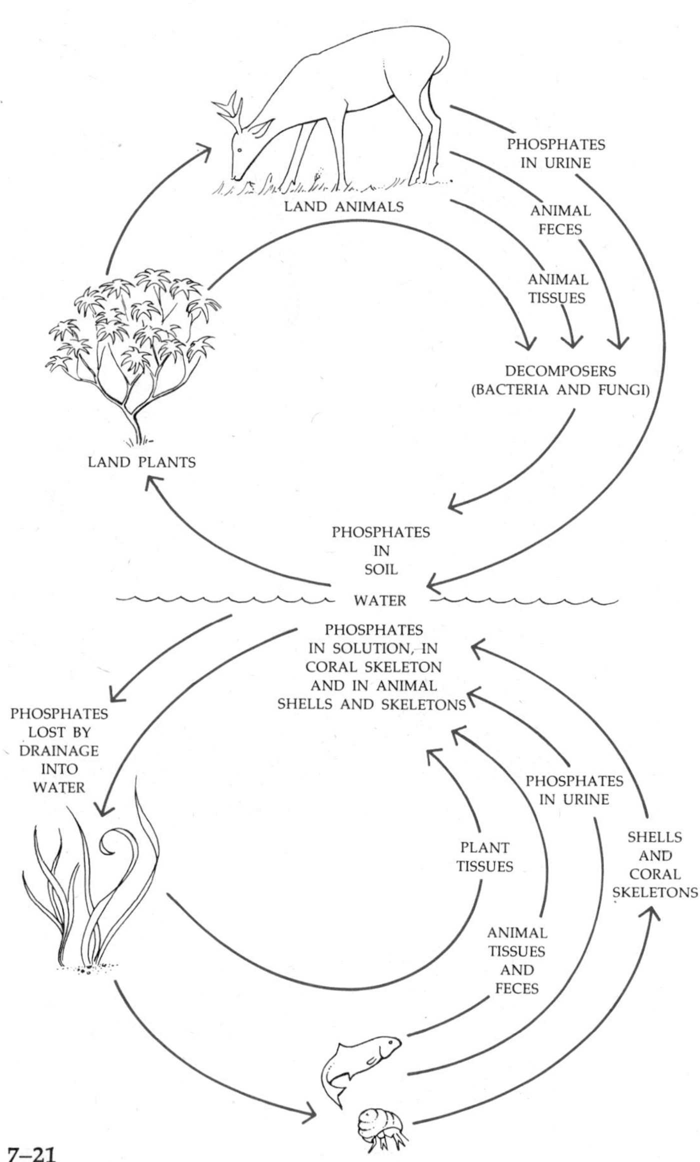

7–21
The phosphorus cycle.

Concentration of Elements

Many of the elements needed by living systems are present in the tissues of these systems in far higher concentrations than in the surrounding air, soil, or water. This concentration of elements comes about as a result of selective uptake of such compounds by living cells, amplified by the channeling effects of the ecological pyramids of biomass and energy. Under natural circumstances, this concentration effect is usually valuable; animals usually have a greater requirement for minerals than do plants because so much of the biomass of plants is cellulose. In man-made environments, however, the results can be surprising and even catastrophic. We have an example in strontium-90, a byproduct of the testing of atomic weapons, which we shall discuss in Section 9.

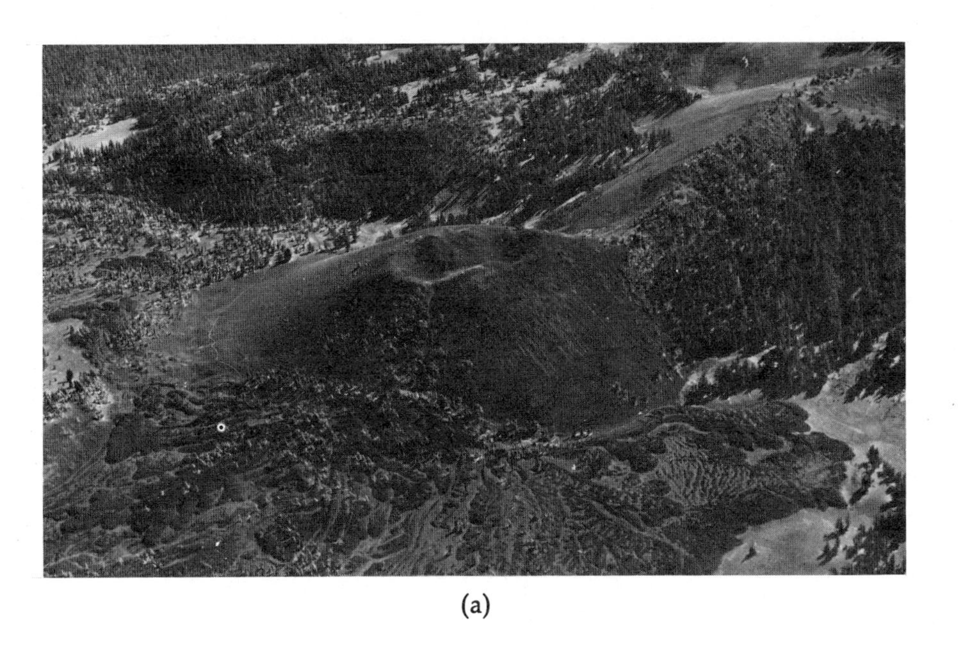

(a)

(b)

7–22
(a) *Yapoah Crater, a volcanic cinder cone east of the Cascade Mountains in central Oregon. Succession leading to the establishment of climax forest on such a cone may take centuries and may often be interrupted by further volcanic activity long before it is complete.* (b) *An early stage of succession on a rocky slope. Lichens have begun to accumulate soil, and a bladder fern has sprung up in a small crevice.*

ECOLOGICAL SUCCESSION

If land is laid bare as the result of a landslide, erosion, excavation, the eruption of a volcano, the retreat of a glacier, the rising of a new island from the sea, or some other such phenomenon, it will, if the environment is not too harsh, slowly become covered with vegetation and its accompanying animal life. The vegetation that initially colonizes the bare land is usually replaced in the course of time by a second type, which gradually crowds out the first and which itself eventually may be replaced. In the northeastern United States, where these stages of replacement have been studied by ecologists over a number of years, it has been found that such communities of living organisms replace one another in a predictable and orderly sequence; this process of community change is known as *ecological succession*.

The process of ecological succession is carried out by the living organisms themselves. Each temporary community changes the local conditions of temperature, light, humidity, soil content, and other abiotic factors and so sets up favorable conditions for the next temporary community. When the site has been modified as much as possible (here the limits are set by the environment), succession ceases. The final community is known as the *climax community*. In a sense, the climax community, too, is in a state of constant change; individual organisms die, and their places are taken by new individuals. However, these individuals will probably be of the same sort and of the same species as the ones being replaced. Further changes in the character of the community will take place only if the community is disturbed or if the climate undergoes a drastic change (as in the Ice Ages).

Primary Succession

The occupation by plants of an area not previously covered by vegetation is known as *primary succession.* The first stage in primary succession is often the formation of soil. Rocks are broken down by weathering processes, such as freezing and thawing or heating and cooling, which cause substances in the rocks to expand and contract, thus splitting the rocks apart. Water and wind exert a scouring action that breaks the fragmented rock into smaller particles, often carrying the fragments great distances. Water enters between the particles, and soluble materials such as rock salt dissolve in the water. Water in combination with carbon dioxide from the air forms a mild acid which dissolves substances that will not dissolve in water alone. Chemical reactions begin to take place that contribute to the disintegration of the rock. Soon, if other conditions such as light and temperature permit, bacteria, fungi, and then small plants begin to gain a foothold. Growing roots split rock particles, and the disintegrating bodies of the plants and those of the animals associated with them add to the accumulating material. Finally, the larger plants move in, anchoring the soil in place with their root systems, and a new community has begun.

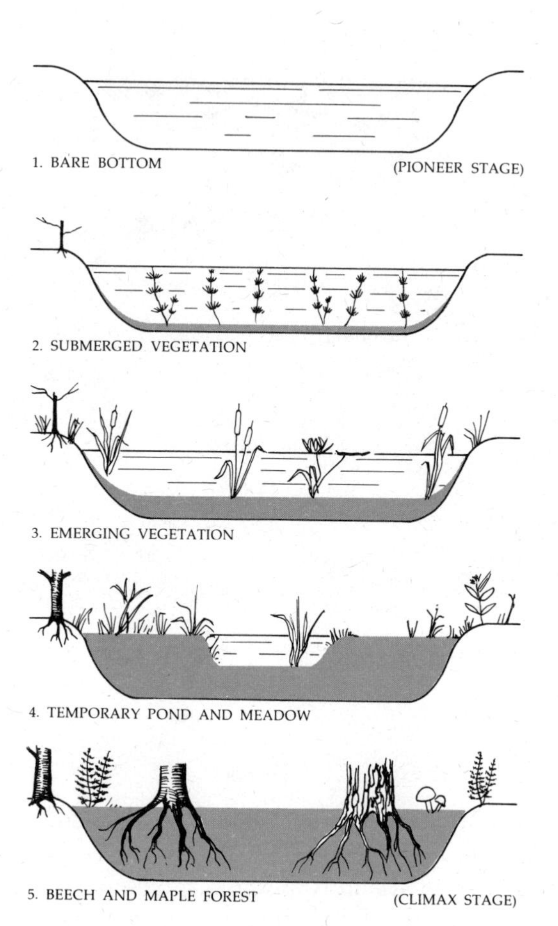

7–23

An example of primary succession: a pond at the south end of Lake Michigan gradually becomes woodland. Succession begins (stage 2) when bottom-rooted plants start to replace the plankton. The plants and their debris trap the silt and slowly fill up the pond (stage 3), which becomes a bog. Then water-loving grasses and rushes bind the soil of the bog together, until the part of the land which once was a pond has become a meadow (stage 4). Finally, the meadow passes through additional stages to become the climax forest.

Secondary Succession

When succession occurs on a site previously occupied by vegetation, it is known as *secondary succession*. If a farmer abandons a plowed field, for example, it is bombarded by the seeds of numerous plants and is captured by those which can germinate most quickly. In an open field, these are the plants that can survive the sunlight and drying winds—weeds and grasses and such trees as cedars, white pines, poplars, and birch. For a while, these plants rule the forest, but eventually they eliminate themselves because their seedlings cannot compete in the shade cast by the parent trees. Some seedlings can thrive in partial shade, however—for

SOIL PROFILES

On the surface of the soil, in most natural communities, there is an accumulation of litter—leaves and other plant parts, feces, the carcasses of insects and other animals—from which soluble organic materials leach down into the underlying soil. The soil itself typically has three layers: the A horizon, the B horizon, and the C horizon.

The A horizon, or topsoil, is the zone of maximum organic accumulation (humus). The B horizon, or subsoil, consists of weathered soil which either may have formed from the rock below or may have been imported by wind, water, or ice. This layer contains some soluble organic matter, leached down from the topsoil. It also contains a larger amount of clay, which is composed of extremely small particles, sifted down from the upper layer, that tend to stick together. The C horizon is made up of loose rock, which extends down to the bedrock beneath it.

Soils have characteristic profiles that depend on the conditions under which they were formed. The soil of the northeastern United States, because of extensive glaciation, contains no residual soil (soil formed from the bedrock beneath it) but consists entirely of sedimentary (imported) soil. The grasslands of the Midwest, on the other hand, are covered deeply by residual soil. Much of the land under cultivation in these areas has lost its humus-rich A horizon as a result of erosion. Such soils are sometimes called "hardpan" because of their high clay content.

Prairie soil. Note how the roots of the grasses bind the topsoil.

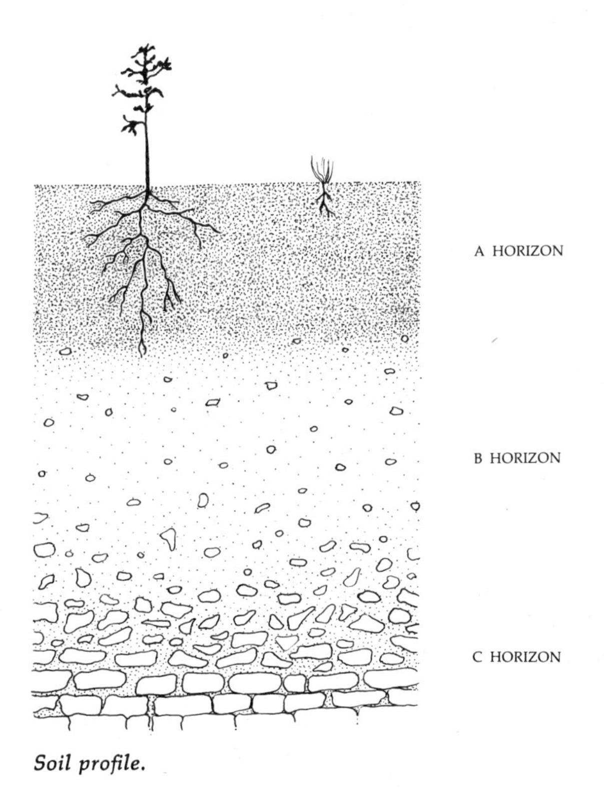

Soil profile.

(a)

(b)

7–24
(a) *When fire sweeps through a forest, secondary succession—with regeneration from nearby unburned stands of vegetation—is initiated. Some plants produce sprouts from the stumps; others seed abundantly on the burn. In one group of pines, the closed-cone pines, the cones do not open to release their seeds until they have been burned.* (b) *Sugar pines in the southern Sierra Nevada of California. With the control of forest fires, these trees are being replaced by others, such as the incense cedar* (*the two trees to the right*).

instance, oaks, red maples, white ash, and tulip trees. As the forest matures, these trees grow tall, and finally they shut out so much light that even *their* seedlings cannot grow. Eventually, the only young trees that can grow in the forest are those that can survive in the dimmest light, such as the hemlock, beech, and sugar maple, and these ultimately take over the forest. Nothing else can compete with them in the conditions that have been established. This is the climax forest.

Forests in all stages of succession can be found throughout the United States. Secondary successions typically occur where land has been cultivated and abandoned or following fires or extensive lumbering activities.

The Climax Community

Whether succession is primary or secondary, the final stages and the climax community will be the same in a given area. The physical characteristics of the environment determine the nature of the climax community. In regions where these are particularly unfavorable, such as the tundra, the process of succession involves relatively few stages and the climax community is correspondingly simple. Where physical conditions are not limiting, the climax community is rich and diverse.

(a)

(b)

7–25
(a) *The seedling of a red maple rising above needles of white pine. Mature white pines filter the light, so that their own seedlings cannot survive and only those tolerant of shade, such as maple and oak seedlings, can gain a foothold.* (b) *Seedling trees of balsam fir growing up under and replacing quaking aspen in northern Minnesota—a stage in forest succession leading to a climax community of white spruce and balsam fir.*

Immature and Mature Ecosystems

As ecosystems pass through the various stages of succession, one type of plant and animal community is replaced by another, distinctly different type. Although the changes that take place as the ecosystem matures differ in detail as to the exact species involved and the rates of change, they all have certain results in common.

First, there is an increase in total biomass. Compare, for example, a recently abandoned field, which is an immature ecosystem, with a deciduous forest, which is a more mature one.

Second, there is an increase in the number of species. This increase has several important consequences. One is a greater stratification of the plant life; in the mature forest, for example, the successive arrangement of different types of plants at different levels ensures the maximum utilization of all intensities and wavelengths of light. Another result of this increase in the number of species is that food webs are more complex. As a consequence, they will be less subject to drastic fluctuations. As the ecosystem increases in diversity, it approaches a state of dynamic equilibrium—homeostasis—comparable to the homeostasis of complex organisms.

SUMMARY

Ecology is the study of the relationships of organisms with each other and with their abiotic environment. An ecosystem is a particular community of plants and animals and the environment in which they live. It is characterized by (1) a flow of energy from the sun through autotrophs to heterotrophs, and (2) a cycling of minerals and other inorganic materials.

Within an ecosystem, there are trophic (feeding) levels. All ecosystems have at least two such levels: autotrophs, which are plants or photosynthetic algae, and herbivores, which are usually animals. The autotrophs, the primary producers in the ecosystem, convert a small proportion (in the neighborhood of 1 percent) of the sun's energy into chemical energy. The herbivores, which eat the autotrophs, are the primary consumers. A carnivore which eats the herbivore is a secondary consumer, and so on. About 10 percent of the energy transferred at each trophic level is stored in body tissue. There are seldom more than five links in a food chain.

The movements of water, carbon, nitrogen, and minerals through ecosystems are known as biogeochemical cycles. In such cycles, inorganic materials from the air, water, or soil are taken up by primary producers, passed on to consumers, and eventually transferred to decomposers, chiefly bacteria and fungi. The decomposers break down dead and discarded organic material and return it to the soil or water in a form in which it can be used again by the primary producers.

The cycling of nitrogen from the soil, through the bodies of plants and animals, and back to the soil again is known as the nitrogen cycle. It involves several stages. Nitrogen reaches the soil in the form of organic material of plant and animal origin. This material is decomposed by soil organisms. Ammonification, the breakdown of nitrogen-containing molecules to ammonia (NH_3) or ammonium (NH_4^+), is carried out by certain soil bacteria and fungi. Nitrification is the oxidation of ammonia or ammonium to form nitrites and nitrates; these steps are carried out by two different types of bacteria. Nitrogen enters plants almost entirely in the form of nitrates. Within the plant, nitrates are reduced to ammonium with a carbon-containing group. Nitrogen-containing organic compounds are eventually returned to the soil, principally through death and decay, completing the nitrogen cycle.

Nitrogen is lost from the soil by harvest, erosion, fire, leaching, and denitrification. Nitrogen is increased in the soil by nitrogen fixation, which is the incorporation of elemental nitrogen into organic components. Biological nitrogen fixation is carried out entirely by microorganisms, including blue-green algae, or by a symbiotic association of bacteria (rhizobia) and leguminous plants.

Ecological succession is an orderly sequence of changes in the type of vegetation and other organisms on a particular site. It results from modifications produced by the living organisms themselves and culminates in the establishment of a climax community, which is the characteristic community for that area.

Maturation in ecosystems is accompanied by increases in biomass and in number of species. The mature system is characterized by greater stability than in the immature system.

QUESTIONS

1. Plants cannot live in soils in which their roots do not get oxygen. Why not?

2. What biological characteristics might you expect to find in a day-neutral plant?

CHAPTER 7–3

The Community in Action: Interactions of Organisms

The plants, animals, and other organisms that live in a particular area make up a *community*, and a group of individuals of one species within a community is a *population*. The composition of a community is determined by a variety of factors, some of which are physical, such as rainfall and the type of soil, and some of which are determined by the interactions of the living things within the community. This chapter describes the many different types of interactions of the populations in a community and their effects.

COMPETITION

Competition occurs when two different species in a community both require some resource which is not unlimitedly available. Competition is often for food, but it may be for light, water, nesting space, or some other necessity. If two species in a community are in direct competition for a limited resource, only one species will survive in that community. This principle of competitive exclusion is sometimes referred to as Gause's principle, after the Russian biologist G. F. Gause.

Gause formulated his principle on the basis of a number of laboratory experiments. His simplest, now classic experiment involved laboratory cultures of two species of paramecia, *Paramecium aurelia* and *Paramecium caudatum*. When the two species were grown under identical conditions in separate containers, *P. aurelia* grew much more rapidly than *P. caudatum*, indicating that the former used the available food supply more efficiently than the latter. When the two were grown together, the former rapidly outmultiplied the latter, which soon died out.

A similar simple experiment was made with two species of clover. Either could grow well alone, it was found, but when the two were mixed together, one soon eliminated the other. The reason was simple: the successful species held its leaves slightly higher than those of its competitor and so, once the cultures grew dense enough, completely overshadowed it.

Numerous observations of natural communities have confirmed Gause's principle. When two species with similar requirements are found in the same community, close inspection usually reveals that they are not actually in direct competition.

A study of barnacles provides an example. Before barnacles, which are a kind of crustacean, change from their immature, larval forms, in which they are free-swimming, into their adult, feeding forms, they cement themselves to a rock and secrete a shell. In Scotland, one species of barnacle, *Chthamalus stellatus*, occurs in the high part of the intertidal seashore and another barnacle, *Balanus balanoides*, occurs lower down. Although young *Chthamalus* often attach to the rock in the lower, *Balanus*, zone after their short period of drifting in the plankton, no adults are ever found there.

The history of a population of barnacles can be recorded very accurately by holding a pane of glass over a patch of

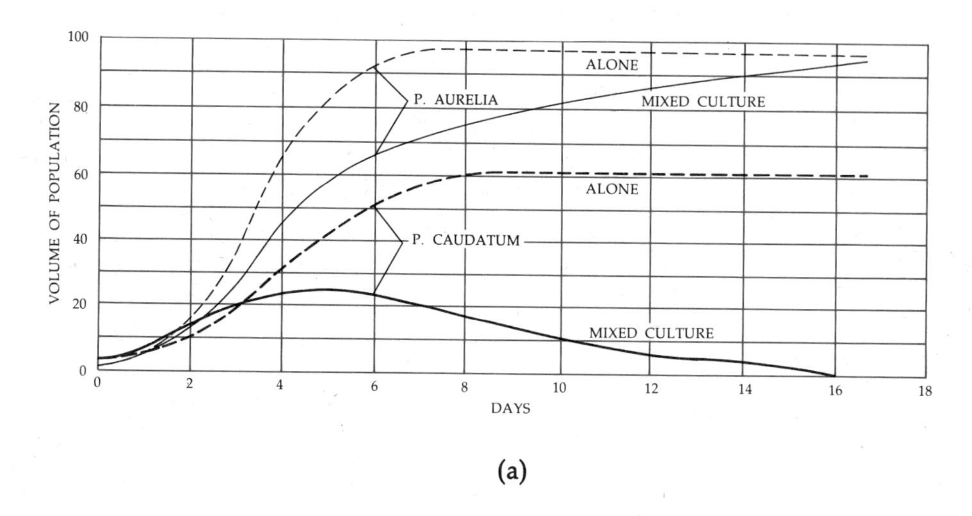

(a)

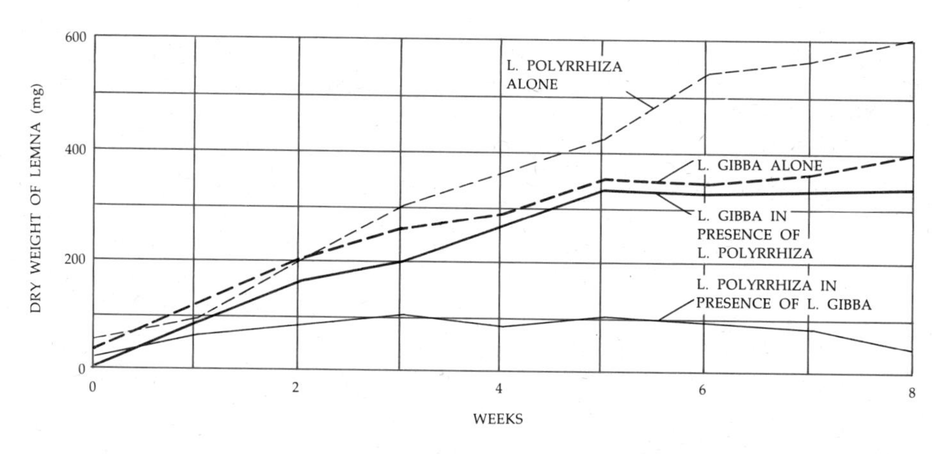

(b)

7–26

(a) *Results of Gause's experiment with two species of* Paramecium *demonstrate Gause's principle that if two species are in competition for the same resource—in this case, food—one eliminates the other.* Paramecium caudatum *and* Paramecium aurelia *were first grown separately under controlled conditions and with constant food supply. As you can see,* P. aurelia *grew much more rapidly than* P. caudatum, *indicating that it uses available food supplies more efficiently. When the two protozoans were grown together,* P. aurelia *rapidly outmultiplied* P. caudatum, *and the latter was eliminated.*

(b) *A similar experiment with two species of floating duckweed, tiny angiosperms found in ponds and lakes. One species,* Lemna polyrrhiza, *grows more rapidly in pure culture than the other species,* Lemna gibba. *But* L. gibba *has tiny air-filled sacs which, like pontoons, float it on the surface, and so it shades the other species, making it a victor in the competition for light.*

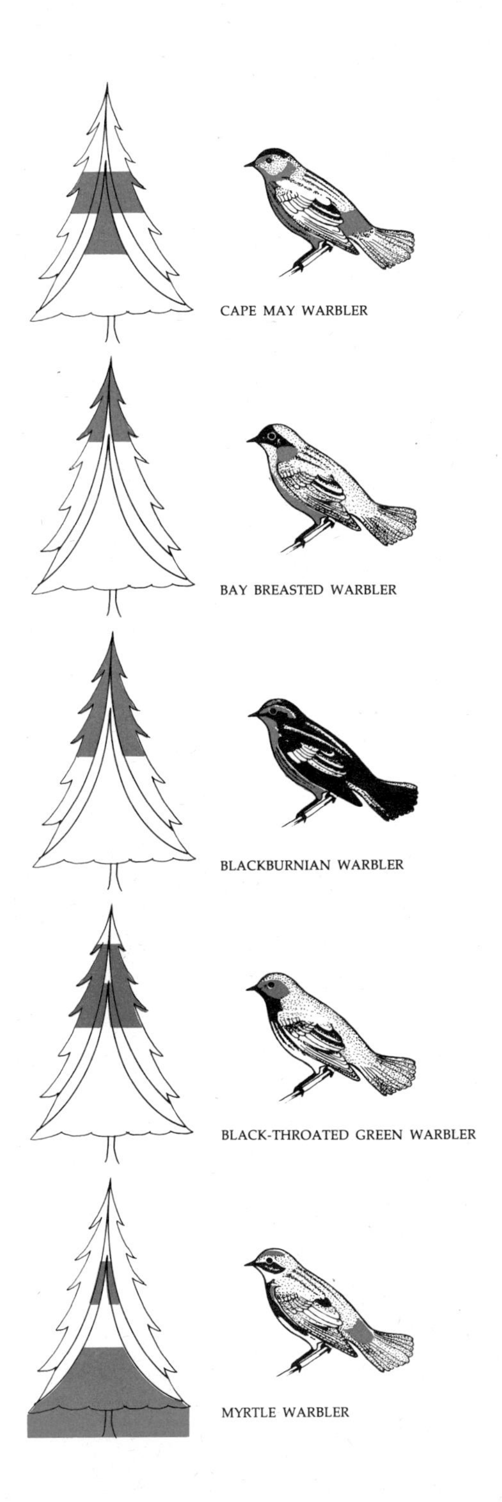

7–27
A demonstration of Gause's principle: the feeding zones in a spruce tree of five species of North American warblers. The colored areas in the tree indicate where each species spends at least half its feeding time. In this way, all five species feed in the same trees without competing.

barnacles and noting with glass-marking ink each spot where a barnacle is. Once attached, barnacles remain fixed, so that by returning later, one can check exactly which barnacles have died and which new ones have arrived.

By doing this, the investigator was able to show that in the lower zone, *Balanus,* which grows faster, ousts *Chthamalus* by crowding it off the rocks or growing over it. When *Chthamalus* was isolated from contact with *Balanus,* it lived with no difficulty in the lower zone, showing that the reason for the restriction of *Chthamalus* to the high shore levels was its competition with *Balanus.*

Gause's Principle and the Empty Niche

An interesting corollary to Gause's principle can be found in the fact that in some communities, because some resources are not being utilized, there is a "vacancy," or an empty ecological niche. Sometimes these "vacancies" are recognized only after they are filled. For example, after the Suez Canal was opened in 1869, 15 Red Sea species of fish colonized the Mediterranean, and some have become quite abundant. These species apparently filled vacant niches, because there seemed to be no decrease in the frequency of any of the Mediterranean species. No Mediterranean fish succeeded in entering the Red Sea, however, indicating that there were no vacancies available in the Red Sea, or at least none which could be filled by Mediterranean fish. The concept of the vacant niche is valuable in terms of evolutionary theory, as we shall see in Section 8.

SYMBIOSIS

Symbiosis ("living together") is a close association between two organisms of different species which is beneficial to both. We have already mentioned several examples of symbiosis. One is the association between a ruminant animal, such as a cow, and the bacteria that inhabit its stomach and break down cellulose. *Trichonympha,* a protist which is the source of the spectacular cross section of cilia shown in Figure 2–10, has a similar association with wood termites. Symbiotic algae, such as those found in the coral reef, provide their host with the products of photosynthesis and receive protection from this relationship. Symbiotic combinations of bacteria and root tissues can fix nitrogen although neither can do this alone.

A classic example of symbiosis is provided by the lichens, which are part alga, part fungus. The body of the lichen is composed largely of fungal mycelium, and held within this mycelium are numerous photosynthetic algal cells. The two organisms together form a closely integrated unit which can grow under conditions where neither the fungus nor the alga alone could survive. Lichens occur from arid desert regions to the Arctic; they grow on bare soil, tree trunks, sunbaked rocks, and windswept alpine peaks all over the world. They are often the first colonists of bare rocky areas.

PARASITES AND HOSTS

Parasitism is a close relationship between two organisms of different species that is beneficial to only one of them. Nearly all diseases in organisms (except some in higher animals, such as heart disease, which is linked to a prolonged life-span) are caused by parasites, chiefly microorganisms. Such diseases are special cases of predator-prey relationships. In these instances, the predator is always considerably smaller than the prey, and the prey is known as the host, which implies a gracious acceptance and is therefore somewhat misleading.

Large animals may sometimes be parasites. The hagfish, which sucks out the flesh of other fish, might be classified

7–28

One of the most ecologically important symbioses is the mycorrhiza *("fungus-root"), a relationship between certain fungi and the roots of plants. In the experiment whose results are shown in this picture, both groups of white pine seedlings were raised in a sterile nutrient solution for two months. Those on the left were transplanted directly to prairie soil. Those on the right were placed for two weeks in forest soil, where mycorrhizae developed, before they were planted in the prairie. The fungi apparently convert minerals in the soil to a form more readily usable by plants, and the plant roots seem to secrete organic compounds the fungi can utilize.*

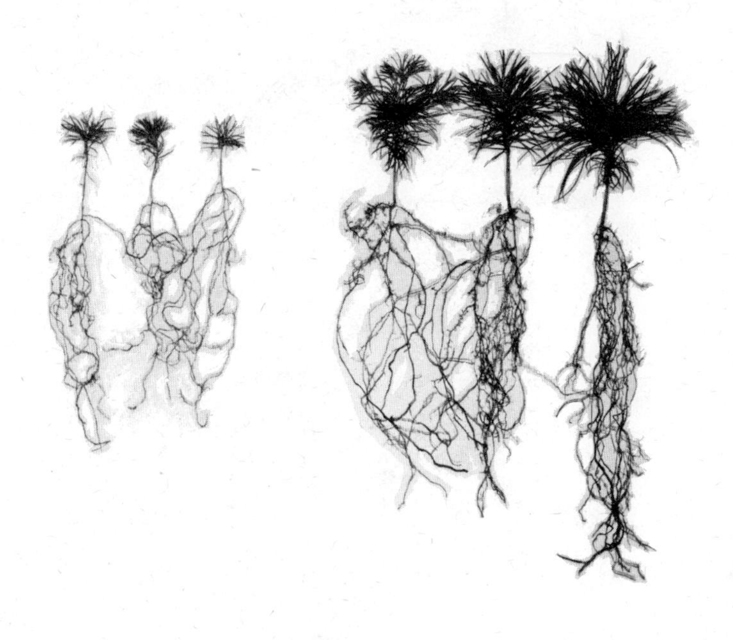

(a)

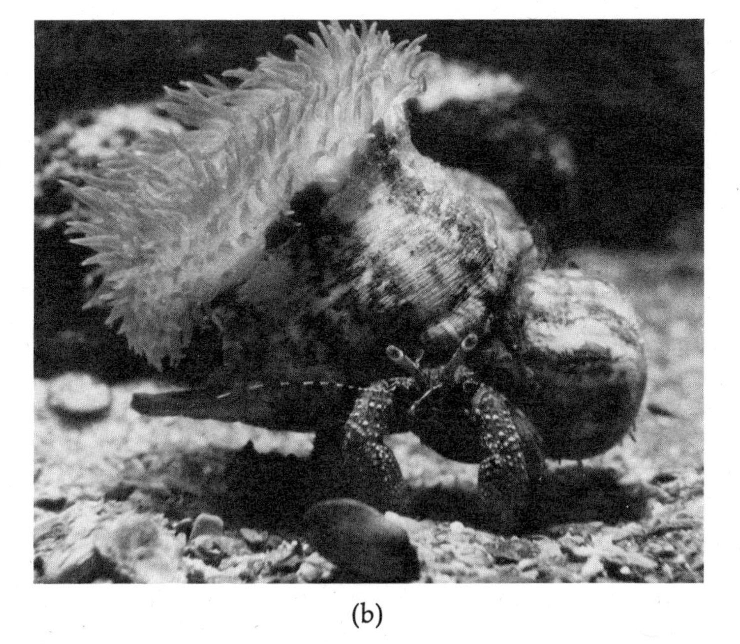

(b)

(c)

7–29
Symbioses. (a) *One kind of small fish, known as cleaner fish, is permitted to approach larger fish with impunity because they feed off the algae, fungi, and other microorganisms on the fish's body. The fish recognize the cleaners by their distinctive markings. There are other groups of fish that, by closely resembling the cleaners, get close enough to the large fish to take large scraps of food from them.* (b) *Sea anemone on the shell of a hermit crab. The crab enjoys the protection of the coelenterate and the sea anemone gains mobility—and so a wider feeding range—from its association with the crab.* (c) *The small bird seated on the rhinoceros's back, eats its ticks serving both itself and the rhino. The accompanying egrets feed on insects stirred up by the rhino's passage. The rhino was probably the unicorn of mythology. Its "horn" is actually a solid mat of heavy hair.*

either as a predator or as a parasite. The tapeworm, often yards long, is indisputably a parasite.

Most plants and animals in a natural community support hundreds of parasites of many species—in fact, perhaps millions, if one were to count the viruses. Parasites do not usually kill their host, and they almost never wipe out entire populations. We know this from our own experience. Bacterial disease is fatal only when the bacteria find themselves a particularly favorable place to multiply, such as in the blind pouch of an infected appendix, in an open wound, or when the host's resistance is reduced. Most bacteria are usually harmless, and some are helpful, such as those in the stomachs of the herbivores and those—*Escherichia coli*, for example—in our own intestinal tracts. Modern antibiotics often cause nausea and diarrhea because they destroy the useful and protective bacteria of the intestinal tract. And of all the viral diseases of man, only one, rabies, is regularly fatal if unchecked.

In animal and plant communities, diseases are most likely to wipe out the very young, the very old, and the disabled—either directly or, often, indirectly by making them more susceptible to other predators or to the effects of climate. Most infectious diseases tend to spread more rapidly when populations are crowded and to subside as the population thins out. They thus serve as a check on population density in much the same way that food shortage does.

It is logical that a parasite-caused disease should not be too virulent or too efficient. If a parasite were to kill all the hosts for which it is adapted, it, too, would perish. This principle is particularly well illustrated by the changes that have taken place in the myxoma virus of rabbits in Australia. The myxoma virus causes only a mild disease in the South American rabbit, its normal host, but a rapidly serious one, laboratory studies have shown, in the European rabbit, which has had no previous evolutionary experience with this disease. To end the scourge of introduced rabbits in Australia, myxoma-infected rabbits were set free on the continent. At first the effects were spectacular, and the rabbit population steadily declined, yielding a share of pastureland once more to the sheep herds on which much of the economy of the country depends. But then occasional rabbits began to survive, and their litters also were resistant. Now some 95 percent of the rabbits survive an attack of myxoma virus. A double process of selection took place. The virus, as originally introduced, was so rapidly fatal that often a rabbit died before it could be bitten by a mosquito and thereby infect another rabbit; the virus strain then died with the rabbit. Strains less drastic in their effects, on the other hand, had a better chance of survival since they had a greater opportunity to spread to a new host. (After the initial infection, the rabbit is immune to the virus, just as human beings usually become immune to mumps or measles after one infection.) So, first, selection began to work in favor of a less virulent strain, as proved by tests of the

7–30

In South Africa, a campaign was waged against hippopotamuses. Unlovely and unloved, they cluttered up rivers, were "useless," and so were shot on sight. The result was a rapid increase in a debilitating disease called schistosomiasis, caused by a snail-carried parasite that attacks the human liver. Without the hippos, which served as natural bulldozers, the rivers silted up, the snails proliferated, and schistosomiasis has become as great a hazard in some parts of Africa as malaria was 50 years ago.

Australian virus on European rabbits in the laboratory. Almost simultaneously, resistant rabbits began to appear, as proved by tests of the South American virus on Australian rabbits. Now the two are reaching a peaceful equilibrium, like most hosts and parasites.

PREDATOR-PREY RELATIONSHIPS

As we saw in the previous chapter, the various trophic levels of an ecosystem are linked by predator-prey relationships. Animals prey on plants, and other animals prey on the herbivores. In complex ecosystems, predator-prey relationships tend to regulate numbers of organisms but, unlike competition, seldom result in the elimination of a species. Almost all predators in a complex system prey on a wide variety of other organisms. When one species begins to decrease in number, its members are preyed upon less frequently and the predators turn their attention to one increasing in number.

Predation is not necessarily bad for the population as a whole. Wolves, for instance, have great difficulty overtaking healthy adult caribou or even healthy calves. A study of Isle Royale, an island on Lake Superior, showed that in some seasons more than 50 percent of the kills that wolves made were caribou with lung disease, although the incidence of such individuals in the population was less than 2 percent. (Human hunters, however, with their superior weapons and their desire for a "prize" specimen, are more likely to injure or destroy strong, well-adapted animals.)

Early in the 1900s, animal lovers undertook to protect a population of mule deer on the Kaibab Plateau in Arizona. There were about 4,000 deer living there at the time in a healthy, stable population, but it appeared that there was ample food for more (actually, the carrying capacity of the plateau at that time had been estimated at about 30,000 deer). Why not protect these gentle deer by exterminating the bloodthirsty animals that preyed upon them, it was argued. Hunters were encouraged to kill hundreds of natural predators—mountain lions, coyotes, and wolves. By 1918, the deer population had increased to 40,000, and the food plants were beginning to show damage. By 1920, some fawns were starving. In 1923, the deer population was up 100,000 and the forage on the range was so damaged that 60,000 deer died of starvation during the next two winters. The population then dropped to about 10,000, where it has remained stabilized.

7–31
The lemming is an inhabitant of the tundra, a biome in which there are relatively few species. In such areas, fluctuations either in the supply of food or in the number of predators can cause a single population to increase or decrease very rapidly. In lemmings, these population cycles result in spectacular mass migrations every three or four years.

In very simple ecosystems, such as the desert, the tundra, or areas of human agriculture, predator and prey numbers may fluctuate widely. When prey is abundant, the population of the predator increases. The population of the prey is then reduced drastically in number, and many of the predators, lacking other prey, starve to death. The prey then increases in number, and the cycle begins again.

(a)

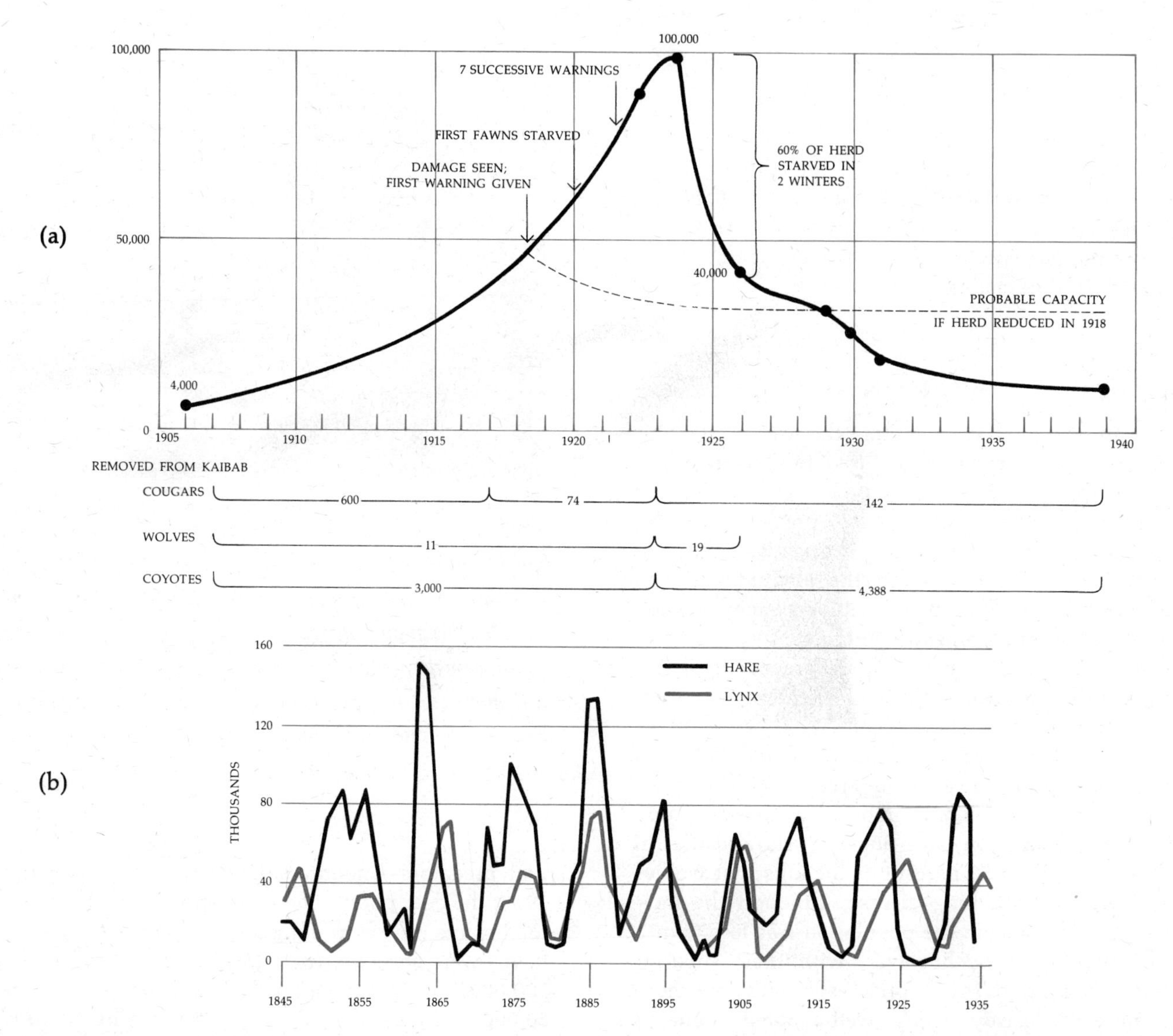

(b)

7–32

Predator-prey relationships: (a) *The effect of the removal of natural predators on the deer population of the Kaibab Plateau in Arizona. In* 1907, *a bounty was placed on cougars, wolves, and coyotes—all natural predators of the deer. In* 1918, *when the deer population had increased tenfold, investigators warned that the range was being overcrowded. More than half the herd eventually starved.* (b) *The number of lynx and snowshoe hare pelts received yearly by the Hudson Bay Company over a period of almost* 100 *years, indicating a pattern of* 10-*year oscillations in population density. The lynx reaches a population peak every* 9 *or* 10 *years, and these peaks are followed in each case by several years of sharp decline. The snowshoe hare follows the same cycle, with a peak abundance generally preceding that of the lynx by a year or more.*

(a)

(b)

7–33

The introduction of a new species in a community that does not support any natural predators for that species may sometimes have disastrous results. When prickly-pear cactus was brought to Australia from South America, it soon escaped from the garden of the gentleman who had imported it and spread into fields and pastureland until more than 30 million acres were so densely covered with prickly pears that they were valueless (a). *The cactus then began to take over the rest of Australia at the rate of about a million acres a year and was not brought under control until a natural predator was imported. This was a South American moth* (Cactoblastis cactorum) (b), *whose caterpillars live only on the cactus. Now only an occasional cactus and a few moths can be found.*

DEFENSES

Natural selection favors the most efficient predator, and so its genes are most likely to remain in the gene pool. Thus, over the course of the generations, the big cats become swifter and more cunning, the necks of the giraffes become progressively longer, and the grosbeaks' bills grow thicker and stronger. Natural selection also favors the prey most able to survive, and under strong evolutionary pressures from predators, plants and animals have developed some interesting and, to the human eye, extremely ingenious defenses.

Natural Defenses in Plants

Some natural defenses in plants are structural, such as the sharp-toothed edges of the holly leaf, the thorn of the rosebush, the spine of the cactus, and the sting of the nettle. Plants also produce a number of chemical substances for which the plant itself has no physiological use and which appear to function as defenses against leaf-eating insects and other predators. These chemicals include many substances presently useful to man, among them digitalis, quinine, castor oil, peppercorns, and other spices; some substances of more questionable value, such as nicotine, caffeine, and morphine; and others, such as the active principles in marijuana, mescaline, and peyote, whose desirability for the animal world is a matter of debate at many institutions of higher learning. Plants appear to have been waging such chemical warfare long before the coming of man, and it is possible that some small leafhopper was the first of all animals to have its mind expanded in a psychedelic experience.

Natural Defenses in Animals

Natural defenses in animals include structural adaptations and the capacity to produce noxious chemicals, both of which are seen in plants, and behavioral strategies as well. Some animals are formidably armored—for example, the armadillo, the porcupine, and the sea urchin. Some have ingenious behavioral devices. The armadillo and the pill bug

7–34
Natural defenses. (a) *Spines, such as in this flowering pincushion cactus, protect plants from herbivores.* (b) *Long-eared owl confronting an intruder.* (c) *Box turtle in retreat.* (d) *Skunks, experts in chemical warfare, are conspicuously marked. Many would-be predators learn to avoid second encounters.*

roll themselves up in tight armor-plated balls. Cephalopods of some species vanish, jet-propelled, leaving behind only an ink cloud. Many lizards have brightly colored tails that break off when they are attacked and, bright-colored and wriggling, divert the predator from the prey—which is escaping, tailless but with all its vital organs intact. Hermit crabs often carry sea anemones as passengers on their shells. The anemone protects the crab from predators while enjoying the benefits of mobility (and perhaps some stray morsels of food). When the crab changes shells, it often coaxes its sea anemone from its old home to the new one.

Grazing animals feed in herds, and birds, when threatened, tend to tighten their flock. It is almost impossible for a hawk to catch a bird that is within the flock since the hawk will be subjected to a constant bombardment of other bird bodies moving at higher speed.

The periodical cicadas protect one another in a quite different way. By emerging all together from their well-hidden pupal casings only after a long interval, they greatly reduce the likelihood that a natural predator dependent on cicadas will be present; most predators simply cannot wait that long.

Some birds emit alarm cries to warn the flock of an approaching predator. This seeming altruism has been interpreted by some scientists as running counter to theories of natural selection within species since the bird that calls attention to itself in this way is more likely to be lost—and its genes with it—to the predator than the bird that is not so altruistic. Other scientists point out, however, that since the bird is nearly always closely related genetically to the other members of its own flock, the genotype it saves, while perhaps not its own, is very similar. Therefore, genotypes dictating warning behavior would be more likely to be returned to the gene pool than those that do not. A somewhat similar but simpler situation is seen in the worker bee that dies on stinging an intruder; in so doing, she saves her hive of sisters.

Birds also take risks to save their young. "Injury feigning" is a common device among many types of ground-nesting birds. The parent bird flutters on the ground as though crippled, uttering piteous cries but always managing to stay just beyond reach of the intruder. Some scientists prefer to call this behavior "distraction display" to emphasize that the bird does not know what it is doing, although this latter term implies that the bird is being distracting, which it is equally unlikely to know. Regardless of the bird's intention, anyone who has seen this display cannot fail to marvel at the perfection of the performance, shaped as stringently and delicately by the unarguable forces of evolution as a bone is shaped, or a beak, or a feather.

Concealment and Camouflage

Hiding is one of the chief means of escape from predators. The young of many birds and of some other vertebrates respond to the warning cries of their parents by "freezing" or by running to cover. Small mammals often have nests or burrows or makeshift residences in hollow logs or beneath tree roots in which they conceal themselves from their enemies.

Protective coloration is common. Mice, lizards, and arthropods that live on the sand are often light-colored, and such light-colored species, if removed from their home territories, will immediately try to return to them. Snails that live on mottled backgrounds are often banded. Grass snakes are grass-colored, as are many of the insects that live among the grasses.

Some animals are countershaded for camouflage. The next time you pass a fish market, look at the specimens laid out on view. Fish are nearly always darker on the top than on the bottom. Countershading reduces the contrast between the shaded and unshaded areas of the body when the sun is shining on the organism from overhead. A fish was once found—the Nile catfish—that was reverse-countershaded; that is, its dorsal surface was light, and its ventral surface dark. The selective theory of camouflage was momentarily threatened, but scientific order was restored when

7–35

Camouflage. (a) *Common tree frog clinging to trunk of a white oak.* (b) *Walkingstick. Unlike the praying mantis, which it somewhat resembles, this wingless insect is a vegetarian.* (c) *Female pheasant crouching on nest in which she is incubating eggs.*

it was discovered that the Nile catfish characteristically swims upside down.

Other organisms hide by looking like something else. One type of insect, the treehopper, looks like a thorn; another, the walkingstick, like a twig. Young larvae of swallowtail butterflies look like bird droppings. In order for such disguises to work, animals must behave appropriately. *Biston betularia,* the peppered moth, lies very flat and motionless on its tree trunk, so that its colors blend with those of the tree. Some moths that sit exposed on the bark of trees even habitually orient themselves so that the dark markings on their wings lie parallel to the dark cracks in the bark. Some desert succulent plants look like smooth stones, revealing their vegetable nature only once a year when they flower.

Some insects manage through warning coloration to frighten off their would-be predators. Large spots that look like eyes are commonly found on the backs of butterflies or the bodies of caterpillars, where they will suddenly appear when the insect spreads its wings or arches its body. Small

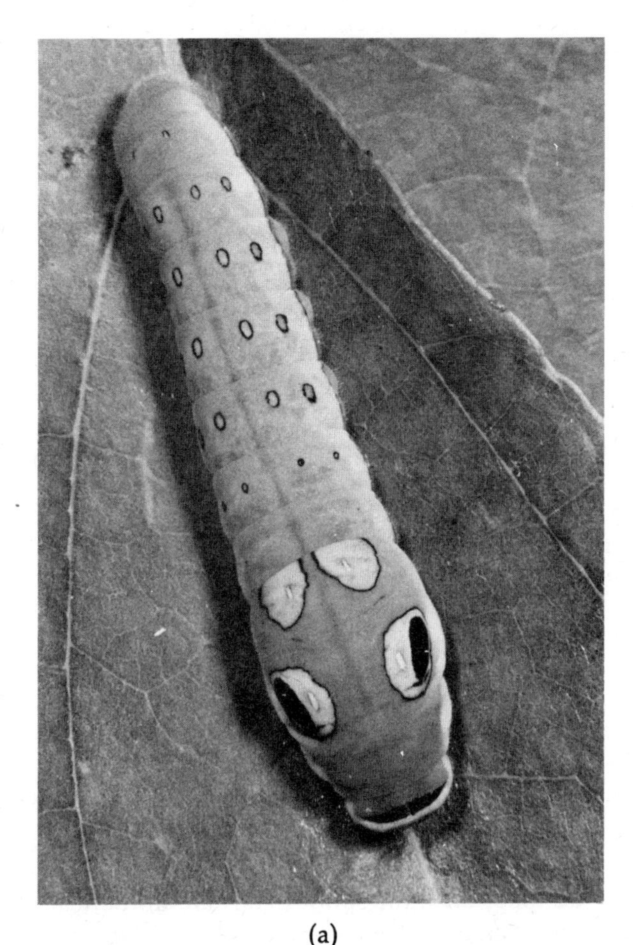

(a)

(b)

7–36
Many species have evolved surface markings that look like eyespots. Some of these eyespots apparently frighten would-be predators. Others deflect attack from the head to some less vulnerable part of the body. (a) *The startled expression of this caterpillar of the spicebush swallowtail butterfly is only skin deep. The head is actually at the opposite, less conspicuous end.* (b) *A predator striking at the "eye" on the tail of this butterfly fish would probably encounter empty water as the fish darted away.*

birds that have been reported to flee at the sight are probably birds that themselves are likely to be the prey of larger, large-eyed birds, such as owls or hawks. Smaller eyespots, while probably not frightening, seem to have the effect of deflecting the point of attack away from the head. Examination of wounded butterflies has shown that if a part of the wings bears beak marks or is missing, it is most often the part that contains the eyespots. One investigator has tested and confirmed this conclusion by painting eyespots on the wings of living insects, releasing them, and later recapturing them for examination.

On Being Obnoxious

Some animals defend themselves, or their species at least, by having a disagreeable taste, odor, or spray, often derived from distasteful chemicals in the plants they eat. Such animals are, generally, carefully avoided. The monarch butterfly, for instance, tastes bad to birds; it flies very slowly, which helps a would-be predator to recognize its highly characteristic colors. Louis Leakey, the well-known anthro-

(a)

(b)

7–37

(a) *Flowers of the milkweed family contain poisons damaging to the heart and other tissues of animals. Monarch butterflies, which are immune to these poisons, pollinate the milkweed flower, drink the poison-containing nectar and concentrate it in their own bodies, thus detering their own would-be predators.* (b) *The monarch has numerous mimics. Model and mimic characteristically fly slower, thus giving potential predators opportunity to recognize their warning colors.*

pologist, has hypothesized that primitive man may have survived, like the monarch butterfly, by not tasting very good to other animals. This hypothesis is supported by the fact that man, who is relatively defenseless compared with other vertebrates his size, apparently shared water holes and other habitats with large carnivores.

Obviously, tasting bad, while useful, is not an ideal defense from the point of view of the individual, since making this fact known may demand a certain amount of personal sacrifice. (Actually, birds drop the monarch butterfly after the first bite, but they often inflict fatal injury in the process.) Obnoxious sprays and odors have the advantage of warding off predators before they harm the prey. To man, the most familiar of such animals is the skunk, which advertises its malodorous threat by its distinctive coloration. Many insects and other arthropods have developed defense secretions. In some millipedes, for example, the secretion oozes out of a gland onto the surface of the animal's body. Other arthropods are able to spray their secretion over a distance, in some instances even aiming it precisely at the attacker. The caterpillar *Schizura concinna,* whose single spray gland opens ventrally just behind its head, directs the spray simply by aiming its front end. In the beetle *Eleodes longicollis,* the spray glands are in the rear, and the beetle, when disturbed, does a quick headstand and spreads a secretion from its abdominal tip. The soldiers of certain termite species possess a pointed cephalic nozzle from which their defensive spray is ejected. This spray not only can incapacitate a small predator but also acts as an attractant to summon more troops. In the whip scorpion, two glands open at the tip of a short knob that moves like a gun turret. Many arthropods possess a number of glands but discharge only from those closest to the point of attack, thus saving ammunition and gaining efficiency. The secretions usually act as topical irritants, especially to the mouth, nose, and eyes of the predator. Because of their relatively permeable skin, frogs and toads, common predators of arthropods, are sensitive to these irritants over their entire bodies.

(a)

7–38

(a) *The beetle* Eleodes longicollis *has glands in its abdomen which secrete a foul-smelling liquid. When disturbed, it stands on its head and sprays the liquid at the potential predator.*
(b) Eleodes longicollis *is on the left. On the right is* Megasida obliterata *which, as you can see, somewhat resembles* Eleodes longicollis *and emphasizes this resemblance by also standing on its head. However,* Megasida obliterata *has no similar glands, no noxious secretion, and no spray.* (c) *A grasshopper mouse which has solved the problem of how to eat* Eleodes longicollis. *The mouse drives the posterior end of the beetle into the ground and eats it head first. It would probably eat* Megasida obliterata *the same way.*

(b)

(c)

Vertebrates learn quickly to recognize an obnoxious prey. Blue jays, for instance, having once been sprayed by a walkingstick, will remain aloof from it even when the insect is presented two or three weeks later. Predators also may learn to counter these defenses. Grasshopper mice feed on beetles like *Eleodes longicollis* by jamming them butt-end into the earth, so that their chemical arsenal is harmlessly expended in the soil, and then eating them head first.

Müllerian Mimicry: Advertising

For animals that have a highly effective protective device, such as a sting, a revolting smell, or a poisonous or bad-tasting secretion, it is advantageous to advertise. The more inconspicuous or rare such an animal is, the larger the proportion of individuals that must be sacrificed before the bird or other predator learns to avoid it. Müllerian mimics, named after F. Müller, who first described the phenomenon, are groups of insects that, although not closely related phylogenetically, all have effective obnoxious defenses and all resemble one another. Bees, wasps, and hornets probably offer the most familiar example; even if we cannot tell which is which, we recognize them immediately as stinging insects and keep a respectful distance. Similarly, large numbers of bad-tasting butterflies are look-alikes. Müllerian mimicry is adaptive for all species involved because each prospers from a predator's experience with another.

Batesian Mimicry: Deception

Batesian mimicry, first described by the British naturalist H. W. Bates in 1862, is deceptive mimicry. In Batesian mimicry, the mimic fools its predator by resembling a stinging or bad-tasting "model" which the predator has learned to avoid, although the mimic itself is innocuous. Some species of harmless flies resemble bees or hornets, and many species of butterflies resemble monarchs or other unpalatable butterflies or moths.

Laboratory experiments have clearly demonstrated Batesian mimicry in operation. Jane Brower, working at Oxford, made artificial models by dipping mealworms in a solution of quinine, to give them a bitter taste, and then marking each one with a band of green cellulose paint. Other mealworms, which had first been dipped in distilled water, were painted green like the models, so as to produce mimics, and still others were painted orange to indicate another species. These colors were chosen deliberately: orange is a warning color, since it is clearly distinguishable, and green is usually found in species that are not repellent and therefore do not wish to advertise.

The painted mealworms were fed to caged starlings, which ordinarily eat mealworms voraciously. Each of the nine birds tested received models and mimics in varying proportions. After initial tasting and violent rejection, the models were generally recognized by their appearance and avoided. In consequence, their mimics were protected also. Even when mimics made up as much as 60 percent of the green-banded worms, 80 percent of the mimics escaped the predators.

Batesian mimicry obviously works to the advantage only of the mimic. The model, on the other hand, suffers from attacks not only from inexperienced predators but from predators who have had their first experience with mimic rather than with model. The mimetic pattern will be at its greatest advantage if the mimic is rare, that is, less likely to be encountered than the model, and also if the mimic makes its seasonal appearance after the model, thus reducing its chances of being encountered first. However, if the model is sufficiently distasteful—as in the case of the quinine-soaked mealworm—it may protect mimics even if the latter are very common.

CONCLUSIONS

In this chapter, we have tried to show some of the ways in which interactions of organisms shape the community and affect the characteristics of its inhabitants. In each of these cases, we have spoken about the relationships between two populations. Obviously, community interactions involve many different species, all linked together in what has aptly been called the "web of life." The sum total of forces acting upon even one population is so complex that it can probably never be understood; however, isolated examples, such as we have given here, afford at least a glimpse of the forces at work which create the great diversity of living things.

SUMMARY

Populations within a community interact in a variety of ways, including competition, predator-prey relationships, and parasite-host associations.

Competition is a principal determinant of the survival of individual species within a community. Gause's principle, the principle of competitive exclusion, states that if two species are in competition for a limited resource, one species will eliminate the other.

Symbiosis ("living together") is a term used to describe a close relationship between two different species in which both species benefit. An example is the lichen, which is the result of a symbiotic association between an alga and a fungus and which grows where neither organism can grow alone.

Parasitism is a close relationship between species in which only one species benefits. Nearly all diseases in organisms are caused by parasites. Most parasites do not kill their host, and they almost never wipe out entire populations.

The trophic levels of an ecosystem are linked by predator-prey associations. These associations exert a regulating force on the size of populations and also have profound evolutionary effects on the various species involved.

Plants and animals have developed a variety of defenses against predation. These include "armor" and other forms of physical protection, as seen in cacti, armadillos, turtles, and numerous other organisms, and chemical weapons, such as the plant poisons and noxious secretions of insects. Many organisms are camouflaged.

Some insects have come to resemble organisms of other species either to advertise an effective protective device that they possess in common with the other species (Müllerian mimicry) or to "pretend" they possess such a device when they actually do not (Batesian mimicry).

All these associations help to determine the character of the community and of the organisms within it.

QUESTION

Introducing a new species into a community can have a number of possible effects. Name some of these possible consequences both to the community and to the species. What types of studies should be made before the importing of an "alien" organism? Some states and many countries have laws restricting such importations. Has your own state adopted any such laws? Are they, in your opinion, ecologically sound?

CHAPTER 7–4

The Community in Action: Social Behavior

Social behavior includes all interactions of individuals of the same species. Like the spine of a cactus and the armor of an armadillo, patterns of social behavior took form as a consequence of evolutionary pressures and have survival value for the species as a whole.

INSECT SOCIETIES

Of all animal organizations, the insect societies are probably the best understood, in terms both of their evolution and of the interplay of forces that keep them together. Like many other animal societies, they are based on the family unit, and the way in which the mother-egg relationship has gradually developed over the course of millennia into a tightly knit large group is indicated by observations of the behavior of existing species. In fact, the type and degree of care provided to the young insect by the wasp or bee are coming to be used as a taxonomic characteristic for determining relationships between the many species, just as slight differences in the shape of the body or the color of the wing might be used.

Subsocial Bees

Although we think of all bees as living in communal hives, most of the species are actually solitary, as are most of the wasp species, which are closely related. Among the solitary species, the female bee builds a small nest, either burrowing into soil or wood or piling up earth or plant material, lays her eggs in it, stocks it with a mixture of honey and pollen (which is the protein source for the larvae), seals it off, and leaves it.

In subsocial species, the mother returns to feed the young larvae; similar behavior is shown by many of the wasps. In one type of subsocial bee (*Allodape pringlei*), the eggs are set in a curved row, like a spiral staircase, winding up the inside wall of a hollow stem. After hatching, the larvae hold themselves in place by short, plump "arms." The mother drops food into the hollow stem for the larvae, and they hold the food in these special fleshy protuberances and feed themselves.

In another type of subsocial bee (those belonging to the genus *Halictus*), the mother forms a crude comb of 16 to 20 cells underground, packs in food continuously for some time, and finally closes off the chamber. She then guards the entrance so long as she survives, which may be until the young bees emerge. In some of these species, the emergent young extend the old comb and, in turn, remain to watch over their own offspring. This small community is usually destroyed during the winter, and in the spring, each female that has survived founds a new colony.

Bumblebees

The bumblebee is a representative of the next stage of socialization. As with the *Halictus* bees, the bumblebees (*Bombidae*) must found their colony anew each spring. Every bumblebee you see in flight in the early spring is a queen and the potential founder of a colony. When she finds a suitable nest site, she constructs two cells from the wax which exudes from the surface of her abdomen. One cell she fills with nectar and pollen from her foraging trips. In the second cell, she lays a group of eggs, usually about eight. She then caps the egg cell with wax and settles down on it like a broody hen. In some three to five days, depending on the species, the eggs hatch; the mother then feeds them on the nectar and pollen. About seven days after hatching, the larvae spin cocoons. The mother continues to guard the cocoons, and she also constructs additional cells, laying eggs in each; in her spare time she forages for nectar. The larvae within the cocoons pupate for about two weeks, and then, helped by the mother, the damp, soft-bodied, pale-colored young bees crawl out of their cocoons. These workers are all females. In two or three days, the young bees (called callows) become hardened, develop the "furry" coat and bright colors of their mother, and go to work, helping to gather nectar and pollen and to care for each successive brood of younger sisters.

With the emergence of the first group of workers, a big step in the socialization process is observed. Now the mother devotes almost all her time to egg laying. The workers do not lay eggs of their own but rather enlarge the nest, which gradually assumes the form of a rough comb, and spend the rest of their time foraging for the insatiable and ever-growing brood. In other words, the workers care for their sisters (and, eventually, their brothers) rather than try to raise families of their own. As a consequence of this "altruistic" behavior, bumblebees are referred to as truly social rather than merely as subsocial.

The nectar is carried in a special honey stomach of the bee, where it is processed by enzymes. It is then regurgitated into a wax container in hives, where it evaporates into honey. The pollen is transported simultaneously in pollen baskets formed of long stiff hairs on the bees' hind legs. The returning worker scrapes the pollen off her hind legs by means of her middle legs and into one of the pollen bins in the hive.

The colony usually numbers only a few hundred bees, but it may grow as large as a thousand or so. Toward the end of the season, the production of young males and young queens begins, and the young queens mate. The workers, males, and old queens die at the end of the summer, and the young queens scatter and hibernate, emerging in the spring to found a new colony.

Honeybees

The honeybee colony, which usually has a population of 30,000 to 40,000 workers, differs from that of the bumblebee and many other social bees or wasps in that it survives the winter. Like other bees, the isolated honeybee cannot fly if the temperature falls below 10°C (50°F) and cannot walk if the temperature is below 7°C (45°F). Within the hive, bees maintain their temperature by clustering together in a dense ball; the lower the temperature, the denser the cluster. The clustered bees produce heat by constant muscular movements of their wings, legs, and abdomens. In very cold weather, the bees inside the cluster keep moving toward the center, while those in the core of the cluster move to the colder outside periphery. The entire cluster moves slowly about on the combs, eating the stored honey from the combs as it moves.

Egg laying begins early in the year, in January or February, with each egg deposited in a separate wax cell. The white, grublike larvae which hatch from the eggs are fed by the nurse workers; each larval bee eats about 1,300 meals a day. After the larva has grown until it fills the cell, a matter of about six days, the nurses cover the cell with a wax lid, sealing it in. It pupates for about 12 days, after which an adult emerges. The adult rests for a day or two and then begins successive phases of employment. She is first a nurse, bringing honey and pollen from storage cells to the queen, drones, and larvae. This occupation usually lasts about a week, but it may be extended or shortened, depending on the conditions of the community. Then she begins to produce wax, which is exuded from the abdomen, passed forward by the hind legs to the front legs, chewed

7–39
Honeybee. The antennae, attached to the head by a ball-and-socket joint, are actually "smellers" rather than feelers. The large compound eyes cannot see red (which is black, or colorless, to them) but can see ultraviolet, which is colorless to human eyes. The sting is at the tip of the abdomen.

7–40
Workers store honey in the wax cells of the comb. The honey is made from nectar, processed by special enzymes in the workers' bodies.

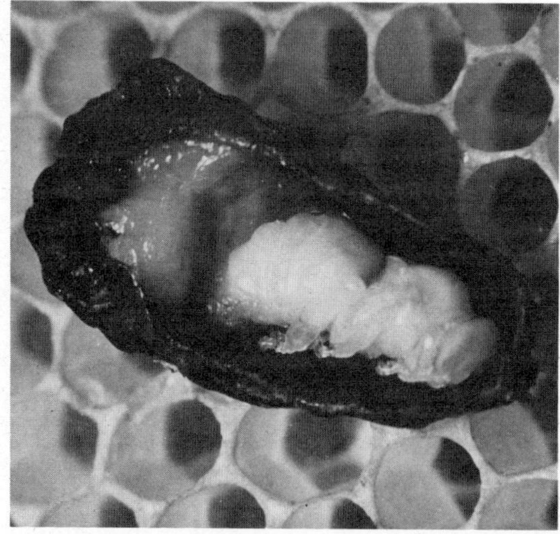

7–41
Royal nymph. Nourished on royal jelly, such larvae are the future queens.

7–42
Another task of the worker is airconditioning. Here a worker beats her wings to cool the hive.

thoroughly, and then used to enlarge the comb. These houseworking bees also remove sick or dead comrades from the hive, clean emptied cells for reuse, and serve as guards at the hive entrance. During this period, they make brief trips outside, seemingly to become familiar with the immediate neighborhood of the hive. In the third and final phase of their existence, the worker bee forages for honey and nectar. The life-span of a worker is usually only about six weeks.

Each hive has only one adult queen. The queen caste is genetically the same as the worker caste. All the differences between the two castes depend on the substance fed the queen-to-be in the larval stage and on the hormonal influences she, in turn, exerts upon her subjects. For the first two days of life, all bee larvae are fed "brood food," a white paste produced in the glands of young workers and secreted from the mouth. Thereafter, worker and drone larvae are fed honey and pollen, whereas the larva being made ready for queenhood is fed only the glandular secretions (hence known as royal jelly) all during its larval stage. Attempts have been made to identify the substance in royal jelly that confers queenhood, but so far these have not been successful.

Queens are raised in special cells which are larger than the ordinary cells and are oddly shaped, somewhat resembling an empty peanut shell. If a hive loses its queen, workers notice her absence very quickly, sometimes in only 15 minutes, and become quite agitated. Very shortly, they begin enlarging worker cells to form emergency queen cells. The larvae in the enlarged cells are then fed exclusively on royal jelly until a queen develops. Any larva so treated will become a queen.

The queen bee is inseminated only once in her lifetime, and the supply of sperm that she receives at that time is stored in her body and can be used to fertilize each egg as she produces it. Every fertilized egg develops into a female; the workers, all of which are sterile females, are required in large numbers for the economy of the hive. From time to time, however, males are needed to fertilize new queens, either because the hive has become too big and is dividing or because the old queen is nearing the end of her life. At these times, the queen bee produces eggs without releasing sperm, so that some bees develop parthenogenetically, that is, from unfertilized eggs. These are males. Female bees have 32 chromosomes, the diploid number, but the males have only 16, the haploid number.

The queen influences her subjects by means of chemicals called *pheromones*, which when secreted by an animal exert their effects on other individuals of the same species. One of the pheromones, known as queen's substance, inhibits ovarian development in the worker bees and prevents them from becoming queens or producing rival queens. The pheromone passes through the workers of the hive orally. As the workers meet, they often exchange the contents of their stomachs. Studies in which queen substance has been tagged with a radioactive label have shown that as a result of this activity, the pheromone travels through the hive with remarkable rapidity. Within only half an hour after removal of the queen, the shortage of queen's substance is already noticed and the hive begins to grow restless. It is difficult to understand how a single queen can produce enough pheromone to influence the entire hive of as many as 40,000 workers or more, as well as tend to her stupendous egg-laying chore, but it has been suggested that after the pheromone is passed among the workers, it is fed back to the queen in a reduced form and she need simply oxidize it to reactivate it.

In the spring, when the nectar supplies are at their peak, so many new broods are raised that the group separates into two colonies. The new colony is always founded by the old queen, who leaves the hive taking about half of the workers with her. This helps to ensure survival of the new colony since this queen is of proven fertility. The group stays together in a swarm for a few days, gathered around the queen, after which the swarm either will settle in some suitable hollow tree or other shelter found by its scouts or will be picked up by a beekeeper and transferred to an empty hive, which it will then furnish with wax.

In the meantime, in the old hive, new queens have begun to develop, often even before the old queen has left, and ovarian development has begun in some of the workers, some of which lay eggs which, since they are unfertilized, develop into males or drones. After the old queen leaves the hive, a new young queen emerges, and any other developing queens are destroyed. The young queen then goes on her nuptial flight, exuding a pheromone that entices the drones of the colony to follow. She mates only on this one occasion and then returns to the hive to settle down to a life devoted to egg production. During this single mating, she receives enough sperm to last her entire life, which may be some five to seven years. The sperm cells are stored in her spermatheca and are released, one at a time, to fertilize each egg as it is being laid. The queen usually lays unfertilized eggs only in the spring, at the time males are required to inseminate the new queens, so the release of the spermatozoa from the spermatheca is apparently under the control of her internal or external environment. Participation in the nuptial flight is the only service to the hive performed by the drones. Since they are unable to feed themselves, they become an increasing liability to the hive. As nectar supplies decrease in the fall, they are stung to death or driven out to starve by their sisters.

Termites

Termite societies are by far the most ancient communities in the world. Man has existed as a social animal for less than 1 million years, ants for about 100 million years, and termites for something like 200 million years. There are five different families and almost 2,000 species of termites. All species are social; some termitaries contain as many as 3 million individuals.

Termite societies are organized by a rigid caste system in which differences of occupation are clearly reflected in body size and shape. The largest member of the colony is the queen, which may in some species reach several inches in length, far larger than any member of her "court." Huge, grublike, and almost shapeless, she serves as an egg factory, being constantly fed by the workers and laying as many as 10,000 eggs a day. Even if she were able to move, her swollen body could not pass through the narrow chambers of the termite colony, so she spends her entire existence, often 10 or 15 years, sealed in her nuptial chamber. In many species, the king, the sole reproductive male of the colony, is sealed in the chamber with her.

7–43

In the rigidly structured termite society, the different occupations of the several castes are reflected in anatomical specializations. Shown here are three workers and, in the center, a soldier, distinguished by its larger size and massive jaws.

The two largest termite castes are the soldiers and the workers. The soldiers, defenders of the community, possess large, strong mandibles and also, in some species, a syringelike beak from which they can eject a thick, irritating liquid. In many species, the headparts of the soldiers have become so massive that they are no longer able to feed themselves and must be taken care of by the workers, who also tend the royal couple, feed the young, and maintain the many-chambered colony.

The eggs laid by the queen develop into nymphs. All nymphs are the same through the first stages. If the colony is short of workers, the nymphs will develop, in their final molt, into workers; if soldiers are needed, a proportion of

7–44
With her enormous abdomen, the largest member of the termite colony is the queen. Her entire adult existence, which may last for 15 years or more, is spent producing eggs. Constantly fed by the workers, she lays as many as 10,000 eggs a day!

7–45
Swarming termites: winged males and queens on the nuptial flight.

nymphs will become soldiers. (On the other hand, if there are too many soldiers, the workers will destroy some of them.) If the king or queen should die, a nymph will develop into a reproductive form and assume the burdens of the royal chamber. At specific periods of the year, depending on the species and the climate, termite colonies produce winged, reproductive males and females. These leave the colony on a brief nuptial flight, form pairs, lose their wings, and dig a burrow in which they will close themselves forever from the outside, thereby founding a new colony.

Termites lead most of their lives in hermetically sealed, warm, damp burrows, and many species perish of dessication if exposed to the drying effects of light and air outside their burrows. This is why, incidentally, they can do so much damage when they infest a building, since they will hollow out entire beams without exposing themselves. The common destructive form prevalent in the United States has one important point of vulnerability: a need to return to the soil, apparently for moisture or nutrients.

OTHER SOCIAL ORGANIZATIONS

Many aggregations of organisms are based solely on environmental conditions—humidity, shelter, temperature, availability of food—and so do not represent true social behavior. In these groups each animal responds to environmental stimuli as if it were alone. However, in other aggregations, stimuli are exchanged among members of the group, and these serve both to hold the group together and to determine its responses to the environment.

Animal Families

Many animal groups are based on family relationships. Often these associations are temporary, the young remaining with their parents only until they reach maturity. Sometimes, however, the generations stay together, and the family unit enlarges into a herd or flock. Some of the most stable flocks are matriarchies, such as flocks of sheep or of red deer, in which one of the oldest females leads a

7–46
Birds typically have nuclear families—mother, father, and one or more offspring. In some species males and females mate for life; in others, the union dissolves either after mating or when the young are able to take care of themselves. (a) *Father Emperor penguin with offspring. The single infant will spend its first several weeks riding about on its mother's or father's large flat feet, nestled under a warm fold of abdominal skin.* (b) *Young bearded tits. Gaping in nesting birds is a strong stimulus to the parents, who will literally exhaust themselves feeding their importunate brood. Note the mouth markings in these nestlings. Birds of this species will feed only young with these markings.*

(a)

(b)

group consisting of other females and of young. In these cases, the males usually congregate separately—the stag line—except at breeding time.

Other groups are dominated by males. In some of these, only one male of reproductive age is allowed in any one unit, which usually keeps the groups small. The young bachelors live outside the group, waiting for a chance to drive off or destroy the older male, and a male that is not sufficiently strong or aggressive is doomed to a life of celibacy.

In communities such as those of the howler monkeys, pair bonds are not formed and females in heat are unjealously shared by the males of the troop. Care of the young, although devolving primarily on the mother, is shared by the adults of the group.

Flocks and Herds

Animals as diverse as bats, birds, locusts, antelope, bison, fish, and gnats tend to move always in flocks or herds. If

7–47
An Alaskan fur seal harem; the bull is in the upper right corner of the picture. In this species, the family consists of a single male and a variable number of females and young, and the group stays together only during the summer months. The dominant male is always fully grown, since only the adult bulls are large enough and strong enough to secure a beach territory and round up a harem for the summer; the younger males live by themselves in nearby "bachelor quarters." The females arrive at the breeding grounds in June, about a month after the males, who have been busy establishing their territories. There is a period of almost constant conflict while each bull herds as many females as he can into his own territory. In a very short time, the females give birth to pups conceived during the preceding summer. There is then a period of mating, and for the rest of the summer the bull will continue to guard his harem, neither eating nor leaving his territory for several months. In the fall, the males and females separate, the females and young going south and the bulls moving to the Gulf of Alaska. In the spring, the bull will gather another family together again, but his harem may consist of an entirely new group of females.

7–48
Like most primates, macaques are organized in bands or tribes. When the infants are very young, the mothers usually have primary responsibility for their care, but as they grow older, care is shared by other members of the group. Here a juvenile whose place with his mother has been usurped by a new baby is comforted by one of the adult males of the tribe.

you watch birds feeding in a field, for example, they will all move off together when one of them takes off. Even very young minnows, less than 10 millimeters in length, tend to orient themselves in parallel rows for swimming. Starlings and other birds fly in flocks that may number a thousand or more, swooping and diving together with great precision. The simultaneous, coordinated behavior shown by these animals is sometimes referred to as "infectious" behavior, since it appears as though an individual cannot help "catching" the behavior of the group. A common example of infectious behavior in humans is yawning. Besides yawning, alarm, courtship and copulation, and feeding behavior all tend to spread infectiously in animal societies. Schooling in fish is an extreme example of social organization based on such imitative behavior.

7–49
Birds which fly in flocks position themselves to take advantage of windcurrents produced by other members of the group. The leader, who has to work the hardest, frequently changes places with another member of the flock.

Advantages of Group Life

Group life affords the individual protection against predators. Large groups are unlikely to be attacked; even birds as savage as the hawk nearly always select a solitary flyer or a straggler. Fish that form large schools are less likely to be attacked by larger fish. It is believed that the barracuda, for instance, will mistake a large school of very small fish for one large fish and so leave it alone. Sometimes the sheer numbers of individuals in a large group are confusing to the hunter. *Daphnia,* which are small crustaceans, are common aquarium food, as fish fanciers know. If fish are fed many daphnids all at once, a smaller number are eaten than if a few daphnids are presented at one time; apparently the fish get confused by the conflicting stimuli. In a group, if one member senses danger, its actions warn the rest; in some groups of birds and of mammals, certain individuals appear to act as sentinels.

There are other advantages. Small warm-blooded animals, such as mice, provide each other with warmth, and groups of aquatic animals can sometimes survive in a toxic environment more successfully than isolated animals since their combined metabolic processes can serve to detoxify it.

Although the survival value of these social organizations is clearly evident, little is known of the stimuli which keep these societies together. Recent studies indicate that pheromones, already shown to exert such powerful influences in insect societies, may also have an important role in vertebrate social behavior.

SOCIAL DOMINANCE

Animal societies can be organized in a variety of ways. Often, as we have mentioned, they will be led or directed by a senior female or a senior male. One type of social dominance that has been studied in some detail is what is called the *pecking order* in chickens. A pecking order is established whenever a flock of hens is kept together over any period of time. In any one flock, one hen usually dominates all the others; she can peck any other hen without being pecked in return. A second hen can peck all hens but the first one; a third, all hens but the first two; and so on through the flock, down to the unfortunate pullet that is pecked by all and can peck none in return. Hens that rank high in pecking order have privileges such as first chance at the food trough, the roost, and the nest boxes. As a consequence, they can usually be recognized at sight by their sleek appearance and confident demeanor. Low-ranking hens tend to look dowdy and unpreened and to hover timidly on the fringes of the group. A hen experimentally moved from flock to flock may have different ranks in each of several different flocks. In a stable flock, the old-timers tend to occupy the top ranks and any new arrival usually ranks low and has to peck her way up.

Pecking orders reduce the breeding population. Cocks and hens low in the pecking order copulate much less frequently than socially superior chickens. Similar observations have been made in baboons. When a female baboon comes into heat, juvenile males and the less dominant adults copulate with her. It is only when she is actually ovulating (as indicated by a marked swelling of her genital area), however, that the most dominant males mate with her, thus greatly increasing their chances of fathering her offspring.

TERRITORIALITY

Another means by which animals organize their societies is through the establishment of territories. Territoriality in some form is seen in animals as widely diverse as crickets, howler monkeys, fur seals, dragonflies, red deer, beaver, prairie dogs, many types of lizards, and a large number of species of birds and fish.

Territoriality was first recognized by an English amateur naturalist and bird watcher, Eliot Howard, who observed that the spring songs of male birds served not only a courtship function but also to warn other males of the same species away from the terrain that the prospective father had selected for his own. In general, a territory is established by a male and is an area in which he will not permit other males of the same species to intrude. Courtship of the female, nest building, raising of the young, and often feeding are carried out within this territory. Possession of a territory assures a mating pair of a monopoly of food and nesting materials in the area and of a safe place to carry on all the activities associated with reproduction and care of the young. Some pairs carry out all their domestic activities within the territory. Others perform the mating and nesting activities in the territories, which are defended vigorously by the males, but do their food gathering on a nearby communal feeding ground, where the birds congregate amicably together. A third type of territory functions only for courtship and mating, as in the bower of the bowerbird, the arena of the prairie chicken, or the stamping ground of the antelope. In these territories, the male prances, struts, and postures—but very rarely fights—while the females look on and eventually indicate their choice of a mate by entering his territory. Males which

7–50
Many birds proclaim their territories by singing but the ruffed grouse drums. The noise, which sounds like a distant outboard motor, is produced by rapid beating of the wings, as shown here.

have not been able to secure a territory for themselves are not able to reproduce; in fact, there is evidence from studies of some territorial animals, such as the Australian magpie, that adults who do not secure territories do not mature sexually.

Thus, in effect, territoriality ranges from the actual monopoly of a physical area to the symbolic territoriality of the stamping ground.

The Flexibility of Territories

As older animals die, younger ones will contend for their places, keeping the breeding population stabilized. If the range is enlarged, more animals can gain access to the breeding community. This is illustrated by observations of muskrats, a species which has been particularly well studied because of its economic importance. Muskrats make their burrows along the edges of streams or around any other body of shallow water close to a supply of grass or grains. All breeding takes place within the burrows, from which nonbreeding populations are excluded. If the population increases, more animals are driven out. On the other hand, if freezing, drought, or trapping reduces the population, outsiders are allowed in and breeding increases. Limited hunting or trapping of the animals does not decrease the population; similarly, control of other predators does not increase the population. The only way to increase the muskrat population is to increase the carrying capacity of the breeding area. This was clearly demonstrated when the construction of dams, dikes, and canals in the Saskatchewan River diverted water into marshes suitable for muskrat habitation. In four years, the number of muskrats in the area increased from 1,000 to 200,000.

In some territorial animals, the size of the territory varies according to the food supply. Among Scottish red grouse, for instance, an area that contained 40 territories in one year in which the vegetation growth was ample had contained only 16 two years earlier, when the growth was poor.

7–51
Male fiddler crab signals with his large claw to attract a female to his territory—a hole in the sand—and warn off other males.

Territory Boundaries

A territory may be an area of plain, a corner of a small wood, or a few feet at the bottom of a pond. Sometimes, the territory consists of little more than the nest itself and the immediate area around it. For the male bitterling, a small fish, the territory is an area immediately surrounding a freshwater mussel. The bitterling admits only egg-laden females into his territory, then leads them to the mussel, where they lay their eggs within its gills. The male then injects his sperm while swimming over the siphon of the mussel.

Territories vary greatly in size. Among birds, for instance, the golden eagle defends a territory of 93 square kilometers, or about 37 square miles; the European robin, a territory of about 6 square kilometers; and the king penguin, a territory of only ½ square meter. Individuals of the same species often have territories of somewhat different sizes, depending partly upon the density of food and shelter in a given area, partly upon population pressure, and partly upon the aggressiveness of the individual.

Even though they may be invisible, the boundaries of almost all territories are clearly defined by the territory owner. With birds, for example, it is not the mere proximity of another bird of the same species that elicits aggression but his presence within a part of a particular area. The territory owner patrols his territory by flying from tree to tree. He will ignore a nearby rival outside his territory, but he will fly off to attack a more distant one that has crossed the border. Animals of other species are generally ignored unless they are prey or predators.

Ordinarily, the territory is established by the male, who waits for the female; but in some cases, such as that of the red-necked phalarope, the female establishes and defends the territory while the drab male defends the nest, incubates the eggs, and takes care of the young. Some birds, such as the American robin and the European swallow, tend to return to the same territory year after year, often with the same mate. Similarly, bull seals year after year occupy the same territory, on which they keep their harems, while immature bulls occupy bachelor areas on the edges of the breeding grounds.

Sometimes territory is the only or the primary bond between partners. Males and females of the South European green lizard, for example, defend their territory against members of the same sex only. Thus the most powerful male and the most powerful female are likely to occupy the most desirable natural dwelling place and so to breed together more frequently than with other partners. Storks and night herons, similarly, are attached to a nesting place rather than to a mate, and if the mate of the previous year does not return the following year, he or she is readily exchanged for a new incumbent.

Territorial behavior not only limits the effective breeding population but it tends to ensure that those who do breed are the best adapted. In this sense, territoriality is an extremely important evolutionary tool for shaping the course of development of the species.

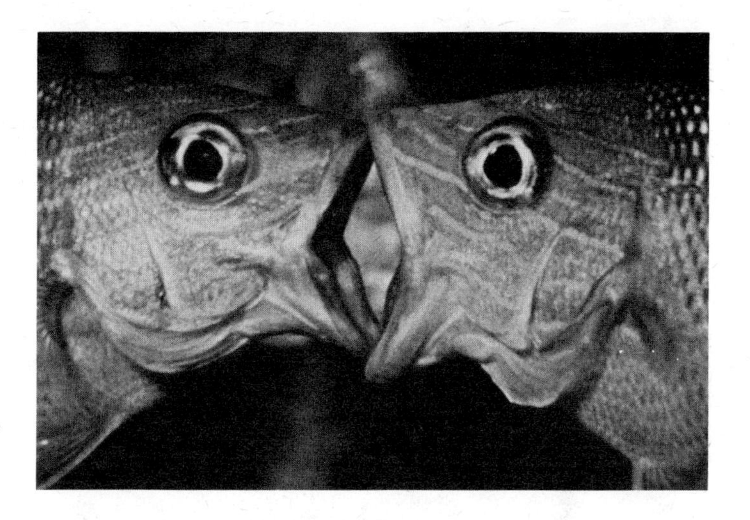

7–52

A conflict between two male grunts on their territorial borderline. Each grasps the other by the lips, and the loser is the one who lets go first. Although defeated, the loser is allowed to swim away unharmed.

AGGRESSION IN ANIMALS

Social conventions such as pecking orders and territoriality greatly reduce conflict among members of the same species. During the period when a pecking order is being established, frequent and sometimes bloody battles may ensue, but once rank is fixed in the group, a mere raising or lowering of the head is sufficient to acknowledge the dominance or submission of one hen in relation to another. Life then proceeds in harmony. If a flock is disrupted, the entire pecking order must be reestablished, and the subsequent disorganization results in more fighting, less eating, and less tending to the essential business, from the poultry dealer's point of view, of growth and egg laying.

Similarly, once an animal has taken possession of a territory, he is virtually undefeatable on it. Among territory owners, prancing, posturing, scent marking, and singing and other types of calls usually suffice to dispel intruders because of the great psychological disadvantage of the latter.

(a)

(b)

(c)

7–53

Among animals of the same species, fighting rarely results in death or even injury to either of the combatants. (a) *The combat dance of Pacific rattlesnakes.* (b) *Kangaroo rats attack one another with their hind legs. The battle ends when one gives up and trots off.* (c) *Two male ibex engaged in a territorial dispute for a courtship arena.*

If a male stickleback is placed in a test tube and moved into the territory of a rival male, he visibly wilts, his posture becoming less and less aggressive the further within the territory he is transported. Similarly, a male cichlid will dart toward a rival male within his territory, but as he chases the rival back into his own territory, he will begin to swim more and more slowly, his caudal fins seemingly working harder and harder, just as if he were making his way against a current which increases in strength the further he pushes into the other male's home ground. The fish know just where the boundaries are and, after chasing each other back and forth across them, will usually end up with each one trembling and victorious on his own side of the truce line.

Moreover, competition is often confined to particular parts of the day. Among the red grouse of Scotland, the males crow and threaten only very early in the morning when the weather is good. This ceremony may become so threatening that weaker members of the group leave the moor, often starving as a consequence or being killed by predators. Once the early-morning contest is over, however, the remaining birds flock amiably together and feed side by side for the rest of the day. The cock's crow at dawn, the dawn chorus of songbirds, the massed maneuvers of starlings at their roosts at the end of the day, and the night choirs of frogs are all examples of this phenomenon.

When fighting does occur among animals of the same species, it is often ritualized in such a way that both animals remain unharmed. For instance, male iguanas of the Galapagos Islands fight by pushing their heads against one another; the one which drops to its belly in submission is no longer attacked. Male cichlid fish of one species first display, presenting themselves head on and then side on, with their dorsal fins erected, and then beat water at each other with their tails. If this does not bring about a decision, each grasps the other by its thick strong lips; they then pull and push with great force until one lets go and, unharmed but defeated, swims away. Rattlesnakes, which could kill each other with a single bite, never bite when they fight but glide along side by side, each pushing its head against the head of the other, trying to push it to the ground in a form of Indian wrestling. Many antlered animals, such as stags of the fallow deer, which have long and vicious horns, follow an equally careful ceremony and attack only when they are facing each other, so that their horns are used only for dueling and not for goring.

These proscriptions against intraspecific killing "make sense" biologically. It is to the advantage of the species, in terms of its survival, not to waste its strength in needless bloodshed, and therefore genetic traits that channel aggressions and reduce conflict will be selected under evolutionary pressures. Man, whose instincts are apparently much less strong than those of other animals, is one of the few species that regularly kills its own kind.

SUMMARY

Social behavior involves interactions of individuals of the same species. Among the most complex societies are those of insects. These societies are matriarchies, centering around the care of the queen and the raising of the brood. The behavior of members of the society and even their physical characteristics, as evidenced by the very diverse members of termite society, are determined by chemical substances –pheromones—exchanged among members of the society.

Many animal social groups, especially among birds and mammals, are organized around the necessity of feeding the young and are temporary in nature, dissolving as the young approach maturity. Other animal social groups, such as those found among many birds, many of the hoofed mammals, and most primates, are more permanent in nature.

Two of the ways in which vertebrate societies are organized are social dominance and territoriality. Socially inferior animals—those low in the pecking order—reproduce less frequently than their social superiors and are often psychological castrates. They are also the first to starve if food is limited or to be driven out if shelter is limited.

Territories may be "real"—that is, they may be actual areas of land containing food and nesting material to support a mating pair and young—or they may be symbolic, such as a stamping ground. In either case, the only animals that breed are those which have territories. Animals without territories provide replacements for territory owners and a reserve for population expansion if the range or food supply of the population is increased.

Acts of aggression are restricted in animal societies by social dominance, territoriality, and other forms of social behavior. Intraspecific killing is found only rarely among animals other than man.

QUESTIONS

1. In what ways are human societies different from insect societies? How are they similar?

2. What behavioral conventions limit intraspecific aggression in man? Do any foster it?

GLOSSARY

BIOGEOCHEMICAL CYCLE: The circulation of inorganic substances throughout an ecosystem, encompassing its geological components, such as the atmosphere, soil, and water systems, and its biological components, that is, the producers, consumers, and decomposers of organic matter.

BIOMASS: The total weight of all organisms in a particular habitat or area.

BIOME (**by**-ohm): A worldwide complex of similar natural communities, characterized by distinctive vegetation and climate; for example, the grassland areas collectively form the grassland biome, and the tropical rain forests form the tropical rain forest biome.

CARNIVORE: A flesh-eating organism.

CHAPARRAL: A community of small trees and shrubby plants especially adapted to rainy winters and dry summers. Such communities usually border a desert and are widely distributed in southern California.

CLIMAX COMMUNITY: The final or stable community in a successional series, which is more or less in equilibrium with existing environmental conditions and is composed of a definite group of plant and animal species.

COMMUNITY: The organisms inhabiting a common environment and interacting with one another.

COMPETITION: The effect of a common demand by two or more organisms on a limited supply of food, water, light, minerals, etc.

DISTRACTION DISPLAY: A form of defensive behavior in animals with young which diverts the attention of a predator from the nesting site or litter.

ECOLOGICAL DOMINANT: A species which, by virtue of size, number, or habits, exerts a controlling influence on its environment and, as a result, determines what other kinds of organisms exist in that ecosystem.

ECOLOGICAL NICHE: The role of a particular species in the activities of the ecosystem of which it is a part.

ECOLOGICAL SUCCESSION: The process by which a community goes through a number of temporary developmental stages, each characterized by a different species composition, before reaching a stable condition (climax community).

ECOLOGY [Gk. *oikos*, house, + *logos*, a discussion]: The overall pattern of relationships of organisms to one another and to their environment in an ecosystem.

ECOSYSTEM: All organisms in a community plus the associated environmental factors with which they interact.

EPILIMNION [Gk. *epi*, upon, + *limnion*, small lake]: The layer of water that overlies the thermocline of a lake and that is subject to the action of wind; in contrast to hypolimnion.

EPIPHYTE [Gk. *epi*, upon, + *phyton*, plant]: A plant that grows upon another plant but is not parasitic upon it; a characteristic plant of tropical rain forests.

FOOD CHAIN, FOOD WEB: A chain of organisms existing in any natural community such that each link in the chain feeds on the one below and is eaten by the one above; there are seldom more than five or six links in a chain, with plants, bacteria, and other scavenging forms on the bottom and the largest carnivores on the top.

GAUSE'S PRINCIPLE [after G. F. Gause, German geneticist]: The statement that any two species with the same ecological requirements cannot coexist in the same locality and that occupation of that space will go to the species which is more efficient in utilizing the available resources.

GROUND WATER: Water in the zone of saturation where all openings in rocks and soil are filled, the upper surface of which forms the water table.

HERBIVORE: A plant-eating organism.

HOST: An organism on or in which a parasite lives.

HYDROSPHERE: The water on the surface of the earth.

HYPOLIMNION [Gk. *hypo,* under, + *limnion,* small lake]: The layer of stagnant water below the thermocline of a lake which maintains a uniform temperature except during periods of overturn; in contrast to epilimnion.

LATERITE [L. *later,* brick, tile]: A red iron-containing soil found in the tropics.

LEACHING: The removal of minerals and other elements from soil by the downward movement of water.

LITHOSPHERE: The solid part of the earth, including crust, mantle, and core.

MIMICRY: The resemblance in some conspicuous aspect of physical appearance of a species to one or more other species. In Batesian mimicry, the mimic is a harmless species which resembles another species that is protected from predators by an unpleasant taste, having the ability to sting, or the like. Müllerian mimicry is a physical resemblance among several species, all of whose members possess effective and similar defensive weapons; examples include bees, wasps, and stinging flies and some groups of unpalatable butterflies.

MYCORRHIZA (my-ka-**rise**-a) [Gk. *mykēs,* fungus, + *rhiza,* root]: A symbiotic association of certain fungi and the roots of plants.

PARASITE [Gk. *para,* beside, + *sitos,* food]: An organism which lives on or in an organism of a different species and derives its nutrients from that organism.

PECKING ORDER: The dominance hierarchy in poultry flocks in which each individual according to its rank can peck a number of subordinate fowl with impunity and must submit to pecking by a number of superiors without retaliation.

PHEROMONE: A substance secreted by an animal that influences the behavior or morphological development or both of other animals of the same species.

POPULATION: In ecology, any group of individuals of one species.

PRIMARY SUCCESSION: Ecological succession that develops in a previously unvegetated region.

SAVANNA: A tropical grassland; a transitional area between the tropical rain forest and the desert, with annual rainfall less than that of the rain forest but greater than that of the desert and with seasonal drought. It is characterized by sparse vegetation consisting of grasses, mostly perennial herbs, and scattered growths of trees.

SECONDARY SUCCESSION: Ecological succession that develops in a region previously occupied by vegetation.

SOCIAL DOMINANCE: Hierarchical arrangement of animals in a group of the same species.

SYMBIOSIS (sim-by-**o**-sis) [Gk. *syn,* with, + *bios,* life]: The living together in close association of individuals of two or more species in such a way that each is benefited.

TAIGA: A northern region characterized by persistent winter cover, with precipitation occurring chiefly in summer, with vegetation consisting of coniferous forests and perennial herbs, and with many lakes, bogs, and marshes.

TERRITORY: An area or space occupied by an individual or a group and defended (usually successfully) against trespassers; the site of breeding, nesting, food gathering, or any combination of these activities. Territoriality is common among vertebrates.

THERMOCLINE: The layer of water in a lake or other body of water between the epilimnion and the hypolimnion which is characterized by a gradually decreasing temperature with increasing depth, at the rate of 1°C per meter of depth.

TROPHIC LEVEL [Gk. *trophos,* feeder]: A step in the movement of biomass or energy through an ecosystem.

TUNDRA: A treeless subarctic region with a ground layer of mosses and lichens, various grasses, and a few perennials and with permafrost (a layer of permanently frozen soil) a few inches below the surface.

WATER TABLE: The upper limit of permanently saturated soil; the top layer of the ground water.

SUGGESTIONS FOR FURTHER READING

BATES, MARSTON: *The Forest and the Sea,* Vintage Books, Random House, Inc., New York, 1965.*

A compelling and wide-ranging study of the economy of nature and the ecology of man, beautifully written.

BERRILL, N. J.: *The Living Tide,* Fawcett Publications, Inc., Greenwich, Conn., 1964.*

An informal account, by a marine biologist, of the extraordinary variety of living things to be observed along the coasts of our continent. Dr. Berrill's writing is delightful.

"The Biosphere," *Scientific American,* September, 1970.*

This issue of the Scientific American *is devoted entirely to energy flow and biogeochemical cycles and their relationship to current human problems of food consumption and pollution.*

CARSON, RACHEL: *The Sea around Us,* New American Library, Inc., New York, 1954.*

Miss Carson was a rare combination of scientist and poet. This deservedly popular book traces the history of the formation of the oceans, describes their role in the origin of life and in its evolution, and discusses their present-day importance to human life.

FARB, PETER: *Face of North America,* Harper & Row, Publishers, Inc., New York, 1963.

An interesting and comprehensive account for a general audience of the natural history of the continent—the interplay of geological change, climate, and plant and animal life.

KORMONDY, E. J.: *Concepts of Ecology,* Prentice-Hall, Inc., Englewood Cliffs, N.J., 1969.

A good short modern text.

KRUTCH, J. W.: *Desert Year,* The Viking Press, Inc., New York, 1960.*

A description by one of the best contemporary American nature writers of the animal and plant life of the American desert.

LAYCOCK, GEORGE: *The Alien Animals: The Story of Imported Wildlife,* Natural History Press, Garden City, N.Y., 1966.

The sometimes funny, sometimes tragic, often bizarre consequences of moving plants and animals into new ecological situations.

ODUM, EUGENE P., and HOWARD T. ODUM: *Fundamentals of Ecology,* 3d ed., W. B. Saunders Company, Philadelphia, 1971.

A reliable general textbook of ecology.

RILEY, DENIS, and ANTHONY YOUNG: *World Vegetation,* Cambridge University Press, Cambridge, England, 1966.*

This useful booklet not only desribes the major terrestrial biomes of the world but also explains why these particular kinds of plants are found under the prevailing environmental conditions.

STORER, JOHN H.: *The Web of Life,* New American Library, Inc., New York, 1966.*

One of the first books ever written on ecology for the layman. In its simple presentation of the interdependence of living things, it remains a classic.

* Available in paperback.

SECTION 8

Population Biology

8–1
As Darwin observed, variations exist among the members of any given species.

CHAPTER 8–1

The Modern Theory of Evolution: Variation in Natural Populations

Although it is now more than a hundred years since the first publication of *The Origin of Species,* Darwin's original concept of how evolution comes about still provides the basic framework for our understanding of the process. His concept rests on four premises:

1. Like begets like—in other words, there is stability in the process of reproduction.
2. In any given population, there are variations among individual organisms, and some of these variations are inheritable.
3. In every species, the number of individuals that survive to reproductive age is very small compared with the number produced.
4. Which individuals will survive and reproduce and which will not is determined, in general, not by chance but by reason of these variations. Those with favorable variations will pass their characteristics on to the next and future generations in greater numbers.

Modern evolutionary theory conceives of evolution as Darwin conceived of it with the important addition of an understanding of the mechanisms of inheritance. Twentieth-century genetics answers two questions that Darwin was never able to resolve: (1) why genetic traits are not "blended out" but can disappear and reappear (like whiteness in pea flowers) and (2) how the variations appear on which natural selection acts. This combination of evolutionary theory and genetics is known as the synthetic theory of evolution. (Here "synthetic" does not mean artificial, which is the connotation it has for us in these days of man-made fabrics and artificial colors and flavors, but has its original meaning of the putting together of two or more different elements.)

The branch of genetics that emerged from this synthesis of Darwinian evolution and Mendelian principles is known as *population genetics.* A population, for the geneticist, is an interbreeding group of organisms. For instance, all the fish of one particular species in a pond are a population, and so are all the fruit flies in one bottle. The population is defined and united by its gene pool, which is simply the sum of all the alleles of all the genes of all the individuals in a population.

In this view, the individual is only a temporary vessel, holding a small portion of the gene pool for a short time, testing a particular combination. If the individual has a favorable combination of genes, his genes are more likely to be returned to the pool through his progeny and so will be present in an increased proportion in the next generation. If the combination is not favorable, his contribution to the gene pool will be reduced or perhaps eliminated. Thus the final criterion—the measurement—of an individual's fitness is the number of surviving offspring he produces, which, of course, determines his contribution to the gene pool of future generations. Evolution is the end product of such accumulated changes in the gene pool.

8–2
Even in populations in which the individuals appear almost identical, such as the members of this group of gannets, we know that variations exist because individuals have no difficulty in identifying their own mates or offspring. To gannets, all people probably look alike.

LIKE BEGETS LIKE

Like begets like, we know now, because of the remarkable precision with which the DNA is transmitted from cell to cell, so that the DNA in the sperm or eggs of any individual is a true copy of the DNA that that individual received from its father or mother. In fact, this mechanism of copying serves not only to link man to his immediate ancestors but to link all living things to one another. For instance, human cells have some of the enzymes found in protists and even in prokaryotes, which indicates that certain sequences of DNA have been copied over and over for billions of years.

This constancy is, of course, essential to the survival of the individual organisms of which the population is composed. By and large, an organism, the product of tens of thousands of years of evolution, is well in tune with its environment, and unless the environment changes, wide variations in structure or function are almost always doomed to failure.

However, if evolution is to occur, variations among individuals must be present. Such variations make it possible for species to change under changing circumstances and provide the raw material on which natural selection acts. Population genetics thus becomes the study of this perpetual tug-of-war between the forces of constancy and the forces of change.

VARIABILITY IN NATURAL POPULATIONS

Darwin's awareness of the existence of variations among individuals in a population was based largely on his observations as a naturalist, both in England and during his voyage on the *Beagle.* He was also well aware of the extent to which variations among individuals could be exploited by what he called "artificial selection" to produce different and often very exotic breeds of domestic animals. (Pigeon fanciers, for example, were especially common at that time, and Darwin remarked upon the extraordinary strains of birds being produced by their efforts.) He was not an experimenter, merely a constant and thoughtful observer. Twentieth-century population biologists have amply confirmed by experimental means Darwin's concepts of the extent of heritable variations within a natural population.

Bristle Number in Drosophila

In one group of studies, for example, the extent of latent variability in a natural population was demonstrated in the laboratory by experiments in which a population of *Drosophila melanogaster* was selected for an easily ascertainable hereditary trait: the number of bristles on the ventral surface of the fourth and fifth abdominal segments. In the starting stock, the average number of bristles was 36. Two selection groups were run, one for increase of bristles and one for decrease. In every generation, individuals with the fewest bristles were selected and crossbred, and so were individuals with the highest number of bristles. Selection for low bristle number resulted in a drop after 30 generations to an average of 30 bristles. In the high-bristle-number line, progress was at first rapid and steady. In 21 generations, bristle number rose steadily from 36 to an average of 56. No new genetic material had been introduced. It was apparent that within the single population, a very wide range of possible variation existed and that this had become manifest by selection pressures. Subsequent experiments in *Drosophila* and other organisms have shown that the choice of bristle number was not a mere fortunate accident. No matter what single characteristic is selected for breeding experiments, all reveal a comparable range of natural variability.

There is a second part to the bristle-number story. The low-bristle-number line soon died out owing to sterility. Presumably, changes in factors affecting fertility had also taken place during selection. When sterility became severe in the high-bristle line, a mass culture was started; members of the high-bristle line were permitted to interbreed without selection. The average number of bristles fell sharply, and in five generations went down to 40. Thereafter, as this line continued to breed without selection, the bristle number fluctuated up and down, usually between 40 and 45, which still was higher than the original 36. At generation 24, selection for high bristle number was begun again for a portion of this line. The previous high bristle number was regained, and this time there was no loss in

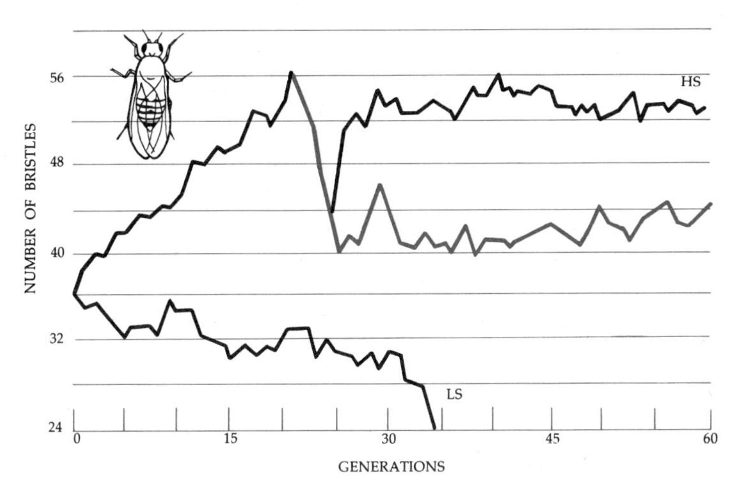

8–3

The results of an experiment with Drosophila melanogaster, *demonstrating the extent of latent variability in a natural population. From a single parental stock, one group was selected for an increase in the number of bristles on the ventral surface (HS, high selection line) and one for a decrease in the bristle number (LS, low selection line). As you can see, the HS line rapidly reached a peak of 56, but then the stock began to become sterile. Selection was abandoned at generation 21 and begun again at generation 24. This time, the previous high bristle number was regained and there was no apparent loss in reproductive capacity. Note that after generation 24 the stock interbreeding without selection was also continued, as indicated by the line of color. After 60 generations it had 45 bristles. The LS line died out owing to sterility.*

reproductive capacity. Apparently, the genotype had rearranged and reintegrated itself so that the genes controlling bristle number were present in more favorable combinations.

Mapping studies have shown that bristle number is controlled by a large number of genes, at least one on every chromosome and sometimes several at different sites on the same chromosome. Selection for bristle type, therefore, although the trait itself would appear to be neither useful nor harmful, in some way disrupted the entire genotype. Livestock breeders are well aware of this consequence of artificial selection. Loss of fertility is a major problem in virtually all circumstances in which animals have been purposely inbred for particular traits. This result emphasizes the fact that in natural selection it is the entire phenotype that is selected rather than certain isolated traits, as is often the case in artificial selection.

Variability in Gene Products

Molecular biology has put forward another method for studying latent variability. Genes, as we saw in Section 3, dictate proteins. Experimenters from the University of Chicago ground up fruit flies and extracted proteins from them. From these proteins they selected eighteen different enzymes which they were able to isolate on the basis of their enzymatic activity. Then they analyzed each group separately to see if it was composed of a single protein or different proteins. (The method they used was electrophoresis—the same method used by Pauling to separate the proteins of normal and sickle-cell hemoglobin.) Of the eighteen enzyme groups studied in this way, nine were composed of identical proteins; in other words, the genes that had produced these proteins were the same throughout the entire group of fruitflies studied. Nine of the groups, however, contained structurally different enzymes. Therefore without any direct analysis of the genes themselves, the investigators were able to conclude that among the fruitflies studied, there were two or more alleles of the gene responsible for that enzyme. In fact, seven of the eighteen enzyme groups analyzed (39 percent of those studied) contained relatively common variants, often several different ones. In one group of genes, there were as many as six slightly different structural forms; that is, six alleles for that gene were shown to exist in the species as a whole. Each population tested was heterozygous for almost a third of the genes tested. Each individual, it was estimated, was probably heterozygous for about 12 percent of its genes.

Similar studies in human populations, using an accessible tissue such as blood or placenta, indicate that, in any given group, most genes are represented by at least two alleles and most individuals are heterozygous for at least 16 percent of their genes.

THE ORIGIN OF VARIATIONS

Mutation

Hereditary variations originate as mutations, that is, sudden changes in a genetic character. In terms of molecular genetics, a mutation can be defined as an alteration in the sequence of nucleotides in a DNA molecule which results in a change in the instructions for the assembly of a particular protein. The alteration may be very small; as we saw in Chapter 3–6, the change in a single base pair is responsible for the abnormal hemoglobin associated with sickle-cell anemia.

Mutations, as we noted previously, can be produced by exposure to x-rays, ultraviolet, and other agents. Most occur "spontaneously"—meaning simply that we do not know the reasons for them.

The Role of Recombination

By the process of meiosis, variations arising from mutations are reshuffled, worked into the gene pool, and so brought into new combinations with other genes, eventually giving rise to new phenotypes upon which natural selection operates. The effect of a particular allele will vary, depending both on the other allele and on the rest of the genotype.

In many organisms, sexual combinations can occur only when the organisms are of different mating strains, as in *Chlamydomonas*, or of different sexes, as in the higher animals. Even among those invertebrates which are hermaphrodites, such as earthworms, an individual seldom fertilizes its own eggs. Some plants, such as the holly, have male flowers on one tree and female on the other, ensur-

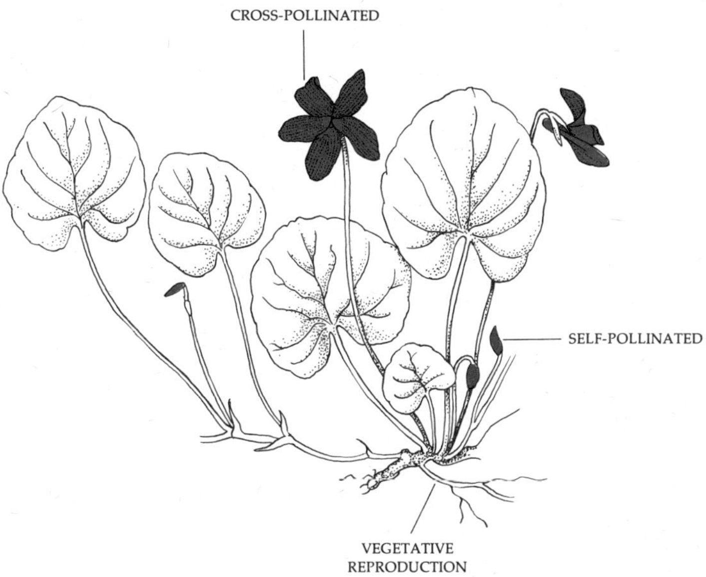

8–4
Plants such as the violet shown here often reproduce both sexually and asexually. The larger flowers are cross-pollinated by insects, and the windborne seeds are carried some distance from the parent plant. The smaller flowers, closer to the ground, are self-pollinated and never open. Seeds from these flowers drop close to the parent plant and produce plants that are genetically similar to the parent and so, presumably, apt to grow successfully near the parent. Creeping underground stems also eventually produce a new series of genetically identical plants right next to the parent (vegetative reproduction).

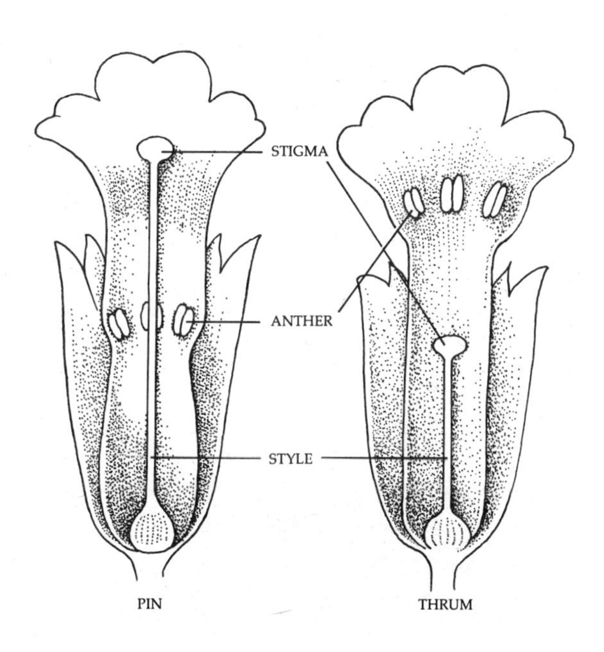

8–5
Cross sections of a pin type and a thrum type of a primrose. Notice that the anthers of the pin flower and the stigma of the thrum flower are both situated about halfway up the corolla tube and that the pin stigma is level with the thrum anthers. An insect foraging for nectar in these plants would collect pollen in different areas of his body, so that thrum pollen would be deposited on pin stigmas, and vice versa.

ing cross-fertilization. Others have genes for self-sterility. These are genes that prevent the pollen from fertilizing the ovule if it falls on the stigma of the same plant. Typically, such a gene has multiple alleles—s^1, s^2, s^3, and so on. A pollen grain carrying the allele s^1 cannot fertilize a plant with an s^1 allele, and one with an s^1/s^2 genotype cannot fertilize any plant with either of those alleles, and so forth.

In one population of about 500 evening primrose plants, 37 different self-sterility alleles were found, and it has been estimated that there are more than 200 alleles for self-sterility in red clover. As a consequence, the chances of a particular pollen grain fertilizing an ovule are great if it has a rare self-sterility allele (which will nearly always be accompanied by a rare genotype) and slight if it has a common self-sterility allele. Such a system strongly encourages variability in a population.

Pin and Thrum

In many species of primroses, cross-pollination is ensured by another mechanism. In these plants, two types of flowers occur in approximately unchanging proportions in every population. In one of these (known as pin), the style is long, raising the stigma high in the flower, and the stamens are short, so that the anthers are located low in the body of the flower. In the other form (thrum), the stigma is low and the anthers are high (see Figure 8–5). A pollinator that

8–6

By artificial selection, it is possible to produce "purebreds," such as this afghan hound on the left. However, the introduction of a new genotype—such as that of the less exotic dog on the right—results in the rapid introduction of variations into the lineage, a fact of which dog breeders are only too well aware. Similarly, although usually to a less striking degree, gene flow into natural populations is likely to diminish the effects of natural selection.

moves from primrose to primrose will receive pollen from one type on the part of its body most likely to come into contact with the stigma of the other type. In this way, cross-pollination is promoted.

In addition to the anatomical differences between the two forms, there are other physiological differences that promote cross-pollination of the two types. For instance, the cells of pin stigma react against pin pollen grains, inhibiting the formation of pollen tubes, but these cells promote the growth of tubes from thrum pollen.

The existence of these elaborate systems is an indication that genetic recombination and the patterns of variation it helps to maintain have a survival value for the species.

Gene Flow

Another source of variability in populations is gene flow. Gene flow is simply the introduction of new genes into a population. It is usually the result of the immigration into a local population of individuals which, although of the same species, have a slightly different gene pool.

Gene flow, as we shall see, by increasing variability, tends to retard the process of natural selection.

SUMMARY

Evolution is the product of natural selection acting on random heritable variations. Natural populations can be shown, by a variety of methods, to harbor a wide spectrum of latent variations.

Variations are introduced into the gene pool in the form of mutations. They are brought into new genetic combinations by meiosis and fertilization. Both plants and animals have developed mechanisms to prevent self-fertilization. These include mating strains, sexes, self-sterility alleles (in plants), and special anatomical adaptations, such as the pin and thrum types among species of primrose.

New genes are also introduced into local populations by the immigration of individuals of the same species but with slightly different genotypes. This process is known as gene flow.

Mutation, recombination, and gene flow act to promote variations among the individuals in a population.

QUESTION

Compare self-sterility genes among primroses with incest taboos among humans.

CHAPTER 8–2

Darwin and Mendel and Hardy and Weinberg

Darwin, as we saw in the preceding chapter, correctly perceived the existence of heritable variations in natural populations and their role in evolution. However, the theories of inheritance at that time were unable to explain how the small heritable variations, so important to evolutionary theory, were not blended in, like a drop of colored water in a beaker, but managed to persist, like a single blue marble in a bag of different colored ones. In fact, troubled by this very question, Darwin somewhat modified his thinking in later editions of *The Origin of Species* to move toward a more Lamarckian viewpoint, accepting Lamarck's premise that acquired characteristics may also be inherited. For similar reasons, some of Darwin's followers were prepared, at the beginning of the twentieth century, to place far too great an emphasis on the role of mutation in evolution. Their concept, which is sometimes called the "hopeful monster" theory, is that new strains of organisms appeared virtually overnight as a result of sudden changes in the hereditary material.

Mendel's great contribution was to demonstrate that traits—even when they seem to disappear—are actually preserved, "hidden" in the heterozygote. This Mendelian principle is one of the foundations of the synthetic theory of evolution. Its application to population biology is known as the *Hardy-Weinberg law,* after an English mathematician (Hardy) and a German physician (Weinberg).

According to the Hardy-Weinberg law, the proportions both of different alleles and therefore of different genotypes will remain the same in a sexually reproducing population, generation after generation, if (1) the population is large, (2) mating is random, (3) no mutations occur, (4) there is no gene flow, and (5) there is no natural selection. The equilibrium is expressed mathematically in terms of the relative frequencies of the two alleles: $p^2 + 2pq + q^2 = 1$, a formula which we shall explain in the pages to follow.

What the Hardy-Weinberg law says, in effect, is that sexual recombination does not *by itself* change the overall composition of the gene pool. A dominant allele, for instance, does not tend to increase in a population (a concept that was not understood even by many biologists in the early twentieth century) and a recessive one does not tend to decrease. In nature, however, conditions 2 to 5 are rarely met. Mating, as we shall see in the next chapter, is usually nonrandom, and mutations, immigration, and natural selection occur continuously. In short, we know that changes do occur in the frequencies of alleles in natural populations.

Why is it, then, that the Hardy-Weinberg law is of such central significance in modern evolutionary theory—as central, really, as Mendel's principles are to the theory of heredity? Its value to the population biologist is that it offers a model against which to compare actual gene frequencies in a natural population and thus provides a means of evaluating the effects of the forces acting to produce genetic change. In this chapter, we shall discuss the basis of this important principle and some of its applications.

THE HARDY–WEINBERG EQUILIBRIUM

The Hardy-Weinberg formula for equilibrium under random mating ($p^2 + 2pq + q^2 = 1$) is based both on Mendel and on the mathematics of probability. How do the laws of chance operate in random mating? Suppose, for instance, that in a large group of female mice, one-fourth of the females are brown. The chance of picking a brown female, at random, is therefore one in four, or $1/4$. In the group of male mice, one-fourth are also brown, and so the chance of picking a male is one in four, or $1/4$. But what are your chances of picking a brown female and a brown male, at random, at the same time? There are 4×4, or 16, possible combinations of male and female mice. (You can easily work this out in a diagram.) Therefore, your chances of picking a brown female and brown male together are one in sixteen. According to the laws of chance, multiplying the probability of one event (picking a brown female) by the probability of another event (picking a brown male) will give you the probability of the two events occurring simultaneously (picking a brown female and a brown male): $1/4 \times 1/4 = 1/16$.

Hardy-Weinberg in Mendelian Terms

To demonstrate the operation of the Hardy-Weinberg principle, let us begin by crossbreeding two populations, one of which is homozygous *AA* and one of which is homozygous *aa*. (These are the sort of populations with which Mendel began his studies.) All the first filial generation will be heterozygous (*Aa*). When members of this generation interbreed (*Aa* × *Aa*), the typical Mendelian ratio results; one-fourth of the population will be *AA*, one-fourth *Aa*, one-fourth *aA*, and one-fourth *aa*:

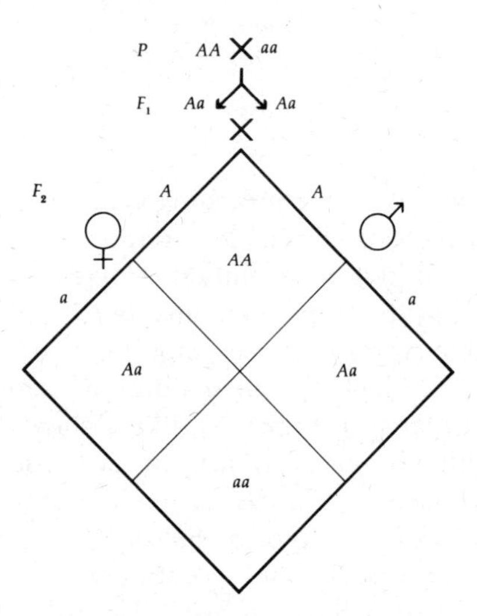

Remember that in these Punnett squares, like those in Section 3, each square represents a group of individuals rather than a single individual.

What will happen to these genotype proportions in the next generation? Let us consider first all the different kinds of crosses that will produce *AA* in the next generation. The first combination, female *AA* × male *AA*, will produce all *AA*s, and the chances of this occurring are $1/4 \times 1/4$, or $1/16$:

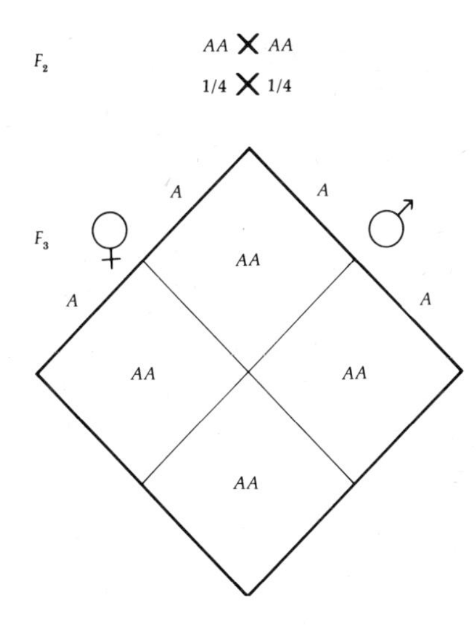

Female AA × male Aa (or aA) will produce some AAs. The chances of these matings occurring are ¼ × ½, or ⅛, but only one-half of the offspring will be AAs, so the chances of getting AA from these crossings are ¼ × ½ × ½, or $\frac{1}{16}$:

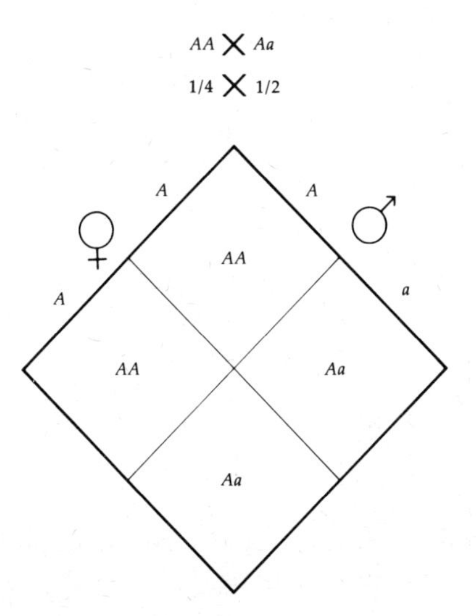

The third combination that will produce some AAs is female Aa × male AA. Here again, the chance of the offspring being AA is $\frac{1}{16}$:

Aa × AA

1/2 × 1/4

A A a A

AA Aa AA Aa

The fourth combination that will produce AAs is Aa × Aa. There is one chance in four that such a mating will occur and one chance in four that it will produce an AA, again $\frac{1}{16}$:

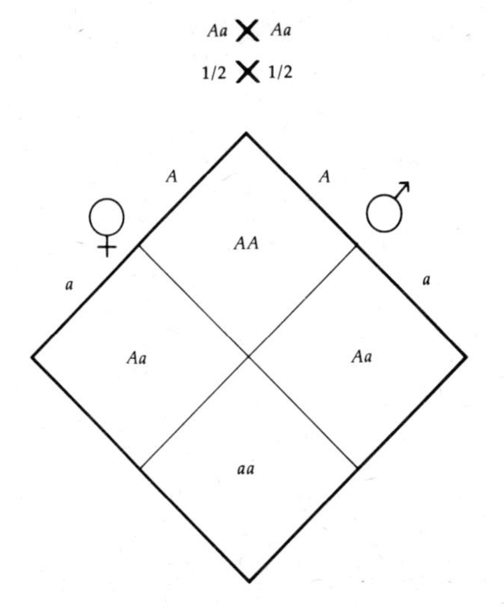

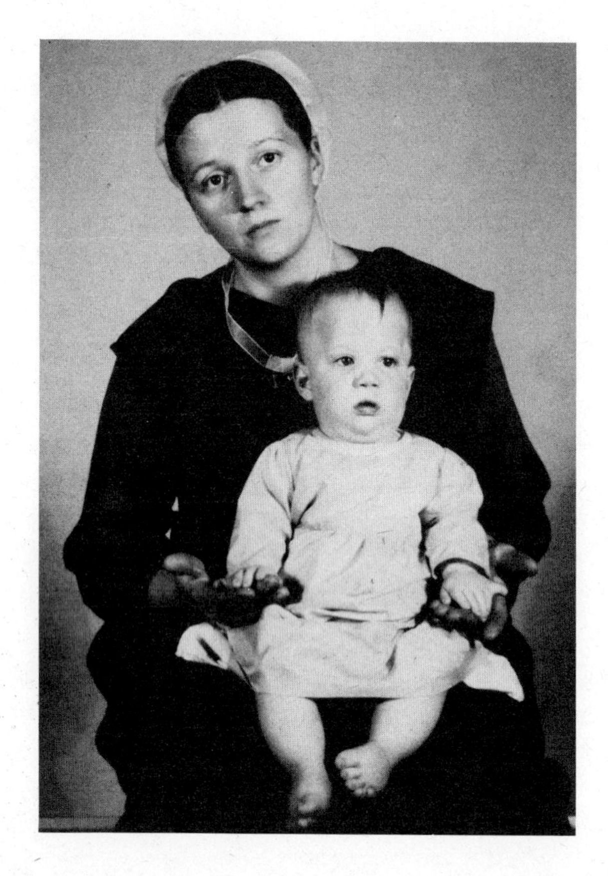

8–7

Among the Old Order Amish, a group founded by only three couples some 200 years ago, there is an unusually high frequency of a rare allele which, in its homozygous state, results in extra fingers and dwarfism. This Amish child is a six-fingered dwarf.

Now add these up. The chances of producing an *AA* in the third generation are $\frac{1}{16}$ plus $\frac{1}{16}$ plus $\frac{1}{16}$ plus $\frac{1}{16}$, or $\frac{4}{16}$ ($= \frac{1}{4}$), the same as in the F_2 generation. In other words, the probability is that the proportion of dominants in the third generation will be just the same as the proportion in the second generation. And, in fact, the proportions of *Aa*, *aA*, and *aa* will remain the same also, as you can readily demonstrate in this same way. In short, in a large population in which there is random mating, in the absence of forces that change the proportion of genes (conditions 3 to 5), the original frequencies of the two alleles will be retained from generation to generation.

Hardy-Weinberg in Algebraic Terms

For studies in population genetics, the Hardy-Weinberg principle is usually stated in algebraic terms, with the fractions that we used as examples before expressed as decimals. In the case of a gene for which there are two alleles in the gene pool, the frequency (represented by the symbol p) of one gene and the frequency (q) of the other must together equal the frequency of the whole, or 1; $p + q = 1$. Or, to put it another way, $1 - p = q$ and $1 - q = p$. This is equivalent to saying that if there are only two alleles, *A* and *a*, for instance, of a particular gene and if half (0.50) of the alleles in the gene pool are *A*, the other half (0.50) has to be *a*. Similarly, if 99 out of 100 (0.99) are *A*, 0.01 are *a*.

Then how do we find the relative proportions of individuals who are *AA*, *Aa*, and *aa*? As we established previously, these proportions can be calculated by multiplying the frequency of *A* (male) by that of *A* (female), *A* (male) times *a* (female), *a* (male) times *A* (female), and *a* (male) times *a* (female). We can express these multiplications in algebraic terms as $p^2 + 2pq + q^2$—which equals, as you probably know, $(p + q)^2$. Using this expression, we can see that if half (0.50) of the gene pool is *A* and half *a*, the proportion of *AA* will be 0.25, the proportion of *Aa* will be 0.50, and the proportion of *aa* will be 0.25, which is exactly what

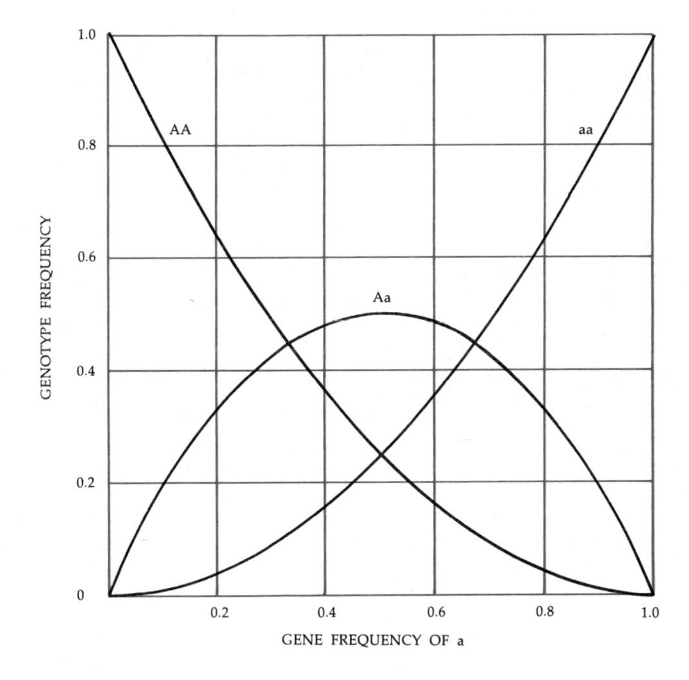

8–8

The relationship between the frequency of allele a *in the population and the frequency of the genotypes* AA, Aa, *and* aa. *Naturally, the more* AAs *there are, the lower the frequency of* a. *Because of the interrelationship of* AA, Aa, *and* aa, *a change in the frequency of either allele results in a corresponding and symmetrical change in the frequencies of the other allele and of the genotypes.*

Mendel said—although in slightly different terms—in the first place.

Now let us look at a gene pool in which p (the frequency of A) equals 0.8 and q (the frequency of a) equals 0.2. The genotypic frequencies are AA equals 0.64, Aa equals 0.32, and aa equals 0.04, providing that mating takes place at random:

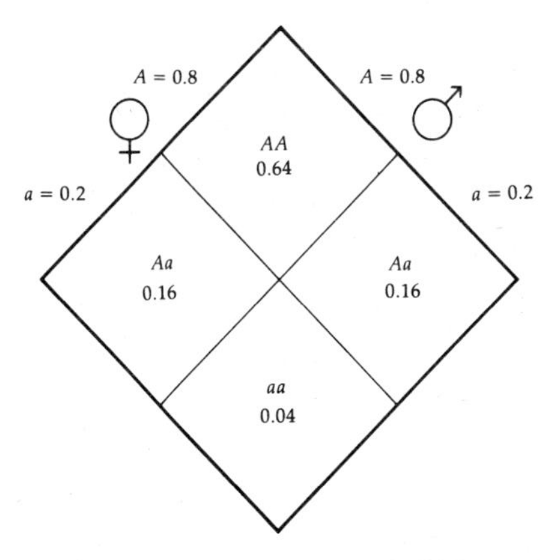

The resulting population will thus be made up of 64 percent *AA*s, 32 percent *Aa*s, and 4 percent *aa*s. And, as we have just proved, the percentages will tend to remain in this equilibrium.

APPLICATIONS OF THE HARDY-WEINBERG PRINCIPLE

By applying the Hardy-Weinberg formula, population biologists are able to study and compare the frequencies of individual alleles in populations. Differences in frequencies of particular alleles in comparable populations provide an indication, for example, of the adaptive value of a particular trait under varying conditions. Changes in allele frequency over periods of time provide a mean for monitoring the operation of natural selection.

8–9
Hereditary variations, the raw material for evolution, also may be the source of congenital diseases. This little girl, age 1½, has phenylketonuria (PKU), which is caused by a defect in an enzyme that breaks down the amino acid phenylalanine. Accumulated phenylalanine may cause brain damage; about one-fourth of the patients with PKU are in mental institutions. This child has just had a blood test, part of a treatment program that includes a diet low in phenylalanine.

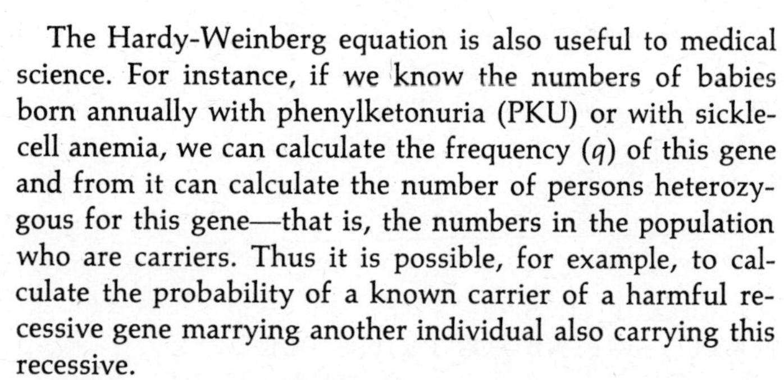

The Hardy-Weinberg equation is also useful to medical science. For instance, if we know the numbers of babies born annually with phenylketonuria (PKU) or with sickle-cell anemia, we can calculate the frequency (q) of this gene and from it can calculate the number of persons heterozygous for this gene—that is, the numbers in the population who are carriers. Thus it is possible, for example, to calculate the probability of a known carrier of a harmful recessive gene marrying another individual also carrying this recessive.

For example, the incidence of PKU in the general population is estimated to be 1 in 15,000. Thus q^2, the proportion of homozygous recessives, is 1/15,000, or 0.000067. The frequency of the allele for PKU is q, which equals $\sqrt{0.000067}$, or about 0.0082. If $q = 0.0082$, p equals $1.00 - q$, or 0.9918, $pq = 0.0081$, and $2pq =$ about 0.016. In other words, in the population under study, 16 in every 1,000 persons are heterozygous carriers for the PKU gene.

Multiple Alleles

So far we have talked about alleles only in terms of pairs, but actually there may be many mutant forms of the same gene. For instance, in the case of the gene for red eyes carried on the female chromosome in *Drosophila* (see Chapter 3–4), there are a variety of intermediate forms known by names such as apricot, buff, eosin, and honey, as well as the mutant that results in white eyes. More than 15 alleles have been recognized, and probably more could be detected if there were ways of distinguishing finer gradations of color.

The Hardy-Weinberg formula applies equally well when there are more than two alleles of the same gene, although the calculations then become more complex. The expression $(p + q + r)^2$, for example, represents the genotypic frequencies of a three-allele gene, with r representing the frequency in the gene pool of the third allele.

SOME IMPLICATIONS OF THE HARDY-WEINBERG EQUILIBRIUM

Protection of the Recessive Gene

Natural selection acts on the phenotype. As a consequence, a recessive gene is exposed to natural selection only in an individual homozygous for the gene. The importance of this fact becomes evident when we apply the Hardy-Weinberg formula:

Frequency of allele a *in gene pool*	*Genotype frequencies* AA	Aa	aa	*Percentage of* a *in heterozygotes*
0.9	0.01	0.18	0.81	10
0.1	0.81	0.18	0.01	90
0.01	0.9801	0.0198	0.0001	99

As you can see, the lower the frequency of allele *a*, the smaller the proportion of it exposed in the homozygotes becomes, and the removal of the allele by natural selection slows down accordingly. This result is of special interest to students of eugenics. For instance, in the case of genetic disorder manifest only in the homozygous recessive (like PKU) with a frequency of about 0.01 in the human population, *aa* individuals make up 0.0001 of the population (1 child for every 10,000 born). You can calculate that it would take 100 generations, roughly 2,500 years, of a program of sterilization of defective homozygote individuals to halve the gene frequency and reduce the number to 1 in 40,000. If a dominant gene is lethal, however, it will be removed from a population in one generation unless it is replaced by new mutations.

Importance of Heterozygote Superiority

If the allele is a rare one, as we can see in the table, it is present largely in the heterozygous state. This fact has an interesting consequence: if the heterozygote (*Aa*) is in some way superior to either homozygote (*AA* or *aa*), the proportion of the recessive allele—even though it is harmful in the homozygous state—may increase in the population.

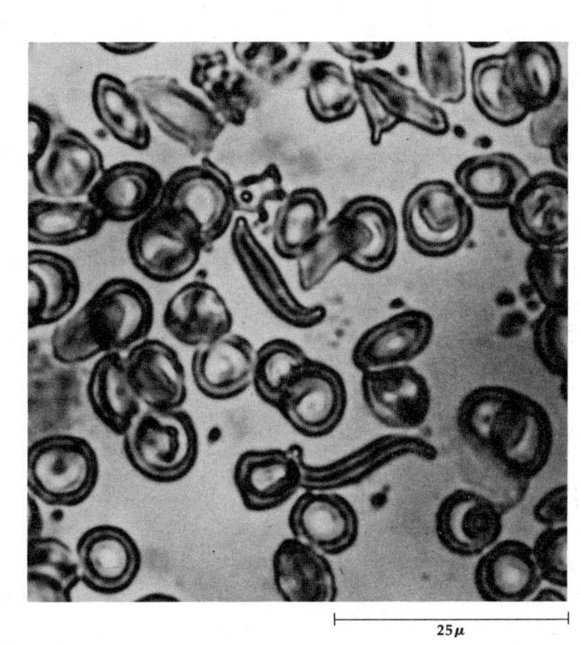

8–10

Photomicrograph showing normal red blood cells and sickle cells.

One of the most dramatic instances of this phenomenon, known as heterozygote superiority, is found in association with sickle-cell anemia. Individuals homozygous for sickling almost never live to maturity. Therefore, almost every time one sickling gene encounters another, two sickling genes are removed from the population. At one time, it was thought that the sickling gene was maintained in the population by a steady influx of new mutations. Yet in some African tribes, heterozygotes for sickling number as high as 45 percent, and to replace the loss of sickle genes by mutations alone would require a rate of about 1 mutant per 100 genes. This is about five thousand times greater than any other known mutation rate in man.

In the search for an alternative explanation, it was discovered that the heterozygote is maintained because it confers a selective advantage. In many African tribes, malaria is one of the leading causes of illness and death, especially

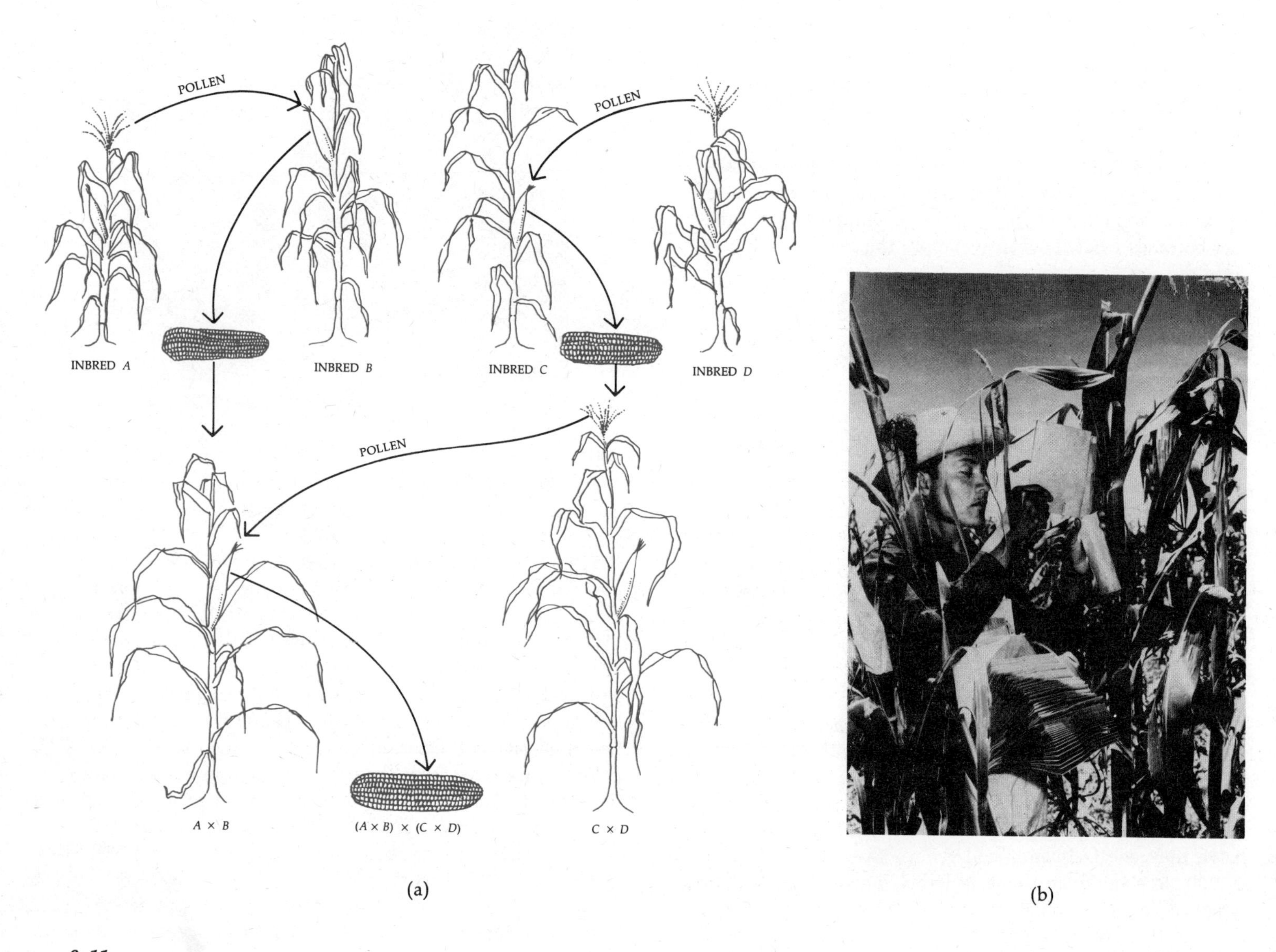

8–11

(a) *Hybrid corn is derived by first crossing inbred strains* A *with* B *and* C *with* D *and then crossing the resulting single-cross plants to produce double-cross seed for planting. The increased size and hardiness of the hybrid are probably due to its increased heterozygosity.* (b) *Collecting pollen for hybridization.*

among young children. Studies of the incidence of malaria among young children show that susceptibility to malaria is significantly lower in individuals heterozygous for sickling. Moreover, women who carry the sickling gene are more fertile.

Among blacks in the United States, only about 9 percent are heterozygous for the sickling allele. Since no more than half of this loss of the sickling allele can be explained by the black-white admixture in America, the conclusion is that once selection pressure for the heterozygote is relaxed, the mutant will tend to be eliminated slowly from the population.

Other genes that are harmful in the homozygous state appear to be maintained in the population by heterozygote superiority. For example, for reasons that are not known, women who are carriers of hemophilia (heterozygotes) have a fertility rate about 20 percent greater than that of the rest of the human female population.

Outbreeding

Heterozygote superiority is apparently also responsible for the phenomenon known as *hybrid vigor*, a striking example of which is found in hybrid corn. The development of hybrid corn caused a revolutionary improvement in the corn crop of the United States in the 1930s, because of the increased size and hardiness of the plants, derived by crossing two different varieties to produce the seeds for each planting. Hybrid vigor is apparently the result of the fact that the plants are heterozygous at far more loci than most natural varieties. In addition to possible benefits from heterozygote superiority, such hybrids are also less likely to be homozygous for deleterious recessives.

Most human cultures encourage outbreeding, at least to some extent. Many pretechnological societies, for example, demand that a young man choose a wife from another village rather than from his own community, and virtually all cultures have strong prohibitions against incest. Such prohibitions are particularly interesting when you consider that intermarriages of brother and sister or father and daughter would tend to keep property or power within the family and so could be socially and economically advantageous. Followers of Freud would maintain that incest is forbidden by strong psychological taboos, as reflected in the Oedipus myth. More pragmatic biologists hold that the prohibition stems from the observed ill effects of incestuous matings and has been merely reinforced by cultural restraints, which have now become deeply rooted in our subconscious.

THE IMPORTANCE OF POPULATION SIZE

As we stated previously, the Hardy-Weinberg law holds true only if the population is large. This qualification is necessary because the Hardy-Weinberg equilibrium depends on the laws of probability. These laws—the laws of chance—apply equally as well to flipping coins, rolling dice, betting at roulette, and taking a poll to see who is going to win an election. In flipping coins, it is possible for heads to show up five times in a row, but on the average, heads will show up half the time and tails half the time; the more times the coin is flipped, the more closely the expected frequencies of half (0.50) and half (0.50) are approached. Similarly, in a presidential poll, it is possible that the first five people interviewed will be Republicans, and it is not until a sufficiently large sample of the voting population is queried that accurate predictions become possible. In a small population, as in a small sample, chance plays a large role.

Genetic Drift

Consider, for example, an allele that is present in 2 percent of the individuals ($q = 0.02$). In a population of a million, 20,000 representatives of this particular allele would be present in the gene pool. But in a population of 50, only one individual would carry this gene. If this individual should fail to mate or should be destroyed by chance before leaving offspring, this allele would be completely lost. Similarly, if 10 of the 50 without this allele were lost, the frequency would jump from 1 in 50 to 1 in 40. This phenomenon is known as *genetic drift*.

The Founder Principle

A small population that branches off from a larger one may or may not be representative of the larger population from which it was derived. Some rare alleles may be overrepresented (like the Republicans in the poll) or may be lost completely. As a consequence, even when and if the small population increases in size, it will have a different genetic composition—a different gene pool—from that of the parent group. This phenomenon is known as the *founder principle.* In a small population, there is likely to be less variation in the population, because of the loss of rare alleles either by genetic drift or by sampling error, and hence a decrease in heterozygosity. Thus, even though matings may be at random, the mates will be more closely related to each other—more similar genetically—with a consequent increase in homozygosity.

An example of the founder principle is found in the Old Order Amish of Lancaster, Pennsylvania. Among these people, there is an unprecedented frequency of a gene which, in the homozygous state, causes a combination of dwarfism and polydactylism (extra fingers). Since the group was founded in the early 1770s, some 61 cases of this rare congenital deformity have been reported, about as many as in all the rest of the world's population. Approximately 13 percent of the persons in the group, which numbers some 17,000, are estimated to carry this rare mutant gene.

The entire colony, which has kept in virtual isolation from the rest of the world, is descended from three couples. By chance, one of the six must have been a carrier of this gene. Although inbreeding does not cause disease or deformity, it may lead to its manifestation if such alleles are already present.

SUMMARY

The Hardy-Weinberg law is a basic principle of population genetics. It states that: The proportions of different alleles in a population will remain the same if (1) the population is large, (2) mating is random, (3) no mutations occur, (4) there is no gene flow, and (5) there is no natural selection. The mathematical expression of the Hardy-Weinberg equilibrium is $(p + q)^2$, which represents the proportions of genotypes in the population when p equals the frequency of one allele and q equals the frequency of an alternate allele. Thus the homozygotes for one allele are represented by p^2, the homozygotes for the second allele by q^2, and the heterozygotes by $2pq$.

Analysis of the Hardy-Weinberg equation demonstrates that in the case of a rare allele, the allele is present largely in the heterozygous state. If the allele is recessive—that is, if its effects are not manifest in the phenotype—it is therefore protected against selection pressures. Selection for the heterozygote tends to increase the proportion of a rare allele in the population even though the allele is harmful in the homozygous state.

The Hardy-Weinberg equilibrium operates only if the population is large enough. In a small population, certain alleles tend to get lost or are overrepresented by chance; this phenomenon is known as genetic drift. A small population which has branched off from a larger population may not be a representative sample of the larger population; this phenomenon is known as the founder principle.

QUESTIONS

1. In terms of molecular genetics, what is a possible explanation for the many different eye colors in *Drosophila?*

2. Short, stubby fingers (brachydactylism) are the phenotypic expressions of a dominant gene. Since the gene for brachydactylism is dominant, why doesn't everybody have short, stubby fingers? Explain your answer.

3. What is the difference between gene flow and genetic drift? How do they each affect the Hardy-Weinberg equilibrium?

CHAPTER 8–3

Adaptation: The Direction of Evolution

In the last two chapters, we discussed the extent of heritable variations among individuals in populations and the ways in which these variations are introduced into the populations and maintained there, thus providing the raw material for natural selection. As we have seen, the forms that the variations take are entirely accidental. They become significant only as they are acted upon by a large group of environmental factors which, in combination, result in what Darwin called natural selection. Selection in no way causes the individual variations. However, by continuously eliminating particular genotypes from a population, it channels variations already in the population. Selection thus results in the nonrandom reproduction of genotypes.

It is important to remember that when we talk about selection, we are describing events that have already taken place. Only after we find that the proportion of a certain allele or a certain genetic trait is higher in a particular generation than it was in a previous generation can we say that selection has occurred.

The result of selection is the adjustment of populations to their environment, or *adaptation*. Adaptation results in the development of differences between groups of organisms. Evolution, as we shall see in the next chapter, is the consequence of adaptation.

NATURAL SELECTION

Selection is always in operation. Darwin's original concept of the process of evolution was based upon the observation that comparatively few organisms in most populations survive and reproduce. However, selection also continues to take place even in the most rapidly expanding populations. Even, for example, if there is a complete lack of competition for food or some other vital resource, some individuals will be more fertile than others and hence leave more representatives of their own genotype.

Three general types of selection operate within populations: stabilizing, disruptive, and directional. *Stabilizing selection*, a process which goes on at all times in all populations, is the continual elimination of extreme individuals. Most mutant forms are probably immediately weeded out in this way, many of them in the zygote or the embryo. The second type of selection, *disruptive selection*, is said to have occurred when two extreme types in a population increase at the expense of intermediate forms. (The experiment on bristle number in fruit flies is an example of artificial disruptive selection.) The third type, which is the one we shall be most concerned with, is *directional selection*. Directional selection results in the gradual replacement of

(a)

(b)

(c)

(d)

8–12
Adaptations. (a) *Moles, which seldom venture above ground, have small, almost sightless eyes. For sensing their prey—consisting largely of worms and other underground invertebrates—they rely almost entirely on their fleshy noses. Notice also the large shovel-like forepaws.* (b) *In a behavioral adaptation, the bittern conceals itself from potential enemies by standing absolutely still among the reeds which are its usual habitat, and pointing its large slender bill directly upward.* (c) *Tarsiers, primitive primates which live entirely in trees, have hands and feet with enlarged skin pads, specialized for grasping. Their huge eyes are related to their nocturnal existence.* (d) *Lamarck believed that giraffes' necks grew longer generation after generation as a result of their reaching for food in the higher branches. According to modern evolutionary theory, giraffes genetically endowed with slightly longer necks will have a slight advantage over their fellows which will be reflected in their leaving an increased number of offspring.*

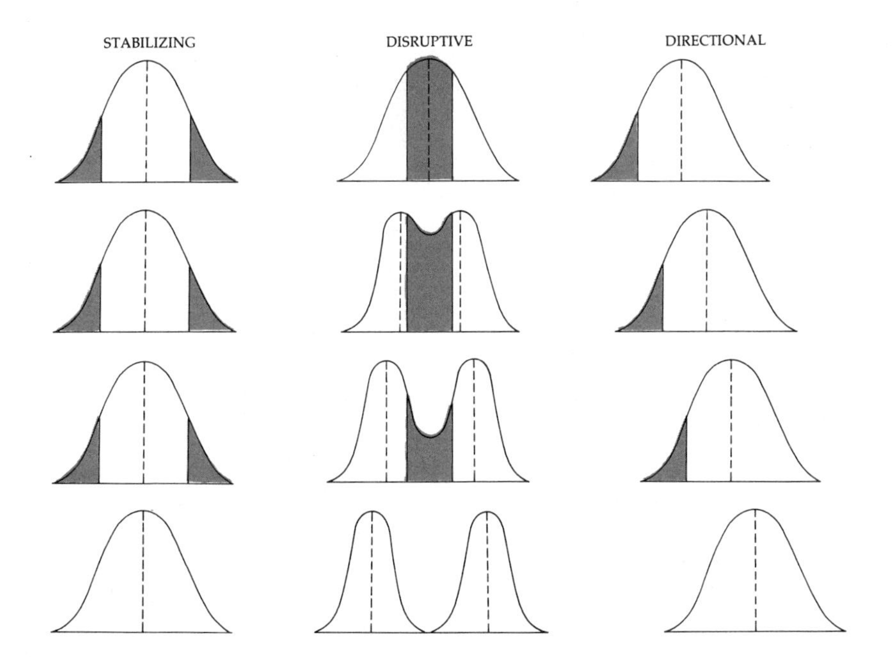

8–13
The three different types of natural selection. Stabilizing selection involves the elimination of extremes. In disruptive selection, intermediate forms are eliminated, producing two divergent populations. Directional selection, which is the gradual elimination of one phenotype in favor of another, produces adaptive change.

one allele by another in the gene pool. These replacements manifest themselves by changes in the anatomy, physiology, and behavior of the individual phenotypes on which the selection process operates.

THE FORCES OF NATURAL SELECTION

As we have noted in previous chapters, the face of our planet and the conditions of life upon it have undergone constant change. Some 3 billion years ago, oxygen began to accumulate in the atmosphere, rendering many forms of life obsolete and opening the way for the evolution of new types of organisms. Similarly, the transition from a warm, moist climate to an arid one some 250 million years ago favored the selection of plants which did not require free water for fertilization and in which the dormant embryos were protected within seeds. The glaciations of the recent Ice Age led to the extinction of many of the large mammals and strongly influenced the evolutionary course of most of the organisms, including man, that have survived.

The interactions of different types of organisms have also affected the course of their evolution. At the beginning of Section 5, we traced the coevolution of flowers and pollinators. In Section 7, we saw some of the results of predator-prey adaptations—camouflage, mimicry, and other plant and animal defenses. These examples of adaptive change seem to confirm, at least to some extent, the phrase "survival of the fittest," often used in describing the Darwinian

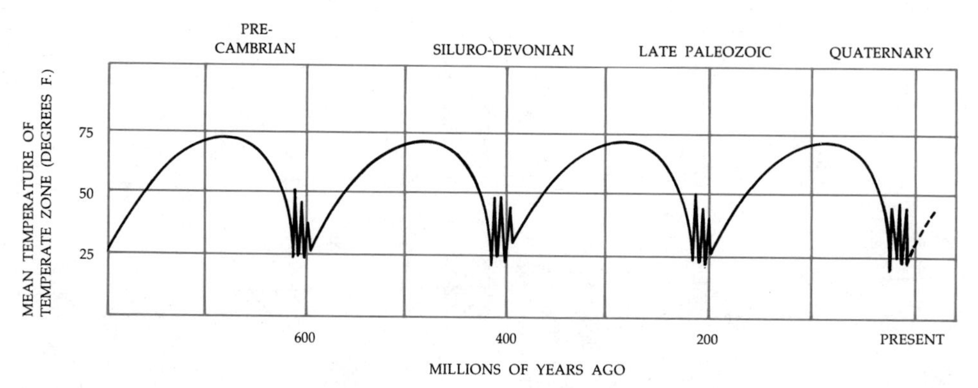

8–14

Temperature cycles deduced from the fossil record reveal the climatic patterns of the past. The mean temperature of the earth is normally much higher than it is today. The long warm cycles are interrupted by Ice Ages (jagged peaks) at intervals of approximately 200 *million years. The temperature changes in the immediate geological future should follow the pattern of the broken line. As you can see, we are now in the process of recovering from a period of extreme cold which, climatologists predict, may be repeated in another* 10,000 *to* 15,000 *years. These changing climatic patterns and, in particular, the extensive glaciations of the Ice Ages, have been strong selective forces.*

theory. In the early twentieth century, the doctrine of survival of the fittest in natural populations was used by some businessmen to defend ruthless competitive tactics in industry on the grounds that they were merely following a "law" of nature. This business philosophy was referred to by some as social Darwinism.

However, in actual fact, very little in the process of adaptive change fits the concept of "nature red in tooth and claw." One species of duckweed floating gently above another and casting a shadow is an apt model of the struggle for survival.

Differential Fertility

One of the strongest forces in directional selection is relative fertility. The factor or factors that make a particular protozoan, plant, or animal able to produce more surviving offspring than another of the same species undoubtedly vary from individual to individual but are clearly potent selective forces in natural populations.

Nonrandom Mating

Another strong selective pressure in nonrandom mating. As Hardy and Weinberg pointed out, alleles tend to remain in equilibrium in the population *if* mating occurs at random. However, nonrandom mating is the rule rather than the exception in most natural populations. The bright colors of flowers, the striking markings of many birds and fishes, and complex patterns of mating behavior all serve as inducement to nonrandom matings. Territoriality and dominance hierarchies, by excluding certain individuals from the breeding population, also ensure that some individuals make more contributions to the gene pool than others.

Among most insects and vertebrates, males do the courting and females do the choosing; therefore, the selection process of nonrandom mating affects males more than females. Females are usually quick to reject suitors who deviate in any way from the "ideal." For instance, female fruit flies will not mate with a white-eyed male if a red-eyed male is present. Clearly, nonrandom mating often results in stabilizing selection, but this is not necessarily the case. The selection pressures which led to the evolution of the elaborate plumage and huge tails of certain birds or the giant antlers of some deer or the long, intricate mating rituals of many birds and insects appear to be related almost entirely to feminine conquest.

8–15
Among most of the higher animals, including the higher invertebrates, the female of the species does the choosing. The female will usually select only a male clearly of her own species; even a slight deviation, such as a difference in eye color in a male fruitfly, is grounds for rejection. As a consequence, special anatomical and behavioral characteristics identifying males of particular species have been under heavy selection pressure. The huge antlers of the now extinct Irish elk (a) *presumably served mainly for courtship and territorial displays. The spectacular tail* (b) *of the otherwise quite modestly endowed male lyre bird* (c) *serves a similar function.*

(a)

(b)

(c)

8–16
The size of the appendages in a particular type of animal can often be correlated with the climate in which it lives. The fennec (a) *of the North African desert has large ears which help it to dissipate body heat. The red fox* (b) *of the eastern United States has ears of intermediate size, and the Arctic fox* (c) *has relatively small ears. Similar correlations between size, weight, color, etc., and environment can also sometimes be made among animals of a single species living over an extended geographic range.*

NATURAL SELECTION AND ADAPTATION

Adaptation, in population biology, is the adjustment of populations to their environment as a consequence of the operation of natural selection. Support for the concept that small variations among groups of organisms can be correlated with variations in their immediate environments has come from observations of members of the same species in different climates. These correlations are known as clines. Small mammals, for example, tend to have a smaller body size in the warmer parts of the range of the species and a larger body size in the cooler parts; in cooler regions, a larger body size is advantageous because it is easier for larger animals to conserve heat. On the other hand, tails, ears, bills, and other extremities of animals are relatively longer in the warmer part of a species range than in the cooler parts; such adaptations are correlated with the usefulness of the extremities in radiating heat.

Biologists studying the incubation time of gypsy moths in various parts of the United States found that an increase in length of incubation could be correlated with how far north the species occurred. Those farther north incubated for longer periods, thus ensuring that the adults did not emerge until the cold weather was safely past. Similarly, plants growing in the south often have slightly different requirements for flowering or for breaking dormancy than the same plants growing in the north, although they all may belong to the same species.

DISCONTINUOUS VARIATIONS

Adaptation always represents a temporary balance of the numerous demands made on an organism by the environment. In the case of continuous variations—variations representing differences of degree, such as larger, longer, browner—the phenotype represents a compromise between demands. Other traits are discontinuous; that is, they show up as sharp, clear-cut differences among groups in populations. This phenomenon is known as *polymorphism.* Studies of polymorphism in populations indicate that it is maintained by balances among different selective pressures.

Color and Banding in Snails

One of the best studied examples of polymorphism is found among land snails of the genus *Cepaea.* In one species (*Cepaea nemoralis*), for instance, the shell of the snail may be yellow, brown, or any shade from pale fawn through pink and orange to red. The lip of the shell may be black or dark brown (normally) or pink or white (rarely), and up to five black or dark or dark-brown longitudinal bands may decorate it. (See Figure 8–17.) Fossil evidence shows that these different types of shells have coexisted since before the Neolithic period.

Studies among English colonies of *Cepaea nemoralis* have revealed selective forces at work in some of the colonies. An important enemy of the snail is the song thrush. Song thrushes select snails from the colonies and take them to nearby rocks, where they break them open, eating the soft parts and leaving the shells. By comparing the proportions of types of shells around the thrush "anvils" with the proportions in the nearby colony, the investigators have

(a)

(b)

8–17

(a) *Polymorphism in land snails.* (b) *An "anvil," where song thrushes break land snails open in order to obtain the soft, edible parts. From the evidence left by the empty shells, investigators have been able to show that in areas where the background is fairly uniform, unbanded snails have a survival advantage over the banded type. Conversely, in colonies of snails living on dark, mottled backgrounds (such as woodland floors), banded snails are preyed upon less frequently.*

been able to show correlations between the types of snails seized by the thrushes and the habitats of the snails. For instance, of 560 individuals taken from a small bog near Oxford, 296 (52.8 percent) were unbanded; whereas of 863 broken shells collected from around the rocks, only 377 (43.7 percent) were banded.* In other words, in bogs, where the background is fairly uniform, unbanded snails are less likely to be preyed upon than banded ones.

Studies of a wide variety of colonies have confirmed these correlations. In uniform environments, a higher proportion of snails is unbanded, whereas in rough, tangled habitats, such as woodland floors, far more tend to be banded. Similarly, the greenest communities have the highest proportion of yellow shells, but among snails living on dark backgrounds, the yellow shells are much more visible and are clearly disadvantageous, judging from the evidence conveniently assembled by thrushes.

Many of the snail colonies studied were at distances so great from one another that the possibility of immigration between populations could be entirely ruled out. How, then, are the differences in shell types maintained in the face of such strong selection? One would expect populations living on uniform backgrounds to be composed only of unbanded snails and colonies on dark mottled backgrounds to lose all their yellow-shelled individuals. The answer to this problem is not fully worked out, but it seems to involve the fact that there are physiological factors which are correlated with the particular shell patterns and which form a part of the same group of genes that control color and banding. Experiments have shown, for instance, that unbanded snails (especially yellow ones) are more heat-resistant and cold-resistant than banded snails. In other words, linked with color and banding are other strong selection pressures that are also at work, and these may maintain the genetic variations.

* This difference (about 10 percent) may seem too small to be significant, but actually the number of shells counted was sufficiently large that there is only 1 chance in 1,000 that this difference could have occurred by chance.

Blood Groups in Man

The human blood types A, B, AB, and O are familiar examples of polymorphism. Apparently, the three alleles associated with these blood types are a part of our ancestral legacy, since the same blood types are also found in other primates. For a number of years after the discovery of blood type, it was assumed that they were all neutral in terms of their selective value and, as a consequence, were all maintained in invariant frequencies in the human population. Now geneticists are becoming more and more reluctant to dismiss any gene as "neutral," and information is accumulating that may be used as evidence against this concept of neutrality. It is known, for example, that persons with type A blood run a relatively higher risk of cancer of the stomach and of pernicious anemia. Men with type B blood are about 2 inches taller, on the average, than those with other blood types. Persons with type O blood have a higher risk of duodenal ulcers. Furthermore, there are irregular geographic distributions of the A, B, AB, and O groups, and many geneticists are inclined to believe that these reflect some selective force favoring particular blood groups under particular conditions.

EVOLUTION IN A MAN-MADE ENVIRONMENT

Darwin believed that evolution is such a slow process that it can never be observed directly. Within recent years, however, the effects of human civilization have been producing new and extremely strong selection pressures, leading to what is probably an unprecedented rate of directional selection. One of the most striking and best studied examples is that of *Biston betularia*, the peppered moth. These moths fly at night and spend their days at rest. They were well known to naturalists of the nineteenth century, who remarked that they were usually found on lichen-covered trees and rocks. Against this background, their light coloring made them practically invisible. Until 1845, all reported

(a)

(b)

8–18

Biston betularia, *the peppered moth, and its melanic form,* carbonaria, *at rest on a lichen-covered tree trunk in unpolluted countryside* (a) *and on a soot-covered tree trunk near Birmingham, England* (b). *A striking example of an evolutionary change resulting from a drastic environmental change, the black form of the moth began to appear in the latter part of the nineteenth century as the English countryside became increasingly polluted by industrial smoke.*

specimens of *Biston betularia* had been light-colored, but in that year one black moth of this species was captured at the growing industrial center of Manchester.

With the increasing industrialization of England, smoke particles began to pollute the foliage in the vicinity of industrial towns, killing the lichens and leaving the tree trunks bare. In heavily polluted districts, the trunks and even the rocks and the ground became black. During this period, more and more black *Biston betularia* were found. This mutant black form spread through the population until black moths made up 99 percent of the Manchester population. Where did the black moths come from? Eventually, it was demonstrated that the black color was the result of a rare recurring mutation.

Why were they increasing so rapidly? H. B. D. Kettlewell was among those who hypothesized that the color of the moths protected them from predators, notably birds. In the face of a number of strongly opposed entomologists—all of whom claimed they had never seen a bird eat a *Biston betularia* of any color—he set out to prove his hypothesis. He marked a sample of moths of both colors, carefully putting a spot on the underside of the wings, where it could not be seen by a predator. Then he released known numbers of marked individuals into a bird reserve near Birmingham, an industrial area where 90 percent of the local population consisted of black moths, and another sample into an unpolluted Dorset countryside where no black moths ordinarily occurred. He returned at night with light traps to recapture his marked moths. From the area around Birmingham, he recovered 40 percent of the black moths but only 19 percent of the light ones. From the area around Dorset, 6 percent of the black moths and 12.5 percent of the light moths were retaken.

To clinch the argument, Kettlewell placed moths on tree trunks in both locations, focused hidden movie cameras on them, and was able to record birds actually selecting and eating the moths, which they do with such rapidity that it is not surprising that it was not previously observed. Near Birmingham, when equal numbers of dark and light moths were available, the birds seized 43 light-colored moths and only 15 black ones; near Dorset, they took 164 black moths but only 26 light-colored forms.

Until recently, only a few of the light-colored populations could be found, and these were far from industrial centers. Because of the prevailing westerly wind in England, the moths to the east of industrial towns tended to be of the black variety right up to the east coast of England, and the few light-colored populations were concentrated in the west, where lichens still grow. A similar tendency for dark-colored forms to replace light-colored forms has been found among some 70 other moth species in England and some 100 species of moths in the Pittsburgh area of Pennsylvania.

Most recently, strong controls have been instituted in Great Britain on particulate content of smoke, and the heavy soot pollution has begun to decrease. The light-colored moths are already increasing in proportion to the black forms, but it is not yet known whether a complete reversal either in pollution or in adaptive coloration will come about.

Insecticide Resistance

The development of resistance to insecticides by flies, cockroaches, and similar unwelcome visitors is a phenomenon only too familiar to the general public. Insecticide resistance in a population develops gradually over a number of generations, but from man's point of view, this development is remarkably rapid. For instance, during World War II, DDT was highly effective in controlling body lice, but by the time of the Korean War, it was virtually useless. In fact, within two years after DDT was first used, resistant strains of many types of insects had developed independently in different parts of the world.

A particularly striking example of insecticide resistance has been found in the scale insects (*Coccidae*) that attack citrus trees in California. In the early 1900s, a concentration of hydrocyanic gas sufficient to kill nearly 100 percent of the insects was applied to orange groves at regular intervals with great success. By 1914, orange growers near Corona began to notice that the standard dose of the fumigant was no longer sufficient to destroy one type of scale insect, the red scale. A concentration of the gas that had left less than 1 in 100 survivors in the nonresistant strain left 22 survivors out of 100 in the resistant strain. By crossing resistant and nonresistant strains, it was possible to show that the two differed in a single gene. The mechanism for this resistance is not known, but one group of experiments suggests that the chief point of difference is that the resistant individual can keep its spiracles closed for 30 minutes under unfavorable conditions whereas the nonresistant insect can do so for only 60 seconds.

Flax Mimics

Another evolutionary change resulting from man-made selection pressures can be found in a weed of the mustard family. This weed (*Camelina sativa*) has developed races capable of growing in the flax fields of northern Europe, where flax is still an important commercial crop.

The parental form of the weed is a branched bushy annual with small white flowers, small seed pods, and rather small rounded seeds; this type is commonly found on roadsides and in open fields. Even if it encroaches on the flax fields, it does not survive there for more than a generation because its seeds are so different in size and shape from the large flat seeds of flax that they are automatically discarded with the chaff when the flax seeds are purified by winnowing and so are not replanted.

However, in areas where flax is commonly grown, new forms of *Camelina* have evolved. The plants are tall and unbranched and thus able to compete for sunlight with the tall, unbranched, closely growing flax plants. More important, they have much larger seeds than the races of mustard weed found along the roadside. These seeds, although they are not of the same shape as flax seeds, are not separated by the winnowing machines and so are resown each year by the flax growers. This new race, the flax mimic, produces fewer seeds in each seed capsule than do wild strains, and since the plant is unbranched, there are fewer seed capsules. On the other hand, the seeds enjoy the uncommon luxury of being sown in a carefully prepared field.

SUMMARY

Natural selection is the nonrandom reproduction of genotypes. It channels variations by continuously eliminating particular genotypes. There are three general types of selection: stabilizing, disruptive, and directional. The result of selection is the adjustment of populations to their environment, a phenomenon known as adaptation. Forces that channel the direction of selection include changes in climate and other inorganic factors and predator-prey, symbiotic, and parasitic relationships with other species in the community, leading to patterns of coevolution. Within species, differential fertility and nonrandom mating can exert powerful selective pressures.

Clines are correlations between subpopulations of organisms of the same species and differences in their environment.

Polymorphism is the existence within a population of discontinuous traits, such as shell colors in snails or blood groups in man. The persistence of polymorphism in populations is believed to reflect the existence of different but approximately equal selective pressures for the particular characteristics.

Environmental changes brought about by human technology have produced strong selective pressures, resulting in rapid adaptive change in affected populations. Examples include industrial melanism in moths, insecticide resistance in insects, and mimicry among weeds.

QUESTIONS

1. What selection pressures are likely to be in operation for the human species at the present time?

2. Cave-dwelling species of fish are often blind. How would Lamarck have explained this phenomenon? How would modern evolutionary theory explain it?

CHAPTER 8–4

The Origin of Species

Evolution is the accumulation of adaptive changes in populations. Since this process of adaptive change is continuous, it is difficult to identify the moment when evolution can be said to have taken place. However, the emergence of a new species clearly marks a milestone along the evolutionary pathway, and how new species—different *kinds* of organisms—come into existence has always been a central problem in evolutionary theory. Although Darwin did not ever deal directly with this question, his recognition of its central importance is reflected in the title of his major work, which is also the title of this chapter.

What is a species? In the words of Terrell H. Hamilton, a leading young population biologist: "A species may be envisioned as an isolated pool of genes flowing through space and time, constantly adapting to changes in its environment as well as to the new environments encountered by its extension into other geographic regions."*

Speciation is the formation of one or more new species. It is now generally agreed that a number of factors are involved in speciation, of which the three most important are (1) genetic isolation of a population, (2) the availability of a new ecological niche, and (3) time.

How do we decide when a new species has come into being? The moment at which one species splits to become two contemporary species is difficult to pinpoint in time—although, as we shall see, it now seems to have been observed in the laboratory. There is no doubt, however, that one species has become two when the two populations can coexist in the same natural environment without interbreeding, and for the purposes of this discussion, this will be the working definition of a species.

* Terrell H. Hamilton, *Process and Pattern in Evolution*, The Macmillan Company, New York, 1967.

DARWIN'S FINCHES

Admirers of Charles Darwin find it particularly appropriate that one of the best examples of speciation is provided by the finches observed by Darwin on his voyage to the Galapagos Islands. All the Galapagos finches are believed to have arisen from one common ancestral group—perhaps a single pair—transported from the South American mainland, some 600 miles away. As we saw in Chapter 8–2, a small population is apt to be nonrepresentative of its larger, parent population and, as a consequence, lacks the stability of the parent population and so is more susceptible to rapid change. The small size of the founding group was probably an important factor in the remarkable changes that subsequently took place. How they got there is, of course, not known, but it is probable that they were the survivors of some particularly severe storm. (Every year, for instance, some American birds and insects appear on the coasts of Ireland and England after having been blown across the North Atlantic.)

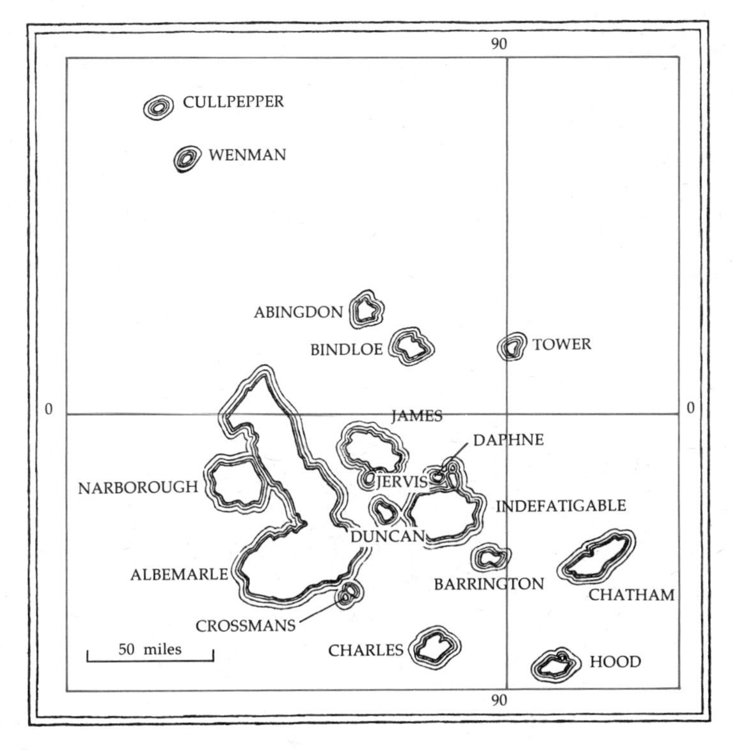

8–19

The Galapagos Islands, some 600 miles west of the coast of Ecuador, have been called "a living laboratory of evolution." Species and subspecies of plants and animals that have been found nowhere else in the world inhabit these islands. "One is astonished," wrote Charles Darwin in 1837, "at the amount of creative force . . . displayed on these small, barren, and rocky islands. . . ."

We do have an idea of what greeted these unwilling adventurers. The Galapagos archipelago consists of 13 main volcanic islands, with many smaller islets and rocks. On some of the islands, craters rise to heights of 3 or 4 thousand feet. The islands were pushed up from the sea more than a million years ago, and most of them are still covered with black basaltic lava.

The major vegetation is a dreary grayish-brown thornbush, making up miles and miles of dense leafless thicket, and a few tall tree cactuses—"what we might imagine the cultivated parts of the Infernal regions to be," young Charles Darwin wrote in his diary. Inland and high up on the larger islands, the land is more humid, and there one can find rich black soil and tall trees covered with ferns, orchids, lichens, and mosses, kept damp by a mist that gathers around the volcanic peaks. During the rainy season, the area is dotted by sparkling shallow crater lakes. The vegetation may have differed when the finches first arrived, but clearly plants preceded the birds or the birds would have been unable to survive.

The differences between the lush vegetation of the South American coast and the much harsher environment of the Galapagos undoubtedly put the new arrivals under strong selective pressures.

Thirteen different species of finches live on the archipelago, and an additional species lives on Cocos Island, several hundred miles to the north. The ancestral type is believed to have been a ground finch, and six of the Galapagos finches are ground finches. Four species of ground finches live together on most of the islands. Three of them eat seeds and differ from one another mainly in the size of their beaks, which in turn, of course, influences the size of seeds that they eat. The fourth lives largely on the prickly pear and has a much longer and more pointed beak. The two other species of ground finch are usually found only on outlying islands, where some supplement their diet with cactus.

In addition to the ground finches, there are six species of tree finches, also differing from one another mainly in beak size and shape. One has a parrotlike beak, suited to its diet of buds and fruit. Four have insect-eating beaks, each adapted to a different size range of insects. The sixth and most remarkable of the insect eaters is the woodpecker finch, which has a beak like a woodpecker's but lacks the woodpecker's long prying tongue and so carries about a twig or cactus spine to dislodge insects from crevices in the bark. (See Figure 8–20.)

8–20
(a) *Three of the 13 different species of Darwin's finches. There are six species of ground finch (of which three are shown here), six species of tree finch, and one warblerlike species—all derived from a single ancestral species. Except for the warbler finch, which resembles a warbler more than a finch, the species look very much alike; the birds are all small and dusky-brown or blackish, with stubby tails. The differences between them lie mainly in their bills, which vary from small, thin beaks to huge, thick ones.* (1) *The small ground finch* (Geospiza fuliginosa) *and* (2) *the medium ground finch* (Geospiza fortis) *are both seed eaters.* G. fortis, *with a somewhat larger beak than* G. fuliginosa, *is able to crack larger seeds.* (3) *The cactus ground finch* (Geospiza scandens) *lives on cactus blooms and fruit. Notice that its beak is much larger and more pointed than those of the other two ground finch species.* (b) *The woodpecker finch* (Camarhynchus pallidus) *is a rare phenomenon in the bird world because it is a tool user. Like the true woodpecker, it feeds on grubs which it digs out of trees; but lacking the woodpecker's chisel beak and long barbed tongue, it resorts to an artificial probe, a twig or cactus spine, to dislodge the grub. The woodpecker finch shown has selected a cactus spine which it inserts into the grub hole. The bird has succeeded in prying out the grub, which it eats. If the pick selected turns out to be an efficient tool, the bird will carry it from tree to tree in its search for grubs.*

The thirteenth species of finch is hardly a finch at all. Classical taxonomists, using all ordinary standards of external appearance and behavior, would classify it as a warbler, but its internal anatomy and other characteristics clearly place it among the finches, and there is general agreement that it, too, is a descendant of the common ancestors.

As students of the Galapagos finches reconstruct the story, the original founding party from the mainland probably made its landfall on one of the larger islands, gained a foothold, and hung on. From time to time, members of the group were carried to other islands. Each founder group that survived must have remained isolated long enough to reconstruct a new gene pool. This might have been a matter of at least 10,000 years for each of the species, although since there are 13 main islands, several species were probably evolving at the same time.

Finches are not particularly good fliers; if they had been, they might not have speciated. Shore plants, for example, even when found on widely separated islands, are likely to be of the same species since most of these plants are well able to survive long ocean voyages. To speciate, a population must remain genetically isolated; otherwise, its emerging genetic differences will be swamped, swallowed back up into the original gene pool. Once speciation has occurred, however, the different species can live together and maintain their genetic identities. Because the isolated groups of finches became adapted to different ecological niches—in particular, to the utilization of different food resources—they were able to coexist once reunited. (Remember that we do not know how many populations were lost as a result of inability either to adapt rapidly enough or to compete with established groups.) As many as 10 species of finch now can be found together on some of the larger islands.

The Galapagos, in short, offered a situation in which the three criteria for speciation could readily be met: genetic isolation, a number of unoccupied ecological niches, and because of the geographical distances involved, time.

GENETIC ISOLATION

Because of the importance of genetic isolation in speciation, remote islands provide many other examples of evolution in progress. The tortoises of the Galapagos, for which the archipelago is named, are different from mainland tortoises and seem to offer an example of speciation in process. The races, or subspecies, of Galapagos tortoises are so different from each other that, as Darwin observed, the sailors who frequent the Galapagos can tell by the appearance of a tortoise which of the islands it came from. Perhaps in another several thousand years, they will evolve into distinct species. Unfortunately, this experiment will probably never take place; the huge, lumbering tortoises are virtually defenseless, and their numbers have been seriously reduced. For years, sailors stopped regularly at the Galapagos to provision their ships, carrying away two or three hundred tortoises each time. The tortoises, known to be able to live for as long as a year without food or water, were stacked in the hold as a welcome source of fresh meat. In recent years, the tortoises have been protected against man, but they are still threatened by introduced mammals which have gone wild. Donkeys and goats compete with them for food, and wild pigs, dogs, and rats prey upon the eggs and the young.

Most animals on remote islands, including the Galapagos tortoises, have no fear of man. Many island visitors have reported their feelings of delight at having wild birds alight casually on their shoulders or at being able to reach out and stroke a sunning sea lion or walk unnoticed through a crêche of nesting penguins. Nature moves slowly, and by the time mechanisms of fear of man and defense against his predations have evolved, hunters will have reaped a rich harvest of meat, fur, and feathers and many island animals will have gone forever. One species of Darwin's finches has already disappeared.

Hundreds of examples of island speciation have now been studied, particularly on the Pacific islands. Here one can sometimes trace the repeated ocean voyages of a genus of land snails, for example, from New Guinea, Australia, or the Philippines eastward to New Caledonia, the Samoas, Tahiti, the Marquesas, the Hawaiian Islands, or the Galapagos or westward from the coast of South America across the same island chain. Some of these organisms show only the beginnings of species formation, like the Galapagos tortoises, and some are so changed that their genealogy has become almost impossible to trace.

Other Types of Geographic Barriers

What is an island? For a plant, it may be a mist-veiled mountaintop, and for a fish, a freshwater lake. A forest grove may be an island for a small mammal. A few meters of dry ground can isolate two populations of snails. Islands of this sort usually form by the creation of barriers between formerly contiguous geographic zones. The Isthmus of Panama, for instance, has repeatedly submerged and reemerged in the course of geological time. With each new emergence, the Atlantic and Pacific Oceans became "islands," populations of marine organisms were isolated, and some new species formed. Then, when the oceans joined again (with the submergence of the Isthmus), the continents separated and became, in turn, the "islands."

There are a number of natural geographic barriers in the United States. The Grand Canyon and Upper Colorado River, for example, although not barriers for birds, are very effective at separating populations of small land animals. Several kinds of pocket gophers, wood mice, pocket mice, and field voles reach their northern limits at the south rim of the Canyon; there are related but different species that inhabit the plateau of the other rim. Two such subspecies of squirrel can easily be distinguished. The Abert squirrel, which lives south of the Grand Canyon, has a grayish tail, a white belly, and great tufted ears. The Kaibab squirrel, found only on the north rim, has a black belly and a pure white tail. These squirrels are very much the same; both kinds obviously had a common ancestor in the distant past, before the Canyon formed and became a barrier.

Similarly, San Francisco Bay is an important barrier for some mammals. Of 24 species of small mammals studied in the area, 11 occur both north and south of the bay, whereas 8 are found only in the north and 5 only in the south. What constitutes a barrier for some kinds of animals is not a barrier for others.

Another example of the effects of geographic barriers is provided by the *Oeneis* butterfly, which is found only on the tops of the Rocky Mountains in Colorado and the White Mountains of New England and along the cool coast of Labrador.

Recently a study was made comparing populations of wood frogs (*Rana sylvatica*) from the Colorado-Wyoming area with those from southern Canada. Presumably the more southerly populations represent a group that was displaced southward by the most recent glaciation, about 10,000 years ago, and the northern populations represent survivors of the parent group. The groups differ in several physical characteristics, including average leg length and markings, and the mating calls of the southern group tend to be lower-pitched. Most important, in cross-fertilization experiments, 99 percent of the eggs failed to develop, indicating that the southern group has become a new species.

Geographic Races

Every widespread species that has been carefully studied has been found to contain geographically representative populations that differ from each other to a greater or lesser extent. A species composed of subspecies of this sort is obviously particularly susceptible to speciation if geographic barriers arise, as with the wood frog.

Because of the work of John Moore of Columbia University, the leopard frog (*Rana pipiens*) has become one of the best-known examples of subspeciation. Leopard frogs are found in North America as far north as Quebec and as far south as Mexico. Moore has studied and compared a large number of characteristics in 29 separate populations of this species. He has found that there are marked

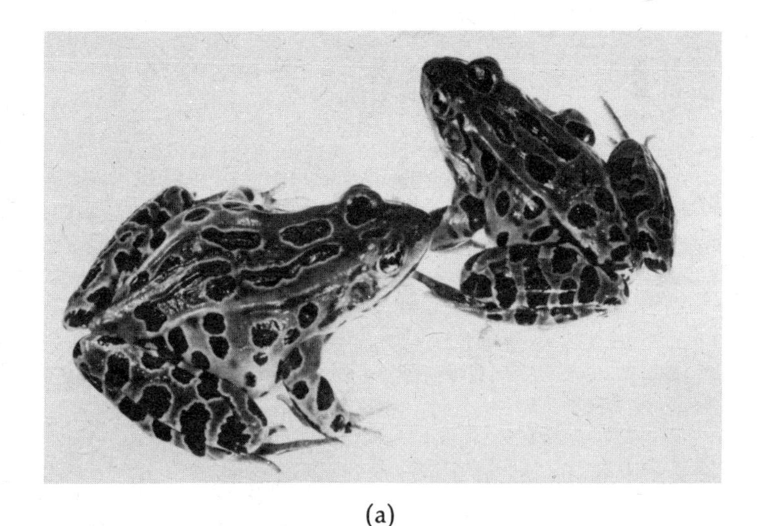

(a)

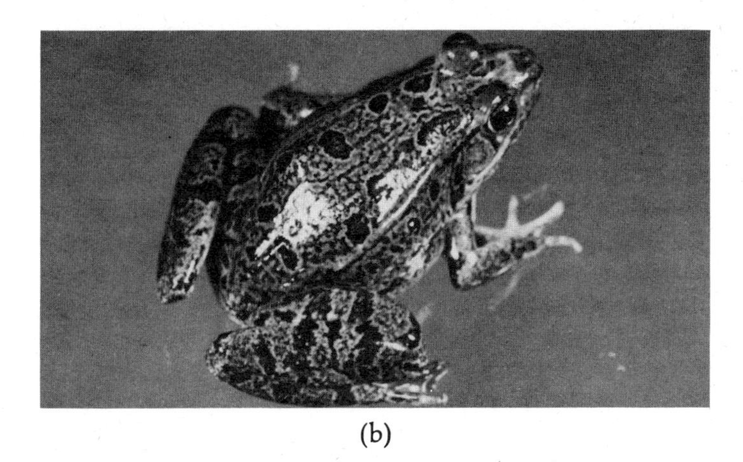

(b)

(c)

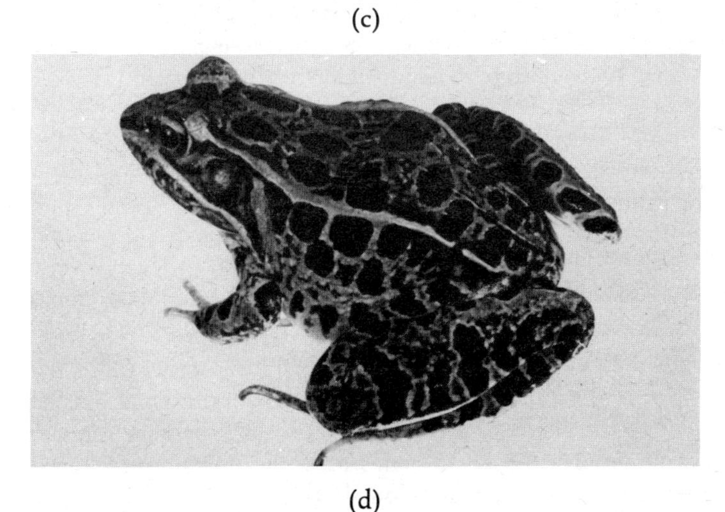

(d)

8–21
A well-known example of subspeciation is offered by the leopard frog (Rana pipiens). *These frogs are found as far north as Quebec and as far south as Mexico. The subspecies shown are from Quebec* (a), *western Texas* (b), *Louisiana* (c), *and southern Florida* (d).

variations, as might be expected, from population to population. Crossbreeding of individuals from different populations produces normal offspring when parents are drawn from populations that are geographically adjacent, such as central and southern Florida, or that lie at roughly the same latitude, such as Texas and central Florida. However, the greater the north-south gap separating the home populations of the parents, the greater also is the proportion of defective, often inviable, offspring. In Texas-Vermont hybrids, for example, mortality among the developing embryos may reach as high as 100 percent. Yet because there is a gene flow among adjacent populations, *Rana pipiens* is still one species. One can see, however, that if the intermediate, bridging populations were eliminated, separate species would evolve quite quickly, as they have done with *Rana sylvatica.*

(a)

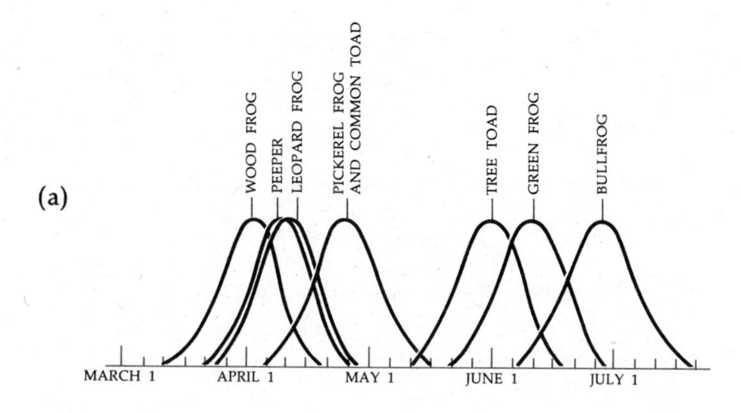

(b)

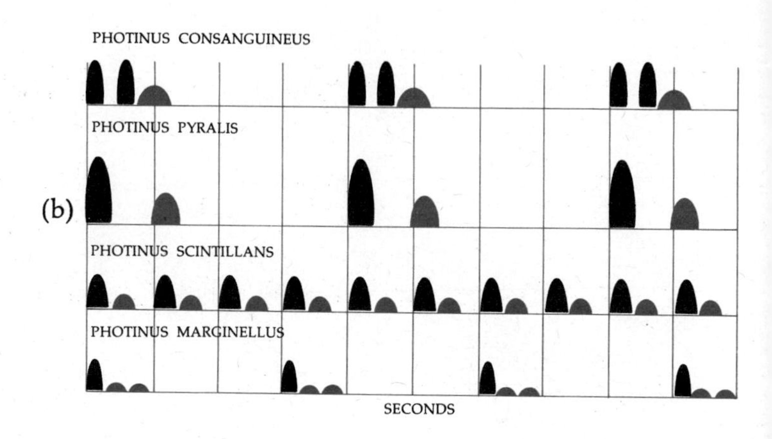

8–22

Genetic isolating mechanisms. (a) *Mating timetable for various frogs and toads that live near Ithaca, N. Y. In the two cases where two different species have mating seasons that coincide, the breeding sites differ. Peepers prefer woodland ponds and shallow water; leopard frogs breed in swamps; pickerel frogs mate in upland streams and ponds; and common toads use any ditch or puddle.* (b) *Flashing patterns of various species of fireflies found in Delaware. Each division on the horizontal lines represents one second, and the height of each curve represents the brilliance of the flash. The black curves are male flashes, and the colored ones are the female responses. The male flashes first and the female answers, returning the species-specific signal. Firefly flashes differ not only in duration, intensity, and timing, but also in color.*

Behavioral Isolation

A recent experiment indicates that new species of animals may be able to form without geographic isolation if there is genetic isolation. Theodosius Dobzhansky and a coworker at the Rockefeller University took two strains of *Drosophila* that interbred freely and "tagged" each of them by inbreeding for recessive mutations. If the offspring were the product of matings between two individuals of the same strain, they would show the mutation (*aa*); if they were hybrids, then they would be phenotypically wild-type (*Aa*). The matings were observed with a hand lens. The hybrids, which were vigorous and fertile, were removed from the population. Within about 60 to 70 generations of this strong artificial selection against the hybrids, clear-cut evidence of behavioral isolation began to appear. A female of one strain exposed to males of both strains would tend always to choose a male from her own strain rather than from the other one. Eventually a high degree of isolation between the two strains was achieved, on the basis of the sexual behavior of the female alone.

Thus, for the first time, the crucial step in the establishment of a new species—the development of its independence from other gene pools—was observed as it took place.

Polyploidy

Polyploidy is another means by which genetic isolation is produced without geographic isolation. A haploid cell, you will recall, has one set of chromosomes ($1n$). A diploid cell has two sets ($2n$), one from each parent. A cell that has more than two sets of chromosomes (or an individual made up of such cells) is known as a polyploid.

Polyploidy originates when the chromosomes replicate and divide but the cell does not. A polyploid cell may have three, four, five, or more sets of chromosomes.

8–23
Polyploidy can be induced in plants by treating dividing cells with a chemical called colchicine which prevents the development of the spindle and so blocks the separation of the chromosomes during mitosis. Because polyploidy may increase the size of a plant or the number of its floral parts, many cultivated plants are polyploids, such as the dahlia shown here.

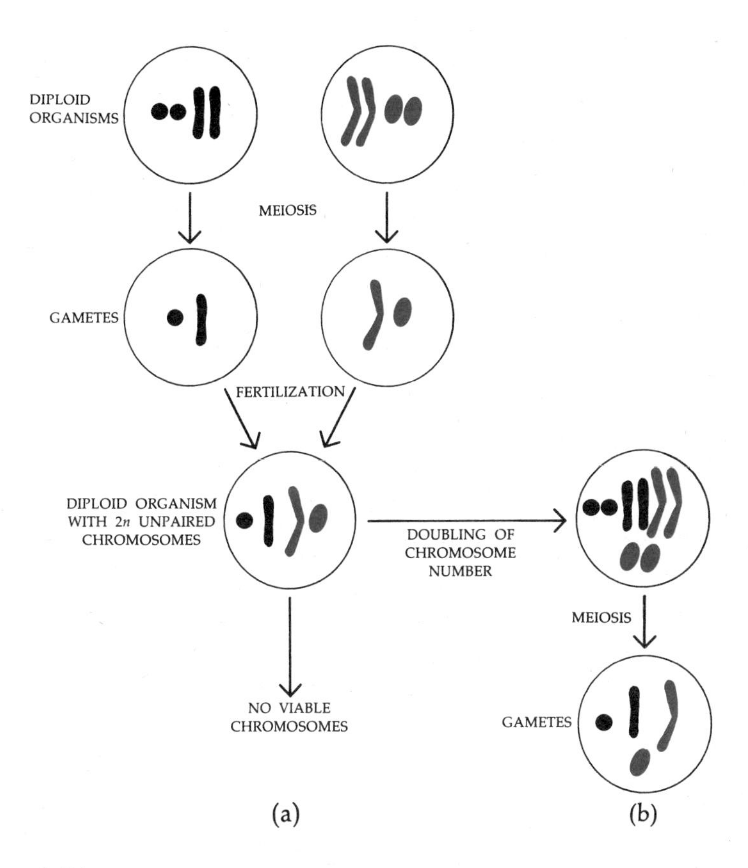

8–24
A hybrid organism (such as a mule) produced from two 1*n gametes can grow normally because mitosis is normal, but it cannot reproduce because the chromosomes cannot pair for meiosis* (a). *However, a hybrid organism produced from two* 2*n gametes can both divide normally mitotically and undergo meiosis. Since each chromosome will have a partner, the chromosomes can pair at meiosis* (b).

Polyploidy rarely occurs in animals but is very frequent in plants. Among the largest group of plants, the angiosperms, about half of the approximately 300,000 known species are polyploids.

A polyploid plant may reproduce vegetatively, or a flower may form from a part of the plant that is polyploid. Cells that are $4n$ or $6n$ (or any even multiple of n) can undergo meiosis because each chromosome has a suitable partner. However, they cannot produce fertile hybrids with the $1n$ parent type (such a hybrid would be $3n$). They can, however, produce viable offspring with other polyploids. In effect, these polyploid types are genetically isolated from their parent strains.

Polyploidy also makes it possible for hybrid crosses to occur in plants. The man-made hybrid *Raphanobrassica*, formed from the radish and the cabbage, provides an example. Some years ago, an enterprising plant geneticist was struck by the commercial possibilities of such a hybrid since, in the one plant, the roots are edible and, in the other, the leaves. In each plant, the haploid (1*n*) number is 9. A cross between the two produced a plant with nine radish chromosomes and nine cabbage chromosomes, which was, like the mule, a sturdy hybrid but infertile. But since the chromosomes could not pair at meiosis, an artificial tetraploid (4*n*) was developed in which the somatic cells had 18 radish chromosomes and 18 cabbage chromosomes, making it possible for the gamete to receive a full complement (9) of each. As their originator had hopefully anticipated, the hybrids were both sturdy and fertile. Their only drawback: they had a root like that of a cabbage and a head like that of a radish!

Many important domestic crops, including wheat, are hybrid polyploids.

THE AVAILABLE NICHE

The second major factor required for speciation, as we noted earlier, is an available ecological niche, a place for a new species in the working community. Again, one of the best illustrations is the Galapagos finches. On these deserted islands, there was not only a niche for finches but also niches for warblers and woodpeckers, and so the immigrant finches became "warblers" and "woodpeckers."

TIME

Evidence of the length of time required for a new species to develop is based largely on what is known about the ages of various islands. The British Isles, for instance, have been separated from the continent of Europe only since the last glaciers retreated, about 10,000 years ago. During this time Britain has not developed any distinctly new species, so that a period of 10,000 years or so seems inadequate, in most cases, although it was apparently sufficient for *Rana sylvatica*. On the other hand, islands whose age is probably a few million years seem to have many unique species. So between 10,000 and a few million years seems necessary.

ADAPTIVE RADIATION

Major evolutionary change appears to take place in sudden bursts, rather than slowly and steadily. Such bursts characteristically take the form of *adaptive radiation*, which is the diversification of a group of organisms that share a common ancestor (often, itself, newly evolved) to fill all the ecological space available. Adaptive radiation is believed by some population biologists to be the major pattern of evolution. Darwin's finches are an example of adaptive radiation on a miniscale, the marsupials of Australia (Figure 8–25) on a larger one.

The fossil record contains many examples of adaptive radiation. For example, some 425 million years ago, the first terrestrial plants and animals suddenly emerged. A large number of different populations apparently began the same move at about the same time. Most undoubtedly perished, but a few survived and then differentiated into the many different types of terrestrial plants and animals existing today. A similar, even more rapid burst of evolution gave rise to the birds. Apparently, according to the fossil record, a number of populations of reptiles must have taken to climbing and gliding at about the same time. The shift was made very rapidly from the land to the air, and although these were intermediate populations, gliding cold-blooded reptiles undoubtedly existed at one time. Then they disappeared rapidly and permanently. The mammals similarly burst forth on the evolutionary scene, with many different kinds appearing simultaneously in the fossil record.

What is necessary for these great evolutionary moves to take place? First, there must be an ample store of genetic variation in the gene pool. Second, a genetic upheaval must occur. A population that is well in tune with its environment will probably not undergo the sudden changes that lead to evolutionary breakthroughs. A drastic genetic change is most likely to occur in a small population under environmental stress.

There are important limits to what can occur. Evolution must proceed by certain "logical" steps. Water, which is

"WOODCHUCK" *Wombat*

"BEAR" *Koala Bear*

"RAT" *Kangaroo Rat*

"DOG" *Tasmanian Wolf*

8–25
In Australia, marsupials fill the niches occupied by placental mammals on other continents. Here the marsupials are labeled with the common names of their placental counterparts.

(a)

(b)

8–26

Although porpoises (a) *are evolutionary descendents of primitive mammals, their shapes closely resemble those of large, primitive ocean dwelling fish, as exemplified by the sand shark* (b). *Organisms that come to occupy similar ecological niches as a result of adaptive radiation often come to resemble one another.*

800 times heavier than air, can support an animal like a jellyfish and can provide sufficient resistance for ciliary motion, but a jellyfish could never become terrestrial. Animals needed to develop strong supporting skeletons and strong muscles before they could invade the land. A third necessary condition for evolution is, therefore, preadaptations, the existence of structures that could become modified to other uses.

For example, the lungs and lobes of the lungfish made it possible for animals to begin the invasion of land once the plants had prepared the way for them. These structures developed in the lungfish not as an adaptation to terrestrial life but as an adaptation to conditions of drought. The lungs and lobes had been in existence for millions of years. Similarly, although birds and reptiles are very different in their modes of life, very little modification of the basic plan of the reptile skeleton was necessary to produce the skeleton of the bird; for the bird, feathers were the big evolutionary breakthrough. The famous fossil *Archaeopteryx* (Figure 4–36) is essentially a reptile with feathers. With insects, the rapid development of a great multitude of species depended upon the existence of appendages that could be modified into a large number of specialized structures.

The fourth condition necessary for an evolutionary breakthrough to occur is that a new adaptive zone must be available. The land was once such an adaptive zone, and the air another. Once an adaptive zone becomes available, divergent lines of ancestral stock try to cross the barrier until one of them makes it into the new zone. The shift either is rapid or fails completely. Once the shift is made, radiation takes place until the zone is filled. Other bursts of adaptive radiation appear to have taken place with the subsiding of the glaciers, when essentially new land was opened for habitation. For Darwin's finches, the underoccupied Galapagos Islands represented a new adaptive zone.

Convergent Evolution

Organisms that move into similar ecological niches as a result of adaptive radiation often come to resemble one another. The Australian placentals, so strikingly similar to their mammalian counterparts, are a good example. The whales, a group which includes the dolphin and porpoises,

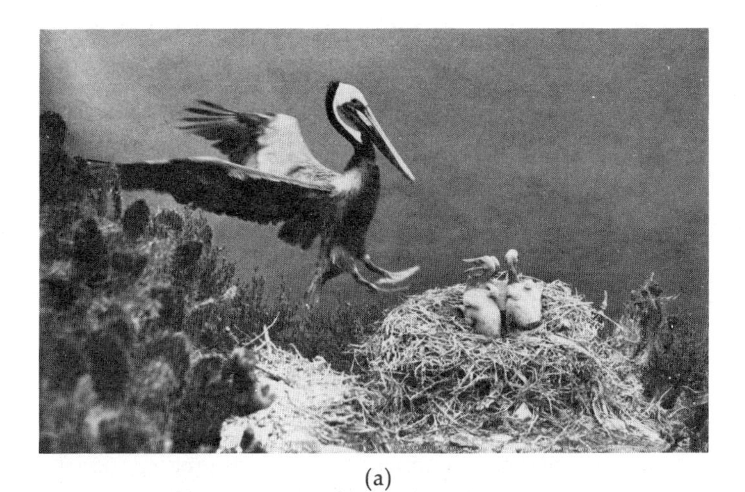

(a)

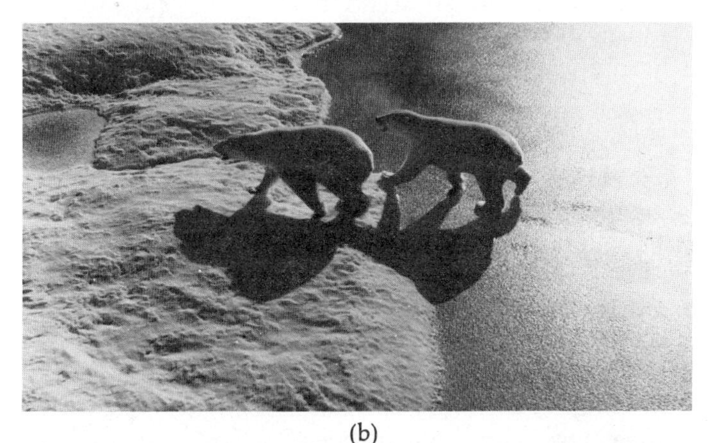

(b)

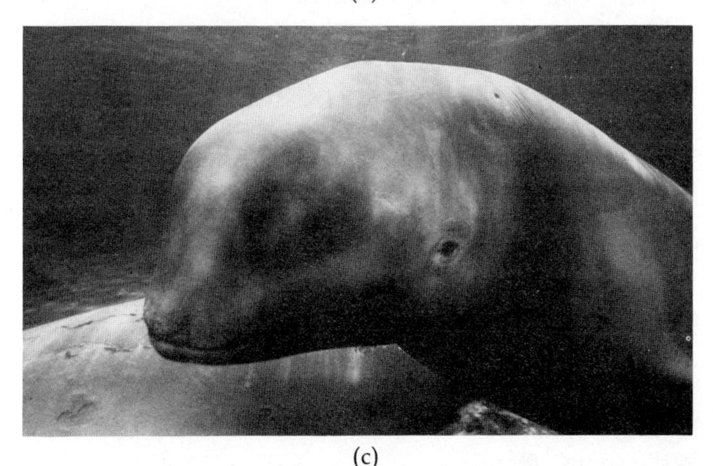

(c)

8–27
Endangered species. (a) *The California brown pelican is a victim of DDT pollution, apparently the consequence of the dumping of wastes from an insecticide-manufacturing plant into the coastal waters. DDT affects the birds' reproductive capacities; no pelican chicks have been successfully raised in California for the last two years.* (b) *Polar bears, prized as trophies, are now being hunted from the air. They are driven across the ice until exhausted, and then shot by sportsmen.* (c) *In Captain Ahab's time, a whaling boat averaged a whale a month; today a whale is killed every* 12 *minutes by the modern whale factory fleets of the Soviet Union and Japan. Blue, humpback and gray whales, now nominally protected, are close to extinction, and the number of whales of other species, such as the beluga whale shown here, is dropping rapidly. But Russian and Japanese spokesmen maintain that whale meat is an essential source of proteins for their populations.*

are very similar in many exterior features to sharks and other large fish; although their fin conceals the remnants of a vertebrate hand, they are warm-blooded, like their land-dwelling ancestors, and they have lungs rather than gills. (Gills are clearly more useful than lungs in aquatic respiration—another example of the fact that evolutionary processes work only with what is already available.) Similarly, two families of plants invaded the desert, giving rise to the cacti and the euphorbs. Both evolved large fleshy stems with water-storage tissues and protective spines and appear superficially similar. However, their quite different flowers reveal their widely separate evolutionary origins.

EXTINCTION

Only a small fraction of all the species that have ever lived are presently in existence—certainly less than $1/10$ of 1 percent, perhaps less than $1/1000$ of 1 percent. Little is known about the processes of extinction, although because the removal of old species usually leaves an ecological vacancy to be filled by new ones, extinction is also one of the essential features of evolution. Environmental changes (especially cold and drought), competition, and predation are probably major forces in the elimination of existing species, just as they are major forces in the formation of new ones. Recently man, partly by predation but more by disruption of the natural environment, has replaced climatic change as the chief destroyer of organisms. Also, he has introduced a new pattern in evolution: there is, apparently, no ecological replacement of some of the niches made available by this human-mediated extinction. However, based on the history of evolution it seems likely that man, too, will eventually become extinct. We do not, as yet, have any indication as to what new species, if any, will take his place.

SUMMARY

Three major factors required for speciation are genetic isolation of a population, the availability of a new ecological niche, and time. Probably between 10,000 and a few million years are required for speciation to occur. Genetic isolation is often achieved by geographical isolation, as occurs on islands. Apparently, genetic isolation can occur by behavioral isolation. Polyploidy is another means of genetic isolation and is particularly common among plants. It makes possible new hybrid combinations.

Evolution often takes place by adaptive radiation, the diversification of species to fill the ecological niches in a new adaptive zone. Requirements for adaptive radiation include the existence of a population with much latent variation, the existence within this population of appropriate preadaptations, and the availability of a new adaptive zone.

In the past, extinction has made way for new species. Man is now the dominant force in the extinction of species. We do not yet know what species, if any, will fill the niches now being vacated.

QUESTION

In your opinion, why is there at present only one species in the genus *Homo*?

GLOSSARY

ADAPTATION: 1. The adjustment of populations to their environment by the operation of natural selection. 2. A peculiarity of structure, physiology, or behavior of an organism which especially aids in fitting the organism to its particular environment.

ADAPTIVE RADIATION: The evolution from a relatively primitive and unspecialized type of organism to several divergent forms specialized to fit numerous distinct and diverse ways of life, as occurred in Darwin's finches or the marsupials of Australia.

CLINE [Gk. *klinein*, to lean]: A correlation between a series of gradual differences within a species with differences in climate or other geographical factors.

CONTINUOUS VARIATION: A variation in a particular trait, such as size, that is characterized by small gradations of difference within a population.

DIRECTIONAL SELECTION: The general shift in a population in favor of a particular characteristic.

DISCONTINUOUS VARIATION: A variation in a particular trait, such as coloration, that is characterized by sharply defined differences within a population.

DISRUPTIVE SELECTION: The increase of two or more extreme types in a population at the expense of intermediate types.

GENE POOL: All the alleles of all the genes in a population.

GENETIC ISOLATION: The inhibition of gene exchange between different species by morphological, behavioral, or physiological mechanisms.

HARDY-WEINBERG LAW: The mathematical expression of the relationship between relative frequencies of two or more alleles in a population; it demonstrates that the frequencies of alleles in a gene pool are not changed by the process of sexual recombination.

HETEROZYGOTE SUPERIORITY: The greater fitness of a heterozygote as compared with the two homozygotes.

HYBRID: The offspring of two parents that differ in one or more heritable characters; the offspring of two different varieties or of two different species.

INBREEDING: The breeding of plants or animals which are closely related genetically. In plants it is usually brought about by repeated self-pollination.

LATENT VARIABILITY: The potential range of variation in a particular genetic trait already present in a natural population.

MUTATION: An inheritable change of a gene from one allelic form to another.

OUTBREEDING: The breeding of plants or animals which are not closely related genetically.

POLYMORPHISM [Gk. *polys,* many, + *morphē,* form]: The occurrence together of two or more distinct forms of a species.

POLYPLOIDY [Gk. *ploos,* fold or times]: The possession of more than two complete sets of chromosomes per cell.

POPULATION: In genetic terms, an interbreeding group of organisms.

RECOMBINATION: The appearance of gene combinations in progeny which differ, as a result of the sexual process, from the combinations present in the parents.

SELECTION: The nonrandom reproduction of genotypes, resulting from the increased frequency of certain genes in the population and the decreased frequency of others.

SPECIATION: The emergence in a single population of smaller groups sufficiently different in genetic constitution to be capable of coexistence in the same environment without interbreeding.

SPECIES: [L. *species,* kind, sort]: A group of animals or plants that actually (or potentially) interbreed and are reproductively isolated from all other such groups.

STABILIZING SELECTION: The continual elimination of extreme types from the population as a result of their severe incompatibility with the environment.

SUGGESTIONS FOR FURTHER READING

CARLQUIST, SHERWIN: *Island Life: A Natural History of the Islands of the World,* The Natural History Press, Garden City, New York, 1965.

An exploration of the nature of island life and of the intricate and unexpected evolutionary patterns found in island plants and animals.

DOBZHANSKY, THEODOSIUS: *Genetics and the Origin of Species,* 3d ed., Columbia University Press, New York, 1951.*

A presentation of the basic themes of population genetics by one of the foremost leaders in studies in Drosophila.

FISHER, RONALD A.: *The Genetical Theory of Natural Selection,* 2d ed., Dover Publications, Inc., New York, 1958.

First published in 1929, this is one of the crucial books in the new synthetic theory of evolution and remains one of the best expositions of the subject for the interested student.

HAMILTON, TERRELL H.: *Process and Pattern in Evolution,* The Macmillan Company, New York, 1967.*

This short book, designed as a supplementary text, is outstanding for its clarity of definition and presentation of evolutionary concepts.

HUTCHINSON, G. EVELYN: *The Ecological Theater and the Evolutionary Play,* Yale University Press, New Haven, Conn., 1965.

By one of the great modern experts on freshwater ecology, this is a charming and sophisticated collection of essays on the influence of environment in evolution—and also on an astonishing variety of other subjects.

LACK, DAVID: *Darwin's Finches,* Harper & Row, Publishers, Inc., New York, 1961.*

This short, readable book, first published in 1947, gives a marvelous account both of the Galapagos, its finches, and other inhabitants and of the general process of evolution.

MAYR, E.: *Animal Species and Evolution,* Harvard University Press, Cambridge, Mass., 1963.

A masterly, authoritative, and illuminating statement of contemporary thinking about species—how they arise and their role as units of evolution.

SMITH, JOHN MAYNARD: *The Theory of Evolution,* 2d ed., Penguin Books, Inc., Baltimore, 1966.

Written for the general public, this book is notable for its many concrete examples of evolution, past and present.

STEBBINS, G. LEDYARD: *Processes of Organic Evolution,* Prentice-Hall, Inc., Englewood Cliffs, N.J., 1966.*

A brief review of the entire field by one of its outstanding practitioners.

* Available in paperback.

PICTURE ESSAY

Across North America*

Sea otters in a bed of brown algae off Kirkof Point, Alaska.

* By a leisurely route, proceeding irregularly from west to east.

On the Olympic peninsula in Washington. As much as 12 feet of rain fall annually from the moisture-laden Pacific westerlies. The result is a temperate rain forest, rich in many species of trees, including the Douglas fir, which reaches 18 feet in diameter, huge red cedars, western hemlocks, and Sitka spruces which grow to heights of 300 feet. The roots of even the largest trees penetrate only about 3 feet into the water-soaked ground, and as a consequence, uprooting is common. The rotten logs of the fallen giants provide nurseries for seedlings which send their roots around their flanks to the soil below. The forest floor and the trunks and limbs of the trees are carpeted with more than 70 species of mosses.

(a)

A community is made up of many populations of interdependent organisms. The redwood community along the coast of northern California and southern Oregon is dominated by the redwood (a), Sequoia sempervirens, *whose distribution is determined by the coastal fog that provides the required moisture during the long dry summers. Little light reaches the floor of the redwood forest, which is thickly carpeted with mosses and needles. Among the plants that can grow in this damp, shady, acid soil are the redwood sorrel* (b) *and the swordferns* (c). *These coastal redwoods and the giant sequoias,* Sequoiadendron giganteum, *which are found only in widely scattered groves on the west slope of the Sierra Nevada, are among the last survivors of a redwood empire that, 25 million years ago, stretched all the way across the northern hemisphere.*

(b)

(c)

The Sierra Nevada is a single vast block of the earth's crust, 400 miles long, tilted toward the southwest, so that its Pacific face forms a gentle slope. Its northeastern edge, by contrast, is a jagged crest, the High Sierra, the topmost point of which is Mt. Whitney (14,495 feet).

Yosemite Valley, a deep trough gouged out by glaciers, lies east of the High Sierra. In the spring, waterfalls, fed by the melting snows of the mountain peaks, leap a thousand feet and more down sheer canyon walls.

In the foothills of southern California, neighboring states, and northern Mexico, there is a heavy rainfall in the winter, followed by a long dry summer and autumn. The native vegetation is chaparral. Mule deer live in the chaparral during the spring growing season, moving out to cooler regions during the summer. The resident vertebrates—brush rabbits, wood rats, lizards, and brown towhees—are generally small and dull-colored, matching the dull-colored, leathery-leaved shrubs. Ponderosa pine is growing on the heights. To the east, chaparral grades into desert.

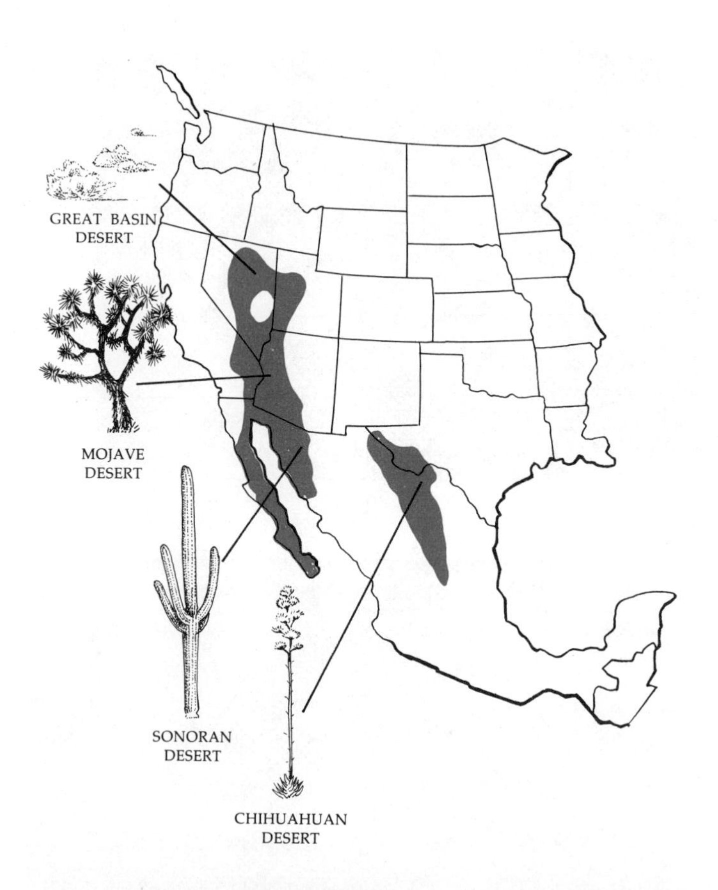

Joshua tree in the desert.

The North American deserts extend for nearly 500,000 miles. The Sonoran stretches from southern California to western Arizona and down into Mexico. It is dominated by the giant saguaro cactus, often as much as 50 feet high, with a wide-spreading network of shallow roots. Water is stored in a thickened stem which expands, accordionlike, after a rainfall.

To the southeast is the Chihuahuan desert whose principal plant form is the agave, or century plant, a monocot.

North of the Sonoran is the Mojave, whose characteristic plant is the grotesque Joshua tree, named by early Mormon colonists who thought that its strange, awkward form resembled a bearded patriarch gesticulating in prayer. The Mojave contains Death Valley, the lowest point on the continent (280 feet below sea level), only 80 miles from Mt. Whitney.

The Mojave blends into the Great Basin, a cold desert, bounded by the Sierra Nevada to the west and the Rockies to the east. It is the largest and bleakest of the American deserts. Here the dominant plant form is sage brush.

Most amphibians require two or three months to develop their adult form, but the desert spadefoot, conceived in a fleeting rain puddle, has been observed to progress from egg to tadpole to a small, hopping toad in only 9½ days. As an adult, the spadefoot can lie dormant for 9 or 10 months, concealed in a burrow dug with its spadelike hindfeet and lined with a gelatinlike substance of its own secretion.

Desert scorpion.

Jackrabbit. Why would you expect these large ears in a desert species?

Following the brief spring rains, the desert may be carpeted with flowers. The seeds of some desert annuals germinate only after sufficient rain has fallen to leach away inhibitory chemicals in the seed coat, an environmental cue that enough moisture is present for the plant to complete a hasty cycle from seed to flower to seed again.

Texas horned lizard, camouflaged.

Zabriskie Point, Death Valley, California.

Extending 2,200 miles from the Yukon south to New Mexico, the Rockies, the "backbone of the continent," are a discontinuous range of mountains and highlands, interrupted by rolling valleys and plateaus. They reach their greatest width in Colorado (300 miles) and also their greatest heights; here there are 54 mountains more than 14,000 feet high, 250 between 13,000 and 14,000 feet. This is Castle Creek, near Aspen, Colorado.

Tundra on top of Pecos Baldy Peak, northern New Mexico. Such mountain tundra is comparable in many respects with that found hundreds of miles to the north in the arctic. Here, however, forested slopes are found within a few hundred feet. The change from biome to biome is more rapid with elevation than with latitude.

The biomes of the tundra and taiga form two circumpolar belts around the northern hemisphere, bounded on the north by the polar icecap, the remains of the Pleistocene glaciers. The characteristic vegetation of the tundra consists of lichens, grasses, sedges and dwarf willows, all small, stunted and compact, resistant to the wind and snow. The tundra blends southward into the coniferous trees of the taiga, mainly pine, spruce and fir, which are intermingled with some deciduous trees, such as birch, poplar, and alder. A thick layer of needles and dead twigs which decompose slowly covers the ground.

Caribou in Yukon territory. Most grazing animals such as these move from the tundra to the taiga after the first, early snowfalls cover the sparse vegetation.

Two white-tailed ptarmigans. One of the few birds that never leave the tundra, the ptarmigan changes its coat three times a year, following photoperiodic cues. They are brown in summer, gray in autumn.

Arctic terns, like many other species, nest in the tundra, taking advantage of the long hours of daylight to gather food for their voracious young. They winter in the Antarctic, following migration routes of 8,000 to 10,000 miles. Within three months, the birds from this still unhatched clutch must be ready for this voyage.

Timber wolf, predator of the taiga.

Moose, near Banff, Alberta. Many areas of the taiga are wet and even marshy in the summer, as a result of the melting snow. (Taiga, in Russian, means "swamp forest.") Through the long winter, however, the water of the tundra and taiga is locked in the form of ice.

Between the Rocky Mountains and the Appalachians to the east, is the great "heartland" of America. About 350 million years ago, much of it was a vast inland sea. Now it is almost entirely grasslands and forest. To the west, in the rainshadow of the Rockies, the grass is short and tough but the air gathers and releases moisture from the Gulf of Mexico and the Great Lakes as it moves eastward, and the grasses become increasingly longer and more luxuriant, blending on their eastern border with the deciduous forest.

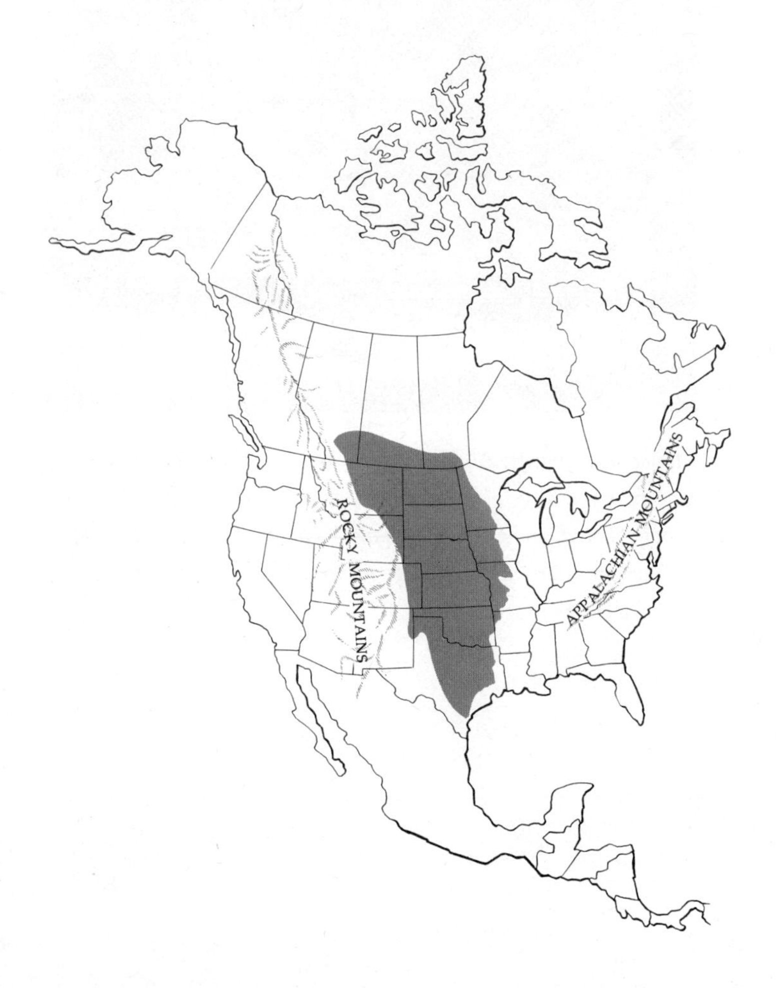

Buffalo were at one time the native herbivore of the short-grass prairie. Their place has now been taken largely by beef cattle.

The land that was previously long-grass prairie is now almost completely covered by wheat and cornfields.

Prairie hawk.

Prairie dogs, natives of the short-grass prairie, live in large communities which are complex both in their underground structures of tunnels and burrows and in their social organization. This town member has sighted possible danger —presumably the photographer—and is emitting a high, whistle-like cry of alarm.

Texas cowhands.

The eastern third of the continent is forestland, although much of this natural forest has been taken over for agriculture and for the great metropolitan centers of the east coast. Moving southward, the coniferous forest of the taiga gradually fills with deciduous trees. These mixed northern forests give way to deciduous forest, a biome characteristic of an area which receives at least 15 inches of rainfall, much of it in the summer months (in contrast to the chaparral) and a growing season of at least three months.

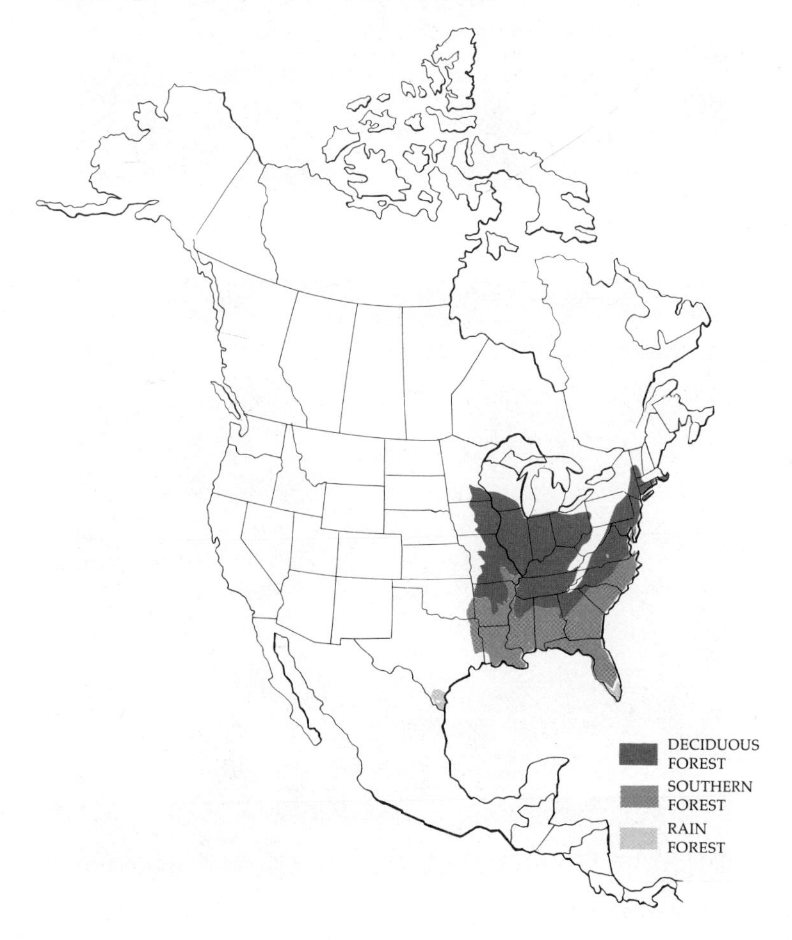

The Great Lakes, which lie near the northern border of the deciduous forest, were carved out of the face of the continent by the last glaciation. Here is a bay along the north shore of Lake Superior as seen from the lighthouse on Split Rock.

Young deer drinking.

Rainy spring day in the Adirondacks. These trees are a mixture of conifers, largely spruce, and deciduous species, including maple and birch. The many lakes of upstate New York are also of glacial origin.

White-footed mouse. It has been gnawing on the deer antler, a source of minerals.

Saw-whet owl catching white-footed mouse.

The deciduous forest in the southern Appalachians is one of the richest temperate zone forests in the world, containing many kinds of hardwoods. The Appalachians, which extend from Newfoundland 1,500 miles southward to central Alabama, are worn and ancient mountains with few peaks above the timberline. This photograph was taken in autumn.

Badgers.

Young raccoons.

Young stand of oak and hickory in Virginia. These trees are typical of the southern and lowland areas of the deciduous forest.

The perennial herbs of the deciduous forest flower in early spring before the leaves have grown back on the trees. Among the common harbingers of spring is the hepatica.

The southern forest is dominated by yellow pines—loblolly, slash, and longleaf—mixed with broadleafed species such as maples, elms, cottonwoods, and a variety of oaks.

Secondary succession. A young stand of loblolly pine is taking over an abandoned field in Wake County, North Carolina. The plow furrows are still clearly visible.

Bobcat, a predator of the southern forest.

The opossum is the only marsupial of North America. The babies are only about one inch long when they are born; in order to survive they must make their way to the mother's pouch.

Freshwater swamp, with waterlilies and cattails.

In two small areas on the continent—at the tip of Florida and a small section of southeastern Texas—are found the only examples in North America of the tropical forest biome. Here, where there is little seasonal change in temperature and rainfall, is a rich dense vegetation of broadleafed evergreens. Here, in the Florida Everglades, an egret takes flight. The trees are mangroves.

During the Ice Ages, the great weight of the glaciers bore down on the flexible crust of the earth and, along the northeast coast, tilted it downward as much as 1,200 feet in some areas, submerging the coastal plain. Mountains and ridges near the shore were transformed into the numerous islands that dot the coast of eastern Maine and Nova Scotia and the inland mountains were hacked away by the cold seas to form bluffs and escarpments. The coast line is highly irregular because of the many valleys now taken over by the sea.

Within this long egg case, composed of a parchmentlike material, hundreds of eggs are being deposited. When the baby whelks hatch, they will break through thin areas in the individual capsules and emerge as miniature copies of their parents.

Diamondback terrapin in salt water marsh.

(Above) *From the tip of Cape Cod southward is one continuous coastal plain, a broad ribbon of sand formed by millenia of weathering and decay of rocks.*

Laughing gulls (right) *in summer plumage over the Atlantic.*

Southern coast with cattails.

SECTION 9

The Evolution of Man and the Ecological Crisis

9–1
Is the chimpanzee following man's evolutionary route? Chimpanzees are just now making the transition from an arboreal existence to life in the savanna and, according to recent field studies, they use a number and variety of simple tools. Perhaps most striking, they appear to share with us some complex emotional and social attributes once regarded as exclusively human.

SECTION 9

The Evolution of Man and the Ecological Crisis

Man is a member of the Kingdom Animalia. As such, he is a highly organized collection of eukaryotic cells without walls and without chloroplasts. He has a nervous system, and he moves by means of contractile fibers. He is a heterotroph, relying either directly or indirectly on green plants as his source of energy, of carbon compounds, and of nitrogen. And like all other living things, man is a part of an ecosystem, on which he is dependent and which he influences by his activities.

Man is a member of the phylum Chordata, animals which have, during at least some stage of their existence, a notochord, pharyngeal gill slits, and a hollow nerve cord on the dorsal side. He belongs to the largest subphylum of the Chordata, the Vertebrata, in which the notochord, in the course of development, is replaced by cartilage or bone, forming the segmented vertebral column, the backbone. Members of this phylum have a skull surrounding a well-developed brain and, usually, a tail. The Vertebrata are divided into a number of classes, including fish, amphibians, reptiles, birds, and mammals. Man is a member of the class Mammalia. Like other living organisms, he is a product of his evolutionary history and can be understood completely only in terms of this history.

EVOLUTION OF THE MAMMALS

According to the fossil record, the first mammals arose from a primitive reptilian stock about 200 million years ago, at about the time of the first dinosaurs. Our information about these mammals is very slight, consisting of only a few fragments of skulls and some rare teeth and jaws. From these scraps of evidence, we believe that the first mammals were about the size of a rat or mouse. They had sharp teeth, indicating that they were flesh or insect eaters, but since they were too small to attack other vertebrates, they are assumed to have lived mostly on insects and worms, supplementing their diet with tender buds, fruit, perhaps eggs, and probably whatever else they could find. Although they were tiny and seemingly helpless compared with the reptiles that were their contemporaries, they had several advantages. The most crucial perhaps is that they were warm-blooded, or to put it more precisely, homeothermic. As a consequence of their warm-bloodedness, these early mammals were able to be more active over long periods of time and more alert than their reptilian contemporaries. Also, and perhaps most important at the beginning, their high and constant body temperature enabled them to be active at twilight and throughout the night, as they most likely were, when the reptiles were rendered inactive by the cold.

For about 80 million years, these small animals led their secretive, probably nocturnal existences in a land dominated by carnivorous reptiles. Then suddenly, as geologic time is measured, the dinosaurs disappeared. Their disappearance occurred at a time when, geologists believe, there was a drop in the average temperature and, perhaps more im-

9–2
Among the great reptiles of North America some 90 million years ago were the duck-billed dinosaurs, such as Edmontosaurus *on the right, and the hooded* Corythosaurus *and crested* Parasaurolophus *in the swamp on the left. Their contemporaries included the heavily armored* Palaeoscincus *in the center and the ostrich-like* Struthiomimus *in the background. All of these are representatives of the peak of dinosaur evolution which was reached shortly before the extinction of the entire group.*

portant, a marked increase in seasonal temperature fluctuations. Small reptiles are able to regulate their temperatures to a large degree by their behavior—sunning themselves or burrowing or hibernating—but the giant reptiles, because of their great bulk and consequent lower surface-to-volume ratio (remember?), would have had much more of a problem both in unloading heat and in warming up again to a physiologically fully operational temperature. Perhaps it is significant that the warm, essentially unchanging climate under which the dinosaurs evolved and thrived lasted 130 million years. Perhaps these ideal conditions—from a reptilian point of view—led to such a high degree of stabilizing selection that the dinosaur populations no longer contained enough genetic variability to undergo a major adaptive change. In any case, by the end of the Cretaceous period the very large reptiles had disappeared forever, and about 75 million years ago an explosive radiation of the mammals began.

As we mentioned in Section 4, there are three groups of mammals: (1) the egg-laying mammals, or monotremes, such as the duck-billed platypus; (2) the marsupials, examples of which are the kangaroo and the opossum; and (3) the placentals, by far the largest group. Among the placentals are a number of different lines, or orders. The one whose members most closely resemble the earliest mammals is the order of insectivores, which includes moles, hedgehogs, and shrews. Other familiar placental orders are the carnivores, ranging in size from the now-extinct saber-toothed tiger down to small, weasel-like creatures; the herbivores, or ungulates, grazing animals from which most of our domesticated farm animals are descended; rodents; rabbits and hares; bats, the only flying mammals; whales; and the primates, the order to which man belongs.

In the tertiary period—the "Age of Mammals"—there existed, as far as we know, a total of 32 different orders of mammals and some 15,000 species. Today, there are only 18 orders, with some 4,500 species. The greatest wave of extinction took place in North America only a few thousand years ago and involved most of the large mammals that inhabited the continent, including the horse (which did not return to North America until the coming of the Europeans in comparatively modern times), the ground sloths, giant armadillos, camels, mastodons, and mammoths. No adequate explanation has ever been given for the disappearance of these animals. It was apparently not due to climatic changes; the last of the glaciations was over by that time. The only major new element in the environment was man. Did his activities in some way upset ecological balances and so lead to these extinctions, as such activities are doing even today? At the present time, it is simply not known.

9–3

A tree shrew, a small nocturnal animal which is believed to resemble the early mammals and also the earliest of the primates. Notice the relatively unspecialized five-digited feet, partially grasping the trunk.

Evolution of the Primates

The primate ancestor was a shrewlike mammal. Primate evolution began when a group of these small insectivores took to the trees. Adaptation to arboreal life profoundly influenced the evolution of the primate hand, preadapting it to the holding of tools. It also resulted in a shift from dependence on smell, the dominant sense of most mammalian lines, to a dependence on sight, far more useful to tree dwellers. Most of all, the degree of judgment and rapidity of decision required for jumping from limb to limb must have placed a high selective value on intelligence. The primates are the most intelligent of the mammalian orders.

The Coming of Man

During the Miocene epoch, some 20 million years ago, climatic changes produced new environments, gradually converting the tropical forest into savanna with scattered clumps of trees. Some of the primates retreated as the forest retreated. Others—those ancestral to the great apes and to ourselves—adapted to this new environment. About 14 million years ago, according to the fossil record, a group of animals appeared that were distinct from their ape contemporaries. This group, perhaps because they were driven out of the groves by the more powerful ape species, moved from the wooded savanna out into the open grasslands, where they became fully bipedal (two-legged). Bipedalism freed their hands for carrying food and other objects, throwing rocks and other weapons, and using tools. They became omnivorous, eating both meat and vegetables, unlike other apes, which eat a mostly vegetable diet, supplemented occasionally by eggs and small or infant animals which they chance across. Like most of the other apes (apparently excepting the orangutan), man seems to have been gregarious, moving in small nomadic bands. At some point the human female underwent an important biological change: she became sexually receptive all the time. Other

9–4
Some close relatives of man. (a) *Gibbons are the smallest and, apparently, the least intelligent of apes. Modern gibbons, like their ancestors of the tropical forest, spend most of their time in the treetops.* (b) *Instead of jumping and scrambling through the trees as monkeys do, apes swing hand over hand. As a consequence, the hand became preadapted for manipulating objects and the body became more upright, facing forward and outward—a preadaptation for bipedalism. The modern chimpanzee sleeps in trees but spends much of his waking time on the ground, where he scampers on all fours.* (c) *Gorillas have largely abandoned the trees, except for some of the smaller animals that sleep in the lower branches. They usually walk on all fours, carrying their weight on the second joints of their fingers and the soles of their feet, but they can stand and walk erect, freeing their hands for carrying.*

(b)

(a)

(c)

female mammals, including the modern apes, are sexually receptive only during estrus, when they are "in heat." This development strengthened the pair bonds between male and female and is apparently the biological basis of the human family. Perhaps most important of all, man developed the capacity for speech. Many other animals can recognize and respond to verbal cues, but apparently man alone has the capacity to formulate those symbols we call "words," use them for the storage of information, and articulate them.

About 1.5 million years ago, the first of the four major cold phases of the last Ice Age began. In each of these four glaciations, sheets of ice as much as a mile thick in some regions spread out from the poles, scraped their way over much of the continents—reaching as far south as the Ohio and Missouri Rivers in North America and covering Scandinavia, most of Great Britain, northern Germany, and northern Russia—and then receded again. We are living at the end of the fourth glaciation, which began its retreat only some 20,000 years ago. During these periods of violent climatic changes, living organisms were placed under extraordinary evolutionary pressures. Plant and animal populations had to move, to change, or to become extinct. In the interglacial periods, during which the average temperatures were warmer than those of today, the tropical forests and their inhabitants spread up through today's temperate zones. During the periods of glaciation, only animals of the northern tundra could survive in these same locations. Rhinoceroses, great herds of horses, large bears, and lions roamed Europe in the interglacial periods, and in North America, as the fossil record shows, there were mammoths, camels, horses, lions, saber-toothed cats, and great ground sloths, one species as large as an elephant. In the colder periods, reindeer ranged as far south as southern France, and during the warmer periods, the hippopotamus reached England.

It was during this extraordinarily difficult time that modern man came into being. His rate of evolution, as judged by changes in cranial capacity, was more rapid than that

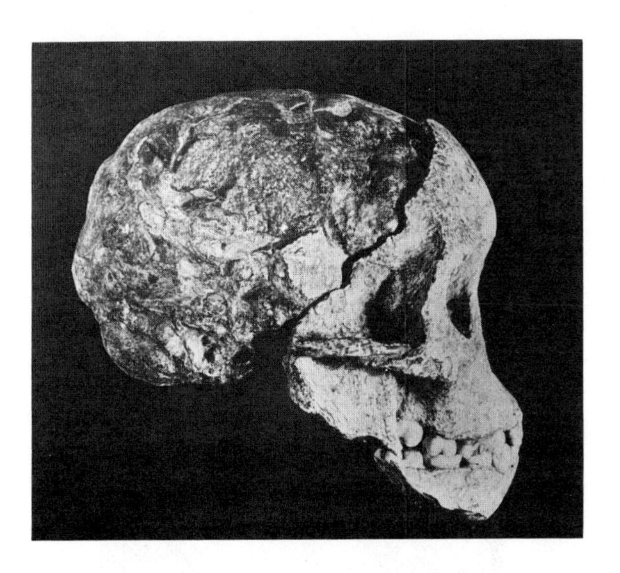

9–5

The skull and a reconstruction drawing of a humanlike child who lived more than 2 million years ago in Africa. Australopithecus africanus, *"the southern ape of Africa," walked upright, made and used tools, and was probably ancestral to modern man.*

9–6
Neanderthal man, a member of the genus Homo, *lived about 100,000 years ago. He was short, probably muscular, and hunted a wide range of game, including mammoths and wooly rhinoceroses. Recent Neanderthal findings include graves with gifts of food and flowers and the skeleton of a man so crippled he must have been cared for since birth, both of which suggest that Neanderthal, for all his primitive appearance (as we reconstruct him), was far more "civilized" than previously suspected—capable not only of compassion and concern but of abstract ideas, such as the concept of life after death, and so also probably of speech.*

9–7
Cro-Magnon man, who closely resembled modern man, replaced Neanderthal about 35,000 years ago. He, too, was a hunter. His graceful, imaginative cave drawings—such as this one of a galloping wild boar—have now been found in many sites throughout Europe. They seem to have played a part in ceremonies associated with the hunt.

known for any other species during any other comparable time interval. Cro-Magnon man, who made his appearance about 35,000 B.C., is physically indistinguishable from modern man, and we have no reason to believe that he was any less intelligent. The history of man from the last glaciation until now is primarily a story of his cultural evolution. Like his biological evolution, man's cultural evolution has taken place at an ever-increasing pace until, at this moment in time, these changes pose problems which, in the view of some observers, threaten the survival of this new and dominant species.

THE AGRICULTURAL REVOLUTION

The most important single event in the evolution of human culture, it is generally agreed, was the development of agriculture. Bound to the soil, men stopped leading their wandering, hazardous existence, settled in ever-enlarging communities, multiplied, and prospered. Also, a given area under agriculture can support many times as many people as the same area used for hunting. The agricultural revolution seems to have been the direct cause of our present population crisis. About 25,000 years ago there were perhaps 3 million people. By the close of the Pleistocene era, some 10,000 years ago, the human population, then probably numbering a little more than 5 million, spread over the entire world.

Agriculture appears to have begun simultaneously about 8,000 B.C. in several different centers in both the Old World and the New. From the Old World sites, we have fossil imprints of cultivated wheat and barley, remains of domesticated goats, sheep, and cattle, and pottery vessels, stone bowls, and mortars. The farmers of the New World grew corn, pumpkins, squash, gourds, and cotton. The potato, the sweet potato, the peanut, and the tomato are also New World crops. The one domesticated animal found in the ancient farms of both the Old World and New, is the dog, man's oldest and still most faithful companion.

Urbanization

With a more stable food supply and the greater physical safety of community living, man began rapidly to increase his numbers. By 4000 B.C., about 6,000 years ago, the population had increased enormously, to more than 86 million, and by the time of Christ, it is estimated, there were 133 million people. In other words, the population increased by more than 25 times between 10,000 and 2,000 years ago. Agriculture not only increased the food supply, it also brought people together in groups, which, as the population increased, grew larger. During the 5,000 years in which agriculture was developing, more and more new centers grew up, and by 3,000 B.C. all the plains of Europe south of the Scandinavian mountains were inhabited by people who lived in villages of a more or less permanent character. Because a single person could produce more than enough food for his own family, many people were freed from the land and began to pursue other occupations.

When man was a hunter and food gatherer, 2 square miles, on an average, were required to provide enough for one family to eat, and so population had to be of very low density. Following the development of agriculture, increasing concentrations of people in urban centers facilitated the increasingly rapid exchange of ideas. Since 1650, when the world population reached 500 million, the development of science, technology, and industrialization has brought about further profound changes in the life of man and in his relationship to nature.

The Conquest of Death

One of the most important changes brought about in the last several hundred years is the reduction in the death rate, due partly to the discovery in the twentieth century of the new "wonder drugs" but even more to the earlier and less spectacular development of public health measures, such as clean water supplies, pasteurization of milk, and sanitary means for sewage disposal.

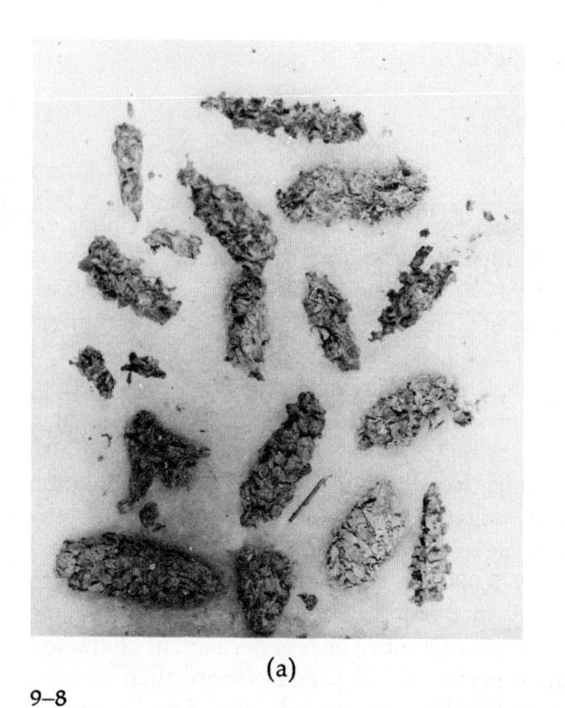

(a)

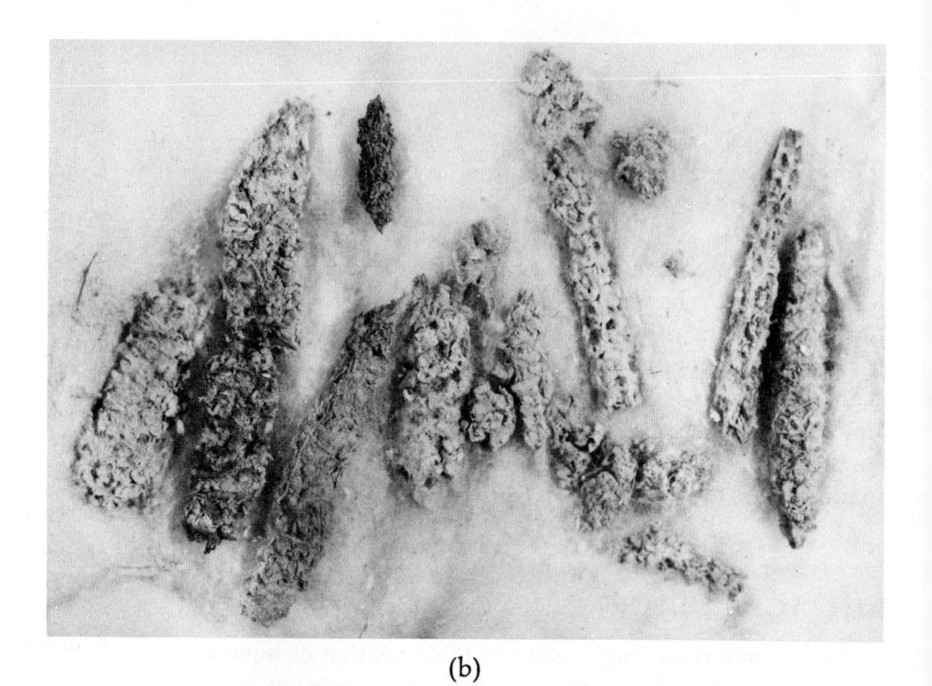

(b)

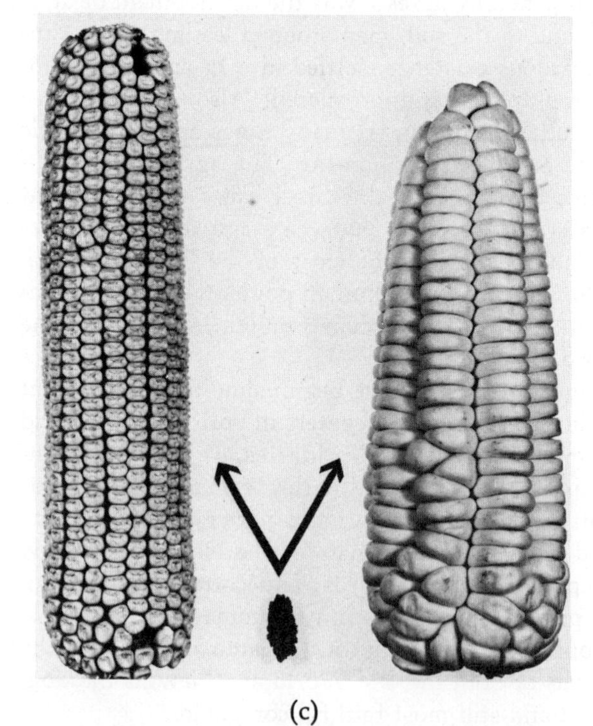

(c)

9–8
Agriculture has resulted in marked changes in many food crops. (a) *Fossil cobs of wild corn from the San Marcos Cave in the Tehuacan Valley of Mexico. The corn cobs, shown actual size, have been found among archeological remains dating from* 5200 *to* 3400 B.C. (b) *Fossil cobs of early cultivated corn, from* 3400 *to* 2300 B.C. *These are much larger, probably as a result of irrigation and of weeding of competing plants. Like the wild cobs, these early cultivated ones are attached to the plant by only a fragile stalk, and each kernel is surrounded by bracts (chaff).* (c) *A fossil cob, about* 2 *to* 3 *centimeters long, from a cave in New Mexico compared with an ear of modern corn-belt corn (left) and large-seeded Peruvian flour corn (right).*

(a)

(b)

9–9

(a) Miniature painting of a plague hospital of the Middle Ages, showing the infected swellings (buboes) characteristic of the disease which is caused by a bacterium transmittted to man by fleas from infected rats. The bubonic plague caused the death of an estimated 25,000,000 people, one-fourth of the population of Europe. (b) Macabre fifteenth-century woodcut of the Dance of Death, reflecting the spirit of that grim age.

An infant born 300 years ago had little better than a fifty-fifty chance of surviving its first year, whereas in many countries today, infant mortality is less than 2 percent. (Incidentally, the United States ranks comparatively low in its control of infant mortality: fifteenth among modern nations. This low figure is a reflection of the level of medical care received by the poor, particularly in the cities.)

Judging from the fossils, Neanderthal man, once he reached adulthood, might be expected to live to be about 35 years old. A man reaching adulthood in fourteenth-century England had about the same life expectancy. In the United States the average life expectancy, which has increased 13 years since 1900, now has reached about 67 years for men and about 74 for women.

It would seem that these accomplishments, of which modern civilization is justly proud, have also resulted directly in the most pressing problem of our time—the explosive growth of the human population.

THE POPULATION EXPLOSION

In 1972, there were approximately 3.8 billion people on our planet. This is an almost incomprehensibly large figure. Imagine, for instance, that all the people in the world stood in one long single line. The line would reach 1¼ million miles. It could stretch to the moon and back twice or could circle the earth more than 50 times at the equator.

Moreover, the rate of increase of this enormous population is unprecedented. The human family, according to the Population Reference Bureau, is presently gaining an average of 1,400,000 members a week, 199,104 a day, 8,296 an hour, and 138 a minute. If it continues growing at its present rate of 2 percent a year, the world population will reach 4 billion in 1975, 5 billion in 1986, and 6 billion in 1995. Around 1650, when there were some 500 million people in the world, the doubling rate of the population was approximately 1,000 years. By 1850, however, the population had already doubled, indicating a doubling rate of 200

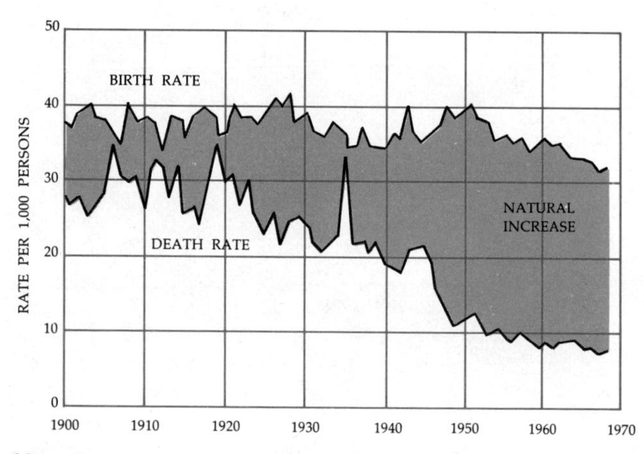

9–10

In many tropical countries, the death rate has fallen rapidly since 1940, resulting in a rapid growth of the population. The drop in death rate is the result of increased medical services, control of malaria by DDT, and the availability of new anti-bacterial drugs, especially the antibiotics. Note that the birthrate has also begun to fall. The data shown here are from Ceylon.

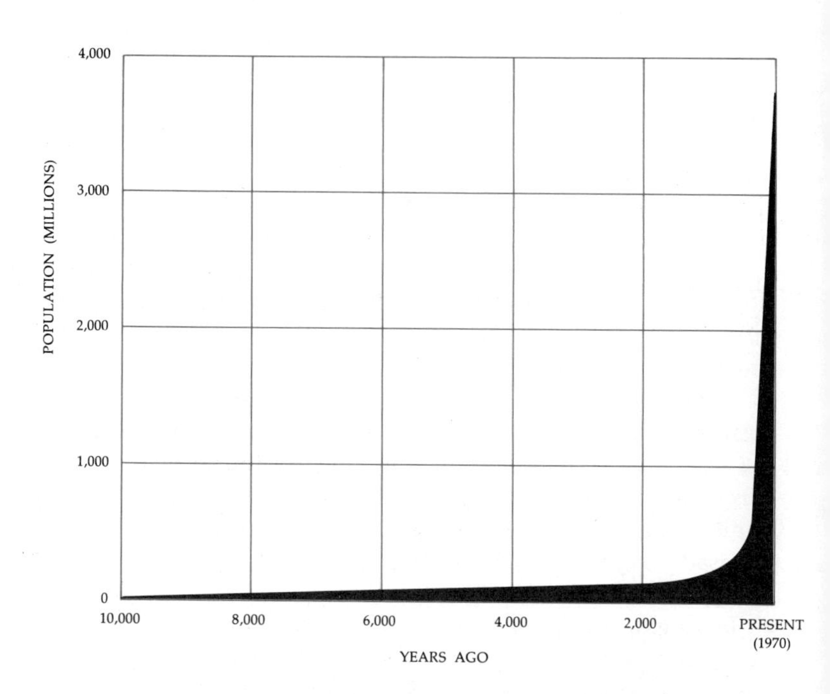

9–11

Growth of the human population.

years. And by 1930, only 80 years later, there were 2 billion people! The doubling rate of the human population is now estimated at approximately 37 years. It took all of man's history—perhaps a million years—for the human population to reach its first billion, which occurred shortly before 1850. At the present rate, by about 2005, well within your lifetimes, the population will be double what it is today—more than 7 billion people.

Population is not increasing at a uniform rate in all parts of the world. In many of the countries of Latin America, Asia, and Africa—the so-called "underdeveloped" countries—doubling times generally range from about 20 to 35 years. According to figures released by the Population Reference Bureau, some examples are: India and Nigeria, 28 years; United Arab Republic, 24 years; Pakistan, 23 years; Brazil and Rhodesia, 22 years; Mexico and the Philippines, 20 years; and El Salvador, 19 years. For the "developed" countries, doubling times are from 50 to 200 years. For example, the population of the United States will double in 70 years, if current estimates hold; that of the United Kingdom will double in about 140 years; those of West Germany and Italy in 117 years; France in 70 years; and the Soviet Union and Japan in 63 years. This unequal distribution of growth makes the problem more difficult. First, births are occurring at the greatest rates in precisely those areas where the new arrivals have the least chance of an adequate diet, good housing, schools, medical care, or future occupations. Moreover, because the affluent citizens of "developed" countries are not constantly reminded of the soaring population rate, they may feel that it is not their problem. Yet a child born to a middle-class American will consume, in his lifetime, a far greater amount of the limited resources of the world—more than twice the amount of food, for instance—than a child born in an underdeveloped country. In the words of Jean Mayer, chairman of the National Council on Hunger and Malnutrition in the United States:

> Rich people occupy much more space, consume more of each natural resource, disturb the ecology more, and create more land, air, water, chemical, thermal and radioactive pollution than poor people. So it can be argued that from many viewpoints it is even more urgent to control the numbers of the rich than it is to control the numbers of the poor.

POPULATION CONTROL

What is going to stop the population growth? One possible answer to this question—a tragic one—is famine. Of the world's 3.8 billion people, at least 2 billion are inadequately nourished now. Although there are good reasons to believe that food production can be increased—a possibility that we shall discuss later in this chapter—it is clear that food production can never be increased as rapidly as the population is increasing at the present time. As Thomas Robert Malthus pointed out in 1798, population grows exponentially (2, 4, 8, 16, 32), whereas food production, if it increases at all, increases linearly (1, 2, 3, 4, 5, 6). (Malthus strongly influenced Darwin's formulation of the principles of natural selection.) Famine could kill enough people, presumably, to keep the population in check. It could also, as any number of recent studies stress, leave indelible scars on any infants that survive in terms not only of physical deprivation but also of lowered intelligence and mental capabilities. Such a generation would be hard-pressed to find leaders to guide it out of the ruin that a great famine would leave behind.

Another possibility, of course, is war. There is already a sharp division between "haves" and "have-nots." Among middle-aged citizens of the United States, one of the most pressing medical problems is obesity; at the same time half of the children of the world go to bed hungry. Yet the United States government rents 20 million acres from our farmers so that they will not grow food on them, and a number of other countries, including Australia, Canada, and Argentina, are involved in similar efforts to restrict production—at a time of worldwide hunger. The "have" nations, moreover, with their industrial economies and high consumer demands, are rapidly depleting the world's natural resources—minerals, timber, fuels—at the expense of the "have-nots." All known reserves of natural gas, for example, will be entirely used up by 1985 at the present rate of consumption. The United States, with 6 percent of the world's population, accounts for almost 35 percent of the world's yearly consumption of fossil fuels.

9–12

The biome of modern man. An aerial view of Oakland, California. More than 70 percent of the population of the United States now live in cities or their suburbs.

As competition for these resources becomes more bitter and as the problems of hunger and crowding and unemployment become more pressing, war will become increasingly difficult to avoid. In a large-scale war, it is difficult to see how the use of nuclear weapons could be prevented. A nuclear war would, of course, solve the human population problem—perhaps permanently.

Birth-control Programs

A third possibility is a drastic reduction in the birthrate, which would limit the population to somewhere near its present size, that is, zero population growth. This solution also has its problems, but the problems, although difficult, are slight compared with those previously considered. (The late Maurice Chevalier, when asked how he liked being seventy-five years old, replied that he liked it fine, considering the alternative.)

One of the problems is that an ideal method of birth control is not yet available. The Pill (see Chapter 6–1), which is the most reliable of all the methods except sterilization, carries with it certain hazards, although these hazards seem to be less than those associated with the pregnancies that it prevents. Also, the Pill is relatively expensive and must be taken on schedule, which interjects problems of human carelessness. By far the most effective means of birth control and the safest, biologically speaking, is vasectomy of the male, a simple operation which prevents the sperm from reaching the seminal fluid. (See page 318.) Vasectomy is growing in popularity but is not yet widely accepted. There are three reasons apparently. One is a lack of information about the procedure; many men, according to a recent poll, believe, quite erroneously, that vasectomy interferes in some way with sexual performance or gratification. Second, even among men who do not plan to father additional children, there are some individuals to whom the idea of being sterile poses a psychological threat. Third—and this is the most realistic objection—there is at present no guarantee that a vasectomy can be reversed (although it often can be) and fertility restored. Therefore, many men wish to reserve the possibility of producing additional children in case they are divorced or their present children die. (Storage of frozen sperm in sperm banks has been suggested as a solution to this last objection.) Undoubtedly, better methods of birth control will eventually be found, but there may not be time to wait.

However, even if an ideal method of birth control were developed and made available on a worldwide basis, we have no reason to believe that any significant reduction in population growth would immediately follow. Among many peoples of the world, including many Americans, opposition to birth control is deeply rooted in tradition, religious belief, or social custom. The domestic role assigned to women, with its consequent ideal of the "good mother," has contributed to this resistance since large families offer women status, an occupation, and a means of demonstrating success. And many families simply do not wish to limit the number of their children and resent the regulation of what seems to them to be a personal affair. There has been a tendency among middle-class Americans, and among the middle class of other Western nations, to regard family planning as primarily a way of controlling the numbers of the poor. In the words of a black delegate to a recent meeting on population control, "You've got Planned Parenthood ladies calling on us twice a day, but nobody is knocking on doors in suburbia and the rural areas."

Most importantly, there seems to be a correlation between economic deprivation and high birthrates. As we have seen, it is in the "underdeveloped" countries—the countries that can least afford it—that the greatest rates of population growth are found today. In India, for example—a country in which large numbers of people are now starving—the annual rate of population increase is 2½ percent. The total population passed the 555-million mark sometime during 1970, according to the Population Reference Bureau and, at this rate, will reach a billion between 1990 and 1995. Yet, according to statistics gathered in one of the family-planning programs now under way in India, most Indian women do not seek help in birth control until they

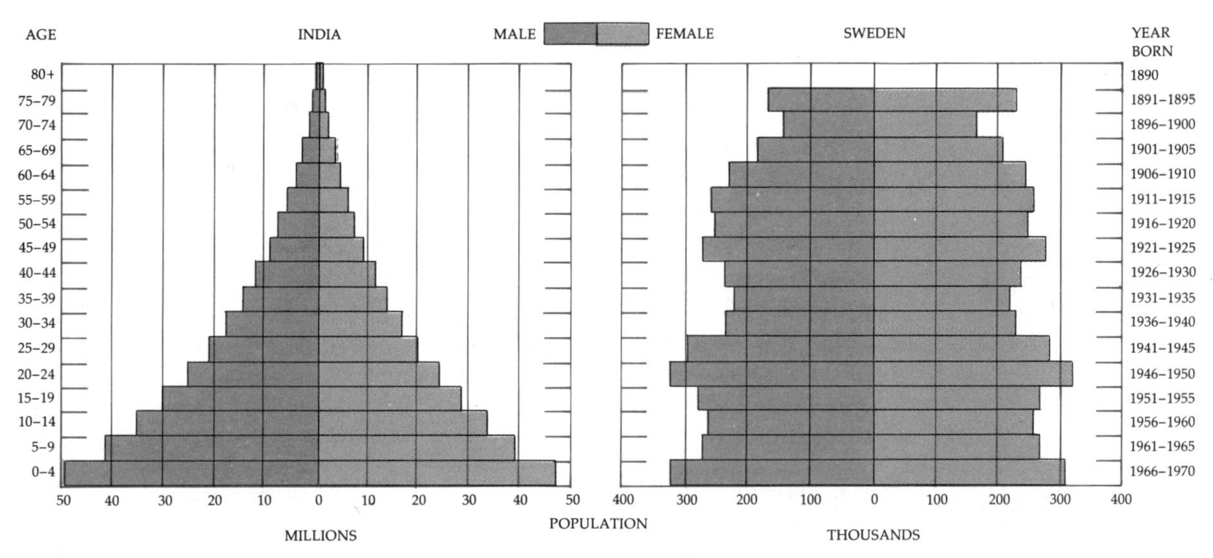

9–13

Charting populations by age and sex permits predictions about future growth rate. In India, for example, nearly 45 percent of the population is under fifteen years of age. Even if these young men and women limit their family size enough to produce only enough children to replace themselves (which means cutting the current birthrate in half), population growth will not level off until about the year 2040—and at a level of well over a billion. Sweden, by contrast, whose population has grown very slowly in recent years, will remain at the same rate unless the birthrate is dramatically increased.

have three or four children; about one-third of all sterilized Indian parents, in fact, have six living children. The desire for large families is deeply rooted in the Indian culture. Children are seen as an economic advantage. They earn their cost before they leave home, and in Indian tradition, they provide security for their parents in old age. Indians know that there is strong probability that at least half of one's children will die before maturity and that there is a likelihood that all of them will. Two recent studies of life expectancy in India show that with the high death rate among children, it is necessary for a mother to bear five children if the couple is to be 95 percent certain that one son will survive the father's sixty-fifth birthday.

Birth Rates, Death Rates and Social Security

Despite the fact that the most recent phase of the population increase is related to a reduction in death rates, particularly from infectious diseases, many experts believe that further reductions in death rates and a general increase in the standard of living will reduce the rate of population growth.

Barry Commoner of Washington University in St. Louis, who has long been in the front ranks of our environmental protectors, is of the opinion that population pressures can be significantly reduced by reducing death rates, particularly among infants. Wherever income rises and death rates drop below a critical point, birthrates begin to fall too, he points out. This was true historically in industrialized nations, and it is becoming true in countries where living standards are just beginning to rise.

China offers an example of a country with a falling birthrate. The birthrate in rural Chinese compounds has now been reduced to 16 per thousand (about what it is in the United States). On being interviewed, both men and

POPULATION CONTROL, STERILIZATION, AND IGNORANCE*

We recently submitted a questionnaire to students and faculty at Cornell University designed to test attitudes and preferences concerning family size and contraceptive technique. The 1,059 respondents (74 percent males) were a mixed lot who represented the physical and biological sciences, humanities, and social sciences and who included faculty (294), graduate students (174), upperclassmen (264) and freshmen (327). Given the level of education of the sample, the results were unexpected in several respects.

First, although there was general agreement (84 percent) on the desirability of limiting family size, a substantial majority (65 percent) said it wanted three children (39 percent) or more (26 percent). Only 30 percent favored two children, and a mere 5 percent expressed preference for one or none. Choice was in no major way affected by age, sex, marital status, parenthood, or professional specialty. Even the respondents whom we expected to be most concerned about the population crisis (for example, graduate students and young faculty in biology) included a minimum of 50 percent with a desire for three children or more.

As regards contraception, about one-half favored "the pill" over all other available means as a way both to space children (53 percent) and to maintain family size at its desired limit (50 percent). Other contraceptive appliances such as condoms, diaphragms, and intrauterine devices were each given top preference by no more than 13 percent of the sample. Voluntary sterilization, either of man or woman, was judged as decidedly undesirable. Only 6 percent opted in favor of vasectomy as the preferred form of contraception once full family size had been achieved. . . . It is of interest in this connection that the consequences of sterilization are not generally understood. For example, asked whether vasectomy would abolish the ability to ejaculate, nearly half the respondents (49 percent) confessed to ignorance or expressed either certainty or probability that emission would no longer accompany orgasm. Biology students scored no better than nonbiologists, and graduate students, even after marriage and parenthood, seemed to be no better informed than freshmen. The only exceptional group was the biology faculty, but, even there, 30 percent were either misinformed or uninformed on this point. . . .

We are bothered by these results. . . . What are we to make of the educated youth growing up among us that is either unconcerned about population growth or, at the very least, unable or unwilling to apply to itself the simple arithmetic of compound interest? And what, if any, are the prospects for improved sex education when ignorance about the reproductive system is widespread even among those who should know best? Thomas Eisner, Ari van Tienhoven, Frank Rosenblatt, Cornell University

* Excerpted from an editorial in *Science*, **167**, Jan. 23, 1970.

women stated that they are voluntarily reducing the size of their families because their old-age security, which children used to provide, is now guaranteed and also because the women are actively employed outside the home.

Japan, too, has a falling birthrate. In a period of about ten years, Japan cut its birthrate almost in half—from 30.2 per thousand in 1947 to 17.6 per thousand in 1959 and it is now being maintained at about this level. No new methods of birth control were introduced (abortion has been a major means of control), and there has been very little expenditure of government funds for public education or propaganda.

A similar voluntary movement seems to be under way among young people in this nation, many of whom plan to have only one or two children of their own and adopt others if they want a larger family. In fact, some, relieved of social and psychological pressures by the population explosion, find themselves free to choose not to have any. The effect on the birthrate of these sentiments will be known over the next 10 years.

9–14
Families from East Pakistan, now Bangladesh, the eighth largest nation in the world (population, 75 million) seeking refuge in India (the second largest). Despite poverty, famine, and a shockingly high rate of infant mortality (15 percent in East Pakistan)—or perhaps because of them—birth rates remain high and population in these countries continues to climb.

OUR HUNGRY PLANET

The most serious of the immediate problems posed by the rapid growth in the population and the widening gap between the "haves" and "have-nots" is chronic starvation. Famines have occurred in the past, but these have characteristically been short-term, localized events, usually the direct effects of particular natural or man-made disasters. This is the first time in human history that a large percentage of the population faces the prospect of never having enough to eat.

Malnutrition takes two forms. There may be a shortage of total calories, in which case the individual cannot meet his energy requirements and must draw on the reserve supplies of his own tissues until these, too, are exhausted. The second form of malnutrition, which may exist where the first type does not, involves a shortage of essential food elements, usually proteins and often vitamins. These are required for building and replenishing tissues and so are particularly important for children, who, unfortunately, are those most likely to be deprived of protein while receiving the dangerous comfort of enough calories. Examples include the "sugar babies" of the Caribbean, who are seldom seen without their piece of sugar cane, and the children of our own cities whose diet is made up largely of candy bars and soft drinks–"empty calories," in the words of the nutritionists.

Protein shortages are more difficult to remedy than shortages in overall calories. Meat is a major source of protein in diets of the "developed" countries, as it seems to have been since man first became a hunter, but meat is now a luxury available in adequate amounts only to the relatively affluent. The reason for this is easily understood in terms of the losses at every trophic level which we discussed in Chapter 7–2. For example, 100,000 pounds of corn can produce, as a rough simplification, 10,000 pounds of beef, which, in turn, produces 1,000 pounds of humanity. The same 100,000 pounds of corn eaten as corn bread, tortillas, couscous, or polenta will provide 10,000 pounds of humanity.

9–15
(a) *A child with kwashiorkor, a West African word that means "the sickness a child develops when another child is born." The skin discolorations, swollen extremities, apathy, and weakness are typical. The syndrome, which is the result of a deficiency in protein but not in total calories, usually develops after a child is weaned. Children with kwashiorkor can be found in the United States as well as in "under-developed" countries.* (b) *Child dying of starvation, caused by deficiency in total calories.*

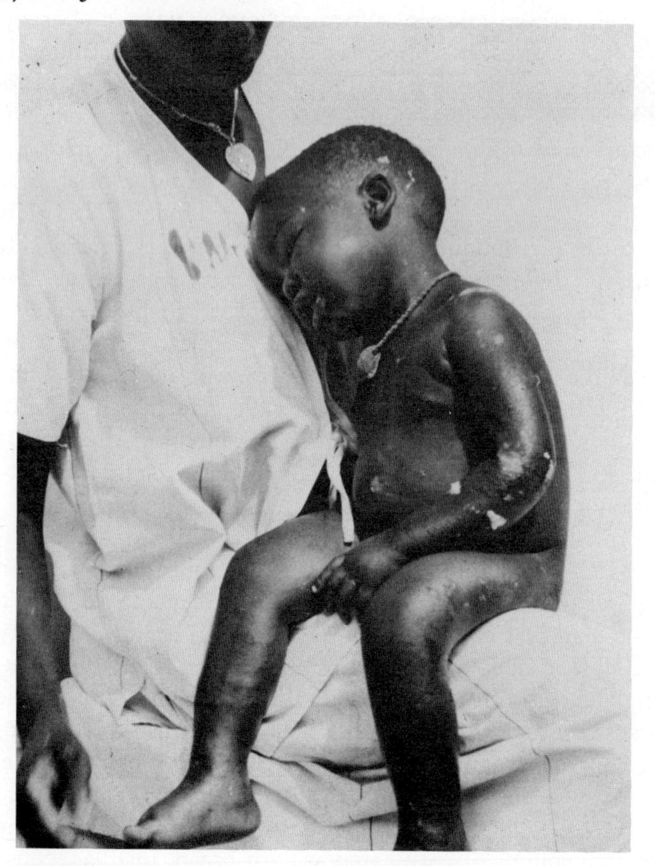

(a)

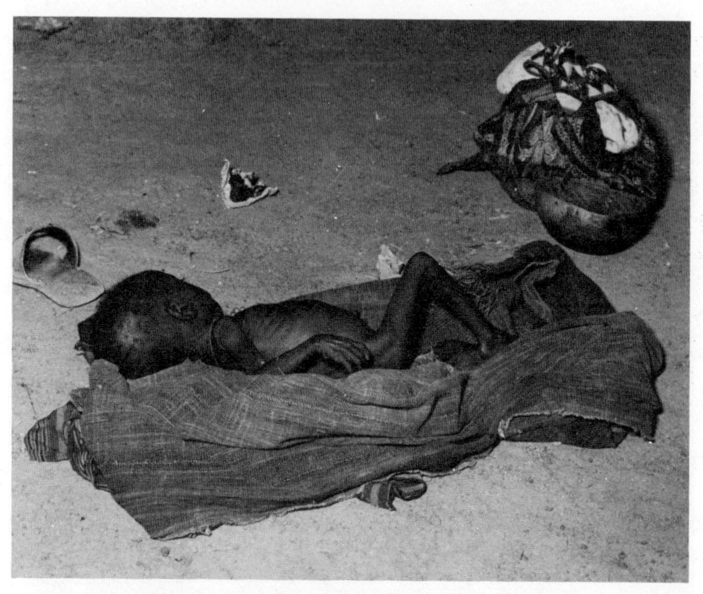

(b)

Therefore, most of the proteins for most of the populations of the "underdeveloped" countries come from the grains—actually, the seeds—of grasses, chiefly rice, wheat, and corn. As a consequence, fluctuations in harvests of these crops have immediate and devastating effects.

In the United States, one-half of the harvested agricultural land is planted with feed crops. Livestock in this country are being fed about 20 million tons of protein, primarily from sources that could be used for human consumption: 89 percent of our corn crop, 87 percent of our oat crop, etc. Of this amount, only one-tenth—about 2 million tons—is retrieved as protein for human consumption. The amount "lost"—18 million tons—is equivalent to 90 percent of the yearly world protein deficit—enough to provide 12 grams of protein a day for every person in the world!*

* These statistics are from Frances Moore Lappé, *Diet for a Small Planet*, Ballantine Books, Inc., New York, 1971. According to her calculations, an adult of average size needs about 43 grams of protein a day.

9–16

The protein conversion ratio is simply the number of pounds of protein fed to livestock to produce 1 pound of protein for human consumption. Beef, as you can see, is by far the most wasteful protein source. (Actually cattle do not require cereal grains or other proteins; all the proteins they need can be provided by the cellulose-digesting bacteria in their rumen. High-protein food is given cattle to fatten or "finish" them before marketing; marbling the meat with fat improves its taste, although not its nutritional value, and enables cattle farmers to command a higher price for the livestock.)

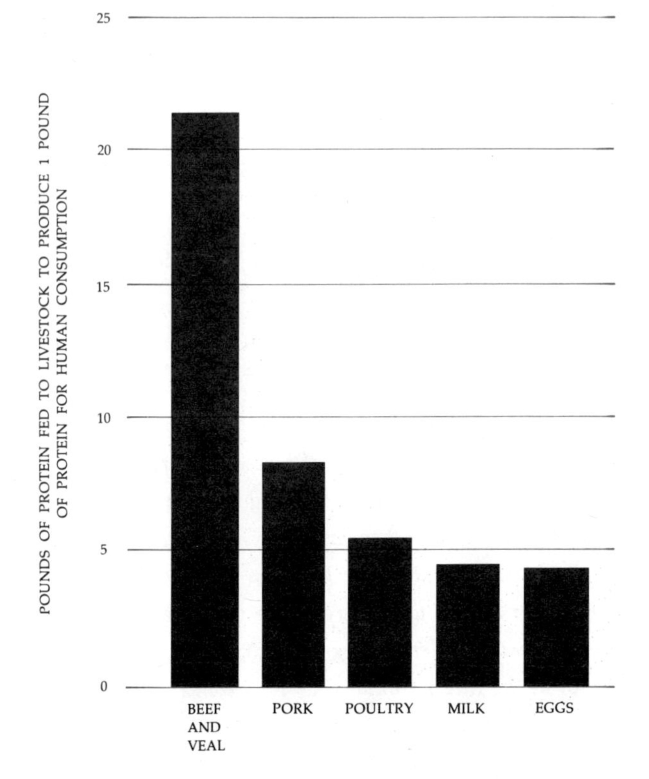

9–17

Off the western coast of South America, the Humboldt Current moves surface water away from the coastal slopes and cold water, rich in nutrients, is brought to the surface in upwellings. As a consequence, these waters abound with fish. In 1968, the United States imported 700,000 tons of fish from Chile and Peru for use as feed for North American livestock. This amount would have provided more than enough protein for the entire population of Peru. This family of five —a little boy is sleeping under the paper at right—was photographed in Lima, Peru.

Protein and Brain Development

The effects of protein deprivation are particularly serious in the young. As we saw in Section 4, the brain is proportionately extremely large in the embryo, and its growth continues to be very rapid during the period after birth. During the first 3 years of a child's life, his body grows to only 20 percent of its adult size, but the brain during this same period reaches 80 percent of its adult size. And, apparently, if it does not grow during this period, it cannot catch up later. More than half of the dry weight of the brain is protein. Studies of rats show that in infant rats who have been deprived of adequate protein, the brain size is smaller and so, often, is the cranium, the portion of the skull that holds the brain. There are indications that these same effects can be seen in undernourished children, and it is clear that intellectual function is impaired. For instance, in a study of slum children in Santiago, Chile, only 51 percent of malnourished children reached the normal range of mental development, as compared with 95 percent of children from the same background who received a supplemented diet.

The Green Revolution

The most successful attempt to date to expand food production has involved the development by selective breeding of new strains of grain plants which produce higher yields. Under the sponsorship of the Rockefeller Foundation, several new strains of wheat have been developed in Mexico which, when heavily fertilized, can produce as much as twice the normal yield per acre. Moreover, these new strains have much less exacting photoperiod requirements than the older ones, so that two or even three crops can be grown in a single year. Using these new strains, Mexico has increased its wheat production sixfold in the last 20 years. More recently, these new strains have been introduced into India and Pakistan, and in 5 years these countries have increased their wheat harvests by more than 60 percent. Forty million acres are now planted with these new seeds, and for the first time in recent history, food

(a)

(b)

9–18

(a) *Norman Borlaug, who was awarded the Nobel Peace Prize in 1970. Borlaug was the leader of a research project sponsored by the Rockefeller Foundation under which new strains of wheat were developed in Mexico. This new wheat, a cross between a native wheat and a Japanese dwarf wheat, has shorter, stiffer stalks, which are much less likely to topple even when heavily laden with grain.* (b) *Field workers weeding a rice plot at the International Rice Research Institute in the Philippines. New strains of rice have been developed at the Institute with the support of the Rockefeller and Ford Foundations.*

production is increasing faster than the birthrate in these two countries. Yet, despite this hopeful progress, there are more hungry people in the world today than ever before.

New Food Sources

A number of other new methods of increasing food production are being studied. Many of these seem promising, although none has yet proved itself on a mass basis. One possibility is to grow algae in large quantities as a source of animal fodder. Another is to grow microorganisms—probably bacteria or yeasts—on simple hydrocarbons, such as butane or methane (see page 66), plus inorganic minerals as a source of food for domestic animals or humans. These organisms not only are good energy sources but are high in the essential amino acids. The problems are largely technological; methods are needed to grow the microorganisms on a large scale, to provide them with oxygen and prevent contamination, to harvest them, and to process them in some form acceptable for consumption. Present national interest in prolonged space flight may provide the impetus for the solution of some of these technical problems and also may make a diet of algae and bacteria seem more glamorous.

A particularly promising approach to increasing the world's harvests, and one which also needs a technological solution, is the desalination of water for the irrigation of the desert. About one-third of the world is made up of desert, and a surprisingly large proportion of this desert borders the ocean. Experiments with freshwater irrigation have proved that the desert, with its warm constant daytime temperatures and abundant sunlight with high light intensity, can be extremely productive in agriculture. Desalination of water is technically feasible but at present is very expensive in relation to the amount of water needed. It has been estimated that 200 gallons of water a day are required to grow the 2,500 calories needed by one person for one day. (Remember that more than 98 percent of the water taken up by a plant—corn, for instance—is transpired.) The solution may lie in the development of facilities using nuclear energy combined with means to reduce water requirements. The latter might include, for example, the selection of plants with a tolerance for sodium, so that slightly salty water could be used, and the construction of asphalt floors under the desert floors or plastic greenhouses over them to reduce water loss.

9–19
Beet fields in the Negev Desert, Israel.

Plankton, which has been often mentioned as a potential food source, no longer seems so promising in this regard. In the open ocean, where the greatest total quantity of plankton is found, light is adequate only very near the surface of the water and phosphorus and other minerals are scarce. As a consequence, the growth is very thin, which seems to make harvesting a technical impossibility. Also, removing large quantities of plankton from the ocean would further threaten the already waning supplies of fish, which depend directly or indirectly on this first trophic level.

A possibility which may or may not be practical but which is interesting to think about involves the importing of animals from Africa to fill unoccupied ecological niches

9–20
Some ecologists have suggested importing animals from Africa to fill vacant ecological niches in the United States. One candidate for import is the eland, which can subsist on the shrubby vegetation of the chaparral. Elands are reported to be gentler and easier to manage than domestic cattle, to grow faster and larger, and to produce excellent meat.

9–21
Abandoned cars under approach to Brooklyn Bridge, New York City.

in North America. According to this line of reasoning, these ecological niches have been vacant since the mysterious but apparently man-linked extinction of a number of large animals some 10,000 years ago. Some of these animals, such as one species of giant sloth and the American camel, were browsers which lived on tough leaves of plants of the chaparral, such as creosote bushes, yucca, and agave. Certain animals of the African savanna—the eland, for instance, which is domesticated easily—are able to subsist on these plants and might be used to convert them to protein for human consumption. There are 1 million square miles of such land in 11 Western states and Mexico.

How much further we can increase grain yields by the introduction of new strains and whether or not any of the other means now being considered will be successful are simply not known at this time. It does seem reasonable to expect that food production can be increased somewhat. It is even more clear that the production of new human beings can be expected to continue to increase, and as Malthus pointed out 200 years ago, the latter can always exceed the former. So actually, the effect of all our present efforts may simply be to buy time—time in which, if the world is very lucky, the population problem can be brought under control.

POLLUTION

Another of the major threats posed by the cultural evolution of man is pollution. In one sense, the pollution problem is as old as life itself. All living creatures produce metabolic wastes, which are discarded into the air, the water, or the land. When man was a wanderer, he simply moved to another campsite, and his excrement and debris were taken up by the land, broken down, and recycled through living systems. When people began to pack together in cities, waste disposal became a major problem. The solution of this problem, largely as a result of advances in the engineering of sanitation facilities—such as sewers

(a)

(b)

(c)

(d)

—and of public health laws, brought about tremendous improvements in the livability of urban centers and striking decreases in mortality rates from infectious diseases. As a result of these efficient and largely silent and invisible systems, coupled with our general affluence, we have learned to throw things away in increasing numbers. The problem now is to find an "away" that can accommodate the growing mass of material to be discarded. "Away" no longer usually means an alley or the heads of our neighbors, as it did in medieval times, but it often means the river running past the next town or the offshore waters of our oceans.

The pollution problem imposed by our increasing numbers is made far more difficult by our advancing technology. To replace containers made of clay, wood, or skin, we have perfected the aluminum beer can, the disposable bottle, and the plastic bag. In addition to our personal wastes, the industries which provide our necessities and luxuries spew their waste products into the air and waters at an ever-increasing rate.

9–22
(a) *St. Louis,* (b) *Dallas,* (c) *New York and* (d) *Minneapolis.*

9–23

Water pollution. (a) *One of the victims of the Santa Barbara oil slick. In* 1971, *ocean experts began to warn that pollution of ocean waters was beginning to kill the phytoplankton on which the "great chain of life" in the sea depends.* (b) *Another washday miracle. In many areas where fresh water cycles through wells and cesspools, the groundwater is foaming with detergents.* (c) *Estuaries and marshes, where fresh and salt-water meet, not only are rich in crustaceans and shellfish but are the spawning grounds and shelter the young of many of the important food fish of the oceans. This polluted marsh in Walnut Creek, California, was rescued by a water reclamation program.* (d) *Eutrophication—meaning, ironically, "good nourishment"—is a result of the pouring into lakes and ponds of excess phosphates and nitrates, in the form of fertilizers, detergents, and sewage. The result is an explosive overgrowth of algae, as in this scene along Chicago's lakefront, which then die and are decomposed by bacteria whose respiratory activities consume so much oxygen that the fish and other aquatic animals can no longer survive. The Great Lakes, because they are located in the midst of large farming areas, have been particularly affected.*

The pollution problem is the most soluble of our environmental crises—which does not necessarily mean that it will be quickly resolved. It must be made profitable to recycle inorganic materials, in the same way that it is "profitable" for the ecosystem to reuse its natural materials. Such recycling not only will prevent the accumulation of wastes but will retard the squandering of our resources—such as the thousands of acres of timberland turned into newsprint and other paper products each year. Also, it must be made profitable for industry to control its wastes and the wastes of its finished products, such as the exhausts of automobiles. Or, to put it more harshly, it is necessary to make it unprofitable not to. These problems are largely matters of legislation; however, unfortunately, much of this legislation will be unpopular not only with producers but with many consumers.

Insecticides

In addition to the pollutants which are by-products of our daily life and industry and which accidentally poison the living things in our environment, we have systematic programs for the development and use of materials with which we deliberately poison particular living organisms in our attempts to control our ecosystem. Chief among these poisons are insecticides.

DDT, the most widely used of the insecticides, was first introduced on a large scale during World War II and proved to be a spectacular success at curing body lice and, therefore, typhus. It soon became a major worldwide weapon against insect-borne diseases. Look again at the decline in the death rate in Ceylon (Figure 9–10). In 1945, it was 22 per thousand. DDT was introduced in 1946, with the consequent rapid control of the malaria-carrying mosquitoes. The death rate dropped 34 percent between 1946 and 1947 and was down to 10 per thousand by 1954. By 1969, it was 8 per thousand. Most of these lives were saved by DDT. And now that DDT is no longer being sent to Ceylon, malaria is beginning to appear again.

DDT is an efficient insect killer. In the nervous systems of both vertebrates and invertebrates, transmissions across synaptic junctions occur as the result of the discharge of certain chemicals, chief among which is acetylcholine. Acetylcholine, which acts swiftly to stimulate the adjacent neuron, is rapidly destroyed by an enzyme, cholinesterase. This enzymatic action is indispensable to the smooth functioning of the nervous system. DDT destroys cholinesterase in insects, causing their nervous system to run wild, and so kills them; the amount required is extremely small, far too little to have any effect on other animals.

The problem with DDT is that it persists and it accumulates. Prior to World War II, most insecticides were organic compounds based on natural substances which have coevolved with the insects and which protect the plants from them. Nicotine compounds are an example. Such substances break down relatively quickly. DDT and other synthetic insecticides retain their activity far longer, however. And once farmers and gardeners came to recognize the increase in yields of crop plants, they began to apply it routinely, on fixed schedules, and DDT began to accumulate in the soil and water. A Long Island marsh that had been sprayed for 20 years to control mosquitoes was recently found, for example, to contain up to 32 pounds of DDT per acre. It is now estimated that more than a billion pounds of DDT is circulating in the biosphere at the present time; even if DDT and related insecticides were never to be used again—from this second onward—they would be with us for years to come.

Insecticides such as DDT, because they persist, accumulate along food chains. In 1949, 1954, and 1957, Clear Lake, California, was sprayed to control midges, a small, harmless insect that was an annoyance to vacationers. The pesticide used was DDD, a close chemical relation of DDT, less toxic but equally persistent. The amount present in the lake after each spraying was about 0.02 part per million, and after 2 weeks, in all three instances, no DDD was detectable in the water at all. Before 1950, Clear Lake had been a nesting ground for about 1,000 pairs of western grebes (ducklike birds that eat small fish and other aquatic organisms).

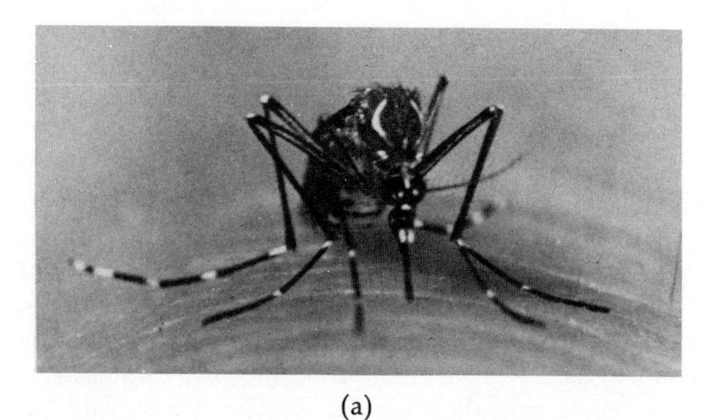

(a)

(b)

(c)

(d)

9–24

(a) *A female of the species* Aedes aegypti (*the male lives on nectar*). *If she draws blood from a person with malaria, she will probably pick up cells of* Plasmodium, *a parasitic protist that causes malaria. Worldwide use of DDT has virtually wiped out this species in many areas of human habitation and so eliminated malaria, a major cause of death in many tropical and subtropical countries.* (b) *An eaglet and an egg which will never hatch, photographed in a nest near the Muskegon River in Michigan. DDT causes a bird's liver to break down the hormones that mobilize calcium at the time of egg production, resulting in thin-shelled, fragile eggs, like the one shown here. Birds at the top of food chains, such as the osprey, peregrine falcon, and bald eagle, are principal victims.* (c) *Boll weevil, one of the agricultural pests controlled by DDT.* (d) *Fish killed by insecticides in a farm pond in Wake County, North Carolina.*

(a)

(b)

9–25

(a) *Alternatives to DDT and other insecticides are the use of natural predators, such as the ladybug shown here eating its way through a feast of aphids, and synthetic pheromones, which, by simulating the chemical stimuli emanated by female insects at mating time, lure the males to their death.* (b) *Huge feathery antennae of many male insects, such as the Cecropia moth shown here, are capable of detecting these chemicals in very small amounts and at great distances.*

From 1950 to 1961, no young were produced at all; in 1962, 5 years after the last spraying, one hatched. It was not until 1969 that they once more began to reproduce successfully. Studies undertaken to determine the concentration of DDD along the various trophic levels revealed that the concentration in the plankton was about 250 times as much as that originally present in the water. The concentration in frogs was 2,000 times the original concentration; in sunfish, about 12,000 times; and in grebes, as high as 80,000 times. The reason that no DDD could be detected in the water immediately after spraying became all too clear: it had been taken up completely by the living organisms of the lake. In the meantime, by 1957, the midge and about 150 other species of insects had developed some immunity to the pesticide.

In birds, it is now known, DDT, DDD, and related pesticides collect in the liver and induce the production of enzymes that break down various hormones and so interfere with reproduction. A number of species of predatory birds, including the bald eagle, the peregrine falcon, and the brown pelican, are now under threat of extinction because of pesticides.

The effect of DDT on other animals is not yet known. Virtually all populations of vertebrate animals around the world now have appreciable amounts of DDT in their tissues. Even animals in the Arctic, where DDT has almost never been used, show traces of it, the result of airborne distribution. At present, human tissues contain, on the average, about 11 parts per million of DDT and related chemicals. (Seven parts per million is the statutory maximum for meat transported in interstate commerce.) DDT tends to concentrate in fat. Eventually, we may have accumulated so much DDT in our fat that it will be dangerous to go on a reducing diet that would release the stored DDT into our systems. It also accumulates in milk; milk samples taken from the breasts of American women have been found to contain 0.1 to 0.2 part per million, which is several times more DDT than the federal government allows in dairy milk meant for human consumption.

Wide-scale routine applications of DDT have now been discontinued in most areas of the United States, and its use has been completely outlawed in some states. New means for controlling insect pests are being sought, though none as effective and inexpensive as DDT has yet been found.

One might hope that the lessons learned from DDT will persist in our environment as long as the chemical itself. This seems unlikely, however. For the last 9 years, we have been destroying the crops and forests of the South Vietnamese (our allies) with herbicides whose long-term effects are completely unknown; more than 5 million acres now have been sprayed—about one-seventh of the total land area—10 percent of it cropland. As a consequence largely of the protest of the scientific community, our spraying program in Vietnam has been discontinued although it is still being carried out, to a lesser extent, by the South Vietnamese. At present, however, we are bulldozing some 1,000 acres a day, returning rich timberland to a stage of early succession, an enterprise that will bring about far-reaching ecologic and economic damage.

We are, at present, adding hundreds, probably thousands, of new synthetic chemicals to our environment every year in the form of insecticides, fungicides, weed killers, food preservatives and additives, and industrial wastes. Are we prepared to predict their pathway through the biosphere?

One more example. Strontium is an element closely related to calcium. Like calcium, it can be taken up by grasses through their foliage as well as through the soil, eaten by dairy cows, concentrated in milk, which is largely consumed by children, and deposited in growing bone tissues and teeth. Strontium-90, a radioactive form of strontium and a by-product of the testing of atomic weapons, follows exactly the same route. As a consequence of atomic tests in the 1950s, by 1959 the bones of children in North America and Europe averaged an estimated 2.6 micromicrocuries of strontium-90 per gram of bone calcium, compared with 0.4 micromicrocurie per gram of calcium in the bones of adults. This amount of radioactivity has not been proved dangerous, but exposure to radioactive elements is known to cause leukemia, bone cancers, and genetic abnormalities and to generally shorten the life-span; and the minimum exposure that can produce such effects is not established. The half-life of strontium-90 is 28 years; in other words, it takes 28 years for half of the element to lose its radioactivity, so this exposure still continues.

The channeling of strontium-90 along the calcium food chain might have been anticipated. Other results were less predictable. For example, in the Arctic, only slight radiation exposure from the atomic tests was expected because the amount of fallout that reaches the ground at the poles is much less than it is in the temperate parts of the United States. However, Eskimos in the Arctic were discovered to have amounts of radioactivity in their bodies which were much higher than those found in the inhabitants of temperate regions. The link in the chain was the lichens. Since they have no real roots, lichens absorb their minerals directly from the air, and so they absorbed a large amount of direct fallout, little of which had had time to decay and none of which was dissipated by absorption into the soil. In the winter, caribou live almost exclusively on lichens, and at the top of the food chain, Eskimos live largely on caribou.

The point of all these stories is, of course, that the world has now become, in many ways, a single ecosystem, that

CATS TO BORNEO

Some years ago the World Health Organization instituted a program of spraying DDT widely on the tropical island of Borneo in an effort to control the mosquitoes. This program proved very effective in destroying mosquitoes. It also destroyed wasps, however. These wasps preyed on caterpillars, which, in turn, ate the thatched roofs of the natives' houses. So as a direct consequence of the mosquito-control programs, the roofs collapsed inside the huts.

Meanwhile, a program of spraying with DDT was initiated to kill houseflies. Up to that point, the houseflies had been controlled by small lizards—geckoes—which crawled over the ceilings and walls of the houses. When the flies were poisoned with DDT, the geckoes, which continued to eat them, also began to die. Falling to the floor of the huts, they were eaten by the house cats, which in turn began to die. Eventually, this led to an enormous increase in the rat population, which not only began to invade the houses and consume the food but also spread plague among the native population. Finally, a program of parachuting cats into remote Bornean villages was instituted in an effort to restore the ecological balance that was so badly disturbed by a seemingly reasonable program of spraying houseflies with DDT.

we are having profound effects on this ecosystem, and that we do not know what these effects or their consequences may be.

AGRICULTURE AND THE SOIL

In a natural ecosystem, as we have seen, organisms interact with one another and with their physical environment. As a result of these interactions, carbon, nitrogen, phosphates, and other materials are cycled and recycled through the ecosystem. In terrestrial ecosystems, one of the most important elements in this recycling, and certainly the most complex, is the soil. Natural soils contain a variety of minerals, which are taken up by plant and animal bodies and then returned in discarded leaves and branches, animal excreta, and decomposing plant and animal bodies. In addition to these nonliving components, the soil contains a vast array of living creatures, a teeming underground world of animal life, of plant roots, of fungi, bacteria, and protozoas. The insects, earthworms, and other invertebrates and the microorganisms that live in the soil decompose the organic material, converting it to a form in which it can be used by plants and also, as a consequence, turning the soil and aerating it.

When the land is turned to agriculture, the subtle relationships of this soil world are destroyed. The minerals are removed from the land when the crop plants are harvested or consumed by cattle and other herbivores. Insecticides and fungicides and herbicides kill the microorganisms of the soil. Plows rip up the roots that hold the soil together. One of the prime examples of the destructive consequences of agriculture is to be found in the tropical forests. The tropical rain forest is by far the richest, in terms of varieties of organisms, and the most luxuriant, in terms of density, of all biomes. Unlike the deciduous forests, which are typically carpeted with leaves and other debris, the floor of the tropical forest is bare; in this warm, damp climate, the pace of life is so swift that everything that falls to the ground

(a)

(b)

9–26
(a) *The prairie soils were once so bound together by the roots of grasses that they could not be cultivated until special plowing instruments were developed. But once the land was clear and the top soil became depleted by intensive agriculture or overgrazing, the soil became vulnerable to wind erosion. This photograph was taken in Texas County, Oklahoma, in 1937; John Steinbeck's novel,* The Grapes of Wrath, *described the emigration of dust bowl farmers to California.* (b) *Water erosion and an abandoned farmstead in California.*

that is not immediately devoured is quickly decomposed and reused. In fact, some recent studies of tropical soils indicate that perhaps there is no reservoir of minerals in the inorganic portion of the soil and that the only reservoir is the bodies of the soil microorganisms.

Areas where deciduous forests have stood are characteristically fertile and so good for farming; in fact, such has been the fate of most of the deciduous forests of the United States. One might suppose that the tropical forest would also provide rich farmland, but this is not the case. Once the natural community of plants is removed and the soil plowed, the ground may become almost as hard as cement. This red, hard soil has a special name, "laterite"; it is useful for making bricks for building. The lifeless laterite soil will not hold any minerals; they are leached right through it with the tropical rainfall. For centuries, the farmers of the damp tropics have cleared patches of rain forest by slashing and burning, have planted crops for two or three seasons until the ground hardened, and have then moved on, leaving the wound in the forest to heal itself. Now, however, under the tremendous pressures of the population growth in these countries, more and more of the tropical forest is being destroyed and reduced to laterite. It is estimated that by the year 2000 almost none of the original tropical forest will remain. This is a double tragedy: Most modern plant and animal life is believed to have had its beginnings in these forests, and yet we know little about them. Moreover, they are clearly the richest and most interesting of all biomes, and yet they have been little studied by ecologists and, once lost, can never be reconstructed. If they were contributing substantially toward feeding the hungry peoples of the world, one might be more easily reconciled to the loss, but because of our present lack of knowledge, they are not nor is it likely that they will be in the foreseeable future.

In the temperate zones, also, agriculture has left cruel scars on the land. When the roots and other organic materials that bind the soil together are ripped up by the plow, the land becomes much more vulnerable to the effects of the weather. Wind and water erode the soil, and the topsoil,

(a)

(b)

(c)

9–27

When the rich flora of the tropical jungle (a) *is slashed and burned for agriculture,* (b) *the soil often turns hard and infertile.* (c) *A clue to the formation of these infertile soils may be this network of roots and fungi lifted off a decomposing leaf in the Amazon forest. Mycorrhizae (associations between fungi and the roots of higher plants) have recently been found to play a crucial role in mineral cycling. In fact, current research indicates that in tropical soils, the chief reservoir of minerals is not in the soils themselves but in the soil fungi. Minerals are passed directly from the fungi to living root cells. In this way, little mineral leaks into the soil and leaching in undisturbed soils is at a minimum. In tilled tropical soils, on the other hand, the mycorrhizae are destroyed and the soils are rapidly depleted by leaching.*

9–28
Wheatfield U.S.A.

which contains the humus, is lost. Minerals are leached from the soil and washed away, eventually to the sea.

Fertilizers

In order to combat the effects of the constant removal of nutrients from the soil which occurs with agriculture, more and more massive amounts of fertilizers must be applied. The modern use of fertilizers has often made possible spectacular increases in crop yield; one of the great advantages of the new dwarf strains of wheat and rice is their capacity to respond to large amounts of fertilizers. However, fertilizers have also produced some disastrous side effects. It is likely that runoff of fertilizer from adjacent agricultural areas was the chief factor in the pollution and subsequent "death" of Lake Erie and the loss of its once sizable fishing industry.

One of the chief requirements of plants, as we have seen, is for nitrogen, which must enter the plant in the form of an organic compound, almost always as a nitrate (NO_3^-). When nitrates are applied to crops in large amounts, plants may concentrate them in considerably greater quantities than are normal. Nitrates are relatively innocuous for humans, but nitrite (NO_2^-) compounds, always present in small amounts when nitrates are in solution, are not. Particularly in infants, high concentrations of nitrites interfere with the capacity of the red blood cells to carry oxygen, causing serious disease (called methemoglobinemia) and sometimes death. The evidence is mounting that some crops of commercially available spinach have such a high nitrite concentration as to be dangerous to infants if eaten in quantity. In several communities in the heavily farmed San Joaquin Valley of California, nitrites from fertilizers have reached the groundwater and so are taken up again in wells for drinking purposes. The level of nitrites in the water is now so high that physicians in these communities routinely advise that infants drink bottled water until they are about six years old. And the use of nitrates has been steadily increasing. In the last 25 years, although the production of food in the United States has increased in proportion to population growth, the production of nitrogen fertilizers is up 1,050 percent.

THE IMPOVERISHMENT OF THE ECOSYSTEM

How is it that our once sturdy little planet, on which life has persisted and even flourished for more than 3 billion years, has now become so fragile? The answer to this question can be found in a reexamination of the concept of mature and immature ecosystems (Chapter 7–2). Any ecosystem under the "control" of man—and now man's influence is extending over the entire biosphere—is returned rapidly to an immature stage, a stage at which there are few species and therefore little complexity of interaction. A mature, complex ecosystem, with its complicated food webs, has many built-in checks and balances. Individual members of the plant and animal community may be sick or die, but the ecosystem itself is healthy and species tend to endure and in relatively stable numbers. Under agriculture, plants do not grow in complex communities, as

9–29
Off-the-road vehicles, such as dune buggies and trail bikes, are extending the destructive influence of man into the last wild corners of the continent. This wolf was shot from a snowmobile. Other, less fortunate animals are often run to exhaustion and left to die of starvation.

they do in a forest, but in pure stands. A cornfield, for example, has little inherent stability. If not constantly guarded by man, it will be immediately overrun with insects and weeds. It is for this reason that insecticides and herbicides play such a large and, indeed, indispensable role in our modern life. The susceptibility of modern crops to predators and parasites was tragically illustrated by the great potato famine of Ireland, which was caused by a fungus infection. The famine of 1845–1847 was responsible for more than 1 million deaths from starvation and initiated large-scale emigration from Ireland to the United States; within a decade, the population of Ireland dropped from 8 million to 4 million. Virtually the entire Irish potato crop was wiped out in a single week in the summer of 1846. Some modern plant geneticists are warning that the new strains of wheat and rice, which promise to contribute so importantly toward feeding the growing populations, will be particularly susceptible, because of their genetic uniformity and widespread distribution, to such disasters.

The selection of certain plants and animals and the destruction of their enemies are part of a deliberate program designed for mankind's prosperity. At the same time, we are participating in accidental programs of selection and destruction. House mice, rats, fleas, houseflies, and cockroaches are remarkably well suited to living in man-made environments, and so such organisms are flourishing. On the other hand, many species are becoming extinct. Some, as we have mentioned, are being destroyed by DDT. Others, including some species of great whales, crocodiles, alligators, and many of the large cats are being hunted to death. Still other species are threatened with extinction because of disturbances of their customary feeding and breeding sites. Not only has the agricultural revolution changed the relative numbers of different species; it has caused the wholesale extinction of many animals and plants. Some of our concern about endangered species can be dismissed as sentimental or esthetic; the world has learned to get along without the saber-toothed tiger and the dodo. However, there still remains the question of how many strands we can destroy in the web of life before the fabric of our own existence becomes threatened.

In short, we are creating a world that is more and more susceptible to damage, while at the same time, as a consequence of our increasing numbers and technology, we are amplifying our capacities to affect it. Although we can trace the biological roots of these phenomena, biology does not hold any ready solutions. There is an analogy here with the dilemma of twentieth century physics. Through the gentle genius of scientists like Albert Einstein and Neils Bohr, physics created the knowledge that led to the atomic bomb, but physics could not tell us whether or not to explode it over Hiroshima. Biology has not created its equivalent of the atomic bomb but, like physics, it has created knowledge for which it cannot dictate the uses. Biology, for instance, can tell us the details of the reproductive process; medical science, building on biological knowledge, can develop new means for limiting the birth rate; but biology cannot determine whether, by whom, or under what circumstances these new methods should be applied. Biology can tell us how an insecticide, amplified by its passage through the food chain, can strike a nesting tern, never for a moment its target; but in a world in which malaria and starvation are major causes of death, biology cannot decide the extent to which we should use insecticides to kill mosquitoes and to protect crop plants. Biology can tell us why animal protein is so much more expensive—in terms of human lives—than vegetable protein. Faced with the choice, few of us would rather eat a steak than share a potful of rice and beans with a group of hungry children. But we do not know how to share our excess with the millions of undernourished people.

In the words of Barry Commoner:

> "These are *value* judgments; they are determined not by scientific principle, but by the value we place on economic advantage and on human life. . . . These are matters of morality, of social and political judgments. In a democracy they belong not in the hands of 'experts,' but in the hands of the people and their elected representatives."*

What are we, then, to do—not as experts, or biologists, or even students of biology, but just as ordinary individual human beings, inhabitants of a small planet?

We can always, of course, do nothing. Some contend that man is merely following his nature in his exploitation of his environment and of other living things, including his fellow man, and that this course will lead him inevitably to destruction. Others believe that things are so bad already that there is little hope in trying to do anything now; unfortunately, the gloomy portents of some authorities in the field, although they may not have had that intention, lend credence to this point of view. This attitude—a fashionable one in some circles—may, of course, be right. If more than 99 percent of all species are extinct, the odds are that *Homo sapiens* will follow this same route, and perhaps we should not be too sentimental about it, or too concerned about which other species may fill our ecological niche. Yet, although it may be the correct answer, few of us can live comfortably with it.

For those of us who do elect to take action, there are, as Commoner again reminds us, two types of action open to us. The first is individual action. We can try to limit our personal consumption and our personal waste—pick up our litter, clean up our automobile exhausts, learn to live with midges, and gypsy moths, and with wash that is not whiter than white, limit the size of our families, and even try to change our personal diet so that there is a little left over for someone else. These actions may be important not only for their own sakes but because they serve to foster a new climate in which new ideas can take root.

The second type of action is political and social and—in the broadest sense—revolutionary. Nations—not just our own, but all nations—are dedicated, by the fact of their existence, to promoting the welfare of their own citizens, and within nations, governments find it necessary to support the interests of particular groups who play vital roles in the national economy. We are living in a country and at a time in which growth has been the proud criteria of progress for many decades: growth in population, in power consumption, in paved roads, in numbers of cars per family, and above all, in gross national product, our total production of goods and services. If things are to change, it is necessary to adopt a new standard of values in which something much less measurable—such as quality of life not only for ourselves but for others—replaces our consumer's yardstick. Some authorities believe that this can take place only

* (*The Closing Circle* Alfred A. Knopf, New York, 1971)

under conditions of much more rigid political, economic, and even psychological control. Others point to present changes—an apparent decrease in birthrates, liberalization of abortion laws in a few states, and widespread concern about pollution and conservation—and see in them evidence that Americans may be prepared to deal with these problems voluntarily and perhaps even effectively. It is too early to tell. Recent events in our history have shown, however, that governments do respond, although slowly, to massive change in the climate of public opinion. If we really do want a change, not only in our immediate environment but also in our political and social priorities, this desire will manifest itself in a large number of ways.

Finally, to return to biology, we should remember that man, although a member of the Kingdom Animalia, is different in certain important respects from other animals. Because of these differences, it is largely irrelevant whether man is naturally aggressive or passive, exploitative or generous. Unlike any of the other animals, man can make deliberate choices about his own course of behavior.

In the age of the dinosaurs, the earliest primates survived, it would appear, largely by their wits; now, if man is to survive the monsters of his own creating, he will have to do it, again, by the contents of his own skull. For within the mind of man—that complex collection of neurons and synapses—resides the uniquely human capacity to accumulate knowledge, to plan with foresight, and so to act with enlightened self-interest.

SUMMARY

Man, like all living organisms, is a product of evolution. He belongs to the phylum Chordata and the subphylum Vertebrata, which includes also the fish, amphibians, reptiles, birds, and other mammals. Most mammals are placental. Because of their high constant body temperature, they are far more alert and intelligent than the reptilian line from which they evolved.

The line of apes from which man eventually evolved first became distinct from the other primates about 14 million years ago. Evolutionary trends among this group included the change from four-legged to two-legged walking, the use of the hands for carrying and for manipulation of objects, the making of tools, the development of speech, and a great increase in cranial capacity and, presumably, in intelligence.

The last rapid steps in the evolution of man took place in the last 1.5 million years, during the Ice Ages. Physical evolution was completed about 35,000 years ago. Cro-Magnon man, who made his appearance at that time, was clearly a modern man, biologically speaking. Human evolution since that time has been cultural.

The most important event in the cultural evolution of man was the advent of agriculture, about 10,000 years ago. Agriculture both ensured a more dependable food supply and permitted men to live together in communities, which hastened the spread of ideas and technology and, as a consequence, the pace of cultural change.

About 8000 B.C., when agriculture had its beginning, there were probably about 5 million people in the world. By 4000 B.C., the population had increased to 86 million; by the time of Christ, there were 133 million people in the world; and by 1650, there were 500 million.

In the last few hundred years, the population has been increasing at a faster and faster rate. By 1972, there were approximately 3.8 billion people. The doubling rate of the human population is now estimated at approximately 37 years, which means there will probably be about 7 billion people in the world by the close of the century. Because the capacity of the planet to support its human population is limited—although these limits have not been defined—attempts are under way to control the population by voluntary means rather than by the more ancient though highly effective means of war, pestilence, or famine. One obstacle to the reduction in births is the unavailability of an ideal contraceptive method. A much more stubborn obstacle is the resistance, for various religious, sociological, psychological reasons, of many individuals and groups to birth control. Many experts believe that only by improving health and living standards in the "have-not" countries can significant reductions in birthrate be achieved.

The most serious present consequence of the rapid increase in world population is hunger. Malnutrition may involve either a shortage of total calories or a protein deficit, or both. It is estimated that more than half of the persons now alive suffer from some form of malnutrition. Effects of protein deprivation are particularly serious in infants; protein shortages early in life permanently impair not only physical but mental development.

The development of new strains of wheat and rice has greatly increased food yields in Mexico, India, and Pakistan, and for the first time in recent history, food production is now increasing more rapidly than the birthrate in these countries. Yet despite this progress, there are more hungry people in the world today than ever before.

New methods of food production are being explored. These include the cultivation on a large scale of algae, bacteria, and yeasts which can be used for animal fodder or processed for human consumption. Another approach involves the desalination of seawater for irrigation of the desert.

A major consequence of man's recent cultural evolution has been pollution of air, water, and land by industrial and agricultural waste products and by the by-products of human existence. Steps toward the solution of this problem are the development of means and facilities for recycling waste products and the imposition of strict regulations on both private individuals and industry.

In addition to accidental and careless pollution, in recent years we have deliberately added a number of new synthetic chemicals to the biosphere. Our experience with DDT, the most widely used and consequently the most studied of such chemicals, provides an example of the possible effects of this form of pollution. DDT is degraded very slowly and so persists, in contrast to organic products which are broken down rapidly under natural circumstances. Moreover, it accumulates along the food chain and so poses a particular threat to organisms of the higher trophic levels, such as birds of prey and man. Even if the use of DDT were discontinued immediately—which it will not be because of its importance in the control of malaria—its effects would persist in the ecosystem for at least another 20 years.

The extensive agriculture needed to grow food for the expanding human population has harsh effects upon the soil. The soils of tropical forests rapidly become infertile once the land is cleared for farming. They form a cement-like material known as laterite. Soils of the temperate zone can be kept fertile by massive applications of fertilizers. However, the nitrites which accumulate as a result of such fertilization programs are now also creating pollution problems in some heavily farmed areas.

In terms of its future effect on the living things on our planet, the most serious result of man's rise to ecological dominance may prove to be his simplification of the biosphere. Farming efficiency favors growing particular crops in large stands, and at the same time, as a result of agriculture, urbanization, and pollution, the number of species is being reduced. As a consequence, the biosphere is coming more and more to resemble an immature ecosystem, one in which there is little stability and in which the living components of the system are much more susceptible to the insults of the physical components of the environment and, undoubtedly, to man-made threats as well.

Evolution moves relentlessly in a single direction. We cannot regain a previous kind of relationship between man and nature any more than we can resurrect the dodo. But perhaps if we learn to use, increase, and respect our knowledge of our biosphere, to learn from our experiences, and to recognize that the future is already upon us, we may preserve some of what still remains.

QUESTIONS

1. One of the pieces of evidence supporting the conclusion that the first mammals were homeothermic is that, according to fossil specimens, they had a bony palate separating the nose and mouth. The hard palate makes it possible to breathe and eat at the same time (see Figure 6–57). Why would such a capacity be associated with warm-bloodedness?

2. Suppose the birthrate were to drop immediately to the so-called "replacement level"—two children per couple. Assuming that the great majority of children are born to women between 15 and 30 years of age, what would happen to the population of a country with a triangular population profile (Figure 9–13)? One with a barrel-shaped profile? One whose profile is an inverted pyramid?

3. In an editorial entitled "Food, Overpopulation, and Irresponsibility" that appeared recently in a scientific journal, the authors concluded: "Because it creates a vicious cycle that compounds human suffering at a high rate, the provision of food to the malnourished nations of the world that cannot, or will not, take very substantial measures to control their own reproductive rates is inhuman, immoral and irresponsible." What is your opinion?

4. Buckminster Fuller, the architect, has said, "Pollution is resources we are not harvesting." Give some examples to support this statement. Do you agree with his definition?

SUGGESTIONS FOR FURTHER READING

BAKER, HERBERT G.: *Plants and Civilization,* Wadsworth Publishing Company, Inc., Belmont, Calif., 1965.*

An introduction to the study of plants in relation to man, this book illustrates the profound influence of plants on man's economic, cultural, and political history.

"The Biosphere," *Scientific American,* September, 1970.*

This issue of the Scientific American *is devoted entirely to the "thin film of living matter at the surface of the earth" and the recycling processes by which it maintains itself.*

BRESLER, JACK B. (ed.): *Human Ecology. Collected Readings.* Addison-Wesley, Reading, Mass., 1966.

A collection of articles, by experts in many different fields, on the wide variety of physical and biological factors that have influenced the evolution of man and that affect his present existence. The subjects range from biological clocks to insanity among lemmings (a cautionary tale) to the search for life on Mars.

CARSON, RACHEL: *Silent Spring,* Houghton Mifflin Company, Boston, 1962.*

This is the book that awakened the nation to the dangers of pesticides. Once seen as highly controversial, it is now regarded as a classic of modern ecology.

COMMONER, BARRY: *The Closing Circle,* Alfred A. Knopf, Inc., New York, 1971.

The author traces the relationship between the present ecological crisis and technological and social factors in our society. His title, The Closing Circle, *refers to his conviction that if we are to survive, "we must learn how to restore to nature the wealth we borrow from it."*

EHRLICH, PAUL, and ANNE EHRLICH: *Population/Resources/Environment,* W. H. Freeman and Company, San Francisco, 1970.

Required, although not cheerful, reading for all concerned with the ecological crisis. The Ehrlichs present a wealth of useful data, even though you may not agree with all their conclusions.

LAPPE, FRANCES MOORE: *Diet for a Small Planet,* Ballantine Books, Inc., New York, 1971.*

If the world is lucky, this will be one of the most important books of 1971. Mrs. Lappé clearly establishes the feasibility of eating "low on the food chain," thereby greatly increasing the availability of proteins to other human populations. Includes practical suggestions and many recipes.

NELSON, GIDEON E., and JAMES D. RAY, JR.: *Biologic Readings for Today's Students,* John Wiley & Sons, Inc., New York, 1971.

A group of readings selected particularly for use in introductory biology on the basis of their readability, their interest, and their relevance to today's problem and to progress in medicine and other related fields.

SAUER, CARL ORTWIN: *Land and Life,* University of California Press, Berkeley, 1967.

Carl Sauer was concerned with the forces that relate human life to the web of plant and animal life sustaining it and with man's role in changing the face of the earth. These essays, written over the author's lifetime, range from studies of prehistoric man to descriptions of pioneer life in America.

SHEPARD, PAUL and DANIEL MC KINLEY (eds.): *The Subversive Science: Essays toward an Ecology of Man,* Houghton Mifflin Company, Boston, 1969.*

A very readable collection of essays by some noted authors on various aspects of human populations and their relation to the environment.

VAN LAWICK-GOODALL, JANE: *In the Shadow of Man,* Houghton-Mifflin Company, Boston, 1971.

An absorbing personal account of eleven years spent observing the complex social organization of a single chimpanzee community in Tanzania.

* Available in paperback.

APPENDIX A

Metric Table

	Fundamental Unit	*Quantity*	*Numerical Value*	*Symbol*	*English Equivalent*
Length	meter			m	39.4 inches
		kilometer	1000 m	km	.62137 miles
		centimeter	.01 m	cm	.3937 inches
		millimeter	.001 m	mm	
		micron	.000001 m	μ	
		nanometer (millimicron)	.000000001 m	nm (mμ)	
		Ångstrom	.0000000001 m	Å	
Mass	gram			g	3.527 ounces
		kilogram	1000 g	kg	2.2 pounds
		milligram	.001 g	mg	
		microgram	.000001 g	μg	
Time	second			sec	
		millisecond	.001 sec	msec	
		microsecond	.000001 sec	μsec	
Volume (solids)	cubic meter			m^3	1.308 cubic yards
		cubic centimeter	.000001 m^3	cm^3	.061 cubic inches
		cubic millimeter	.000000001 m^3	mm^3	
Volume (liquids)	liter			l	1.06 quarts
		milliliter	.001 liter	ml	
		microliter	.000001 liter	μl	

APPENDIX B

Temperature Conversion Scale

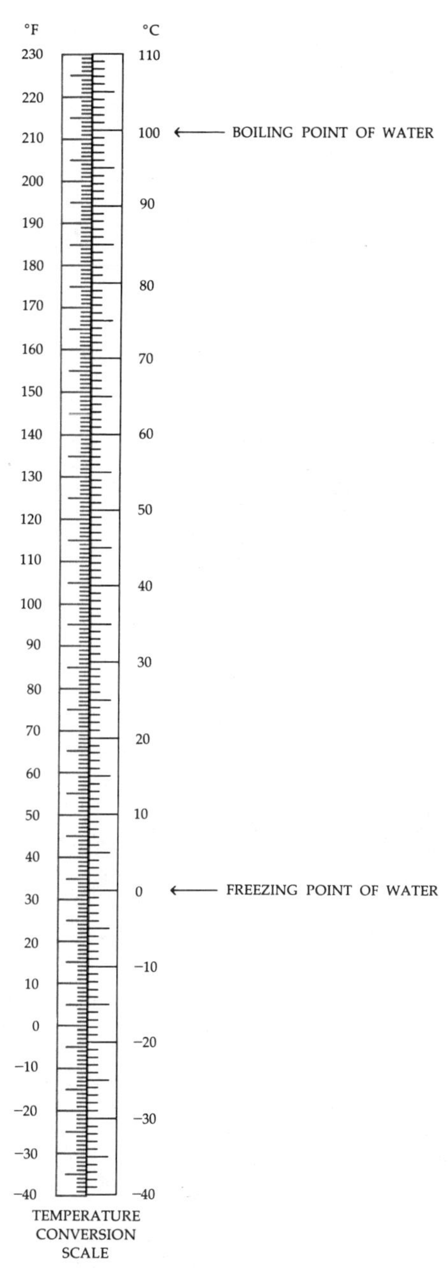

TEMPERATURE CONVERSION SCALE

FOR CONVERSION OF FAHRENHEIT TO CENTIGRADE, THE FOLLOWING FORMULA CAN BE USED:

$$°C = \frac{5}{9}(°F - 32)$$

FOR CONVERSION OF CENTIGRADE TO FAHRENHEIT, THE FOLLOWING FORMULA CAN BE USED:

$$°F = \frac{9}{5}°C + 32$$

APPENDIX C

Classification of Organisms

There are several ways to classify organisms. The one presented here follows the overall scheme described in Section 4, in which organisms are divided into five major groups, or kingdoms: Monera, Protista, Fungi, Plantae, and Animalia. The following classification is based closely upon that of Whittaker (in *Science 163*:150–160).

The chief taxonomic divisions are kingdom, phylum, class, order, family, genus, species. This classification includes all of the major phyla. Certain classes and orders, particularly those mentioned in this book, are also included, but the listing is far from complete. The number of species given for each group is the estimated number of living species described and named.

KINGDOM MONERA

Prokaryotic cells which lack a nuclear envelope, plastids and mitochondria, and 9-plus-2 flagella. Monera are unicellular but sometimes aggregate into filaments or other superficially multicellular bodies. Their predominant mode of nutrition is absorption, but some groups are photosynthetic or chemosynthetic. Reproduction is primarily asexual, by fission or budding, but portions of DNA molecules may also be exchanged between cells under certain circumstances. Those that move, move by simple flagella or glide.

About 1,600 species of bacteria and bacterialike organisms are recognized at present, but doubtless thousands more await discovery. Classification of Monera is based largely upon metabolic features and is not comparable to that of eukaryotes. One group, the class Rickettsiae—very small bacterialike organisms—occurs widely as parasites in arthropods, and may contain tens or even hundreds of thousands of species, depending upon the criteria used; they have not been included in the estimate given. No satisfactory classification of the Monera has yet been proposed. One of the groups included is the phylum Cyanophyta, the blue-green algae, which contain chlorophyll *a* and are photosynthetic, and of which there are about 200 nonsymbiotic distinct species.

KINGDOM PROTISTA

Eukaryotic unicellular organisms in which the cells are sometimes aggregated into filaments or other superficially multicellular bodies. Their modes of nutrition include ingestion, photosynthesis, and sometimes absorption. True sexuality is present in some phyla. They move by 9-plus-2 flagella or pseudopods or are nonmotile.

PHYLUM

PHYLUM PROTOZOA: microscopic, unicellular or simple colonial animals; usually with distinct nucleus and cytoplasm; reproduction usually asexual by mitotic division; classified by type of locomotion. About 30,000 species.

CLASS

Class Mastigophora: protozoans with flagella, including a number of symbiotic forms such as *Trichonympha* and *Trypanosoma,* the cause of sleeping sickness.

Class Sarcodina: protozoans with pseudopodia, such as amoebas. No stiffening pellicle outside cell walls.

Class Ciliophora: protozoans with cilia, including *Paramecium* and *Stentor.*

Class Sporozoa: parasitic protozoans; usually without locomotive organs during a major part of their life cycle. Includes *Plasmodium,* several species of which cause malaria.

PHYLUM

PHYLUM CHRYSOPHYTA: golden algae and diatoms. Autotrophic organisms with chlorophylls *a* and *c* and the accessory pigment fucoxanthin. Food stored as the carbohydrate leucosin or as large oil droplets. Cell walls consisting mainly of pectic compounds, sometimes heavily impregnated with siliceous materials. Some 6,000 to 10,000 living species.

CLASS

Class Bacillariophyceae: diatoms. Chrysophyta with double siliceous shells, the two halves of which fit together like a pillbox. They are sometimes motile by the secretion of mucilage fibrils along a specialized groove, the raphe. There are many extinct and 5,000 to 9,000 living species.

Class Chrysophyceae: golden algae. A diverse group of organisms including flagellated, amoeboid, and nonmotile forms, some naked and others with a cell wall that may be ornamented with siliceous scales. At least 1,000 species.

PHYLUM

PHYLUM PYRROPHYTA: golden-brown algae. Autotrophic organisms with chlorophylls *a* and *c*. Food is stored as starch. Cell walls contain cellulose. The same phylum contains some 1,100 species, mostly biflagellated organisms, of which the great majority belong to the following class:

CLASS

Class Dinophyceae: the dinoflagellates. Pyrrophyta with lateral flagella, one of which beats in a groove that encircles the organism. They probably have no form of sexual reproduction and their mitosis is unlike that in any other organism. There are more than 1,000 species.

PHYLUM

PHYLUM EUGLENOPHYTA: the euglenoids. Autotrophic (or sometimes secondarily heterotrophic) organisms with chlorophylls *a* and *b*. They store food as paramylon, an unusual carbohydrate. Euglenoids usually have a single apical flagellum and a contractile vacuole. The flexible cell wall (pellicle) is rich in proteins. Sexual reproduction is unknown. Euglenoids occur mostly in fresh water. There are some 450 species.

PHYLUM GYMNOMYCOTA: the slime molds. Heterotrophic amoeboid organisms that mostly lack a cell wall but form sporangia at some stage in their life cycle. Predominant mode of nutrition is by ingestion. There are three classes:

CLASS

Class Myxomycetes: the plasmodial slime molds. Slime molds with multinucleate plasmodium which creeps along as a mass and eventually differentiates into sporangia, each of which is multinucleate and eventually gives rise to many spores. About 450 species.

Class Acrasiomycetes: cellular slime molds. Slime molds in which there are separate amoebas which eventually swarm together to form a mass but retain their identity within this mass, which eventually differentiates into a compound sporangium. Seven genera and about 26 species.

Class Protostelidomycetes: In this recently discovered group, the amoebas may remain separate or mass, but each one eventually differentiates into a simple stalked sporangium with one or two spores at its apex. Five genera and more than a dozen species.

KINGDOM FUNGI

Eukaryotic unicellular or multinucleate organisms in which the nuclei occur in a basically continuous mycelium; this mycelium becomes septate (partitioned off) in certain groups and at certain stages of the life cycle. They are heterotrophic, with nutrition by absorption. Reproductive cycles often include both sexual and asexual phases. All are provisionally included in a single phylum, Mycota. There are some 100,000 valid species of fungi to which names have been given, but at least twice this many more probably await discovery, and many of the described ones will be eventually found to have been named two or more times; this is particularly so for fungi that may be classified both as ascomycetes and as members of the Fungi Imperfecti.

CLASS

Class Oomycetes: Mostly aquatic fungi with motile cells characteristic of certain stages of the life cycle, their cell walls are composed of glucose polymers including cellulose. There are several hundred species.

Class Zygomycetes: terrestrial fungi, such as black bread mold, with the hyphae septate only during the formation of reproductive bodies; chitin predominant in the cell walls. The class includes several hundred species.

Class Ascomycetes: terrestrial and aquatic fungi, including *Neurospora*, powdery mildews, morels and truffles. The hyphae are septate but the septa perforated; complete septa cut off the reproductive bodies, such as spores or gametangia. Chitin is predominant in the cell walls. Sexual reproduction involves the formation of a characteristic cell, the ascus, in which meiosis takes place and within which spores are formed. The hyphae in many ascomycetes are packed together into complex "fruiting bodies" known as ascocarps. Yeasts are unicellular ascomycetes that reproduce asexually by budding. There are about 30,000 species.

Class Basidiomycetes: terrestrial fungi, including the mushrooms and toadstools, with the hyphae septate but the septa perforated; complete septa cut off reproductive bodies, such as spores or gametangia. Chitin is predominant in the cell walls. Sexual reproduction involves formation of basidia, in which meiosis takes place and on which the spores are borne. There is often complex differentiation of "tissues" within their basidiocarps. There are some 25,000 species.

Fungi Imperfecti: Mainly fungi with the characteristics of Ascomycetes but in which the sexual cycle has not been observed; a few probably belong to other classes. The Fungi Imperfecti are classified by their asexual spore-bearing organs. There are some 25,000 species, including a *Penicillium*, the original source of penicillin, fungi which cause athlete's foot and other skin diseases, and many of the molds which give cheese, such as Roquefort and Camembert, their special flavor.

Lichens: Logically regarded as another class of fungi, the lichens are mostly Ascomycetes symbiotic with unicellular algae which multiply within their densely packed hyphae. A very few lichens appear to involve Basidiomycetes. About 17,000 described species.

KINGDOM PLANTAE

Multicellular eukaryotes and related unicellular forms. Principal mode of nutrition by photosynthesis. Their photosynthetic pigment is chlorophyll *a*, with chlorophyll *b* and a number of carotenoids serving as accessory pigments. The cell walls contain cellulose as a matrix. They are primarily nonmotile organisms attached to a substrate. There is considerable differentiation of organs and tissues in the more specialized forms. Their reproduction is primarily sexual with alternating gametophytic and sporophytic phases; the gametophyte phase has been progressively reduced in the course of evolution. The plants have clearly evolved on at least three occasions; the red algae and brown algae have an origin separate from the rest.

PHYLUM

PHYLUM RHODOPHYTA: red algae. Primarily marine plants characterized by the presence of chlorophyll *a* and pigments known as phycobilins. Their carbohydrate reserve is a special type of starch (Floridean). No motile cells are present at any stage in the complex life cycle. The plant body is built up of closely packed filaments in a gelatinous matrix and is not differentiated into leaves, roots, and stem. It lacks specialized conducting cells. There are some 4,000 species.

PHYLUM PHAEOPHYTA: brown algae. Multicellular marine plants characterized by the presence of chlorophyll *a* and *c* and the pigment fucoxanthin. Their food reserve is a carbohydrate called laminarin. Motile cells are biflagellate, with one forward flagellum of the tinsel type and one trailing one of the whiplash type. A considerable amount of differentiation is found in some of the kelps, with specialized conducting cells for transporting photosynthate to dimly lighted regions of the plant present in some genera. There is however no differentiation into leaves, roots, and stem, as in the land plants. There are about 1,100 species.

PHYLUM CHLOROPHYTA: green algae. Unicellular or multicellular plants characterized by chlorophylls *a* and *b* and various carotenoids. The carbohydrate food reserve is starch. Motile cells have two whiplash flagella at the apical end. True multicellular genera do not exhibit complex patterns of differentiation. Multicellularity has arisen at least three times, and quite possibly more often. There are about 7,000 known species and possibly many more.

PHYLUM BRYOPHYTA: mosses, hornworts, and liverworts. Multicellular plants with the photosynthetic pigments and food reserves similar to those of the green algae. They have gametangia with a multicellular sterile jacket one cell layer thick. The sperm are biflagellate and motile. Gametophytes and sporophytes both exhibit complex multicellular patterns of development, but the conducting tissues are usually completely absent and not well differentiated when present. Most of the photosynthesis in these primarily terrestrial plants is carried out by the gametophyte, upon which the sporophyte is initially dependent. There are more than 23,500 species.

CLASS

Class Hepaticae: the liverworts. The gametophytes are either thallose (not differentiated into roots, leaves and stems) or leafy, and the sporophytes relatively simple in construction. There are about 9,000 species.

Class Antherocerotae: the hornworts. The gametophytes are thallose. The sporophyte grows from a basal meristem for as long as conditions are favorable. Stomata are present on the sporophyte. There are about 100 species.

Class Musci: the mosses. The gametophytes are leafy. Sporophytes have complex patterns of spore discharge. Stomata are present on the sporophyte. There are about 14,500 species.

PHYLUM

PHYLUM TRACHEOPHYTA: the vascular plants. Terrestrial plants with complex differentiation of organs into leaves, roots, and stem. The only motile cells are the male gametes of some species, which are propelled by many cilia. The vascular plants have well-developed strands of conducting tissue for the transport of water and organic materials. The main trends of evolution in the vascular plants involve a progressive reduction in the gametophyte, which is green and free-living in ferns but heterotrophic and more or less enclosed by sporophytic tissue in the others; the loss of multicellular gametangia and motile sperm; and the evolution of the seed. The phylum includes the following subphyla with living representatives:

SUBPHYLUM

SUBPHYLUM LYCOPHYTINA: the lycophytes. Homosporous and heterosporous vascular plants with microphylls; extremely diverse in appearance. All lycophytes have motile sperm. There are five genera and about 1,000 species.

SUBPHYLUM SPENOPHYTINA: the horsetails. Homosporous vascular plants with jointed stems marked by conspicuous nodes and elevated siliceous ribs and sporangia borne in a strobilus at the apex of the stem. Leaves are scalelike. Sperm are motile. Although now thought to have evolved from a megaphyll, the leaves of the horsetails are structurally indistinguishable from microphylls. There is one genus, *Equisetum*, with about two dozen living species.

SUBPHYLUM PTEROPHYTINA: ferns, gymnosperms, and flowering plants. Although diverse, these groups possess in common the megaphyll, which in certain genera has become much reduced. There are about 260,000 species.

CLASS

Class Filicineae: the ferns. They are mostly homosporous although some are heterosporous. The gametophyte is more or less free-living and usually photosynthetic. Multicellular gametangia and free-swimming sperm are present. There are about 11,000 species.

Class Coniferinae: the conifers. Seed plants with active cambial growth and simple leaves, in which the ovules are not enclosed and the sperm are not flagellated. There are some 50 genera and about 550 species, the most familiar group of gymnosperms.

Class Cycadinae: cycads. Seed plants with sluggish cambial growth and pinnately compound, palmlike or fernlike leaves. The ovules are not enclosed. The sperm are flagellated and motile, but are carried to the vicinity of the ovule in a pollen tube. Cycads are gymnosperms. There are nine genera and about 100 species.

Class Ginkgoinae: ginkgo. Seed plants with active cambial growth and fan-shaped leaves with open dichotomous venation. The ovules are not enclosed and are fleshy at maturity. Sperm are carried to the vicinity of the ovule in a pollen tube, but are flagellated and motile. They are gymnosperms. There is one species only.

Class Angiospermae: the flowering plants. Seed plants in which the ovules are enclosed in a carpel (in all but a very few genera), and the seeds at maturity are borne within fruits. They are extremely diverse vegetatively but characterized by the flower, which is basically insect-pollinated. Other modes of pollination, such as wind pollination, have been derived in a number of different lines. The gametophytes are much reduced, with the female gametophyte often consisting of only eight cells or nuclei at maturity. Double fertilization involving two of the three nuclei from the mature microgametophyte gives rise to the zygote and to the primary endosperm nucleus; the former becomes the embryo and the latter a special nutritive tissue, the endosperm. There are about 250,000 species.

SUBCLASS

Subclass Dicotyledonae: the dicots. Flower parts are usually in fours or fives; leaf venation is usually netlike, pinnate, or palmate; there is true secondary growth with vascular cambium commonly present; there are two cotyledons; and the vascular bundles in the stem are in a ring. There are about 190,000 species.

Subclass Monocotyledonae: the monocots. Flower parts are usually in threes, leaf venation is usually parallel, true secondary growth is not present, there is one cotyledon, and vascular bundles in the stem are scattered. There are about 60,000 species.

KINGDOM ANIMALIA

Eukaryotic multicellular organisms. Their principal mode of nutrition is by ingestion. Many animals are motile, and they generally lack the rigid cell walls characteristic of plants. Considerable cellular migration and reorganization of tissues often occurring during the course of embryology. Their reproduction is primarily sexual, with male and female diploid organisms producing haploid gametes which fuse to form the zygote. There are more than a million described species of animals, but the actual number may be closer to 10 million.

SUBKINGDOM PARAZOA

Includes the sponges, the only major group of multicelled animals with no digestive cavity.

PHYLUM

PHYLUM PORIFERA: the sponges. Simple multicellular animals, largely marine, with stiff skeletons, and bodies perforated by many pores that admit water containing food particles. About 4,200 species.

SUBKINGDOM METAZOA

Animals whose bodies consist of many cells, as distinct from the unicellular Protozoa.

PHYLUM

PHYLUM COELENTERATA: the coelenterates. Animals with radially symmetrical, "two-layered" bodies of a jellylike consistency. Reproduction is asexual or sexual. They are the only organisms with cnidoblasts, special stingy cells. All are aquatic and most are marine. About 9,600 species.

CLASS

Class Hydrozoa: Hydra, Obelia and other hydra-like animals. Often colonial, and often having a regular alternation of asexual and sexual generations. Polyp form dominant.

Class Scyphozoa: the marine jellyfishes or "cup animals," including *Aurelia*. Medusa form dominant; true muscle cells.

Class Anthozoa: sea anemones ("flower animals") and colonial corals. No medusa stage.

PHYLUM

PHYLUM CTENOPHORA: the comb jellies and sea walnuts. They are free-swimming, often almost spherical animals. They are translucent, gelatinous, delicately colored and often bioluminescent. They possess eight bands of cilia, for locomotion. About 80 species.

PHYLUM PLATYHELMINTHES: the flatworms. Bilaterally symmetrical with three tissue layers. The gut, when present, has only one opening. No coelom or circulatory system; complex hermaphroditic reproductive system; excretion by means of special (flame) cells. About 15,000 species.

CLASS

Class Turbellaria: the planaria and other non-parasitic flatworms. Ciliated, carnivorous, have ocelli ("eye spots").

Class Trematoda: the flukes. Parasitic flatworms with a digestive tract.

Class Cestoidea: the tapeworms. Parasitic flatworms with no digestive tract; absorb nourishment through body surface.

PHYLUM

PHYLUM NEMERTEA: the proboscis or ribbon worms. Non-parasitic; usually marine; have a tubelike gut with mouth and anus, and a protrusible proboscis armed with a hook for capturing prey; simple circulatory and reproductive systems. About 550 species.

PHYLUM NEMATODA: the roundworms and nematodes. Includes minute free-living forms, such as vinegar eels, and plant and animal parasites, such as hookworms. A fairly large phylum; characterized by elongated, cylindrical, bilaterally symmetrical bodies. About 80,000 species.

PHYLUM ACANTHOCEPHALA: the spiny-headed worms. Parasitic worms with no digestive tract and a head armed with many recurved spines. About 300 species.

PHYLUM CHAETOGNATHA: arrow worms. Free-swimming planktonic marine worms; have a coelom, a complete digestive tract and a mouth with strong sickle-shaped hooks on each side. About 50 species.

PHYLUM NEMATOMORPHA: the horsehair worms. Extremely slender, brown or black worms up to 3 feet long; adults are free-living, but the larvae are parasitic in insects. About 250 species.

PHYLUM ROTIFERA: miscroscopic, wormlike or spherical animals, commonly called "wheel animalcules." Have complete digestive tract, flame cells, and a circle of cilia on the head, the beating of which suggests a wheel; males minute and either degenerate or unknown in many species. About 1,500 species.

PHYLUM GASTROTRICHA: microscopic, wormlike animals moving by longitudinal bands of cilia. About 140 species.

PHYLUM BRYOZOA: "moss" animals. Microscopic aquatic organisms; characterized by a U-shaped row of ciliated tentacles, with which they feed; usually fixed and in branching colonies; superficially resemble hydroid coelenterates but are much more complex; have anus and coelum; retain larva in special brood pouch. About 4,000 species.

PHYLUM BRACHIOPODA: the lamp shells. Marine animals with two hard shells (one dorsal and one ventral), superficially like a clam; obtain food by means of ciliated tentacles; fixed by a stalk or one shell in adult life. About 260 living species; 3,000 extinct.

PHYLUM PHORONIDEA: sedentary, elongated, wormlike animals that secrete and live in a leathery tube; U-shaped digestive tract and a ring of ciliated tentacles with which they feed. Marine. About 15 species.

PHYLUM ANNELIDA: ringed or segmented worms, usually with well-developed coelom (fluid-filled cavity within mesoderm), one-way digestive tract, head, and circulatory system; appendages nonjointed, when present; have nephridia, well-defined nervous system. About 7,000 species.

CLASS

Class Archiannelida: small simple, probably primitive, marine worms. About 35 species.

Class Polychaeta: mainly marine worms, such as *Nereis.* Have a distinct head with palps and tentacles and many bristled lobose appendages; parapodia often brightly colored. About 4,000 species.

Class Oligochaeta: soil, fresh water and marine annelids, including the earthworm (*Lumbricus*). Scanty bristles and usually a poorly differentiated head. About 2,500 species.

Class Hirudinea: leeches. Posterior sucker and usually an anterior sucker surrounding the mouth; fresh water, marine and terrestrial; either free-living or parasitic. About 300 species.

PHYLUM

PHYLUM MOLLUSCA: unsegmented animals, with a head, a mantle and a muscular foot, variously modified; mostly aquatic; soft-bodied, often with one or more hard shells; all but *Pelecypoda,* the bivalves, have a radula (rasplike organ used for scraping or marine drilling); three-chambered heart. About 100,000 species.

CLASS

Class Amphineura: the chitons. Simplest type of mollusks; elongated body covered with mantle in which are imbedded eight dorsal shell plates. About 700 species.

Class Pelecypoda: two-shelled, bivalve mollusks, including clams, oysters, mussels, scallops. Usually have a hatchet-shaped foot and no distinct head. Generally sessile. About 15,000 species.

Class Scaphopoda: the tooth or tusk shells; includes *Dentalium.* Marine mollusks with a conical tubular shell. About 350 species.

Class Gastropoda: asymmetrical mollusks including snails, whelks, slugs. Usually with a spiral shell and a head with one or two pairs of tentacles. About 80,000 species.

Class Cephalopoda: octopus, squid, *Nautilus.* "Headfoot" with eight or ten arms or many tentacles; mouth with two horny jaws; well-developed eyes and nervous system; shell is external (*Nautilus*), internal (squid) or absent (octopus); all except *Nautilus* have ink gland. About 400 species.

PHYLUM

PHYLUM ARTHROPODA: the largest phylum in the animal kingdom. Segmented animals; paired jointed appendages, hard jointed exoskeleton; growth involves molting; complete digestive tract; coelom reduced; no nephridia; brain dorsal; nerve cord ventral; paired ganglia in each segment. About 765,257 species.

CLASS

Class Merostomata: horseshoe crabs. Aquatic; book gills; "living fossils." 5 species.

Class Crustacea: lobsters, crabs, crayfishes, shrimps. Mostly aquatic; two pairs of antennae; one pair of mandibles, and typically two pairs of maxillae; thoracic segment with appendages; abdominal segments with or without appendages. About 25,000 species.

Class Arachnida: spiders, mites, scorpions. Most members terrestrial, air-breathing; usually 4 or 5 pairs of legs; first pair of appendages used for grasping; have chelicerae (pincers or fangs) instead of jaws or antennae. About 30,000 species.

Class Onychophora: simple, terrestrial arthropods. All belong to one genus, *Peripatus;* have many short unjointed pairs of legs. About 73 species.

Class Insecta: insects. Most are terrestrial; most breathe air by means of trachea; one pair of antennae, three pairs of legs; three distinct parts of the body (head, thorax and abdomen); most have 2 pairs of wings. Include bees, ants, beetles, butterflies, fleas, lice, bugs, flies, etc. About 700,000 species.

Class Chilopoda: the centipedes. 15 to 173 trunk segments, each with one pair of jointed appendages. About 2,000 species.

Class Diplopoda: the millipedes. Abdomen with 20 to 100 segments, each with two pairs of appendages. About 7,000 species.

PHYLUM

PHYLUM ECHINODERMATA: radially symmetrical in adult stage; well-developed coelom formed from enteric pouches; endoskeleton of calcareous ossicles and spines; unique water vascular system; tube feet; marine. About 5,700 species.

CLASS

Class Crinoidea: sea lilies and feather stars. Sessile animals, often having a jointed stalk for attachment; ten arms bearing many slender lateral branches. Most species are fossils.

Class Asteroidea: starfishes. Have five to fifty arms; oral surface directed downward. Two to four tube feet.

Class Ophiuroidea: brittle stars, serpent stars. Greatly elongated, highly flexible slender arms; rapid horizontal locomotion.

Class Echinoidea: sea urchins, sand dollars. Skeletal plates form rigid test which bears many movable spines.

Class Holothuroidea: sea cucumbers. Sausage-shaped or wormlike elongated body.

PHYLUM

PHYLUM CHORDATA: animals having at some stage a notochord, pharyngeal gill slits, and a hollow nerve cord on the dorsal side. About 44,794 species.

SUBPHYLUM

SUBPHYLUM HEMICHORDATA: small group of marine animals, including the acorn, or proboscis worms. Wormlike animals with a notochordlike structure in the head end; gill slits; solid nerve cord. About 91 species.

SUBPHYLUM TUNICATA: tunicates or ascidians. Saclike; adults usually sessile, often forming branching colonies; marine; feed by ciliary currents; have gill slits, reduced nervous system, no notochord. Larva active with well developed nervous system and notochord. About 1,600 species.

SUBPHYLUM CEPHALOCHORDATA: lancelets. Small subphylum containing only *Amphioxus* and related forms. Somewhat fishlike marine animals with a permanent notochord the whole length of the body; nerve cord; pharynx with gill slits; no cartilage or bone. About 13 species.

SUBPHYLUM VERTEBRATA: the vertebrates. Most important subphylum of Chordata. Notochord replaced by cartilage or bone, forming the segmented vertebral column or backbone; skull surrounding a well-developed brain; usually a tail. About 43,090 species.

CLASS

Class Agnatha: lampreys and hagfishes. Eel-like aquatic vertebrates without limbs; jawless sucking mouth; single nostril; no bone, scales or fins.*

Class Chondrichthyes: sharks, rays, skates, and other cartilaginous fish. Without air bladders; have complicated copulatory organs, scales. Almost exclusively marine.*

Class Osteichthyes: the bony fish, including nearly all modern freshwater fish, such as sturgeon, trout, perch, anglerfish, lungfish and some almost extinct groups. Usually with an air bladder or (rarely) a lung.*

Class Amphibia: salamanders, frogs, and toads. Usually breathing by gills in the larval stage and by lungs in the adult stage; incomplete double circulation; skin usually naked; the limbs are legs. The first vertebrates to inhabit the land, and ancestors of the reptiles; eggs unprotected by a shell and embryonic membranes. About 2,000 species.

Class Reptilia: turtles, lizards, snakes, crocodiles; includes extinct species such as the dinosaurs. Breathe by lungs; incomplete double circulation; skin usually covered with scales; the four limbs are legs (absent in snakes); cold-blooded; mostly live and reproduce on land though some are aquatic; embryo enclosed in an egg shell and has a protective amnion and an allantois. About 5,000 species.

Class Aves: the birds. Warm-blooded animals with complete double circulation; skin covered with feathers; the forelimbs are wings. Includes the extinct *Archeopteryx*. About 8,590 species.

Class Mammalia: the mammals. Warm-blooded animals with complete double circulation. Distinguished by skin usually covered with hair; young nourished with milk secreted by the mother; four limbs, usually legs (forelimbs sometimes arms, wings, or fins); diaphragm used in respiration; lower jaw made up of a single pair of bones; three bones in each middle ear connecting ear drum and inner ear; 7 vertebrae in neck. About 4,500 species.

* The total number of species of fishes is estimated to be about 23,000.

SUBCLASS

Subclass Prototheria: the monotremes. Oviparous (egg-laying) mammals with imperfect temperature regulation. Only two living species are the duckbill platypus and spiny anteater of Australia and New Guinea.

Subclass Metatheria: the marsupials, including kangaroos, opossums, and others. Viviparous mammals without a placenta (or with a poorly developed one); the young are born in an undeveloped state and are carried in an external pouch of the mother for some time after birth. Found chiefly in Australia.

Subclass Eutheria: mammals with a well-developed placenta. Comprises the great majority of living mammals. The principal orders of Eutheria are the following:

Order

Insectivora: shrews, moles, hedgehogs, etc.
Edentata: toothless mammals—anteaters, sloths, armadillos, etc.
Rodentia: the rodents—rats, mice, squirrels, etc.
Artiodactyla: even-toed ungulates—cattle, deer, camels, hippopotamuses, etc.
Perissodactyla: odd-toed, hoofed mammals—horses, zebras, rhinoceroses, etc.
Proboscidea: elephants
Lagomorpha: rabbits and hares
Sirenia: the manatee, dugong, and sea cows. Large aquatic mammals with the forelimbs finlike, the hind limbs absent.
Carnivora: carnivorous animals—cats, dogs, bears, weasels, seals, etc.
Cetacea: the whales, dolphins, and porpoises. Aquatic mammals with the forelimbs fins, the hind limbs absent.
Chiroptera: the bats. Aerial mammals with the forelimbs wings.
Primates: the lemurs, monkeys, apes, and man.

ILLUSTRATION ACKNOWLEDGMENTS

1 Radio Times Hulton Picture Library

2 Burndy Library

3 National Park Service

4 By permission of the Royal College of Surgeons of England

5 A. W. Ambler, from National Audubon Society

1–1 Mount Wilson and Palomar Observatories

1–2 National Aeronautics and Space Administration

1–3 Photograph by S. Jónasson. R. Anderson, *et al.*, *Science*, **148:** 1179–1190, 28 May 1965. Copyright 1965 by The American Association for the Advancement of Science.

1–4 Cyril Ponnamperuma

1–5 Eric V. Gravé

1–6 E. S. Barghoorn and J. W. Schopf, *Science*, **152:** 758–763, 6 May 1966. Copyright 1966 by The American Association for the Advancement of Science.

1–7 J. Pickett-Heaps

1–8 George I. Schwartz

1–9 Oscar Erpenstein, from National Audubon Society

2–1 George I. Schwartz

2–2 (b) Eric V. Gravé

2–3 The Bettmann Archive, Inc.

Page 30 Micrograph by W. A. Jensen, from *Cell Ultrastructure* by William A. Jensen and Roderic B. Park. © 1967 by Wadsworth Publishing Company, Belmont, California. Reprinted by permission of the publisher.

Page 31 (*left*) W. A. Jensen, from *Cell Ultrastructure* by William A. Jensen and Roderic B. Park. © 1967 by Wadsworth Publishing Company, Belmont, California. Reprinted by permission of the publisher. (*right*) L. M. Biedler

2–5 Micrograph by George Palade

2–7 Micrograph by Eric Gravé; drawing after A. Jurand and G. G. Selman, *Anatomy of* Paramecium aurelia, St. Martin's Press, New York, 1969.

2–8 A. Jurand and G. G. Selman, *Anatomy of* Paramecium aurelia, St. Martin's Press, New York, 1969.

2–10 (b), (c) A. Jurand and G. G. Selman, *Anatomy of* Paramecium aurelia, St. Martin's Press, New York, 1969.

2–12 E. H. Newcomb

2–14 K. R. Porter

2–16 (a) G. Decker

2–17 Grant M. Haist, from National Audubon Society

2–18 Esso Research and Engineering Co.

2–19 Laurence Pringle, Photo Researchers, Inc.

2–20 Jack Dermid

2–21 Jack Dermid

2–24 Associated Photographers, courtesy of Calgon Corporation.

2–29 K. R. Porter

2–30 (b) R. D. Preston; (c) Eric V. Gravé

2–31 John H. Gerard

2–35 B. E. Juniper

2–39 (b) California Institute of Technology

2–40 (a) Eila Kairinen and Emil Bernstein, The Gillette Company Research Institute; (b) K. R. Porter

Page 82b Micrograph by L. A. Staehelin

Page 82c K. R. Porter

Page 82d (*left*) Micrograph by Ursula Goodenough; (*right*) micrograph by J. Pickett-Heaps

Page 82e Micrograph by C. S. Raine, Albert Einstein College of Medicine

Page 82f (*left*), (*top right*) A. Jurand and G. G. Selman, *Anatomy of* Paramecium aurelia, St. Martin's Press, New York, 1969. (*bottom right*) A. V. Grimstone

Page 82g Micrographs by K. R. Porter

Page 82h (*left*) K. R. Porter; (*right*) J. Heslop-Harrison

Page 82i K. R. Porter

2–43 National Aeronautics and Space Administration

2–44 (a) Jack Dermid; (b) Russ Kinne, Photo Researchers, Inc.; (c) Y. Haneda; (d) Australian News and Information Bureau

2–45 Yerkes Regional Primate Research Center

2–52 K. R. Porter

2–57 (a), (b) L. K. Shumway

3–1 Kurt Hirschhorn, M.D., from *Birth Defects: Original Article Series*, Vol. IV, No. 4 *Guide to Human Chromosome Defects* by Audrey Redding and Kurt Hirschhorn, M.D., The National Foundation.

3–2 D. Branton, from *Cell Ultrastructure* by William A. Jensen and Roderic B. Park. © 1967 by Wadsworth Publishing Company, Inc., Belmont, California. Reprinted by permission of the publisher.

3–6 Howard Towner

3–7 (a) S. Inoué, *Polarization Optical Studies of the Mitotic Spindle, I. The Demonstration of Spindle Fibers in Living Cells, Chromosome Bd. 5:* 487–500, Springer, 1953. (b) After E. J. DuPraw, *Cell and Molecular Biology*, Academic Press, New York, 1968, page 593. (c) Etienne de Harven

3–8 Eric V. Gravé

3–9 James Cronshaw

3–16 The Bettmann Archive, Inc.

Page 144 (*left*) Gary Laurish; (*center*), (*right*) Samples courtesy of Patricia Farnsworth, Ph.D., Barnard College; scanning electron micrographs by Irene Piscopo

3–22 From John A. Moore, *Heredity and Development*, Oxford University Press, Cambridge, 1963.

3–23 C. G. G. J. van Steenis

3–24 Edmund B. Gerard

3–26 F. B. Hutt, *Journal of Genetics* **22:** 126, 1930

3–32 Bernard John

3–34 From James D. Watson, *Molecular Biology of the Gene*, W. A. Benjamin, Inc., Menlo Park, Calif., 1965.

3–35 B. P. Kaufmann

3–38 The Bettmann Archive, Inc.

3–46 (a) Lee D. Simon; (b) from James D. Watson, *Molecular Biology of the Gene*, W. A. Benjamin, Inc., Menlo Park, Calif., 1965. (c) A. K. Kleinschmidt, D. Lang, D. Jacherts and R. K. Zahn, *Biochim. Biophys. Acta*, **61:** 857–864, 1962, Figure 1.

3–47 From *The Double Helix* by James D. Watson. Copyright © 1968 by James D. Watson. Reprinted by permission of Atheneum Publishers, New York.

3–48 Photograph by A. C. Barrington-Brown

3–51 John Cairns

3–55 (b) Hans Ris

3–57 O. L. Miller, Jr. and Barbara R. Beatty, Biology Division, Oak Ridge National Laboratory

3–60 (b) Jack Griffith

3–61 George T. Rudkin

4–1 Lynwood M. Chace, from National Audubon Society

4–3 (a) J. F. M. Hoeniger, from *General Microbiology*, **40:** 29–42, 1965. (b) Charles C. Brinton, Jr.

Page 208 J. D. Almeida and A. F. Howatson, *Journal of Cell Biology*, **16:** 16, 1963.

Page 209 (*top right*) Photograph by Frederick A. Murphy, National Communicable Disease Center; (*bottom*) E. Boy de la Tour

4–4 Micrograph by Norma J. Lang

4–5 (a) Eric V. Gravé; (b) Ingmar Holmasen; (c) Franco Rossi, Engis Equipment Company; (d) E. B. Small and D. S. Marszalek, *Science*, 163: 7 March 1969. Copyright 1969 by The American Society for the Advancement of Science. (e) Eric V. Gravé

4–6 (a) James Oschman; (b) Leonard Muscatine

4–7 (a) Jack Dermid; (b) C. Bracker; (c) L. West

4–8 (a) Douglas P. Wilson; (b) Don Longenecker

4–9 R. W. Hoshaw

4–11 © Field Museum of Natural History, painting by Charles R. Knight

4–12 (a), (b) Ross E. Hutchins; (c), (d), (e) Jack Dermid

4–13 Myron C. Ledbetter

4–14 William M. Harlow

4–15 (a) Russ Kinne, Photo Researchers, Inc.; (b) Hugh Spencer, from National Audubon Society; (c) Karl H. Maslowski, from National Audubon Society; (d) Jeanne White, from National Audubon Society; (e) Jack Dermid

4–18 (a) Hugh Spencer, from National Audubon Society; (b) Douglas P. Wilson; (c), (d) Russ Kinne, Photo Researchers, Inc.

4–21 (a) James V. McConnell; (b) Douglas P. Wilson; (c) Armed Forces Institute of Pathology Photograph; (d) U. S. Department of Agriculture; (e) M. Wimmer-Mizzaro

4–23 From W. D. Russell-Hunter, *A Biology of Lower Invertebrates*, The Macmillan Company, New York, 1968, page 86.

4–26 (a) Alvin E. Staffan; (b) Queensland Tourist Bureau; (c), (d) Douglas P. Wilson

4–27 Douglas P. Wilson

4–29 (a), (b) Jack Dermid; (c) Alvin E. Staffan; (d) L. West

4–30 (a) Ross E. Hutchins

4–31 (a) Hugh Spencer, from National Audubon Society; (b) Hal H. Harrison, from National Audubon Society; (c)

John H. Gerard; (d) Bernard Gluck, from National Audubon Society

4–33 Donald H. Fritts

4–34 Russ Kinne, Photo Researchers, Inc.

4–35 Jack Dermid

4–36 Courtesy of the American Museum of Natural History

4–37 Roberts Rugh, *Guide to Vertebrate Development*, Burgess Publishing Company, Minneapolis, 1971.

5–1 Jack Dermid

5–3 Jack Dermid

5–4 Gary Laurish

5–7 (a), (b) L. West; (c) Edward S. Ross; (d) Claude Schoepf; (e) Edward S. Ross; (f) Luoma, from National Audubon Society; (g) Edward S. Ross

5–8 (a) J. Heslop-Harrison; (b), (c), (d) Patrick Echlin

5–11 L. West

5–15 (a) Lynwood M. Chace, from National Audubon Society; (b), (c) Jack Dermid

5–16 Peter M. Ray, from *The Living Plant*, Holt, Rinehart and Winston, Inc., New York, 1963.

5–19 H. R. Chen

5–21 (a), (b) J. P. Nitsch

5–22 S. H. Wittwer

5–23 J. E. Varner

5–26 Jack Dermid

5–29 Marvin Winter

5–34 (a) J. Heslop-Harrison; (b) B. E. Juniper

5–36 (b) L. West

5–39 Jack Dermid

5–41 (a) Martin H. Zimmermann; (b) George A. Schaefers

6–1 Ken Heyman

6–4 David M. Phillips

6–12 Hippie, Ken Heyman; elephant seal, Al Lowry, Photo Researchers, Inc.; peacock, Jack Dermid; lion, A. W. Ambler, from National Audubon Society

Page 330a Roberts Rugh and Landrum B. Shettles, M.D., *From Conception to Birth: The Drama of Life's Beginnings*, Harper & Row, Publishers, Inc., New York, 1971.

Page 330b Carnegie Institution of Washington

Page 330c Carnegie Institution of Washington

Page 330d Carnegie Institution of Washington

Page 330e Carnegie Institution of Washington

Page 330f Fetus at 39 days, H. Nishimura, from Rugh and Shettles, *op. cit.*; fetus at 42 days, Rugh and Shettles, *op. cit.*; fetus at 55 days, E. Ludwig, from Rugh and Shettles, *op. cit.*; fetus at 2 months, Rugh and Shettles, *op. cit.*; fetal skeleton, Rugh and Shettles, *op. cit.*

Page 330g Fetus at 68 days, E. Ludwig, from Rugh and Shettles, *op. cit.*; fetus at 10 weeks, Rugh and Shettles, *op. cit.*

Page 330h Carnegie Institution of Washington

Page 330i Rugh and Shettles, *op. cit.*

Page 330j Rugh and Shettles, *op. cit.*

Page 330k Rugh and Shettles, *op. cit.*

Page 330l (*left*), (*center*) Rugh and Shettles, *op. cit.*; (*right*) E. Wilcox

6–17 K. B. Roper

6–19 K. R. Porter

6–20 (b) David H. Hubel

6–21 (a) K. R. Porter; (c) E. R. Lewis

6–25 Marilyn Silverstone, Magnum Photos

6–28 (b) E. R. Lewis

6–30 The Bettmann Archive, Inc.

6–33 Photograph by Jack Dermid

6–36 K. R. Porter

6–37 Oscar Auerbach

6–41 Jack Dermid

6–43 S. Baty

6–45 (a) Nicholas Mrosovsky; (b) Leonard Lee Rue, III, from National Audubon Society; (c) Sdeuard C. Bisserôt

Table 6–4 (Page 369) From C. Ladd Prosser and Frank A. Brown, Jr., *Comparative Animal Physiology*, 2d ed., W. B. Saunders Company, Philadelphia, 1961, page 158.

6–46 UNESCO/Dominique Roger

6–47 John H. Gerard

6–48 Russ Kinne, Photo Researchers, Inc.

6–54 (a) K. R. Porter; (b) L. M. Biedler

6–56 L. M. Biedler

Page 386 U. S. Department of Agriculture

6–60 Eric V. Gravé

6–61 Medical Audio Visual Department, Walter Reed Army Institute of Research

6–65 From Lewis Carroll, *Alice's Adventures in Wonderland*, illustrated by John Tenniel

6–66 Owner: Sally Curtis

6–67 Bob Leatherman, from National Audubon Society

6–70 José M. Delgado, Yale School of Medicine

6–73 David H. Hubel

6–81 Jerre Levy, Colwyn Trevarthen, and Roger W. Sperry, *Brain: Journal of Neurology*, 1972.

6–82 Frank Lotz Miller, Black Star

7–1 © Elihu Blotnick, BBM Associates

7–6 Karl Weidmann

7–7 J. M. Conrader, from National Audubon Society

7–8 From Peter Farb, *Face of North*

America: The Natural History of the Continent, Harper & Row, Publishers, Inc., New York, 1963, page 11.

7–9 (a) P. J. Webber; (b) U. S. Fish and Wildlife Service

7–10 Alan Root

7–11 Ingmar Holmasen

7–13 (a) Jack Dermid; (b) Jeanne White, from National Audubon Society

7–14 Ron Church, Photo Researchers, Inc.

7–16 (a) Alvin E. Staffan, from National Audubon Society; (b) Jack Dermid; (c) Sixten Jonsson; (d) R. D. Estes

7–20 (a) The Nitragin Company, Inc.; (b) P. J. Dart; (c) R. R. Herbert, R. D. Holsten, and R. W. F. Hardy, E. I. duPont de Nemours and Co., Inc.

7–22 (a) U. S. Forest Service; (b) L. West

Page 453 Photograph by Robert H. Wright, from National Audubon Society

7–24 (a) Jack Dermid; (b) U. S. Forest Service

7–25 (a) Jack Dermid; (b) U. S. Forest Service

7–28 S. A. Wilde

7–29 (a) Hermann Kacher; (b) Nat Fain/Marineland of Florida; (c) Jen and Des Bartlett, Photo Researchers, Inc.

7–30 Emil Schulthess

7–31 R. J. Brooks

7–33 (a) John Mann, Australian Department of Lands; (b) C. L. Hogue and R. S. Lasebeer, *Natural History Magazine*, **75**: May 1966.

7–34 (a) Jack Dermid; (b) Lewis Walker; (c) Jack Dermid; (d) John H. Gerard

7–35 (a) Jack Dermid; (b) Ross E. Hutchins; (c) Eric Hosking

7–36 (a) L. West; (b) Marineland of Florida

7–37 (a) Alvin E. Staffan; (b) Ross E. Hutchins

7–38 Thomas Eisner

7–39 Russ Kinne, Photo Researchers, Inc.

7–40 Edwin Way Teale

7–41 Edwin Way Teale

7–42 Edwin Way Teale

7–43 Ross E. Hutchins

7–44 Ross E. Hutchins

7–45 Ross E. Hutchins

7–46 (a) U. S. Navy; (b) Eric Hosking

7–47 V. B. Scheffer, U. S. Fish and Wildlife Service

7–48 Wisconsin Regional Primate Research Center

7–49 Joe Van Wormer, Photo Researchers, Inc.

7–50 Hope Buyukmihci, from National Audubon Society

7–51 Alvin E. Staffan, from National Audubon Society

7–42 Hermann Kacher

7–53 (a) G. E. Kirkpatrick, Frank Lane; (b) Willis Peterson; (c) Uzi Paz

8–1 Baron Wolman

8–2 Eric Hosking

8–6 Rapho Guillumette Pictures

8–7 Victor A. McKusick

8–9 March of Dimes

8–10 Walter Dawn, from National Audubon Society

8–11 (b) Food and Agricultural Organization, United Nations

8–12 (a) Leonard Lee Rue, III, from National Audubon Society; (b) Jack Dermid; (c) Douglas Fisher, Frank Lane; (d) Sally Anne Thompson, Scala Fine Arts Publishers, Inc.

8–15 (a) © Field Museum of Natural History, painting by Charles R. Knight; (b), (c) Harold J. Pollock

8–16 (a) Ron Garrison, San Diego Zoo; (b) courtesy of the Pennsylvania Game Commission; (c) Leonard Lee Rue, III, from National Audubon Society

8–17 P. M. Sheppard

8–18 H. B. D. Kettlewell

8–20 I. Eibl-Eibesfeldt

8–21 John A. Moore

8–23 A. W. Ambler, from National Audubon Society

8–25 "Woodchuck": Ron Garrison, San Diego Zoo; "rat," "bear," "dog": Australian News and Information Bureau

8–26 Russ Kinne, Photo Researchers, Inc.

8–27 (a) L. W. Walker, National Audubon Society; (b) Canadian Government Travel Bureau; (c) The New York Times/Jack Manning

Page 535a Alaska Department of Fish and Game

Page 535b (*left*) Virgil Argo; (*right*) Save-the-Redwoods League

Page 535c (*top left*) Walter Knight, Jepson Herbarium; (*bottom left*) U. S. Forest Service; (*right*) National Park Service

Page 535d (*left*) Jack Dermid; (*bottom right*) Dennis Stock, Magnum Photos

Page 535e (*left*), (*top right*) Jack Dermid; (*bottom right*) Willis Peterson

Page 535f (*top left*) Standard Oil Company (N. J.); (*bottom left*) John H. Gerard; (*right*) Alexander Lowry, Photo Researchers, Inc.

Page 535g (*left*) Art Bilsten, from National Audubon Society; (*right*) U. S. Forest Service

Page 535h Photographs courtesy of Canadian Government Travel Bureau

Page 535i (*top left*) Eric Hosking; (*bottom left*) John H. Gerard; (*right*) courtesy of Department of Indian Affairs and Northern Development, Canadian Government

Page 535j Photograph by Steven C. Wilson

Page 535k (*top left*) Ken Heyman; (*bottom left*) Miloslav Posepny, from *Animals of Many Lands* by Hanns Reich Verlag, München. Used with the permission of Hill & Wang, Inc. (*top right*) Bill Gabriel, from National Audubon Society; (*bottom right*) Fritz Henle, Photo Researchers, Inc.

Page 535l Photograph by Erling Larsen, Frederic Lewis Photography

Page 535m (*top*) D. Mohrhardt; (*bottom left*) Grant M. Haist, from National Audubon Society; (*bottom right*) Jack Dermid

Page 535n (*left*) Jack Dermid; (*top right*) G. Ronald Austing; (*bottom right*) Wisconsin Conservation Department

Page 535o (*left*) Photographs by Jack Dermid; (*right*) U. S. Forest Service

Page 535p (*left*), (*top right*) Jack Dermid; (*bottom right*) Paul A. Moore

Page 353q (*left*) Jack Dermid; (*right*) Max Hunn, from National Audubon Society

Page 535r (*left*) George Daniell, Photo Researchers, Inc.; (*top right*) Jack Dermid; (*bottom right*) Marineland of Florida

Page 535s (*top*) Mary M. Thacher, Photo Researchers, Inc.; (*bottom*) Jack Dermid

Page 535t Jack Dermid

9–1 Russ Kinne, Photo Researchers, Inc.

9–2 © Field Museum of Natural History, painting by Charles R. Knight

9–3 San Diego Zoo

9–4 Ron Garrison, San Diego Zoo

9–5 Courtesy of the American Museum of Natural History

9–6 Courtesy of the American Museum of Natural History

9–7 Courtesy of the American Museum of Natural History

9–8 Paul C. Mangelsdorf, Richard S. MacNeish, and Walton C. Galinat, *Science*, **143:** 538–545, 7 February 1964. Copyright 1964 by the American Association for the Advancement of Science.

9–9 The Bettmann Archive, Inc.

9–12 Sunderland Aerial Photographs, Oakland, California

9–14 Wide World Photos, Inc.

9–15 (a) Food and Agricultural Organization; (b) P. Pittet/Food and Agricultural Organization; United Nations

9–17 Wide World Photos, Inc.

9–18 (a) Wide World Photos, Inc.; (b) International Rice Research Institute

9–19 Louis Goldman, Rapho Guillumette Pictures

9–20 Karl H. Maslowski, Photo Researchers, Inc.

9–21 F. B. Grunzweig, Photo Researchers, Inc.

9–22 (a) Wide World Photos; (b) Dallas Morning News; (c) National Air Pollution Control Administration; (d) Minneapolis Tribune

9–23 (a) Susan Landor; (b) S. Collins, Photo Researchers, Inc.; (c) Phiz Mozesson; (d) John Hendry, Jr., from National Audubon Society

9–24 (a) S. A. Smith, The London School of Hygiene and Tropical Medicine; (b) Robert Harrington, Michigan Department of Natural Resources; (c) U. S. Department of Agriculture; (d) courtesy of Federal Water Quality Administration, U. S. Department of the Interior

9–25 (a) Jerome Wexler, from National Audubon Society; (b) Lynwood Chace, from National Audubon Society

9–26 U. S. Department of Agriculture

9–27 (a), (b) Walter Dawn, from National Audubon Society; (c) F. W. Went

9–28 U. S. Department of Agriculture

9–29 Rohn Engh

Page 571 Ken Heyman

INDEX

Page numbers in **boldface** indicate major text references; **underlined boldface** refers to glossary entries; *italic* refers to illustration captions or table entries; and regular type indicates incidental text references.